SENSOR NETWORK OPERATIONS

SENSOR NETWORK OPERATIONS

Edited by

Shashi Phoha

Thomas LaPorta

Christopher Griffin

IEEE PRESS

WILEY-INTERSCIENCE

A JOHN WILEY & SONS, INC PUBLICATION

Published by John Wiley & Sons, Inc.
Published simultaneously in Canada.

For general information on our other products and services or for technical support, please contact our Customer Care Department within the United States at (800) 762-2974, outside the United States at (317) 572-3993 or fax (317) 572-4002.

Wiley also publishes its books in a variety of electronic formats. Some content that appears in print may not be available in electronic formats. For more information about Wiley products, visit our web site at www.wiley.com.

Library of Congress Cataloging-in-Publication Data is available.

ISBN-13 978-0-471-71976-2
ISBN-10 0-471-71976-5

Printed in the United States of America

10 9 8 7 6 5 4 3 2 1

To Him in whose presence to present oneself is to find oneself.

—*Shashi Phoha*

I dedicate this book to Lisa, Abigail, and Sophia.

—*Thomas F. LaPorta*

For Amy, my darling wife, who put up with me during this project. I love you more than you can imagine.

—*CHG*

CONTENTS

PREFACE

In recent years, interest in sensor networks has evolved from a subject solely of research to one of deployed systems. As new applications have arisen that use sensor networks, the demands on these networks have grown. One of the main challenges with these networks is how to manage missions from application requirements to network operation. By their nature, most sensor networks are mission specific. They therefore place specific requirements on network systems that in turn dictate the use of certain algorithms.

In this book we present the state-of-the art on methods for sensor network operations. This includes algorithms for configuring networks to meet mission requirements and applications in the real world.

FOCUS

The focus of this book is on sensor network operations, starting from translating mission specifications into operational algorithms and applications. The technologies presented are general in terms of the types of transducers and data being gathered.

INTENDED AUDIENCE

This book is intended for researchers and practitioners interested in any aspect of sensor network operations. For researchers, this book serves as a starting point into research in areas such as algorithms for sensor placement, power management, routing, or applications of sensor networks. It may serve as a comprehensive tutorial or be used as the basis for new research projects. For practitioners, this book contains many algorithms, protocols, and experiments that will prove useful in the design and deployment of mission-specific sensor networks.

ORGANIZATION

This book is organized into three parts. The first part provides the motivation and overview of the book and gives a background on sensor platforms and mission-oriented sensor networks. The second part of the book is arranged into seven chapters and presents many algorithms for controlling sensor networks. This includes deployment and localization, mobility management, low layer protocols, routing, power management, data gathering and dissemination, and security. The third part of the book presents several sensor network applications and illustrates how sensor networks may be used.

SHASHI PHOHA
THOMAS LAPORTA
CHRISTOPHER GRIFFIN

January 2006

CONTRIBUTORS

Gordon B. Agnew, University of Waterloo, Ontario, Canada

Anish Arora, Ohio State University, Computer Science & Engineering, Columbus, Ohio

Mahesh Arumugam, Michigan State University, East Lansing, Michigan

Saurabh Bagchi, Purdue University, West Lafayette, Indiana

Kiran S. Balagani, Pennsylvania State University, Mechanical Engineering, State College, Pennsylvania

N. Balakrishnan, Indian Institute of Science, Supercomputing Research Center, Bangalore, Karnataka, India

Pierre Baldi, Center for Pervasive Communications and Computing, Irvine, California

Steve Beck, University of Tennessee, Knoxville, Tennessee

Pratik K. Biswas, Pennsylvania State University, State College, Pennsylvania

Hasan Çam, Arizona State University, Computer Science and Engineering, Tempe, Arizona

Guohong Cao, Pennsylvania State University, State College, Pennsylvania

Krishnendu Chakrabarty, Duke University, Electrical and Computer Engineering, Durham, North Carolina

Chang Wen Chen, Florida Institute of Technology, Electrical Computer Engineering, Melbourne, Florida

Wook Choi, University of Texas at Arlington, Computer Science and Engineering, Arlington, Texas

Kaviraj Chopra, University of Arizona, Electrical and Computer Engineering, Tucson, Arizona

Song Ci, University of Massachusetts, Boston, Massachusetts

Sajal K. Das, University of Texas at Arlington, Computer Science and Engineering, Arlington, Texas

Sridhar Dasika, University of Arizona, Electrical and Computer Engineering, Tucson, Arizona

Hakan Deliç, Bogazici University, Bebek, Istanbul, Turkey

Murat Demirbas, Ohio State University, Computer Science & Engineering, Columbus, Ohio

Roberto Di Pietro, Università, di Roma, Dipartimento di Informatica, Roma, Italy

Tassos Dimitriou, Athens Information Technology, Athens, Greece

Cem Ersoy, Bogazici University, Bebek, Istanbul, Turkey

Mohamed Gouda, University of Texas at Austin, Computer Sciences, Austin, Texas

Stanley Grant, California Institute for Telecommunications and Information Technology, La Jolla, California

Eric Grele, Pennsylvania State University, State College, Pennsylvania

Christopher Griffin, Pennsylvania State University, State College, Pennsylvania

Bechir Hamdaoui, University of Wisconsin, Madison, Wisconsin

Sharif Hamid, University of Nebraska–Lincoln, Computer and Electronics Engineering, Omaha, Nebraska

Kristin Herlugson, Washington State University, Pullman, Washington

Anh Tuan Hoang, National University of Singapore, Electrical & Computer Engineering, Singapore

Alireza Hodjat, UCLA, Electrical Engineering, Los Angeles, California

Fei Hu, Rochester Institute of Technology, Computer Engineering Department, Rochester, New York

Tzonelih Hwang, National Cheng Kung University, Department of Computer Science and Information Engineering, Tainan, Taiwan

S. S. Iyengar, Louisiana State University, Computer Science Department, Baton Rouge, Louisiana

Ekta Jain, University of Texas at Arlington, Electrical Engineering, Arlington, Texas

Xiang Ji, Pennsylvania State University, Department of Computer Science and Engineering, Universtiy Park, Pennsylvania

Holger Junker, Wearable Computing Lab, EE, Zurich, Switzerland

Raja Jurdak, University of California, Irvine, Information and Computer Science, Irvine, California

G. Kesidis, Pennsylvania State University, State College, Pennsylvania

John Koch, Pennsylvania State University, State College, Pennsylvania

Ioannis Krontiris, Athens Information Technology, Athens, Greece

Sandeep Kulkarni, Michigan State University, East Lansing, Michigan

Thomas LaPorta, Pennsylvania State University, State College, Pennsylvania

Bennett Lau, Purdue University, West Lafayette, Indiana

Ha V. Le, Vietnam National University, Computer Science Department, Hanoi, Vietnam

Duke Lee, UC Berkeley, Electrical Engineering and Computer Science, Berkeley, California

Myung J. Lee, City College of New York, Electrical Engineering, New York, New York

Tian-Fu Lee, National Cheng Kung University, Department of Computer Science and Information Engineering, Tainan, Taiwan

Zhiyuan Li, Purdue University, West Lafayette, Indiana

Jie Lian, University of Waterloo, Ontario, Canada

Qilian Liang, University of Texas at Arlington, Electrical Engineering, Arlington, Texas

R. Logananthraj, University of Louisiana at Lafayette, Center for Advanced Computer Studies, Lafayette, Louisiana

Cristina Videira Lopes, University of California, Irvine, Chemical Engineering and Materials Science, Irvine, California

Yung-Hsiang Lu, Purdue University, West Lafayette, Indiana

Stefan Lucks, NEC Europe Ltd., Network Laboratories, Heidelberg, Germany and Mannheim University, Computer Science, Mannheim, Germany

Paul Lukowicz, Electronics Laboratory, EE, Zurich, Switzerland

Bharat Madan, Pennsylvania State University, State College, Pennsylvania

Nipoon Malhotra, Purdue University, West Lafayette, Indiana

Luigi V. Mancini, University of Rome, Department of Information, Rome, Italy

Carter May, Rochester Institute of Technology, Computer Engineering Department, Rochester, New York

Alessandro Mei, University of Rome, Department of Information, Rome, Italy

Mehul Motani, Institute for Infocomm Research, Singapore

Amit U. Nadgar, Pennsylvania State University, Mechanical Engineering, State College, Pennsylvania

Kshirasagar Naik, University of Waterloo, Ontario, Canada

Prashant Nair, Arizona State University, Computer Science and Engineering, Tempe, Arizona

Fotios Nikakis, Athens Information Technology, Athens, Greece

Krishna Nuli, University of Nebraska—Lincoln, Computer and Electronics Engineering, Omaha, Nebraska

Ertan Onur, Bogazici University, Bebek, Istanbul, Turkey

Suat Ozdemir, Arizona State University, Computer Science and Engineering, Tempe, Arizona

Symeon Papavassiliou, New Jersey Institute of Technology, Electrical and Computer Engineering, Newark, New Jersey

Shashi Phoha, Pennsylvania State University, University Park, Pennsylvania

Vir V. Phoha, Louisiana Tech University, Computer Science, Ruston, Louisiana

Raviraj Prasad, University of Nebraska—Lincoln, Computer and Electronics Engineering, Omaha, Nebraska

Hairong Qi, University of Tennessee, Knoxville, Tennessee

Parmeswaran Ramanathan, University of Wisconsin, Madison, Wisconsin

Vaithiyam Ramesh, Rochester Institute of Technology, Rochester, New York

Hamid Sharif, University of Nebraska, Lincoln, Nebraska

Asok Ray, Pennsylvania State University, Mechanical Engineering, University Park, Pennsylvania

Sandip Roy, Washington State University, Pullman, Washington

Harshavardhan Sabbineni, Duke University, Electrical and Computer Engineering, Durham, North Carolina

Ali Saberi, Washington State University, Pullman, Washington

H. Ozgur Sanli, Arizona State University, Computer Science and Engineering, Tempe, Arizona

Vrudhula Sarma, University of Arizona, Electrical and Computer Engineering, Tucson, Arizona

Guna Seetharaman, Air Force Institute of Technology, Department of Electrical and Computer Engineering, Wright Patterson AFB, Ohio

Sivakumar Sellumuthu, Wayne State University, Computer Science, Detroit, Michigan

Sengupta Raja, UC Berkeley, Electrical Engineering and Computer Science, Berkeley, California

Kewei Sha, Wayne State University, Computer Science, Detroit, Michigan

Weisong Shi, Wayne State University, Computer Science, Detroit, Michigan

Waqaas Siddiqui, Rochester Institute of Technology, Rochester, New York

Yuldi Tirta, Purdue University, West Lafayette, Indiana

Gerhard Troester, Wearable Computing Lab, EE, Zurich, Switzerland

Pravin Varaiya, UC Berkeley, Civil Engineering, Berkeley, California

Raviteja Varanasi, Pennsylvania State University, Mechanical Engineering, State College, Pennsylvania

Ingrid Verbauwhede, UCLA, Electrical Engineering, Los Angeles, California

Xiaoling Wang, University of Tennessee, Knoxville, Tennessee

André Weimerskirch, Ruhr-University Bochum, Communication Security, Bochum, Germany

Hsiang-An Wen, National Cheng Kung University, Department of Computer Science and Information Engineering, Tainan, Taiwan

Dirk Westhoff, NEC Europe Ltd., Network Laboratories, Heidelberg, Germany and Mannheim University, Computer Science, Mannheim, Germany

Min Wu, University of Missouri–Columbia, Electrical Engineering, Columbia, Missouri

Jie Yang, New Jersey Institute of Technology, Electrical and Computer Engineering, Newark, New Jersey

Bin Yao, Pennsylvania State University, State College, Pennsylvania

Erik Zenner, NEC Europe Ltd., Network Laboratories, Heidelberg, Germany and Mannheim University, Computer Science, Mannheim, Germany

Hongyuan Zha, Pennsylvania State University, Department of Computer Science and Engineering, State College, Pennsylvania

Jianliang Zheng, City College of New York, Electrical Engineering, New York, New York

Jin Zhu, New Jersey Institute of Technology, Electrical and Computer Engineering, Newark, New Jersey

I

SENSOR NETWORK
OPERATIONS OVERVIEW

1

OVERVIEW OF MISSION-ORIENTED SENSOR NETWORKS

1.1 INTRODUCTION

Sensor networks represent a new frontier in technology that holds the promise of unprecedented levels of autonomy in the execution of complex dynamic missions by harnessing the power of many inexpensive electromechanical microdevices. Miniature sensing and computational devices, often embedded in wireless electromechanical platforms, are being developed to interact directly with the physical world. Spanning time and space, and cognizant of a common mission, they monitor changes in the operational environment and collaborate to actuate distributed tasks in dynamic and uncertain environments. Dispersed over a hostile battlefield, these devices may self-organize to act as numerous eyes and ears of soldiers surveying the field from a safe distance. Embedded in unmanned air vehicles, they may monitor bio/chemical plumes in the atmosphere or handle hazardous materials on the ground. Mobile robots with embedded sensor systems explore the surface of Mars; and integrated systems of undersea robots are being designed to hunt for mines in shallow water and to develop high fidelity now casts and forecasts of the ocean through time–space coordinated sampling. Sensor networks are expected to play an important role in transportation management and safety and in medical applications. More commonplace applications include fine-grain monitoring of indoor environments, buildings, and home appliances. In general, the next phase of automation calls on networks of sensors to take on the dull, dirty, and dangerous functions of human interest, accomplishing them with the perception and adaptation of humans, in collaboration with humans. As a system of interacting sensor nodes, a sensor network is a human-engineered, complex dynamic system that must combine its understanding of the physical world with its computational and control functions and operate with constrained resources. As a distributed dynamic system, these tiny distributed devices must collectively comprehend the time evolution of physical and operational phenomena and predict their effects on mission execution and then actuate control actions that execute common high-level mission goals.

Sensor Network Operations, Edited by Phoha, LaPorta, and Griffi

Copyright © 2006 The Institute of Electrical and Electronics

This book presents new advances for engineering and operating sensor networks to meet specified mission goals. Prior to deployment, these mission-oriented sensor networks (MoSNs) need to be endowed with distributed high-level representations of mission specifications that can be dynamically executed by harnessing the collective powers of distributed sensor/actuator nodes in unknown or uncertain environments. Collaborative intelligent inference is necessary to circumvent limitations of sensor data, communications, and equipment faults. Emergent behaviors and phase transitions must be modeled, predicted, and controlled.

1.2 TRENDS IN SENSOR DEVELOPMENT

Shashi Phoha and Thomas LaPorta

Sensors of physical phenomena with integrated servomechanisms have been commonplace throughout the latter half of the twentieth century controlling thermostats and valves, monitoring flow or adapting to changes in pressure or stress, and providing alarms for fire or flooding. As dynamic systems, they have been expected to perform these and many other localized isolated tasks with precision and reliability. These applications relied on statically positioned sensors designed to operate independently for long periods of time (months to years) with nonrenewable power supplies. Traditional sensor technology was characterized by large transducers, highly capable processing platforms, and complex signal and data processing software. These characteristics limited the types of applications that could make use of sensor technology. Sensor technology has matured resulting in smaller and more efficient transducers, processing platforms, and communication modules. In addition, the communications capabilities of sensors have greatly improved to allow large-scale networks of sensors to be formed. These advancements have paved the way for a much broader set of applications of sensors.

The present-day demands on sensor networks entail comprehensive perception of locally sensed changes in the physics of the environment and adaptive time–space coordinated control of individual servomechanisms in support of a common mission [1]. The state-of-the-art in sensor technology now supports extremely small sensors that may be highly mobile and power efficient and are equipped with sufficient computing capabilities to run distributed algorithms to manage their motion, process data, and form and manage networks. As a result, algorithms for managing sensor networks, and the applications that use them, have grown increasingly complex. In this book we provide a collection of studies that represent a comprehensive treatment of the current state of research with respect to sensor networks. We provide a brief overview of the state-of-the art in sensor platforms and algorithms dealing with the computational infrastructure issues for sensor networks.

1.2.1 Sensor Platforms

Sensor platforms are comprised of four main components: transducers, a hardware computing platform, an operating system, and communication modules. The transducers are responsible for monitoring an area of interest and gathering data. The computing platform and operati responsible for processing and formatting data received from the application that is analyzing data from the sensor field.

These modules also run control algorithms to move sensors, form networks, aggregate data, and perform security functions. The hardware computing platform typically consists of a central processing unit (CPU), memory, and input/output (I/O) ports. The operating system runs on the computing hardware and is used to provide a software interface to the hardware and to provide a degree of programmability. The communication module has two functions. First, it provides an interface for the transducers to transmit their gathered data into the computing platform. Second, it is used to transmit data back to a server where it is analyzed along with data received from other sensors. This module may include multiple I/O interfaces. Today, wireless interfaces are becoming the dominant communication technology for sensor networks because of their ease of deployment and reduced cost.

There are a tremendous number of research efforts on transducer technologies that are beyond the scope of this book. Overviews of some major efforts can be found at http://www.cens.ucla.edu/ and http://www.el.utwente.nl/tt/. This research has resulted in miniature transducers for sensing many types of phenomena, thus placing the onus on the designers of computing platforms, operating systems, and communication modules to reduce the size and cost of their components so that entire sensor packages are small. In the following subsections we review the prevailing technology in the hardware computing platform, operating systems, and communication modules.

Computing Hardware   By far, the most popular processing platform for small sensor devices is based on the so-called Mote hardware that was developed at the University of California at Berkeley. This family of hardware platform has been productized by Crossbow (www.xbow.com) as the MICA product line. The platforms are characterized by small size, power efficiency, and very limited CPU and memory capabilities when compared to conventional processing platforms, such as desktop personal computers (PCs). However, despite their limitations, they provide a highly capable system for developing sensor applications in a form factor that may be used in many harsh, inaccessible environments.

The MICA 2 Mote weighs 0.7 ounces and is $58 \times 32 \times 7$ mm, making it ideal for deployment for many applications that require very small sensors. The MICA 2 has 128 kbytes of program memory and 512 kbytes of memory for storing samples. The MICA 2 has a 10-bit analog-to-digital (A/D) converter, so it can store over 100,000 samples in its memory. The MPR400CB processor board runs the communications and networking protocols simultaneously with application software. The MICA 2 has a 51-pin expansion connector to allow it to interface with many types of external transducers. It also supports several internal transducer cards. It draws 8 mA while active, and less than 15 μA while in sleep mode. We will discuss the radio capabilities of the MICA 2 in the next subsection, but it is designed to be deployed in large-scale sensor networks of over 1000 nodes. If a smaller platform is required, the MICA 2DOT has capabilities similar to the MICA 2, but has a form factor of approximately the size of a quarter, or a thickness of 6 mm and a diameter of 25 mm. The major difference between the MICA 2 and the MICA 2DOT is that the MICA 2DOT offers far fewer I/O connections. It has 18 pins for connecting external peripherals. It is clear from the description of the hardware computing platform, that while the presence of a CPU and memory allows for many types of algorithm to execute, they must be specially designed to account for the hardware limitations. We illustrate this point with three examples. First, consider security. In most Internet environments, security is provided using encryption using either DES or AES. The DES algorithm requires about 20 kbytes of memory to store the program if written in C, and approximately another 20 kbytes of memory to store the variable used during its run-time operation. Therefore, it would occupy almost one-third of the available memory on the platform, making it infeasible to use. Solutions include using hardware support for encryption/decryption or using simpler algorithms. These choices present trade-offs in terms of hardware complexity, power consumption, and overall strength of security. Second, consider routing. In an Internet environment, routing is performed using proactive protocols that exchange link state or distance vectors, requiring large tables to be stored in individual routers. These tables included next-hop routes for all destinations. In a large sensor network, if sensor nodes forward data for each other, these tables will become prohibitively large to store on memory-limited sensor nodes. Therefore, new routing algorithms and protocols must be developed.

Finally, the operating system itself is typically several megabytes on a conventional computing platform. Given the limited memory on a Mote, new operating systems must be defined, as discussed in the next subsection. Another example of very simple sensor nodes are RFID tags, which are often passive devices with no power or computing capabilities. Active badges, such as those developed in the iBadge project (http://nesl.ee.ucla.edu/projects/ibadge/) at UCLA are another example. The iBadge is 2.3 ounces and has a lifetime of over 4 h. It uses BlueTooth for radio communication and has on-board localization and speech processing capabilities.

In addition to simple end devices, much more capable sensor computing platforms exist. These are typically used as gateways to aggregate traffic from simple sensors to a backbone network or operate in controlled environments with persistent power supplies. One example is the Crossbow Stargate XScale Network Interface and Single Board Computer. The Stargate runs the Linux operating system and provides USB and PCMCIA and Ethernet interfaces. Another example is the Sensoria (http://www.sensoria.com/) sGate. Like the Stargate, the sGate runs Linux. It has a 32-bit 300-MIPs processor. Essentially, these are general-purpose processors that can perform complex functions to support security, routing, and data processing.

Operating Systems   Operating systems for sensor nodes must be very lightweight and occupy only a small amount of memory. Because sensor applications have many common characteristics, the operating system design can be very specialized. The operating systems most commonly used across a wide range of sensor platforms is the TinyOS, which was developed as part of the Smart Dust project at Berkeley, the same project that led to the Mote. While the Mote has been productized by Crossbow, the TinyOS is maintained as open source by the research group at Berkeley and has a very large user community. Details of the TinyOS, the source code, and a list of platforms that support its use are available at http://webs.cs.berkely.edu/tos.

The TinyOS is designed to support event-driven applications. In addition, it supports concurrency so that many events may be monitored simultaneously. These two characteristics are the most important user features of the OS. It is designed to run with minimal support from hardware, thus enabling sensor computing platforms to use simple, low-power devices. TinyOS supports programming in a language very similar to C. More capable sensor nodes, such as the Stargate and sGate discussed above, often use off-the-shelf operating systems, such as Linux.

Communication Modules   As stated earlier, the communication modules of sensor platforms support both reading data from transducers and communication links that are used to form a network for passing sensor data back to a server for processing. Wireless is the most popular media for sensor networks. Much research is still ongoing to determine the best wireless communications technology and low layer access protocols to be used in sensor networks. Considerations include transmission range, power consumption, bandwidth, and traffic types to be supported. Whereas many sensor applications of disparate type have migrated to the Mote platform and TinyOS for a computing platform, because these applications have vastly different data transmission requirements, several radio technologies are still under consideration. For example, many sensor applications assume that sensors will be densely deployed and that low-bit-rate telemetry or event reporting will be transmitted across the network. For these applications, a low-power, low-bit-rate radio suffices because sensors may relay traffic for each other, and not much data is being transmitted. On the other hand, sensor networks that support applications that include the transmission of images or video streams when an event of interest is detected, must support the transmission of high-bit-rate, bursty data. These sensors require the use of radios more typical of wireless local area networks. Because power consumption of wireless transmission may be high, the radio interfaces tend to be more specialized with respect to applications than the computing hardware platform or operating system.

The MICA 2 sensors use radios that operate in the ISM band, specifically at 868, 916, 315, or 315 MHz. Depending on the model, between 4 and 50 channels are supported on a single platform. Data is transmitted at 38.4 kbaud using Manchester encoding. These radios work at low power, 25 to 27 mA, for transmitting at maximum power, 8 to 10 mA to receive, and less than 1 μA while in sleep mode. Their outdoor transmission range is 500 to 1000 ft. One ongoing research effort to produce a much lower power radio is the PicoRadio project at Berkeley. Details can be found at http://bwrc.eecs.berkeley/Research/PicoNet. The goal of this project is to produce a radio that costs less than 50 cents and draws less than 5 nJ per correctly transmitted bit. In fact, the goal is to design the overall node to be so low power that it can scavenge energy from the environment through vibrations or other means. A second direction for radio advancements for sensor networks is through the 802.15.4 standard. Ember (http://www.ember.com/) has a commercially available version of a radio designed for sensor networks based on this technology. The radio is 7×7 mm, has a range of 75 m, and supports 128-bit AES encryption. The radio operates in the 2.4 GHz ISM band and supports up to 16 channels with 5-MHz spacing per channel. Data is transmitted at 250 kbps using OQPSK Direct Sequence Spread Spectrum. The power consumption is similar to that of the MICA 2 radio 20.7 mA to transmit, 19.7 mA to receive, and 0.5 μA while idle. Other wireless interfaces are also popular in sensor networks, including well-known standards such as BlueTooth and 802.11.

Sensor Platform Summary   As we have discussed, a very small form factor for sensor nodes is critical for many applications. To meet these requirements great innovations have been applied to transducers, computing hardware, operating system, and communication design. These systems are now commercially available from several companies. With the ability to support more complex applications, more complex algorithms to support these applications are required to run in the sensor nodes. Even with the advances in sensor platform technology, the resulting platforms are still quite limited compared to desktop and server computing platforms. For this reason, much research is ongoing in designing and implementing these algorithms with high efficiency.

1.2.2 Sensor Network Algorithms

Many algorithms and protocols execute in sensor nodes to fulfill the mission of the networked sensing system. These algorithms must first enable dispersed sensors to form a network, determine their locations, and reconfigure or perhaps move to reposition so that the system may fulfill its mission. They must allow sensor nodes to efficiently gather data, access transmission media, communicate information to distant nodes, and disseminate information that has been learned. Depending on the type of application, different levels of security must be provided to protect the integrity and privacy of the data being gathered and disseminated. Finally, all of these algorithms must be designed with power efficiency in mind.

1.3 MISSION-ORIENTED SENSOR NETWORKS: DYNAMIC SYSTEMS PERSPECTIVE

Shashi Phoha

For executing complex time-critical missions, a sensor network may be viewed as a distributed dynamic system with dispersed interacting smart sensing and actuation devices that may be embedded in mobile or stationary platforms. A sensor network operates on an infrastructure for sensing, computation, and communications, through which it perceives the time evolution of physical dynamic processes in its operational environment. A mission-oriented sensor network (MoSN) is such a dynamic system that has also been endowed with a high-level description of the goals of a specific mission. The MoSN nodes accept inputs from interacting nodes for situation awareness and participate in individual or cluster-wide dynamic adaptation to meet mission goals. Advances in integrated wireless communications, fast servocontrolled sensors/actuators, and micro- and nanotechnologies, have enabled large-scale integration of inexpensive computational and sensing devices that can be spatially dispersed for distributed monitoring of physical phenomena. With intelligent mechanisms for self-organization and adaptation, the sensor network can take on many functions of human interest with the perception and adaptation of humans. The interactive nonlinear and multi-time-frame dynamics of the resulting systems can approach the complexity of biological systems.

Part II of the book covers recent research developments relating to the computational, communications, and networking designs of MoSNs that provide an adaptive infrastructure for dependable data collection for real-time control and actuation. In harnessing the true potential of networked sensors, a perceptive infrastructure is needed that adapts to the dynamics of the mission. The infrastructure enables these dynamically self-reconfigurable and introspective networks of possibly mobile sensor nodes to be capable of understanding and interpreting mission objectives and adapting to the dynamics of harsh and often unknown physical environments. These tiny distributed devices must collectively comprehend the time evolution of physical phenomena and their effect on mission execution to close the distributed feedback control loop.

Part III of the book presents a wide range of pragmatic applications that are enabled by sensor networks. Multiple types of sensors are involved: acoustic, video, wearable context-sensitive sensor nodes, and even multimodal sensor nodes. A system of wearable sensors is described for context recognition in human subjects. An unmanned underwater sensor

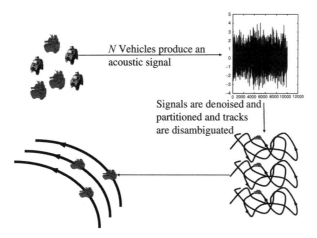

Figure 1.1 High-level behavior recognition and prediction using a distributed sensor network.

network is designed for autonomous undersea mine hunting operations. An experimental sensor network deployed in the dessert for tracking vehicles shows significant performance degradation due to environmental noise. Algorithms are developed for its autonomous adaptation to environmental noise. Soft-sensing techniques are presented to robustly operate networked robotic sensors to autonomously detect and mitigate effects of emerging software failures, like memory leak or mutex lock that result in erratic behavior of the system. The chapters in this part depict the broad potential of sensor networks to achieve the next level of automation in pragmatic applications.

The book addresses major research issues for designing and operating MoSNs. The development of a perceptive infrastructure for dependable data collection for human interpretation is the first concern. In harnessing the true potential of networked sensors, however, this is only the first step. In order to autonomously execute complex adaptive missions while comprehending and adapting to the dynamics of harsh and often unknown physical environments, these tiny distributed devices must collectively comprehend the time evolution of physical phenomena and their effect on mission execution and activate action to close the distributed feedback control loop. To thus endow the esprit de corps on isolated computational electromechanical devices, much more is needed. For example, in Figure 1.1, the acoustic signals emanating from a set of target vehicles in a noisy environment may be denoised and collaboratively processed by a network of acoustic sensors, using dynamic space–time clustering and beam-forming techniques [1–3]. Signal partitioning may be used to determine and predict the individual random mobility patterns of each targeted vehicle. However, a higher level of comprehension of mission goals is needed if the sensor network is called upon to understand and predict coordinated movement in formation, a behavior that may be of significant more interest to the execution of the mission. If the sensor network must act as the eyes and ears of humans, allowing them to stay at a safe distance from a dangerous battlefield, it must dependably comprehend the criticality of its sensor perceptions and responses to mission execution and convey these proficiently to humans for time-critical interaction. There is simply no time for humans to receive and analyze a data sheet plotting locations, speed, and direction of movements of individual vehicles and to infer and deter movement in formation.

The design and operation of sensor networks calls for the confluence of computational sciences with physical sciences and with decision and control sciences [4]. Physical sciences model the nonlinear dynamics of physical phenomena. Sensor networks, as distributed dynamic systems, must comprehend and predict the effects of emerging phenomena on mission execution and actuate control actions to successfully execute mission specifications. Prior to deployment, sensor networks need to be endowed with distributed high-level representations of mission specifications that can be dynamically executed by harnessing the collective powers of distributed sensor/actuator nodes in unknown or uncertain environments. The first phase of this research is presented in this book. Research challenges still abound. Advances in symbolic dynamics are needed to identify atomic physical events in sensor data that capture the causal dynamics of the underlying nonlinear processes and abstract event sequences that associate the time evolution of these processes to mission specifications at various levels of fidelity. Advances in nonlinear dynamic systems for nonlinear modeling and control of distributed multi-time-scale processes are needed to enable individual sensors to comprehend the higher level dynamics and respond to global changes. Collaborative intelligent inference is necessary to circumvent limitations of sensor data, communications, and equipment faults. Emergent behaviors and phase transitions need to be modeled, predicted, and controlled. These dynamically self-reconfigurable and introspective networks of mobile sensor nodes must be capable of understanding and interpreting mission objectives and adapting their behaviors. Sensor networking technology as a true extension of ourselves as the eyes and ears in the field calls for a collective intelligence that comprehends the distributed images and sounds to ascertain and enable executable action and actuation.

REFERENCES

1. I. F. Akyildiz, W. Su, Y. Sankarasubramaniam, and E. Cayirci, "A survey on sensor networks," *IEEE Communications Magazine*, vol. 40, no. 8, pp. 102–114, 2002.

2. S. Phoha, N. Jacobson, and D. Friedlander, "Sensor network based localization and target tracking through hybridization in the operational domains of beamforming and dynamic space-time clustering," in *Proceedings of the 2003 Global Communications Conference*, San Francisco, Dec. 1–5, 2003.

3. K. Yao, R. E. Hudson, C. W. Reed, D. Chen, and F. Lorenzelli, "Blind beamforming on a randomly distributed sensor array system," *IEEE Journal on Selected Areas in Communications*, vol. 16, pp. 1555–1657, 1998.

4. S. Phoha, "Guest editorial: Mission-oriented sensor networks," *IEEE Transactions on Mobile Computing*, vol. 3, no. 3, pp. 209–210, 2004.

II

SENSOR NETWORK DESIGN AND OPERATIONS

2

SENSOR DEPLOYMENT, SELF-ORGANIZATION, AND LOCALIZATION

2.1 INTRODUCTION

A key attribute of sensor networks is to be able to self-form, that is, when randomly deployed to be able to organize into an efficient network capable of gathering data in a useful and efficient manner. Often, gathering data in a useful manner requires that the exact location of a sensor be known. This requires that sensors be able to determine their location. This location information is often reused for other purposes. For example, once sensors know their location, and that of their neighbors, redundant sensors can be powered down to save energy. Likewise, low-energy communication paths may be established between nodes. Coverage holes in the sensor network may be uncovered and, through mobility, healed. In this chapter we address issues of network formation, including localization.

Sensor positioning problems are a critical area of research for sensor network operations. Sensor networks are useless if their configurations are not robust to power degradation or they are prone to breach by the very objects they are designed to detect. Many distributed algorithms rely on sensors with accurate knowledge of their position. While this can be achieved by providing each sensor with a Global Positioning System (GPS) unit, this is not always possible or desirable. Hence, internal localization algorithms are required. This chapter explores the issues of sensor placement for robust and scalable target detection and sensor node localization over large distances.

Section 2.2 by Sabbineni and Chakrabarty describes a fully distributed algorithm for exploiting redundancy in sensor networks to maintain connectivity and coverage in response to power degradation. When active nodes fail due to energy depletion or other reasons such as wearout, SCARE replaces them appropriately with inactive nodes.

Section 2.3 by Ji and Zha studies some situations where most existing sensor positioning methods tend to fail to perform well. It then explores the idea of using dimensionality reduction to estimate sensors coordinates in space; a distributed sensor positioning method based on multidimensional scaling technique is proposed.

Sensor Network Operations, Edited by Phoha, LaPorta, and Griffin
Copyright © 2006 The Institute of Electrical and Electronics Engineers, Inc.

The location estimation or localization problem in wireless sensor networks is to locate the sensor nodes based on ranging device measurements of the distances between node pairs. A distance is *censored* when the ranging devices are unreliable and the distance between transmitting and receiving nodes is large. Section 2.4 by Lee, Varaiya, and Sengupta compares several approaches for estimating censored distances with a proposed strategy called trigonometric k clustering.

Section 2.5 by Onur, Ersoy, and Deliç cedilla considers the sensing coverage area of surveillance wireless sensor networks. The sensing coverage is determined by applying the Neyman–Pearson detection model and defining the breach probability on a grid-modeled field. Weakest breach paths are determined using Dijkstra's algorithm.

The discussions in this chapter enhance the state of the art in sensor network operations by presenting solutions to the problems of sensor placement and localization. These results can be used to prolong the life of deployed sensor networks, enhance the quality of service of perimeter networks, and provide introspection necessary for sensor localization in highly distributed networks.

2.2   SCARE: A SCALABLE SELF-CONFIGURATION AND ADAPTIVE RECONFIGURATION SCHEME FOR DENSE SENSOR NETWORKS

Harshavardhan Sabbineni and Krishnendu Chakrabarty

We present a distributed self-configuration and adaptive reconfiguration scheme for dense sensor networks. The proposed algorithm, termed self-configuration and adaptive reconfiguration (SCARE), distributes the set of nodes in the network into subsets of coordinator and noncoordinator nodes. Redundancy is exploited not only to maintain the coverage and connectivity provided by sensor deployment but also to prolong the network lifetime. When active nodes fail due to energy depletion or other reasons such as wearout, SCARE replaces them appropriately with inactive nodes. Simulation results demonstrate that SCARE outperforms the previously proposed Span method in terms of coverage, energy usage, and the average delay per message.

2.2.1   Background Information

Advances in miniaturization of microelectronic and mechanical structures (MEMS) have led to battery-powered sensor nodes that have sensing, communication, and processing capabilities [1, 2]. Wireless sensor networks are networks of large numbers of such sensor nodes. Example applications of such sensor networks include the monitoring of wildfires, inventory tracking, assembly line monitoring, and target tracking in military systems. Upon deployment in a remote or a hostile location, sensor nodes might fail with time due to loss of battery power, an enemy attack, or a change in environmental conditions. The replacement of each failed sensor node with a new sensor node is expensive and often infeasible, and it is therefore undesirable. Hence in such cases, a large number of redundant sensor nodes are deployed with the expectation that these nodes will be used later when some other nodes fail. The self-configuration of a large number of sensor nodes requires a distributed solution. In this section, we present a scalable self-configuration and an adaptive reconfiguration (SCARE) algorithm for distributed sensor networks.

An effective self-configuration scheme should have the following characteristics. It should be completely distributed and localized because a centralized solution is often not scalable for wireless sensor networks. It should be simple without excessive message overhead because sensor nodes typically have limited energy resources. It should be energy-efficient and require only a small number of nodes to stay awake and perform multihop routing, and it should keep the other nodes in a sleep state.

We propose a solution that meets the above design requirements. We present a distributed self-configuration scheme that distributes the set of nodes in the sensor network into subsets of coordinator nodes and noncoordinator nodes. While coordinator nodes stay awake, provide coverage, and perform multihop routing in the network, noncoordinator nodes go to sleep. When nodes fail, SCARE adaptively reconfigures the network by selecting appropriate noncoordinator nodes to become coordinators and take over the role of failed coordinators. This scheme only needs local topology information and uses simple data structures in its implementation.

2.2.2 Relevant Prior Work

A number of topology management algorithms have been proposed for ad hoc and sensor networks [3–6]. While the connectivity problem has been studied in considerable detail for wireless ad hoc networks, less attention has been devoted to the problem of balancing connectivity and coverage. The GAF scheme [4] uses geographic location information of the sensor nodes, and it divides the network into fixed-size virtual square grids. GAF identifies redundant nodes within each virtual grid and switches off their radios to achieve energy savings. In contrast, SCARE achieves energy savings by selectively powering down some of the nodes that are within the sensing radius of a coordinator. A coverage-preserving node scheduling scheme is described in [7] that extends the LEACH [8] protocol to achieve energy savings. In this scheme, nodes advertise their position information in each round. Each node evaluates its eligibility to switch itself off by calculating its sensing area and comparing it with its neighbors's. If a node's sensing area is embraced by a union set of its neighbors's, then it turns itself off. To prevent blind spots in coverage due to several eligible nodes switching themselves off simultaneously, a back-off-based scheduling is used. After the back-off interval has elapsed, nodes broadcast a status advertisement message to let other nodes know about their on/off status. Thus, each node broadcasts two messages in this scheme. In contrast, SCARE needs fewer than two messages per node on average during its operation. The scheme in [7] also utilizes location information of the nodes for its operation. SCARE only needs an estimate of the distance between the nodes.

The STEM scheme described in [6] trades off latency for energy savings by putting nodes aggressively to sleep and waking them up only when there is data to forward. It uses a second radio operating at a lower duty cycle for transmitting periodic beacons to wake up nodes when there is data to forward. SCARE does not use a separate paging channel for self-configuration. Nevertheless, SCARE can integrate well with STEM to achieve significant energy savings.

In AFECA [9], nodes listen to the channel for transmissions. AFECA conservatively tries to keep nodes awake when there are not too many neighbors in its radio range. In order to deduce this information, each node has to listen to transmissions that are not meant for it. In SCARE, however, nodes listen at only periodic intervals in order to determine their states.

The PAMAS [10] multiaccess protocol saves power by switching off the radio of a node when it is not transmitting or receiving. This method saves power when idle listening consumes significantly less energy compared to message reception.

The Span approach [5] appears to be the most closely related to SCARE. Span attempts to save energy by switching off redundant nodes without losing the connectivity of the network. Nodes make decisions based on their local topology information. However, SCARE differs from Span in that it uses distance estimates to determine the state of a node. Span uses a communication mechanism to obtain this information. Since Span was developed for ad hoc networks, its main focus is on ensuring network connectivity through energy-efficient topology management. It is not directed toward ensuring the sensing coverage of a given region. SCARE also differs from Span in that, in addition to ensuring network connectivity and low-energy self-configuration, it attempts to provide a high level of sensing coverage.

A TDMA-based self-organization scheme for sensor networks is presented in [11]. Each node uses a superframe, similar to a TDMA frame, to schedule different time slots for different neighbors. However, this scheme does not take advantage of the redundancy inherent in wireless sensor networks to power off some nodes.

SCARE utilizes a localization scheme for periodic transmission of beacon signals and for the synchronization of the clock signals of sensor nodes. A number of such localization schemes have been proposed in the literature for sensor networks [1–3]. These schemes use a special set of nodes, called the reference nodes, that transmit beacon signals to let the sensor nodes self-estimate their position. The approach in [12] is based on the received signal strength from the reference nodes to carry out location estimation of the sensor nodes. It is shown that despite fading and mobility, a small window average is sufficient to do location estimation.

Traditionally, GPS [13] receivers are used to estimate positions of the nodes in mobile ad hoc networks. However, their high cost and the need for more precise location estimates make them unsuitable for sensor networks. It is expensive to add GPS capability to each device in dense sensor networks.

In [14], a scheme is presented to estimate the relative location of nodes using only a few GPS-enabled nodes. It uses the received signal strength information (RSSI) as the ranging method. Whitehouse and Culler [15] use an ad hoc localization technique called *Calamari* in combination with a calibration scheme to calculate distance between two nodes using a fusion of radio-frequency (RF)-based RSSI and acoustic time of flight (TOF). Acoustic ranging [16] can also be used to get fine-grained position estimates of nodes.

Finally, several clustering techniques have been proposed in the ad hoc networking literature. Vaidya et al. [17] propose a scheme that attempts to find maximal cliques in the physical topology and use a three-pass algorithm to find the clusters. Although this scheme finds a connected set of clusters, it consumes a significant amount of energy during clustering and cannot be directly applied to sensor networks. The adaptive clustering scheme proposed in [18] uses node identifications (IDs) to build two-hop clusters in a deterministic manner. SCARE differs from this scheme in two ways. First, the main goal of SCARE is to use distance information to power down redundant sensor nodes, whereas in [18] node IDs are used to provide better QoS guarantees by clustering nodes. Second, in [18], as in [17], energy efficiency is a secondary concern. In [19], clustering schemes for both static and mobile networks are proposed. However, there is no provisioning for switching off redundant nodes in these schemes. Thus, [19] cannot be directly applied to sensor networks.

On the other hand, SCARE is specifically designed for sensor networks to take advantage of their inherent redundancy.

2.2.3 Outline of SCARE

SCARE is a decentralized algorithm that distributes all the nodes in the network into subsets of coordinator nodes and noncoordinator nodes. While the coordinator nodes stay awake and provide coverage and perform multihop routing in the network, noncoordinator nodes go to sleep. Noncoordinator nodes wake up periodically to check if they should become coordinators to replace failed coordinators.

SCARE achieves four desirable goals. First, it saves energy by selecting only a small number of nodes as coordinators and putting other nodes to sleep. Second, it uses only local topology information for coordinator election and hence is highly scalable. Third, it provides nearly as much sensing coverage compared to the coverage obtained if all the nodes are awake. Finally, it preserves network connectivity by using a protocol based on CHECK and CHECK_REPLY messages. We next describe a basic scheme for self-configuration. The basic scheme will subsequently be extended to prevent network partitions.

Basic Scheme   In self-configuration based on SCARE, each node starts by generating a random number with uniform probability between 0 and 1. A node becomes eligible to be a coordinator if the random number thus generated is greater than a threshold (say 0.9). Therefore, a very small percentage of the nodes actually become coordinators. Other nodes just wait and listen. The threshold value can be preset depending on the application. A higher value for the threshold results in a small number of initial coordinator nodes. This has the effect of delaying the convergence of the self-configuration algorithm, but it might result in a better selection of coordinator nodes. On the other hand, a low value for the threshold implies that a high number of coordinator nodes are selected randomly in the beginning. This hastens the convergence of the protocol although a larger number of coordinator nodes may be selected.

A node that is eligible to be a coordinator waits for a random amount of time before declaring itself to be a coordinator by broadcasting a HELLO message. This wait time, for example, can be chosen from a uniform distribution of values between T and NT where T is a preset slot time and N is the number of neighbors of the node that are coordinators. Initially N can be chosen to be a constant, for example, 6. This prevents excessive contention on the wireless channel that might result if all the nodes decide to become coordinators at once.

Upon receipt of a HELLO message, a sensor node compares its distance from the sender C of the HELLO message to its sensing range s. A node within a distance s from a coordinator immediately becomes a noncoordinator node and stores the ID of the node that sent the HELLO message in its local cache. A node that is at a distance greater than s from C but within transmission range r becomes eligible to be a coordinator node. This is shown in the Figure 2.1. The shaded region in the figure represents the sensing range of the node C. The outer circle represents the transmission range of the sensor node. Here, we assume that the sensing radius is smaller than the transmission radius. This is often the case for sensors in a sensor node [20].

While SCARE assumes the presence of an appropriate localization mechanism [1–3], exact distance calculations are not necessary. We show later that a moderate error in distance estimation has little effect on the outcome of the self-configuration procedure.

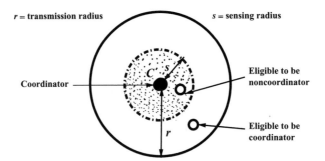

Figure 2.1 Sensing and transmission radii of a node.

Network Partitioning Problem The basic scheme described above can sometimes result in a partitioning of the network; see Fig. 2.2*a*. Here, coordinator node *F* makes node *A* a non-coordinator. However, coordinator node *D* can communicate with *F* only through *A*. This can potentially result in the partitioning of the network if coordinator (active) node *D* is unable to reach any other active nodes. As a result, network connectivity cannot be guaranteed in this situation. In Figure 2.2*b*, *G* and *K* are coordinator nodes and *B* and *C* are noncoordinator nodes. This situation again results in network partitioning as nodes *G* and *K* cannot reach each other.

To prevent such situations, we extend the basic SCARE scheme outlined above, which results in a more effective technique for self-configuration. In the basic scheme, if there is a network partition, a node might never receive a HELLO message. This results in the node waiting eternally for a HELLO message, which results in wastage of energy. Hence, we choose a time-out value T_{off} after which the nodes that are still undecided about their state can become eligible to become coordinator nodes. The time-out value can be chosen based on the probability threshold discussed in Section 2.2.3. A lower value for the threshold means that the procedure converges quickly and needs a lower T_{off} value and vice versa.

To prevent the network partitioning that occurs due to the pathological cases shown in Figure 2.2, a node that initially receives a HELLO message from a coordinator node does not become a noncoordinator immediately and go to the sleep state. Instead, it continues to listen for messages from other coordinator nodes and remains in the "eligible to be a

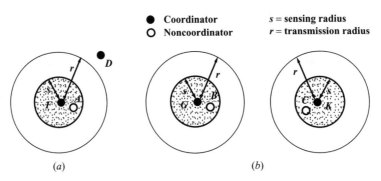

Figure 2.2 Network partitions in the basic scheme: (*a*) first scenario and (*b*) second scenario.

noncoordinator" (ETNC) state. A sensor node that is in the ETNC state can become a coordinator node in two cases:

1. If it can connect two neighboring coordinator nodes[1] that cannot reach each other in one or two hops. It can deduce this information from the HELLO messages it received earlier. As shown in Figure 2.2a, node A, which is in the ETNC state, receives HELLO messages from node F and node D and decides to become a coordinator; this eliminates the partition.
2. If it can connect two neighboring coordinator nodes that cannot reach each other in one or two hops via a node in the ETNC state. As shown in Figure 2.2b, nodes B and C, that are in the ETNC state, receive HELLO messages from nodes G and K, respectively, and decide to become coordinators as there is no match between the node lists of G and K; this eliminates the partition.

To achieve this, each ETNC node sends a CHECK message. This CHECK message contains the neighbor list of the coordinator node that caused this node to be in the ETNC state. Intuitively, this case is more likely to occur if there are few coordinators in the vicinity and less likely if there are more coordinator neighbors. Any ETNC node that receives this CHECK message replies with a CHECK_REPLY message and becomes a coordinator if there is no node common to the neighbor lists of both the nodes. Upon receipt of the CHECK_REPLY message, the node that sent the CHECK message also becomes a coordinator. In Figure 2.2b, noncoordinator node B sends a CHECK message and gets a CHECK_REPLY message from node C. Both nodes B and C, therefore, become coordinator nodes. This procedure removes the network partition.

If the HELLO message is received from the same partition, and the node lists contained in the HELLO message do not have any common neighbors with the node lists the node received from other HELLO messages, then the node goes from the ETNC state to the "eligible to be coordinator" (ETC) state. This removes partitions if the HELLO messages are from different partitions. If they are from the same partitions, then the node connects the two coordinator nodes.

To prevent oscillations during the selection of coordinators, we enforce the condition that once a node becomes a coordinator, it continues to remain a coordinator until it is unable to provide any service. This strategy is used despite the fact that this coordinator might become redundant later during self-configuration. This penalty is reasonable since it occurs infrequently, especially in contrast to the energy needed to select an optimum number of coordinators. As the density of nodes increases, the fraction of noncoordinator nodes increases, and this leads to more energy savings. Due to its distributed nature, SCARE has a slightly larger number of coordinators than the minimum number necessary for coverage and connectivity. This also happens due to the randomness involved in the distributed selection of coordinator nodes.

After self-configuration, each coordinator periodically broadcasts a HELLO message along with a list of its one-hop neighbors that are coordinators. Noncoordinator nodes listen to these messages. Noncoordinator nodes also maintain a timer to keep track of the coordinator node that made them a noncoordinator. If this timer goes off, a noncoordinator

---

[1] A neighboring node lies within the node's transmission radius.

node assumes that the corresponding coordinator node has failed and goes into an undecided state. This results in noncoordinator nodes becoming eligible to become coordinators.

SCARE can also be applied to mobile sensor networks. A node that has moved to a new location is treated in the same way as the appearance of a new node at that location. It sets itself to the undecided state and listens to the network until either the timer T_{off} goes off or it receives a HELLO message. Similarly, when a node moves away from one location, this is treated as a node failure by its neighbors. Failure of noncoordinator nodes does not result in any change in the topology. However, the movement of coordinator nodes is detected by the noncoordinator nodes, and this makes them eligible to subsequently become coordinators.

2.2.4 Details of SCARE

A set of control rules governs the state of the sensor node, while a set of defer rules decide when a node should postpone its decision. Timeout rules specify the time after which sensor nodes should make a decision.

A sensor node executing the SCARE procedure can be in one of the following states: coordinator (C), noncoordinator (NC), eligible to be a coordinator (ETC), eligible to be a noncoordinator (ETNC), and undecided (U) (Fig. 2.3). The ETC and ETNC states are temporary and exist only during the T_{setup} period explained below. There are seven timeout values in SCARE:

1. T_{off} Time after which a node that is in the undecided state about its state becomes eligible to be a coordinator and goes into the ETC state.

2. T_{rand} Time for which the sensor node that is in ETC state waits before becoming a coordinator. It then sends a HELLO message along with all its coordinator neighbors that it has identified.

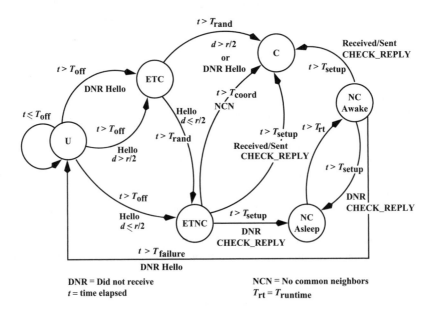

Figure 2.3   State diagram of SCARE.

3. $T_{runtime}$ After every $T_{runtime}$ units of time, all noncoordinator nodes wakeup and listen.

4. T_{setup} Time interval for which the noncoordinator nodes wake up and listen, after which they go to sleep if they still remain noncoordinators. This is also the period during which beacon messages are sent to synchronize the nodes.

5. T_{coord} Time interval during which only the coordinators send HELLO messages. This occurs at the beginning of the T_{setup} period.

6. $T_{noncoord}$ Time interval during which only the noncoordinators send messages. This is the latter part of the T_{setup} period. This period starts immediately after the T_{coord} period ends.

7. $T_{failure}$ A noncoordinator node waits for time $T_{failure}$ for the HELLO messages from the coordinator node that made it the noncoordinator. If no HELLO message is received within this time interval, it decides that the corresponding coordinator node has failed and sets its state to undecided.

Next we describe the type of messages in more detail. There are three types of messages in SCARE:

- HELLO   These messages are sent by coordinators. They also contain a list of the one-hop coordinator neighbors of the sender node.
- CHECK   These messages are periodically sent by the noncoordinator nodes. They are used to remove the potential network partitions. Each CHECK message also contains of list of coordinator neighbors of the node that made it the noncoordinator.
- CHECK_REPLY   Upon receipt of a CHECK message, noncoordinator node compares the coordinator neighbor list included in the CHECK message with the neighbor list of the node that made it a noncoordinator. If there are no common entries in the two lists, it sends a CHECK_REPLY message. Thus, SCARE adopts a conservative strategy in creating paths in the network and prevent partitions. A noncoordinator node becomes a coordinator node if two coordinators at the end of the T_{coord} period cannot reach each other within one or two hops.

Recall that we used r to denote the transmission radius of a node. Similarly, recall that s is the sensing radius of a node. The control rules that decide the state of the sensor node are as follows:

1. A sensor node that generates a random number between 0 and 1, and greater than a threshold, becomes a coordinator.

2. A sensor node that lies at a distance between s and r of a coordinator node becomes eligible to become a coordinator node and goes into the ETC state.

3. A sensor node that lies at a distance at most s from a coordinator node becomes eligible to become a noncoordinator node and goes into the ETNC state.

4. A sensor node that is in ETNC state listens to the HELLO messages sent by the coordinator nodes for the T_{coord} period. From this list of coordinator nodes contained in the HELLO messages, if it determines that two coordinator nodes do not have a common neighbor that is a coordinator, this node becomes a coordinator at the end of the T_{coord} period. On the other hand, if there are common neighbors in the node lists, then the node stays in the ETNC state.

5. A sensor node that is in the ETNC state at the end of T_{coord} period broadcasts a CHECK message. This message contains a list of the coordinator neighbors of the node that caused it to go to the ETNC state.

6. A sensor node that receives a CHECK message compares the list of neighbors in the CHECK message with its neighbor list. If there is no match between the two lists, it transmits a CHECK_REPLY message to the sender of the CHECK message.

7. Upon receipt of a CHECK_REPLY to its CHECK message, a sender node that is in the ETNC state becomes a coordinator node. The node that sent the CHECK_REPLY also becomes a coordinator.

8. A sensor node that is in the ETNC state and does not satisfy conditions 4 and 5 becomes a noncoordinator node at the end of the setup period.

9. A sensor node that is in the ETC state becomes a coordinator node after the T_{coord} period if it does not become a noncoordinator node due to the selection of some other coordinator node.

10. A sensor node with data to send can opt to become a coordinator for as long as it has data to transmit.

The defer rules for SCARE are as follows:

1. If a node becomes eligible to be a coordinator, it listens for T_{rand} period.
2. If a node becomes eligible to be a noncoordinator at the end of the T_{coord} period, it listens for time $T_{noncoord}$ period.

The timeout rules are as follows:

1. A sensor node at the end of the T_{rand} period broadcasts a HELLO message.
2. A sensor node at the end of the T_{setup} period becomes a noncoordinator if it is still eligible to be a noncoordinator.
3. A sensor node at the end of the T_{coord} becomes a coordinator if it is still eligible to become a coordinator.
4. A sensor node wakes up and listens to the medium after the timer $T_{runtime}$ expires.
5. After its T_{off} timer expires, a sensor node becomes eligible to become a coordinator if it is still undecided about its state.

A state diagram for the SCARE algorithm is shown in Figure 2.3. The distance estimate is denoted by d, and we set $s = r/2$ in this figure. The timeout values in SCARE are application-dependent, and they need to be tuned specific to the application. For example, the T_{off} value that triggers the state transition from an undecided state to an ETNC state depends on the radio range of the specific sensor used in the sensor network.

Time Relationships   The relationships between T_{off}, $T_{runtime}$, T_{setup}, T_{coord}, and $T_{noncoord}$ are as follows:

1. $T_{off} < T_{coord} < T_{setup}$.
2. $T_{coord} < T_{setup}$ and $T_{non-coord} < T_{setup}$.

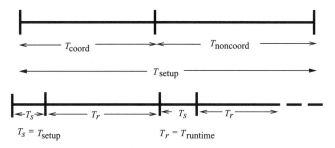

Figure 2.4 Illustration of the relationships between the time intervals.

These relationships are illustrated in Figure 2.4.

Figure 2.5 shows the result of applying SCARE to an example sensor network with 100 randomly deployed nodes in a 100-m × 100-m grid. The sensor nodes have a radio range of 25 m. Timeout values of $T_{failure}$ of 3 s, T_{coord} of 3 s, $T_{noncoord}$ of 2 s, T_{setup} of 5 s, and $T_{runtime}$ of 95 s are used. SCARE selects 32 nodes as coordinators and the rest are designated as noncoordinators.

Ensuring Network Connectivity We next discuss how SCARE prevents network partitioning. Let S be a set of nodes containing the partial set of coordinators that are connected and the associated nodes in the ETNC state. Each coordinator in set S can reach any other coordinator in set S in a finite number of hops. Let X denote the region enclosing the nodes present in set S. Now consider a node not in set S. Any node not present in S can lead to the following scenarios. We use the notation P_A to represent the area within the transmission range of node A.

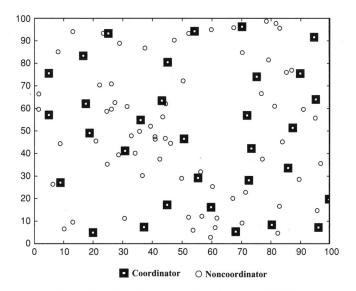

Figure 2.5 Result of self-configuration using SCARE.

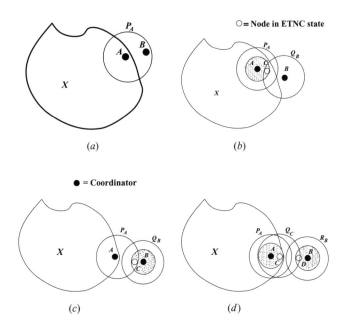

Figure 2.6   Illustration of how network partitioning is prevented in SCARE.

1. Coordinator B outside the region X but within the transmission range of the coordinator A in region X as shown in Figure 2.6a. In this case, both the coordinators can reach each other and the set $S = S \cup \{B\}$ and the region X expands to include the coordinator B.

2. Coordinator B is outside the transmission range of the coordinator A but is within the transmission range of ETNC node C; see Figure 2.6b. However, as node C listens to the HELLO messages from both coordinator nodes A and B, it becomes a coordinator if there is no other path from A to B by becoming a coordinator. Now this reduces to (case 1) with coordinators C and B within reach of each other. C becomes a coordinator, and the region X expands to include the coordinator B, that is, $S = S \cup \{B\}$.

3. Coordinator B is outside the transmission range of coordinator A. However, node C in ETNC state due to node B is within the reach of coordinator A as shown in Figure 2.6c. Node C listens to HELLO messages from A and B, and it becomes a coordinator. Now, A and C are within reach of each other, and this reduces to case 2; hence $S = S \cup \{C\}$. By a similar procedure, node B is also included.

4. Coordinator B and coordinator A cannot reach each other as shown in Figure 2.6d. However, nodes C and D that are in ETNC state can reach other. Node C and node D send and receive CHECK and CHECK_REPLY messages and become coordinators if there is no other path from node A to B. Once C becomes a coordinator, coordinator C in region X and coordinator D outside region X are within reach of each other. This reduces to case 2 and $S = S \cup \{D\}$. Region X expands to include node B and node D.

5. A node F that is outside the reach of either a coordinator or a node in ETNC state in region X. In this case, as the region X expands to include more nodes, node F falls into one of the above categories and as a consequence becomes connected with the nodes present in region X.

We have therefore shown that network partitioning can never arise during self-configuration.

Message Complexity The total number of control messages, referred to as message complexity in SCARE, can be determined as follows: Suppose N is the total number of nodes in the network. Let N_c be the number of coordinator nodes selected. The number of noncoordinator nodes in the network is then simply $N - N_c$. Each coordinator node sends a HELLO message and each noncoordinator node sends a CHECK message. Let Δ be the average number of coordinator neighbors of a noncoordinator node. A noncoordinator node sends a CHECK_REPLY message in response to a CHECK message if and only if there is no match between the coordinator neighbor lists of the noncoordinator nodes. In Span, each noncoordinator node sends one message and each coordinator node sends two messages. Therefore, the number of messages sent in each T_{period} interval is $N + N_c$.

Consider two noncoordinator nodes A and B. For every node in the coordinator neighbor list of A, let α be the probability that this node is present in the coordinator neighbor list of node B. The probability that there are is no match is then $(1 - \alpha)^\Delta$. Thus the expected number of CHECK_REPLY messages is $(1 - \alpha)^\Delta (N - N_c)$. The total expected number of control messages sent in SCARE is therefore $(1 - \alpha)^\Delta (N - N_c) + N_c + (N - N_c) \approx N$ for sufficiently large Δ in dense sensor networks. This is clearly less than the $N + N_c$ messages needed in Span. The size of each message in SCARE is almost equal to the size of each message in Span since the almost same information is contained in both sets of messages.

2.2.5 Optimal Centralized Algorithm

In this section, we develop a provably optimal centralized algorithm that selects a smallest number of nodes to maximize coverage yet maintains network connectivity. This optimal algorithm is compared to SCARE in order to evaluate the effectiveness of the distributed algorithm.

We model the sensor network as a graph G, and use this model to develop an algorithm MAXCOVSPAN that generates a spanning subgraph of G. In addition, G provides the maximum coverage among all spanning subgraps of G, where the nodes in the spanning subgraph correspond to the active sensor nodes in the network. The results provided by the centralized procedure MAXCOVSPAN can then be compared with the results obtained using the distributed SCARE procedure.

Recall that SCARE selects nodes as coordinators on the basis of a distance metric. MAXCOVSPAN also uses the distance between nodes to include nodes in a spanning subgraph such that the coverage is maximized.

Problem Statement Find a spanning subgraph of G that provides the maximum coverage. The vertices in G correspond to the sensor nodes. If two nodes are within radio range of each other, an edge is included in G between the corresponding vertices. The weight of this edge denotes the distance between the two sensor nodes. The algorithm is described in terms of the following rules that are applied to G.

Initialization

Rule 0: Color all vertices white.

Rule 1: Start with an arbitrary node. Call this node Current and color it black.

Selection

Rule 2: Pick an adjacent vertex that is connected by an edge to the Current vertex of maximum weight and color this node black. Color all other neighbors of the Current node gray. Call the vertex that has most recently been colored black as Current.

Rule 3: If the vertex belonging to the longest edge is already colored black, follow rule 4; otherwise repeat rule 2.

Rule 4: If there are still white vertices, pick a gray vertex that has most white neighbor vertices and call it Current.

Termination

Rule 5: Repeat rules 2, 3, and 4 until all the vertices are colored either black or gray.

Theorem 2.2.1   The algorithm MAXCOVSPAN runs in $O(n^2)$ time for a graph with n vertices.

PROOF   At each time instant, one vertex is colored either black or gray. There are n vertices in the graph. However, we need to check for the remaining gray nodes that have white nodes as their neighbors. This takes $O(n)$ time as we might have to check all the n nodes in the worst case. Hence, it takes a total of $O(n^2)$ time to complete the algorithm.   ■

Theorem 2.2.2   MAXCOVSPAN always generates a spanning subgraph.

PROOF   It suffices to show that at the end of the algorithm, each node is colored either gray or black. This can be shown as follows. According to rule 4, if a gray node has white vertices as neighbors, then it is colored black and all its neighbors except the neighboring vertex belonging to the longest edge are colored gray. A black node has all its neighbors colored black or gray according to rule 2. This completes the proof of the theorem.   ■

Theorem 2.2.3   The spanning subgraph G' generated by MAXCOVSPAN provides the highest coverage among all spanning subgraphs of G that have the same number of nodes as G'.

PROOF   In order to avoid case-by-case analysis, we prove this theorem using mathematical induction. Suppose $G = (V, E)$ is the graph corresponding to the sensor network. Let $P_i = (V_i, E_i)$ denote the partial (incomplete) spanning subgraph of size i generated by MAXCOVSPAN. Let $Cov(P_i)$ denote the coverage obtained with P_i. Consider the base case $P_1 = (v_1, \phi)$, where v_1 is any node selected at random. $Cov(P_1)$ is the maximum as all nodes have the same sensing range, and the coverage provided by any node is the same. Next we assume that that the coverage of P_n is the maximum among all partial spanning subgraphs of size n. The coverage provided by a partial connected spanning subgraph of size $n + 1$ is given by $Cov(P_{n+1}) = Cov(P_n) \bigcup Cov(v_2)$ where v_2 is the node added to the partial spanning subgraph of size n. For $Cov(P_{n+1})$ to be maximum, v_2 needs to have minimum overlap with P_n. This is ensured by MAXCOVSPAN. The algorithm selects the node that is farthest from the partial spanning subgraph, and this results in the coverage of the new selected node to have minimum overlap with the partial spanning subgraph of size *n*. From this observation, it follows that $\text{Cov}(P_{n+1})$ is maximum. Hence by the principle of mathematical induction, we have shown that MAXCOVSPAN generates a connected spanning subgraph that provides maximum coverage. ∎

Coverage Comparisons Figure 2.7 shows the variation of coverage with the total number of nodes for three scenarios: all nodes awake, MAXCOVSPAN, and SCARE. We assume that the nodes are placed randomly on a 100-m × 100-m grid. We assume a sensing range of 12.5 m and a transmission range of 25 m. We vary the number of nodes from 50 to 300, and the results are averaged over 100 runs. The results show that the distributed SCARE procedure performs nearly as well as the centralized MAXCOVSPAN procedure, and for a large rnumber of deployed nodes, both these methods perform nearly as well as the scheme of keeping all nodes awake.

2.2.6 Performance Evaluation

To better understand the performance issues in SCARE, we use simulations to determine the effectiveness of SCARE in terms of coverage, connectedness, and network lifetime. We compare SCARE to Span in the simulations. Finally, we examine the impact of distance estimation errors on the effectiveness of SCARE.

Each sensor node is assumed to have a radio range of 25 m. The bandwidth of the radio is assumed to be 20 kbps. The sensor characteristics are given in Table 2.1 [21].

Simulation Methodology We have developed a simulator in Java to evaluate the performance of SCARE. Our simulator uses geographic forwarding as the routing layer and

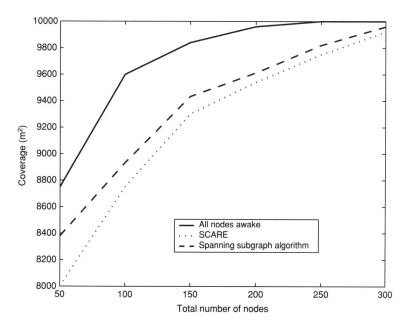

Figure 2.7 Coverage versus number of nodes.

Table 2.1   Radio Characteristics [21].

Radio Mode	Power Consumption (mW)
Transmit (T_x)	14.88
Receive (R_x)	12.50
Idle	12.36
Sleep	0.016

IEEE 802.11 [22] as the MAC layer. Each sensor node that receives a packet forwards it to the neighbor coordinator node that is closest to the destination. If no neighboring coordinator node is closer to the destination than the node itself, the packet cannot be forwarded and it is dropped. SCARE runs on top of IEEE 802.11 MAC and below the routing layer to help coordinate the forwarding of packets from source to destination.

We use a grid size of 100 m × 100 m, and sensor nodes with radios having a nominal radio range of 25 m and a bandwidth of 20 kbps. Initially, nodes are randomly deployed in the grid with the only condition that the nodes form a connected network. We simulate different node densities by increasing the number of nodes and keeping the grid size constant. To study the effect of increase in the number of nodes on SCARE, we simulate 50, 100, 150, 200, 250, and 300 nodes in our simulations. The results presented in this section are averaged over 100 simulation runs.

In the remainder of this section, we compare SCARE with Span and show that SCARE selects a smaller number of coordinators compared to Span and thus provides significant energy savings. To study the effect of SCARE coordinator selection on packet loss rate, we used a constant bit rate (CBR) traffic. However, to more closely understand the effectiveness of SCARE, we separate the nodes that generate traffic from the nodes that execute SCARE and participate in multihop forwarding of packets. Sources and destinations of traffic are placed outside the simulated region, and these nodes do not execute the SCARE procedure. A total of 10 source nodes and 10 destination nodes are used in our simulations. Each source node selects a random destination from the 10 destination nodes and sends a CBR traffic of 10 kbps to it.

To study the effect of mobility on SCARE, we use a random way-point model [23]. In this model, each node randomly selects a destination location in the sensor field and starts moving toward it with a velocity randomly chosen from a uniform distribution. Once it reaches the destination, it waits for a certain predetermined pause time, before it starts moving again. The pause time determines the degree of mobility in this model. We simulated five different pause times of 0, 100, 200, 500, and 1000 s and a velocity range of 0 to 10 m/s. A pause time of 1000 s corresponds to the stationary sensor network while a pause time of 0 s corresponds to high mobility. We used $T_{failure} = 3$ s, $T_{coord} = 3$ s, $T_{noncoord} = 2$ s, $T_{setup} = 5$ s, and $T_{runtime} = 95$ s in our simulations.

Although SCARE relies on a localization scheme, we do not simulate it in our simulator for simplicity. Instead, we make use of the geographic locations of sensor nodes provided by our simulator to aid SCARE in deciding the state of each sensor node. However, since the message overhead due to SCARE is negligible, only one message per node, we believe that this does not affect the results significantly.

Simulation Results   In this subsubsection, we first evaluate the coverage provided by SCARE. We define coverage as the total sensing area spanned by all the coordinator nodes.

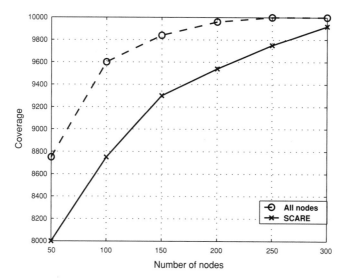

Figure 2.8 Coverage obtained with SCARE compared to the case when all nodes are awake.

We assume that noncoordinator nodes turn off their sensors. Although SCARE does not provide complete coverage due to the random deployment gaps in the sensing range of the coordinators, its coverage is very close to the maximum coverage. Yet, SCARE selects only a few nodes as coordinators to provide this coverage, thus achieving considerable energy savings. Therefore, SCARE efficiently trades off minimum loss in coverage with a tremendous gain in energy savings.

In Figure 2.8, we show the coverage versus the number of deployed nodes for SCARE. Recall that coverage is measured by the total sensing area spanned by all the coordinator nodes. We also show the coverage when SCARE is not run and all nodes are kept awake. As expected, the coverage obtained with SCARE is slightly less than the coverage obtained if all nodes have their sensors and radios turned on. However, the coverage produced by SCARE becomes comparable to the best-case coverage as the number of nodes increases. In these simulations, the grid size is kept constant, hence an increase in the number of nodes represents an increase in the node density.

We next compare the number of coordinators selected in Span with the corresponding number for SCARE. As shown in Figure 2.9, the number of coordinators selected by SCARE is much less than in Span. (For 50 nodes, Span selects fewer coordinators, but the coverage is too low.) SCARE selects a smaller number of coordinators yet provides nearly the same coverage.

Figure 2.10 shows the coverage obtained by using SCARE and Span. SCARE tends to provide better coverage than Span for a range of values for the number of nodes below 100. Both provide similar coverage as the number of nodes increases beyond a threshold.

Figure 2.11 shows the fraction of nodes selected as coordinators with an increase in the number of nodes. SCARE selects a small fraction of nodes as coordinators with increase in node density. Hence, compared to Span, more energy savings are obtained with SCARE for dense sensor networks.

Figure 2.12 compares the number of coordinators selected by SCARE compared to the ideal number of coordinators needed for the square tiling configuration discussed in

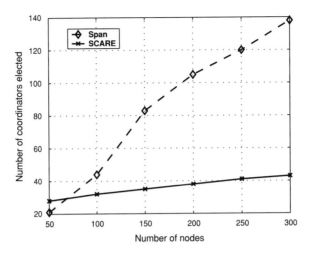

Figure 2.9   Number of coordinators selected with an increase in nodes.

Section 2.2.5. SCARE selects almost the same number of coordinators as in the ideal case. This behavior is different from the behavior of SCARE in Figure 2.9 as here the nodes are placed in a regular fashion and not randomly deployed. Random deployment results in SCARE selecting more nodes as coordinators to cover the entire grid and still maintain connectivity. Any self-configuration algorithm should have minimal control message overhead. In Figure 2.13, we compare the number of control messages used by SCARE and Span for the self-configuration. SCARE uses a smaller number of control messages compared to Span because it takes advantage of the random initialization of the nodes. This leads to a partial configuration of the network; hence SCARE uses fewer number of control messages to achieve self-configuration.

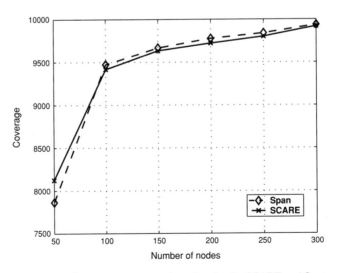

Figure 2.10   Coverage versus number of nodes for SCARE and Span.

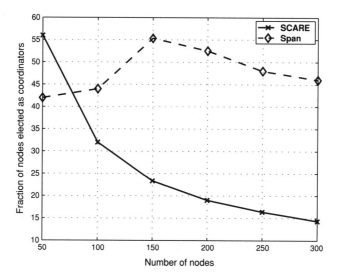

Figure 2.11   Fraction of nodes selected as coordinators in SCARE and Span.

Figure 2.14 shows the effects of mobility on packet loss rate for both Span and SCARE. Nodes follow the random way-point model described in the previous subsubsection. Packet loss rate is calculated as the ratio of the number of lost packets to the number of packets actually sent. We note that the packet loss rates for both these methods are comparable.

Figure 2.15 shows the fraction of surviving nodes as a function of simulation time for both SCARE and Span. SCARE uses fewer control messages and consumes less energy for self-configuration and reconfiguration of the network. The number of surviving nodes falls below 80% at 765 for SCARE compared to 700 for Span.

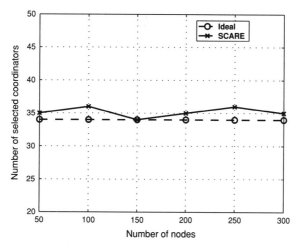

Figure 2.12   Coordinators selected in SCARE versus an ideal number of coordinators selected based on square tiling.

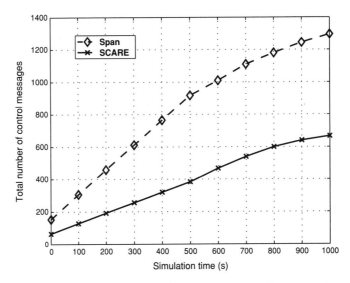

Figure 2.13   Number of control messages used for self-configuration.

Effect of Location Estimation Error on Results

We next investigate how errors in distance estimation affect the performance of SCARE. Since nodes use distance estimation only to determine their eligibility to go to the sleep state, we do not expect SCARE to be significantly affected because of moderate errors in distance estimates.

To measure this feature of SCARE quantitatively, we ran simulations by introducing artificial errors in distance estimation. We modeled such errors by shifting the location of each node by a random amount in the range $[x \pm e, y \pm e]$, where e is either 10 or 20% of the radio range of a node and $[x, y]$ is the location of a sensor node. Nodes use these

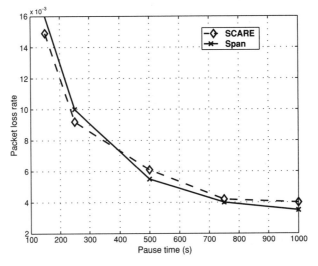

Figure 2.14   Packet loss rate as a function of pause time.

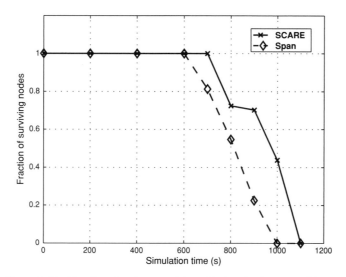

Figure 2.15 Fraction of nodes remaining with time for Span and SCARE.

artificial locations rather than their real location to estimate the distance between them and a coordinator node. We refer to this scheme as either SCARE-10 or SCARE-20.

Figure 2.16 shows the results of these simulations. The simulations using SCARE-10 and SCARE-20 are based on incorrect estimation of the distance from the coordinators by the nodes. Consequently, the number of coordinators is different from the case when there is no error. However, the increase in the number of coordinators is negligible, while the decrease in coverage is found to be minimal. In the case of SCARE-10, the increase is only 3% for a small number of nodes and negligible ($<0.2\%$) for a large number of nodes.

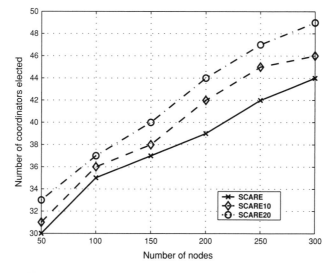

Figure 2.16 Effect of error in distance estimation on SCARE.

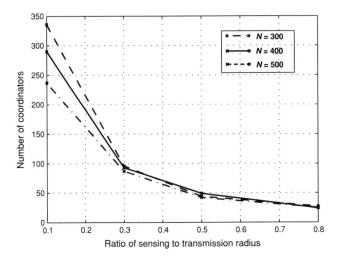

Figure 2.17   Number of coordinators selected versus s/r.

The decrease in coverage was found to be at most 0.7%. As can be seen from the graph in Figure 2.16, the results are similar for the SCARE-20 case.

In all the simulation results shown above, the sensing radius has been taken to be one-half of the transmission radius. We now examine the effect of varying the sensing radius (s) as a fraction of the transmission radius (r) of a node. Figure 2.17 shows that the number of coordinators selected by SCARE as the ratio of sensing radius to the transmission radius is varied. The number of coordinators selected drops rapidly as the ratio increases. As expected, the coverage increases with an increase in s/r; see Figure 2.18. At the s/r value of 0.3, we obtain almost 93% coverage with only 25% nodes selected as coordinators.

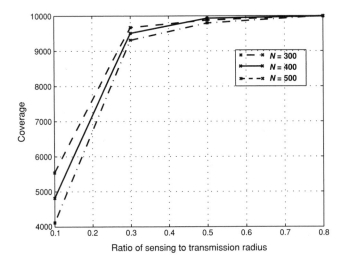

Figure 2.18   Coverage obtained versus s/r.

In the absence of calibrated data for the timeout parameters, we repeated the above set of experiments for different values of the parameters. The details are not listed here due to reasons of conciseness. We obtained similar experimental results in all cases.

2.2.7 Conclusion

We have presented a new scalable algorithm, termed SCARE, for self-configuration and adaptive reconfiguration in dense sensor networks. The proposed approach distributes the set of nodes into subsets of coordinator and noncoordinator nodes. It exploits the redundancy inherent in these networks to maintain coverage in the presence of node failure, as well as to prolong the network lifetime. We have presented a novel node replacement strategy that allows SCARE to use noncoordinator nodes to replace coordinator nodes that fail. We have presented simulation results to highlight the advantages of SCARE over a previously proposed topology management scheme Span for ad hoc networks.

2.3 ROBUST SENSOR POSITIONING IN WIRELESS AD HOC SENSOR NETWORKS

Xiang Ji and Hongyuan Zha

Wireless ad hoc sensor networks are being developed to collect data across the area of deployment. It is necessary to identify the position of each sensor to stamp the collected data or facilitate communication protocols. Most existing localization algorithms make use of trilateration or multilateration based on range measurements obtained from the received signal strength indicator (RSSI), the time of arrival (TOA), the time difference of arrival (TDoA), and angle of arrival (AoA). In this section, we first study some situations that most existing sensor positioning methods fail to perform well. An example of such situations is when the topology of a sensor network is anisotropic. We propose a distributed sensor positioning method with an estimation–comparison–correction paradigam to address these conditions. In detail, multidimensional scaling (MDS) technique is applied to recovering a series of local maps for adjacent sensors in two- (or three-) dimensional space. These maps are then stitched together to form a global map marking all sensors' locations. Then, the estimated positions of anchor sensors are compared with their physical positions, to calculate adjustment parameters, with which sensors' locations are calibrated by iterative process of estimation, comparison and correction. The method avoids measurement errors caused by anisotropic network topology and complex terrain, which previous research fail to address. We also study the application of MDS to centralized position estimation and present some related heuristic principles for collecting pairwise distances between sensors.

2.3.1 Background Information

Wireless ad hoc sensor networks have recently attracted much interest in the wireless research community as a fundamentally new tool for a wide range of monitoring and data-gathering applications. Many applications with sensor networks are proposed, such as habitat monitoring [24–27], health care [28, 29], battle-field surveillance and enemy tracking [30, 31], and environment observation and forecasting [32–34]. A general setup of a wireless sensor network consists of a large number of sensors randomly and densely deployed in a certain area. Each compact sensor usually is capable of sensing, processing data at a small scale, and communicating through omnidirectional radio signal [28, 35]. Because omnidirection radio signal attenuates with a distance, only sensors within a certain range can communicate with each other. This range is called hop distance R. Wireless sensor networks significantly differ from classical networks on their strict limitations on energy consumption, the simplicity of the processing power of nodes, and possibly high environmental dynamics.

Determining the physical positions of sensors is a fundamental and crucial problem in wireless ad hoc sensor network operation for several important reasons. We briefly list two of them in the following: First, in order to use the data collected by sensors, it is often necessary to have their position information stamped. For example, in order to detect and track objects with sensor networks, the physical position of each sensor should be known in advance for identifying the positions of detected objects. In addition, many communication protocols of sensor networks are built on the knowledge of the geographic positions of sensors [36–38]. However, in most cases, sensors are deployed without their position information known in advance, and there is no supporting infrastructure available to locate them after deployment. Although it is possible to find the position of each sensor in a wireless sensor network with the aid of Global Positioning System (GPS) installed in all sensors, it is not practical to use GPS due to its high power consumption, expensive price, and line-of-sight conditions. Thus, it is necessary to find an alternative approach to identify the position of each sensor in wireless sensor networks after deployment. In the general model of wireless ad hoc sensor network, there are usually some landmarks or nodes, called *anchor* nodes, whose position information is known, within the area to facilitate locating all sensors in a sensor network.

In the section, we first analyze some challenges of the sensor positioning problem in real applications. The conditions that most existing sensor positioning methods fail to perform well are the anisotropic topology of the sensor networks and complex terrain where the sensor networks are deployed. Moreover, cumulative measurement error is a constant problem of some existing sensor positioning methods [37, 39, 40]. In order to accurately position sensors in anisotropic network and complex terrain and avoid the problem of cumulative errors, we propose a distributed method that consists of iterations of estimation–comparison–correction. In detail, a series of local maps of adjacent sensors along the route from an anchor (starting anchor) to another anchor (ending anchor) are computed. We apply MDS, a technique that has been successfully used to capture the intercorrelation of high dimensional data at low dimension in social science, to compute the local maps (or relative positions) of adjacent sensors with high errortolerance. These local maps are then pieced together to get the approximation of the physical positions of the sensor nodes. Because the position information of the starting anchor is known, with the stitched maps, the position of the ending anchor can be estimated, which may be different from the true position of the ending anchor. When aligning the calculated position and the true position of ending anchor, the positions of sensors in the stitched maps will approximate their true positions effectively. With very few anchors (≥ 3), our approach usually generates more accurate sensor position information in a network with anisotropic topology and complex terrain than any other positioning method. The method is also efficient in eliminating cumulative measurement errors. At last, we also study the position estimation based on MDS in centralized paradigm and propose several heuristic models to efficiently collect pairwise distances in a wireless sensor network.

The focus of the section is on the position estimation algorithm instead of communication protocol details. Our methods are illustrated with two-dimensional networks, and they can be easily extended to three-dimensional cases.

2.3.2 Previous Work

There have been many efforts to solve the sensor positioning problem. They mainly fall into one of the following four classes or the combinations of them. The first class of methods improve the accuracy of distance estimation with different signal techniques. The RSSI technique was employed to measure the power of the signal at the receiver. Relatively low accuracy is achieved in this way. However, because of its simplicity, it is widely used in previous research. Later, ToA and TDoA are used by Savvides et al. [41] and Priyantha et al. [42] to reduce the errors of range estimation, but these methods require each sensor node being equipped with a CPU with powerful computation capability. Recently, Niculescu et al. use AoA to measure the positions of sensors [43]. The AoA sensing requires each sensor node installed with an antenna array or ultrasound receivers.

The second class of positioning methods relies on a large amount of sensor nodes with positions densely distributed in a sensor network [36, 44, 45]. These nodes with positions known, which are also named as *beacons* or *anchor* nodes, are arranged in a grid across the network to estimate other nodes' positions nearby them.

The third class of methods employ distance vector exchange to find the distances from the nonanchor nodes to the anchor nodes. Based on these distances, each node can estimate its position by performing a trilateration or multilateration [39, 41]. The performance of the algorithms is deteriorated by range estimation errors and inaccurate distance measures, which are caused by complex terrain and anisotropic topology of the sensor network. Savarese [46] tried to improve the above approach by iteratively computing. However, this method adds a large number of communication costs into the algorithm and still cannot generate good position estimation in many circumstances. Moreover, the accuracy of this class of algorithms relies on the average hop distance estimation, and it tends to deteriorate when the topology of the sensor network is anisotropic. For example, in Figure 2.19, sensors

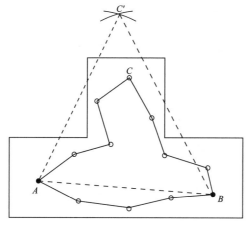

Figure 2.19 Sensor network in nonsquare area.

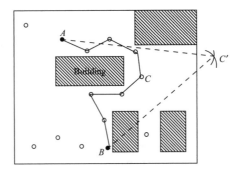

Figure 2.20 Sensor network deployed in a square area with obstacles.

are deployed in a *T-shaped* area, instead of a square area, which is assumed and used as the fundamental condition by most existing research works. A and B are two anchors, A may estimate hop distance with the distance of AB and hop count in the route from A to B. If A and B estimate their distances to C with the estimated hop distance, the estimated distances will be increased a lot by error. A similar situation happens to the case in Figure 2.20, in which sensors are deployed in a square area. But there are some buildings that are marked by dark rectangle areas, and sensors cannot access them. Thus, the routes between a pair of sensors are detoured severely by the buildings in the square area, and the estimated distances of AC and BC are increased significantly. Another example is illustrated by Figure 2.21, where sensors are deployed on a square area with deep grass or bush on the left part and clear ground on the right. The complexity of and terrain leads to different signal attenuation factors and hop distances in the field.

The last class of methods [37, 39, 40] locally calculate maps of adjacent nodes with trilateration or multilateration and piece them together to estimate the nodes' physical or relative positions. The performance of these algorithms relies heavily on the average hop distance estimation and suffers from the cumulative range error during the map stitching.

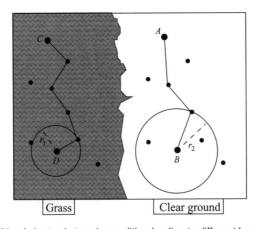

Figure 2.21 Anisotropic terrain condition leading to different hop distances.

2.3.3 Challenges

Considering the real sensor network application scenario, there are several challenges in designing positioning algorithm. Since a large number (up to thousands) of sensors are usually used when they are densely deployed across a given area, we hope to achieve good position estimation as well as keep the hardware design of sensors simple and cheap. In many circumstances it is impossible to get a large number of anchor nodes deployed densely and uniformly to assist position estimation of nonanchor nodes. Thus, it is desirable to design a sensor positioning method that is able to generate accurate position estimation with as few anchors as possible. Third, sensors may be deployed in battle fields or urban areas with complex terrain and vegetation (Figs. 2.19 and 2.20). The sensor network may be badly anisotropic (Fig. 2.21). However, most existing research explored sensor positioning algorithms based on isotropic network topology in a square area. Neither their algorithms nor their experimental environment dealt with sensor network with anisotropic topology such as shown in Figures 2.19, 2.20, and 2.21. Finally, most of the previous methods estimate an average hop distance and broadcast it to the whole network. In many cases, sensors may be deployed on an area with anisotropic vegetation and terrain condition (Fig. 2.21). Thus, sensors at different portions of the area can have different hop distances, and a uniform hop distance in calculation will lead to nonnegligible errors [39,41,46] and serious cumulative errors [37,40].

2.3.4 Calculating Relative Positions with Multidimensional Scaling

Multidimensional scaling (MDS), a technique widely used for the analysis of dissimilarity of data on a set of objects, can disclose the structure in the data [47–50]. We use it as a data-analytic approach to discover the dimensions that underlie the judgments of distance and model data in geometric space. The main advantage in using the MDS for position estimation is that it can always generate high accurate position estimation even based on limited and error-prone distance information. There are several varieties of MDS. We focus on classical MDS and the iterative optimization of MDS, the basic idea of which is to assume that the dissimilarity of data are distances and then deduce their coordinates. More details about comprehensive and intuitive explanation of MDS are available in [47–50].

Classical Multidimensional Scaling We will use $T = [t_{ij}]_{2 \times n}$ to record the physical positions of the n sensors deployed, each in two-dimensional space. The term $d_{ij}(T)$ represents the Euclidean distance between sensor i and j based on their positions in T, and we have

$$d_{ij} = \left[\sum_{a=1}^{2} (t_{ai} - t_{aj})^2 \right]^{1/2} \tag{2.1}$$

The collected distance between node i and j is denoted as δ_{ij}. If we ignore the errors in distance measurement, δ_{ij} is equal to $d_{ij}(T)$. We will discuss the error effects to position estimation caused by the differences between δ_{ij} and $d_{ij}(T)$ later. The expression $X = [x_{ij}]_{2 \times n}$ denotes the estimated positions of the set of n sensor nodes in two-dimensional (2D) space.

If all pairwise distances of sensors in T are collected, we can use the classical multidimensional scaling algorithm to estimate the positions of sensors:

1. Compute the matrix of squared distance D^2, where $D = [d_{ij}]_{n \times n}$.
2. Compute the matrix J with $J = I - e(e^T/n)$, where $e = (1, 1, \ldots, 1)$.
3. Apply double centering to this matrix with $H = -\frac{1}{2} J D^2 J$.
4. Compute the eigen-decomposition $H = U V U^T$.
5. Suppose we want to get the i dimensions of the solution ($i = 2$ in the 2D case), we denote the matrix of largest i eigenvalues by V_i and U_i the first i columns of U. The coordinate matrix of classical scaling is $X = U_i V_i^{1/2}$.

Iterative Multidimensional Scaling   In many situation, the distances between some pairs of sensors in the local area are not available. When this happens, the iterative MDS is employed to compute the relative coordinates of adjacent sensors. It is an iterative algorithm based on multivariate optimization for sensor position estimation in 2D space. Since only a portion of the pairwise distances are available, δ_{ij} is undefined for some i, j. In order to assist computation, we define weights w_{ij} with value 1 if δ_{ij} is known and 0 if δ_{ij} is unknown and assume

$$\delta_{ij} = d_{ij}(T)$$

in the following induction, where X is randomly initialized as $X^{[0]}$ and will be updated into $X^{[1]}, X^{[2]}, X^{[3]} \ldots$ to approximate T with our iterative algorithm.

We hope to find the position matrix X to approximate T by minimizing

$$\sigma(X) = \sum_{i<j} w_{ij} \left[d_{ij}(X) - \delta_{ij} \right]^2 \tag{2.2}$$

This is a quadratic function without containts. The minimum value of such functions is reached when its gradient is equal to 0. The update formula for the iterative algorithm is thus induced to

$$X = V^{-1} \left[\frac{w_{ij}\delta_{ij}}{d_{ij}(T)} A_{ij} \right] T \tag{2.3}$$

where A_{ij} is a matrix with $a_{ii} = a_{jj} = 1$, $a_{ij} = a_{ji} = -1$, and all other elements zeros, and $V = \sum_{i<j} w_{ij} A_{ij}$. If V^{-1} does not exist, we replace it with the Moore–Penrose inverse of V given by $V^- = (V + 11')^{-1} - n^{-2}11'$ [47].

We summarize the iteration steps as:

1. Initialize $X^{[0]}$ as random start configuration, set $T = X^{[0]}$ and $k = 0$, and compute $\sigma(X^{[0]})$.
2. Increase the k by 1.
3. Compute $X^{[k]}$ with the above update formula and $\sigma(X^{[k]})$.
4. If $\sigma(X^{[k-1]}) - \sigma(X^{[k]}) < \epsilon$, which is a small positive constant, then stop; Otherwise set $T = X^{[k]}$ and go to step 2.

The ϵ is an empirical threshold based on accuracy requirement. We usually set it as 5% o
the average hop distance. This algorithm generates the relative positions of sensor nodes in
$X^{[k]}$.

2.3.5 Distributed Sensor Positioning Method Based on MDS

In this sensor positioning method, the above MDS techniques are used in a distributed
manner by estimating a local map for each group of adjacent sensors, and then these maps
are stitched together. In this section, the details of distributed sensor positioning method are
presented.

Hop Distance and Ranging Estimation   Here we employ the widely used distance
measurement model of received signal strength indication (RSSI). A circle centered in
a sensor node bounds the maximal range for the direct communication of the sensor's
radio signal, which is called the *hop distance*. Nodes within one hop distance can directly
communicate with each other, while nodes that are in more than one hop away relay messages
through some media node in hop-by-hop fashion. The power of the radio signal attenuates
exponentially with distance, and this property enables the receiver to estimate the distance
to the sender by measuring the attenuation of radio signal strength from the sender to
the receiver. For example, there are four sensor nodes *A, B, C,* and *D* in Figure 2.22.
Hop distance is r_h. The distance between *A* and *D*, r_{ad}, can be inferred from *A*'s signal
strength at the position of *D* and r_h. It is necessary to point out that some other distance
measure approaches, such as TOA, TDOA, AoA, and ultrasound, can also be applied here.
They even generate more accurate distance measure than RSSI, but they need very complex
hardware equipped in each sensor. In the section, we intend to use RSSI and simple hardware
configuration to achieve competitive performance.

Aligning Relative Positions to Physical Positions   After the pairwise distances of
a group of adjacent sensors are estimated, their relative positions (or a local map) can be
calculated with the MDS techniques. Since we hope to compute the physical positions of all
sensors with our distributed positioning method, it is necessary to align the relative positions
to physical positions with the aid of sensors with known positions. It is known that, in an

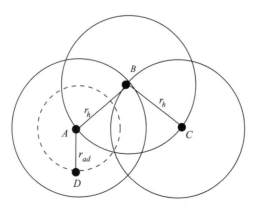

Figure 2.22   Hop distance and signal strength.

adjacent group of sensors, at least three sensors' physical positions are required in order to identify the physical positions of remaining nodes in the group in the 2D case. Thus, each group of adjacent sensors must contain at least three nodes with physical positions known, which can be anchors or nodes with physical positions calculated previously.

The alignment usually includes shift, rotation, and reflection of coordinates. $R = [r_{ij}]_{2 \times n} = (R_1, R_2, \ldots, R_n)$ denotes the relative positions of the set of n sensor nodes in two-dimensional space. $T = [t_{ij}]_{2 \times n} = (T_1, T_2, \ldots, T_n)$ denotes the true positions of the set of n sensor nodes in two-dimensional space. In the following explanation, we assume the nodes 1, 2, and 3 are anchors. A vector R_i may be shifted to $R_i^{(1)}$ by $R_i^{(1)} = R_i + X$, where $X = R_i^{(1)} - R_i$. It may be rotated counterclockwise through an angle α to $R_i^{(2)} = Q_1 R_i$, where

$$Q_1 = \begin{bmatrix} \cos(\alpha) & -\sin(\alpha) \\ \sin(\alpha) & \cos(\alpha) \end{bmatrix}$$

It may also be reflected across a line

$$S = \begin{bmatrix} \cos(\beta/2) \\ \sin(\beta/2) \end{bmatrix}$$

to $R_i^{(3)} = Q_2 R_i$, where

$$Q_2 = \begin{bmatrix} \cos(\beta) & \sin(\beta) \\ \sin(\beta) & -\cos(\beta) \end{bmatrix}$$

Before alignment, we only know R and three or more anchor sensors' physical positions T_1, T_2, T_3. Based on them, we computer $T_4, T_5, \ldots, T_n$. Based on the above rules, we have

$$(T_1 - T_1, T_2 - T_1, T_3 - T_1) = Q_1 Q_2 (R_1 - R_1, R_2 - R_1, R_3 - R_1) \tag{2.4}$$

With $R_1, R_2, R_3, T_1, T_2,$ and T_3 known, we can compute

$$Q = Q_1 Q_2 = (T_1 - T_1, T_2 - T_1, T_3 - T_1)/(R_1 - R_1, R_2 - R_1, R_3 - R_1) \tag{2.5}$$

Then $(T_4, T_5, \ldots, T_n)$ can be calculated with

$$(T_4 - T_1, T_5 - T_1, \ldots, T_n - T_1) = Q(R_4 - R_1, R_5 - R_1, \ldots, R_n - R_1), \tag{2.6}$$

$$(T_4, T_5, \ldots, T_n) = Q(R_4 - R_1, R_5 - R_1, \ldots, R_n - R_1) + (T_1, T_1, \ldots, T_1). \tag{2.7}$$

Distributed Physical Position Estimation   An anchor node called *starting anchor* initializes flooding to the whole network. When other anchor nodes, called *ending anchors*, get the flooding message, they pass their positions back to the starting anchor along with the reverse routes from starting anchor to each of them. Then the starting anchor knows the positions of ending anchors and routes to each of them. Starting anchor estimates the positions of those sensors that are on these routes and one hop away from it. Figure 2.23

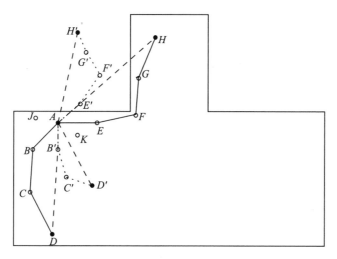

Figure 2.23 Position estimation for a neighborhood.

illustrates the procedure: A is the starting anchor, D and G are the ending anchors. Anchor A knows the positions of D and H as well as the routes to them, which are (A, B, C, D) and (A, E, F, G, H), respectively. Anchor A estimates that the position of B is B' on dashed line AD and the position of E is E' on dashed line AH. Anchor A also estimates the average hop distances in the direction of AD and AH, respectively. With the collection of pairwise distances among neighboring nodes by RSSI sensing, MDS can be performed to calculate the local map (or the relative positions) for neighboring sensor nodes. In Figure 2.23, the relative positions of neighboring nodes A, B, E, J, K are calculated by A. Through aligning the relative positions of A, B, E with their physical positions, the physical positions of J, K can be calculated as well. In the same way, localized mapping and alignment are performed for sensor nodes along a route from the starting anchor to an ending anchor. Figure 2.24 illustrates the procedure of propagated position estimation from starting anchor to ending anchor.

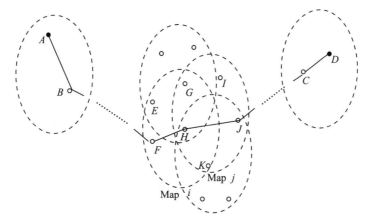

Figure 2.24 Propagation of position estimation.

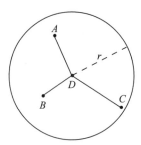

Figure 2.25 Classical MDS.

In Figure 2.24, A is the starting anchor and D is the ending anchor. The remaining nodes are along the route of flooding from A to D, and each local map is represented with a dash ellipse. Map i contains adjacent sensors E, F, G, H, K. Since the physical positions of E, F, G are calculated previously, the physical positions of H, K can be computed with the above MDS and alignment techniques. Then H, K, I, J, and G are adjacent sensors and build map j to further estimate I and J's positions. Figure 2.25 illustrates four adjacent sensors A, B, C, and D; r is the hop distance; A, B, and C are nodes with positions known; D collects the position of A, B, and C, and then calculates their pairwise distances; D also has its distances to A, B, C, respectively. Thus, D can perform a classical MDS to compute the local map (or relative positions of the four sensors).

Figure 2.26 illustrates an example of six adjacent sensors A, B, C, D, E, and F; r is the hop distance; sensors A, B, C, and D know their positions, and sensors E and F do not know their positions; E collects the position of A, B and its distances to them. Then E relays this information to F; and F collects the positions of C, D and its distances to them. Thus, F can compute the pairwise distances of the six sensors except the distances of AF, BF, CE, and DE. Term F can perform an iterative MDS to compute the local map (or the relative positions of the six sensors).

Then, positions of all nodes around a route from a starting anchor to an ending anchor and the ending anchor itself can be estimated. For example, in Figure 2.23, the estimated position of nodes E, F, G are E', F', G', respectively. With the physical position of G known in advance, we can compare G' and G and align them if they are not equal (rotate $\angle G'AG$ with A as center and then scale AG' to AG). We can also apply the same alignment to the coordinates of all sensors along the route, such as E' and F'. In general, the positions

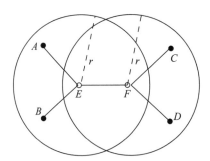

Figure 2.26 Iterative MDS.

of E' and F' are effectively corrected and approximated to their true positions, respectively. The above position estimation procedures are executed iteratively on a route from a starting anchor to an ending anchor until estimated positions converge. Our experimental results indicate that this procedure usually generate highly accurate position estimation for sensors along a route. Then, those nodes with positions accurately estimated are viewed as anchor nodes, and they initialize other position estimation for sensors along different routes. The estimation method can be performed on different portion of sensors in an ad hoc sensor network simultaneously until all sensors know their positions.

2.3.6 Centralized Sensor Positioning Method Based on MDS

Multidimensional scaling can also be used to estimate all sensors' relative positions through one centralized computation. In order to collect some of pairwise distances among sensors, we select a number of source sensors, and they initialize flooding to the whole network to estimate some of the pairwise distances. These estimated distances are then transmitted to a computer or sensor for centralized computation of MDS. The paradigm of centralized computation is supported by sensor system design [35] or fly-over base station, and has been used by Doherty et al. [51] in their sensor position algorithm. The details on flooding and pairwise distances collection is presented next.

Pairwise Distance Collection Usually, a network of sensors are randomly, densely distributed. They are sufficiently connected in a general ad hoc sensor network model. The key operation in pairwise distance collection is to select several sensor nodes flooding across the network. An anchor node is selected as *source sensor* to initialize a broadcast containing its ID, position, and hop count equal to 0. Each of its one-hop neighbors hears the broadcast, appends its ID to the message, increases the hop count by 1, and then rebroadcasts it. Every other node that hears the broadcast but did not hear the previous broadcasts with lower hop count will append its ID, increase the hop count by 1, and then rebroadcast. The process continues until all nodes in the sensor network get the message broadcasted by the original source node. Each node that is far away from the source node usually keeps a route from source node to it. An example broadcast is illustrated in Figure 2.27, where node S initializes a flooding and the average hop distance is r. Each route found is indicated with connected

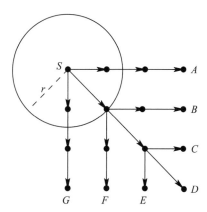

Figure 2.27 Routes of a flooding initialized by node *S*.

arrow lines. Nodes A, B, C, D, E, F, G each keep the corresponding route information from node S to them, respectively. The distance of any pair of nodes on one of the routes can be calculated by multiplying the average hop distance by the number of hop counts between them on the route. Usually, a source node's broadcast only collects the pairwise distances of nodes for which the route information is available. When another anchor node hears the broadcast, it uses the information in the received message to induce the average hop distance. The anchor node is then selected as a new source node, and it initializes another broadcast later to collect more pairwise distances as well as publish the average hop distance. Similarly, we can select some other nodes as source nodes to broadcast.

For n sensors in a sensor network, there are $n(n-1)/2$ pairwise distances in total. Our experimental results indicate that a source node flooding usually collects 3 to 8% of all pairwise distances depending on the relative position of the source node in the network, the connection degree of nodes, and hop distance. The above iterative MDS will generate position estimation with various accuracies depending on the percentage of the pairwise distances collected to all pairwise distances. Usually, more than 10% pairwise distance is needed for a good position estimation. Thus, a certain number of source nodes (anchor nodes or nonanchor nodes) should be selected to initialize flooding. However, the total number of pairwise distances collected does not increase linearly with the number of source nodes, since there are a lot of overlaps among the sets of collected flooding routes by different source nodes.

We hope to initialize with as less source sensors flooding as possible and collect as many pairwise distances as possible. This requires that broadcast from each source sensor can collect relatively more pairwise distances and the overlap among sets of pairwise distances collected by every source node's broadcast should be small. Figure 2.28 illustrates a typical network of sensors (dots) and flooding routes (lines) from a source node (the triangle in the left-top corner). There are 400 sensor nodes, average hop distance 1.2, and 3.1% sensor pairwise distances collected in the network. We take a heuristic analysis of source nodes selection with approximating the topology of a network of sensors with grids. An approximated topology of a sensor network with 37 sensor nodes is plotted in Figure 2.30.

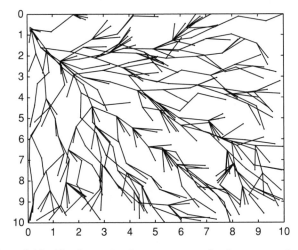

Figure 2.28   Flooding routes from a source node of a sensor network.

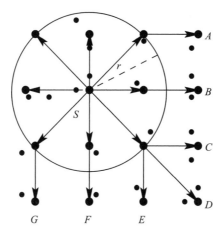

Figure 2.29 Broadcast initialized by node *S* collects 24 pairwise distances.

Node *S* initializes the broadcast and the circle centered with *S* represents the range of signal. Some routes marked by arrow lines are selected to connect 16 nodes, while other routes are omitted. These selected routes contain relatively more nodes than other routes.

Based on the route information in nodes *A*, *B*, *C*, *D*, *E*, *F*, *G*, we can induce 34 pairwise distances. Based on the grid model, we have the following observations:

1. A flooding initialized by a source node located at the border of the network usually collects more pairwise distance than that of a source node located at the center of the network.
2. Flooding initialized by source nodes geodesically far away from each other tends to generate pairwise distance sets with less overlap.

We illustrate the above principles with Figure 2.29 and Figure 2.31. If Figure 2.30 is the first flooding, the flooding in Figure 2.29 is the second flooding in the network and collects and

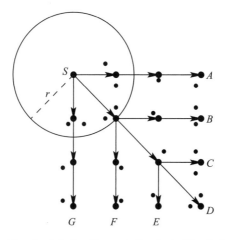

Figure 2.30 Flooding initialized by node *S* collects 34 pairwise distances.

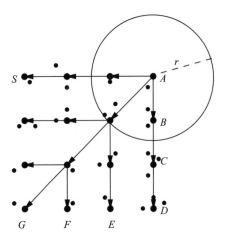

Figure 2.31 Broadcast initialized by node *A* collects 34 pairwise distances.

extra 24 pairwise distances, while the flooding in Figure 2.31 only collects extra 6 pairwise distances.

2.3.7 Simulations

We simulated our proposed distributed and centralized positioning methods with 2 ns. In order to exam the performance of our distributed positioning method, different sensor deployment strategies are considered to model anisotropic network topology and complex terrain. The first strategies is that 400 nodes are randomly placed in a square region, and the average node connection degree is 8.5. The second strategies is that 400 nodes are randomly placed in the *T*-shaped region as in Figure 2.19. The average node connection degree is 7.4. The third strategy is that 400 nodes are randomly placed in a square region, and the region is equally divided into four nonoverlapped square regions. Sensors have different hop distances. The average connection degrees in different small square regions are 6.2, 7.4, 8.9, and 13.4. All of above numbers are determined by randomly generated data sets. We also consider the errors of neighboring sensor distance estimation with RSSI. The measurement error is in the range 0 to 50% of the average hop distance, uniformly distributed.

We measure the performance of the algorithm with mean error, which has been widely used in previous research works:

$$\text{Error} = \frac{\sum_{i=m+1}^{n} \|x_{\text{est}}^i - x_{\text{real}}^i\|_2}{(n - m) \times (\text{radio} - \text{range})} \tag{2.8}$$

where n and m are the total number of sensors and the number of anchors, respectively. A low error means good performance of the method.

Considering the limited pages, we only present some of our results. Figures 2.32, 2.33, and 2.34 are the experimental results with distributed method. Simulation results show that our distributed sensor positioning method outperforms previous research methods. The similarity of the performance in Figures 2.33 and 2.34 indicates that our algorithm is robust in position estimation under different network topologies and terrain conditions. We also

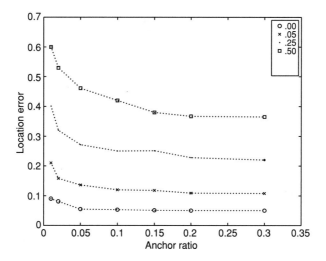

Figure 2.32 Errors when applying the distributed method to sensors in square region.

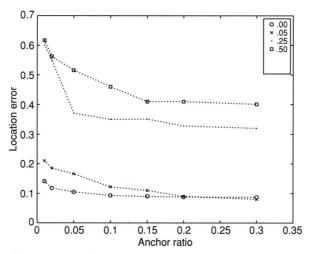

Figure 2.33 Errors when applying the distributed method to sensors in *T* sharp region.

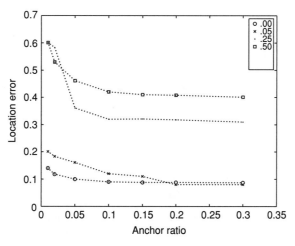

Figure 2.34 Errors when applying the distributed method to sensors in a region with different signal attenuation factors.

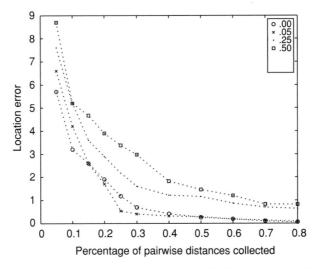

Figure 2.35 Errors when varying the percentage of collected pairwise distances.

use the centralized MDS positioning method to estimate randomly deployed sensor nodes on a square area, in which the available pairwise distances are varying. The experimental results are presented in Figure 2.35. The results indicate that more pairwise distances will bring better performance of our algorithm. In order to demonstrate the scheme we proposed for source nodes selection, we compare the number of collected pairwise distances based on random source nodes selection and our selection scheme. The results are shown in Figure 2.36 and indicates that our source nodes selection scheme is efficient in pairwise distance collection.

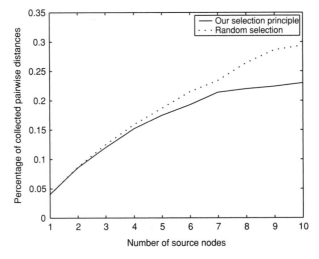

Figure 2.36 Percentage of collected pairwise distances when increasing the number of source nodes for flooding.

2.3.8   Conclusion and Future Work

We address shortcomings, which are caused by anisotropic network topology and complex terrain, of existing sensor positioning methods. Then, we explore the idea of using multidimensional scaling (MDS) technique to compute relative positions of sensors in a wireless sensor network. A distributed sensor positioning method based on MDS is proposed to get the accurate position estimation and reduce error cumulation. The method works in the manner of estimation–comparison–correction. Comparing with other positioning methods, with very few anchors, our approach can accurately estimate the sensors' positions in networks with anisotropic topology and complex terrain as well as eliminate measurement error cumulation. We also study the position estimation based on MDS in a centralized paradigm. Experimental results indicate that our distributed method for sensor position estimation is very effective and efficient.

However, we did not analyze the communication costs during the operation of our methods yet. We are doing experiments related to message complexity and power consumption. Results will be reported very soon. We would also like to investigate the positioning problem for mobile sensors in wireless ad hoc sensor networks.

2.4   TRIGONOMETRIC *K* CLUSTERING (TKC) FOR CENSORED DISTANCE ESTIMATION

Duke Lee, Pravin Varaiya, and Raja Sengupta

The location estimation or localization problem in wireless sensor networks is to locate the sensor nodes based on ranging devices measurements of the distances between node pairs. These ranging measurements typically become unreliable when the distance is large; we say that such distances are *censored*. The section compares several approaches for estimating censored distances with a strategy proposed here called trigonometric *k* clustering, or TKC.

2.4.1   Background Information

Recent advances in microelectronic and mechanical structures (MEMS) sensors and low-power radios have inspired proposals to deploy wireless sensor networks for monitoring a spatially distributed physical process such as the environment in a building or the traffic on a highway. In these applications, the network comprises a set of nodes, each consisting of one or more sensors, a processor, and a radio. The sensor measurements are locally processed and forwarded to a central place. To understand the received data, it is necessary to associate the measurements from each node with its physical location.

A significant literature is devoted to localization in wireless sensor networks. Bulusu et al. proposed localization using a grid of landmarks [52]. Niculescu and Nath [53, 54] introduced distributive censored distance estimation with an information exchange strategy similar to the distant vector routing algorithm. Doherty et al. [51] developed localization based on convex optimization. Simic and Sastry [55] devised a simple, distributive method for localization by restricting the connectivity area to a rectangular box. Savvides et al. [56], Savarese and Langendoen [57], and Whitehouse [58] among others called for iterative refinement of position estimates using gradient descent algorithms. Nguyen and Sinopoli [59] proposed the use learning theory to cope with noisy observations.

In some situations the locations of the nodes may not be known or may be too costly to determine manually. However, the nodes may have ranging devices used to measure the distances between node pairs. One common way is to use the strength of a signal received by a node to estimate its distance from the transmitter. Another way is to measure the sound wave's time of flight to infer the distance between transceivers by transmitting ultrasonic wave.

However, measurements from these ranging devices are unreliable when the distance between transmitting and receiving nodes is large. We then say that the distance is censored. In these cases, indirectly estimating the censored distance is a better option. For example, if there are three nodes, A, B, and C, and the distances between A and B and between B and C are not censored but the distance between A and C is censored, it may be better to estimate the latter as the sum of the two previous distances.

This section focuses on algorithms for censored distance estimation, also called multihop distance estimation. These algorithms estimate censored distances using distance measurements of neighboring node pairs. Three algorithms, DV-hop, DV-distance, and Euclidean method, are discussed in [54]. We expand on the discussion of censored distance estimation and introduce new concepts based on trigonometric constraints. In particular, we introduce a new algorithm, called *trigonometric k clustering*, or TKC. Furthermore, we will compare the performance of various censored distance estimation algorithms using data from a testbed that uses Berkeley Motes [60] and generated using both radio signal strength and ultrasonic wave flight time.

System Model and Notations   An *anchor node*, or an *anchor*, can measure its position reliably. A *floating node* is a nonanchor node. We estimate location of *target nodes* in M-dimensional physical space, $\mathcal{Y}$; V is the set of all nodes; and A is the set of all anchors. The position of node i is $p(i) \in \mathcal{Y}$; its measurement is $\tilde{p}(i)$; and the measurement error is $e(i) = \tilde{p}(i) - p(i)$. The distance between i and j is $d(i, j)$; its measurement is $\tilde{d}(i, j)$; and its measurement error is $e(i, j) = \tilde{d}(i, j) - d(i, j)$. The neighbors of node i, $N(i)$, is the set of all nodes when the distance from i can be measured reliably. The distance between node i and j is censored if the distance cannot be measured reliably. The following summarizes the notation for the measurements.

$$\tilde{p}_i = \begin{cases} p_i + e_i & \forall i \in A \\ \text{NULL} & \text{otherwise} \end{cases}$$

$$\tilde{d}(i, j) = \begin{cases} d(i, j) + e(i, j) & \forall j \in N_i \\ \text{NULL} & \text{otherwise} \end{cases}$$

$$\hat{d}(i, j) = \begin{cases} \| \tilde{p}_i - \tilde{p}_j \| & \forall i, j \in A \\ \text{NULL} & \text{otherwise} \end{cases}$$

where $\delta(i, j)$ is a distance estimate of $d(i, j)$ obtained from measurements $\{\tilde{d}(i, j) | i, j \in V\}$ and $\{\hat{d}(i, j) | i, j \in A\}$. We assume that $\delta(i, j) = \delta(j, i), \forall i, j \in V$. The network topology is the graph (V, E) where E is the set of distance estimations between nodes labeled by $\delta(i, j)$: An edge exists between node i and node j if $\delta(i, j) \neq \text{NULL}$.

2.4.2 Distance Measurements

Ranging technologies are ways of measuring the distance between a pair of nodes. We investigate four popular ranging technologies—network connectivity, radio signal strength, RF time of flight, and acoustic time of flight; these differ in their range, accuracy, directionality, and response to obstacles. We conclude this section with a strategy to combine various measurements.

Network Connectivity   Network connectivity can be used to approximate the distance between a pair of nodes. For instance, the range of an omnidirectional communication device can be simplified as a circle centered at the device with radius RANGE; this yields,

$$d(i, j) \begin{cases} \leq \text{RANGE} & \text{if} (i, j) \in E \\ > \text{RANGE} & \text{otherwise} \end{cases}$$

Network connectivity information can be readily obtained from on-board radios. However, modeling range accurately is difficult due to obstacles and atmospheric conditions. In fact, the range can be quite irregular in shape as shown in [58]. Second, the range may be too coarse for reliable distance estimation—an IEEE 802.11 radio, for example, can communicate over 1-km distance in an open area; in this case, the connectivity information is ineffective for finer-grained localization.

Signal Strength   A receiver can estimate its distance from a transmitter by measuring the signal attenuation. Similar to the network connectivity-based solutions, on-board RF-based communication devices can be readily used. However, inaccuracies can result from unpredictable power attenuation due to multipath, interference, fading, and shadowing. This is seen in Figure 2.37, a log–log plot of received signal strength between a pair of Cisco IEEE 802.11b wireless cards with respect to the distance between the cards. The measurements were taken from a slow-moving vehicle with respect to a fixed transmitter in an open area at the UC Berkeley Richmond Field Station. Furthermore, the direction and altitude of the antenna can affect signal strength greatly. According to [56], raising the antenna 1.5 m above ground can increase the radio transmission range from 20 to 100 m.

RF Time of Flight   The Global Positioning System (GPS) [61], a widely available technology for localization, uses RF time difference of arrival from four GPS satellites synchronized with atomic clocks. GPS technology is scalable and has a resolution of 2 to 3 m with an error of up to 10 to 20 m. However, there are three main concerns when using GPS technology in wireless sensor networks. First, each GPS receiver needs line-of-sight connections to at least four GPS satellites. Thus localization solutions solely based on GPS readings may not work well indoors or near tall buildings. Second, each receiver needs an accurate clock—an inaccuracy of 1 Ms corresponds to a 300-m error. Third, GPS receivers are still too costly for many sensor network applications and consume a great deal of power. A typical GPS receiver costs around $100 and consumes power in the order of watts [52].

However, with single-chip GPS solutions, such as [62], cost and power consumption are expected to decrease dramatically. Furthermore, if the mobility of sensors is limited, sensors can further reduce the impact of the high-energy consumption by running localization algorithms sparingly. This leads us to believe that the GPS solution can be viable in the

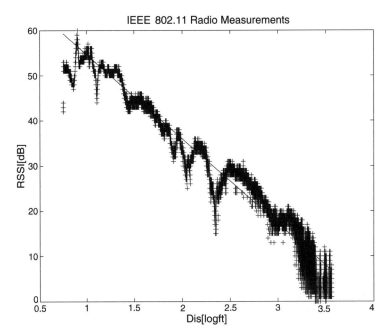

Figure 2.37 Signal-strength-based ranging.

future for outdoor sensor network applications with moderate accuracy requirement for position estimation.

Acoustic Time of Flight Ranging technologies using acoustic time of flight is used in MIT's Cricket [42], ActiveBat [63], UCLA's AHLoS [56, 64], and UC Berkeley's Motes [58]. This technology is robust against fluctuating received signal strength. Savvides et al. [56] observed error less than 2 cm up to a range of 3 m. Data collected by Whitehouse [65] also showed reliability of the measurement—less than a 10-cm error up to the distance of 5 m in the best case. The error varied somewhat between calibration locations and measurement locations (Section 2.4.5). However, compared to RF technologies, acoustic time-of-flight technologies are more susceptible to atmospheric conditions, shorter in range, higher in reflection coefficients, and less able to penetrate into solid objects such as walls.

Furthermore, acoustic time-of-flight technology needs time synchronization between transceivers. Because a global time synchronization is difficult to achieve, most wireless sensor network systems including [42], [63], [56], and [58] use an RF signal as a time synchronizing signal. A transmitter sends an RF signal and an acoustic signal at the same time; and a receiver measures the time difference of arrival of the two signals. For this to work, the algorithm must correctly identify which acoustic signal corresponds to which RF signal.

Calibration Calibration is one major challenge of ranging in wireless sensor networks. As mentioned in [66], an uncalibrated wireless communication radio can transmit twice the power of another radio. In addition, the characteristics of acoustic and RF signals vary significantly depending on the terrain. As noted earlier, raising the antenna 1.5 m above ground can increase the radio transmission range from 20 to 100 m.

Furthermore, the cost of each wireless sensor must be kept at a minimum; and in many applications on-site calibration is difficult due to the inaccessibility of the terrain. One approach is to predict the calibration coefficient from simulated settings. Whitehouse [58] advocates calibration of transmission and reception gains for each transceiver by linear regression using all transmission and reception information.

Distance Estimation   According to the model in Section 2.4.1, we have up to three measurements for $d(i, j)$: $\hat{d}(i, j)$, $\tilde{d}(i, j)$, and $\tilde{d}(j, i)$. If the joint probability density function f is known, the maximum-likelihood estimate of $\delta(i, j)$ of $d(i, j)$ is

$$\delta(i, j) = \arg\max_{d(i,j)} f[\hat{d}(i, j), \tilde{d}(i, j), \tilde{d}(j, i)|d(i, j)] \tag{2.9}$$

Suppose that $e(i)$, $e(i, j)$, and $e(j, i)$ are independent Gaussian random variables, that is, $e(i) = N(0, \sigma_p)$, $e(i, j) = N(0, \sigma_{dij})$, and $e(i, i) = N(0, \sigma_{dji})$. Then VAR$[\hat{d}(i, j)]$ is VAR$[e(i)]$ + VAR$[e_j]$. Thus, $\hat{d}(i, j) = N[d(i, j), 2\sigma_p]$. The maximum-likelihood estimate for $d(i, j)$ is $\delta(i, j)$ such that

$$\frac{\partial}{\partial \delta(i, j)} \ln f(\hat{d}(i, j), \tilde{d}(i, j), \tilde{d}(j, i)|\delta(i, j)) = 0$$

This reduces to

$$\frac{\hat{d}(i, j) - \delta(i, j)}{2\sigma_p} + \frac{\tilde{d}(i, j) - \delta(i, j)}{\sigma_{dij}} + \frac{\tilde{d}(j, i) - \delta(i, j)}{\sigma_{dji}} = 0$$

And we get an explicit expression for $\delta(i, j)$:

$$\delta(i, j) = \frac{\sigma_{dij}\sigma_{dji}\hat{d}(i, j) + 2\sigma_p\sigma_{dji}\tilde{d}(i, j) + 2\sigma_p\sigma_{dij}\tilde{d}(j, i)}{\sigma_{dij}\sigma_{dji} + 2\sigma_p\sigma_{dij} + 2\sigma_p\sigma_{dji}}$$

2.4.3 Censored Distance Estimation

A ranging device for localization is useful within a limited range. Distance estimates outside this range are censored. The following example demonstrates the importance of censored distance estimation. Suppose the (i, j)th element of matrix (2.10) is the estimated distance between nodes i and j, $1 \le i, j \le 4$. The censored distances are marked NULL in the matrix. If we ignore the knowledge that $\delta(1, 2)$ is censored, it would be impossible to distinguish between the two configurations of Figure 2.38 using the distance estimates (2.10):

$$\begin{bmatrix} 0 & \text{NULL} & \delta(1, 3) & \delta(1, 4) \\ \text{NULL} & 0 & \delta(2, 3) & \delta(2, 4) \\ \delta(3, 1) & \delta(3, 2) & 0 & \delta(3, 4) \\ \delta(4, 1) & \delta(4, 2) & \delta(4, 3) & 0 \end{bmatrix} \tag{2.10}$$

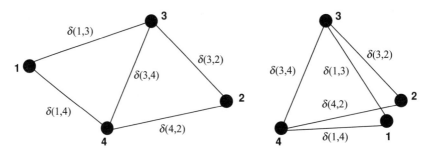

Figure 2.38   Two acceptable configurations specified by (2.10).

Simple Substitution Method   A censored distance between a pair of nodes can be replaced by a value based on prior knowledge about topology and ranging technologies. We call this the simple substitution method. For instance, we can replace censored distances with an average of $\{d(i, j)|d(i, j) > \text{RANGE}\}$. Suppose $\delta(i, j) = [(\{d(i, j)|d(i, j) > \text{RANGE}\}]$ and $\delta(i, j) < d(1, 2)$. We estimate location of nodes in two-dimensional physical space $(M = 2)$. If all measurements are accurate, node placements that satisfies the censored distance estimate $\delta(i, j)$ and (2.10) requires a three-dimensional space as in Figure 2.39. Projecting this three-dimensional placement onto the two-dimensional physical space reduces estimated distances between nodes: In the figure, node 1 is projected closer to node 3 and node 4 than indicated by $\delta(1, 3)$ and $\delta(1, 4)$, respectively. One can do better by incorporating more observations from neighboring nodes, as we discuss next.

Shortest-Hop Method   In a network with evenly distributed nodes, the shortest-hop count in (V, E) multiplied by an estimated hop length can effectively estimate the distance between nodes. One challenge is to restrict the connectivity to achieve a good resolution while maintaining a set of core neighbors. To see the reason for restricting connectivity, consider a fully connected network. In that network, connectivity information is ineffective in discriminating between nearby nodes from far-away nodes. Another challenge is to estimate the average hop length. DV-hop [53] is a distributed algorithm, wherein anchors

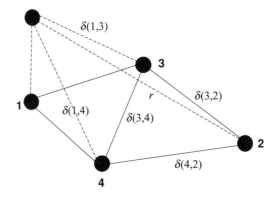

Figure 2.39   Inaccuracies in simple substitution.

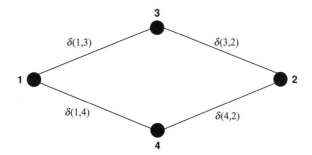

Figure 2.40   Preference of negative error for shortest-path algorithm.

estimate the hop length by averaging hop length to other anchors; and nodes obtain average hop length from their nearest anchor.

Shortest-Path Method   For dense networks with relatively accurate ranging devices, the shortest-path length [along the graph (V, E)] between any two nodes can be used to approximate the distance between them. Because the shortest path length is greater than or equal to the straight-line distance between them, there is a tendency for the shortest-path method to overestimate the distance. However, this tendency is partly balanced by the algorithm's tendency toward negative error when minimizing over various paths. To illustrate, the shortest path estimate of $d(1, 2)$ in Figure 2.40 is

$$\delta(1, 2) = \begin{cases} \delta(1, 3) + \delta(3, 2) & \text{if } \delta(1, 3) + \delta(3, 2) < \delta(1, 4) + \delta(4, 2) \\ \delta(1, 4) + \delta(4, 2) & \text{otherwise} \end{cases}$$

If $\delta(1, 3), \delta(3, 2), \delta(1, 4)$, and $\delta(4, 2)$ are unbiased, the shortest-path algorithm tends to underestimate path lengths because it always picks the shorter of the two equal-length paths.

Lemma 2.4.1   If $\delta(i, j)$ is an unbiased estimate of $d(i, j)$ for all $i, j \in V$, the expected value of the output of the shortest-path algorithm is less than or equal to the shortest-path length.

PROOF   We denote $p_k = [k_1, k_2, \ldots, k_{M_k}]$ to be kth path between two fixed nodes. Since $\min : \mathcal{R}^2 \to \mathcal{R}$ is a concave function, Jensen's inequality implies $E\{\min_k[\sum_l \delta(k_l, k_{l+1})]\} \le \min_k\{E[\sum_l \delta(k_l, k_{l+1})]\}$. But since $E[\delta(i, j)] = d(i, j)$, we have $E\{\min_k[\sum_l \delta(k_l, k_{l+1})]\} \le \min_k[\sum_l d(k_l, k_{l+1})]$.   ■

2.4.4   Trigonometric Censored Distance Estimation

We discuss strategies to estimate censored distances using trigonometric constraints from two adjoining triangles. We focus on techniques on two-dimensional physical space. As shown in Figure 2.38, $\delta(1, 3), \delta(1, 4), \delta(3, 2), \delta(4, 2)$, and $\delta(3, 4)$ in (2.10) form two adjoining triangles $\Delta(\delta(3, 2), \delta(4, 2), \delta(3, 4))$ and $\Delta(\delta(3, 4), \delta(1, 3), \delta(1, 4))$. Using the law of cosine, $\delta(1, 2)$ can be identified up to the ambiguity shown in the figure. This trigonometric constraint of two adjoining triangles is the basis for the trigonometric censored distance estimation. To generalize, we relabel nodes 1, 2, 3, and 4 of Figure 2.38 as nodes i, j, k, and l, respectively. Using the new notation, the localization problem is restated: find $\delta(i, j)$ given l and k, such that $\delta(i, k)$, $\delta(k, j)$, $\delta(i, l)$, $\delta(l, j)$, and $\delta(k, l)$ are defined. The following equations can be derived using the law of cosine:

$$\theta_{jlk} = \cos^{-1} \left[\frac{\delta(k, j)^2 + \delta(k, l)^2 - \delta(l, j)^2}{2\delta(k, j)\delta(k, l)} \right]$$

$$\theta_{ilk} = \cos^{-1} \left[\frac{\delta(i, k)^2 + \delta(k, l)^2 - \delta(i, l)^2}{2\delta(i, k)\delta(k, l)} \right]$$

$$\theta_{ilj}^{\min} = \min(|\theta_{ilk} - \theta_{jlk}|, |\theta_{ilk} + \theta_{jlk}|)$$

$$\theta_{ilj}^{\max} = \max(|\theta_{ilk} - \theta_{jlk}|, |\theta_{ilk} + \theta_{jlk}|)$$

$$\delta(i, j)^{\min} = \sqrt{\delta(i, k)^2 + \delta(k, j)^2 - 2\delta(i, k)\delta(k, j) \cos\left(\theta_{jlk}^{\min}\right)}$$

$$\delta(i, j)^{\max} = \sqrt{\delta(i, k)^2 + \delta(k, j)^2 - 2\delta(i, k)\delta(k, j) \cos\left(\theta_{jlk}^{\max}\right)}$$

These equations produce two random variables, $\delta(i, j)^{\min}$ and $\delta(i, j)^{\max}$, corresponding to the two possible configurations as shown in Figure 2.38. One of the two possible configurations corresponds to the desired estimate. The following algorithms, called trigonometric resolution methods, decide between $\delta(i, j)^{\min}$ and $\delta(i, j)^{\max}$ by attempting to find the true configuration.

Trigonometric Resolution Methods   Trigonometric resolution methods resolve the ambiguity between the two configurations shown in Figure 2.38.

Random Resolution Method   The random resolution method chooses between the two configurations at random. This method is simple and does not give preference to $\delta(i, j)^{\min}$ or $\delta(i, j)^{\max}$. In a well-connected network where multiple pairs of node l and node k that satisfy such configuration can be found for each node pair i and j, the random resolution method often produces more accurate results than those of more sophisticated algorithms with a bias toward $\delta(i, j)^{\min}$ or $\delta(i, j)^{\max}$.

Threshold Resolution Method   The threshold resolution method uses the knowledge of the ranging device's ability to measure the distance between a pair of nodes as a function of the distance between them. Suppose that $P(d)$ represents the probability that a pair of nodes at distance d from each other can measure the distance between them. Using $P(d)$ we can choose between $\delta(i, j)^{\min}$ and $\delta(i, j)^{\max}$ by

$$\delta(i, j) = \begin{cases} \delta(i, j)^{\max} & \text{if } P(\delta(i, j)^{\min}) > P(\delta(i, j)^{\max}) \\ \delta(i, j)^{\min} & \text{otherwise} \end{cases}$$

Using the concept above, we can devise a simple filtering method by assuming a value for RANGE: $\delta(i, j)^{\max} >$ RANGE implies that $\delta(i, j) = \delta(i, j)^{\min}$.

Euclidean Method   The Euclidean method of Niculescu and Nath [53] is based on the observation that choosing between the ambiguities is equivalent to choosing between one of two partitions of the plane separated by the line connecting node l and node k; then node j is likely to be in the same partition as node i if a majority of N_j is connected to node i. More specifically,

$$\delta(i, j) = \begin{cases} \delta(i, j)^{\max} & \text{if } 2|N_i \cap N_j| < |(N_i \cup N_j) - (N_i \cap N_j)| \\ \delta(i, j)^{\min} & \text{otherwise} \end{cases}$$

We found from our experiments that the Euclidean method is robust when distance measurements are accurate. Euclidean method slightly outperforms the random resolution method in sparse networks and slightly underperforms the random resolution method in dense networks.

Degenerate Triangle Method Suppose $\Delta(\delta(k, j), \delta(l, j), \delta(k, l))$ or $\Delta(\delta(k, l), \delta(i, k),$ $\delta(i, l))$ is a degenerate or an approximately degenerate triangle. If $\Delta(\delta(k, j), \delta(l, j), \delta(k, l))$ or $\Delta(\delta(k, l), \delta(i, k), \delta(i, l))$ approximately form a line, then the triangle is degenerate. From this, we can estimate the distance without ambiguity. If $\delta(j, k)$, $\delta(j, l)$, and $\delta(l, k)$ forms a line then

if $(\text{abs}(\delta(l, k) - (\delta(l, j) + \delta(j, k))) \approx 0)$
$\quad \delta(i, j) = \min((\delta(l, j) + \delta(l, i)), (\delta(j, k) + \delta(i, k)))$
elseif $(\text{abs}(\delta(l, j) - (\delta(l, k) + \delta(j, k))) \approx 0)$
$\quad \theta = \pi - a\cos((\delta(l, k)^2 + \delta(i, k)^2 - \delta(l, i)^2)/(2 * \delta(i, k) * \delta(l, k)))$
$\quad \delta(i, j) = \sqrt{\delta(i, k)^2 + \delta(j, k)^2 - 2 * \delta(i, k) * \delta(j, k) * \cos(\theta)}$
else
$\quad \theta = \pi - a\cos((\delta(l, k)^2 + \delta(l, i)^2 - \delta(i, k)^2)/(2 * \delta(l, i) * \delta(l, k)))$
$\quad \delta(i, j) = \sqrt{\delta(l, i)^2 + \delta(l, j)^2 - 2 * \delta(l, i) * \delta(l, j) * \cos(\theta)}$

Multiple Trigonometric Resolution Methods Let $T(i, j)$ denote a set of all (l, k) for each unknown $\delta(i, j)$ for which $\delta(i, k)$, $\delta(k, j)$, $\delta(i, l)$, $\delta(l, j)$, and $\delta(k, l)$ exist. Multiple trigonometric resolution methods estimate $d(i, j)$ given $\{\delta(i, k), \delta(k, j), \delta(i, l), \delta(l, j),$ and $\delta(k, l)|(l, k) \in T(i, j)\}$.

Averaging We can average estimates obtained from trigonometric resolution algorithms to obtains $\delta(i, j)$. Assuming that estimates obtained from trigonometric resolution methods correspond to the true configuration, we hope to obtain a better estimate by averaging. However, we have no guarantee that each estimate is obtained from the true configuration. The next algorithm, trigonometric k clustering, tries to correct the wrong guesses on the true configuration by a refinement process.

Trigonometric k Clustering Trigonometric k clustering (TKC) obtains $\delta(i, j)$ by refining the choices between $\delta(i, j)^{\min}_{(l,k)}$ and $\delta(i, j)^{\max}_{(l,k)}$, $\forall(l, k) \in T(i, j)$ using a variant of the k-clustering algorithm. The intuition is based on the assumption that every pair contains a random variable that corresponds to the true configuration. With small estimation errors, each pair is likely to contain an element that is close to the unknown $d(i, j)$. TKC attempts to identify clusters among the pairs, such that a cluster is formed near $d(i, j)$. The following is the pseudocode for the variant of the k-clustering algorithm used by TKC. We start out with an estimate on the distance that may be obtained from any censored distance estimation algorithm.

1. We form a cluster by selecting one from each pair closer to the current estimate for $d(i, j)$.
2. If there is no change in the membership of the cluster, exit; otherwise, continue.
3. We set the mean of cluster as the new estimate for $d(i, j)$, and go back to the first step.

This algorithm is guaranteed to converge since (2.11) is monotonically decreasing, and there are only finitely many values for (2.11):

$$\sum_{(l,k)\in T(i,j)} \left[\delta(i, j)_{(l,k)} - \frac{1}{|T(i, j)|} \sum_{(l,k)\in T(i,j)} \delta(i, j)_{(l,k)} \right]^2 \qquad (2.11)$$

2.4.5 Experimental Results

All ranging data used below are collected by Whitehouse and presented in his study [65]. The experimental data are calibrated using a method developed by Whitehouse [66]. From the analysis of the experimental data we attempt to gain an insight into the performance of ranging devices and behavior of censored distance estimation algorithms. Furthermore, we establish disErr (2.12) as a performance metric for censored distance estimation algorithms:

$$\text{disErr} = \frac{1}{|V|^2} \sum_{i,j\in V} \| d(i, j) - \delta(i, j) \| \qquad (2.12)$$

To investigate the usefulness of disErr (2.12) as a performance measure of censored distance estimation algorithms, we used distance estimates obtained from each algorithm to obtain position estimates. In the analysis, disErr (2.12) is compared with posErr (2.13), a sum of position errors from a position estimation. To obtain position estimates, we used multidimensional scaling, or MDS [67].

$$\text{posErr} = \frac{1}{|V|} \sum_{i\in V} \| p(i) - \tilde{p}(i) \| \qquad (2.13)$$

Using the experimental data, we compare the performance of shortest-hop (Section 2.4.3), shortest-path (Section 2.4.3), and trigonometric resolution methods (Section 2.4.4). We denote shortest hop as Hop and the shortest-path algorithm as Path. As for the trigonometric resolution methods, we focus on four configurations: MEuclid, MRand, TKCEuclid, and TKCRand. MEuclid takes the mean of results from the Euclidean method (Section 2.4.4). TKCRand applies the trigonometric k clustering (Section 2.4.4) starting from an initial estimate obtained from MRand.

Signal-Strength-Based Ranging   We discuss two experiments carried out in an open, grassy field at the west gate of the UC Berkeley campus and in an office space in the east wing of the Intel Lab at Berkeley. Information about the experiment can be found at [68]. Berkeley Mica Motes [69] were placed on a 4×4 grid with 3-m spacing in the field and a

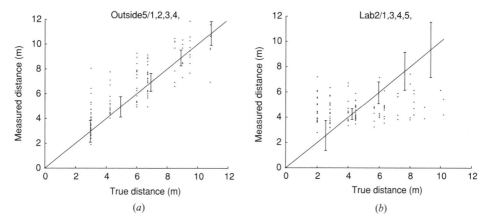

Figure 2.41 Accuracy of signal strength ranging: (*a*) outside and (*b*) lab.

3×6 grid with 2-m spacing in the lab. They were intentionally placed carelessly to achieve random antenna orientation. Using ChipCon CC1000 radios on-board on the Mica Motes, each node took turns transmitting 10 messages, and the signal strengths of transmitted signals were collected at each node. The experiments were repeated five times; each time the motes were shuffled around on the grid.

We used four of five experiments to find a mapping from received signal strength to distance and used the remaining experiment for measurements. Figure 2.41 is a scatter plot of distance measurements against their true distances. Outside5/1,2,3,4 means that experiment 1, 2, 3, and 4 were used to derive coefficients for distance measurements in experiment 5. In the figure, points below the true distance 45° line are underestimates; and points above the line represent overestimates. The standard deviations of errors are plotted as brackets around the true distance line. As seen by the large standard deviations, the distance measurements from the ChipCon CC1000 radio are not very accurate. The lab measurements are more inaccurate.

Table 2.2 summarizes the results for signal-strength-based distance estimation technique using ChipCon radios. The posErr (2.13) is obtained from MDS using four corner nodes at (0,0), (0,9), (9,0), and (9,9) as anchor nodes. With respect to posErr (2.13), Path and TKCRand perform equally well outdoors. However, Path does poorly indoors compared to TKCRand, when ranging errors are increased. Figure 2.42*a* shows placements of nodes in Outside5/1,2,3,4 obtained by MDS and TKCRand. As shown in Figure 2.42*b* the position

Table 2.2   Chipcon Radio Signal-Strength-Based Ranging Experimental Results

Measured/Trained	Hop	Path	MRand	MEuclid	TKCRand	TKCEuclid
			dis Err			
Outside5/1,2,3,4	0.090	0.056	0.058	0.070	0.058	0.068
Lab2/1,3,4,5	0.102	0.094	0.095	0.096	0.094	0.096
			posErr			
Outside5/1,2,3,4	3.063	1.281	1.499	1.747	1.193	1.720
Lab2/1,3,4,5	3.003	2.935	3.023	2.529	2.488	2.529

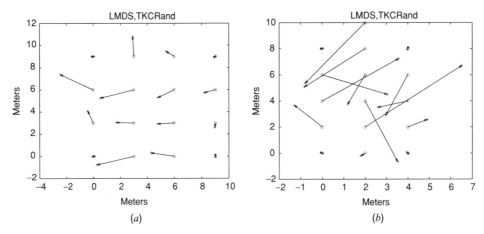

Figure 2.42   Position estimation based on signal strength ranging: (*a*) outside and (*b*) lab.

estimates obtained indoors are terribly inaccurate. In fact, MDS/TKCRand does worse than a simple position estimation algorithm that places all floating nodes at (2, 4), the center of all anchors.

Ultrasonic-Based Ranging   The ranging data were collected in three different environments: grass, grasscups, and pavement. In the grasscups environment, nodes were placed on top of cups on the grass to increase the range area. The topologies were the same for all three experiments. We focus on four configurations, namely Grass Grasscups, Grass Pavement, Grasscups Grass, and Grasscups Pavement. Grass Grasscups were a set of distance measurements taken in the grass environment with calibration coefficients obtained from the grasscups environment. As shown in Figure 2.43, the accuracy of the acoustic-based estimation is significantly better than that of signal-strength-based ranging from the ChipCon CC1000 radio.

Table 2.3 summarizes the results for time difference of arrival based on a distance estimation technique using concurrent ultrasonic and RF signals. The posErr in Table 2.3 is obtained from MDS using the censored distance estimation algorithms with four corner nodes at (3.08, 2.88), (0.45, 0.11), (4.24, 3.95), and (4.08, 1.22) as anchor nodes. TKCRand, TKCEuclid, and Path all perform well with respect to posErr in the grasscups environment. However, TKCEuclid and Path do not do as well in the grass environment compared with TKCRand. This is due to increased error in the grasscups environment compared to the grass environment. In this example, TKCRand is more robust against ranging errors. Figure 2.44 shows placements obtained by TKCRand.

Figure 2.45 is a histogram of posErr divided by the corresponding disErr from all experimental results. It is apparent that there is a correlation between disErr and posErr. We make the cautious claim that disErr is a good performance metric for a censored distance estimation algorithm. The ultimate performance metric for a censored distance estimation algorithm, however, needs to be determined from a metric based on positional error. The choice of the localization algorithm and the choice of the performance metric for position estimation can affect the preference for the censored distance estimation performance metric.

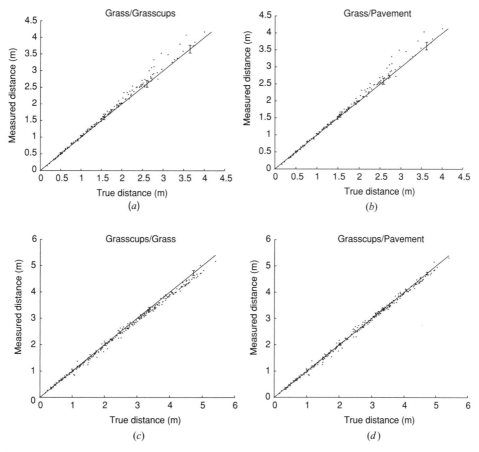

Figure 2.43 Accuracy of acoustic ranging: (*a*) grass trained on grasscups, (*b*) grass trained on pavement, (*c*) grasscups trained on grass, and (*d*) grass trained on pavement.

Table 2.3 Acoustic-Based Ranging Experimental Results

Measured/Trained	Hop	Path	MRand	MEuclid	TKCRand	TKCEuclid
			dis Err			
Grass/Pavement	0.020	0.003	0.017	0.017	0.004	0.008
Grass/Grasscups	0.020	0.005	0.014	0.017	0.005	0.009
Grasscups/Pavement	0.038	0.002	0.003	0.008	0.002	0.002
Grasscups/Grass	0.038	0.003	0.004	0.009	0.003	0.003
			posErr			
Grass/Pavement	0.679	0.091	0.479	0.638	0.072	0.405
Grass/Grasscups	0.695	0.090	0.468	0.642	0.088	0.432
Grasscups/Pavement	1.430	0.052	0.069	0.108	0.045	0.045
Grasscups/Grass	1.555	0.067	0.066	0.105	0.066	0.066

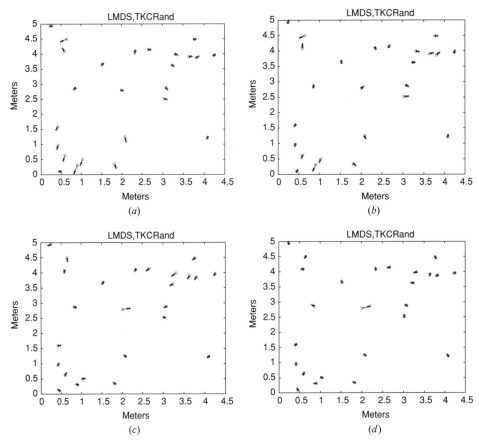

Figure 2.44 Position estimation based on acoustic ranging: (*a*) grass trained on grasscups, (*b*) grass trained on pavement, (*c*) grasscups trained on grass, and (*d*) grasscups trained on pavement.

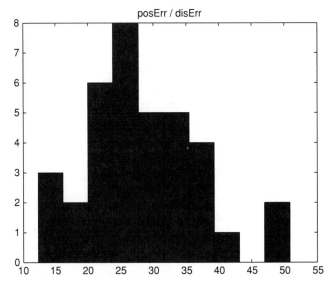

Figure 2.45 disErr as a metric for censored distance estimation algorithm: histogram of posErr/disErr.

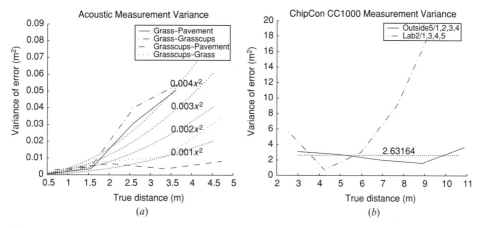

Figure 2.46 Modeling of ranging error: Variance of ranging error versus distance: (*a*) acoustic and (*b*) signal strength.

2.4.6 Simulation Results

Our simulation models are based on the experimental results in the previous section. In the case of acoustic ranging, we model the ranging error on a Gaussian random variable with variance proportional to the square of the distance. Given a distance estimate $\delta(i, j)$, the variance of error DVAR is modeled as

$$\text{DVAR} = \alpha \times \delta^2(i, j) \tag{2.14}$$

To obtain the scaling constant α, we fit square curves to the sample variance curve obtained from acoustic ranging data, as shown in Figure 2.46.

As for the signal strength data, we assume the variance to be constant over the distance measurements up to 11 m. We assume that the ranging area is roughly symmetrical. Figure 2.47 plots the probability of reliable distance measurement as a function of distance. For

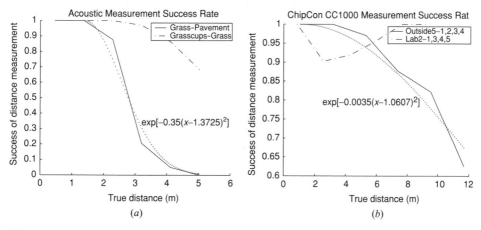

Figure 2.47 Modeling of ranging area: Successful ranging versus distance: (*a*) acoustic and (*b*) signal strength.

Table 2.4   Signal-Strength-Based Ranging Simulation Results

	Hop	Path	MRand	MEuclid	TKCRand	TKCEuclid
			dis Err			
$8 \times 8gvc$	0.0265	0.0296	0.0099	0.0206	0.0059	0.0078
$9\text{-}5 \times 9\text{-}5g1c$	0.0169	0.0081	0.0050	0.0055	0.0049	0.0049
$10 \times 10g3c$	0.0228	0.0575	0.0745	0.0636	0.0168	0.0091
			posErr			
$8 \times 8gvc$	2.7226	7.0072	0.8169	1.3477	0.5107	0.6285
$9\text{-}5 \times 9\text{-}5g1c$	2.9364	0.6592	0.4691	0.5613	0.4750	0.4495
$10 \times 10g3c$	2.2554	4.5291	7.3757	3.2778	1.6357	1.0763

the simulations, we approximate the success rate curve by an exponential curve, as shown in Figure 2.47. We censor distances greater than 12 m because of the lack of experimental data.

As explained in more detail below, we find that TKC is a reliable censored distance estimation algorithm. TKCEuclid tends to perform slightly better than TKCRand if ranging measurements are accurate or network is sparse. When network connectivity is high, TKCRand seems to outperform all other algorithms. The performance of Path is not consistent because its tendency to overestimate (path curvatures) and its tendency to underestimate (preference of negative error) need to be balanced. Low network connectivity, uniformly placed nodes, and large ranging error all increase the comparative performance of Hop against other algorithms. However, the performance of Hop fell below Path in most of the cases.

Signal-Strength-Based Ranging Simulation   Table 2.4 summarizes results from simulation using the signal strength ranging model. The first experiment, $8 \times 8vc$ (Figs. 2.48 and 2.49), is run on a 8×8 grid with varying spacing with four anchor nodes at (0,0), (0,11.2), (11.2,0), and (11.2,11.2). This is a well-connected network, as all nodes are reachable in two hops indicated by Figure 2.48a—only one estimate of distance is evaluated using Hop. Both TKCEuclid and TKCRand perform well, as plotted in Figure 2.49. TKCRand performs slightly better than TKCEuclid, since the network is well-connected and ranging error is relatively high. The performance of Path is poor due to the large variance of ranging error. The preference for negative error is illustrated in Figure 2.48b.

The topology for the second experiment, $9\text{-}5 \times 9\text{-}5g1c$ (Figs. 2.50 and 2.51), is a 9×9 grid with a 5×5 hole in the middle. Anchors are located at (0, 0), (0, 8), (8, 0), and (8, 8). Path greatly underestimates the distances. Trigonometric-constraints-based algorithms outperform path-based algorithms due to the dominance of ranging error and uneven topology.

The third experiment, $10 \times 10g3c$ (Figs. 2.52 and 2.53), is on a 10×10 grid with 3-m spacing. The anchors are located at (0, 0), (0, 27), (27, 0), and (27,27). Path performs less well due to increased underestimation caused by increased size of the network. TKCEuclid performs better than TKCRand, since the network is not as well-connected as in the first example.

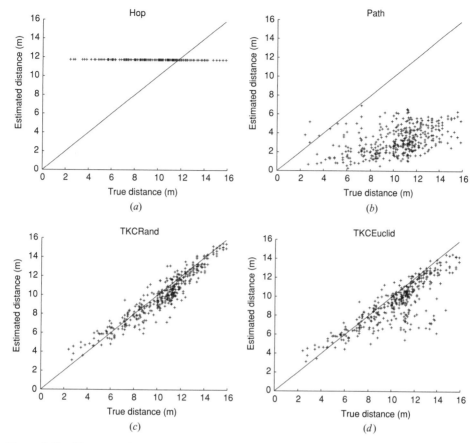

Figure 2.48 Distance estimations (8 × 8*gvc*: 8 × 8, variable spacing, ChipCon CC1000): (*a*) shortest hop, (*b*) shortest path, (*c*) TKC random, and (*d*) TKC Euclidean.

Acoustic-Based Ranging Simulation   Table 2.5 summarizes the results from simulations using acoustic ranging model. In the first experiment, $8 \times 8va$ (Figs. 2.54 and 2.55), nodes are placed on a 8×8 grid with varied spacing. The anchors are located at $(0, 0)$, $(0, 5.6)$, $(5.6, 0)$, and $(5.6, 5.6)$. In this example, TKCEuclid and TKCRand perform well (Fig. 2.55) thanks to the superiority of acoustic ranging compared to signal-strength-based ranging. Path performs better than with signal-strength-based ranging because the error from its preference of negative error is reduced. However, TKCEuclid and TKCRand both outperform Path because the uneven density of nodes negatively affects the performance of Path.

The second experiment (Figs. 2.56 and 2.57), denoted by $10 \times 10g1a$, nodes are placed on 10×10 grid with 1-m spacing with anchors at $(0,0)$, $(0,8)$, $(8,0)$, and $(8,8)$. In this example, TKCEuclid is clearly the best, as plotted in Figure 2.57. The Euclidean method is superior to the random resolution method due to accuracy of measurements. Poorer performance of Path comes from its preference for negative error as shown in Figure 2.56. As the number of hops increases, error due to the negative bias also increases.

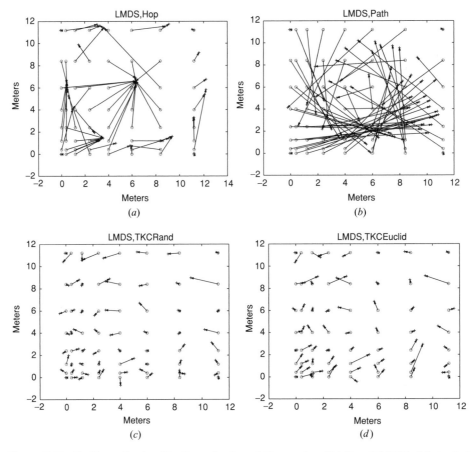

Figure 2.49 Position estimates ($8 \times 8gvc$: 8×8, variable spacing, ChipCon CC1000): (*a*) shortest hop, (*b*) shortest path, (*c*) TKC random, and (*d*) TKC Euclidean.

2.4.7 Conclusion

Censored distance estimation is an important part of location estimation. In this section, we explored existing algorithms and further developed trigonometric-constraint-based censored distance estimation algorithms. Using experimental data from received signal-strength-based ranging and acoustic-based ranging technologies, we compared the performance of censored distance estimation algorithms and developed a sound simulation model. We found that the trigonometric k-clustering method is a robust and reliable censored distance estimation algorithm.

2.5 SENSING COVERAGE AND BREACH PATHS IN SURVEILLANCE WIRELESS SENSOR NETWORKS

Ertan Onur, Cem Ersoy, and Hakan Deliç

In this section, the sensing coverage area of surveillance wireless sensor networks is considered. Sensing coverage is determined by applying the Neyman–Pearson detection model

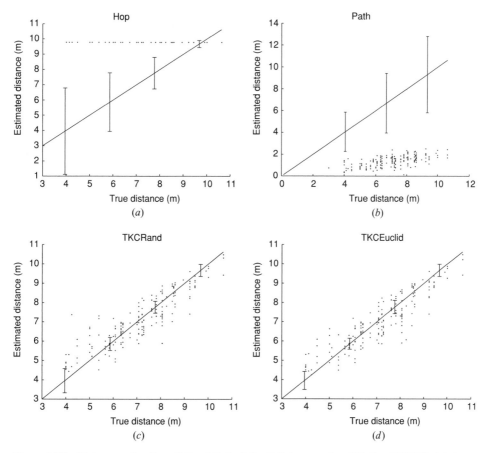

Figure 2.50   Distance estimations (9-5 × 9-5 *g*1 *c*: 9-5 × 9-5, 1-m spacing, ChipCon CC1000): (*a*) short-est hop, (*b*) shortest path, (*c*) TKC random, and (*d*) TKC Euclidean.

and defining the breach probability on a grid-modeled field. Using a graph model for the perimeter, Dijkstra's shortest-path algorithm is used to find the weakest breach path. The breach probability is linked to parameters such as the false alarm rate, size of the data record, and the signal-to-noise ratio. Consequently, the required number of sensor nodes and the surveillance performance of the network are determined. The false alarm rate and the field width turn out to be the two most influential parameters on the breach probability.

2.5.1 Background Information

Wireless sensor networks (WSN) are appropriate tools to monitor an area for surveillance. The primary challenges in building a surveillance wireless sensor network (SWSN) pertain to the decisions to be considered while deploying the sensors. These decisions may consist of communication and sensing range of sensor nodes and density of the SWSN, deployment strategy to be applied (random, regular, planned, etc.), and sink deployment. Depending on the range and the number of sensors, the sensing coverage area of the SWSN may contain

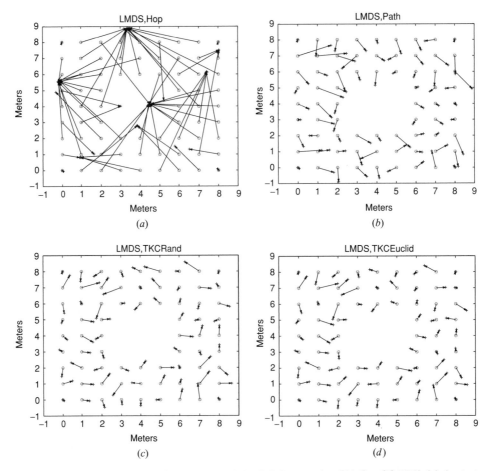

Figure 2.51 Position estimates (9-5 × 9-5g1c: 9-5 × 9-5, 1-m spacing, ChipCon CC1000): (a) shortest hop, (b) shortest path, (c) TKC random, and (d) TKC Euclidean.

breach paths. The probability that a target traverses the region through the breach path gives precious insight about the level of security provided by the SWSN. Thus, it is the aim of this section to analyze the probability of the weakest breach path and draw important inferences regarding the sensing and deployment parameters in an SWSN.

The sensing and communication ranges of some propriety devices are listed in [70]. For example, the sensing range of the Berkeley motes acoustic sensor, HMC1002 magnometer sensor and through-beam type of photoelectric sensor are nearly 1, 5, and 10 m, respectively. The communication range of the Berkeley motes MPR300, MPR400CB, and MPR520A are 30, 150, and 300 m, respectively. The ratio of the communication and sensing ranges shows that the network must be densely deployed. The high redundancy level of the network necessitates energy conservation schemes.

The effect of sensor deployment on the performance of target detection is considered in [71], where the authors propose a measure of goodness of deployment, namely the path exposure, which is the likelihood of detecting a target that traverses the region using a given path. The unauthorized traversal problem is defined, and an incremental sensor deployment

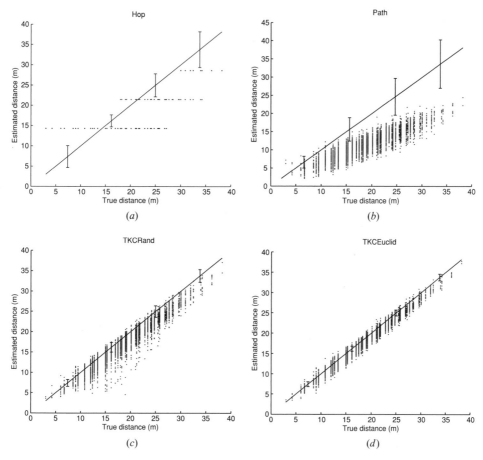

Figure 2.52 Distance estimations ($10 \times 10g3c$: 10×10, 3-m spacing, ChipCon CC1000): (*a*) shortest hop, (*b*) shortest path, (*c*) TKC random, and (*d*) TKC Euclidean.

strategy is proposed. Zou and Chakrabarty propose a virtual force algorithm to increase the coverage after an initial random deployment of sensors [72]. The problem is stated as maximizing the coverage area within a cluster in cluster-based sensor networks subject to a given number of sensors. In both studies, the area to be monitored is a rectangular field. However, most of the time, the area under surveillance is irregular in shape. Considering the perimeter security applications, the field to be monitored is usually narrow and long. Therefore, nonuniform deployment must also be considered.

An incremental sensor deployment strategy is proposed in [73], where there are no prior models of the static environment, and all of the sensors are identical and are able to communicate with a remote base station. The proposed algorithm runs to maximize the coverage area while maintaining full line-of-sight connectivity, and it is shown to produce similar coverage results as the model-based algorithms. The authors analyze the trade-offs in sensor network infrastructure in [74], where continuous update and phenomenon-driven application-level scenarios are analyzed by considering accuracy, latency, energy efficiency, good-put (ratio of total packet count received by observer to the total packet count sent by all

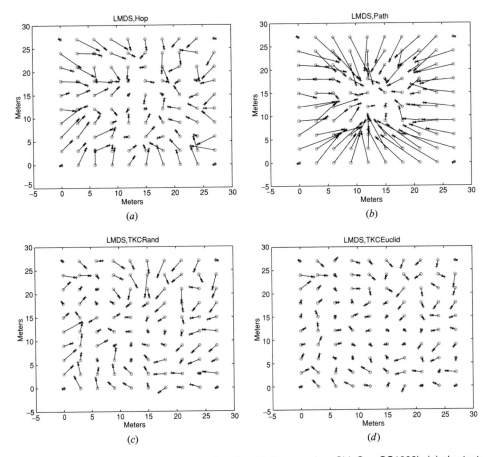

Figure 2.53 Position estimates ($10 \times 10g3c$: 10×10, 3-m spacing, ChipCon CC1000): (*a*) shortest hop, (*b*) shortest path, (*c*) TKC random, and (*d*) TKC Euclidean.

sensors), and scalability as the performance measures. It turns out that there is no appreciable difference between grid-type deployment and random deployment; yet, biasing density to target movement pattern increases accuracy. However, for fields that are irregular in shape, rigorous analysis is required to reach a stronger conclusion about the effects of random and deterministic deployment strategies.

Table 2.5 Acoustic-Based Ranging Simulation Results

	Hop	Path	MRand	MEuclid	TKCRand	TKCEuclid
			dis Err			
$8 \times 8gva$	0.0091	0.0023	0.0143	0.0151	0.0009	0.0009
$10 \times 10g1a$	0.0069	0.0017	0.0242	0.0452	0.0006	0.0010
			posErr			
$8 \times 8gva$	0.7824	0.1616	0.7632	1.3367	0.0611	0.0638
$10 \times 10g1a$	0.8646	0.1919	2.2035	2.5948	0.1176	0.1713

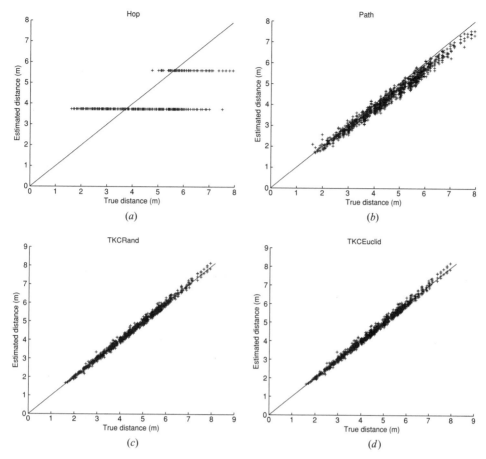

Figure 2.54 Distance estimations ($8 \times 8gva$: 8×8, variable spacing, acoustic): (*a*) shortest hop, (*b*) shortest path, (*c*) TKC random, and (*d*) TKC Euclidean.

In [75], Megerian et al. introduce the exposure concept as the ability to observe a target moving in a sensor field. By expressing the sensibility of a sensor in a generic form, the field intensity is defined as the sum of the active sensor sensibilities. The exposure is then defined as the integral of the intensities (involving all sensors or just the closest one) on the points in a path in the sensor field. Next, they develop a method to calculate the minimum exposure path between any two points in a sensor field. However, some important questions are left unanswered. It is not clear what the threshold value of the minimum exposure has to be to determine the required number of sensor nodes. Determining the threshold becomes too complex when different types of sensors are utilized.

The research efforts summarized above are all based on a generic sensing model that merely says that the detection performance of a sensor is inversely proportional to some power of the sensor-to-target distance. A superficial understanding of the sensor operation that just takes into account the path loss neglects such crucial parameters as the false alarm rate and the number of data processed per sensor decision, as well as the noise phenomenon with an acting signal-to-noise ratio (SNR) at the receiving end. This shortcoming leads to

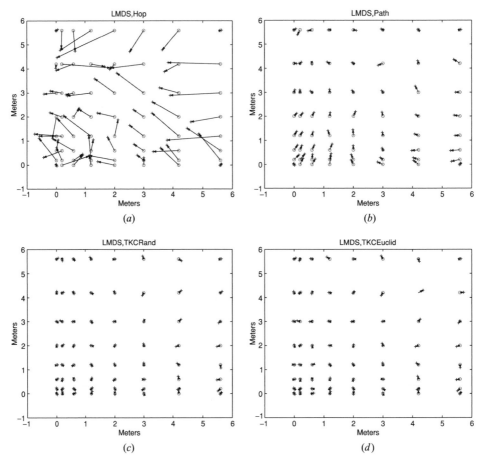

Figure 2.55 Position estimates (8 × 8*gva*: 8 × 8, variable spacing, acoustic): (*a*) shortest hop, (*b*) shortest path, (*c*) TKC random, and (*d*) TKC Euclidean.

the inability to establish a link with other deployment-critical issues such as the required number of sensor nodes for a specified sensing coverage level and energy efficiency. Another common modeling flaw is to assume the same sensing capability for all sensor nodes. This is clearly not possible since different propagation and noise conditions will imply nonidentical detection capabilities, even if one supposes that the same type of sensors are deployed. The exposure calculations and deployment strategies described above become "extremely difficult" if the sensor types are of different characteristics [75]. The need for a unifying sensing model is evident.

He et al. report that sensor nodes generate false alarms at a nonnegligible rate when an SWSN is run in an energy-efficient manner [76]. This observation further suggests that the sensing model must include the false alarms in its formulation. Motivated by the desire to gain more insight about the impact of the parameters listed in the preceding paragraph, we employ in this section the Neyman–Pearson dector for each sensor, which ties performance to a maximum allowable false alarm rate, the size of the data set collected by the sensors

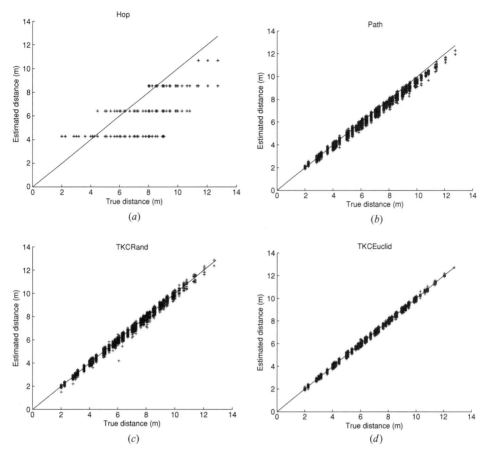

Figure 2.56   Distance estimations ($10 \times 10g1a$: 10×10, 1-m spacing, acoustic): (*a*) shortest hop, (*b*) shortest path, (*c*) TKC random, and (*d*) TKC Euclidean.

at each stage of the decision process, and the signal-to-noise ratio. We define the weakest breach path as the one that has the lowest end-to-end detection probability, a quantity that is defined in a precise manner in the next section. We then proceed to find the breach path through Dijkstra's shortest-path algorithm by assigning the negative logarithms of the miss probabilities as weights of the grid points. With NP dector and the associated use of Dijkstra's algorithm, we study the breach probability as a function of all parameters for both uniformly and normally distributed random sensor deployment. One of the notable outcomes is the evaluation of the relationship between the field shape and the required number of sensor nodes.

In the next section we describe the weakest breach path problem and present how to find the sensing coverage using Neyman–Pearson detection. Dijkstra's shortest-path algorithm is proposed as a solution to this problem by defining a grid-based field model. After presenting the problem formally, the results are analyzed in Section 2.5.3. After the results and discussions, we draw our conclusions in Section 2.5.4.

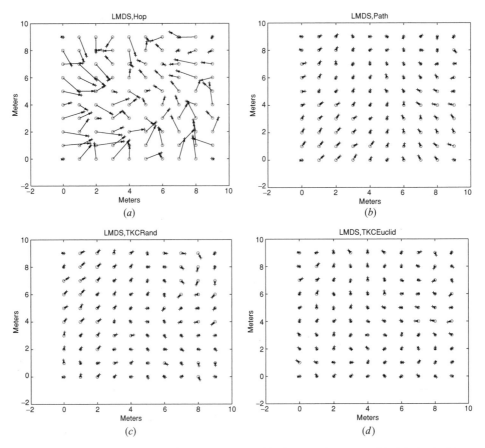

Figure 2.57   Position estimations ($10 \times 10g1a$: 10×10, 1-m spacing, acoustic): (*a*) shortest hop, (*b*) shortest path, (*c*) TKC random, and (*d*) TKC Euclidean.

2.5.2   Weakest Breach Path Problem

In an SWSN, the region to be monitored may be a large perimeter that might be several kilometers. Before deploying sensors in the field, the perimeter may have to be segmented in order to deal with the complexity. Segmentation can be done according to the environmental properties of the perimeter such as altitude and topography. In this section, we work with a single segment.

The security level of an SWSN can be described with the breach probability, which can be defined as the miss probability of an unauthorized target passing through the field. We define the weakest breach path problem as finding the breach probability of the weakest path in an SWSN. To calculate the breach probability, one needs to calculate the sensing coverage of the field in terms of the detection probabilities.

To simplify the formulations, we model the field as a cross-connected grid. A sample field model is presented in Figure 2.58. The field model consists of the grid points, starting point, and the destination point.

The aim of the target is to breach through the field from the starting point, which represents the insecure side, to the destination point, which represents the secure side. The horizontal

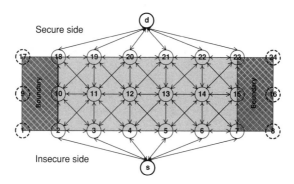

Figure 2.58 Sample field model constructed to find the breach path for the length is 5 m, width is 2 m, boundary is 1 m, and the grid size is 1 m ($N = 8, M = 3$).

axis is divided into $N - 1$ and the vertical axis is divided into $M - 1$ equal parts. In this grid-based field model along the y axis, we add boundary regions to the two sides of the field. Thus, there are NM grid points plus the starting and destination points. To simplify the notation, instead of using two-dimensional grid point indices (x_v, y_v) where $x_v = 0, 1, \ldots, N - 1$ and $y_v = 0, 1, \ldots, M - 1$, we utilize one-dimensional grid point index v, which is calculated as $v = y_v N + x_v + 1$. For the starting point, $v = 0$, and for the destination point, $v = NM + 1$. To represent the connections of the grid points, which a target uses to proceed through the field, the connection matrix $\mathbf{C} = [c_{v,w}]$ is defined as

$$
c_{v,w} = \begin{cases}
1 & \text{if } 0 < v, w < NM + 1 \text{ and } (x_v - x_w, y_v - y_w) \in D \\
1 & \text{if } v = 0 \text{ and } y_w = 0 \\
1 & \text{if } w = N \times M + 1 \text{ and } y_v = M - 1 \\
0 & \text{otherwise}
\end{cases}
\tag{2.15}
$$

where $\mathbf{C}$ is $(NM + 2) \times (NM + 2)$, and $D = \{\{-1, 0, 1\} \times \{-1, 0, 1\}\} - \{(0, 0)\}$, which is the set of possible difference tuples of the two-dimensional grid point indices excluding the condition that $v = w$. The first condition of the partial function of connection matrix $\mathbf{C}$ states that each grid point (excluding starting and destination points) is connected to the grid points, which are either one hop away or cross-diagonal. The second condition states that the starting point is connected to all of the initial horizontal grid points of the field. The third condition says that all of the final horizontal grid points are connected to the destination point. Otherwise, the two grid points are not connected.

Neyman–Pearson Detection Model   Using the field model described above, the detection probabilities are to be computed for each grid point to find the breach probability. The optimal decision rule that maximizes the detection probability subject to a maximum allowable false alarm rate α is given by the Neyman–Pearson formulation [77]. Two hypotheses that represent the presence and absence of a target are set up. The Neyman–Pearson (NP) detector computes the likelihood ratio of the respective probability density functions and compares it against a threshold that is designed such that a specified false alarm constraint is satisfied. Note that NP detector is also used in [78], which introduces the co-grid method and follows a different context than the breach path problem. However, the NP detector is not combined with the path loss model unlike what follows here, and parametric links with detection performance are not established in [78].

Suppose that passive signal reception takes place in the presence of additive white Gaussian noise (AWGN) with zero mean and variance σ_n^2, as well as path loss with propagation exponent η. The symbol power at the target is ψ, and the signal-to-noise power ratio (SNR) is defined as $\gamma = \psi/\sigma_n^2$. Each breach decision is based on the processing of L data samples. We assume that the data are collected fast enough so that the Euclidean distance d_{vi} between the grid point v and sensor node i remains about constant throughout the observation epoch. Then, given a false alarm rate α, the detection probability of a target at grid point v by sensor i is [77]

$$p_{vi} = 1 - \Phi\left(\Phi^{-1}(1-\alpha) - \sqrt{L\gamma_{vi}}\right) \tag{2.16}$$

where $\Phi(x)$ is the cumulative distribution function of the zero mean, unit variance Gaussian random variable at point x, and

$$\gamma_{vi} = \gamma A d_{vi}^{-\eta} \tag{2.17}$$

represents the signal-to-noise ratio at the sensor node i, with A accounting for factors such as the antenna gains, transmission frequency, as well as propagation losses. Active sensing can be accommodated by properly adjusting the constant A.

Because the NP detector ensures that

$$\lim_{d_{vi} \to \infty} p_{vi} = \alpha,$$

instead of using p_{vi}, we introduce the measure

$$p_{vi}^* = \begin{cases} p_{vi} & \text{if } p_{vi} \geq p_t, \\ 0 & \text{otherwise,} \end{cases} \tag{2.18}$$

where $p_t \in (0.5, 1)$ is the threshold probability that represents the confidence level of the sensor. That is, the sensor decisions are deemed sufficiently reliable only at those d_{vi} distances where $p_{vi} > p_t$. Depending on the application and the false alarm requirement, typically $p_t \geq 0.9$. Note that p_{vi}^* is not a probability measure, but we shall nevertheless treat it as one in the ensuing calculations.

The detection probability p_v at any grid point v is defined as

$$p_v = 1 - \prod_{i=1}^{R}(1 - p_{vi}^*) \tag{2.19}$$

where R is the number of sensor nodes deployed in the field. The miss probabilities of the starting and destination points are one, that is $p_0 = 0$ and $p_{NM+1} = 0$. More clearly, these points are not monitored because they are not in the sensing coverage area. The boundary regions are not taken into consideration.

The weakest breach path problem can now be defined as finding the permutation of a subset of grid points $V = [v_0, v_1, \ldots, v_k]$ with which a target traverses from the starting

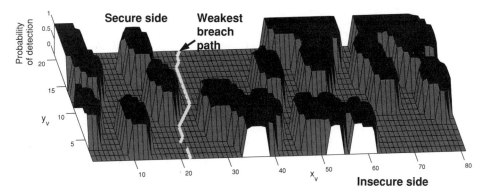

Figure 2.59 A sample sensing coverage and breach path where the field is 70 × 20 m, the boundary is 5-m wide, and the grid size is 1 m ($N = 81$, $M = 21$, $L = 100$, $R = 30$, $\alpha = 0.1$, $\eta = 5$, $\gamma = 30$ dB.).

point to the destination point with the least probability of being detected where $v_0 = 0$ is the starting point and $v_k = NM + 1$ is the destination point. The nodes v_{j-1} and v_j, $j = 0, 1, \ldots, k$, are connected to each other where $c_{v_{j-1}, v_j} = 1$. Here we can define the breach probability P of the weakest breach path V as

$$P = \prod_{v_j \in V} (1 - p_{v_j}) \tag{2.20}$$

where p_{v_j} is the detection probability associated with the grid point $v_j \in V$, and it is defined as in Eq. 2.19. A sample sensing coverage and breach path is shown in Figure 2.59. Using the two-dimensional field model and adding the detection probability as the third axis, we obtain hills and valleys of detection probabilities. The weakest breach path problem can be informally defined as finding the path which follows the valleys and through which the target does not have to climb hills so much.

In order to solve the weakest breach path problem, Dijkstra's shortest path algorithm [80] can be used. The detection probabilities associated with the grid points cannot be directly used as weights of the grid points, and consequently, they must be transformed to a new measure d_v. Specifically, we assign the negative logarithms of the miss probabilities, defined as

$$d_v = -\log(1 - p_v) \tag{2.21}$$

as weights of the grid points. This algorithm finds the path with the smallest negative logarithm value that turns out to be the largest breach probability. A similar application of Dijkstra's algorithm can be found in [81] for a network with decision fusion, where the sensor detection is not NP optimal.

Using Dijkstra's algorithm, the breach probability can be defined as the inverse transformation of the weight d_{NM+1} of the destination point which is

$$P = 10^{-d_{NM+1}}. \tag{2.22}$$

The found path, V can be used to calculate the breach probability in Eq. 2.20 that is equal to the value computed in Eq. 2.22.

Table 2.6   Parameter Values Used in the Simulations for the LCFA and HFCA Scenarios

Parameter	LCFA	HCFA
Length	20 m	100 m
Width	5 m	10 m
Boundary	10 m	10 m
Grid size	1 m	1 m
N	41	121
M	6	11
α	0.1	0.01
η	5	3
γ	30 dB	30 dB
L	100	100
p_t	0.9	0.9
R	17	31

In the next section, the impact of various parameters on the breach probability is investigated. The effect of the field shape on the breach probability is also analyzed, and a method for computing the required number of sensor nodes is provided.

2.5.3   Breach Probability Analysis

In this section, two SWSN scenarios are considered:

- *Low-Cost False Alarm (LCFA):*   For example, a house or a factory is to be monitored for intrusion detection. In this scenario, the cost of false alarms is relatively low.
- *High-Cost False Alarm (HCFA):*   In this class of applications, the financial and personnel cost of a false alarm is significantly higher compared to LCFA. For example, the perimeter security of some mission-critical place such as an embassy or nuclear reactor is to be provided by deploying an SWSN to monitor unauthorized access. The cost of a false alarm might involve the transportation of special forces and/or personnel of related government agencies to the embassy/museum, as well as, the evacuation of residents in the surrounding area.

The false alarm rate is set to 0.01 and 0.1 for HCFA and LCFA, respectively. The other parameter values are listed in Table 2.6. The grid size is taken as 1 m to be able to assume that the detection probabilities of targets on adjacent grid points are independent. These models can be considered as the building blocks that may be used to cover larger fields. The results are the averages of 50 runs.

Effects of Parameters on the Breach Probability   The breach probability P is quite sensitive to the false alarm rate α. As shown in Figure 2.60a for the LCFA scenario and in Figure 2.60b for the HCFA scenario, as α increases, the SWSN allows more false alarms. Because α reflects the tolerance level to false alarm errors, the NP detection probability and the detection probability p_v of the targets at grid point v both increase in α. Consequently, the breach probability decreases.

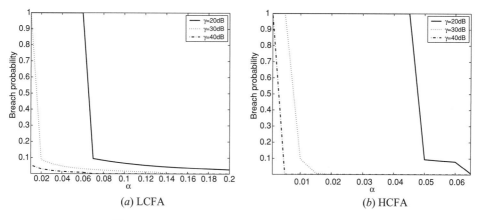

Figure 2.60 The effect of α on the breach probability.

For a given α and γ pair, there is an upper bound on the path-loss exponent for which a breach probability requirement can be met. When the false alarm rate is high as in LCFA, cluttered and obstructed environments are still successfully monitored by the network. For instance, Figure 2.61a suggests that $\gamma = 20$ dB is sufficient for $\eta = 4.5$. On the other hand, with tight control of the false alarms, the sensors must be carefully positioned to have line-of-sight (see Fig. 2.61b).

As the signal-to-noise ratio γ increases, the detection performance improves (see Fig. 2.62), and the breach probability decreases. Depending on the path-loss exponent, $\gamma = 10$ dB yields minimal breach probability for both LCFA and HCFA. Note that η and γ display a duality in that if one is fixed, the performance breaks down when the other parameter is below or above some value. For example, for $\gamma = 30$ dB, $P \to 1$ as soon as η exceeds 5.5 in LCFA (Fig. 2.61a). Similarly, for the same scenario, breach detection becomes impossible once $\gamma < 6$ dB if $\eta = 4$ (Fig. 2.62a). The deterioration is somewhat more graceful for HCFA.

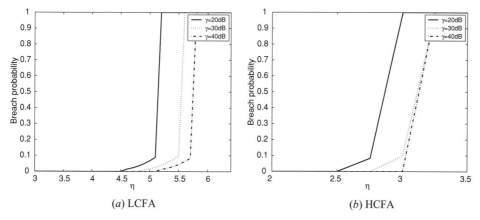

Figure 2.61 The effect of η on the breach probability.

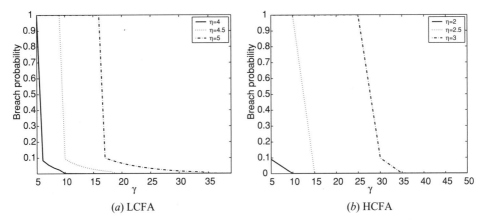

Figure 2.62 The effect of γ on the breach probability.

Figure 2.63b depicts that a data record of 60 and 115 samples per breach decision is sufficient for LCFA and HCFA, respectively, if $P \approx 0.1$ is good enough. In general, more data samples per breach decision are required if a low false alarm rate is desired. However, note that L grows asymptotically to the same quantity for both LCFA and HCFA as $P \rightarrow 0$. For active sensors, restrictions on energy consumption may prohibit collecting too many samples.

Determining the Required Number of Sensor Nodes While analyzing the required number of sensor nodes for a given breach probability, we consider two cases of random deployment. In the first case, we assume that the sensor nodes are uniformly distributed along both the vertical and horizontal axes. In the second case, the sensor nodes are deployed uniformly along the horizontal axis and normally distributed along the vertical axis with mean $M/2$ and a standard deviation of 10% of the width of the field. In the simulations, the sensor nodes that are deployed outside the field are not included in the computations of the detection probabilities.

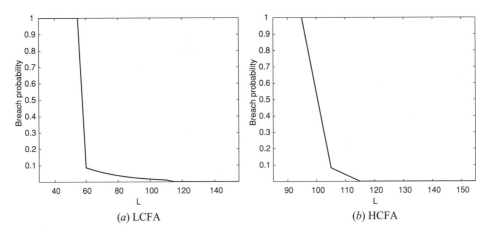

Figure 2.63 The effect of L on the breach probability.

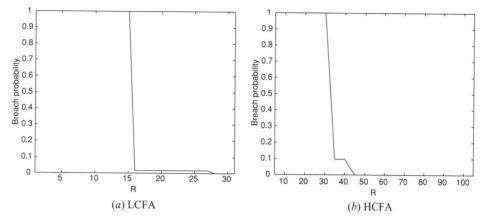

Figure 2.64 The effect of the number of sensor nodes on the breach probability for $y_v \sim$ Uniform$(0, M - 1)$.

Considering uniformly distributed y-axis scheme, the required number of sensor nodes for a given breach probability is plotted in Figure 2.64. A breach probability of 0.01 can be achieved by utilizing 16 sensor nodes for LCFA, and 45 for HCFA. Exchanging the false alarm rates to $\alpha = 0.01$ for LCFA and $\alpha = 0.1$ for HCFA, the requirement becomes 28 and 30 sensor nodes, respectively. The rapid decrease in the breach probability at $R = 16$ in Figure 2.64a can be justified by the fact that most of the grid points are covered with high detection probabilities (saturated) for $R = 15$, and adding one more sensor node decreases the breach probability drastically. Once the saturation is reached, placing more sensors in the field has marginal effect.

Analyzing Figure 2.65, the above-mentioned saturation is seen more clearly for the normally distributed y-axis scheme. For this kind of deployment, since the sensor node may fall outside the field, the breach probability decreases slower compared to the uniformly distributed y-axis scheme.

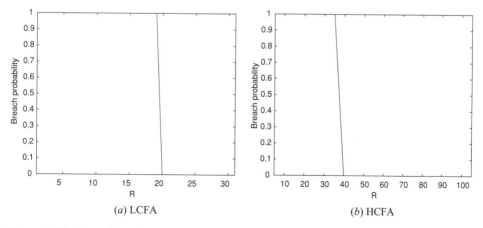

Figure 2.65 The effect of the number of sensor nodes on the breach probability for $y_v \sim$ Normal$(M/2, N/10)$.

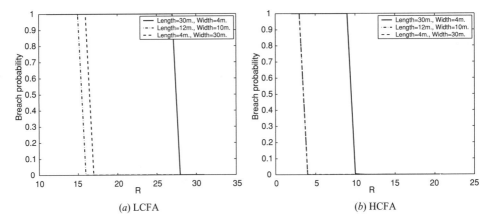

Figure 2.66   The effect of the field shape on breach probability for $y_v \sim$ Uniform(0, $M - 1$).

Effect of Field Shape on the Breach Probability   Depending on the application, the field shape of the grid model may vary. In Figures 2.66 and 2.67, the effect of the field shape on the breach probability is depicted considering uniformly and normally distributed y-axis schemes, respectively. For a given number of sensor nodes, the breach probability is larger for narrow and long fields compared to the thick and short fields. For example, when uniform random deployment on both axes are considered, with 20 sensor nodes, it is possible to provide a breach probability below 0.01 for a field where the length is 4 m and the width is 30 m. However, with the same number of sensor nodes the breach probability turns out to be around one for the field where the length is 30 m and width is 4 m.

In Tables 2.7 and 2.8, different grid sizes are simulated for LCFA and HCFA, respectively, and the required number of sensor nodes are tabulated for $P \leq 0.01$. As the size of the grid becomes shorter and thicker, the required number of sensor nodes decreases. For the LCFA scenario, as the field is shortened and widened, the difference between the required number of sensor nodes for the uniformly and normally distributed y_v schemes decreases. However, the largest difference is obtained for the fields where the width is the smallest. The normally

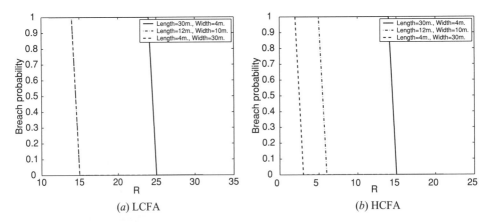

Figure 2.67   The effect of the field shape on breach probability for $y_v \sim$ Normal($M/2$, $N/10$).

Table 2.7 Effect of Field Shape on Required Number of Sensor Nodes for a Breach Probability of 0.01 for LCFA Scenario

Length (m)	Width (m)	$y_v \sim \text{Uniform}(0, M-1)$	$y_v \sim \text{Normal}(M/2, N/10)$
40	3	16	20
30	4	11	15
24	5	11	13
20	6	7	13
15	8	4	3
12	10	4	3
10	12	3	3
8	15	3	2

distributed y_v scheme is more determining of the required number of sensor nodes, because it produces a deployment where many sensor nodes are placed around the center line of the field along the horizontal axis. This deployment scheme produces a well-secured barrier in the middle of the field.

2.5.4 Conclusions

In this section, we employ the Neyman–Pearson detector to find the sensing coverage area of the surveillance wireless sensor networks. In order to find the breach path, we apply Dijkstra's shortest path algorithm by using the negative log of the miss probabilities as the grid point weights. By defining the breach probability as the miss probability of the weakest breach path, the false alarm rate constraint has a significant impact on the breach probability, as well as the required number of sensor nodes for a given breach probability level. For fields where the signal attenuates faster with distance, large SNR levels are needed. Upon analyzing the effect of the field shape on breach probability, it is concluded that the differences between the breach probabilities of uniformly and normally distributed y-axis schemes are larger for narrower fields. Furthermore, the width of the field has a noticeable impact on the breach probability.

The model and results developed herein give clues that link false alarms to energy efficiency. Enforcing a low false alarm rate to avoid unnecessary response costs implies either a larger data-set (L) and hence a greater battery consumption, or a denser sensor network,

Table 2.8 Effect of Field Shape on Required Number of Sensor Nodes for a Breach Probability of 0.01 for HCFA Scenario

Length (m)	Width (m)	$y_v \sim \text{Uniform}(0, M-1)$	$y_v \sim \text{Normal}(M/2, N/10)$
40	3	10	13
30	4	4	3
24	5	4	3
20	6	4	3
15	8	3	2
12	10	2	2
10	12	2	2
8	15	2	2

which increases the deployment cost. Similar qualitative and/or quantitative inferences about the relationships between various other parameters can also be made. Wireless sensor networks are prone to failures. Furthermore, the sensor nodes die due to their limited energy resources. Therefore, the failures of sensor nodes must be modelled and incorporated into the breach path calculations in the future. Simulating the reliability of the network throughout the entire life of the wireless sensor network is also required. Lastly, especially for the perimeter surveillance applications, the obstacles in the environment play a critical role in terms of sensing and must be incorporated in the field model.

REFERENCES

1. D. Estrin, R. Govindan, J. Heidemann, and S. Kumar, "Next century challenges: Mobile networking for smart dust," in *Proceedings of the MOBICOM*, Seattle, 1999, pp. 271–278.

2. G. J. Pottie and W. Kaiser, "Wireless sensor networks," *Communications of the ACM*, vol. 43, pp. 51–58, 2000.

3. K. M. Alzoubi, P. J. Wan, and O. Frieder, "Distributed heuristics for connected dominating sets in wireless ad hoc networks," *Journal of Communications and Networks*, vol. 4, pp. 1–8, 2002.

4. J. H. Y. Xu and D. Estrin, "Geography informed energy conservation for ad hoc routing," in *Proceedings of the MOBICOM*, Rome, 2001, pp. 70–84.

5. B. Chen, K. Jamieson, H. Balakrishnan, and R. Morris, "Span: An energy-efficient co-ordination algorithm for topology maintenance in ad hoc wireless networks," in *Proceedings of the MOBICOM*, Rome, 2001, pp. 85–96.

6. C. Schurgers, V. Tsiatsis, and M. B. Srivastava, "Stem: Topology management for energy-efficient sensor networks," in *Proceedings of the IEEE Aero Conference*, Big Sky Montana, 2002, pp. 135–145.

7. D. Tian and N. D. Georganas, "A coverage-preserving node scheduling scheme for large wireless sensor networks," in *Proceedings of the WSNA*, Atlanta, 2002, pp. 32–41.

8. W. R. Heinzelman, A. Chandrakasan, and H. Balakrishnan, "Energy-efficient communication protocol for wireless microsensor networksw. r. heinzelman, a. chandrakasan and h. balakrishnan," in *Proceedings of the HICSS*, Maui, Hawaii, 2000, pp. 3005–3014.

9. J. H. Y. Xu and D. Estrin, "Adaptive energy conservating routing for multihop ad hoc routing," Technical Report 527, USC/ISI, 2000.

10. C. S. Raghavendra and S. Singh, "Pamas: Power-aware multi-access protocol with signaling for ad hoc networks," *ACM Communications Review*, vol. 28, pp. 5–26, 1998.

11. K. Sohrabi and G. J. Pottie, "Performance of a novel self-organization protocol for wireless ad hoc sensor networks," in *Proceedings of the IEEE VTC*, 1999, pp. 1222–1226.

12. P. Bergamo and G. Mazzini, "Localization in sensor networks with fading and mobility," in *Proceedings of the PIMRC*, Lisbon, Portugal, 2002.

13. B. Hoffman-Wellenhof, H. Lichteneger, and J. Collins, *Global Positioning System: Theory and Practice*, 4th ed., Vienna, Austria: Springer-Verlag, 1997.

14. N. Patwari and R. J. O'Dea, "Relative location in wireless networks," in *Proceedings of the IEEE VTC*, vol. 2, 1991, pp. 1149–1153.

15. K. Whitehouse and D. Culler, "Calibration as parameter estimation in sensor networks," in *Proceedings of the WSNA*, Atlanta, 2002, pp. 59–67.

16. L. Girod and D. Estrin, "Robust range estimation for localization in ad hoc sensor networks," Technical Report CS-TR-2000XX, University of California, Los Angeles, 2000.

17. N. H. Vaidya, P. H. Krishna, M. Chatterjee, and D. K. Pradhan, "A cluster-based approach for routing in dynamic networks," *ACM Computer Communications Review*, vol. 27, pp. 49–65, Apr. 1997.

18. M. Gerla and J. T. Tsai, "Multicluster, mobile, multimedia radio network," *Wireless Networks*, vol. 1, pp. 255–265, Oct. 1995.

19. S. Basagni, "Distributed clustering for ad hoc networks," in *Proceedings of the ISPAN*, Fremantle, Australia, 1999, pp. 310–315.

20. "Wireless integrated network systems," http://wins.rsc.rockwell.com.

21. "Ash transceiver's designers guide," http://www.rfm.com.

22. *Wireless LAN Medium Access Control and Physical Layer Specifications*, IEEE 802.11 Standard (IEEE LAN MAN Standards Committee), Aug. 1999.

23. J. Broch, D. Maltz, D. Johnson, Y. Su, and J. Jetcheva, "A performance comparison of multi-hop wireless ad hoc network routing protocols," in *Proceedings of the MOBICOM*, Dallas, 1998, pp. 85–97.

24. E. Biagioni and K. Bridges, "The application of remote sensor technology to assist the recovery of rare and endangered species," *International Journal of High Performance Computing Applications*, vol. 16, no. 3, Aug. 2002, special issue on Distributed Sensor Networks.

25. A. Cerpa, J. Elson, L. G. D. Estrin, M. Hamilton, and J. Zhao, "Habitat monitoring: Application driver for wireless communications technology," in *Proceedings of the ACM SIGCOMM Workshop on Data Communications*, San Francisco, 2001.

26. A. Mainwaring, J. Polastre, R. Szewczyk, D. Culler, and J. Anderson, "Wireless sensor networks for habitat monitoring," in *Proceedings of ACM International Workshop on Wireless Sensor Networks and Applications*, Atlanta, Sept. 2002.

27. H. Wang, J. Elson, L. Girod, D. Estrin, and K. Yao, "Target classification and localization in habitat monitoring," in *Proceedings of the IEEE ICASSP*, Hong Kong, Apr. 2003.

28. I. F. Akyildiz, W. Su, Y. Sankarasubramaniam, and E. Cayirci, "A survey on sensor networks," *IEEE Communications Magazine*, vol. 40, no. 8, pp. 102–114, 2002.

29. L. Schwiebert, S. K. S. Gupta, and J. Weinmann, "Research challenges in wireless networks of biomedical sensors," in *Proceedings of Mobile Computing and Networking*, Rome, 2001, pp. 151–165.

30. P. Boettcher, J. A. Sherman, and G. A. Shaw, "Target localization using acoustic time-difference of arrival in distributed sensor networks," in *Proceedings of the SPIE Forty-Seventh Annual Meeting*, Seattle, 2002.

31. E. Howden, "Networked sensors for the objective force," in *Proceedings of SPIE Forty-Seventh Annual Meeting*, Seattle, 2002.

32. "Alert systems," http://www.alertsystems.org/.

33. "Corie observations," http://www.ccalmr.ogi.edu/CORIE/.

34. D. C. Steere, A. Baptista, D. McNamee, C. Pu, and J. Walpole, "Research challenges in environmental observation and forecasting systems," in *Proceedings of the Sixth Annual International Conference on Mobile Computing and Networking*, Boston, 2000, pp. 292–299.

35. J. Hill, R. Szewczyk, A. Woo, S. Holar, D. Culler, and K. S. J.Pister, "System architecture directions for networked sensors," *Operating Systems Review*, vol. 34, pp. 93–104, 2000.

36. N. Bulusu, J. Heidemann, and D. Estrin, "Gps-less lowcost outdoor localization for very small devices," *IEEE Personal Communications*, vol. 7, no. 5, pp. 28–34, 2000.

37. S. Capkun, M. Hamdi, and J.-P. Hubaux, "Gps-free positioning in mobile ad-hoc networks," in *Hawaii International Conference on System Sciences (HICSS-34)*, Maui, HI, Jan. 2001, pp. 3841–3490.

38. F. Ye, H. Luo, J. Cheng, S. Lu, and L. Zhang, "A two-tier data dissemination model for large-scale wireless sensor networks," in *MOBICOM*, Atlanta, 2002.

39. D. Niculescu and B. Nath, "Ad-hoc positioning system," in *Proceedings of IEEE GlobeCom*, San Antonio, Texas, Nov. 2001.

40. C. Savarese, J. Rabaey, and J. Beutel, "Locationing in distributed ad-hoc wireless sensor networks," in *IEEE International Conference on Acoustics, Speech, and Signal Processing (ICASSP)*, Salt Lake City, UT, May 2001, pp. 2037–2040.

41. A. Savvides, C.-C. Han, and M. Srivastava, "Dynamic fine-grained localization in ad-hoc networks of sensors," in *Seventh ACM International Conference on Mobile Computing and Networking (Mobicom)*, Rome, Italy, July 2001, pp. 166–179.

42. N. Priyantha, A. Chakraborty, and H. Padmanabhan, "The cricket location support system," in *Sixth ACM MOBICOM*, Boston, MA, Aug. 2000.

43. D. Niculescu and B. Nath, "Ad hoc positioning system (aps) using aoa," in *INFOCOM*, Apr. 2003.

44. N. Bulusu, J. Heidemann, and D. Estrin, "Adapive beacon placement," in *Proceedings of the Twenty-First International Conference on Distributed Computing Systems (ICDCS-21)*, Phoeniz, AZ, Apr. 2001.

45. N. Bulusu, J. Heidemann, V. Bychkovskiy, and D. Estrin, "Density-adaptive beacon placement algorithms for localization in ad hoc wireless networks," in *IEEE Infocom 2002*, New York, June 2002.

46. A. Savarese, "Robust positioning algorithms for distributed ad hoc wireless sensor networks," Master's thesis, University of California, Berkeley, 2002.

47. I. Borg and P. Groenen, *Modern Multidimensional Scaling Theroy and Applications*, Berlin: Springer, 1997.

48. P. Green, F. Caromone, and S. Smith, *Multidimensional Scaling: Concepts and Applications*, Newton, MA: Allyn and Bacon, 1989.

49. N. Metropolis, A. Rosenbluth, M. Rosenbluth, A. Teller, and E. Teller, "Equations of state calculations by fast computing machines," *Journal of Chemical Physics*, vol. 21, pp. 1087–1092, 1953.

50. S. Schiffman, M. Reynolods, and F. Young, *Introduction to Multidimensional Scaling*, New York: Academic, 1981.

51. L. Doherty, K. Pister, and L. E. Ghaoui, "Convex position estimation in wireless sensor networks," in *IEEE Infocom 2001*, Anchorage, AK, Apr. 2001.

52. N. Bulusu, J. Heidemann, and D. Estrin, "Gps-less low cost outdoor localization for very small devices," *IEEE Personal Communications Magazine*, vol. 7, no. 5, pp. 28–34, Oct. 2000. Available: http://lecs.cs.ucla.edu/~bulusu/papers/Bulusu00a.html.

53. D. Niculescu and B. Nath, "Ad hoc positioning system (aps)," in *IEEE GLOBECOM*, Nov. 2001.

54. D. Niculescu and B. Nath, "Dv based positioning in ad hoc networks," *Kluwer Journal of Telecommunication Systems*, pp. 267–280, 2003.

55. S. N. Simic and S. Sastry, "Distributed localization in wireless ad hoc networks," Technical Report UCB/ERL M02/26, University of California at Berkeley, 2002.

56. A. Savvides, C.-C. Han, and M. B. Strivastava, "Dynamic fine-grained localization in ad-hoc networks of sensors," in *Proceedings of Mobile Computing and Networking*, Rome, 2001, pp. 166–179.

57. J. R. Chris Savares and K. Langendoen, "Robust positioning algorithms for distributed ad-hoc wireless sensor networks," paper presented at the USENIX Technical Annual Conference, June 2002.

58. K. Whitehouse, "The design of calamari: An ad-hoc localization system for sensor networks," Master's thesis, University of California at Berkeley, 2002.

59. M. I. J. XuanLong Nguyen and B. Sinopoli, "A kernel-based learning approach to ad hoc sensor network localization," *ACM Trans. Sensor Networks*, vol. 1, no. 1, pp. 134–152, 2005.

60. R. H. K. J. M. Kahn and K. S. J. Pister, "Mobile networking for smart dust," paper presented at the ACM/IEEE International Conference on Mobile Computing and Networking (MobiCom 99), Seattle, Aug. 17–19, 1999.

61. B. Hofmann-Wellenhof, H. Lichtenegger, and J. Collins, *Global Positioning System: Theory and Practice*, Berlin: Springer-Verlag, 1992.

62. "Sychip gps module," http://www.sychip.com/gps-module.html, June 2005.

63. A. Ward, A. Jones, and A. Hopper, "A new location technique for the active office," *IEEE Personal Communications*, vol. 4, no. 5, pp. 42–47, Oct. 1997.

64. L. Girod and D. Estrin, "Robust range estimation using acoustic and multimodal sensing," in *IEEE/RSJ International Conference on Intelligent Robots and Systems (IROS)*, Maui, HI, Oct. 2001.

65. K. Whitehouse, paper submitted at the USENIX Technical Annual Conference, June 2002, submitted to Sensys.

66. K. Whitehouse and D. Culler, "Macro-calibration in sensor/actuator networks," *Mobile Networks and Applications Journal (MONET)*, special issue on Wireless Sensor Networks, vol. 8, no. 4, pp. 463–472, June 2003.

67. J. B. Kruskal, "Multidimensional scaling," *Quantitative Applications in the Social Science*, SAGE Publications Thousand Oaks, CA, 1978.

68. K. Whitehouse, http://www.cs.berkeley.edu/ kamin/calamari/index.html, June 2005.

69. A. Woo, "Mica sensor board," http://today.cs.berkeley.edu/tos/hardware/hardware.html, June 2005.

70. H. Zhang and C.-J. Hou, "On deriving the upper bound of α-lifetime for large sensor networks," Technical Report UIUCDCS-R-2004-2410, Department of Computer Science, University of Illinois at Urbana-Champaign, 2004.

71. T. Clouqueur, V. Phipatanasuphorn, P. Ramanathan, and K. K. Saluja, "Sensor deployment strategy for detection of targets traversing a region," *Mobile Networks and Applications*, vol. 8, no. 4, pp. 453–461, Aug. 2003.

72. Y. Zou and K. Chakrabarty, "Sensor deployment and target localization based on virtual forces," in *Proceedings of the IEEE INFOCOM*, San Francisco, Apr. 2003, pp. 1293–1303.

73. A. Howard, M. J. Mataric, and G. S. Sukhatme, "An incremental self-deployment algorithm for mobile sensor networks," *Autonomous Robots*, vol. 13, no. 2, pp. 113–126, Sept. 2002.

74. S. Tilak, N. B. Abu-Ghazaleh, and W. Heinzelman, "Infrastructure tradeoffs for sensor networks," in *Proceedings of the First ACM International Workshop on Wireless Sensor Networks and Applications*, Atlanta, 2002, pp. 49–58.

75. S. Megerian, F. Koushanfar, G. Qu, G. Veltri, and M. Potkonjak, "Exposure in wireless sensor networks: Theory and practical solutions," *Wireless Networks*, vol. 8, no. 5, pp. 443–454, Sept. 2002.

76. T. He, S. Krishnamurthy, T. A. J. A. Stankovic, L. Luo, R. Stoleru, T. Yan, and L. Gu, "Energy-efficient surveillance system using wireless sensor networks," in *Proceedings of the Second International Conference on Mobile Systems, Applications, and Services*, Boston, June 2004, pp. 270–283.

77. D. Kazakos and P. Papantoni-Kazakos, *Detection and Estimation*, New York: Computer Science Press, 1990.

78. G. Xing, C. Lu, R. Pless, and J. A. O'Sullivan, "Co-grid: An efficient coverage maintenance protocol for distributed sensor networks," in *Proceedings of the 3rd International Symposium on Information Processing in Sensor Networks*, Berkeley, Apr. 2004, pp. 414–423.

79. T. S. Rappaport, *Wireless Communications: Principles and Practice*, Upper Saddle River, NJ: Prentice-Hall, 1996.

80. M. A. Weiss, *Data Structures and Algorithm Analysis in C++*, 2nd ed., Reading, MA: Addison-Wesley, 1999.

81. V. Phipatanasuphorn and P. Ramanathan, "Vulnerability of sensor networks to unauthorized traversal and monitoring," *IEEE Transactions on Computers*, vol. 53, no. 3, pp. 364–369, Mar. 2004.

3

PURPOSEFUL MOBILITY
AND NAVIGATION

3.1 INTRODUCTION

Mobility in wireless communications networks is an external stimulus that must be considered in the design of the network and refers to the mobility of the user community utilizing the network. When mobility is so enforced on a sensor network, either by an independent mobile platform in which a sensor node is embedded or by the environment, we call it passive mobility. Most recent literature deals with this kind of mobility, generally modeled as random mobility, and the network protocols are designed to adapt to this uncontrolled repositioning of nodes. In contrast, mobile sensor networks may be designed to incorporate purposeful mobility that enables the dynamic placement of sensors in critical locations to maintain connectivity, conserve resources, or to support space–time critical mission goals with limited resources.

Sensor networks generally observe systems that are too complex to be simulated by computer models based directly on their physics. In operating a sensor network, it is assumed that the underlying physical processes evolve at different time scales. In the fast time scale, that is, over short periods of time, the positions of the sensors may be considered to be static, and the system may be assumed to be an ergodic, discrete-event Markov process.

In the slowly varying time scale, that is, over longer periods of time, the sensor positions may need to shift due to passive mobility, failures, or changes in the operational environment. In this case, the assumption is that the time scale for assessing changes in the network topology is much longer than the time scale for changes in the operational dynamical variables. The system may therefore be considered to be in semiequilibrium for time-critical observations of short duration so its dynamics can be determined from data samples whose duration is long compared to changes in the dynamical variables but short compared to changes in sensor positioning.

Sensor Network Operations, Edited by Phoha, LaPorta, and Griffin
Copyright © 2006 The Institute of Electrical and Electronics Engineers, Inc.

The sections in this chapter develop algorithms and protocols to utilize controlled mobility to enhance the ability of a sensor network to execute its mission in this state of semiequilibrium and to adapt to changes in the slow time scale. In Section 3.2, Tirta et al. develop resource-efficient algorithms for adaptation to passive mobility by introducing controllable mobile nodes that can reposition themselves to mitigate the negative effects of passive mobility. Section 3.3, by Cao et al., addresses sensor operations with purposeful mobility in the presence of failures to best harness the power of the network to monitor its environment. Section 3.4, by Roy et al., develops control and actuation mechanisms to achieve movement in formation with collision avoidance by taking advantage of the multiple directions of motion available to each sensor node. Section 3.5, by Lian et al., addresses life extension of a statically deployed data collection sensor network by nonuniform sensor deployment strategies and also by introducing a protocol for predictable sink mobility to maximize distributed data collection with limited resources.

3.2 CONTROLLED MOBILITY FOR EFFICIENT DATA GATHERING IN SENSOR NETWORKS WITH PASSIVELY MOBILE NODES

Yuldi Tirta, Bennett Lau, Nipoon Malhotra, Saurabh Bagchi, Zhiyuan Li, and Yung-Hsiang Lu

A large class of sensor networks is used for data collection and aggregation of sensory data about the physical environment. Since sensor nodes are often powered by limited energy sources, such as a battery that may be difficult to replace, energy saving is an important criterion in any activity. Some deployments of sensor networks have *passive mobile* nodes, that is, nodes that are mobile without their own control. For example, a node mounted on an animal for biohabitat monitoring, or a light-weight node dropped into a river for water quality monitoring. Passive mobility makes the activity of data gathering challenging since the positions of the nodes can change arbitrarily. As a result, the nodes may move too far from the data aggregation point, such as a base station, making the data transmission extremely energy intensive. In extreme cases, the nodes may become disconnected from the rest of the network, making them unusable. We propose a sensor network architecture with some nodes capable of *controlled mobility* to solve this problem. Controlled mobility implies the nodes can be moved in a controlled manner in response to commands, with a determined direction, speed, and so forth. We present the different categories of nodes in our architecture and mobility algorithms for the two classes, called *collector* and *locator*, that have controlled mobility. It is well accepted that efficient data gathering benefits from the knowledge of locations of nodes. Passive mobile nodes makes location determination (i.e., localization) a crucial problem. We propose the use of the locators for this through a novel scheme based on triangulation. We provide theoretical and simulation based analysis of the mobility algorithms with respect to the metrics of energy, latency, and buffer space requirements.

3.2.1 Background Information

Sensor networks are a particular class of wireless ad hoc networks in which the nodes have small components for sensors, actuators, and radio frequency (RF) communication components. Sensor nodes are dispersed over the area of interest, called sensor field, and are capable of short-range RF communication (about 100 ft) and contain signal processing engines to manage the communication protocols and for data processing [1]. The individual nodes have a limited processing capacity but are capable of supporting distributed applications through coordinated effort in a network that can include hundreds or even thousands of nodes. Sensor nodes are typically battery-powered. Since replacing batteries is often very difficult, reducing energy consumption is an important design consideration for sensor networks. Because transmitting power is proportional to the square or quadruple of the transmission range, the range of a sensor node is constrained in most deployments.

In sensor networks, the final data gathering and analysis station, called the base station, is sometimes placed far from the sensing nodes. This may be because the sensing field is in a hazardous environment, such as enemy territory, or a physically harsh environment, with high temperature, and the like. It may be impossible to locate a protected, high-end base station with large computational and communication capabilities in such a hazardous environment. Since the transmission range of the individual nodes is limited, the large separation between the sensing region and the base station implies long-distance and multihop transmission has to occur. This architecture is difficult to deploy and maintain with regular sensing nodes acting as relay nodes.

Some nodes in the network may be mobile, either in a controlled or in an uncontrolled manner. Uncontrolled mobility, referred to as *passive mobility* implies the nodes move but not of their own volition. Examples include nodes embedded into animals, nodes carried by human beings who move according to other considerations, and light-weight nodes carried by physical processes such as running water, glaciers, or wind. Nodes with passive mobility make data collection more challenging. Some nodes may move far away from the data aggregation point, such as a base station or a cluster head, and even become disconnected from the rest of the network. With mobility, routing data to the nodes may become inefficient since many sensor network routing protocols rely on position knowledge [2, 3]. Hence, traditionally, mobility has been looked on as an adversary to efficient deployment of sensor networks due to degradation of the topology affecting parameters such as connectivity and coverage [4] or due to increased failure rates of wireless links [5]. We turn this argument on its head and propose to use controlled mobility of certain nodes in the network to our advantage. *Controlled mobility* implies the ability to control the movement of an entity (direction, velocity, etc.) by sending it control commands.

Several studies combine controlled mobility and sensor networks. LaMarca et al. [6] suggested using mobile robots to deploy and calibrate sensors, to detect their failures, and to recharge nodes using radio frequency or infrared signals. They built a prototype of sensor networks for house plants with a mobile robot. Sibley et al. [7] built miniature robots with sensors, called robomotes. These sensors were equipped with wireless communication, odometer, infrared object sensors, and solar cellars. Each robomote is only 47 cm^3. Rybski et al. [8] presented a system for reconnaissance and surveillance using two types of robots: scouts and rangers. A scout is a mobile robot smaller than a soda can. A ranger can carry 10 scouts and launch each 10-scout unit up to a distance of 30 m. These examples demonstrate the practicability of combining sensor networks and robots. On the other hand, they have not fully addressed the issues encountered in large-scale sensor networks with passively mobile nodes. Miller et al. [9] explained that the slow motion of the Mars rover was not due to the limitation of technology; instead, it was mainly the concern of safety and the long communication delay between Mars and Earth. Hence, it is possible to construct sensory robots with controlled movement for data collection.

In this section, we demonstrate the use of controlled mobility to solve the problem of energy-efficient data collection in sensor networks that have a large geographical spread and management of nodes that exhibit passive mobility. Since mobility is an energy-expensive process itself, our solution equips only a small fraction of nodes with the ability for controlled mobility and, for one class of nodes, enables its energy source to be periodically recharged. We propose a network architecture with five classes of nodes. The first class consists of ordinary *sensing nodes* that are dispersed over the sensor field and have the constraints of communication, computation, and energy traditionally associated with sensor nodes. Some of these nodes may exhibit passive mobility, while the rest are stationary. The network is divided into clusters with each sensing node assigned to a cluster. Each cluster has a set of *cluster heads*, only one of which plays that role at a time. The cluster head buffers data from the sensing nodes awaiting further transmission. The cluster heads are off-the-shelf sensor nodes with one addition—they have larger memory for storage of the sensed data. The third class of nodes is the *data collector*. A data collector has the capability for controlled motion and visits the cluster heads according to a predetermined schedule, collecting the sensed data from them and transmitting the data to the base station. The data collectors have the option of occasionally returning to the base station for recharging their energy source. The fourth class of nodes is called *locators*, which are assigned to a cluster. These nodes move within the cluster helping the passively mobile sensing nodes determine their locations and assigning them to the closest cluster head. The locators have a hardware device, such as a Global Positioning System (GPS) receiver, that enables them to determine their own location. The final class of nodes is the *connectors*, which exhibit controlled mobility in order to maintain certain topological properties of the network, such as connectivity and coverage. The role of connectors will form the topic of a separate publication and is not discussed further here.

The goals of our solution using the five classes of nodes are the following:

1. Reduce the distances for data transmission and, therefore, the energy expended in communication of sensed data.
2. Determine the locations of the sensing nodes with passive mobility so that data can be gathered from them efficiently.

An alternate architecture may be proposed that spreads many nodes between the sensing nodes and the base station to act as relay nodes. The data is then communicated through multihop communication from the sensing nodes to the base station. This architecture poses several challenges. The terrain may be such that placement of the intermediate relay nodes is difficult or infeasible. For example, consider marshy lands or water bodies. In such an environment, an unmanned aerial vehicle (UAV) can act as a collector and will likely be easier to deploy. Another challenge is the maintenance of the relay nodes. They will likely be identical to the ordinary sensing nodes since many of them will have to be deployed. Their constraints, including energy drain, will be a matter of concern. The nodes close to the base station will face the funneling effect whereby larger and larger fractions of the network data get funneled through them. This will lead them to drain their energy rapidly.

Our target deployment has a large sensor field and therefore many relay nodes will have to be used. This will make it challenging to provide deterministic bounds on the data latency for the end-to-end transmission of sensed data from the sensing node to the base station. The focus of this section is the controlled mobility algorithms for the data collectors and the locators. The mobility algorithms are evaluated with respect to the energy expended, the buffer usage, and the end-to-end latency of sensed data. A set of mobility algorithms are proposed that differ in their degree of prescience of the network and the parameter that they

Table 3.1 Different Types of Nodes and Their Characteristics

Type	Number	Capability	Mobility	Limitation
Sensor node	Many	Sensing data	Passive or static	Limited memory and energy
Collector	Few	Collect data from heads	Active, high	Energy for long-distance movement
Connector	Some	Improve network	Active, low	Limited energy
Locator	Some	GPS equipped	Active, low	Limited energy
Cluster head	Some	Large data buffer	No	Stationary, limited energy

optimize for. For example, one algorithm assumes knowledge of the size of each cluster and the data rate of the nodes in the cluster. Another algorithm optimizes the energy expended in the motion of the data collectors at the expense of higher data latency. The location determination algorithms are evaluated with respect to the above parameters plus the rate at which the location information is disseminated and the accuracy and precision of the location estimated. The evaluation is performed analytically and through simulation with NS-2 as the simulation environment.

3.2.2 Network Architecture

The sensor network architecture being targeted comprises a large number of sensing nodes that collect information about the physical parameters of their environment. They are embedded in situ in the sensor field and dispersed over a sensor field that has a large geographical spread. The sensor field extends far from the base station to which the sensed data ultimately needs to be communicated back for processing and long-term storage. Some of the sensing nodes may move due to passive mobility.

Apart from the sensing nodes, there are four classes of special nodes introduced in Section 3.2.1—collectors, locators, cluster heads, and connectors. The different classes of nodes and their characteristics are summarized in Table 3.1. A schematic of the network without the four special classes of nodes is shown in Figure 3.1 and a network with the specialized nodes is shown in Figure 3.2. The classes of nodes that are capable of controlled mobility have control interfaces to which commands can be sent for the purpose of directing their motion. The collectors can be mounted on fast aerial vehicles or slower moving surface vehicles.

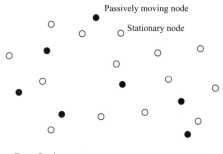

Figure 3.1 When the station is far away from all the sensor nodes, the last hop problem occurs.

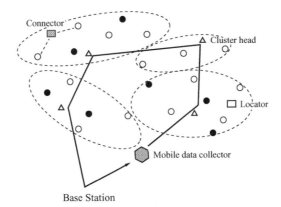

Figure 3.2   Four types of special nodes are added: data collectors, network connectors, locators, and cluster heads.

They have two interfaces—a low-range low-bandwidth RF communication interface for communication with the cluster head and a longer range higher bandwidth GPRS-like communication interface for communication with the base station. The purpose of the collectors is to make all wireless transmission from the sensing nodes and the cluster head short range and energy efficient. In this way, they are analogous to mailmen collecting mail from mailboxes. By using the postal service, people do not have to deliver mail by themselves—they only need to walk short distances to the mailboxes. Each locator is attached to a particular cluster and moves within the cluster for helping the sensing nodes determine their locations. They are equipped with location determination hardware, such as GPS receivers. The cluster heads have larger stable storage compared to the sensing nodes in order to hold the sensed data before handing it over to the collector. Adding cluster heads makes data collection more efficient. The collectors do not have to visit each node; instead, the collectors only need to visit the cluster heads. This is analogous to using mailboxes at the corners of streets to collect mail so that mailmen do not have to visit every house. In this architecture, only the collectors have high mobility and need to travel large distances, possibly at high speeds. Therefore, they have the provision of returning to the base station for recharging their energy source. The locators move within the extent of their respective cluster and are therefore considered to have low mobility. Their energy source may also be charged through the collectors. The locators may be made to return to their cluster head to coincide with the collector visiting the head.

   Let us look at the general-use scenario of the five classes of nodes in a typical deployment of the network. Initially, the nodes are deployed either through precise placement, such as manually placed at precise predetermined and known locations, or through pseudo-random placement, such as light-weight nodes aird-ropped from a moving aerial vehicle. The network is divided into multiple clusters based on geographical boundaries, a cluster head elected in each cluster, and the nodes assigned to different clusters using a high-energy beacon broadcast by each cluster head. Each sensing node collects data about its immediate environment and transmits the data to the cluster head, which is located much nearer to it than the base station. The mobile data collectors visit the cluster heads and collect the data of the entire cluster stored there. The data is then transmitted back by the collector to the base station. Within each cluster, the cluster head role is rotated between the candidate set of nodes, triggered by energy getting drained or for proximity to higher data producing nodes.

The locators are equipped with GPS receivers and move through its cluster helping the cluster heads and the passively mobile sensing nodes determine their location. The location information for the cluster head is used by the collector to decide its movement pattern, while the location information of the sensing nodes is used to assign it to the closest cluster head for efficient gathering of its sensor data.

3.2.3 Mobility Algorithms

Collector Mobility Algorithm   Let us consider initially that the nodes are static and there is a single collector that moves through the network and collects data from the cluster heads. For the moment, we assume that the cluster heads are fixed. We are given n cluster heads that, following the collector's natural traveling path (e.g., based on the geography of the sensor field), are numbered $0, \ldots, n - 1$, such that the collector follows the cycle of $0 \to 1 \to \cdots \to n - 1 \to 0$.

The cluster heads hold sensed data before the collector arrives. Since it can take a long time—possibly hours or days—for the collector to revisit a head, it is important to determine whether the memory in the heads is bounded. If the collector arrives later than the scheduled time, more memory is needed in the heads. Moreover, it takes longer to transmit the data from the heads to the collector. This requires the collector to stay longer at each cluster and further delays the arrival of the next trip. This "positive feedback" may cause the required memory to grow unbounded. The following analysis shows that the memory does not grow indefinitely and hence can reach a stable schedule.

Let r_i be α_i/β_i, for $i \in [0, n - 1]$, where α_i is the sensor data accumulation rate at cluster head i and β_i is the data collection rate of the mobile collector when visiting i. Let d_i be the time for the collector to travel from i to $i + 1$ (modulo n) and D be the sum of d_i. We have the following linear system in terms of variables t_i, which represents the time taken for the collector to collect data from each node i:

$$T = D + \sum_i t_i \tag{3.1}$$

$$t_i \geq r_i T \tag{3.2}$$

$$t_i \geq 0 \tag{3.3}$$

where Eq. (3.2) is from the requirement of $\alpha_i T \leq \beta_i t_i$.

We first study the feasibility of the system defined above. Take the sum of Eq. (3.2) over all i, we have

$$\sum_i t_i \geq \sum_i r_i T \quad \text{or} \quad T - D \geq \sum_i r_i T \tag{3.4}$$

viz.
$$T\left(1 - \sum_i r_i\right) \geq D \tag{3.5}$$

which is possible only if $\sum_i r_i < 1$. To show that $\sum_i r_1 < 1$ is also a sufficient condition for the system to be feasible, we derive the solution for t_i, which minimizes T. (This solution minimizes the required total buffer size for all cluster heads since the required buffer size on each cluster head grows linearly with T. Furthermore, minimizing T also minimizes the latency of data, assuming that the motion of the collector takes much more time than data communication, that is, $D \gg \sum_i t_i$.) Suppose $\sum_i r_i < 1$ is satisfied. We let $T = \tilde{T} \equiv D/(1 - \sum_i r_i)$ and $t_i = \tilde{t}_i \equiv r_i \tilde{T}$. Also, $\tilde{T}$ and $\tilde{t}_i$ satisfy Eqs. (3.2), (3.3), and (3.1) and are hence the solution that minimize T.

Next, we want to see whether the solution for t_i is stable. This is important because occasionally the collector's motion schedule may be delayed unexpectedly due to transient communication slowdown with a certain cluster head or due to changes in the travel conditions (e.g., changing weather). It t_i increases at each delay and does not return to its previous value, then the entire motion schedule may be lengthened without an upper bound, which is an unstable situation. We define the stability in the sense that, if the collector meets an unexpected delay at a node, the system shown above is still feasible and t_i will eventually return back to $\tilde{t}_i$. Let $t_i = \tilde{t}_i \kappa$, where $\kappa > 1$. To show that t_i is still a feasible solution, we observe that $T = D + \kappa \sum_i \tilde{t}_i < \kappa \tilde{T}$. Hence, $t_i = \kappa \tilde{t}_i = \kappa r_i \tilde{T} \geq r_i T$, satisfying Eq. (3.2).

If the collector is delayed unexpectedly by an amount of time L, then each cluster head will accumulate an additional amount, αL, of data, suppose the cluster heads have extra buffer space to store this additional amount and thus avoid data loss. As soon the collector reaches the next cluster head (with delay L), the cluster head empties the buffer in time $\kappa_0 \tilde{t}_i$, for some $\kappa_0 > 1$. The collector then continues to complete the current cycle. When the next cycle begins, the collector no longer needs to spend $\kappa \tilde{t}_i$ at each node i. Instead, it suffices to spend $t_i = r_i(D + \kappa \sum_i \tilde{t}_i)$. The new ratio of t_i over $\tilde{t}_i$ equals

$$\kappa = \frac{D + \kappa_{\text{old}} \sum_i \tilde{t}_i}{D + \sum_i \tilde{t}_i} \tag{3.6}$$

which is less than κ_{old}. Hence, κ is a decreasing positive value. Take the limit on both sides of Eq. (3.6), we see that κ approaches 1. In other words, t_i returns to $\tilde{t}_i$. This proves the stability of the round-robin routine.

Energy Considerations for Buffering at Cluster Head   Let α be the sensing rate and β be the transmission rate. We can use a buffer to store the sensed data during t_1 in Figure 3.3 before transmission. The amount of stored data grows at rate α. During t_2, the transmitter is turned on. The data stored in the buffer decreases at rate $\beta - \alpha$. At the end of t_2, the buffer is empty so the transmitter is turned off again. We assume that $\alpha < \beta$; otherwise, the buffer will definitely overflow. Intuitively, a larger buffer allows the transmitter to stay off longer so more energy can be saved. However, the buffer itself also consumes power so we have to find an appropriate buffer size that causes the most energy savings. Let p_b be the power consumed by the buffer per megabyte and k be the energy overhead to turn on and off the transmitter. The buffer size Q is $(\alpha - \beta)t_1 = \beta t_2$. We can express the value of Q as

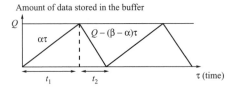

Figure 3.3   Amount of data stored in buffer rises first when transmitter is off and then declines after it is turned on.

$\beta/\alpha(\alpha - \beta)(t_1 + t_2)$. During a period, the additional energy consumed by buffer insertion includes the buffer's energy $Qp_b(t_1 + t_2)$ and the overhead k for power management. The average power is

$$\frac{Qp_b(t_1 + t_2) + k}{t_1 + t_2} = \frac{\beta}{\alpha}(\alpha - \beta)p_b(t_1 + t_2) + \frac{k}{t_1 + t_2}$$

Our earlier study shows that the minimum power occurs when the length of a period $t_1 + t_2$ is $\sqrt{\alpha k/[\beta p_b(\alpha - \beta)]}$ and the buffer size Q is $\sqrt{(\beta k/p_b)(1 - \beta/\alpha)}$ [10].

Different Mobility Algorithms for Collector   The collector moves among the cluster heads using one of three possible schedules—the round-robin schedule, the data-rate-based schedule, and the min-movement schedule. In the *round-robin schedule*, the collector visits each cluster head to collect data in a round-robin manner. In the *data-rate-based schedule*, the collector visits the cluster heads preferentially. The frequency of visiting a cluster head is proportional to the aggregate data rate from all the nodes in the cluster. For example, take four clusters whose aggregate rate of data generation (number of nodes times the data rate of each node) is in the ratio $1 : 2 : 3 : 4$. Define the period for which the collector stays at a cluster head as a *slot* and a consecutive number of slots over which scheduling decisions are made as a *round*. With the data-rate-based schedule, in a round of 10 slots, the collector will visit the cluster heads in the order 1, 2, 3, 4, 4, 3, 2, 4, 4, 3. In the third schedule, called the *min-movement schedule*, the collector visits the cluster heads in the proportion of the aggregate data rate, but also with the goal of optimizing the distance traversed. Thus, in a round of 10 slots, the collector stays at a cluster head for multiple slots continually, the number of slots being calculated as above depending on the aggregate data rate at the cluster. In our simulation, this schedule gives the visit order of the cluster heads as 1, 2, 2, 3, 3, 3, 4, 4, 4, 4. The length of a slot is the time it takes for the cluster head to transfer *all* the data accumulated at the cluster head till the moment the collector arrives.

Analysis of Cluster Head to Collector Communication   The cluster head communicates the data collected from the nodes in the cluster to the collector. In the base case, this communication takes place once the collector reaches the cluster head. This minimizes the transmission energy spent by the cluster head. However, as the analysis in Section 3.2.3 shows, energy is also spent in buffering the data at the cluster head. We explore the possibility of the cluster head transmitting some of the data to the collector as soon as the two are within RF communication distance. In this scheme, the cluster head continues to transmit data to the collector as the collector moves closer. The collector transmits this data to the base station continually. This is possible since the collector uses two different communication interfaces for communicating with the cluster head and the base station—a low-bandwidth RF interface for cluster head communication and a higher bandwidth and longer range interface for communicating with the cluster head (such as GPRS). The scheme has the possible advantage of saving energy expended by the cluster head depending on when it transmits to the collector. It has the decided advantage of reducing the data latency since the collector immediately transmits the received data to the base station.

Let us analyze the distances over which the cluster head should transmit its data to the collector to save the energy. We consider energy saving to be the primary concern and the latency reduction as the consequence. We perform the analysis for one cluster without loss of generality since the cluster-specific parameters (number of nodes and rate of data generation by a node in the cluster) are taken into account in the analysis. Let the number of nodes in the cluster be n and the rate of data generation by each node be ρ. Let the critical distance for the data transmission be θ_{th} such that it is energy efficient for the cluster head to transmit only after the collector is closer than this distance. The time for the collector to traverse distance θ_{th} is $t_{th} = \theta_{th}/v_c$, where v_c is the velocity of the collector. The energy for transmission at the cluster head has two components—one due to the transmission circuitry, which needs to be expended independent of the distance over which the data needs to be sent, and the second component due to the power amplifier whose energy need depends on the distance. The energy due to reception at the collector is due to the receiver circuitry. Let the energy per bit for the two transmission components at the cluster head be E_{tx_elec} and E_{PA} and that for reception at the collector be E_{rx_elec}. Let the energy expended at the power amplifier grow with the square of the transmission distance. Thus, the energy per bit of transmission over distance d is $E_{tx}(d) = E_{tx\_elec} + E_{PA}d^2$. We consider an energy-efficient memory at the cluster head where the unused banks of memory may be turned off. Thus, off-loading some of the data to the collector enables the cluster head to selectively turn off parts of its buffer, thus saving its energy [11–14].

Let the energy spent to keep one bit in a buffer be p_b. The data generated in time t by the nodes in the cluster is $n\rho t$. This is the amount of data buffered at the cluster head. Over the time duration t_{th}, the total energy due to buffering is $\int_0^{t_{tx}} p_b(n\rho t)\,dt$. The amount of data transmitted (and received) while the collector moves over a distance $d\theta$ is $n\rho\,d\theta/v_c$. The energy expended for this data, integrated over the entire distance, is $\int_0^{\theta_{th}}(E_{tx\_elec} + E_{PA}\theta^2 + E_{rx\_elec})n\rho/v_c\,d\theta$. Solving the two definite integrals and equating the two sides [energy due to buffer on left-hand side (LHS) and energy due to RF transmission/reception on the right-hand side (RHS)], we get the critical distance to be

$$p_b\frac{\theta_{th}^2}{2v} = E_{tx\_elec}\theta_{th} + E_{PA}\frac{\theta_{th}^3}{3} + E_{rx\_elec}\theta_{th} \tag{3.7}$$

For the radio used in [15,16], $E_{tx\_elec} = E_{rx\_elec} = 50\,\text{nJ/bit}$ and $E_{PA} = 100\,\text{pJ/bit/m}^2$. From [11], $p_b = 0.012\,\text{W/MB}$. For the collector, we use two different models—one is a fast aerial vehicle of speed 30 m/s, and the second is a robot in our lab of speed 0.1 m/s [17]. The fast collector covers the distance over which RF communication is possible for the sensor nodes in such a short time that it is not worthwhile for the cluster head to begin transmitting. Therefore, we do the calculation for the slow collector. Putting these values in Eq. (3.7), we get a quadratic equation that we solve to get the two roots for θ_{th} of 210.00 and 14.23 m. For all practically available sensor nodes, the RF antennas cannot communicate reliably over 210 m. We take that communication starts at a distance of 100 m (from the spec sheet for Berkeley Mica2 motes [18]) and continues until a distance of 14.23 m. This is explained by the fact that a high distance of separation makes the communication between the cluster head and the collector too energy expensive and, therefore, the cluster head buffers the data. Below the threshold distance, the cluster head has gotten rid of most of the data and has very little data to buffer. Therefore, it is more energy-wise to buffer the data than transmitting it to the collector. The relative energy consumption due to buffering and RF communication is shown in Figure 3.4.

Locator Mobility Algorithm   Let us consider initially that the nodes are static and there is a single locator that moves through the network and aids in the determination of positions of each node in the network. Location determination requires multiple reference points

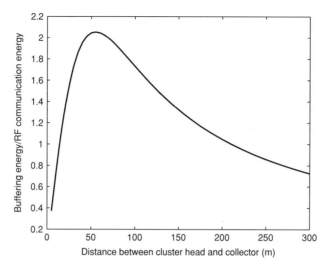

Figure 3.4 Relative energy spent due to buffering at cluster head and RF communication between cluster head and collector.

whose locations are known by, for example, GPS. Equipping many sensor nodes with GPS can be too expensive as well as make the sensing nodes heavy and unwieldy. Therefore, we use mobile locators that move around the sensor nodes as multiple reference points. First, the field is partitioned into *locator cells*, which are disjoint rectangular regions that satisfy a necessary condition (condition 1) and an optional condition (condition 2):

1. The maximum separation of two nodes in the cell is less than the transmission range of the nodes (necessary).
2. The number of nodes in a cell is greater than a threshold, denoted by η_τ (optional).

A cell that satisfies both conditions is called a *complete cell*, and one that satisfies only the necessary condition is called an *incomplete cell*. Note that a cell is different from a cluster. A cell is purely for the purpose of determining locator movement, while a cluster is an aggregate of nodes that have a cluster head that collects and possibly processes data from all the cluster nodes. A cluster comprises multiple cells. A cell is a square region of dimension $<$ (transmission range)$/2\sqrt{2}$.

The given parameters for the sensor network under consideration are the following:

- *Allowable Error in Location Estimation (ε_τ)*   Each node should determine its location with less than ε_τ error on an average. The error arises due to errors in the model correlating physical measurement with distance between one-hop neighbors. This is a requirement imposed on the localization system. The error is given in terms of distance units that the calculated location can differ from the actual location.
- *Velocity of Locator (v_l)*   This is the maximum speed with which the locator can navigate within the cluster.
- *Epoch (ζ_s)*   This is the average duration between successive movements of a passively mobile node.

The locators are mobile and each locator can act as multiple reference points for location estimation by communicating with a sensing node at different time points. The locators are looked upon as mobile entities that roam the sensor field, periodically broadcasting beacon messages with their own locations. An intermediate node when it receives a beacon message, it forwards it after incrementing the hop count. Thus, receiving a beacon message, a passively mobile node can estimate the number of hops it is away from the locator. The sensing nodes collect the beacon messages and after an *appropriate* number of beacon messages perform triangulation to determine their locations. A sensing node knows the number of beacon messages required in order for the location estimation error to be below ε_τ. The error in the estimation (ε) is a function of the number of beacon messages (η_b) and the distance between the node and the locator for each reference point $(\delta_{l,n})$. The distance can be approximated by the product of the number of hops separating the locator and the node $(h_{l,n})$, and the transmission range of each node (r_T). The locator has the same kind of RF communication device as the sensing nodes and, hence, its transmission range is considered identical to that of the sensing nodes.

$$\varepsilon = F(\delta_{l,n}, \eta_b) = F(h_{l,n} \cdot r_T, \eta_b) \tag{3.8}$$

Determination of Localization Error   We wish to determine the number of beacon messages that should be collected by a sensing node to guarantee a desired accuracy in its location estimate. The guarantee will, however, be probabilistic, that is, the guarantee will be stated in terms of a probability that the error in location estimate will be bounded by a desired value. We will start with some background for location estimation. We will present this discussion in terms of a sensing node being surrounded by multiple neighbors each of which knows its own location. The neighbors send beacon messages to the sensing node, which determines the sensing node's location. Our environment where there is a single locator that moves and sends beacon messages from multiple positions is easily mapped to this model. The minimum number of neighbors required for location estimation in an n-dimensional plane is $(n + 1)$. Thus, in our special case of a two-dimensional plane, three neighbors are required, under perfect conditions, namely, where individual distance measurements from a neighbor are completely accurate. Localization with respect to three neighbors gives the following set of equations:

$$
\begin{aligned}
(x_1 - u_x)^2 + (y_1 - u_y)^2 &= r_1^2 \\
(x_2 - u_x)^2 + (y_2 - u_y)^2 &= r_2^2 \\
(x_3 - u_x)^2 + (y_3 - u_y)^2 &= r_3^2
\end{aligned}
\tag{3.9}
$$

In this set of equations, (u_x, u_y) is the position of the sensing node that we intend to locate and (x_i, y_i) is the location of the ith neighbor. Solving Eq. (3.9) we get u_x to be of the following form:

$$u_x = k_1 r_1^2 + k_2 r_2^2 + k_3 r_3^2 + k_4 \tag{3.10}$$

where $k_i \in R$. Similarly, u_y can be expressed in a form like Eq. (3.10). One simple relation for measuring error in u_x and u_y is to differentiate both sides of the equation to get $\Delta u_x = 2k_1 r_1 \, \Delta r_1 + 2k_2 r_2 \, \Delta r_2 + 2k_3 r_3 \, \Delta r_3$. However, this is dependent on topology, and no obvious bounds exist for the error. We choose to work with variances as a measure of the error in estimation. The range measurements are error prone, and this leads to the error in location estimation of the sensing node. If we assume that the errors in range measurements are uncorrelated, then from Eq. (3.10) we get the variance of the estimated location in the following form:

$$\text{Var}(u_x) = k_1^2 \text{Var}(r_1^2) + k_2^2 \text{Var}(r_2^2) + k_3^2 \text{Var}(r_3^2) \tag{3.11}$$

For the purpose of estimating the error, we consider the neighbors to be divided into groups of 3. Triangulation is performed as above for each such group. Each triangulation gives a sample value for u_x and u_y, and the location of the sensing node is finally determined by taking an average of all these u_x and u_y values. For simplicity we consider the number of neighbors to be an integral multiple of 3. Thus, the number of sample values of u_x and u_y is $p = N/3$ where N is the number of neighbors. The expected value of the calculated location (i.e., the sample mean) is the same as the expected value of the population mean, and thus the technique gives an unbiased estimate of the location. The variance of the estimated location is given by

$$\text{Var}(u_{xp}) = \text{Var}(u_x)/p \tag{3.12}$$

Note that $\text{Var}(u_{xp})$ is the *ensemble variation* in space over neighbors that form an *ensemble*, while the variance $\text{Var}(u_x)$ computed in Eq. (3.11) is the time variance. Consider that the measurements of the range vector $[r_1, r_2, r_3]$ represent a stochastic process in time. So the averaging for determination of u_x and u_y is done over time, and therefore the variance in Eq. (3.11) is a time variance. On the other hand, in averaging over $p = N/3$ samples, we are averaging over the neighbors of the sensing node and this is an ensemble average. The corresponding variance is also the ensemble variance. If we assume ergodicity of the mean and the variance, then the time average and variance can be replaced by the ensemble average and variance that we need for our subsequent analysis. Chebyshev's inequality gives that for a random variable X with a distribution having finite mean μ and finite variance σ^2, $P(|X - \mu| \geq t) \leq \sigma^2/t^2, t \geq 0$. Applying this to the random variable u_{xp} we get Probability[Error in location estimate (for the x axis) $\geq$ Error bound in distance units] as follows:

$$P(|u_{xp} - E(u_{xp})| \geq \epsilon) \leq 1/\epsilon^2 \cdot \text{Var}(u_{xp})$$
$$P(|u_{xp} - E(u_x)| \geq \epsilon) \leq 1/(p\epsilon^2) \cdot \text{Var}(u_x) \tag{3.13}$$

Equation (3.13) gives a bound on the probability of error in estimated location exceeding a desired threshold in terms of the number of neighbors. Given a desired accuracy, we can make the probability of error exceeding the desired accuracy to be as small as we like by increasing p, that is, by extension, increasing the number of neighbors. Next, we have to determine the variance in u_x to complete the analysis. In Eq. (3.11), if we take that the errors in the three range measurements r_1, r_2, r_3 are equal, then the variance in u_x can be written as

$$\text{Var}(u_x) = (k_1^2 + k_2^2 + k_3^2) \text{Var}(r^2) \tag{3.14}$$

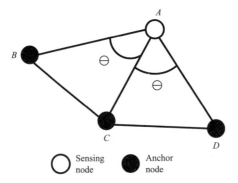

Figure 3.5   Topology of sensing node and three neighboring anchor nodes. Triangulation is done by sensing node using distance measurements from three anchor nodes.

In general, the coefficients k_1, k_2, k_3 are dependent on topology, that is, the relative placements of the neighbors with respect to the sensing node. The upper bound for the sum is infinity (when the three neighbors are collinear), and the lower bound is achieved when the triangles formed by the neighbors with the sensing node are equilateral. We perform an analysis on the topology shown in Figure 3.5. For values of angle θ subtended by the neighbors at the anchor node ($\angle CAD = \angle BAC = \theta$) between $\frac{5}{36}\pi(25°)$ and $\frac{1}{3}\pi(60°)$, it can be shown that $k_1^2 + k_2^2 + k_3^2$ lies close to $0.2/R$ where R is the distance between the neighbor and the sensing node. By our algorithm, the locator sends beacon messages when it is either in the cell or in the adjacent cell to the sensing node and therefore $R \leq r_T$, where r_T is the transmission range.

Next, we compute Var(r^2). Consider the error in range measurements and let the upper bound on the relative error be e. This means that if the actual distance between a neighbor and the sensing node is d, the measurement lies between $d \pm ed$. Assume that the range measurements follow a Gaussian distribution. Therefore, almost all the points (99.74% to be exact) lie within $\mu \pm 3\sigma$. Equating, $\sigma = ed/3$. The upper bound for this is $er_T/3$. Thus, Var(r) $\leq (er_T/3)^2$. Then, Var(r^2) $= E(r^4) - (E(r^2))^2$. The higher order expectations can be calculated using the moment generating function of a Gaussian distribution with mean μ and variance $\sigma^2 M(t) = e^{\mu t + \sigma^2 t^2/2}$. Then $E(r^2)$ and $E(r^4)$ can be derived by differentiating $M(t)$, respectively, twice and four times and evaluating at $t = 0$. Simplifying, we get from Eq. (3.14) the Probability[Error in location estimate (for the x axis) $\geq$ Error bound in distance units] as

$$P(|u_{xp} - E(u_{xp})| \geq \epsilon) \leq 1/(p\epsilon^2) \cdot 0.2e^2 R^3(2.47 \times 10^{-2}e^2 + 4.44 \times 10^{-1}) \quad (3.15)$$

For our environment, we pick $p = 24$ to restrict the error in location estimation in each dimension to 10 m. Note that the above bound can be made much tighter for smaller distances between the neighboring node and the sensing node and for particular topologies. This value of p corresponds to a total of 72 beacon messages by the locator for the nodes in one cell. This means that the number of beacon messages sent out by the locator from each cell (the cell itself and its 8 neighboring cells) is 8.

Once the required number of beacon messages are received by a sensing node, it determines its location. The sensing node then informs the locator of its location. The locator verifies this location using the received signal strength from the node. Then, it assigns the node to the appropriate cluster. In a majority of cases, the node will be assigned to the cluster to which the locator is coupled. Only if the node has moved far away will it be reassigned to a possibly closer cluster head.

Again considering cell i, the locator sends a beacon message while located in cell i. The cells in C_{i_1} are ordered according to the metric (Number of nodes in the cell − Distance of the cell from the current position). The complete and incomplete cells are ordered separately with all the complete cells being ordered above the incomplete cells. The locator picks the next cell to visit in order from the ordered list. Next, the cells in C_{i_1} are ordered, and so on. The movement is stopped when any node in the cell indicates that it has the requisite number of beacon messages.

A constraint for the movement of the locator is that all the η_τ beacon messages must be received by a node before it moves. Let the distance between two cells c_i and c_j be given by $\delta_{i,j}$. The locator has sent η_c beacon messages for cell i and in time t_c. The locator is currently in cell m and the next cell in the ordered list is cell n. The condition that the locator seeks to enforce is $(\eta_\tau - \eta_C) \cdot \delta_{i,j}/v_l < (\zeta_S - t_c)$. If the condition is violated for cell n, the locator moves to another position within the same cell m and sends another beacon message. This probabilistically assures that all beacon messages are received in the time in which it is useful, that is, between two movements of a node.

3.2.4 Experiments and Results

We perform simulations using the NS-2 simulator [19]. The simulation environment is set up as follows. There are four clusters that are separated over a distance such that any two nodes from different clusters are not able to communicate between each other. Each cluster has a cluster head that collects data from its own cluster. Since the cluster head does not have the capability to send the collected data to the base station, there is a mobile data collector that moves to and collects data from each cluster head. The mobile collector then sends the data to the base station for analysis. The large intercluster distance and the separation from the base station underline the need for a mobile collector as opposed to multiple relay nodes between the cluster heads and the base station.

In each simulation, a cluster sends data, at a constant rate, to its own cluster head. The cluster head stores the data in its buffer until the data collector arrives (either physically or within the energy-efficient communicable distance) and collects the data via wireless communication. The data collector follows a predetermined movement schedule to visit each cluster head. The collector periodically goes back to the base station if it does not have enough energy to serve another cluster head. Each cluster is characterized by the position of the cluster head, the number of nodes in the cluster, and the aggregate data rate of the cluster. In our simulation, we consider heterogeneous clusters. For the simulation, the cluster head is considered nonrotating. Since the geographical spread of the cluster is much smaller than the intercluster distance, we believe this will not have much effect on the results. The four clusters with the positions of the cluster heads and the base station is shown in Figure 3.6. The spread of only cluster 2 is shown. The characteristic of the four clusters is summarized in Table 3.2.

The energy model used for the cluster head is the same as for the radio used in [15,16]. The energy has two components—one due to the transmit–receive circuitry and the other due to the power amplifier. The latter component depends on the distance over which the data is transmitted. The amount of energy to send n bits of data over a distance d is given by $n(E_{TxRx} + E_{PA}d^2)$. The bandwidth for the cluster head to collector communication

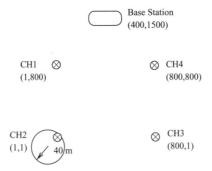

Figure 3.6 Topology of four clusters and base station used for simulation.

follows the value for the Berkeley motes [18] of 38.4 kbps. The collector to base station communication happens using GPRS with a range of over 1 km and a bandwidth of 135 kbps. The GPRS energy model also has the two components. The values are substantially different and taken from the energy for the GSM model [20]. The energy model parameters are summarized in Table 3.3.

Fast and Slow Collector In the simulation we use two collector models—a fast collector and a slow collector. The fast collector is based on the Aerosonde model from NASA [21] and the slow collector is based on a commercial mobile robot called the Palm Pilot Robot Kit (PPRK), which was originally designed at Carnegie Mellon University [22] and is currently used in our lab [17]. These collectors are henceforth referred to as the *fast collector* and the *slow collector*, respectively. The detailed parameters of these two models is shown in Table 3.4. We assume that the slow collector's initial energy is 10 times less than the Aerosonde's to keep the endurance (time between recharges) approximately equal for the two models since for a given travel distance the slow collector consumes approximately one-tenth of the energy of the fast collector. The rationale behind selecting these two models is that the Aerosonde may be used if the data is time-sensitive, such as pertaining to rare event detection by the sensing node, and the mobile robot may be used if there is a budget constraint for the deployment ($40,000 versus $300 per collector in the two cases).

For the simulation, in the slow collector case, the cluster head transfers data to the collector through wireless RF communication for the optimal distance range as calculated in Section 3.2.3. For the fast collector, however, this distance is traversed so fast that we make the simplification that cluster head only transfers data to the collector, once the collector arrives at the cluster head.

Table 3.2 Characteristics of Clusters

Cluster Parameter	1	2	3	4
Cluster head coordinate (m) (Base station is at (400,1500))	(1, 800)	(1, 1)	(800, 1)	(800, 800)
Data rate per cluster node	2 bps	4 bps	6 bps	8 bps
Number of nodes in cluster	500	500	500	500

Table 3.3   **Energy Parameters for Cluster Head and Collector**

Parameter	Value
Cluster head tx/rx circuitry (E_CH_{TxRx})	50 nJ/bit
Cluster head power amplifier (E_CH_{PA})	0.1 nJ/bit/m^2
Collector tx circuitry (E_Coll_{TxRx})	1000 nJ/bit
Collector power amplifier (E_Coll_{PA})	0.008 nJ/bit/m^2

Collector Movement Schedule   Recollect that the collector moves among the cluster heads using one of three possible schedules—the round-robin schedule, the data-rate-based schedule, and the min-movement schedule. In a round of 10 slots, the collector will visit the cluster heads in the order 1, 2, 3, 4, 1, 2, 3, 4, 1, 2 in the round-robin schedule; 1, 2, 3, 4, 4, 3, 2, 4, 4, 3 in the data-rate-based schedule; and 1, 2, 2, 3, 3, 3, 4, 4, 4, 4 in the min-movement schedule. The length of a slot is the time it takes for the cluster head to transfer *all* the data accumulated at the cluster head until the moment the collector arrives (either physically arrives at the cluster head or data transfer through RF communication starts).

We also consider the possibility of variation in the physical conditions of the environment affecting the speed of the collector. The collector may face an obstacle in its path and may have to deviate from part of its precalculated route. Also, the physical conditions, such as wind speed, may change, thereby affecting the collector speed. To take such variations into account, the speed of the collector is varied according to a normal distribution, such that the speed can vary by ±40% of the average speed. This means that the 3σ limit of the normal distribution is taken as ±40% of the average.

Output Parameters   We collect the following output parameters from the simulation of the collector:

- Buffer size at the cluster head
- Data latency for data generated by the sensing nodes
- Time between recharges of the collector

The output parameters are calculated based on 50 rounds of collector movement *after* the initial transient period ends. The initial transient period is characterized by continuous growth in the buffer occupancy at the cluster heads. Each round is defined as 10 slots of movement. The buffer size in bytes is the buffer required at the cluster head for the data generated by the sensing nodes before it is handed over to the collector. Due to the variation in the speed of the collector, the buffer size also varies between rounds. We take

Table 3.4   **Characteristics of Fast and Slow Collector Used in Simulation**

Parameter	Fast Collector (Aerosonde)	Slow Collector (Mobile Robot)
Speed	30 m/s	0.1 m/s
Energy consumed	10 J/s $\simeq$ 0.33 J/m	0.97 J/s $\simeq$ 9.7 J/m
Initial energy	1.44 MJ	144 KJ

Table 3.5   Results for Fast Collector with Three Different Movement Schedules

		Round Robin	Data Rate Based	Min Movement
Buffer size (kbytes)	CH_1	22.43	50.88	44.30
	CH_2	44.93	54.62	80.00
	CH_3	66.53	63.30	107.22
	CH_4	87.92	111.56	124.37
Data latency (s)	CH_1	75.04	181.29	153.79
	CH_2	75.11	93.18	76.99
	CH_3	75.05	62.19	51.36
	CH_4	74.93	46.61	38.55

the maximum buffer size over all the rounds after throwing out the outliers that lie beyond the 2σ range. The maximum buffer size is the relevant metric since the cluster head will have to provision for a buffer of that size to prevent data loss due to overflow. Throwing out the outliers is needed to eliminate the statistical noise and to draw useful conclusions from the results. The data latency is measured as the average of the time gap between the data generated by the sensing node and the data reaching the base station. We approximate the time the data is generated by the sensing node by the time the data reaches the cluster head. The collector returns to the base station when it estimates its energy will run out before it can reach the next cluster head and collect all its data. The third output parameter is the average time between successive visits of the collector to the base station for recharging.

Experiment Set 1: Fast Collector   In the fast collector, we observe the collector has a very high endurance and rarely returns to the base station for recharge. Therefore, the output parameter of average time between recharges is omitted for this set of experiments. The results are presented in Table 3.5.

In the data-rate-based schedule, the buffer size that cluster heads 1, 2, and 3 need is almost similar, but not for cluster head 4. The increasing amount of buffer size in the data-rate-based scenario for cluster heads 1, 2, and 4 compared to the round-robin schedule is because the maximum the number of cluster head(s) that a collector needs to visit before successive visits, to cluster heads 1, 2, and 4 are 9, 4, and 4, respectively. This is higher than the successive visits in the round-robin schedule (3 for all the cluster heads). On the other hand, the gap between successive visits for cluster head 3 is 2, which leads to the decreasing amount of buffer needed. The same reasoning happens in the min-movement scenario since the successive visits to each cluster head are spread farther apart than the round-robin scenario and so is the amount of buffer needed to hold the data. The decrease of average latency in data-rate-based and min-movement compared to round-robin is because both these schedules consider the data rate of each cluster—the higher the data rate, the more often the cluster is visited by the collector. The latency is averaged over the entire data and is therefore improved.

Experiment Set 2: Slow Collector Without Distance Transmission to the Collector   In this set of experiments, we consider the slow collector that waits for the collector to arrive at its location before transferring data to the collector. The buffer size, latency, and time between recharges is shown in Table 3.6.

Table 3.6   Results for Slow Collector with Three Different Movement Schedules under No Distance Transmission to Collector

		Round Robin	Data Rate Based	Min Movement
Buffer size (MBytes)	CH_1	8.60	19.53	16.30
	CH_2	16.75	22.02	29.63
	CH_3	24.29	24.91	37.88
	CH_4	33.62	43.02	46.19
Data latency (h)	CH_1	7.16	17.91	15.11
	CH_2	7.20	9.14	7.55
	CH_3	7.18	6.11	5.03
	CH_4	7.19	4.57	3.77
Time between	CH_1			
recharges (h)	CH_2	43.96	48.52	78.45
	CH_3			
	CH_4			

In this result, the buffer size needed by each cluster is higher compared to the fast collector case in Table 3.5. The reasoning is that the collector speed is much slower such that each cluster produces more data between successive visits, which also increases the data collection time of the collector. The frequency of recharge in the round-robin schedule is higher compared to the other two schedules since the collector has to move each time it has finished collecting all the data at a cluster head. In the data-rate-based schedule, there are two occasions in a round where the collector continues staying at cluster head 4 to retrieve more data. This reduces the movement energy consumption of the collector, thus lengthening the time between recharges. In the min-movement schedule, the collector conserves even more movement energy such that the average time between recharging is even higher.

Experiment Set 3: Slow Collector with Distance Transmission to the Collector
In this scenario, the cluster head starts data transmission when the distance between the collector and itself is 100 m and stops the transmission when the optimum distance d is reached. In our case, as shown in Section 3.2.3, d is 14.23 m. Below the optimum distance, the collector continues to move toward the cluster head, and once it arrives at that cluster head, data transmission resumes again.

Since the cluster head is allowed to start data transmission early, this scenario reduces the average data latency on the three schedules, as shown in Table 3.7 in comparison with Table 3.6. There is also a decrease in buffer requirement for the same reason. The percentage of improvement is shown in the Table 3.8.

In Table 3.8, the min-movement schedule has less improvement due to distance transmission because the collector often stays at a particular cluster head for consecutive slots. The fraction of time over which distance transmission happens in the min-movement schedule in relation to the time the collector spends at a cluster head is smaller than in the other movement schedules. Hence, the relative improvement due to the distance transmission is smaller.

The round-robin schedule helps the recharge interval more than the other schedules. This can be explained by the observation that for cluster 1, the lowest data rate cluster, the collector does not *always* have to physically arrive at the cluster head. In some cases, all

Table 3.7   Results for Slow Collector with Three Different Movement Schedules under Distance Transmission to Collector

		Round Robin	Data Rate Based	Min Movement
Buffer size (Mbytes)	CH_1	7.60	17.33	16.28
	CH_2	15.21	19.46	28.43
	CH_3	21.94	22.27	37.08
	CH_4	29.16	39.33	45.73
Data latency (h)	CH_1	6.37	16.08	14.36
	CH_2	6.37	8.25	7.17
	CH_3	6.35	5.49	4.78
	CH_4	6.34	4.12	3.59
Time between recharges (h)	CH_1			
	CH_2	45.83	49.21	79.83
	CH_3			
	CH_4			

the data is communicated to the collector through distance transmission. In that case, the collector directly goes to the next cluster in its schedule (cluster 2). The distance traversed to go to cluster head 2 is less in this case compared to reaching cluster head 1 and then proceeding to cluster head 2. This is thus energy-efficient and lengthens the time period between recharges of the collector.

Experiment Set 4: Locator   For the locator, we use a commercial mobile robot (PPRK) [17, 22] (same as the one used for the slow collector), equipped with GPS carrying a regular sensor node with capability for short-range RF communication. The locator moves in the cluster with 500 sensing nodes randomly distributed in the cluster spread. All the sensing nodes need to determine their own positions with the help of the locator. The cluster is divided into cells of size $10\,m \times 10\,m$, and the size of the cluster is taken to be a square region of dimension $80\,m \times 80\,m$, which inscribes the circular cluster region. The locator is

Table 3.8   Improvement for Slow Collector with Distance Transmission to Collector

		Round Robin (%)	Data Rate Based (%)	Min Movement (%)
Buffer size	CH_1	11.64	11.24	0.15
	CH_2	9.22	11.65	4.04
	CH_3	9.70	10.61	2.10
	CH_4	13.28	8.56	1.00
Data latency	CH_1	11.01	10.21	5.02
	CH_2	11.57	9.77	5.02
	CH_3	11.56	10.15	5.06
	CH_4	11.84	9.77	4.94
Time between recharges	CH_1			
	CH_2	4.25	1.42	1.76
	CH_3			
	CH_4			

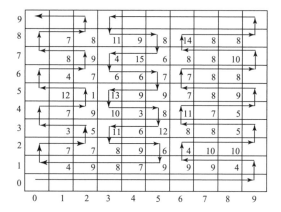

Figure 3.7   Movement of locator within cluster.

capable of broadcasting its own position to the neighboring sensing nodes within a range of 30 m. Thus, a locator in a cell can reach all the nodes in its own cell and in the 8 adjoining square cells (diagonal length $= 2 \cdot \sqrt{2} \cdot 10 < 30$ m. The locator is unaware of the locations of individual sensing nodes. However, it knows the boundary of the cluster.

With this information, the locator moves from one cell to another in an S pattern as shown in Figure 3.7. The number of sensing nodes in each cell is shown by the number in the lower right corner in each cell. During the course of its movement, the locator broadcasts 8 beacon messages in 8 different random positions within each cell. Each beacon message contains the position information of the locator. Since the size of each beacon message is 36 bytes (default packet size in TinyOS [18]) and the locator's bandwidth is 38,400 bps, the locator stays in each position for broadcasting a beacon for 7.5 ms. To simulate a real-world situation, we vary the locator's speed by using a normal distribution, where on an average the locator moves 10 cm/s with a variance of $\pm 40\%$ (same as for the land robot for the slow collector). Each sensing node needs to receive 72 beacon messages to determine its position with error in each dimension of less than 10 m.

We are interested in three output parameters for the locator. The first is the rate at which location information is disseminated in the cluster. This is shown in Figure 3.8. The x

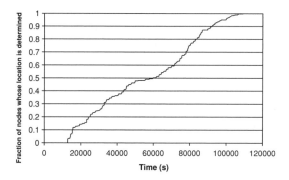

Figure 3.8   Rate of dissemination of location information among sensing nodes in cluster.

axis is the time elapsed since the locator started its movement in the cluster. The y axis is the fraction of the nodes in the cluster that have determined their location. The second parameter of interest is the total time measured from the time the locator starts moving in the cluster such that all the nodes in the cluster know their position. For our simulation, this are 109,359.65 s ($\simeq$ 30.38 h). This can be read off from the curve in Figure 3.8 by considering the time for the y-axis value to be 1.0. This time can be looked upon as the worst-case initial latency when none of the nodes know their location to start with. In the steady state, only a few of the nodes will move and an incremental location determination will be needed. For this reason, we are interested in the third parameter, which is the time for a node to determine its location once the locator arrives in its vicinity. This is given by the time to move among the cell in which the node is located and the 8 adjoining cells and broadcast 8 beacon messages from each cell. For our simulation, this time is 2882.39 s (= 0.8 h). This is also the time for which the node will have to be stationary for the location determination to be useful. This number appears high and is due to the slow speed of the locator (0.1 m/s) compared to the transmission range (30 m). In a network deployment, this will be determined by the estimated pause time in the movement of a node and a locator of an appropriate speed will be chosen.

3.2.5 Conclusion

In this section, we have proposed an architecture for energy-efficient data gathering in large-scale sensor networks. Some sensing nodes in the network may be passively mobile, which makes the problem more challenging. In the network model under consideration, the sensor field has a large geographical spread, and many of the sensing nodes are far away from the base station. We introduce some special classes of nodes some of which have the capability of controlled mobility, that is, mobility of the form that can be directed by sending control signals. The data collectors have the capacity for moving over long distances, possibly at high speeds for collecting data from the cluster heads and communicating the data back to the base station. The cluster heads aggregate the data from the nodes in the cluster and temporarily store them prior to transmission to the collector. To enable efficient data gathering from the passively mobile sensing nodes, they need to be associated with the closest cluster head and, therefore, need to know their locations. This is enabled by GPS-equipped mobile locators that move through the cluster and broadcast beacon messages with location information that helps the sensing nodes determine their position.

In this section, we propose algorithms for motion of the data collectors and the locators. We also propose an extension to the traditional triangulation approach for location determination using mobile locators. The algorithms are analyzed mathematically and simulated to bring out their important characteristics, such as energy consumption, data latency and cluster head buffer requirement (collector algorithm), and convergence time (locator algorithm).

For future work, we plan to consider the movement patterns of the passively mobile sensing nodes in a cluster. We wish to investigate what proactive information broadcast by these passively mobile nodes can benefit the algorithms by making them more informed about the sensor field, such as the density of nodes in a region. We are also investigating the effect of failures of collectors or cluster heads on the data collection. A set of cluster heads can concurrently serve the role. Also cooperation between multiple cluster heads to overcome temporary periods of high data rate is possible. There may be multiple collectors in the network that may work cooperatively in servicing the different cluster heads. Alternately, there may be a set of collectors for emergency data gathering corresponding to rare events being detected by a sensing node. The mobility algorithm needs to take these classes of collector nodes into account.

3.3 PURPOSEFUL MOBILITY IN TACTICAL SENSOR NETWORKS

Guohong Cao, G. Kesidis, Thomas LaPorta, and Bin Yao

Adding mobility to sensor networks can significantly increase the capability of the sensor network by making it resilient to failures, reactive to events, and able to support disparate missions with a common set of sensors. Mobility in sensor networks may be controllable and hence be used to help achieve the network's missions. That is, mobility may be "purposeful" instead of being treated as an uncontrollable external stimulus to which the ad hoc networks must respond. To make use of the purposeful mobility, we propose techniques for mobility-assisted sensing and routing considering the computation complexity, network connectivity, energy consumption of both communications and movement, and the network lifetime. We also define utility functions that can capture the benefits of the movement from the perspective of all missions and maximize the capability of the network.

3.3.1 Background Information

Recent advances [23, 24] in hardware design are enabling low-cost sensors that have sophisticated sensing, communication, and computation capabilities to accomplish multiple, disparate missions [25]. These sensors communicate via radio transmitters/receivers to form a multihop wireless network, that is, a distributed wireless sensor network [26]. Sensor networks can automate information gathering and processing and therefore can support many applications (missions) such as target tracking, perimeter defense, homestead monitoring, and intelligent transportation.

As sensors become widely deployed, multiple missions, each with different requirements, may share common sensors to achieve their goals. Each mission may have its own requirements for the type of data being reported, the sampling rate, accuracy, and location of the sampling. As a single sensor network needs to support different sets of missions under different conditions, the requirement on physical sensor locations becomes dynamic.

For example, in target tracking missions, enough sensors should be deployed along the track of the target, whereas in perimeter defense the requirement is to have adequate sensors along a predescribed perimeter. This dynamic requirement on sensor locations cannot be easily met by deploying a large number of sensors since provisioning for all possible combinations of mission requirements may not be economically feasible. More importantly, precise sensor deployment may not be possible, especially in a hostile environment, where sensors are subject to power depletion, failures, malicious attacks, and may change their physical locations due to external force. Therefore, a fixed sensor network has limitations when applied to support multiple missions or when the network conditions change. In such cases, mobile sensors are essential.

Figure 3.9 illustrates an example of using mobile sensors. Initially, fixed sensors indicated by round dots are randomly distributed. Several mobile sensors represented by squares are also deployed in the network. The mission of the network is to monitor the perimeter of a field and track any target that enters the field. This general mission has specific instances in

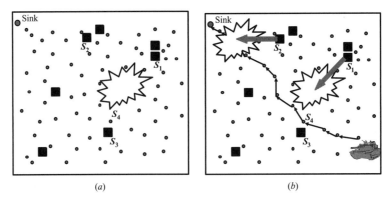

Figure 3.9 Generic sensor network with mobile sensors to satisfy different mission requirements.

airport control, home security, military operations, and the like. During network operation, certain sensors fail due to energy depletion or are destroyed by an external force (fire, bomb), creating *coverage holes* that are not covered by any sensor, as shown in Figure 3.9*a*. Since there are enough sensors along the perimeter of interest to fulfill the initial mission of the sensor network, that is, perimeter monitoring, no adjustment is necessary at this time. Suppose a target enters the field; the mission of the sensor network is to track the target. Since the track taken by the target may be arbitrary, it is desirable to fill the coverage holes. This can be achieved by using the mobile sensors as shown in Figure 3.9*b*. As the target approaches, the network also constructs a data dissemination (routing) path from the sensor monitoring the target to the sink. If some other sensor failure creates a network partition, moving S_2 can fix the routing problem as shown in Figure 3.9*b*.

From this example, we can see that mobility may significantly increase the capability of the sensor network by making it resilient to failures, reactive to events, and able to support disparate missions with a common set of sensors. However, there has not been a great deal of work on adding mobility to sensor networks, referred to as mobile sensor networks. A mobile sensor network is different from an ad hoc network. Although both support mobility, mobility in sensor networks may be controllable and, hence, be used to help achieve the network's missions. That is, mobility may be "purposeful" instead of being treated as an uncontrollable external stimulus to which the ad hoc communication network must respond.

In this section, we propose the following solutions to address purposeful mobility in mobile sensor networks.

- *Mobility-Assisted Sensing* As the mission changes, or to achieve the mission when the network condition (e.g., sensor failure) changes, some mobile sensors must be relocated. The network needs to locate redundant mobile sensors and derive a scenario to move them to the target location considering the computation complexity, movement distance, communication overhead, and the impact of the mobility on other concurrently running missions.

- *Mobility-Assisted Data Dissemination (Routing)* When sensors move in reaction to an event, a mission change, or failure, they may create network partitions, undesirable routes, or cause other disruptions. It is a challenge to invoke sensor mobility to improve network communications considering issues such as network connectivity, energy consumption of both communications and movement, and the network lifetime.

• *Integrated Mobility Management for Sensing and Routing* In addition to moving nodes to fulfill the sensing or communication requirements of a single mission, it is essential to analyze the impact of mobility on all missions sharing the sensors for sensing or communication, and it is a challenge to define utility functions that can capture the benefits of the movement from the perspective of all missions, and maximize the capability of the network.

The rest of this discussion is organized as follows. In Section 3.3.2, we discuss previous work and our solution for mobility-assisted sensing. Section 3.3.3 presents our work on mobility-assisted routing. In Section 3.3.4, we present a problem formulation and the proposed solution for mobility integration. Section 3.3.5 concludes the discussion.

3.3.2 Mobility-Assisted Sensing

In this section, we first give a brief review of the related work on sensor deployment and then present the challenges and the proposed solution on relocating mobile sensors. Finally, we present techniques to find coverage holes, which are used to invoke sensor relocation.

Related Work Since mobility-assisted sensing is still a new area, there is not much work in the literature. The closet work is sensor deployment. Previous work on sensor placement [27–29] largely addressed random and sequential deployment. In a pure random scheme, many more sensors are required than the optimal number to achieve high coverage of a target area. Due to the existence of wind and obstacles, some areas may never be covered no matter how many sensors are dropped. Moreover, during in-building toxic leaks [30], chemical sensors must be placed inside a building from the outside. In these scenarios, it is necessary to make use of mobile sensors [31, 32], which can move to the correct places to provide the required coverage. Based on the work from [33], mobile sensors have already been a reality. Their mobile sensor prototype, called Robomote, is smaller than $0.000047\ m^3$ at a cost of less than $150 in parts. Robomote also has some capability of avoiding obstacles when moved to the designated location.

There have been some research efforts on deploying mobile sensors, but most of them are based on centralized approaches. The work in [34] assumes that a powerful cluster head is able to know the current location and determine the target location of the mobile sensors. However, such a central server may not be available in most cases, and this approach suffers from a single point failure problem. Sensor deployment has also been addressed in the field of robotics [30], where sensors are deployed one by one, utilizing the location information obtained from the previous deployment. This method has strong assumptions on the initial sensor placement in order to guarantee the communication between the deployed and undeployed sensors, and it does not work in case of network partition. Since sensors are deployed one by one, the long deployment time can significantly increase the network initialization time.

In our previous work [31], assuming that all sensors are mobile, we proposed three distributed algorithms for controlling the movement of sensors that are initially randomly placed to get high coverage. The algorithms use Voronoi diagrams to detect coverage holes. In one algorithm, VOR, sensors migrate toward holes. In the second, VEC, sensors move away from each other to achieve a uniform distribution. In the third, minimax, sensors move toward their local center. Although mobile sensors can be used to improve the sensing coverage, their costs may be high. To achieve a good balance between sensor cost and sensor coverage, we proposed a bidding protocol [32] to assist the movement of mobile sensors when a mix of mobile and static sensors are used. In this protocol, mobile sensors act as the hole healing server, the base price of whose service is the size of the hole generated if they leave. Static sensors detect coverage holes and bid mobile sensor based on the hole size. Mobile sensors accept the highest bid and move to heal the hole if the bid is larger than its base price. In this way, mobile sensors always move to heal the largest holes and increase the coverage.

Challenges of Sensor Relocation   The motion capability of sensor nodes can also be used for purposes other than sensor deployment. For example, in case of a sensor failure or node malfunction, other sensors can move to replace the role of the failed node. As an event (i.e., fire, chemical spill, incoming target) occurs, more sensors should relocate to the area of the event to achieve a better coverage. Compared with sensor deployment, *sensor relocation*, which relocates mobile sensors from one place to another place, has many challenges. First, sensor relocation has strict time constraints. Sensor deployment is done before the network is in use, but sensor relocation is on demand and should be finished in a short time. For example, if the sensor monitoring a security-sensitive area dies, another sensor should take the responsibility as soon as possible; otherwise, some security policy may be violated. Second, relocation should not affect other missions currently supported by the sensor network, which means that the relocation should minimize its effect on the current topology. Finally, since physical movement costs much more energy than computation and communication, the moving sensor may suffer. As some nodes die due to low battery power, other nodes need to move again and cost more power. To be fair to each sensor and to prolong the network lifetime, it is important to balance the trade-offs between minimizing the total energy consumption and maximizing the minimum remaining energy of the mobile sensors. Sensor relocation has been mentioned in [35], which focuses on finding the target locations of the mobile sensors based on their current locations and the locations of the sensed events. However, they did not address the challenges of finding the relocation path under time, topology, and energy constraints.

Due to these new challenges, our deployment protocols [31,32] cannot be directly applied for sensor relocation. For example, if the area covered by a failed sensor does not have redundant sensors, moving neighbor sensors may create new holes in that area. To heal these new holes, more sensors need to be involved. This process continues until some area having redundant sensors is reached. During this process, sensors may move back and forth and waste lots of energy. Based on this observation, we propose to first find the location of the redundant sensors and then design an efficient relocation schedule for them to move to the target area (destination).

Finding the Redundant Sensors   Using flooding to find the redundant sensors may significantly increase the message overhead. Techniques based on publisher/subscriber [36] are designed for distributed systems or wired networks and may not be applied to sensor networks due to high overhead. To reduce the message overhead, solutions similar to Two-tier data dissemination (TTDD) [37] can be used. In TTDD, the target field is divided into grids, and each has a grid head. The grid head is responsible for disseminating the sensing data to other grid heads. To find the interested data, the sink floods the query, which will be served by the grid head that has the sensing data. Since the data needs to be flooded to the whole network, although only grid heads, it still has significant overhead.

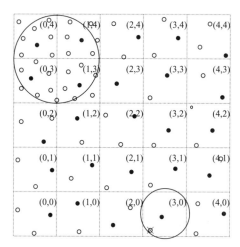

Figure 3.10 System model.

We apply the quorum concept [38–40] to reduce the message overhead. A quorum is a set of grids, and any two quorums must have an intersection grid. If the grid with redundant sensors advertises to sensors in its quorum, any destination grid head can obtain this information by sending a request to the sensors in its quorum. A simple quorum can be constructed by choosing the grids in a row and a column. Suppose N is the number of grids in the network. By using this quorum-based system, the message overhead can be reduced from $O(N)$ to $O(\sqrt{N})$ [38].

By organizing grids as quorums, each advertisement and each request can be sent to a quorum of grids. Due to the intersection property of quorums, there must be a grid that is the intersection of the advertisement and the request. The grid head will be able to match the request to the advertisement. A simple quorum can be constructed by choosing the nodes in a row and a column. Instead of flooding the network with advertisements or requests, the request and the advertisement are only sent to nodes in a row or column. For example, as shown in Figure 3.10, suppose grid (0,3) has redundant sensors, it only sends the advertisement to grids in a row [(0,3), (1,3), (2,3), (3,3), (4,3)] and a column [(0,4), (0,3), (0,2), (0,1), (0,0)]. When grid (3,0) is looking for redundant sensors, it only needs to send a request to grids in a row [(0,0), (1,0), (2,0), (3,0), (4,0)] and a column [(3,4), (3,3), (3,2), (3,1), (3,0)]. The intersection node (0,0) will be able to match the request to the advertisement. Suppose N is the number of grids in the network. By using this quorum-based system, the message overhead can be reduced from $O(N)$ to $O(\sqrt{N})$. Although the message overhead is very low compared to flooding, we can further reduce the message overhead by observing the specialty of our problem.

We can further reduce the message complexity by using the geographic information in sensor networks. For example, we can specify that an advertisement must be sent to grids in one column (advertisement quorum), and a request must be sent to grids in one row (request quorum). Since there is always an intersection grid between any column and row, the grid head of that intersection grid will be able to match the request to the advertisement. Still using the example of Figure 3.10, grids (0,4), (1,4), (0,3), and (1,3) have redundant sensors, while grid (3,0) needs more sensors. The grid head of (1,3) propagates its redundant sensor information through its supply quorum [(1,4), (1,3), (1,2), (1,1), (1,0)]. The grid head in grid (3,0) searches its demand quorum [(0,0), (1,0), (2,0), (3,0), (4,0)]. Grid (1,0) can reply with the information about redundant sensors. Compared to using the quorum in the last example, using grid–quorum cuts the message by half.

Relocating Sensors to the Target Location   After obtaining the information about where the redundant sensors are, the grid head needs to determine how to relocate them. At one extreme, sensors can move to the destination directly. Although this solution can minimize the moving distance, the redeployment time may be long especially when the destination is far away from the source. Furthermore, the sensor moved through a long distance may consume too much energy. If the sensor dies shortly after its movement, this movement is wasted and another sensor has to be found and relocated.

We use a *cascaded movement* to address the problem. The idea is to find some cascading (intermediate) sensors and involve them into the relocation to reduce the delay and balance the power consumption. For example, as shown in Figure 3.11, assume S_0 fails and S_3 is the redundant sensor. S_3 can move to S_2, and S_2 moves to S_1, and S_1 moves to S_0. Since the sensors can first exchange communication messages (i.e., logically move) and ask all relevant sensors to (physically) move at the same time, the relocation time is much shorter. However, the total physical moving distance of this approach may be longer, and it is a challenge to make sure that the sensor coverage is maintained during the sensor movement.

Generally speaking, we have three objectives when determining the relocation schedule: *minimize the relocation time, minimize the total energy consumption*, and *maximize the minimum remaining energy*. Relocation time is mission related and each cascading node has a time constraint. For example, as shown in Figure 3.11, for S_2, if it moves to S_1 before S_3 moves toward S_2, there may appear a new coverage hole around the area covered by S_2. Based on the mission requirement, this may or may not be allowed. From the energy point of view, maximizing the minimum remaining energy at all nodes after the relocation can prolong the network life time since no individual sensor is penalized, but there is a trade-off between minimizing the total energy consumption and maximizing the minimum remaining energy, as illustrated by the two possible paths (going through S_2 and going through S_4) in Figure 3.11.

Previous work on power-aware routing addressed the trade-off based on the current power level [41,42]. When the remaining power is high, minimizing the total energy is more important; otherwise, maximizing the minimum remaining energy is more important since the overuse of individual sensor at this time may deplete their energy and consequently result in disconnections. To achieve a better trade-off, Li et al. [43] designed a *z*-min algorithm, which calculates the route maximizing the minimum remaining power within those paths whose total energy consumption is less than *z* times the minimum energy consumption.

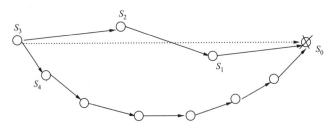

Figure 3.11   Sensor relocation.

However, our problem is much more complicated since cascading sensors also have time constraints.

One solution can be based on minimizing the difference between the total energy consumption and the minimum remaining power. Different from ad hoc routing protocols [41, 42], we may be able to minimize this difference when the sensors are regularly distributed. In this case, dynamic programming techniques can be used to minimize the difference between the total energy consumption and the minimum remaining energy, with the relocation time constraint.

To implement the scheduling algorithm in a distributed way, broadcasting can be used. Each possible cascading sensor broadcasts its decision of the next best sensor to the destination. A sensor can either wait for some amount of time (a system threshold) before broadcasting or make its own decision and rebroadcast the updated version if the previous is wrong. To reduce the frequency of rebroadcasting, geographic information can be used. With this information, the sensors can be sequenced according to their distance to the redundant sensor. A sensor only broadcasts its decision after it receives the decisions from its neighbor sensors, which are located in a search area, which has a high probability to encompass all the cascading nodes.

Sensor Network Diagnosis We will develop techniques to detect coverage holes, which can be used to invoke the sensor relocation mechanisms.

Related Work Exchanging messages between neighbors has the potential of quickly detecting a sensor failure but may suffer from high false positive. Results in system-level diagnosis has been used to address this problem in ad hoc wireless networks [44]. The broadcast nature of the communication network has been exploited to efficiently implement a comparison-based diagnosis protocol. However, every node is eventually notified of the status of every other node, which is not necessary in sensor networks. A less complex approach uses two-tier timeout values [45]. The shorter timeout is used to suspect failure of neighboring nodes, while a longer timeout is used to reduce false positives with input from other neighbors. These techniques fundamentally rely on the participation of all neighbors. If some neighbors of the active node are in the low-power sleep mode, these protocols may not work well.

Low overhead failure detection can be achieved when the topology information is available. In a protocol by Staddon et al. [46], a base station continuously learns the topology of the network. It periodically probes nodes along a preestablished tree structure. If a subtree fails to send back a response, the root node of the subtree is determined to be dead. The status of its children in the subtree is obtained by routing additional probe messages around the dead node. This protocol depends on continuous update of the topology information. Therefore, it may not work well with missions where notification is sent to the sink infrequently, for example, only when an exception occurs.

Techniques for Sensor Network Diagnosis The techniques mentioned in related works do not take dynamic mission requirements into account and focus mainly on the status of the sensors. In reality, the existence of a coverage hole depends on both sensor status and mission requirement. We observe that mission requirement change is initiated by the sink (or other control nodes) and the change can be significant. On the other hand, change in sensor status is a local event and typically has localized impact. We therefore believe two different approaches are necessary: a coverage hole estimation algorithm initiated by the sink whenever mission requirement changes and a sensor status monitoring algorithm that executes continuously but dynamically adapts to mission requirement changes.

COVERAGE HOLE ESTIMATION   We first examine how to quickly estimate the existence and locations of coverage holes given a mission requirement. Each sensor is assumed to know its own location through some localization services, as well as its sensing areas for each parameter of interest to the mission. One important factor to consider is the detection speed. As mobile sensors have finite speed, early detection is critical. Another criterion is a false-positive rate; that is, if a coverage hole is detected, there should be a high probability that the reported hole actually exists.

The detection of a coverage hole relies on aggregate information from multiple sensors. One naive solution is to have the sink collect information from every sensor and perform a local check. This technique can create communication "hot spots" around the sink and therefore is not desirable. More importantly, a false positive may occur if the coverage information from a sensor is lost. We use a *mission-directed data aggregation* technique to address this problem, where the mission requirement information is used to achieve effective data aggregation. Clearly, the sink can partition the network into a set of continuous nonoverlapping areas satisfying the following property: If two areas are adjacent, the parameters being measured in these areas cannot be identical. This information can be used to build an in-network aggregation structure and to efficiently aggregate data from individual sensors. One observation is that we can exploit the sensing result difference of one particular parameter, especially when sensors are identical. Suppose two parameters, p_1 and p_2, need to be monitored in the same area A. The maximum sensing distance of a sensor is d_1 for p_1 and d_2 for p_2, with $d_1 < d_2$. If we can determine there is no coverage hole in A when p_1 is measured, we can safely conclude that there is no coverage hole for p_2 either. Similarly, if there is a coverage hole in A when p_2 is measured, it is guaranteed that the coverage hole will exist when p_1 is measured.

A promising approach to reduce false positives stems from the following observations: Coverage holes and covered areas are nonoverlapping areas, and continuous areas can be effectively aggregated. We can then estimate both coverage holes and covered areas during the same data collection and analysis process, with one of them being more aggressively estimated and the other more conservatively estimated. As a coverage hole and a covered area cannot overlap, we can use the estimation on covered area to reduce false positives on coverage holes.

SENSOR STATUS ANALYSIS   The second approach is to devise a sensor diagnosis technique that can quickly detect a sensor failure and decide whether such failure will result in a new coverage hole. Deciding whether a sensor failure will result in a coverage hole requires information on both sensor status and mission requirements. This can be achieved by using in-network distributed diagnosis that captures distributed mission requirements as well as locally aggregated sensor status. Existing work on network diagnosis [47, 48] can be extended to develop protocols to construct such diagnosis structures and to adapt it according to both mission requirements and sensor status change.

For a given sensor distribution and a set of mission requirements, not every sensor failure will results in a coverage hole. Therefore, the diagnosis technique will concentrate on sensors whose failure results in coverage holes. This technique can result in significant savings in terms of communication and computation overhead.

CO-DESIGN ISSUES   Although the two techniques described above are inherently different, they are used to solve the same problem and need to coexist in the same network. A typical scenario would be as follows. The solution to coverage hole estimation will be executed by the sink whenever the mission requirement changes; the sensor diagnosis technique is executed continuously, but should dynamically adapt to the area of interest to the mission. The coverage hole estimation algorithm needs to collect and analyze information from individual sensors and estimate the coverage holes. During this process, the protocol may need to maintain a certain in-network data structure to perform effective data filtering and aggregation. It is conceivable such infrastructure could be utilized by the sensor diagnosis protocol to quickly adapt its diagnosis and estimation behavior.

3.3.3 Mobility-Assisted Routing

We focus on issues related to mobility for routing. When considering routing, sensors may be cast in one of three roles. First, sensors may be acting solely as relays to transfer data from a source to a sink. In this case they will move to form an energy-efficient route. Second, sensors may be acting solely to gather data, and therefore their movement is dictated by their sensing requirements as discussed in Section 3.3.2. In this case, the routing protocols do not control mobility but must react to it. Third, sensors may be assigned both sensing and relaying responsibilities, simultaneously. In this case, mobility must consider both sensing and routing. We discuss the first two cases in this section and the third case in the next section.

In the first subsection below we discuss mobility algorithms that can minimize the energy consumed due to communication in the relay case and extend it to a scenario in which the sensors are also assigned a sensing mission. In the second subsection we discuss priority-based schemes designed to operate in the presence of high network volatility, the type of which may occur when sensors are highly mobile because of diverse sensing requirements.

Mobility for Routing Using a Distributed Annealing Strategy   Suppose that, as shown in Figure 3.9, certain sensor nodes are assigned target-tracking tasks while others are assigned tasks supporting communication, that is, relaying the tracking data to the data sinks of the network. For the relay nodes, the goals of mobility are to create routing paths between sources and sinks and to maximize the lifetime of the network by moving to positions at which the required transmission power for the tracking data flows is minimized.

Although many routing protocols [41, 42, 49] for ad hoc networks can achieve similar goals, they are not optimized for a different environment that we consider here. Mobile sensor networks are a special kind of ad hoc communication network in which all of the nodes have a communal mission. In some cases, mobility may be controllable and can, therefore, be used to help achieve the network's missions. That is, mobility may be "purposeful" instead of being treated as an uncontrollable external stimulus to which the ad hoc communication network must respond. In other cases, sensor mobility may be random: for example, a group of sensors diffusing through the air, or sensors moving to *scan* a large area [50], where scanning is a special case of mobility for sensing. Even when mobility is controllable, it may be desirable to make it partially random in order to deal with a lack of information locally due to the distributed nature of the network [51].

In this section, we propose a preliminary distributed/decentralized motion decision framework for the relay nodes based on the simulated annealing optimization algorithm (see, e.g., [52]) assuming that nodes can localize their proximal neighbors [53]. Our specific objective is to incrementally find the node positions that minimize the total required transmission power for all the active flows in the network while suitably penalizing for the energy cost of motion in order to find these positions, that is, the mobility energy costs were amortized over the savings in communication power.

More specifically, let $V(x, r)$ be the total power required from the network to transmit the F flows using routes r when the intermediate nodes are in positions x; the optimal choice of routes at position x is

$$R(x) \in \arg \min_{r \in \mathbf{R}(x)} V(x, r)$$

where $\mathbf{R}(x)$ is the set of feasible routes connecting those nodes when in positions x. $R(x)$ is the objective of a distributed routing algorithm (cf. the following subsections) operating at a much faster time scale than that of the motion of the nodes. The amount of power required to maintain a link can be incorporated into the link metrics used for establishing routes.

Under a deterministic greedy mobility strategy, each node moves to a position at which it expects to minimize its energy costs for transmission. This approach may not achieve optimal energy efficiency because local minima may occur. To overcome this characteristic, we restrict motion to a lattice. Then under our annealing motion strategy, node k (currently at x_k) selects a neighboring position z, *at random*, and accepts the move to z according to a "heat bath" probability:

$$\min\{1, \exp(-\beta \Delta_k V(x, z))\}$$

where β is interpreted as inverse annealing temperature. Intuitively, the higher the "temperature," the more random motion that will occur. We chose a lattice over other random mechanisms because it tends to minimize the total energy costs in the network [51].

Given the β parameter, it is possible to tune this algorithm to match the requirements of the pure relaying scenario, or a scenario that includes sensors scanning and relaying simultaneously. When scanning, sensors will move throughout a field to gather information. In these cases, the random motion for routing may coincide with the motion for scanning. When assuming intermediate nodes are solely performing relaying tasks, and stationary nodes are the data sources and sinks, the annealing algorithm can be allowed to "cool" (β increased) to fix the relay nodes in optimal positions. However, if, for example, the tracking nodes themselves move, or the tracking tasking is dynamic, cooling would make the relay network less responsive to this change. Such change is part of more general "volatility" in networking conditions that may be experienced by the relay nodes. We discuss these cases in Section 3.3.4.

We conducted a simulation study on a mobile sensor network, see the results depicted in Figure 3.12. In one set of simulations, no scanning task is set, but in the other, a node moves to scan with equal probability that it makes an "annealing" move for the purposes of relaying data. The figures clearly indicate that tasking scanning resulted in increased power for communication and motion but increased scanning performance where the last figure represents the total number of points visited over a 60-s sliding time window.

Robust (Multipath) Priority-Based Routing   It is natural to assume that because sensors are performing different missions, the data gathered from these sensors will have different requirements in terms of latency when being relayed through the network. Among the flows, high-priority flows include those for tracking traffic, responses to specific queries,

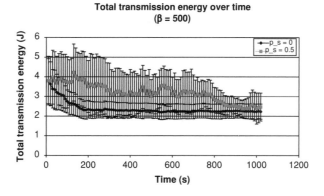

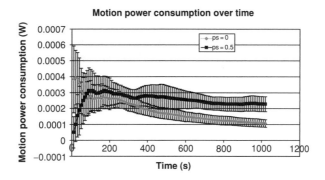

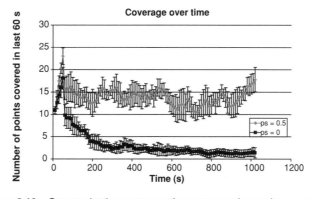

Figure 3.12 Communication power, motion power, and scanning coverage.

the queries themselves, and control (routing) traffic. The goal of the routing protocol for these flows is to minimize energy consumption given a delay constraint. Lower-priority routing flows include passive surveillance traffic and tracking traffic for low-priority targets. The goal of the routing protocol for these flows is to minimize energy consumption. We develop priority-based routing protocols that jointly manage both delay and energy concerns. The principle challenges of such protocols is to reliably route in a highly volatile topology with minimal overhead. Therefore, these protocols will be suitable for cases in which sensors are highly mobile to achieve sensing missions.

Our general approach will attempt to give primary importance to energy efficiency by routing through a good energy path and use priority scheduling to reduce delay for the priority traffic. Nonpriority traffic is not starved under the assumption that priority traffic is bursty and light. Nodes can be notified of the "true" energy resources and delay through each neighbor by a link capacity metric that incorporates the information about queue backlog (related to queuing delay via Little's formula).

We explore suitable routing algorithms based on both "swarm intelligence" and ant-colony meta-heuristics, for example, Ant-Colony-Based Routing Algorithm (ARA) [54] and Termite [55]. ARA consists of three phases: route discovery, route maintenance, and route failure handling. In the route discovery phase, new routes between nodes are discovered with the use of forward-and-backward ants, similar to AntNet. Routes are maintained by subsequent data packets, that is, as the data traverse the network, node pheromone values are modified so that their paths are "reinforced." Also, as in nature, pheromone values decay with time in the absence of such reinforcement. Routing (link) failures, usually caused by node mobility, are detected through missing acknowledgments. When a node detects a routing error, the pheromone value associated with the "missing link" is set to 0. In [56], in addition to forward-and-backward ants, "uniform" ants are introduced to cope with highly mobile nodes.

Both energy and delay issues are considered in [57]. Only delay quantities, however, are considered when computing the pheromone values and forwarding probabilities. The dissipated energy of a node after each ant passes through is calculated by

$$\Delta E_{ij} = \frac{K}{(D_{ij})^2}$$

where K is the amount of energy to transmit the ant over a single unit distance, and D_{ij} is the Euclidean distance between node i and j [58]. The residual node energy at time t is computed by

$$E_i(t) = E_i(t-1) - \sum_j \Delta E_{ij}$$

When a node's energy level simply drops below a prespecified threshold value, the node is removed from the sensor network and alternative routes are found.

The proposed algorithm is based on the Ant mechanism algorithm, and uses energy and delay metrics to perform updates of pheromone levels. We modify the packet header to contain both energy and delay information so that a separate pheromone level will be maintained for each traffic type. Two types of pheromone-based routing algorithms will be developed. In the first framework, packet headers are assumed to have two fields used for routing: one to indicate bottleneck residual energy of a path (to be used for minimizing the energy costs) and the other being a hop count (to minimize delay). In the second framework, packet headers have fields that track the minimum residual energy of the nodes that relay them (as in the first algorithm), and fields that track the cumulative delay based on backlog information of queued packets destined to the packet's source. So, when a packet reaches its destination, it contains the minimum residual energy and the cumulative queuing delay of its route back to its destination.

To reiterate, such pheromone-based approaches [54, 55] are appropriate for highly volatile networking conditions. Such approaches do not exclusively use optimal routes but are highly responsive to changing network topology. Under more "stable" networking conditions, existing protocols like AODV [59] and DSR [60] would yield superior performance/overhead trade-offs, especially when the routing protocol is augmented with *planned* mobility information. Therefore, we will also consider multiprotocol routing in heterogeneous networking environments in which different regions of the network employ the most appropriate routing protocol under the circumstances.

3.3.4 Integrated Mobility for Sensing and Routing

In this section we formulate the problem for integrated mobility. Our goal is to maximize the value generated by the sensor network over time. For example, if multiple missions provide conflicting requirements to the network, higher value is placed on fulfilling the higher priority missions. Likewise, the longer a network remains active, the more value it will generate. Therefore, we develop algorithms and protocols to meet the needs of the composite of the highest value missions while maximizing network lifetime by conserving energy.

We define X_k as a vector representing the location of all sensor in the network during time period k, and t_k as the time that the network remains in this configuration. We define m_j as mission j. The value generated by each sensor i per unit time for configuration k is represented by $v_{i,j}^k(X_k)$. Further, we have

$$v_{i,j}^k(X_k) = u_c(X_k)s_i(m_j, X_k)$$

where $s_i(m_j, X_k)$ is the value of sensor i performing mission j while in configuration k, and $u_c(X_k)$ is either 1 or 0. Then $s_i(m_j, X_k)$, a function specific to each mission, will tend to be higher for more valuable missions when the sensor is optimally placed: As the sensor moves from its optimal position and its accuracy is compromised, or is assigned less critical missions, its value will decrease until it reaches 0. If the sensor is able to communicate its data to the sink in a timely fashion, $u_c(X_k)$ Is 1; otherwise it is 0. This jointly captures the importance of sensing and communicating.

The overall value of the network for configuration k is

$$V_k = t_k \left(\sum_{i,j} v_{i,j}^k(X_k) \right)$$

The overall value of a network is given by $V = \sum_{k=1}^{K} V_k$ where K is the total number of configurations over the lifetime T of the network. To maximize V, we must complete as many missions as possible, which implies conserving energy so that network lifetime is extended. The energy cost of configuration k is

$$C_k = M(X_{k-1} \rightarrow X_k) + E(X_k)t_k$$

where the first term is the cost of moving from configuration $k - 1$, and the second is the cost of sensing and communicating in the new configuration k. Clearly, T and N depend on the C_k. We can see that there is a trade-off between the energy expended to realize a configuration and the energy spent while in the configuration; a critical component to evaluating this trade-off is the time, t_k, spent in the configuration.

There are several interesting challenges to consider when designing algorithms to manage mobility to jointly accommodate sensing and routing. Algorithms that use strict priorities for sensing may not achieve maximum overall value, for example, in cases in which the highest priority task requires exclusive use of sensors, thus allowing no other missions to be accomplished. Sequentially considering missions suffers from possible high latency for relocating sensors. Certain missions may be essential, that is, they must be performed; this must be accounted for when designing algorithms. Algorithms must be carefully designed to account for the impact of location on sensing and relaying; for instance, optimal sensor placement for sensing for one mission may preclude communication, and hence completion of a second mission. When designing these algorithms, we must consider the mobility algorithms discussed in Sections 3.3.2 and 3.3.3 and possible extensions. For example, as discussed in Section 3.3.3, the "temperature" of the annealing algorithm may be modified to make sensors more or less reactive.

3.3.5 Conclusions

Adding mobility to sensor networks can significantly increase the capability of the sensor network by making it resilient to failures, reactive to events, and able to support disparate missions with a common set of sensors. However, there has not been a great deal of work on adding mobility to sensor networks. In this chapter, we addressed three closely intertwined issues to support mobility in sensor networks. First, we proposed solutions to relocate sensors in response to an event or failure, considering the computation complexity, movement distance, relocation time, communication overhead, and the impact of the mobility on other concurrently running missions. Second, we developed mobility assisted routing protocols to improve network communications considering issues such as network connectivity, energy consumption of both communications and movement, and the network lifetime. Finally, we defined utility functions that can capture the benefits of the movement from the perspective of all missions, and maximize the capability of the network.

3.4 FORMATION AND ALIGNMENT OF DISTRIBUTED SENSING AGENTS WITH DOUBLE-INTEGRATOR DYNAMICS AND ACTUATOR SATURATION

Sandip Roy, Ali Saberi, and Kristin Herlugson

In this section, we consider formation and alignment of distributed sensing agents with double-integrator dynamics and saturating actuators. First, we explore the role of the agents' sensing architecture on their ability to complete formation and alignment tasks. We develop necessary and sufficient conditions on the sensing architecture, for completion of formation and alignment tasks using linear dynamic control. We also consider the design of static controllers for the network of agents and find that static control is indeed possible for a large class of sensing architectures. Next, we extend the control strategies developed for completion of formation tasks to simultaneously achieve collision avoidance. In particular, we consider formation stabilization with collision avoidance for sensing agents that move in the plane. The control paradigm that we develop achieves avoidance and formation together, by taking advantage of the multiple directions of motion available to each agent. Our explorations show that collision avoidance can be guaranteed, given some weak constraints on the desired formation and the distance that must be maintained between the agents. Throughout, several examples are developed to motivate our formulation and illustrate our results.

3.4.1 Background Information

A variety of natural and engineered systems comprise networks of communicating agents that seek to perform a task together. In such systems, individual agents have access to partial information about the system's state, from which they attempt to actuate their own dynamics so that the system globally performs the required task. Recently, much effort has been given to developing plausible models for systems of interacting agents and to constructing decentralized controllers for such systems (e.g., [61–65]). These studies vary widely in the tasks completed by the agents (including formation stabilization and collision avoidance), the intrinsic dynamics and actuation of the agents, the communication protocol among the agents, and the structure of the controllers.

Our research efforts are focused on understanding, in as general a manner as possible, the role of the communication/sensing network structure in allowing the network to perform the required task. In this first study, we consider systems with simple but plausible local dynamics (double-integrator dynamics with saturating actuators) and task aims (settling of agents to specific locations or fixed-velocity trajectories in a Euclidean space without collision avoidance, henceforth called *formation stabilization*). Within this simple context, we assume a quite general sensing network architecture,[1] and specify necessary and sufficient conditions on this architecture for the existence of a decentralized dynamic linear time-invariant (LTI) controller that achieves formation stabilization. Using our formulation, we are also able to identify a broad class of sensing architectures for which static decentralized control is possible. While the agent dynamics considered here are limited, we believe that our approach is promising because it clearly extracts the role of the sensing architecture in completing tasks and hence facilitates development of both appropriate sensing architectures and controllers for them. Further, we are able to extend our control design to achieve collision avoidance in addition to stabilization for agents defined in the plane. The goals of our analysis are clearly illustrated with an example. Let us say that three coordinating vehicles seek to locate themselves to the west, east, and south of a target. We aim to achieve this task by controlling the accelerations of the vehicles. Our studies aim to determine the class of observation topologies (ways in which the vehicles observe the target location and/or each others' locations) for which the formation stabilization task can be achieved, without collision among the vehicles.

Throughout our studies, we aim to delineate the connections between our formulation and results and those found in the existing literature on vehicle task dynamics. Broadly, our key contributions to this literature are as follows:

- Our studies consider an arbitrary linear observation topology for the sensing architecture that significantly generalizes the sensing architectures that we have seen in the literature. Of particular interest is the consideration of multiple observations for each agent; we find that multiple observations can sometimes permit stabilization even when a single observation that is an average of these observations does not.
- We consider actuator saturation, which we believe to be realistic in many systems of interest.

---

[1] We feel that our observation architecture is more accurately viewed as a sensing architecture rather than a communication architecture because measurements are assumed to be instantaneous; hence, we will use the term sensing architecture, though our formulation may quite possibly provide good representation for certain communication architectures also.

- We are able to develop explicit necessary and sufficient conditions on the sensing architecture for formation stabilization. This analysis also serves to highlight that the seminal research on decentralized control done by Wang and Davison [66] is central in the study of distributed task dynamics. From this viewpoint, our work buttresses the analysis of [61], by extending the application of [66] to sensing architectures beyond leader–follower ones.
- We show that static stabilizers can be designed for a wide class of sensing architectures, and we explore system performance upon static stabilization through simulations.

3.4.2 Model Formulation

We describe the model of distributed, mobile sensing agents that is studied throughout this section. The model is formulated by first specifying the local dynamics of each agent and then developing a sensing architecture for the agents. A vector representation for the model is also presented.

Local Dynamics   Each of the n agents in our system is modeled as moving in a one-dimensional Euclidean space. We denote the position of agent i by $r_i \in \mathbf{R}$. The position of agent i is governed by the differential equation $\ddot{r}_i = \sigma(u_i)$, where $u_i \in \mathbf{R}$ is a decentralized control input and $\sigma(\ )$ represents (without loss of generality) the standard saturation function. We also sometimes consider double-integrator dynamics without saturation, so that $\ddot{r}_i = u_i$.

One note about our model is of particular importance: In our simulations, we envision each agent as moving in a multidimensional Euclidean space, yet agents in our model are defined as having scalar positions. We can do so without loss of generality because the internal model for the agents in each coordinate direction is decoupled (in particular, a double integrator). Hence, we can simply redefine each agent in a multidimensional system as a set of agents with scalar positions, each of which track the location of the original agent in one coordinate direction. We will discuss shortly how observations in a multidimensional model can be captured using a scalar reformulation.

Sensing Architecture   We define the *sensing architecture* for the system quite generally: Each agent has available one or more linear observations of the positions and velocities of selected agents. Formally, we denote the number of linear observations available to agent i by m_i. The $m_i \times n$ *graph matrix*

$$G_i \triangleq \begin{bmatrix} g_{11}(i) & \cdots & g_{1n}(i) \\ \vdots & & \vdots \\ g_{m_i 1}(i) & \cdots & g_{m_i n}(i) \end{bmatrix}$$

specifies the linear observations that are available to agent i. In particular, the jth ($1 \le j \le m_i$) observation available to agent i is the average

$$\mathbf{a}_{ij} = g_{j1}(i) \begin{bmatrix} r_1 \\ v_1 \end{bmatrix} + \cdots + g_{jn}(i) \begin{bmatrix} r_n \\ v_n \end{bmatrix} \tag{3.16}$$

where $v_i = \dot{r}_i$ is the velocity of agent i. Agent i's m_i observations can be concatenated into a single observation vector:

$$\mathbf{a}_i^T \triangleq \begin{bmatrix} \mathbf{a}_{i1}^T & \vdots & \mathbf{a}_{im_i}^T \end{bmatrix}$$

In vector form, the observation vector for agent i can be written in terms of the state vector as

$$\mathbf{a}_i = C_i \mathbf{x} \tag{3.17}$$

where

$$C_i = \begin{bmatrix} G_i & 0 \\ 0 & G_i \end{bmatrix} \tag{3.18}$$

Sometimes, we find it convenient to append the graph matrices for individual agents into a single matrix. We define the *full graph matrix* for the agents as $G^T = \begin{bmatrix} G_1^T & \cdots & G_n^T \end{bmatrix}$.

A couple notes about our sensing architecture are worthwhile:

- We can represent the graph-Laplacian sensing architecture described in, for example, [62]. Graph-Laplacian observation topologies are applicable when agents know their positions relative to other agents. More specifically, each agent i's observation is assumed to be a average of differences between i's position and other agents' positions. To capture Laplacian observations using our sensing architecture, we constrain each agent to have available a single average (i.e., $m_i = 1$ for all i) and specify the graph matrix entries for agent i as follows:

$$g_{1i}(i) = 1$$
$$g_{1j}(i) = -\frac{1}{|\mathcal{N}_i|}, \qquad j \in \mathcal{N}_i \tag{3.19}$$
$$g_{1j}(i) = 0, \qquad \text{otherwise}$$

where $\mathcal{N}_i$ are the neighbors of agent i (see [62] for details). Note that the full graph matrix for a Laplacian architecture is square, has unity entries on the diagonals, has negative off-diagonal entries, and has row sums of 0.

When we consider Laplacian observation topologies, we will often use a *grounded Laplacian* to represent the sensing architecture. A grounded Laplacian represents a sensing architecture in which the agents associated with each connected component of the full graph matrix have available at least one absolute position measurement of some sort. Mathematically, the full graph matrix has unity diagonal entries and negative off-diagonal entries, but each connected component of the full graph matrix is assumed to have at least one row that sums to a strictly positive value. The difference between a grounded Laplacian architecture and a Laplacian architecture is that each agent's absolute position can be deduced from the observations for the grounded Laplacian architecture but not for the Laplacian architecture. In most applications, it is realistic that absolute positions can be deduced in the frame-of-reference of interest. In, for example, [62], some systems with a Laplacian architecture are shown to converge in a relative frame, which can equivalently be viewed as absolute convergence of state vector differences given a grounded Laplacian architecture. Here, we will explicitly distinguish between these two viewpoints by considering absolute and partial stabilization of our systems.

Our sensing architecture is more general than the graph-Laplacian architecture in that arbitrary combinations of agents' states can be observed, and multiple observations are possible. Consideration of multiple observations is especially important in that it allows comparison of controllers that use averaged measurements with those that use multiple separate measurements. Our analyses show that stabilization is sometimes possible when multiple observations are used, even though it might not be possible when an average of these observations is used.

- When an agent with a vector (multidimensional) position is reformulated as a set of agents with scalar positions, each of these new agents must be viewed as having access to the same information as the original agent. Hence, the graph matrices for these newly defined agents are identical. We note that this formulation allows observations that are arbitrary linear combinations of state variables associated with different coordinate directions.

- Notice that we have structured the model so that the observation architecture is identical for position and velocity measurements (i.e., whenever a particular position average is available, the same velocity average is also available). The stabilization results that we present in the next section do not require identical observation architectures for positions and velocities: in fact, only the sensing architecture for positions is needed to verify stabilization. However, because static stabilization of the model is simplified, we adopt this assumption (which we believe to be quite reasonable in many applications). In some of our results, we will wish to distinguish that only the position observation structure is relevant. For such results, we shall use the term *position sensing architecture* to refer to the fact that the graph structure applies to only positions, while velocity observations may be arbitrary (or nonexistent).

The types of communication topologies that can be captured in our formulation are best illustrated and motivated via several examples.

Example: Vehicle Coordination, String   First, let us return to the vehicle formation example discussed in the introduction. Let us assume that the vehicles move in the plane, and (without loss of generality) that the target is located at the origin. A reasonable assumption is that one vehicle knows its position relative to the target and hence knows its own position. Assuming a string topology, the second vehicle knows its position relative to the first vehicle, and the third vehicle knows its position relative to the second vehicle. To formulate the graph matrices for this example, we define agents to represent the x and y positions of each vehicle. The agents are labeled $1x$, $1y$, $2x$, $2y$, $3x$, and $3y$. The graph matrices for the six agents are as follows:

$$G_{1x} = G_{1y} = \begin{bmatrix} 1 & 0 & 0 & 0 & 0 & 0 \\ 0 & 1 & 0 & 0 & 0 & 0 \end{bmatrix}$$

$$G_{2x} = G_{2y} = \begin{bmatrix} -1 & 0 & 1 & 0 & 0 & 0 \\ 0 & -1 & 0 & 1 & 0 & 0 \end{bmatrix} \tag{3.20}$$

$$G_{3x} = G_{3y} = \begin{bmatrix} 0 & 0 & -1 & 0 & 1 & 0 \\ 0 & 0 & 0 & -1 & 0 & 1 \end{bmatrix}$$

Notice that the sensing architecture for this example is a grounded Laplacian architecture.

Example: Vehicle Coordination Using an Intermediary   Again consider a set of three vehicles in the plane that are seeking to reach a target at the origin. Vehicle 1 knows the x coordinate of the target and hence effectively knows its own position in the x direction. Vehicle 2 knows the y coordinate of the target and hence effectively knows its own position in the y direction. Both vehicles 1 and 2 know their position relative to the intermediary vehicle 3, and vehicle 3 knows its position relative to vehicles 1 and 2. We would like to determine whether or not all three vehicles can be driven to the target.

We can use the following graph matrices to capture the sensing topology described above:

$$G_{1x} = G_{1y} = \begin{bmatrix} 1 & 0 & 0 & 0 & 0 & 0 \\ -1 & 0 & 0 & 0 & 1 & 0 \\ 0 & -1 & 0 & 0 & 0 & 1 \end{bmatrix}$$

$$G_{2x} = G_{2y} = \begin{bmatrix} 0 & 0 & 0 & 1 & 0 & 0 \\ 0 & 0 & -1 & 0 & 1 & 0 \\ 0 & 0 & 0 & -1 & 0 & 1 \end{bmatrix} \tag{3.21}$$

$$G_{3x} = G_{3y} = \begin{bmatrix} -1 & 0 & 0 & 0 & 1 & 0 \\ 0 & -1 & 0 & 0 & 0 & 1 \\ 0 & 0 & -1 & 0 & 1 & 0 \\ 0 & 0 & 0 & -1 & 0 & 1 \end{bmatrix}$$

Example: Measurement Failures   Three aircraft flying along a (straight-line) route are attempting to adhere to a preset fixed-velocity schedule. Normally, each aircraft can measure its own position and velocity and so can converge to its scheduled flight plan. Unfortunately, because of a measurement failure on one of the aircraft, the measurement topology on a particular day is as follows. Aircraft 1 can measure its own position and velocity, as well as its position and velocity relative to aircraft 2 (perhaps through visual inspection). aircraft 3 can measure its own position and velocity. aircraft 2's measurement devices have failed. However, it receives a measurement of aircraft 3's position. Can the three aircraft stabilize to their scheduled flight plans? What if aircraft 2 instead receives a measurement of the position of aircraft 1?

We assume that the aircraft are well-modeled as double integrators. Since each aircraft seeks to converge to a fixed-velocity trajectory, this problem is a formulation–stabilization one. We can again specify the graph matrices for the three aircraft from the description of the sensing topology:

$$G_1 = \begin{bmatrix} 1 & 0 & 0 \\ -1 & 1 & 0 \end{bmatrix}$$

$$G_2 = \begin{bmatrix} 0 & 0 & 1 \end{bmatrix} \tag{3.22}$$

$$G_3 = \begin{bmatrix} 0 & 0 & 1 \end{bmatrix}$$

If aircraft 2 instead receives the location of aircraft 1, then $G_2 = \begin{bmatrix} 1 & 0 & 0 \end{bmatrix}$.

Example: Vehicle Coordination, Leader–Follower Architecture   As in the string of vehicles example, we assume that the vehicles move in the plane and seek a target at the origin. However, we assume a leader–follower sensing architecture among the agents, as described in [61]. In particular, vehicle 1 knows its position relative to the target and hence knows its own position. Vehicles 2 and 3 know their relative positions to vehicle 1. The following graph matrices can be used for this sensing topology:

$$G_{1x} = G_{1y} = \begin{bmatrix} 1 & 0 & 0 & 0 & 0 & 0 \\ 0 & 1 & 0 & 0 & 0 & 0 \end{bmatrix}$$

$$G_{2x} = G_{2y} = \begin{bmatrix} -1 & 0 & 1 & 0 & 0 & 0 \\ 0 & -1 & 0 & 1 & 0 & 0 \end{bmatrix} \tag{3.23}$$

$$G_{3x} = G_{3y} = \begin{bmatrix} -1 & 0 & 0 & 0 & 1 & 0 \\ 0 & -1 & 0 & 0 & 0 & 1 \end{bmatrix}$$

Vector Representation   In state-space form, the dynamics of agent i can be written as

$$\begin{bmatrix} \dot{r}_i \\ \dot{v}_i \end{bmatrix} = \begin{bmatrix} 0 & 1 \\ 0 & 0 \end{bmatrix} \begin{bmatrix} r_i \\ v_i \end{bmatrix} + \begin{bmatrix} 0 \\ 1 \end{bmatrix} \sigma(u_i) \tag{3.24}$$

where $v_i \stackrel{\Delta}{=} \dot{r}_i$ represents the velocity of agent i. It is useful to assemble the dynamics of the n agents into a single state equation:

$$\dot{\mathbf{x}} = \begin{bmatrix} 0 & I_n \\ 0 & 0 \end{bmatrix} \mathbf{x} + \begin{bmatrix} 0 \\ I_n \end{bmatrix} \sigma(\mathbf{u}) \tag{3.25}$$

where

$$\mathbf{x}^T = ([\, r_1 \quad \ldots \quad r_n \mid v_1 \quad \ldots \quad v_n \,])^T$$

and

$$\sigma(\mathbf{u}^T) = ([\, \sigma(\mathbf{u}_1^T) \quad \ldots \quad \sigma(\mathbf{u}_n^T) \,])$$

We also find it useful to define a *position vector* $\mathbf{r}^T = ([\, r_1 \ldots r_n \,])^T$ and a *velocity vector* $\mathbf{v} = \dot{\mathbf{r}}$.

We refer to the system with state dynamics given by (3.25) and observations given by (3.17) as a *double-integrator network with actuator saturation*. We shall also sometimes refer to an analogous system that is not subject to actuator saturation as a *linear double-integrator network*. If the observation topology is generalized so that the graph structure applies to only the position measurements, we shall refer to the system as a *position-sensing double-integrator networks*. (with or without actuator saturation). We shall generally refer to such systems as *double-integrator networks*.

3.4.3 Formation Stabilization

Our aim is to find conditions on the sensing architecture of a double-integrator network such that the agents in the system can perform a task. The necessary and sufficient conditions that we develop below represent a first analysis of the role of the sensing architecture on the achievement of task dynamics; in this first analysis, we restrict ourselves to formation tasks, namely those in which agents converge to specific positions or to fixed-velocity trajectories. Our results are promising because they clearly delineate the structure of the sensing architecture required for stabilization.

We begin our discussion with a formal definition of formation stabilization.

Definition 3.4.1 A double-integrator network can be (semiglobally[2]) formation stabilized to $(\mathbf{r}_0, \mathbf{v}_0)$ if a proper linear time-invariant dynamic controller can be constructed for it, so that the velocity $\dot{r}_i$ is (semiglobally) globally asymptotically convergent to v_{i0} and the relative position $r_i - v_{i0}t$ is (semiglobally) globally asymptotically convergent to r_{i0} for each agent i.

Our definition for formation stabilization is structured to allow for arbitrary fixed-velocity motion and position offset in the asymptote. For the purpose of analysis, it is helpful for us to reformulate the formation stabilization problem in a relative frame in which all velocities and position offsets converge to the origin. The following theorem achieves this reformulation:

Theorem 3.4.1 A double-integrator network can be formation stabilized to $(\mathbf{r}_0, \mathbf{v}_0)$ if and only if it can be formation stabilized to $(\mathbf{0}, \mathbf{0})$.

PROOF   Assume that network can be formation stabilized to $(\mathbf{0}, \mathbf{0})$. Then for every initial position vector and velocity vector, there exists a control signal $\mathbf{u}$ such that the agents converge to the origin. Now let us design a controller that formation stabilizes the network to $(\mathbf{r}_0, \mathbf{v}_0)$. To do so, we can apply the control that formation stabilizes the system to the origin when the initial conditions are computed relative to $\mathbf{r}_0$ and $\mathbf{v}_0$. It is easy to check that the relative position offsets and velocities satisfy the initial differential equation, so that the control input achieves the desired formation. The only remaining detail is to verify that the control input can still be found from the observations using an LTI controller. It is easy to check that the same controller can be used, albeit with an external input that is in general time varying. The argument can be reversed to prove that the condition is necessary and sufficient.   ∎

We are now ready to develop the fundamental necessary and sufficient conditions relating the sensing architecture to formation stabilizability. These conditions are developed by applying decentralized stabilization results for linear systems [66] and for systems with saturating actuators [67].

Theorem 3.4.2 A linear double-integrator network is formation stabilizable to any formation using a proper dynamic linear time-invariant (LTI) controller if and only if there exist vectors $\mathbf{b}_1 \in \text{Ra}(G_1^T), \ldots, \mathbf{b}_n \in \text{Ra}(G_n^T)$ such that $\mathbf{b}_1, \ldots, \mathbf{b}_n$ are linearly independent.

PROOF   From Theorem 3.4.1, we see that formation stabilization to any $(\mathbf{r}_0, \mathbf{v}_0)$ is equivalent to formation stabilization to the origin. We apply the result of [66] to develop conditions for formation stabilization to the origin. Wang and Davison [66] prove that a decentralized system is stabilizable using a linear dynamic controller if and only if the system has no unstable (or marginally stable) *fixed modes*. (We refer the reader to [66] for details on fixed modes.) Hence, we can justify the condition above, by proving that our system has no unstable fixed modes if and only if there exist vectors $\mathbf{b}_1 \in \text{Ra}(G_1^T), \ldots, \mathbf{b}_n \in \text{Ra}(G_n^T)$ such that $\mathbf{b}_1, \ldots, \mathbf{b}_n$ are linearly independent.

---

[2] We define semiglobal to mean that the initial conditions are located in any a priori defined finite set.

It is easy to show (see [66]) that the fixed modes of a decentralized control system are a subset of the modes of the system matrix, in our case $\begin{bmatrix} 0 & I_n \\ 0 & 0 \end{bmatrix}$. The eigenvalues of this system matrix are identically 0, so our system is stabilizable if and only if 0 is not a fixed mode. We can test whether 0 is a fixed mode of the system by using the determinant-based condition of [66], which reduces to the following in our example: The eigenvalue 0 is a fixed mode if and only if

$$\det\left(\begin{bmatrix} 0 & I_n \\ 0 & 0 \end{bmatrix} + \begin{bmatrix} 0 \\ I_n \end{bmatrix} K \begin{bmatrix} C_1 \\ \vdots \\ C_n \end{bmatrix}\right) = 0 \tag{3.26}$$

for all K of the form $\begin{bmatrix} K_1 & 0 & 0 \\ 0 & \ddots & 0 \\ 0 & 0 & K_n \end{bmatrix}$, where each K_i is a real matrix of dimension $1 \times 2m_i$.

To simplify the condition (3.26) further, it is helpful to develop some further notation for the matrices $K_1, \ldots, K_n$. In particular, we write the matrix K_i as follows:

$$K_i = \begin{bmatrix} \mathbf{k}_p(i) & \mathbf{k}_v(i) \end{bmatrix}$$

where each of the four submatrices are length-m_i row vectors. Our subscript notation for these vectors represents that these control gains multiply positions (p) and velocities (v), respectively.

In this notation, the determinant in condition (3.26) can be rewritten as follows:

$$\det\left(\begin{bmatrix} 0 & I_n \\ 0 & 0 \end{bmatrix} + \begin{bmatrix} 0 \\ I_n \end{bmatrix} K \begin{bmatrix} C_1 \\ \vdots \\ C_n \end{bmatrix}\right) = \det\left(\begin{bmatrix} 0 & I_n \\ Q_p & Q_v \end{bmatrix}\right) \tag{3.27}$$

where

$$Q_p = \begin{bmatrix} \mathbf{k}_p(1)(G_1) \\ \vdots \\ \mathbf{k}_p(n)(G_n) \end{bmatrix}$$

and

$$Q_v = \begin{bmatrix} \mathbf{k}_v(1)(G_1) \\ \vdots \\ \mathbf{k}_v(n)(G_n) \end{bmatrix}$$

This determinant is identically zero for all K if and only if the rank of Q_p is less than n for all K, so the stabilizability of our system can be determined by evaluating the rank of Q_p. Now let us prove the necessity and sufficiency of our condition.

Necessity Assume that there is not any set of vectors $\mathbf{b}_1 \in \mathrm{Ra}(G_1^T), \ldots, \mathbf{b}_n \in \mathrm{Ra}(G_n^T)$ such that $\mathbf{b}_1, \ldots, \mathbf{b}_n$ are linearly independent. Then there is row vector $\mathbf{w}$ such that $\sum_{i=1}^n \mathbf{k}_i G_i \neq \mathbf{w}$ for any $\mathbf{k}_1, \ldots, \mathbf{k}_n$. Now consider linear combinations of the rows of Q_p. Such linear combinations can always be written in the form

$$\sum_{i=1}^n \alpha_i \tilde{\mathbf{k}}_p(i) G_i \tag{3.28}$$

Thus, from the assumption, we cannot find a linear combination of the rows that equals the vector $\mathbf{w}$. Hence, the n rows of Q_p do not span the space $\mathbf{R}^n$ and so are not all linearly independent. The matrix Q_p therefore does not have rank n, so 0 is a fixed mode of the system, and the system is not formation stabilizable.

Sufficiency Assume that there is a set of vectors $\mathbf{b}_1 \in \mathrm{Ra}(G_1^T), \ldots, \mathbf{b}_n \in \mathrm{Ra}(G_n^T)$ such that $\mathbf{b}_1, \ldots, \mathbf{b}_n$ are linearly independent. Let $\mathbf{k}_1, \ldots, \mathbf{k}_n$ be the row vectors such that $\mathbf{k}_i G_i = \mathbf{b}_i^T$. Now let us choose the control matrix K in our system as follows: $\tilde{\mathbf{k}}_p(i) = \mathbf{k}_i$. In this case, the matrix Q_p can be written as follows:

$$Q_p = \begin{bmatrix} k_1 G_1 \\ \vdots \\ k_n G_n \end{bmatrix} = \begin{bmatrix} b_1^T \\ \vdots \\ b_n^T \end{bmatrix} \tag{3.29}$$

Hence, the rank of Q_p is n, 0 is not a fixed mode of the system, and the system is formation stabilizable. ∎

By applying the results of [67], we can generalize the above condition for stabilization of linear double-integrator networks to prove semiglobal stabilization of double-integrator networks with input saturation.

Theorem 3.4.3 A double-integrator network with actuator saturation is semiglobally formation stabilizable to any formation using a dynamic LTI controller if and only if there exist vectors $\mathbf{b}_1 \in \mathrm{Ra}(G_1^T), \ldots, \mathbf{b}_n \in \mathrm{Ra}(G_n^T)$ such that $\mathbf{b}_1, \ldots, \mathbf{b}_n$ are linearly independent.

PROOF Again, formation stabilization to any $(\mathbf{r}_o, \mathbf{v}_o)$ is equivalent to formation stabilization to the origin. We now apply the theorem of [67], which states that semiglobal stabilization of a decentralized control system with input saturation can be achieved if and only if

- The eigenvalues of the open-loop system lie in the closed left-half plane.
- All fixed modes of the system when the saturation is disregarded lie in the open left-half plane.

We recognize that the open-loop eigenvalues of the double-integrator network are all zero, and so lie in the closed left-half plane. Hence, the condition of [67] reduces to a check for the presence or absence of fixed modes in the linear closed-loop system. We have already shown that all fixed modes lie in the open left-half plane (OLHP) if and only if there exist vectors

$\mathbf{b}_1 \in \mathrm{Ra}(G_1^T), \ldots, \mathbf{b}_n \in \mathrm{Ra}(G_n^T)$ such that $\mathbf{b}_1, \ldots, \mathbf{b}_n$ are linearly independent. Hence, the theorem is proved. ∎

We mentioned earlier that our formation–stabilization results hold whenever the position observations have the appropriate graph structure, regardless of the velocity measurement topology. Let us formalize this result:

Theorem 3.4.4   A position measurement double-integrator network (with actuator saturation) is (semiglobally) formation stabilizable to any formation using a dynamic LTI controller if and only if there exist vectors $\mathbf{z}_1 \in \mathrm{Ra}(G_1^T), \ldots, \mathbf{z}_n \in \mathrm{Ra}(G_n^T)$ such that $\mathbf{z}_1, \ldots, \mathbf{z}_n$ are linearly independent.

PROOF   The proof of Theorem 3.4.2 makes clear that only the topology of the position measurements play a role in deciding the stabilizability of a double-integrator network. Thus, we can achieve stabilization for a position measurement double-integrator network by disregarding the velocity measurements completely, and hence the same conditions for stabilizability hold. ∎

The remainder of this section is devoted to remarks, connections between our results and those in the literature, and examples.

Remark: Single Observation Case   In the special case in which each agent makes a single position observation and a single velocity observation, the condition for stabilizability is equivalent to simple observability of all closed right-half-plane poles of the open-loop system. Thus, in the single observation case, centralized linear and/or state-space form nonlinear control do not offer any advantage over our decentralized control in terms of stabilizability. This point makes clear the importance of studying the multiple-observation scenario.

Remark: Networks with Many Agents   We stress that conditions required for formation stabilization do not in any way restrict the number of agents in the network or their relative positions upon formation stabilization. That is, a network of any number of agents can be formation stabilized to an arbitrary formation, as long as the appropriate conditions on the full graph matrix G are met. The same holds for the other notions and means for stabilization that we discuss in subsequent sections; it is only for collision avoidance that the details of the desired formation become important. We note that the performance of the controlled network (e.g., the time required for convergence to the desired formation) may have some dependence on the number of agents. We plan to quantify performance in future work.

Connection to [62]   Earlier, we discussed that the Laplacian sensing architecture of [62] is a special case of our sensing architecture. Now we are ready to compare the stabilizability results of [62] with our results, within the context of double-integrator agent dynamics. Given a Laplacian sensing architecture, our condition in fact shows that the system is not stabilizable; this result is expected since the Laplacian architecture can only provide convergence in a relative frame. In the next section, we shall explicitly consider such relative stabilization. For comparison here, let us equivalently assume that relative positions/velocities are being stabilized, so that we can apply our condition to a grounded Laplacian. We can easily check that we are then always able to achieve formation stabilization. It turns out that the same result can be recovered from the simultaneous stabilization formulation of [62] (given double-integrator dynamics), and so the two analyses match. However, we note that, for more general communication topologies, our analysis can provide broader conditions for stabilization than that of [62], since we allow use of different controllers for each agent. Our approach also has the advantage of producing an easy-to-check necessary and sufficient condition for stabilization.

Examples   It is illuminating to apply our stabilizability condition to the examples introduced above.

Example: String of Vehicles   Let us choose vectors $\mathbf{b}_{1x}$, $\mathbf{b}_{2x}$, and $\mathbf{b}_{3x}$ as the first rows of G_{1x}, G_{2x}, and G_{3x}, respectively. Let us also choose $\mathbf{b}_{1y}$, $\mathbf{b}_{2y}$, and $\mathbf{b}_{3y}$ as the second rows of G_{1y}, G_{2y}, and G_{3y}, respectively. It is easy to check that $\mathbf{b}_{1x}$, $\mathbf{b}_{2x}$, $\mathbf{b}_{3x}$, $\mathbf{b}_{1y}$, $\mathbf{b}_{2y}$, and $\mathbf{b}_{3y}$ are linearly independent, and so the vehicles are stabilizable. The result is sensible since vehicle 1 can sense the target position directly, and vehicles 2 and 3 can indirectly sense the position of the target using the vehicle(s) ahead of it in the string. Our analysis of a string of vehicles is particularly interesting in that it shows we can complete task dynamics for non-leader–follower architectures, using the theory of [66]. This result complements the studies of [61] on leader–follower architectures.

Example: Coordination Using an Intermediary   We again expect the vehicle formation to be stabilizable since both the observed target coordinates can be indirectly sensed by the other vehicles, using the sensing architecture. We can verify stabilizability by choosing vectors in the range spaces of the transposed graph matrices, as follows:

$$
\begin{aligned}
\mathbf{b}_{1x}^T &= \begin{bmatrix} 1 & 0 & 0 & 0 & 0 & 0 \end{bmatrix} \\
\mathbf{b}_{1y}^T &= \begin{bmatrix} 0 & -1 & 0 & 0 & 0 & 1 \end{bmatrix} \\
\mathbf{b}_{2x}^T &= \begin{bmatrix} 0 & 0 & -1 & 0 & 1 & 0 \end{bmatrix} \\
\mathbf{b}_{2y}^T &= \begin{bmatrix} 0 & 0 & 0 & 1 & 0 & 0 \end{bmatrix} \\
\mathbf{b}_{3x}^T &= \begin{bmatrix} -1 & 0 & 0 & 0 & 1 & 0 \end{bmatrix} \\
\mathbf{b}_{3y}^T &= \begin{bmatrix} 0 & 0 & 0 & -1 & 0 & 1 \end{bmatrix}
\end{aligned}
\tag{3.30}
$$

It is easy to check that these vectors are linearly independent, and hence that the system is stabilizable.

Consideration of this system leads to an interesting insight on the function of the intermediary agent. We see that stabilization using the intermediary is only possible because this agent can make two observations and has available two actuators. If the intermediary is constrained to make only one observation or can only be actuated in one direction (e.g., if it is constrained to move only on the x axis), then stabilization is not possible.

Example: Measurement Failures   It is straightforward to check that decentralized control is not possible if aircraft 2 has access to the position and velocity of aircraft 3, but is possible if aircraft 2 has access to the position and velocity of aircraft 1. This example is interesting because it highlights the restriction placed on stabilizability by the decentralization of the control. In this example, the full graph matrix G has full rank for both observation topologies considered, and hence we can easily check if the centralized control of the aircraft is possible in either case. However, decentralized control is not possible when aircraft 2 only knows the location of aircraft 3 since there is then no way for aircraft 2 to deduce its own location.

3.4.4 Alignment Stabilization

Sometimes, a network of communicating agents may not require formation stabilization but instead only require that certain combinations of the agents' positions and velocities are convergent. For intance, flocking behaviors may involve only convergence of differences between agents' positions or velocities (e.g., [63]). Similarly, a group of agents seeking a target may only require that their center of mass is located at the target. Also, we may sometimes only be interested in stabilization of a double-integrator network from some initial conditions—in particular, initial conditions that lie in a subspace of R^n. We view both these problems as *alignment stabilization* problems because they concern partial stabilization and hence alignment rather than formation of the agents. As with formation stabilization, we can employ the fixed-mode concept of [66] to develop conditions for alignment stabilization.

We begin with a definition for alignment stabilization:

Definition 3.4.2   A double-integrator network can be aligned with respect to a $\widehat{n} \times n$ weighting matrix Y if a proper linear time-invariant (LTI) dynamic controller can be constructed for it, so that the $Y\mathbf{r}$ and $Y\dot{\mathbf{r}}$ are globally asymptotically convergent to the origin.

One note is needed: We define alignment stabilization in terms of (partial) convergence of the state to the origin. As with formation stabilization, we can study alignment to a fixed point other than the origin. We omit this generalization for the sake of clarity.

The following theorem provides a necessary and sufficient condition on the sensing architecture for alignment stabilization of a linear double-integrator network.

Theorem 3.4.5   A linear double-integrator network can be aligned with respect to Y if and only if there exist $\mathbf{b}_1 \in \mathrm{Ra}(G_1^T), \ldots, \mathbf{b}_n \in \mathrm{Ra}(G_n^T)$ such that the eigenvectors/generalized eigenvectors of $V \overset{\Delta}{=} \begin{bmatrix} \mathbf{b}_1^T \\ \ldots \\ \mathbf{b}_n^T \end{bmatrix}$ that correspond to zero eigenvalues all lie in the null space of Y.

PROOF   For clarity and simplicity, we prove the theorem in the special case that each agent has available only one observation (i.e., $G_1, \ldots, G_n$ are all row vectors). We then outline the generalization to this proof to the case of vector observations, deferring a formal proof to a future publication.

In the scalar-observation case, the condition above reduces to the following: A linear double-integrator network can be aligned with respect to Y if and only if the eigenvectors and generalized eigenvectors of G corresponding to its zero eigenvalues lie in the null space of Y. We prove this condition in several steps:

1. *We characterize the fixed modes of the linear double-integrator network (see [66] for background on fixed modes).* In particular, assume that the network has a full graph matrix

G with $\bar{n}$ zero eigenvalues. Then the network has $2\bar{n}$ fixed modes at the origin. That is, the matrix $A_c = \begin{bmatrix} 0 & I \\ K_1 G & K_2 G \end{bmatrix}$ has $2\bar{n}$ eigenvalues of zero for any diagonal K_1 and K_2. To show this, let us consider any eigenvector/generalized eigenvector $\mathbf{w}$ of G corresponding to a 0 eigenvalue. It is easy to check that the vector $\begin{bmatrix} \mathbf{w} \\ 0 \end{bmatrix}$ is an eigenvector/generalized eigenvector of A_c with eigenvalue 0, regardless of K_1 and K_2. We can also check that $\begin{bmatrix} 0 \\ \mathbf{w} \end{bmatrix}$ is a generalized eigenvector of A_c with eigenvalue zero. Hence, since G has $\bar{n}$ eigenvectors/generalized eigenvectors associated with the zero eigenvalue, the network has at least $2\bar{n}$ fixed modes. To see that the network has no more than $2\bar{n}$ fixed modes, we choose $K_1 = I_n$ and $K_2 = [0]$; then the eigenvalues of A_c are the square roots of the eigenvalues of G, and so A_c has exactly $2\bar{n}$ zero eigenvalues. Notice that we have not only found the number of fixed modes of the network but also specified the eigenvector directions associated with these modes.

2. *We characterize the eigenvalues and eigenvectors of the closed-loop system when decentralized dynamic feedback is used to control the linear double-integrator network.* For our model, the closed-loop system when dynamic feedback is used is given by

$$A_{\mathrm{dc}} = \begin{bmatrix} 0 & I & 0 \\ K_1 G & K_2 G & Q \\ R_1 G & R_2 G & S \end{bmatrix}$$

where R_1, R_2, and S are appropriately dimensioned block diagonal matrices (see [66] for details). It has been shown in [66] that the eigenvalues of A_{dc} that remain fixed regardless of the controller used are identical to the number of fixed modes of the system. Further, it has been shown that the remaining eigenvalues of A_{dc} can be moved to the OLHP through sufficient choice of the control gains. Hence, for the double-integrator network, we can design a controller such that all but $2\bar{n}$ eigenvalues of A_{dc} lie in the OLHP and the remaining $2\bar{n}$ eigenvalues are zero. In fact, we can determine the eigenvectors of A_{dc} associated with these zero eigenvalues. For each eigenvector/generalized eigenvector $\mathbf{w}$ of G, the vectors $\begin{bmatrix} \mathbf{w} \\ 0 \\ 0 \end{bmatrix}$ and $\begin{bmatrix} 0 \\ \mathbf{w} \\ 0 \end{bmatrix}$ are eigenvectors/generalized eigenvectors of A_{dc} corresponding to eigenvalue 0. Hence, we have specified the number of zero modes of A_{dc} and have found that the eigenvector directions associated with these modes remain fixed (as given above) regardless of the controller used.

3. Finally, we can prove the theorem. Assume that we choose a control law such that all the eigenvalues of A_{dc} except the fixed modes at the origin are in the OLHP—we can always do this. Then it is obvious that $Y\mathbf{r}$ is globally asymptotically convergent to the origin if and only if $\begin{bmatrix} Y & 0 & 0 \end{bmatrix}$ is orthogonal to all eigenvectors associated with eigenvalues of A_{dc} at the origin. Considering these eigenvectors, we see that global asymptotic stabilization to the origin is possible if and only if all the eigenvectors of G associated with zero eigenvalues lie in the nullace of Y.

In the vector observation case, proof of the condition's sufficiency is straightforward: We can design a controller that combines the vector observations to generate scalar observations for each agent, and then uses these scalar observations to control the system. The proof of necessity in the vector case is somewhat more complicated because the eigenvectors of A_c and A_{dc} corresponding to zero eigenvalues change direction depending on the controller used. Essentially, necessity is proven by showing that using a dynamic control does not change the class of fixed-mode eigenvectors, and then showing that the possible eigenvector directions guarantee that stabilization is impossible when the condition is not met. We leave the details of this analysis to a future work; we believe strongly that our approach for characterizing alignment (partial stabilization) can be generalized to the class of decentralized control systems discussed in [66] and hope to approach alignment from this perspective in future work.   ∎

Examples of Alignment Stabilization

Laplacian Sensing Topologies   As discussed previously, formation stabilization is not achieved when the graph topology is Laplacian. However, by applying the alignment stabilization condition, we can verify that differences between agent positions/velocities in each connected graph component are indeed stabilizable. This formulation recovers the results of [62] (within the context of double-integrator networks) and brings our work in alignment (no pun intended) with the studies of [63]. It more generally highlights an alternate viewpoint on formation stabilization analysis given a grounded Laplacian topology.

We consider a specific example of alignment stabilization given a Laplacian sensing topology here that is similar to the flocking example of [63]. The graph matrix[3] in our example is

$$
G = \begin{bmatrix}
1 & -\frac{1}{2} & 0 & 0 & -\frac{1}{2} \\
-\frac{1}{2} & 1 & -\frac{1}{2} & 0 & 0 \\
0 & -\frac{1}{2} & 1 & -\frac{1}{2} & 0 \\
0 & 0 & -\frac{1}{2} & 1 & -\frac{1}{2} \\
-\frac{1}{2} & 0 & 0 & -\frac{1}{2} & 1
\end{bmatrix}
\tag{3.31}
$$

The condition above can be used to show alignment stabilization of differences between agents' positions or velocities, for example, with respect to $Y = \begin{bmatrix} 1 & 0 & 0 & -1 & 0 \end{bmatrix}$.

Matching Customer Demand   Say that three automobile manufacturers are producing sedans to meet fixed, but unknown, customer demand for this product. An interesting analysis is to determine whether or not these manufacturers can together produce enough sedans to exactly match the consumer demand. We can view this problem as an alignment stabilization one, as follows. We model the three manufacturers as decentralized agents, whose production outputs r_1, r_2, and r_3, are modeled as double integrators. Each agent clearly has available its own production output as an observation. For convenience, we define a fourth agent that represents the fixed (but unknown) customer demand; this agent necessarily has no observations available and so cannot be actuated. We define r_d to represent this demand. We are concerned with whether $r_1 + r_2 + r_3 - r_d$ can be stabilized. Hence, we are studying an alignment problem.

---

[3] To be precise, in our simulations we assume that agents move in the plane. The graph matrix shown here specifies the sensing architecture in each vector direction.

Using the condition above, we can trivially see that alignment is not possible if the agents only measure their own position. If at least one of the agents measures $r_1 + r_2 + r_3 - r_d$ (e.g., through surveys), then we can check that stabilization is possible. When other observations that use r_d are made, then the alignment may or may not occur; we do not pursue these further here. It would be interesting to study whether or not competing manufacturers such as these in fact use controllers that achieve alignment stabilization.

3.4.5 Existence and Design of Static Stabilizers

Above, we developed necessary and sufficient conditions for the stabilizability of a group of communicating agents with double-integrator dynamics. Next, we give sufficient conditions for the existence of a *static* stabilizing controller[4]. We further discuss several approaches for designing good static stabilizers that take advantage of some special structures in the sensing architecture.

Sufficient Condition for Static Stabilization   The following theorem describes a sufficient condition on the sensing architecture for the existence of a static stabilizing controller.

Theorem 3.4.6   Consider a linear double-integrator network with graph matrix G. Let $\mathcal{K}$ be the class of all block diagonal matrices of the form $\begin{bmatrix} \mathbf{k}_1 & & \\ & \ddots & \\ & & \mathbf{k}_n \end{bmatrix}$, where $\mathbf{k}_i$ is a row vector with m_i entries (recall that m_i is the number of observations available to agent i). Then the double-integrator system has a static stabilizing controller (i.e., a static controller that achieves formation stabilization) if there exists a matrix $K \in \mathcal{K}$ such that the eigenvalues of KG are in the open left-half plane (OLHP).

PROOF   We prove this theorem by constructing a static controller for which the overall system's closed-loop eigenvalues are in the OLHP whenever the eigenvalues of KG are in the OLHP. Based on our earlier development, it is clear that the closed-loop system matrix takes the form

$$A_c = \begin{bmatrix} 0 & I \\ K_1 G & K_2 G \end{bmatrix} \tag{3.32}$$

where K_1 and K_2 are control matrices that are constrained to be in $\mathcal{K}$ but are otherwise arbitrary.

Given the theorem's assumption, it turns out we can guarantee that the eigenvalues of A_c are in the OLHP by choosing the control matrices as $K_1 = K$ and $K_2 = aK$, where a is a sufficiently large positive number. With these control matrices, the closed-loop system matrix becomes

$$A_c = \begin{bmatrix} 0 & I \\ KG & aKG \end{bmatrix} \tag{3.33}$$

---

[4] We consider a controller to be static if the control inputs at each time are linear functions of the concurrent observations (in our case the position and velocity observations).

To show that the *2n* eigenvalues A_c are in the OLHP, we relate these eigenvalues to the *n* eigenvalues of *KG*.

To relate the eigenvalues, we construct the left eigenvectors of A_c. In particular, we consider vectors of the form $\mathbf{w}' = \begin{bmatrix} \mathbf{v}' & \alpha\mathbf{v}' \end{bmatrix}$, where $\mathbf{v}'$ is a left eigenvector of *KG*. Then

$$\mathbf{w}'A_c = \begin{bmatrix} \alpha\rho\mathbf{v}' & \mathbf{v}'(1 + \alpha a\rho) \end{bmatrix} \tag{3.34}$$

where ρ—the eigenvalue of *KG* corresponding to the eigenvector $\mathbf{v}'$—lies in the OLHP. This vector is a left eigenvector of A_c with eigenvalue λ if and only if $\alpha\rho = \lambda$ and $1 + \alpha a\rho = \alpha\lambda$. Substituting $\alpha = \lambda/\rho$ into the second equation and rearranging, we find that $\lambda^2 - a\rho\lambda - \rho = 0$. This quadratic equation yields two solutions, and hence we find that two eigenvalues and eigenvectors of A_c can be specified using each left eigenvector of *KG*. In this way, all *2n* eigenvalues of A_c can be specified in terms of the *n* eigenvalues of *KG*.

Finally, it remains to be shown that the eigenvalues of A_c are negative. Applying the quadratic formula, we find that each eigenvalue λ of A_c can be found from an eigenvalue ρ of *KG*, as

$$\lambda = \frac{a\rho \pm \sqrt{(a\rho)^2 + 4\rho}}{2}.$$

Without loss of generality, let us assume that ρ lies in the second quadrant of the complex plane. (It is easy to check that eigenvalues of A_c corresponding to ρ in the third quadrant are complex conjugates of eigenvalues corresponding to ρ in the second quadrant). We can show that the eigenvalues λ will have negative real parts, using a geometric argument. The notation that we use in this geometric argument is shown in Figure 3.13. In this notation, $\lambda = r \pm q/2$. Since *r* has a negative real part, we can prove that λ has a negative real part by showing that the magnitude of the real part of *q* is smaller than the magnitude of the real part of *r*. To show this, first consider the complex variables *s* and *t*. Using the law of cosines, we can show that the length (in the complex plane) of *s* is less than the length of *t*

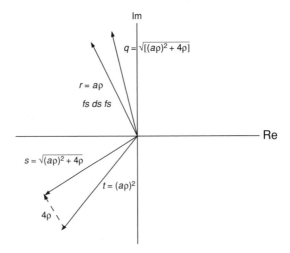

Figure 3.13   We introduce notation used for geometric proof that eigenvalues of A_c are in OLHP.

whenever

$$a > \sqrt{\frac{2}{|\rho| \cos\left(90 - \tan^{-1} \frac{-\text{Re}(\rho)}{\text{Im}(\rho)}\right)}} \qquad (3.35)$$

In this case, the length (magnitude) of q is also less than the magnitude of r. Hence, the magnitude of the real part of q is less than the magnitude of the real part of r (because the phase angle of q is smaller), and the eigenvalue λ has a negative real part. Thus, if we choose

$$a > \max_{\rho} \sqrt{\frac{2}{|\rho| \cos\left(90 - \tan^{-1} \frac{-\text{Re}(\rho)}{\text{Im}(\rho)}\right)}} \qquad (3.36)$$

then all eigenvalues of A_c are guaranteed to be negative. ∎

Our proof demonstrates how to design a stabilizing controller whenever we can find a matrix K such that the eigenvalues of KG are in the OLHP. Since this stabilizing controller uses a high gain on the velocity measurements, we henceforth call it a *high-velocity gain* (HVG) controller.

Using a simple scaling argument, we can show that static stabilization is possible in a semiglobal sense whenever we can find a matrix K such that the eigenvalues of KG are in the OLHP, even if the actuators are subject to saturation. We present this result in the following theorem.

Theorem 3.4.7 Consider a double-integrator network with actuator saturation that has graph matrix G. Let $\mathcal{K}$ be the class of all block diagonal matrices of the form $\begin{bmatrix} \mathbf{k}_1 & & \\ & \ddots & \\ & & \mathbf{k}_n \end{bmatrix}$, where $\mathbf{k}_i$ is a row vector with m_i entries. Then the double-integrator network with actuator saturation has a semiglobal static stabilizing controller (i.e., a static controller that achieves formation stabilization in a semiglobal sense) if there exists a matrix $K \in \mathcal{K}$ such that the eigenvalues of KG are in the open left-half plane (OLHP).

PROOF   In the interest of space, we present an outline of the proof here. Since we are proving semiglobal stabilization, we can assume that the initial system state lies within some finite-sized ball. Notice that, if the double-integrator network were not subject to input saturation, we could find a static stabilizing controller. Further note that, if the static stabilizing controller were applied, the trajectory of the closed-loop system would remain within a larger ball. Say that the closed-loop system matrix for the linear network's stabilizing controller is

$$A_c = \begin{bmatrix} 0 & I \\ KG & aKG \end{bmatrix} \qquad (3.37)$$

Then it is easy to check that the closed-loop system matrix

$$\widehat{A}_c = \begin{bmatrix} 0 & I \\ \frac{KG}{\zeta^2} & \frac{aKG}{\zeta} \end{bmatrix} \qquad (3.38)$$

also represents a static stabilizer for the linear network. Further, the trajectory followed by the state when this new controller is used is identical to the one using the original controller, except that the time axis for the trajectory is scaled by ζ. Hence, we know that the trajectory is bounded within a ball. Thus, if we choose large enough ζ, we can guarantee that the input magnitude is always strictly less than 1 (i.e., that the actuators never saturate), while also guaranteeing stabilization. Such a choice of controller ensures stabilization even when the actuators are subject to saturation, and hence the theorem has been proved. ∎

Design of Static Stabilizers Above, we showed that a static stabilizing controller for our decentralized system can be found, if there exists a control matrix K such that the eigenvalues of KG are in the OLHP. Unfortunately, this condition does not immediately allow us to design the control matrix K or even to identify graph matrices G for which an HVG controller can be constructed since we do not know how to choose a matrix K such that the eigenvalues of KG are in the OLHP.

We discuss approaches for identifying from the graph matrix whether an HVG controller can be constructed and for designing the control matrix K. First, we show (trivially) that HVG controllers can be developed when the graph matrix has positive eigenvalues and give many examples of sensing architectures for which the graph matrix eigenvalues are positive. Second, we discuss some simple variants on the class of graph matrices with positive eigenvalues, for which HVG controllers can also be developed. Third, we use an eigenvalue sensitivity-based argument to show that HVG controllers can be constructed for a very broad class of graph matrices. (For convenience and clarity of presentation, we restrict the sensitivity-based argument to the case where each agent has available only one observation, but note that the generalization to the multiple-observation case is straightforward.) Although this eigenvalue sensitivity-based argument sometimes does not provide good designs (because eigenvalues are guaranteed to be in the OLHP only in a limiting sense), the argument is important because it highlights the broad applicability of static controllers and specifies a systematic method for their design.

Graph Matrices with Positive Eigenvalues If each agent has available one observation (so that the graph matrix G is square) and the eigenvalues of G are strictly positive, then a stabilizing HVG controller can be designed by choosing $K = -I_n$.

The proof is immediate: The eigenvalues of $KG = -G$ are strictly negative, so the condition for the existence of a stabilizing HVG controller is satisfied.

Here are some examples of sensing architectures for which the graph matrix has strictly positive eigenvalues:

- A grounded Laplacian matrix is known to have strictly positive eigenvalues (see, e.g., [68]). Hence, if the sensing architecture can be represented using a grounded Laplacian graph matrix, then a static control matrix $K = -I$ can be used to stabilize the system.
- A wide range of matrices besides Laplacians are also known to have positive eigenvalues. For instance, any strictly *diagonally dominant matrix*—one in which the diagonal entry on each row is larger than the sum of the absolute values of all off-diagonal entries—has positive eigenvalues. Diagonally dominant graph matrices are likely to be observed in systems in which each agent has considerable ability to accurately sense its own position.
- If there is a positive diagonal matrix L such that GL is diagonally dominant, then the eigenvalues of G are known to be positive. In some examples, it may be easy to observe that a scaling of this sort produces a diagonally dominant matrix.

Eigenvalue Sensitivity-Based Controller Design, Scalar Observations Using an eigenvalue sensitivity (perturbation) argument, we show that HVG controllers can be explicitly designed for a very wide class of sensing architectures. In fact, we find that only a certain sequential full-rank condition is required to guarantee the existence of a static stabilizer. While our condition is not a necessary and sufficient one, we believe that it captures most sensing topologies of interest.

The statement of our theorem requires some further notation:

- Recall that we label agents using the integers $1, \ldots, n$. We define an *agent list* $\mathbf{i} = \{i_1, \ldots, i_n\}$ to be an ordered vector of the n agent labels. (For instance, if there are 3 agents, $\mathbf{i} = \{3, 1, 2\}$ is an agent list).
- We define the kth *agent sublist* of the agent list $\mathbf{i}$ to be a vector of the first k agent labels in $\mathbf{i}$, or $\{i_1, \ldots, i_k\}$. We use the notation $\mathbf{i}_{1:k}$ for the kth agent sublist of $\mathbf{i}$.
- We define the kth *subgraph matrix* associated with the agent list to be the $k \times k$ submatrix of the graph matrix corresponding to the agents in the kth agent sublist. More precisely, we define the matrix $D(\mathbf{i}_{1:k})$ to have k rows and n columns. Entry i_w of each row w is assumed to be unity, and all other entries are assumed to be 0. The kth subgraph matrix is given by $D(\mathbf{i}_{1:k})GD(\mathbf{i}_{1:k})'$.

The condition on the graph matrix required for design of an HVG controller using eigenvalue sensitivity arguments is given in the following theorem:

Theorem 3.4.8 If there exists an agent list $\mathbf{i}$ such that the kth subgraph matrix associated with this agent list has full rank for all k, then we can construct a stabilizing HVG controller for the decentralized control system.

PROOF We prove the theorem above by constructing a control matrix K such that the eigenvalues of KG are in the OLHP. More specifically, we construct a sequence of control matrices for which more and more of the eigenvalues of KG are located in the OLHP and hence prove that there exists a control matrix such that all eigenvalues of KG are in the OLHP.

Precisely, we show how to iteratively construct a sequence of control matrices $K(\mathbf{i}_{1:1}), \ldots, K(\mathbf{i}_{1:n})$, such that $K(\mathbf{i}_{1:k})G$ has k eigenvalues in the OLHP. In constructing the control matrices we use the agent list $\mathbf{i}$ for which the assumption in the theorem is satisfied. First, let us define the matrix $K(\mathbf{i}_{1:1})$ to have a single nonzero entry: the diagonal entry corresponding to i_1, or K_{i_1}. Let us choose K_{i_1} to equal $-\text{sgn}(G_{i_1,i_1})$—that is, the negative of the sign of the (nonzero) diagonal entry of G corresponding to agent i_1. Then $K(\mathbf{i}_{1:1})G$ has a single nonzero row, with diagonal entry $-G_{i_1,i_1}\text{sgn}(G_{i_1,i_1})$. Hence, $K(\mathbf{i}_{1:1})G$ has one negative eigenvalue, as well as $n-1$ zero eigenvalues. Note that the one nonzero eigenvalue is simple (nonrepeated).

Next, let us assume that there exists a matrix $K(\mathbf{i}_{1:k})$ with nonzero entries $K_{i_1}, \ldots, K_{i_k}$, such that $K(\mathbf{i}_{1:k})G$ has k simple negative eigenvalues, and $n-k$ zero eigenvalues. Now let us consider a control matrix $K(\mathbf{i}_{1:k+1})$ that is formed by adding a nonzero entry $K_{i_{k+1}}$ (i.e., a nonzero entry corresponding to agent i_{k+1}) to $K(\mathbf{i}_{1:k})$, and think about the eigenvalues of $K(\mathbf{i}_{1:k+1})G$. The matrix $K(\mathbf{i}_{1:k+1})G$ has $k+1$ nonzero rows, and so has at most $k+1$ nonzero eigenvalues. The eigenvalues of $K(\mathbf{i}_{1:k+1})G$ are the eigenvalues of its submatrix corresponding to the kth agent sublist, or $D(\mathbf{i}_{1:k+1})K(\mathbf{i}_{1:k+1})GD(\mathbf{i}_{1:k+1})'$. Notice that this matrix can be constructed by scaling the rows of the $(k+1)$th agent sublist and hence has full rank. Next, note that we can rewrite $D(\mathbf{i}_{1:k+1})K(\mathbf{i}_{1:k+1})GD(\mathbf{i}_{1:k+1})'$ as $D(\mathbf{i}_{1:k+1})K(\mathbf{i}_{1:k})GD(\mathbf{i}_{1:k+1})' + D(\mathbf{i}_{1:k+1})K(\mathbf{i}_{k+1})GD(\mathbf{i}_{1:k+1})'$, where $K(\mathbf{i}_{k+1})$ only has diagonal entry $K_{i_{k+1}}$ nonzero. Thus, we can view $D(\mathbf{i}_{1:k+1})K(\mathbf{i}_{1:k+1})GD(\mathbf{i}_{1:k+1})'$ as a perturbation of $D(\mathbf{i}_{1:k+1})K(\mathbf{i}_{1:k})GD(\mathbf{i}_{1:k+1})'$—which has the same eigenvalues as $K(\mathbf{i}_{1:k})G$—by the row vector $D(\mathbf{i}_{1:k+1})K(\mathbf{i}_{k+1})GD(\mathbf{i}_{1:k+1})'$. Because $D(\mathbf{i}_{1:k+1})K(\mathbf{i}_{1:k+1})GD(\mathbf{i}_{1:k+1})'$ has full rank, we can straightforwardly invoke a standard eigenvalue-sensitivity result to show that $K_{i_{k+1}}$ can be chosen so that all the eigenvalues of $D(\mathbf{i}_{1:k+1})K(\mathbf{i}_{1:k+1})GD(\mathbf{i}_{1:k+1})'$ are simple and negative (please see appendix for the technical details). Essentially, we can choose the sign of the smallest eigenvalue—which is a perturbation of the zero eigenvalue of $D(\mathbf{i}_{1:k+1})K(\mathbf{i}_{1:k})GD(\mathbf{i}_{1:k+1})'$—by choosing the sign of $K_{i_{k+1}}$ properly, and we can ensure that all eigenvalues are positive and simple by choosing small enough $K_{i_{k+1}}$. Thus, we have constructed $K(\mathbf{i}_{1:k+1})$ such that $K(\mathbf{i}_{1:k+1})G$ has $k+1$ simple nonzero eigenvalues. Hence, we have proven the theorem by recursion and have specified broad conditions for the existence and limiting design of static (HVG) controllers. ∎

Examples of Static Control   We develop static controllers for the four examples we introduced earlier. Through simulations, we explore the effect of the static control gains on performance of the closed-loop system.

Example: Coordination Using an Intermediary   The following choice for the static gain matrix K places all eigenvalues of KG in the OLHP and hence permits stabilization with an HVG controller:

$$K = \begin{bmatrix} -1 & 0 & 0 & 0 & 0 & 0 & 0 & 0 & 0 & 0 \\ 0 & 0 & 1 & 0 & 0 & 0 & 0 & 0 & 0 & 0 \\ 0 & 0 & 0 & 0 & 1 & 0 & 0 & 0 & 0 & 0 \\ 0 & 0 & 0 & -1 & 0 & 0 & 0 & 0 & 0 & 0 \\ 0 & 0 & 0 & 0 & 0 & 0 & -1 & 0 & 0 & 0 \\ 0 & 0 & 0 & 0 & 0 & 0 & 0 & 0 & 0 & -1 \end{bmatrix} \tag{3.39}$$

As K is scaled up, we find that the vehicles converge to the target faster and with less overshoot. When the HVG controller parameter a is increased, the agents converge much more slowly but also gain a reduction in overshoot. Figures 3.14, 3.15, and 3.16 show the trajectories of the vehicles and demonstrate the effect of the static control gains on the

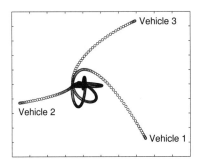

Figure 3.14   Vehicles converging to target: coordination with intermediary.

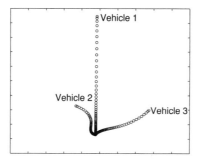

Figure 3.15 Vehicles converging to target. Gain matrix is scaled up by a factor of 5 as compared to Figure 3.14.

performance of the closed-loop system. Also, if a is made sufficiently large, the eigenvalues of A_c move close to the imaginary axis. This is consistent with the vehicles' slow convergence rate for large a.

Example: Vehicle Velocity Alignment, Flocking Finally, we simulate an example, with some similarity to the examples of [63], that demonstrates alignment stabilization. In this example, a controller is designed so that five vehicles with relative position and velocity measurements converge to the same (but initial-condition-dependent) velocity. Figures 3.17 demonstrates the convergence of the velocities in the x direction for two sets of initial velocities. We note the different final velocities achieved by the two simulations.

3.4.6 Collision Avoidance in the Plane

Imagine a pedestrian walking along a crowded thoroughfare. Over the long term, the pedestrian completes a task—that is, she/he moves toward a target or along a trajectory. As the pedestrian engages in this task, however, she must constantly take evasive action to avoid other pedestrians, who are also seeking to complete tasks. In this context, and in the context of many other distributed task dynamics that occur in physical spaces, *collision avoidance*—that is, prevention of collisions between agents through evasive action—is a required component of the task. In this section, we explore collision avoidance among a network of distributed agents with double-integrator dynamics that are seeking to complete a formation task. Like pedestrians walking on a street, agents in our double-integrator

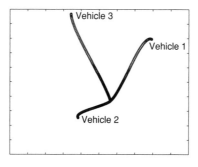

Figure 3.16 Vehicles converging to target when HVG controller parameter a is increased from 1 to 3.

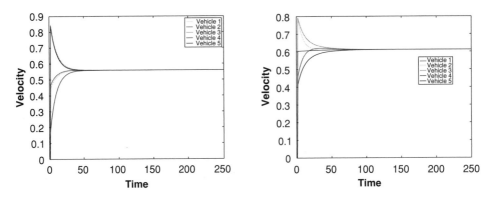

Figure 3.17 We show alignment of *x*-direction velocities in two simulations. Notice that final velocity is different for two simulations.

network achieve collision avoidance by taking advantage of the multiple directions of motion available to them.

We now specifically consider networks of agents that move in the plane and seek to converge to a fixed formation. We develop a control strategy that uses localized sensing information to achieve collision avoidance, in addition to using remote sensing to achieve the formation task. We show that static and dynamic controllers that achieve both formation stabilization and collision avoidance can be designed, under rather general conditions on the desired formation, the size of the region around each agent that must be avoided by the other agents, and the remote sensing topology.

We strongly believe that our study represents a novel viewpoint on collision avoidance in that avoidance is viewed as a localized subtask within the broader formation stabilization task. Our approach is in sharp contrast to the potential function-based approaches of, for example, [63], in which the mechanism for collision avoidance also simultaneously specifies or constrains the final formation.

We recently became aware of the study of [69], which—like our work—seeks to achieve collision avoidance using only local measurements. We believe that our study builds on that of [69] in the following respect: We allow for true formation (not only swarming) tasks, which are achieved through decentralized sensing and control rather than through the use of a set of potential functions. To this end, our collision avoidance mechanism uses a combination of both gyroscopic and repulsive forces, in a manner that leaves the formation dynamics completely unaffected in one direction of motion. This approach has the advantage that the collision avoidance and formation tasks can be decoupled, so that rigorous analysis is possible, even when the task dynamics are not specified by potentials (as in our case).

We believe that our study of collision avoidance is especially pertinent for the engineering of multiagent systems (e.g., coordination of autonomous underwater vehicles), since the task dynamics required for the design are typically specified globally and independently of the collision avoidance mechanism. Our study here shows how such a system can be stabilized to a preset desired formation, while using local sensing information to avoid collisions.

Our Model for Collision Avoidance   In our discussion of collision avoidance, we specialize the double-integrator network model to represent the dynamics of agents moving in the plane, whose accelerations are the control inputs. We also augment the model by specifying collision avoidance requirements and allowing localized sensing measurements that permit collision avoidance. We find it most convenient to develop the augmented model from scratch, and then to relate this model to the double-integrator network in a manner that facilitates analysis of formation stabilization with collision avoidance. We call the new model the *plane double-integrator network* (PDIN).

We consider a network of n agents with double-integrator dynamics, each moving in the Euclidean plane. We denote the x- and y-positions of agent i by r_{ix} and r_{iy}, respectively, and use

$$\mathbf{r}_i \triangleq \begin{bmatrix} r_{ix} \\ r_{iy} \end{bmatrix}^5$$

These n agents aim to complete a formation stabilization task; in particular, they seek to converge to the coordinates $(\bar{r}_{1x}, \bar{r}_{1y}), \ldots, (\bar{r}_{nx}, \bar{r}_{ny})$, respectively. The agents use measurements of each others' positions and velocities to achieve the formation task. We assume that each agent has available two position and two velocity observations, respectively. In particular, each agent i is assumed to have available a vector of position observations of the form $\mathbf{y}_{pi} = C_i \sum_{j=1}^{n} g_{ij} \mathbf{r}_j$, and a vector of velocity observation $\mathbf{y}_{vi} = C_i \sum_{j=1}^{n} g_{ij} \dot{\mathbf{r}}_j$. That is, we assume that each agent has available two position measurements that are averages of its neighbors' positions and two velocity measurements that are averages of its neighbors' velocities. Each of these measurements may be weighted by a *direction-changing matrix* C_i, which models that each agent's observations may be made in a rotated frame of reference. We view the matrix $G = [g_{ij}]$ as specifying a *remote sensing architecture* for the plane double-integrator network since it specifies how each agent's observations depend on the other agents' states.[6] Not surprisingly, we will find that success of the formation task depends on the structure G, as well as the structure of

$$C \triangleq \begin{bmatrix} C_1 & 0 & \cdots \\ & \ddots & 0 \\ \cdots & 0 & C_n \end{bmatrix}$$

Our aim is to design a decentralized controller that places the agents in the desired formation, while providing collision avoidance capability. Collision avoidance is achieved using local measurements that warn each agent of other nearby agents. In particular, we assume that each agent has a circle of radius q about it, called the *repulsion ball* (see Fig. 3.18). *Collision avoidance* is said to be achieved if the agents' repulsion balls never intersect. (We can alternately define collision avoidance to mean that no agent ever enters another agent's repulsion ball, as has been done in [63]. The analysis is only trivially

---

[5] Notice that agents in our new model can have vector state; because we need to consider explicitly, the dynamics of the agents in each coordinate direction, it is convenient for us to maintain vector positions for the agents rather than reformulating each agent as two separate agents.

[6] Two remarks should be made here. First, we note that our notation for the remote sensing architecture differs from the notation for the sensing architecture presented before because agents are considered to have vector states rather than being reformulated as multiple agents with scalar statuses. Second, we note that the analyses described here can be adopted to some more general sensing observations (e.g., with different numbers of measurements for each agent), but we consider this structured formulation for clarity.

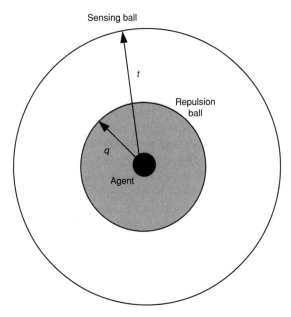

Figure 3.18   Repulsion ball and local sensing ball are illustrated.

different in this case; we use the formulation above because it is a little easier to illustrate our controller graphically.) We also assume that each agent has a circle of radius $t > 2q$ about it, such that the agent can sense its relative position to any other agent in the circle. We call these circles the *local sensing balls* for the agents (see Fig. 3.18). That is, agent i has available the observation $r_i - r_j$, whenever $||r_i - r_j||_2 < t$. The agents can use these *local warning measurements*, as well as the position and velocity measurements, to converge to the desired formation while avoiding collisions.

We view a plane double-integrator network as being specified by the four parameters (G, C, q, t), since these together specify the remote sensing and collision avoidance requirements/capabilities of the network. We succinctly refer to a particular network as PDIN(G, C, q, t).

Formation Stabilization with Collision Avoidance: Static and Dynamic Controllers   We formulate controllers for PDINs that achieve both formation stabilization (i.e., convergence of agents to their desired formation positions) and collision avoidance. In particular, given certain conditions on the remote sensing architecture, the size of the repulsion ball, and the desired formation, we are able to prove the existence of dynamic controllers that achieve both formation stabilization and collision avoidance. Below, we formally state and prove theorems giving sufficient conditions for formation stabilization with collision avoidance. First, however, we describe in words the conditions that allow formation stabilization with guaranteed collision avoidance.

To achieve formation stabilization with collision avoidance for a PDIN, it is requisite that the remote sensing topology permit formation stabilization. Using Theorem 3.4.2, we can easily show that a sufficient (and in fact necessary) condition for formation stabilization using the remote sensing architecture is that G and C have full rank. Our philosophy for concurrently assuring that collisions do not occur is as follows (Fig. 3.19). We find a vector direction along which each agent can move to its formation position without danger of collision. As agents move toward their formation positions using a control based on the remote sensing measurements, we apply a second control that serves to prevent collisions; the novelty in our approach is that this second control is chosen to only change the trajectories of the agents in the vector direction described above. Hence, each agent's motion in the orthogonal direction is not affected by the collision avoidance control, and the component of their positions in this orthogonal direction can be guaranteed to converge to its final value. After some time, the agents are no longer in danger of collision and so can be shown to converge to their formation positions. What is required for this approach to collision avoidance is that a vector direction exists in which the agents can move to their formation positions without danger of collision. This requirement is codified in the definitions below.

Definition 3.4.3 Consider a PDIN(G, C, q, t) whose agents seek to converge to the formation $\bar{\mathbf{r}}$, and consider a vector direction specified by the unit vector (a, b). We shall call this direction *valid* if, when each agent is swept along this direction from its formation position, the agents' repulsion balls do not ever intersect (in a strict sense), as shown in Figure 3.19. From simple geometric arguments, we find that the direction (a, b) is valid

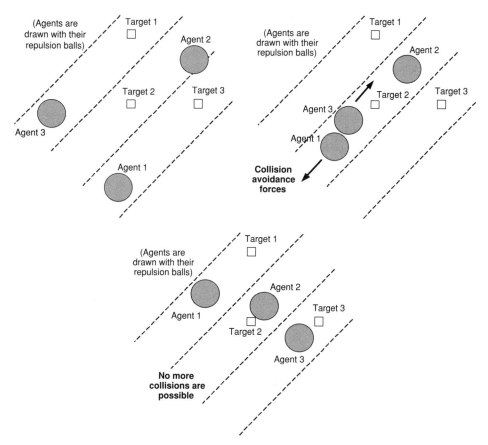

Figure 3.19 Our approach for formation stabilization with collision avoidance is illustrated, using snapshots at three time points.

if and only if $\min_{i,j} | - b(r_{ix} - r_{jx}) + a(r_{iy} - r_{jy})| > 2q$. If $\text{PDIN}(G, C, q, t)$ has at least one valid direction for a formation $\bar{\mathbf{r}}$, that formation is called *valid*.

Theorem 3.4.9 $\text{PDIN}(G, C, q, t)$ has a dynamic time-invariant controller that achieves both formation stabilization to $\bar{\mathbf{r}}$ (i.e., convergence of the agents' positions to $\bar{\mathbf{r}}$) and collision avoidance if G and C have full rank, and $\bar{\mathbf{r}}$ is a valid formation.

PROOF   The strategy we use to prove the theorem is to first design a controller, in the special case that $C = I$ and the valid direction is the x direction $(1, 0)$. We then prove the theorem generally, by converting the general case to the special case proven first. This conversion entails viewing the PDIN in a rotated frame of reference.

To be more precise, we first prove the existence of a dynamic controller for the planar double-integrator network $\text{PDIN}(G, I, q, t)$, when G has full rank and the direction $(1, 0)$ is valid. For convenience, let us define the *minimum coordinate separation f* to denote the minimum distance in the y direction between two agents in the desired formation. Note that $f > 2q$.

Let us separately consider the stabilization of the agents' y positions and x positions to their formation positions. We shall design a controller for the agents' y positions that uses only the remote sensing measurements that are averages of y direction positions and velocities. Then we can view the agents' y positions as being a standard double-integrator network with full graph matrix G. As we showed in our associated article, formation stabilization of this double-integrator network is possible using a decentralized dynamic LTI controller whenever G has full rank. Let us assume that we use this stabilizing controller for the agents' y positions. Then we know for certain that the y positions of the agents will converge to their formation positions.

Next, we develop a decentralized controller that determines the x-direction control inputs from the x-direction observations and the warning measurements. We show that this decentralized controller achieves stabilization of the agents' x positions and simultaneously guarantees collision avoidance among the agents. To do so, note that we can develop a decentralized dynamic LTI controller that achieves convergence of the agents' x positions to their desired formation (in the same manner as for the y-direction positions). Say that the stabilizing decentralized controller is

$$\dot{\mathbf{z}} = A_x \mathbf{z} + B_x \mathbf{y}_x \tag{3.40}$$

$$\mathbf{u}_x = C_x \mathbf{z} + D_x \mathbf{y}_x$$

where $\mathbf{y}_x$ is the x-direction observations (i.e., measurements of x-direction positions and velocities for each agent), $\mathbf{z}$ is the state of the feedback controller, $\mathbf{u}_x$ is the x-direction inputs, and the coefficient matrices are appropriately structured to represent a decentralized controller. To achieve formation stabilization and collision avoidance, we use the following controller:

$$\dot{\mathbf{z}} = A_x \mathbf{z} + B_x \mathbf{y}_x$$

$$\mathbf{u}_x = C_x \mathbf{z} + D_x \mathbf{y}_x + \begin{bmatrix} \sum_{j=1}^n g(\mathbf{r}_1 - \mathbf{r}_j) \\ \vdots \\ \sum_{j=1}^n g(\mathbf{r}_n - \mathbf{r}_j) \end{bmatrix} \tag{3.41}$$

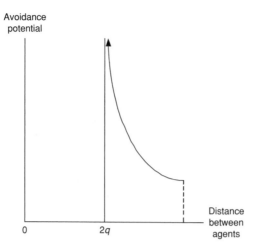

Figure 3.20 Example potential function for collision avoidance is shown.

where the nonlinear input term $g(\ )$, which uses the warning measurements, is described in detail in the next paragraph.[7]

The nonlinear term $g(\mathbf{r}_i - \mathbf{r}_j)$ acts to push away agent i from agent j whenever agent j is too close to agent i. Of course, it can only be nonzero if agent j is in the sensing ball of agent i. In particular, we define the function $g(\mathbf{r}_i - \mathbf{r}_j)$ to be nonzero for $2q \leq ||\mathbf{r}_i - \mathbf{r}_j||_2 \leq \epsilon$ and 0 otherwise, where ϵ is strictly between $2q$ and $\min(f, t)$. Furthermore, we define $g(\mathbf{r}_i - \mathbf{r}_j)$ to take the form $\operatorname{sgn}(r_{ix} - r_{iy})\widehat{g}(||\mathbf{r}_i - \mathbf{r}_j||_2)$, where $\widehat{g}(||\mathbf{r}_i - \mathbf{r}_j||_2)$ is a decreasing function of $||\mathbf{r}_i - \mathbf{r}_j||_2$ in the interval between q and ϵ, and $\int_{t=q}^{\epsilon} g(t)$ is infinite, for any ϵ. An example of a function $g(\ )$ is shown in Figure 3.20.

To prove that this controller achieves collision avoidance, let us imagine that two agents i and j are entering each others' local sensing balls. Now assume that agent i were to follow a path such that its repulsion ball intersected the repulsion ball of agent j. Without loss of generality, assume that $r_{ix} < r_{jx}$ at the time of intersection of the repulsion balls. In this case, agent i would have an infinite velocity in the negative x direction (since the integral of the input acceleration would be infinite over any such trajectory). Thus, the movement of the agent would be away from the repulsion ball, and the agent could not possibly intersect the ball. In the case where more than two agents interact at once, we can still show collision avoidance by showing that at least one agent would be moving away from the others at infinite velocity if the repulsion balls were to collide. Thus, collisions will always be avoided.

Next, let us prove that the x positions of the agents converge to their desired formation positions. To do so, we note that there is a finite time T such that $|r_{iy}(t) - r_{jy}(t)| > \epsilon$ for all $t \geq T$ and for all pairs of agents (i, j), since the y positions of the agents are stable and $(1, 0)$ is a valid direction. Thus, for $t \geq T$, the collision avoidance input is zero for all agents. Hence, the dynamics of the agents' x positions are governed by the stabilizing controller, and the agents converge to the desired formation. Hence, we have shown the existence of a controller that achieves formation stabilization and collision avoidance.

[7] Since the right-hand side of the Closed loop (CL) system is discontinuous, we henceforth assume that solutions to our system are in the sense of Filipov.

We are now ready to prove the theorem in the general case, that is, for a $\text{PDIN}(G, C, q, t)$ whose agents seek to converge to a valid formation $\bar{\mathbf{r}}$, and for which G and C have full rank. Without loss of generality, let us assume that (a, b) is a valid direction. We will develop a controller that stabilizes each agent's *rotated position*

$$\mathbf{s}_i = \begin{bmatrix} a & b \\ -b & a \end{bmatrix} \mathbf{r}_i$$

and *rotated velocity* $\dot{\mathbf{s}}_i$. Stabilization of the rotated positions and velocities imply stabilization of the original positions and velocities, since the two are related by full rank transformations. It is easy to check that the rotated positions and velocities are the positions and velocities of a different PDIN. In the rotated frame, each agent i aims to converge to the position

$$\bar{\mathbf{s}}_i = \begin{bmatrix} a & b \\ -b & a \end{bmatrix} \bar{\mathbf{r}}_i$$

so the network seeks to converge to the formation

$$\bar{\mathbf{s}} \triangleq \left(I_n \otimes \begin{bmatrix} a & b \\ -b & a \end{bmatrix} \right) \bar{\mathbf{r}}$$

Also, each agent has available the observations

$$\mathbf{y}_{pi} = C_i \sum_{j=1}^{n} g_{ij} \mathbf{s}_j \quad \text{and} \quad \mathbf{y}_{vi} = C_i \sum_{j=1}^{n} g_{ij} \dot{\mathbf{s}}_j$$

Hence, in the rotated frame, the agents' positions are governed by

$$\text{PDIN}\left(G, C \left(I_n \otimes \begin{bmatrix} a & b \\ -b & a \end{bmatrix}^{-1} \right), q, t \right)$$

where we implicitly assume that the agents will back-calculate the accelerations in the original frame of reference for implementation. The advantage of reformulating the original PDIN in this new frame of reference is that the new PDIN has a valid direction $(1, 0)$. Also, we note that a controller (that achieves formation stabilization to $\bar{\mathbf{s}}$ and collision avoidance) can be developed for

$$\text{PDIN}\left(G, C \left(I_n \otimes \begin{bmatrix} a & b \\ -b & a \end{bmatrix}^{-1} \right), q, t \right)$$

whenever a controller can be developed for $\text{PDIN}(G, I, q, t)$, because we can simply premultiply (in a decentralized fashion) the measurements of the first PDIN by $C \left(I_n \otimes \begin{bmatrix} a & b \\ -b & a \end{bmatrix}^{-1} \right)^{-1}$ to obtain the measurements for the second PDIN. Hence, from the lemma above, we can show formation stabilization with collision avoidance. Hence, we have converted PDIN(G, C, q, t) to a form from which we can guarantee both formation stabilization and collision avoidance. ∎

Discussion of Collision Avoidance   To illustrate our analyses of formation stabilization with collision avoidance, we make a couple of remarks on the results obtained in the above analysis and present several simulations of collision avoidance.

Existence of a Valid Direction   We have shown that formation stabilization with collision avoidance can be achieved whenever there exists a valid direction—one in which agents can move to their formation positions without possibility of collision with other agents. For purpose of design, we can check the existence of a valid direction by scanning through all possible direction vectors and checking if each is valid (using the test described in Definition 3.4.3). More intuitively, however, we note that the existence of a valid direction is deeply connected with the distances between agents in the formation, the size of the repulsion ball, and the number of agents in the network. More precisely, for a given number of agents, if the distances between the agents are all sufficiently large compared to the radius of the repulsion ball, we can guarantee existence of a valid direction. As one might expect, the required distance between agents in the formation tends to become larger as the repulsion ball becomes larger and as the number of agents increases. We note that existence of valid direction is by no means necessary for achieving collision avoidance; however, we feel that, philosophically, the idea of converging to a valid direction is central to how collision avoidance is achieved: Extra directions of motion are exploited to prevent collision while still working toward the global task. Distributed agents, such as pedestrians on a street, do indeed find collision avoidance difficult when they do not have an open (i.e., valid) direction of motion, as our study suggests.

Other Formation Stabilizers   In our discussion above, we considered a dynamic LTI controller for formation stabilization and overlayed this controller with a secondary controller for collision avoidance. It is worthwhile to note that the condition needed for collision avoidance—that is, the existence of a valid direction—is generally decoupled from the type of controller used for formation stabilization. For instance, if we restrict ourselves to the class of (decentralized) static linear controllers, we can still achieve collision avoidance if we can achieve formation stabilization and show the existence of a valid direction. In this case, our result is only different in the sense that a stronger condition on G is needed to achieve static stabilization.

Different Collision Avoidance Protocols   Another general advantage of our approach is that we can easily replace our proposed collision avoidance mechanism with another, as long as that new protocol exploits the presence of multiple directions of motion to achieve both formation and avoidance. Of particular interest, protocols may need to be tailored to the specifics of the available warning measurements. For instance, if the warning measurements only flag possible collisions and do not give detailed information about the distance between agents, we may still be able to achieve collision avoidance, by enforcing that agents follow curves in the valid direction whenever they sense the presence of other agents. We leave it to future work to formally prove convergence for such controllers, but we demonstrate their application in an example below.

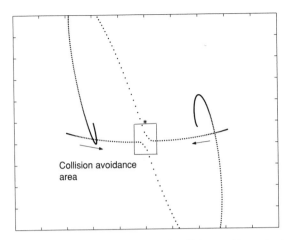

Figure 3.21 Formation stabilization with collision avoidance is shown.

Avoidance of Moving Obstacles   Although we have focused on collision avoidance among controllable agents, our strategy can be adapted to networks with uncontrollable agents (obstacles). Fundamentally, this adaptation is possible because only one agent is required to diverge from its standard trajectory to avoid a collision. Hence, collisions between controllable agents and obstacles can be prevented by moving the controllable agents in a valid direction. Some work is needed to specify the precise conditions needed to guarantee collision avoidance (e.g., we must ensure two obstacles do not converge and that the obstacle does not constantly hinder stabilization). We leave this work for the future.

Simulation   We illustrate our strategy for achieving both collision avoidance and formation stabilization using the examples below. Figure 3.21 shows a direct implementation of the collision avoidance strategy developed in this section. The figure verifies that collision avoidance and formation are both achieved but exposes one difficulty with our approach: The agents' trajectories during collision avoidance tend to have large overshoots because we must make the repulsion acceleration arbitrarily large near the repulsion ball to guarantee that collisions are avoided.

   A simple approach for preventing overshoot after collision avoidance is to apply a braking acceleration as soon as an agent is no longer in danger of collision. We choose this braking acceleration to be equal in magnitude and opposite in direction to the total acceleration applied during collision avoidance. Figure 3.22 shows a simulation of collision avoidance, when a braking force is applied after the collision avoidance maneuver. We note that our proof for formation stabilization with collision avoidance can easily be extended to the case where braking is used.

   Figure 3.23 shows a more advanced approach to collision avoidance. In this example, each agent is guided along a curve in space, as soon it has detected the presence of another agent in its local sensing ball. Curve following can allow the formulation of much more intricate, and optimized, collision avoidance protocols. Some more work is needed, however, to develop curve-following protocols that can be analytically shown to achieve formation and collision avoidance.

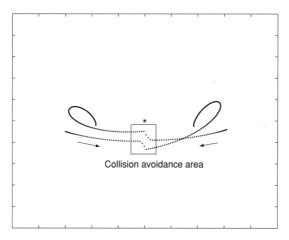

Figure 3.22   Another protocol for collision avoidance is simulated. Here, we have eliminated overshoot after collision avoidance using a braking acceleration.

3.5   MODELING AND ENHANCING THE DATA CAPACITY OF WIRELESS SENSOR NETWORKS

Jie Lian, Kshirasagar Naik, and Gordon B. Agnew

Energy conservation is one of the most important design considerations for battery-powered wireless sensors networks (WSNET). Energy constraint in WSNETs limits the total amount of sensed data (data capacity) received by the sink. The data capacity of WSNETs is significantly affected by deployment of sensors and the sink. A major issue, which has not been adequately addressed so far, is the question of how node deployment governs the data capacity and how to improve the total data capacity of WSNETs by using nonuniform

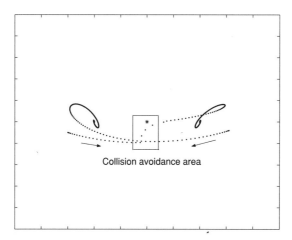

Figure 3.23   In this simulation, we achieve collision avoidance by guiding agents along curves in space, once potential collision is detected.

sensor deployment strategies. We discuss this problem by analyzing the commonly used static model of sensors networks. In the static model, we find that after the lifetime of a sensor network is over, there is a great amount of energy left unused, which can be up to 90% of the total initial energy. This energy waste implies that the potential data capacity can be much larger than the capacity achieved in the static model. To increase the total data capacity, we propose two strategies: a nonuniform energy distribution model and a new routing protocol with a mobile sink. For large and dense WSNETs, both of these strategies can increase the total data capacity by an order of magnitude.

3.5.1 Background Information

A WSNET consists of a set of microsensors deployed within a fixed area. The sensors sense a specific phenomenon in the environment and route the sensed data to a relatively small number of central processing nodes, called sinks. Unlike a mobile ad hoc network (MANET) where bandwidth efficiency and throughput are two important metrics, energy conservation is an important design consideration for WSNETs. This is because all sensors are constrained by battery power. Moreover, since sensor networks generally operate at low data rates, signal interference among neighbors is not much of an issue compared to MANET.

Similar to MANETs, researchers have focused on the medium (media) access control (MAC) and network layer protocols for WSNETs [15, 26, 70–75, 83]. Data aggregation techniques also have been studied in [15, 70, 75, 76]. In fact, all of these focus on increasing the energy efficiency of a WSNET, which is represented by the average energy required to transmit a unit of sensed data to a sink. A commonly used model of WSNETs in those studies is the *static* model in which homogenous sensors are uniformly distributed in the sensed area with one stationary sink. In the static model, we may observe that sensors close to the sink need to forward more data than sensors far away from the sink. Thus, sensors close to the sink exhaust their energy much faster than other sensors. The extreme case occurs with the direct neighbor sensors of the sink, which deplete their energy first. When all neighbors of the sink exhaust their energy, the sink is disconnected from the network, and the lifetime of the network is over. Therefore, the *network data capacity* of a static WSNET, which is defined as the total amount of sensed data received by the sink, is limited and is mainly determined by the total energy in the neighbors of the sink. Meanwhile, when the lifetime of a network is over, there is an unknown amount of energy left unused. Therefore, it is useful to find the data capacity and energy utilization of a WSNET.

The studies listed above for WSNETs emphasize an increasing energy efficiency, and, therefore, prolonging the lifetime of WSNETs. If we assume that the sink receives sensed data in a fixed constant speed, the lifetime of a WSNET can be measured by the network data capacity. However, from the above discussion and the results obtained in this section, in the commonly used static model of WSNETs energy efficiency is not the most important factor in WSNET operation. Instead, we will show in this section that proper deployment (i.e., locations) of sensors and the sink and energy distribution of sensors have a positive impact on the lifetime (or the data capacity) of WSNETs.

The contributions of this section are summarized as follows. First, we will develop a mathematical model of static WSNETs with uniformly distributed, homogenous sensors and a single stationary sink; next, we will analyze its performance. Performance of a WSNET is evaluated using three metrics: *network data capacity*, *energy efficiency*, and *energy utilization.* Unlike the *transport capacity* of a wireless network discussed in the literature [77, 78], *network data capacity* of a sensor network is the total amount of sensed data received by the sink. The main observations from this model are that a significant amount of energy is still left unused after the lifetime of the network is over, and the total data capacity achieved is much smaller than the maximum potential data capacity. For moderate size and large WSNETs, after the lifetime of a network is over, there is a large amount of energy left unused, which can be up to 90% of the total initial energy. These results suggest that the static model is not a good choice for large-scale sensor networks. The main reasons for the inefficient energy utilization are the uniform energy level of all sensors and the stationary sink. Therefore, to increase the total data capacity, we propose a nonuniform energy distribution model and a new routing protocol with mobile sink support. Simulation study shows that these two strategies can improve the total data capacity by an order of magnitude of the data capacity achieved in the static single-sink model.

3.5.2 Basic Assumptions

We focus on large-scale, dense networks with several hundreds to thousands of sensors. For these networks, without any loss of generality, we make the following assumptions:

- The network area is a fixed $W \times H$ rectangular area with width W and height H. N sensors are randomly placed according to Poisson distribution [81, 82] with density λ sensors per unit area. Hence, $N = \lambda WH$.
- All sensors are homogeneous in the sense that all of them have the same amount of initial energy P_s and the same transmission range r_s. Each sensor consumes p_s quantity of energy to transmit one bit of data.
- The locations of the sensors and the sink remain unchanged in the static model.
- The sensed phenomenon is randomly occurred in the network area over time.
- The sink has unlimited energy and the same transmission range r_s as the sensors.
- The energy required to transmit data is much more than the energy consumed by CPU processing, sensing, and data receptions. Thus, only the power consumption of data transmission is considered.

Two sensors or a sensor and the sink are said to be neighbors if they are within the transmission range of each other. The *average degree g* of a sensor is defined as the average number of neighbors of a sensor. Due to the sensors being randomly distributed with Poisson process, we have

$$g = \lambda \pi r_s^2 \tag{3.42}$$

To deploy a sensor network with hundreds of nodes connected with high probability, say, 0.95, the average degree is at least 4 [79–81]. We assume that the required average degree is not less than 5 to ensure connectivity. When a node senses the phenomenon, we assume that the node chooses the shortest path (with minimum number of hop count) to forward data toward the sink. We ignore the route maintenance overhead and the reason is explained in Section 3.5.3. Due to the low data rate in WSNETs, we assume that data collisions at the MAC level are negligible.

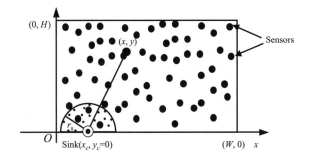

Figure 3.24   Illustration of the SSEP model.

3.5.3 Single Static Sink Edge Placement Model

We consider a static model of WSNETs and analyze its data capacity—and energy utilization. In the model, we consider a WSNET with a single stationary sink located on an edge of the rectangular network area. Also, the sink is at least r_s distance away from the corners of the network area. This model is referred to as the single static sink edge placement (SSEP) model as illustrated in Figure 3.24. We refer to this kind of a sensor network as an SSEP network.

Due to randomly distributed sensors, the data capacity—of an SSEP network is limited by the total initial energy in the neighbors of the sink. Each bit of sensed data in Ψ needs to be forwarded to the sink on a multihop path. Since we assume that the sensed phenomenon is uniformly distributed over time in the network area, the average hop length of all shortest multihop paths can be computed. Using this average hop length, the average energy needed to send one bit of data to the sink can be computed. Therefore, the total energy needed to transmit Ψ bits of sensed data to the sink is easily computed. From this utilized energy and the total initial energy, we can obtain the energy utilization ratio in the SSEP model.

3.5.4 SSEP Model Description

In the SSEP model shown in Figure 3.24, we assume that the sink node is located at coordinate $(x_c, 0)$ and its location never changes. The reason for using this sink placement model is that in many real situations, the sink cannot move into the sensed area due to some adverse conditions in that area, such as the area being close to a volcano, for example. A network in the SSEP model is determined by six parameters: the area width W, the area height H, the average degree g of sensors, the transmission radius r_s, and the location of the sink (x_c, y_c), where $y_c = 0$.

3.5.5 Critical Region and Network Data Capacity

Definition 3.5.1   The *total network data capacity* Ψ of a WSNET is defined as the total amount of sensed data received by the sink. (It does not include other data such as routing overhead.)

The *first critical region*, denoted by V_1, of the sink is defined as the shaded half circle centered at the sink and with a radius r_s in Figure 3.24. It may be noted that V_1 contains an average of $g/2$ (half of the degree) sensors. The sink can only receive data from sensors in V_1. Thus, the lifetime of the network is over if all sensors in V_1 deplete their energy. Since the average total initial energy available in V_1 is $P_s g/2$, where $g = \lambda \pi r_s^2$, and transmission of one bit of data from a sensor in V_1 to the sink consumes p_s amount of energy, the average value of the total network data capacity Ψ, denoted by $\overline{\Psi}$, is given as follows:

$$\overline{\Psi} = \frac{1}{2} P_s g \frac{1}{p_s} = \frac{1}{2} \lambda \pi r_s^2 \frac{P_s}{p_s} \tag{3.43}$$

In an ideal situation where there is no routing overhead, no collision at the MAC level, and the energy consumed in sensing and computation are negligible compared to data transmissions, the average data capacity can be close to $\overline{\Psi}$.

3.5.6 Energy Efficiency of a WSNET in the SSEP Model

To calculate the energy utilization of an SSEP network, we need to know the average amount of energy required to transmit one-bit of sensed data along a forwarding path to the sink. This is done by finding the transmission energy efficiency as follows:

Definition 3.5.2   The *transmission energy efficiency* Λ is defined as the reciprocal of the average amount of energy required to transmit one bit of sensed data along a multihop path from the original sensor to the sink.

A larger value of Λ indicates that less energy is required for data transmission. Since the sensed data should be forwarded to the sink via a multihop path, the smaller the average number of hop counts, the higher the energy efficiency. Therefore, the potential highest energy efficiency can be achieved if sensors use the shortest-hop paths.

To compute Λ, we need to find the average hop count for all shortest-hop paths. However, direct computation of the average hop count is not straightforward for a randomly deployed WSNET. Instead, we use a statistical method as follows. For any given sensor P with many neighbors, P always chooses a neighbor node, which is nearest to the sink, as the next forwarding node. The selected node then follows the same rule to choose the next node, and so on, until the data packet reaches the sink. In general, each forwarding operation reduces the remaining distance from a sensor to the sink by the maximum length, and this length is called the *maximum one-hop progress* toward the sink. Intuitively, a forwarding path established by this rule is generally the shortest-hop path from P to the sink because the network is dense. The hop count of a shortest-hop path from P to the sink is approximately equal to the distance from P to the sink divided by the average maximum one-hop progress. Therefore, the average hop count of all shortest-hop paths is approximately equal to the *average distance* to the sink from all sensors divided by the average maximum one-hop progress. Next, we will derive the expressions for the average distance and the average maximum one-hop progress.

Due to the randomly occurring phenomenon in the sensed area, each shortest-hop path has equal probability of being chosen as a forwarding path. For a large and dense network, the problem is equivalent to finding the average distance from all points in the network area to the sink. Assume that the sink is located at (x_c, y_c). For a given node with coordinate (x, y), the distance d from this node to the sink is given by $d = \sqrt{(x - x_c)^2 + (y - y_c)^2}$. Integrating d over the entire network area, we compute the average value of d, denoted by

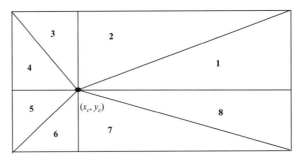

Figure 3.25   Partition of the $W \times H$ area.

$\bar{d}$, as follows:

$$\bar{d} = \frac{1}{WH} \int\limits_{0}^{W} \int\limits_{0}^{H} \sqrt{(x - x_c)^2 + (y - y_c)^2}\, dy\, dx \tag{3.44}$$

It is difficult to solve (3.44) directly. Instead, we consider the double integral in (3.44) in three-dimensional space. Hence, the double integral is a volume enclosed by the function $z = \sqrt{(x - x_c)^2 + (y - y_c)^2}$ and x-y plane with $0 \le x \le W$ and $0 \le y \le H$. The entire $W-H$ area can be partitioned into eight triangles as shown in Figure 3.25. For each triangle, we can compute the volume $V(x, y)$ above the triangle as follows, where x and y denote the lengths of the two nondiagonal edges.

$$V(x, y) = \begin{cases} 0 & \text{if } x = 0 \text{ or } y = 0 \\ \dfrac{x^3}{6}\left(\ln\left(y + \sqrt{x^2 + y^2}\right) - \ln(x) + \dfrac{y}{x^2}\sqrt{x^2 + y^2}\right), & \text{otherwise} \end{cases}$$

The total volume of the double integral is the sum of all volumes of the eight partitioned triangles. Therefore, the solution of (3.44) is the total volume divided by $W \times H$ as follows:

$$\begin{aligned} \bar{d}(W, H, x_c, y_c) = \frac{1}{WH}\, [\,&V(x_c, y_c) + V(y_c, x_c) + V(W - x_c, y_c) + V(y_c, W - x_c) \\ &+ V(x_c, H - y_c) + V(H - y_c, x_c) + V(W - x_c, H - y_c) \\ &+ V(H - y_c, W - x_c)\,]\,, \end{aligned} \tag{3.45}$$

where $\bar{d}(W, H, x_c, y_c)$ denotes the average distance in the configurations of the area with width W, height H, and the sink location (x_c, y_c). Since $y_c = 0$ in the SSEP model, the first four addition terms in (3.45) are 0.

The basic idea behind the average maximum one-hop progress has been illustrated in Figure 3.26. In Figure 3.26, P is a sensor node and Q is the sink. Any node on the arc, which is centered at Q and has a radius $R = d - z$, makes the same forwarding progress $z = d - R$ from P to Q in one hop. When P forwards data to the sink, P always chooses the neighbor node, which makes the maximum progress z, as the next forwarding node. The selected node then follows the same rule to choose the next node, and so on, until the data packet reaches Q. A forwarding path established by this rule is generally the shortest-hop

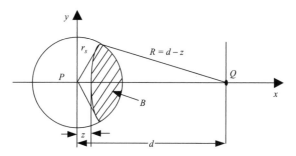

Figure 3.26 One-hop progress form p to Q.

path from P to Q for a dense network. Next we need to find the expected value of the maximum progress z for a WSNET with average degree g.

There is no simple solution available for this problem. A similar, but not identical, one-hop progress problem is discussed in [81]. A solution of the average maximum one-hop progress $\bar{z}$ has been presented in [82], in which the upper and lower bounds of $\bar{z}$ are given. These two bounds are tight enough such that the accuracy of $\bar{z}$ is guaranteed. The average maximum one-hop progress is a function of average degree g, distance d, and radius r_s. If we directly use the solution provided in [15], the model will become too complex to analyze. As discussed in [81], if the ratio of d and r_s is sufficiently large, $\bar{z}$ will be a function of the average degree g only, and the influence of d and r_s is not significant. This result is supported by a simulation study, which is done as follows.

For a particular average degree g, in each round, we randomly generate n nodes based on Poisson distribution with density g within the circle centered at P. Next, we compute the distances from these nodes to Q, choose the node closest to Q, and record its hop progress for this round. We repeat the above three steps for a large number of times and obtain the mean value $\bar{z}$. The experimental results are shown in Figure 3.27 for the value of g ranging from 5 to 80. All distances are normalized to transmission radius r_s, which is treated as unity. The ratios for data series in Figure 3.27 denote the values of d/r_s. All values of $\bar{z}$ are normalized to r_s too. We observe that when both the ratio d/r_s and the average degree are greater than 6, $\bar{z}$ mainly depends on degree g alone. For smaller ratio d/r_s and average degree, the differences among all curves are about 10% of the total $\bar{z}$.

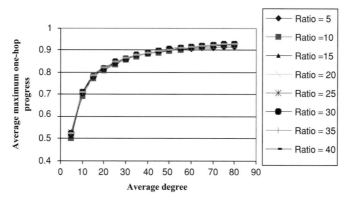

Figure 3.27 Simulation results for the average maximum one-hop progress.

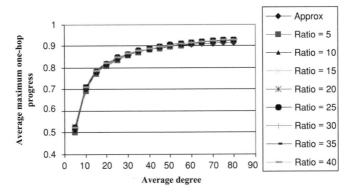

Figure 3.28 Approximated and simulated average maximum one-hop progress.

From the above discussion, an approximate solution for $\bar{z}$ is estimated as follows:

$$\bar{z} = 1 - e^{-1/2}\sqrt{g} - 0.07 - 0.09e^{(5-g)1/3} \tag{3.46}$$

where $\bar{z}$ is normalized to r_s and $g \geq 5$.

The values of $\bar{z}$ obtained from (3.46) and from simulation studies are compared in Figure 3.28. The curve labeled "Approx" denotes the approximated data obtained from (3.46). We observe that the approximated data matches very well with the simulated data within the range $10 \leq d/r_s \leq 40$ and $10 \leq g \leq 60$, which are the most commonly used ranges in real applications. The reader may note that the values of $\bar{z}$ estimated using (3.46) have an error margin of about 10% if $d/r_s \leq 7$.

Equation (3.46) has been validated by performing simulation studies on randomly generated sensors networks. In the experiments, we randomly deploy homogeneous sensors in a $W \times H$ network area with average degree g. We generated networks with different configurations: The ratio of W/H was varied from 1 to 20; the average degree g was varied from 5 to 80; and the sink locations were changed from the middle of the W edge to a corner of the network area. For those networks, we computed the maximum one-hop progress for each node connecting to the sink and then computed the average maximum one-hop progress $\bar{z}$. This simulated average maximum one-hop progress is compared with the calculated progress $\bar{z}$ based on (3.46). From the simulation results (not shown here), we find that the computed values of $\bar{z}$ closely match with the simulated values of $\bar{z}$. The values of $\bar{z}$ estimated using (3.46) have an error margin of approximately 10% for $g \leq 7$. Hence, Equation (3.46) can be used in estimating the value of $\bar{z}$ for a given degree g.

Therefore, the approximated average hop count $\bar{h}$ for all shortest paths is

$$\bar{h} \approx \frac{\bar{d}}{r_s\bar{z}} = \frac{\bar{d}}{r_s(1 - e^{-1/2}\sqrt{g} - 0.07 - 0.09e^{(5-g)1/3})} \tag{3.47}$$

Since sensed data should traverse $\bar{h}$ hops on average to the sink, and transmission for each hop consume p_s amount of energy, the average energy efficiency $\bar{\Lambda}$ is

$$\bar{\Lambda} = \frac{1}{\bar{h}p_s} \tag{3.48}$$

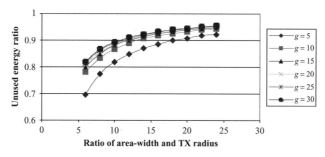

Figure 3.29

3.5.7 Energy Utilization Based on $\overline{\Lambda}$ and $\overline{\Psi}$

After we find the data capacity and the energy efficiency in the SSEP model, the energy utilization can be represented by the unused energy ratio defined as follows.

Definition 3.5.3 The *unused energy ratio* δ of a WSNET is defined as the ratio of the remaining energy available and the total initial energy in the network after all neighbors of the sink deplete their energy.

The total initial energy of a network is NP_s, where N is the total number of sensors and P_s is the initial energy in each sensor. According to (3.43) and (3.48), in an ideal condition, if a perfect routing algorithm is used, $\overline{\Psi}$ bits of data can be sent to the sink, and energy efficiency is $\overline{\Lambda}$. The total energy consumed to transmit $\overline{\Psi}$ bits is $\overline{\Psi}/\overline{\Lambda}$. Therefore, the unused energy ratio δ is

$$\delta = 1 - \frac{\overline{\Psi}}{\overline{\Lambda}NP_s} = 1 - \frac{\overline{h}\pi r_s^{\,2}}{2WH} \tag{3.49}$$

Intuitively, sensors close to the sink forward more data than sensors far away from the sink. Thus, after the lifetime of a WSNET is over, the longer the distance of a sensor from the sink is, the more unused energy left in the sensor. According to (3.49), the higher the energy efficiency $\overline{\Lambda}$, the higher the unused energy ratio δ. Figures 3.29, 3.30, and 3.31 show the result computed from (3.49) for a fixed area A with different H and W.

In these figures, the x axis denotes the ratio of W and the transmission (TX) radius r_s, and the y axis represents the unused energy ratio δ. Each curve is related to a specific average

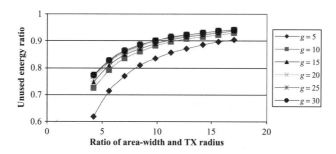

Figure 3.30 Unused energy ratio ($W = H/2$).

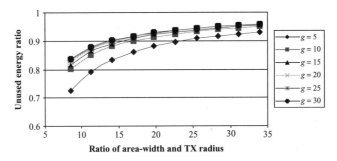

Figure 3.31   Unused energy ratio ($W = 2H$).

degree $g(5 \leq g \leq 30)$. The sink is located at coordinate $(W/2, 0)$. Figures 3.29, 3.30, and 3.31 use $W = \sqrt{A}$, $W = \sqrt{2A}/2$, and $W = \sqrt{2A}$, respectively.

These figures illustrate the waste of energy of SSEP networks. The unused energy ratio δ is mainly determined by W/r_s, height H, and the average degree g. From these figures we conclude that a large amount of energy is wasted in an SSEP network. For example, when $W/r_s = 15$ and nodes have 10 neighbors on an average, about 90% of the total initial energy will be left unused. However, many applications may operate in the range of $W/r_s = 15$. Therefore, SSEP networks cannot efficiently utilize a given amount of total energy.

3.5.8   Simulation Results

We use a routing-level simulator to verify the unused energy ratio predicted by the model above and give the simulation results in Figures 3.32, 3.33, 3.34, and 3.35. In this simulation, networks are configured using the SSEP model. N homogeneous sensors are randomly deployed in a $W \times H$ network area. The sink is located at point $(W/2, 0)$. To make the results reliable, randomly generated sample networks have more than 90% of sensors initially connected to the sink. The sensed events are generated randomly within the network area. Sensors forward data along the shortest-hop paths to the sink. The sensing range of each sensor is half of its transmission range. Each simulation run terminates if 90% of the sensors are disconnected from the sink. Since the number of neighbors of the sink greatly affects the final results, we randomly place exactly $g/2$ nodes within the transmission range of the sink. If a sensor exhausts its energy or is disconnected from the sink, it stops functioning. At the end of all the simulation runs, we calculate the unused energy ratio δ.

Figures 3.32, 3.33, 3.34, and 3.35 illustrate the simulated and predicted unused energy ratios. All these figures can be interpreted in the same way as Figures 3.29, 3.30, and 3.31. Each simulated data point is calculated by taking the average of 40 runs. According to these figures, the predicted data closely matches with the simulation result. Therefore, for most of the commonly used ranges, the SSEP model can be used as a reference model for further analysis.

3.5.9   Discussions of the SSEP Model

For large WSNETs, the SSEP model has a significant drawback in the form of low energy utilization. Multiple sinks can increase the energy utilization to a certain level, but it cannot eliminate the problem when network sizes become larger and larger. There are two reasons

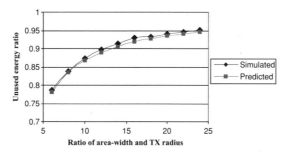

Figure 3.32   Simulated and predicted results ($g = 16$, $W = H$).

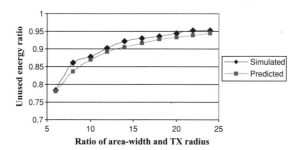

Figure 3.33   Simulated and predicted results ($g = 8$, $W = H$).

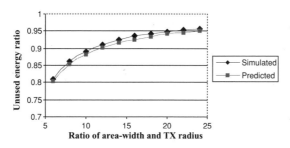

Figure 3.34   Simulated and predicted results ($g = 16$, $W = 2H$).

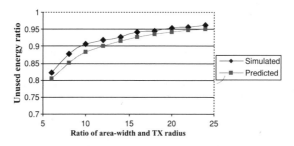

Figure 3.35   Simulated and predicted results ($g = 8$, $W = 2H$).

causing the low energy utilization: (i) the sensors are homogeneous and randomly distributed and (ii) the sink is stationary. Randomly distributed homogeneous sensors lead to uniformly distributed energy in the network area. If a stationary sink is used, the first critical region of the sink is fixed. In general, the closer a sensor is to the sink, the higher the workload is on the sensor to forward data and the faster its energy is exhausted. Hence, the energy in the first critical region will be depleted first. This energy depletion further leads to the sink being disconnected from the remaining network, which results in low data capacity and low energy utilization.

The total data capacity can be improved at two levels: application level and physical level. The idea of *data aggregation* is one example at the application level in which sensors far away from the sink coordinate with each other by using the unutilized energy and send nonduplicated and summarized data to the sink [76]. At the physical level, there are two possible solutions to increase the data capacity. The first one is to use a nonuniform energy distribution strategy. We can either deploy more homogeneous sensors in the area close to the sink or put sensors close to the sink with higher initial energy. In fact, both the techniques focus on the same purpose: Allocate more energy to the area closer to the sink than areas far away from the sink. These two strategies make sensor deployment a difficult task or require nonhomogeneous sensors. The second solution at the physical level is to increase the area of the first critical region of the sink, which can be achieved by using a *mobile sink*. In the subsequent sections, we will discuss these two solutions.

In the previous analyses, we ignored routing overhead. The energy utilization was computed by assuming the availability of a perfect routing algorithm, where sensors always choose the shortest-hop paths, and the cost of computing the shortest paths is negligible. One may argue that if the routing overhead is considered, the unused energy ratio is not as high as the predicted amount. However, such optimal routing algorithms exist as argued below, and the routing overhead involved in the algorithm can be negligible compared with the volume of sensed data.

For example, in the SSEP model the sink can use one round of flooding to establish hop distances for all sensors in the network. During flooding, each sensor can find the smallest hop distance from itself to the sink and record all neighbors with smaller hop distances to the sink. In this way, a forwarding tree rooted at the sink is constructed by using very small routing overhead. Let T denote the time in which the last sensor in the first critical region of the sink depletes its energy (i.e., the sink is disconnected from the network). Before T, it is unlikely that sensors not in the first critical region exhaust their energy. For dense networks, each sensor has many choices to select its next forwarding node. Before time T, one sensor failure, which is not caused by energy depletion, has little impact on the forwarding tree. Since updates of the forwarding tree occurs infrequently, the cost of global topology updates is not significant and can be ignored. Meanwhile, a sensor failure after T need not be considered. Therefore, during the time period T, this routing algorithm is near optimal and the routing overhead is negligible compared to the total sensed data. Even though the routing overhead cannot be omitted in some cases, if the routing overhead is evenly contributed over all sensors, the static model developed still can be used. In this condition, we can assume that sensors in the network have lower useful initial energy.

3.5.10 Nonuniform Energy Distribution Model

We will concentrate on a nonuniform energy distribution model. The main question that this model answers is: For WSNETs with fixed area, uniformly distributed sensors and a fixed total amount of energy, how to allocate energy to sensors such that the total data capacity is close to its maximum potential capacity and unused energy ratio is minimized? In this model, sensors close to the sink have higher initial energy, so the bottleneck of the data capacity from the first critical region is eliminated.

3.5.11 Basic Model

Our discussion of the nonuniform energy distribution is based on the SSEP model. Since the new model uses the same basic configurations as the SSEP model, except nonuniformly (NU) distributed initial energy of sensors, it is referred to as the SSEP-NU model. In the SSEP-NU model, we consider a network with N sensors (uniformly distributed in the area) and fixed total amount of initial energy, denoted by E, for all the sensors. The main objective of this model is to answer the question given above. A typical SSEP-NU network is shown in Figure 3.36.

To simplify our analysis, the $W \times H$ network area is partitioned into four subareas B_1, B_2, B_3, and B_4. For dense networks, data sensed by a node in one subarea will not be forwarded through other areas to the sink in most of the cases. Only nodes very close to the boundary of two adjacent subareas may forward data to nodes in the other area. If the network is sufficiently large, this boundary effect can be ignored. Hence, we consider each subarea separately.

In an ideal condition, the maximum potential data capacity (denoted by Ψ_M) can be achieved when all sensors exhaust their energy at the same time leading to the unused energy ratio 0. Since the network consists of four subareas, we need to know how much data each area contributes. Due to the randomly occurring phenomenon, the data capacity Ψ_{B_n} contributed by B_n, $1 \le n \le 4$, is proportional to its area. Hence, we have

$$\Psi_{B_1} = \Psi_{B_2} = \Psi_M \frac{(W - x_c)H}{2WH} = \Psi_M \frac{W - x_c}{2W} \tag{3.50}$$

$$\Psi_{B_3} = \Psi_{B_4} = \Psi_M \frac{x_c H}{2WH} = \Psi_M \frac{x_c}{2W} \tag{3.51}$$

According to (3.50), B_1 and B_2 generate the same amount of data. However, they consume a different amount of energy since their average path lengths to the sink are not identical. Let $\overline{h}_{B_n}$ denote the average hop path length and $\overline{d}_{B_n}$ the average distance to the sink in area B_n. The average maximum one-hop progress $\overline{z}$ is the same for all subareas since for dense networks $\overline{z}$ depends on the average degree g only. Based on (3.47), the total energy E_{B_1} in

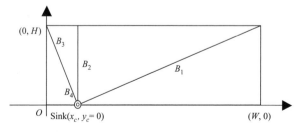

Figure 3.36 Partition of SSEP-NU network.

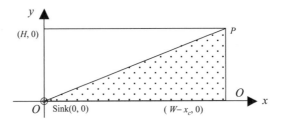

Figure 3.37   Area B_1 after coordinate transformation.

the area B_1 is

$$E_{B_1} = p_s \overline{h_{B_1}} \Psi_{B_1} = p_s \Psi_M \frac{W - x_c}{2W} \frac{\overline{d_{B_1}}}{r_s \overline{z}} \tag{3.52}$$

To compute (3.52), we need to know the value of Ψ_M and $\overline{d_{B_1}}$. Since each data bit traverses $\overline{h}$ hops to the sink on an average, we have

$$\Psi_M = \frac{E}{\overline{h} p_s} \tag{3.53}$$

By using coordinate transformation, area B_1 is shown in a different form in Figure 3.37. For the triangle OPQ shown in Figure 3.37, for simplicity, let $W - x_c = L$. The distance function from an arbitrary point with coordinate (x, y) to the sink is $d = \sqrt{x^2 + y^2}$. Therefore, the average distance $\overline{d_{B_1}}$ from all points in OPQ to the sink is the integration of the distance function over the triangle area OPQ as follows:

$$\overline{d_{B_1}} = \frac{2}{LH} \int_0^L \int_0^{Hx/L} \sqrt{x^2 + y^2}\, dy\, dx \tag{3.54}$$

Combining (3.52), (3.53), and (3.54), we can obtain the value of the total energy E_{B1}. Similarly, we can compute the total energy for other subareas. We will focus on area B_1 since the same steps can be applied to other subareas.

Before allocating E_{B_1} energy to the sensors in B_1, we define the nth *critical region* of the sink in B_1. The first critical region V_1 is already defined in Section 3.5.3. As shown in Figure 3.38, the second critical region V_2 is defined as the region bounded by the x axis, line

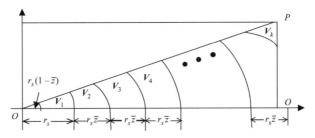

Figure 3.38   Critical regions of the sink in area B_1.

OP, and the arc centered at O with a radius $r_s + r_s\bar{z}$, except the region V_1. Higher critical regions can be defined analogously. Figure 3.38 shows all the critical regions of subarea B_1. In the subsequent sections, we will use V_n to denote both the nth critical region and its area, and the actual meaning is determined by the context.

The main feature of the critical regions is explained like this: On average, a data transmission of a node in the nth critical region will forward the data to a node in the $(n-1)$th critical region. This property holds because of two reasons. First, since nodes forward data by using the shortest-hop paths, for dense networks, the paths are close to straight lines. Second, any two neighboring critical regions are partitioned by the average maximum one-hop progress times transmission radius, $r_s\bar{z}$. Therefore, one transmission will have $r_s\bar{z}$ progress toward the sink on average, and this progress is enough to deliver data to the next critical region. One exception of this feature is V_1. Obviously, nodes in V_1 can send data to the sink directly. This observation indicates that the nodes, with distances to the sink less than $r_s(1 - \bar{z})$, may never be used as forwarding nodes since nodes in V_2 have very low probability to reach these nodes.

Obviously, the total data volume across region V_1 equals to the data generated by the entire area of B_1. The total data volume across region V_n equals to the total data generated by all regions $V_n, V_{n+1}, \cdots, V_k$, where V_k is the final critical region. We define the *data source area* of V_n as $V_n + V_{n+1} + \cdots + V_k$, and denote it by S_n. As the name suggests, the data source area of V_n is the total area in which sensed data needs to be forwarded through V_n. Hence, the energy allocated to V_n is proportional to its S_n. Let $E_{B_1-V_n}$ denote this energy allocated to V_n in B_1. Thus, we have

$$E_{B_1-V_n} = \frac{E_{B_1} S_n}{\sum_{i=1}^{k} S_i} \tag{3.55}$$

Since the computations of V_n and S_n are simple geometric problems, they are ignored here. The same steps can be applied to subareas B_2, B_3, and B_4. In the SSEP-NU model, the closer the critical region to the sink is, the higher the initial energy allocated to the sensors in that region.

3.5.12 Discussion and Application of the SSEP-NU Model

For the SSEP-NU model, there are two possible ways to allocate the computed energy to each critical region: uniform sensor distribution strategy and nonuniform sensor distribution strategy. In the uniform sensor distribution strategy, sensors are still uniformly distributed in the network area, but sensors in different critical regions have different initial energy. Sensors in the same critical region are identical in energy level. This strategy requires a large number of sensor types, which can be up to the total number of critical regions, and this may be impractical in real applications. In the SSEP model, if two sensor types are available, we can only adjust the energy in the first critical region since the first critical region is the bottleneck of the total data capacity. Similarly, if we have m types of sensors, we can adjust the energy in the first $m - 1$ critical regions to solve the bottlenecks created by these levels. In the nonuniform sensor distribution strategy, we assume all sensors to be identical. The basic idea is to put more energy in the area close to the sink, which suggests that the closer the critical region is to the sink, the denser the sensors are deployed in that region. Analysis of the SSEP-NU model depends on the average maximum one-hop progress $\bar{z}$, which is determined by the average degree g. If sensors are deployed unevenly over the network

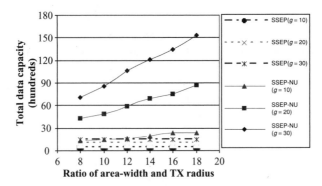

Figure 3.39   Total data capacity ($W = H$).

area, $\bar{z}$ will have a large variance in different areas. When sensors are densely deployed in the network area, the impact on $\bar{z}$ due to the changes in the average degree is not significant. Hence the SSEP-NU model can be used in this environment.

3.5.13   Simulation Results

We developed a simulator to show improvements of the data capacity in the SSEP-NU model. Two allocation strategies are compared in the simulation: One is based on the SSEP model and the other is based on the SSEP-NU model. For the simulation in the SSEP-NU model, we chose a uniform sensor distribution strategy, that is, sensors are uniformly distributed in the network area, but sensors in different critical regions have different initial energy. The simulation environment is the same as the one used in Section 3.5.3, except the initial energy of sensors. For each simulation run, the total energy of all sensors and the network area are fixed. In the SSEP-NU model, given total energy in each critical region by (3.55), we evenly allocate this amount of energy to sensors in this region.

Figures 3.39 to 3.44 show the simulated results for the total data capacity and unused energy ratio in the two energy allocation strategies represented by SSEP and SSEP-NU. Each curve in the figures is related to a certain average degree of the sensors. For all simulations, we set the initial average energy for all sensors equal to 100 and $p_s = 1$. So,

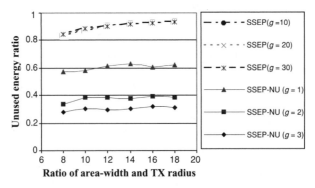

Figure 3.40   Unused energy ratio ($W = H$).

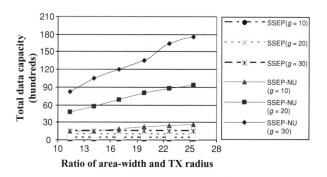

Figure 3.41 Total data capacity ($W = 2H$).

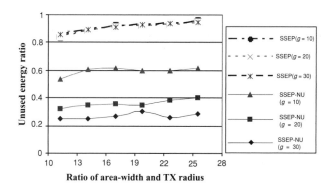

Figure 3.42 Unused energy ratio ($W = 2H$).

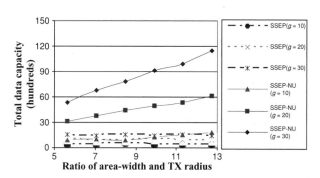

Figure 3.43 Total data capacity ($W = H/2$).

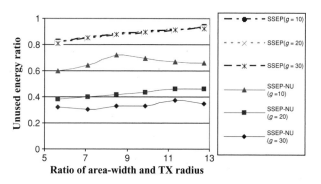

Figure 3.44 Unused energy ratio ($W = H/2$).

the total energy of each simulated network is $100N$, where N is the number of sensors. We varied W and H, but kept the network area to be a constant. Each simulation terminates when more than 50% of total sensors are disconnected from the sink. All data points are computed by taking the average of 10 runs. In these figures, the uniform energy allocation (SSEP) gives a constant total data capacity, which is independent of the network sizes. The total data capacity obtained by the nonuniform energy allocation strategy (SSEP-NU) is proportional to the network size for a fixed average degree. We also observe that for sparse networks ($g = 10$), there is about 60% energy left unused. The main reason is: For large networks, when more and more sensors deplete their energy and the network density decreases below 6, the probability of the sink being disconnected from the network is very high. The energy in the disconnected sensors is considerably large compared to the total initial energy in sparse networks. However, according to the results discussed above, for dense networks, this amount of energy is not significant. Therefore, the higher the average degree is, the lower the unused energy ratio. From the simulation studies, we conclude that for dense and large networks, the total data capacity in an SSEP-NU network can be orders of magnitude of the data capacity in an SSEP network.

3.5.14   Predictable Sink Location Routing Protocol for WSNETs

Another way of increasing the total data capacity is by using a mobile sink instead of a stationary one. Obviously, sink movement makes the sink communicate with more sensors than a stationary sink does. This, in effect, creates a larger "first critical region" and therefore reduces the impact of constrained energy in a smaller first critical region on the data capacity. In this section, we introduce a new routing protocol to improve the total data capacity and avoid high routing overhead introduced by sink mobility.

3.5.15   Predictable Sink Location Routing Protocol

Consider a network with uniformly distributed homogeneous stationary sensors and a single mobile sink. Since sink movement increases the number of sensors with which the sink can communicate, it increases the potential total data capacity of a WSNET. This potential data capacity can be achieved only if the routing overhead incurred due to sink movement is negligible or at least small compared to the total sensed data received by the sink. However, random movement of the sink causes frequent topology updates, which introduces significant routing overhead. To make the impact of routing overhead insignificant, we introduce a simple routing protocol with location awareness support, such as the Global Positioning System (GPS).

The basic idea of this protocol is very simple. Since all sensors are stationary, the main routing overhead comes from unpredictable sink location caused by its movement. If all sensors can predict the location of the sink at any particular time, most topology updates of the sink can be avoided, and the routing overhead will be greatly reduced. Prediction of the sink location requires two conditions to be satisfied: a predictable movement pattern of the sink and synchronization among sensors. If sensors know the predictable movement pattern of the sink and its initial position, they can compute the current position of the sink without waiting for the reception of the position updates from the sink. Obviously, to compute the current location of the sink, sensors and the sink should be synchronized. The former condition can be easily achieved by forcing the sink to move in a predictable pattern. For example, the sink can move along an edge of the network area with a known, constant speed. Synchronization among sensors and the sink is provided by the GPS. This protocol is referred to as the predictable sink location (PSL) routing protocol. A simple description of the PSL protocol is given as follows.

In the PSL protocol, the sink and all sensors are assumed to be synchronized and know their own geographical locations. As the name PSL suggests, sink smovement follows a regular pattern, which can be described by a function of its initial location and current system time T. For instance, assume that the sink starts at location $(x, 0)$, moves along one edge of a rectangular network area with a constant speed v, and moves backward if it reaches a corner. It is easy to use a linear function representing this movement pattern. Then, the sink floods a packet containing its movement pattern and duration to all sensors. The duration field indicates how long this movement pattern lasts. After the expiry of the time indicated in the duration field, the current movement pattern will stop and a new pattern is required. The sink need not use a single pattern during the entire lifetime of the network. Rather, the sink may change its movement pattern at any time by flooding the network with a packet containing a new movement pattern. However, the time interval between pattern changes should be long enough such that the flooding overhead is insignificant. When a sensor sends data to the sink, it can calculate the current sink location by using the movement pattern.

In the PSL protocol, three simple sink movement patterns are used: *edge movement, diagonal movement*, and *rectangular movement*. For the edge movement pattern, the sink moves along the W edge of the network area. When it reaches a corner, it moves backward. Similarly, the sink moves along a diagonal of the network area in the diagonal movement pattern, and it moves backward when it reaches a corner. For the rectangular movement pattern, it moves along the boundary of the network area. In the following, we explain how sensors send data to the sink.

Two potential routing algorithms are available to deliver data from sensors to the sink. One is the geographic random forwarding (GeRaF) algorithm introduced in [82]. In GeRaF, neighbors of each sensor are partitioned into "priority regions" according to their distances to the sink. When a sensor sends data to the sink, it simply broadcasts the data to all neighbors. The neighbors of this sensor in the highest priority region, which are closest to the sink, first contend for the relay position. If a sensor wins the contention, then the winner is selected as the next relay. If no sensor is in the highest priority region, the neighbors in the second highest priority region start a new round of contention until one sensor is selected as the relay. An advantage of GeRaF is that no routing table is required for all sensors, so the routing update cost is avoided. However, GeRaF has several drawbacks. First, for sparse networks, sensors may not find the next relay from its neighbors since sensors never forward packets in the reverse direction away from the sink. The sparser the network is, the higher the probability of sensors not finding the next relay. Second, if the highest nonempty priority region contains more than one node, a collision cannot be avoided. For dense networks, the number of collisions is high.

An alternative way is to use the traditional link state routing algorithm. Each sensor maintains the full topology map of the network and knows the geographical locations of all sensors. The topology and location information can be obtained by using a flooding technique during the initialization stage of the network. Based on the topology map maintained by a sensor n_1, it can select a neighboring sensor n_2, whose hop count to the sink is less than the hop count of n_1. Sensors receiving forwarding data packets follow the same steps as n_1 until the data reaches the sink. After the lifetime of the current movement pattern is over, the sink issues a new pattern, which could be the same as the previous pattern. The advantages of this method are that it avoids the problems involved in GeRaF and guarantees the shortest hop path to be used. The major disadvantage is the cost due to topology maintenance. However, for dense networks, it is unnecessary to immediately flood topology updates when they occur. In this condition, if a sensor cannot overhear its selected neighbor forwarding its packet, it can choose another neighbor with a smaller hop count to the sink to relay its packets. Thus the significant topology updates only happen when a large portion of sensors deplete their energy. At that time, the remaining energy in the network is not significant compared to the total initial energy. For simplicity, we will use the latter method to evaluate the data capacity of WSNETs.

3.5.16 Simulation Results

The purpose of the PSL protocol here is to show how sink movement improves the total data capacity. In this simulation, N homogeneous sensors are randomly deployed in a $W \times H$ network area, and the network contains a single mobile sink. To make the results more reliable, we use randomly generated sample networks where more than 90% of the sensors are initially connected to the sink. The sensed events are generated randomly. The sensors forward data along the shortest hop path to the sink. The sensing range of each sensor is half of its transmission range. Sensors, if exhausting their energy or disconnected from the sink, stop doing anything.

In this simulation, the three sink movement patterns discussed previously are used: *edge movement, diagonal movement*, and *rectangular movement*. For all the three movement patterns, the sink starts its movement from coordinate (0, 0). When the sink moves half distance of the sensor transmission radius, the sink stops moving and stays at the current location for a fixed time period, during which 10 sensed events on average can occur. Then the sink continues to move to the next location and repeats the same behavior. Each simulation run terminates if 50% of the total sensors are disconnected from the sink during one movement cycle. A movement cycle for all patterns is defined as the time needed for the sink to come back to its current location. Initially, each sensor has 100 unit of energy ($P_s/p_s = 100$). All routing overhead is ignored.

The simulation results are shown in Figures 3.45 to 3.52. Each figure has four series of data curves. Series S-Edge, S-Diag, and S-Rect denote the simulation data for PSL protocol in edge movement, diagonal movement and rectangular movement, respectively. The data labeled with P-SSEP denotes the total data capacity calculated in the SSEP model. Figures 3.45 to 3.50 show the simulated data for the total data capacity and their related unused energy ratio for a square network area. Figures 3.51 and 3.52 show the total data capacity with different W and H configurations and $g = 30$ only. We make the following observations from Figures 3.45 to 3.52.

First, the PSL protocol performs best (total data capacity) with the rectangular movement pattern. The next best performance is obtained with the diagonal pattern for different shapes of network areas, such as $W = H$, $W = 2H$, and $W = H/2$. However, the former has lower energy efficiency than the latter, which indicates that the delay of the former is larger than the latter.

Second, the total data capacity (with the edge, diagonal, and rectangular movement pattern) increases when the ratio W/r_s grows. However, the total data capacity of the PSL protocol with edge movement pattern remains unchanged when the ratio H/r_s increases. The total capacities of the PSL protocol with the diagonal and rectangular movement patterns increase along with increasing values of the ratio H/r_s.

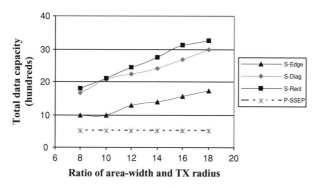

Figure 3.45 Total data capacity ($W = H$, $g = 10$).

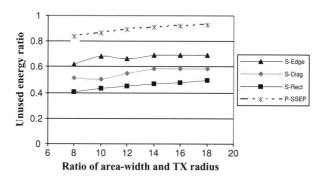

Figure 3.46 Unused energy ratio ($W = H$, $g = 10$).

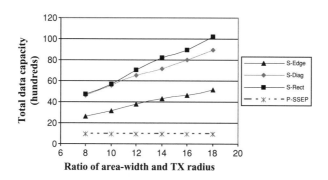

Figure 3.47 Total data capacity ($W = H$, $g = 20$).

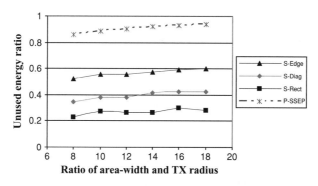

Figure 3.48 Unused energy ratio ($W = H$, $g = 20$).

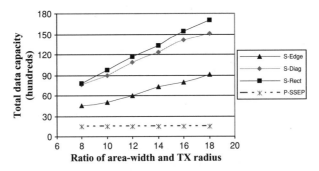

Figure 3.49 Total data capacity ($W = H$, $g = 30$).

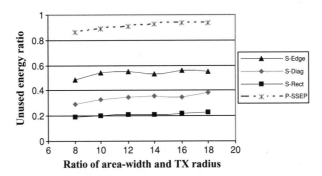

Figure 3.50 Unused energy ratio ($W = H$, $g = 30$).

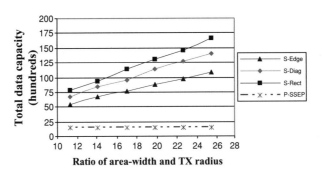

Figure 3.51 Total data capacity ($W = 2H$, $g = 30$).

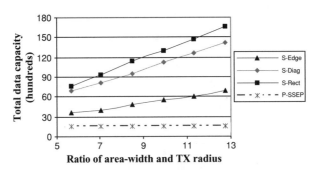

Figure 3.52 Total data capacity ($W = H/2$, $g = 30$).

Our final observation is related to the SSEP-NU model. In Section 3.5.5, the simulation results of the SSEP-NU model are given in Figures 3.39 to 3.44. According to the simulations shown in Section 3.5.5 and this section, we use the same amount of total initial energy and network configurations for all sample networks, so that performance comparison between the SSEP-NU distribution strategy and the PSL in the diagonal and rectangular movement pattern is meaningful. An interesting observation is: For networks with $W = H$ and $W = 2H$, the SSEP-NU distribution strategy and the PSL in rectangular movement pattern generate almost the same amount of total data capacity. For networks with $W = H/2$, the data capacity of the SSEP-NU distribution strategy is much smaller than the PSL with rectangular movement pattern.

The advantage of using the SSEP-NU distribution strategy is that it does not require any additional support, such as positioning and synchronizing signals from the GPS or a moving device to carry the sink. However, its disadvantages are in the form of difficulty in sensor deployment or a need for a large number of sensor types. On the other hand, the PSL protocol is easy and flexible to implement, but it requires expensive devices, such as GPS.

3.5.17 Concluding Remarks

We have defined the SSEP model for static WSNETs. Three performance metrics, namely total network data capacity, transmission energy efficiency, and unused energy ratio, have been defined and discussed for the SSEP model. Due to constrained energy of sensors, unlike in mobile ad hoc networks, the data capacity of WSNETs is limited. When the sink node is static, the energy of sensors in the first critical region is constrained, which leads to the situation where the maximum amount of data received by the sink is even smaller. To transmit this maximum amount of data to the sink, if a perfect routing algorithm is used, there is a great amount of energy left unused after the lifetime of the network is over. For large WSNETs with the SSEP model, this energy waste can be up to 90% of the total initial energy.

It is difficult to greatly increase the maximum data capacity and reduce the unused energy ratio by using efficient routing and MAC algorithms for the SSEP model. However, improvement can be achieved easily by using a nonuniform energy distribution strategy or a mobile sink. Therefore, the SSEP-NU nonuniform energy distribution model has been discussed and a new routing protocol, called predictable sink location (PSL) routing, has been proposed. According to the simulation results, both the SSEP-NU model and the PSL protocol can improve the total data capacity by one order of magnitude of the original SSEP model.

Data capacity is an important design objective in wireless sensor networks. In the static sink model, efficiency of the MAC and routing algorithms becomes of secondary importance. To increase the upper bound of the maximum data capacity, sink mobility, and network deployment such as sink placement, distributions of node positions, and distributions of node energy are options that network designer may consider.

REFERENCES

1. I. F. Akyildiz, W. Su, Y. Sankarasubramaniam, and E. Cayirci, "A survey on sensor networks," *IEEE Communications Magazine*, vol. 40, no. 8, pp. 102–114, Aug. 2002.
2. S. Basagni, I. Chlamtac, V. R. Syrotiuk, and B. A. Woodward, "A distance routing effect algorithm for mobility (DREAM)," in *International Conference on Mobile Computing and Networking*, Dallas, 1998, pp. 76–84.

3. Y. Ko and N. Vaidya, "Location-aided routing (lar) in mobile ad hoc networks," *Wireless Networks*, vol. 6, pp. 307–321, July 2000.

4. S. Cabuk, N. Malhotra, L. Lin, S. Bagchi, and N. Shroff, "Analysis and evaluation of topological and application characteristics of unreliable mobile wireless ad-hoc network," in *Proceedings of the Tenth Pacific Rim Dependable Computing Conference, March, 2004 (PRDC 04)*, Papeete, Tahiti, Mar. 2004.

5. A. Tsirigos and Z. J. Haas, "Multipath routing in the presence of frequent topological changes," *IEEE Communications Magazine*, vol. 39, pp. 132–138, Nov. 2001.

6. A. LaMarca, D. Koizumi, M. Lease, S. Sigurdsson, G. Barriello, W. Brunette, K. Sikorski, and D. Fox, "Making sensor networks practical with robots," Technical Report, IRS-TR-02-004, Intel Research, 2002.

7. G. T. Sibley, M. H. Rahimi, and G. S. Sukhatme, "Robomote: A tiny mobile robot platform for large-scale ad-hoc sensor networks," in *International Conference on Robotics and Automation*, Arlington, VA, 2002, pp. 1143–1148.

8. P. E. Rybski, N. P. Papanikolopoulos, S. A. Stoeter, D. G. Krantz, K. B. Yesin, M. Gini, R. Voyles, D. F. Hougen, B. Nelson, and M. D. Erickson, "Enlisting rangers and scouts for reconnaissance and surveillance," *IEEE Robotics and Automation Magazine*, vol. 7, no. 4, pp. 14–24, Dec. 2000.

9. D. P. Miller, T. S. Hunt, and M. J. Roman, "Experiments and analysis of the role of solar power in limiting mars rover range," in *IEEE/RSJ International Conference on Intelligent Robots and Systems*, Las Vegas, 2003, pp. 317–322.

10. L. Cai and Y.-H. Lu, "Dynamic power management using data buffers," in *Design Automation and Test in Europe*, Paris, 2004, pp. 526–531.

11. L. Cai and Y.-H. Lu, "Dynamic power management using data buffers," in *Design Automation and Test in Europe*, 2004.

12. C. Im, H. Kim, and S. Ha, "Dynamic voltage scheduling technique for low-power multimedia applications using buffers," in *International Symposium on Low Power Electronics and Design*, Huntington Beach, CA, 2001, pp. 34–39.

13. Y.-H. Lu, L. Benini, and G. D. Micheli, "Dynamic frequency scaling with buffer insertion for mixed workloads," *IEEE Transactions on Computer-Aided Design of Integrated Circuits and Systems*, vol. 21, no. 11, pp. 1284–1305, Nov. 2002.

14. Q. Qiu, Q. Wu, and M. Pedram, "Dynamic power management in a mobile multimedia system with guaranteed quality-of-service," in *Design Automation Conference*, Las Vegas, 2001, pp. 834–839.

15. W. Heinzelman, A. Chandrakasan, and H. Balakrishnan, "Energy-efficient communication protocol for wireless microsensor networks," in *Thirty-Third International Conference on System Sciences (HICSS '00)*, Maui, Hawaii, Jan. 2000.

16. S. Lindsey, C. Raghavendra, and K. M. Sivalingam, "Data gathering algorithms in sensor networks using energy metrics," *IEEE Transactions on Parallel and Distributed Systems*, vol. 13, pp. 924–935, 2002.

17. Y. Mei, Y.-H. Lu, C. G. Lee, and Y. C. Hu, "Energy-efficient motion planning for mobile robots," in *International Conference on Robotics and Automation 2004 (ICRA '04)*, New Orleans, 2004.

18. C. Inc., "Mpr/mib mote hardware users manual," http://www.xbow.com/Support/Support_pdf_files/MPR-MIB_Series_User_Manual_7430002105_A.pdf.

19. USC/ISI, "The network simulator—ns-2," http://www.isi.edu/nsnam/ns/.

20. D. Estrin, M. Srivastava, and A. Sayeed, "Wireless sensor networks," Tutorial T5, MOBICOM, Atlanta, 2002.

21. NASA, "Aerosonde unmanned aerial vehicle, manufactured by aerosonde robotic aircraft limited," http://uav.wff.nasa.gov/UAVDetail.cfm?RecordID=Aerosonde, June 2005.

22. G. Reshko, M. Mason, and I. Nourbakhsh, "Rapid prototyping of small robots," Technical Report CMU-RI-TR-02-11, Robotics Institute, Carnegie Mellon University, Pittsburgh, Mar. 2002.

23. I. F. Akyildiz, W. Su, Y. Sankarasubramaniam, and E. Cayirci, "A survey on sensor networks," *IEEE Communications Magazine*, vol. 40, no. 8, pp. 102–114, 2002.

24. G. J. Pottie and W. J. Kaiser, "Wireless integrated network sensors," *Communications of the ACM*, vol. 43, no. 5, pp. 51–58, May 2000.

25. J. M. Kahn, R. H. Katz, and K. S. J. Pister, "Next century challenges, mobile networking for smart dust," in *Proceeding of MOBICOM*, Seattle, Mar. 1999, pp. 271–278.

26. C. Intanagonwiwat, R. Govindan, and D. Estrin, "Directed diffusion: A scalable and robust communication," paper presented at ACM SIGMOBILE, Sixth Annual International Conference on Mobile Computing and Networking (MobiCOM '00), Boston, Aug. 2000.

27. T. Clouqueur, V. Phipatanasuphorn, P. Ramanathan, and K. K. Saluja, "Sensor deployment strategy for target detection," paper presented at the First ACM International Workshop on Wireless Sensor Networks and Applications, Atlanta, 2002.

28. S. Meguerdichian, F. Koushanfar, M. Potkonjak, and M. B. Srivastava, "Coverage problems in wireless ad-hoc sensor network," in *IEEE INFOCOM'01*, Anchorage, AK, Apr. 2001.

29. S. Meguerdichian, F. Koushanfar, G. Qu, and M. Potkonjak, "Exposure in wireless ad-hoc sensor networks," in *MOBICOM*, Rome, 2001.

30. A. Howard, M. J. Mataric, and G. S. Sukhatme, "Mobile sensor networks deployment using potential fields: A distributed, scalable solution to the area coverage problem," paper presented at the *Sixth International Symposium on Distributed Autonomous Robotics Systems*, Fukuoka, Japan, June 2002.

31. G. Wang, G. Cao, and T. L. Porta, "Movement-assisted sensor deployment," in *IEEE INFOCOM*, Hong Kong, Mar. 2004.

32. G. Wang, G. Cao, and T. L. Porta, "A bidding protocol for deploying mobile sensors," in *IEEE International Conference on Network Protocols (ICNP)*, Atlanta, Nov. 2003, pp. 315–324.

33. G. Sibley, M. Rahimi, and G. Sukhatme, "Robomote: A tiny mobile robot platform for large-scale ad-hoc senosr networks," *Proceedings of the International Conference on Robotics and Automation*, Washington, DC, Sept. 2002.

34. Y. Zou and K. Chakrabarty, "Sensor deployment and target localization based on virtual forces," in *INFOCOM*, San Francisco, 2003.

35. Z. Butler and D. Rus, "Event-based motion control for mobile sensor networks," *IEEE Pervasive Computing*, vol. 2, no. 4, pp. 34–43, Oct.–Dec. 2003.

36. P. Eugster, P. Felber, R. Guerraoui, and A. Kermarrec, "The many faces of publish/subscribe," *ACM Computing Surveys*, vol. 35, no. 2, pp. 114–131, June 2003.

37. F. Ye, H. Luo, J. Cheng, S. Lu, and L. Zhang, "A two-tier data dissemination model for large-scale wireless sensor networks," in *ACM MOBICOM*, Atlanta, GA, Sept. 2002, pp. 148–159.

38. S. Cheung, M. Ammar, and M. Ahamad, "The grid protocol: A high performance scheme for maintaining replicated data," *IEEE Transactions on Knowledge and Data Engineering*, vol. 4, no. 6, pp. 582–592, June 1992.

39. H. Garcia and D. Barbara, "How to assign votes in a distributed system," *Journal of ACM*, vol. 32, no. 4, pp. 841–860, May 1985.

40. G. Cao and M. Singhal, "A delay-optimal quorum-based mutual exclusion algorithm for distributed systems," *IEEE Transactions on Parallel and Distributed Systems*, vol. 12, no. 12, pp. 1256–1268, Dec. 2001.

41. K. Kar, M. Kodialam, T. Lakshman, and L. Tassiulas, "Routing for network capacity maximization in energy-constrained ad-hoc networks," in *IEEE infocom*, San Francisco, 2003.

42. S. Singh, M. Woo, and C. S. Raghavendra, "Power-aware routing in mobile ad hoc networks," in *Proceedings of the Fourth Annual International Conference on Mobile Computing and Networking*, Dallas, 1998.

43. Q. Li, J. Aslam, and D. Rus, "Online power-aware routing in wireless ad-hoc networks," in *Proceedings of the Seventh Annual International Conference on Mobile Computing and Networking*, Rome, 2001.

44. S. Chessa and P. Santi, "Comparison based system-level fault diagnosis in ad-hoc networks," in *Proceedings of the Twentieth IEEE Symposium on Reliable Distributed Systems*, New Orleans, Oct. 2001.

45. C.-F. Hsin and M. Liu, "A distributed monitoring mechanism for wireless sensor networks," in *Proceedings of the ACM Workshop on Wireless Security*, Sept. 2002, pp. 57–66.

46. J. Staddon, D. Balfanz, and G. Durfee, "Efficient tracing of failed nodes in sensor networks," in *Proceedings of the First ACM International Workshop on Wireless Sensor Networks and Applications*, Atlanta, Sept. 2002, pp. 122–130.

47. L. Li, M. Thottan, B. Yao, and S. Paul, "Distributed network monitoring with bounded link utilization in IP networks," in *Proceedings of IEEE Infocom*, San Francisco, Mar. 2003.

48. M. Thottan, L. Li, B. Yao, V. S. Mirrokni, and S. Paul, "Distributed network monitoring for evolving IP networks," in *Proceedings of International Conference on Distributed Computing Systems*, Tokyo, Mar. 2004.

49. C.-K. Toh, *Ad Hoc Mobile Wireless Networks*, Upper Saddle River, NJ: Prentice-Hall, 2002.

50. J. Kahn, R. Katz, and K. Pister, "Mobile networking for smart dust," in *Proceedings of the ACM/IEEE International Conference on Mobile Computing and Networking*, Seattle, 1999.

51. G. Kesidis and R. Rao, "Mobility management of ad-hoc sensor networks using distributed annealing," Technical Report CSE03-017, CSF Department, Sept. 2003.

52. G. Kesidis and E. Wong, "Optimal acceptance probability for simulated annealing," *Stochastics and Stochastics Reports*, vol. 29, pp. 221–226, 1990.

53. M. Mauve, J. Widmer, and H. Hartenstein, "A survey on position-based routing in mobile ad-hoc networks," *IEEE Network Magazine*, vol. 15, no. 6, pp. 30–39, Nov. 2001.

54. M. Gunes, U. Sorges, and I. Bouazizi, "ARA—the ant-colony based routing algorithm for MANETs," in *Proceedings of International Conference on Parallel Processing Workshops (ICPPW)*, Vancouver, Aug. 2002.

55. M. Roth and S. Wicker, "Termite: Emergent ad-hoc networking," in *Proceedings of the Second Mediterranean Workshop on Ad-Hoc Networks*, Mahdia, Tunisia, June 2003.

56. J. Baras and H. Mehta, "A probabilistic emergent routing algorithm for mobile ad hoc networks," in *Workshop on Modeling and Optimization in Mobile, Ad Hoc and Wireless Networks*, Sophia-Antipolis, France, Mar. 2003.

57. R. Muraleedharan and L. Osadciw, "Balancing the performance of a sensor network using an ant system," in *Thirty-Seventh Annual Conference on Information Sciences and Systems*, Baltimore, Mar. 2003.

58. M. Mauve, J. Widmer, and H. Hartenstein, "A survey on position-based routing in mobile ad hoc networks," *IEEE Network Magazine*, vol. 15, no. 6, pp. 30–39, Nov. 2001.

59. S. Das, C. Perkins, and E. Royer, "Performance comparison of two on-demand routing protocols for ad hoc networks," in *IEEE Infocom*, 2000, pp. 3–12.

60. D. Johnson and D. Maltz, "Dynamic source routing in ad hoc wireless networks," *Mobile Computing*, Kluwer, 1996, pp. 153–181.

61. D. J. Stilwell and B. E. Bishop, "Platoons of underwater vehicles: Communication, feedback, and decentralized control," *IEEE Control Systems Magazine*, vol. 20, no. 6, Dec. 2000.

62. J. A. Fax and R. M. Murray, "Information flow and cooperative control of vehicle formations," *IEEE Transactions on Automatic Control*, vol. 49, no. 9, pp. 1465–1476, 2004.

63. H. G. T. A. Jadbabaie and G. J. Pappas, "Flocking in fixed and switching networks," *Automatica*, July 2003, submitted for publication.

64. R. O. Saber and R. M. Murray, "Agreement problems in networks with directed graphs and switching topologies," in *IEEE Conference on Decision and Control*, Maui, Hawaii, 2003.

65. C. Reynolds, "Flocks, herds, and schools: A distributed behavioral model," in *SIGGRAPH*, 1987.

66. S. Wang and E. J. Davison, "On the stabilization of decentralized control systems," *IEEE Transactions on Automatic Control*, vol. 18, pp. 473–478, Oct. 1973.

67. A. Saberi, A. A. Stoorvogel, and P. Sannuti, "Decentralized control with input saturation," in *American Control Conference*, Boston, July 2004.

68. C. Godsil and G. Royle, *Algebraic Graph Theory*, New York: Springer-Verlag, 2000.

69. D. E. Chang, S. C. Shadden, J. E. Marsden, and R. Olfati-Saber, "Collision avoidance for multiple agent systems," in *Proceedings of the Forty-Second IEEE Conference on Decision and Control*, Maui, HI, Dec. 2003.

70. W. R. Heinzelman, J. Kulik, and H. Balakrishnan, "Adaptive protocols for information dissemination in wireless sensor networks," in *Proceedings of the ACM MobiCom'99*, Seattle, WA, 1999, pp. 174–185.

71. W. Ye, J. Heidemann, and D. Estrin, "An energy-efficient mac protocol for wireless sensor networks," in *Proceedings of IEEE INFOCOM'02*, New York, 2002, pp. 1567–1576.

72. K. Sohrabi, J. Gao, V. Ailawadhi, and G. Pottie, "Protocols for self-organization of a wireless sensor network," *IEEE Personal Communications*, vol. 7, no. 5, pp. 16–27, 2000.

73. L. Li and J. Y. Halpern, "Minimum-energy mobile wireless networks," in *IEEE International Conference on Communications ICC'01*, Helsinki, Finland, June 2001.

74. A. Woo and D. Culler, "A transmission control scheme for media access in sensor networks," in *Proceedings of ACM MobiCom'01*, Rome, Italy, 2001, pp. 221–235.

75. K. Kalpakis, K. Dasgupta, and P. Namjoshi, "Maximum lifetime data gathering and aggregation in wireless sensor networks," in *Proceedings of IEEE International Conference on Networking*, 2002.

76. B. Krishnamachari, D. Estrin, and S. Wicker, "Modelling data-centric routing in wireless sensor networks," in *Proceedings of IEEE Infocom*, New York, 2002.

77. P. Gupta and P. R. Kumar, "The capacity of wireless networks," *IEEE Transactions on Information Theory*, vol. 46, no. 2, pp. 388–404, Mar. 2000.

78. D. Marco, E. J. Duarte-Melo, M. Liu, and D. L. Neuhoff, "On the many-to-one transport capacity of a dense wireless sensor network and the compressibility of its data," in *Information Processing in Sensor Networks: Second International Workshop*, Palo Alto, CA, April 22–23, 2003.

79. P. Erdös and A. Renyi, "On random graphs I," *Publications Mathematicae*, pp. 290–297, Dec. 1959.

80. H. Dewitt, "The theory random graphs with application to the probabilistic analysis of optimization algorithms," Ph.D. dissertation, Department of Computer Science, UCLA, Los Angeles, 1977.

81. L. Kleinrock and J. Silvester, "Optimum transmission radii for packet radio networks or why six is a magic number," in *Conference Record, National Telecommunications Conference*, Birmingham, AL, 1978, pp. 4.3.2–4.3.5.

82. M. Zorzi and R. R. Rao, "Geographic random forwarding (geraf) for ad hoc and sensor networks: Multihop performance," *IEEE Transactions on Mobile Computing*, vol. 2, no. 4, pp. 337–348, 2003.

83. E. Shih, S. Cho, N. Ickes, R. Min, A. Sinha, A. Wang, and A. Chandrakasan, "Physical layer driven protocol and algorithm design for energy efficient wireless sensor networks," in *Proceedings of ACM MobiCom*, Rome, pp. 272–286.

4

LOWER LAYER ISSUES—MAC, SCHEDULING, AND TRANSMISSION

4.1 INTRODUCTION

In the sensor networks considered in this book, most, if not all, of the communication takes place over wireless media. This provides great flexibility for sensor network deployment because no preexisting communications infrastructure is required before the sensor network is deployed. Many of the hard challenges in wireless sensor networks exist at the lowest layers of the communications protocol stack because it is these layers that are closest to the wireless link and, hence, are responsible for hiding the characteristics of the wireless media from the higher level protocols as much as possible.

Wireless links are inherently shared. The sharing can take place, however, in several dimensions. First, signals on different links can be made orthogonal to each other, that is, close to noninterfering. This can be done by using different frequency bands on each link or using different codes as in code division multiple access (CDMA) systems. These systems require some preconfiguration or real-time signaling to determine the mode of operation of each link. The benefit of these systems is that the performance of each link is independent of the others if the system is operating within its normal conditions. These systems are usually interference limited, that is, can only support a bounded number of users before the signals no longer behave as orthogonal to each other and hence start to interfere and cause communication to degrade.

Second, signals on different links can be shared in time using time division multiple access (TDMA), that is, each user has sole access to a link for a period of time during which no other node will transmit. These systems require some schedule of transmission to be established either before a network is deployed or in real time. If done in real time, the schedule can be established for the duration of a data transfer or for each burst of data. By dividing the time unequally among the nodes sharing the link, QoS differentiation can be provided. For example, sensors transmitting low-resolution data infrequently are given fewer time slots than a node that is transmitting high-resolution video. These systems are limited in capacity by the amount of time they can allocate to each user.

Third, media access can be contention based. In these systems when a node desires to use a link, it must compete with other nodes for the right to transmit. There are many protocols defined to resolve these contentions. These systems have the benefit of being very simple: They are completely distributed and do not require much control information to be passed between nodes. They typically work very well at low loads but struggle at high loads to maintain high link utilizations and provide fairness among nodes.

These low-layer protocols have a profound impact on the overall network. Depending on scheduling, QoS service can be provided. Depending on access technique, network efficiency is impacted. Depending on the protocols defined for resolving contention, power consumption of the nodes is affected. Typically, the design of these protocols is very specific to support a particular application and air interface technology.

In this chapter we present three discussions. In section 4.2, by Kulkarni and Arumugam, a time division system is presented and analyzed. The protocol supports three modes of operation: broadcast, convergecast, and local gossip. The analysis shows that this protocol is superior to contention based-systems. In Section 4.3, by Groves and Fowler, a specific low-layer protocol defined by a standard called 802.15.4 is present and analyzed. This system is designed for low-power personal area networks, and the discussion shows how protocols can be designed to achieve high efficiency for specific uses. In Section 4.4, by Ci, Sharif, Nuli, and Raviraj, an adaptive link layer protocol is presented and analyzed. The purpose of this protocol is to provide flexibility in its operation so that it may be used to support multiple traffic types on a single network.

4.2 SS-TDMA: A SELF-STABILIZING MEDIUM ACCESS CONTROL (MAC) FOR SENSOR NETWORKS

Sandeep S. Kulkarni and Mahesh Arumugam

We focus on the problem of designing a Time Division Multiple Access (TDMA) service for a grid-based sensor network. Such networks are readily found in many applications in the area of monitoring, hazard detection, and so on. We consider three communication patterns, broadcast, convergecast, and local gossip, that occur frequently in these systems. We develop TDMA service that can be customized based on the application requirements and also provide guidance about using this service when the communication pattern is unknown or varies with time. With these customizations, whenever a sensor receives a message, it can forward it to its successors with a small delay. We show that this TDMA service is collision free, whereas existing Carrier Sense Multiple Access (CSMA)-based approaches suffer significant collisions. We also show how this service can be extended to deal with other deployments in a two-dimensional (2D) field, failure of sensors, and sensors that are sleeping as part of a power management scheme. Further, we show that this service can be used in a mobile sensor network that provides localization service.

4.2.1 Background Information

Sensor networks are becoming popular nowadays due to their applications in both academic and industrial environments. Currently, sensor networks are highly useful in applications such as data gathering, active/passive monitoring and tracking of undesirable objects [1], and unattended hazard detection [2–4]. Further, the sensors can be rapidly deployed in the field due to their small size and low cost. However, these sensors are often resource constrained. Specifically, the sensors are constrained by limited power, limited communication distance, and limited sensing capability. Therefore, they need to collaborate with each other to perform a particular task.

Sensor networks can be classified based on the nature of deployment of sensors. One approach for such deployment is random where sensors are distributed with some expected density distribution. Another approach for such deployment is systematic geometric distribution. The results in this section are targeted toward such systematic geometric distribution where sensors are deployed in a (rectangular, hexagonal, or triangular) grid.

One such application that uses such systematic distribution and requires collaborative effort is the DARPA NEST technology demonstration project "A line in the Sand" (LITeS) [1], where a wireless sensor network is used to detect, classify, and track intruders along the deployed field. In this demonstration, sensors are arranged in a thick line (grid). When an intruder (e.g., a person, a soldier, a car, or a heavy vehicle such as a tank) comes into the vicinity of this line or crosses this line, the sensors detect it. Now, to classify the intruder, the sensors that observed the intruder communicate with each other. These sensors have two options for classification: internal or external. In an internal classification, the sensors that detect the intruder communicate with each other. And, in an external classification, the sensors send their observed values to a base station that exfiltrates the data outside the network.

Message Collision and Communication Patterns   One of the important challenges in applications such as LITeS is message collision. Specifically, if a sensor receives two messages simultaneously, then they collide and both messages become incomprehensible. Also, it is difficult for a sensor to know whether a given message reached all its neighbors. This is due to the fact that a message sent by a sensor may collide at one neighbor and be received correctly at another neighbor.

Another important challenge is to deal with the existence of different communication patterns. We consider three commonly occurring communication patterns: *broadcast*, *convergecast*, and *local gossip*. In broadcast, a message is sent to all the sensors in the network. Broadcast is useful when a base station wants to transmit some information (e.g., program capsules for reprogramming the sensors [5, 6], diffusion message to revalidate the time slots allotted to sensors, etc.) to all the sensors in the network. We also consider two other communication patterns, convergecast and local gossip. These communication patterns are based on our experience with LITeS demonstration [1]. Based on the internal classification technique mentioned earlier, we consider the communication pattern, local gossip, where a sensor sends a message to its neighboring sensors within some distance. And, based on the external classification technique mentioned earlier, we consider the communication pattern, convergecast, where a group of sensors send a message to a particular sensor such as the base station.

We present our TDMA service[1] for sensor networks. Our TDMA service ensures collision freedom and fair bandwidth allocation among different sensors. Further, our TDMA service lets one customize the assignment of time slots to different sensors by considering the common communication patterns that occur in the application.

---

[1] A preliminary version of this work appears in [7].

Improper Initializations and State Corruption   In a large sensor network, it is possible that some of the sensors are improperly initialized. Further, if the slots assigned to the sensors are corrupted or the clock skew is higher than expected in some interval, then collisions occur during communication. Hence, starting from such arbitrary states, the TDMA service should be able to recover to states from where the service satisfies its specification. In other words, it is necessary that the TDMA service be fault-tolerant [8,9], that is, starting from an arbitrary state, the system should recover to states from which subsequent communication is collision-free.

Results   We concentrate on designing collision-free communication algorithms for sensor networks. We compare the performance of our TDMA algorithm with collision avoidance protocols applicable in sensor networks. The main results of this section are as follows:

- We present self-stabilizing TDMA (SS-TDMA) service that can be customized for different communication patterns, namely, broadcast, convergecast, and local gossip. Starting from an arbitrary state, SS-TDMA recovers to states from which collision-free communication among sensors is restored. Furthermore, we present simulation results to validate that SS-TDMA is collision-free. We compare SS-TDMA with collision avoidance protocols such as CSMA, and show that they suffer significant number of collisions.
- SS-TDMA can be used in several deployment scenarios. We first consider deployment in a rectangular grid. Then, we extend it to hexagonal/triangular grid. Finally, we show how it can be extended to any geometric distribution provided localization service [10,11] is available.
- We show how slots are assigned to sensors so that the delay in broadcast, convergecast, and local gossip is reduced. Also, under reasonable assumptions, we prove that SS-TDMA is delay-optimal for broadcast. Further, we show how SS-TDMA is used in the context of power management.
- We show that SS-TDMA is fault-tolerant. More specifically, we show that it is possible to reclaim the slots assigned to failed sensors and reassign them to active sensors. Furthermore, we show that SS-TDMA can tolerate errors in the location, that is, even if the sensors are moved slightly from their ideal location, the percentage of collisions is within application requirements. Additionally, we show that mobility is supported in SS-TDMA if localization service is available.
- We outline the middleware architecture of SS-TDMA. Toward this end, we present how our algorithms are implemented as a middleware service. We identify the application programming interfaces (APIs) of our service.

4.2.2 Model and Assumptions

In this section, we present the sensor network model and state our assumptions. The assumptions are in three categories: existence of base station (or exfiltration point), deployment topology of the network, and sensor radio capabilities.

Base Station   We assume that there exists a base station that initiates a diffusing computation to assign initial slots to all sensors. Typically, the base station is more powerful compared to other sensors in the network. Additionally, it has long-range wireless network capability (e.g., IEEE 802.11). If there are multiple base stations in the network, they elect a leader among themselves. Note that this process is independent of the sensor network since the base stations have more powerful radios and use a different wireless network protocol. The elected base station is responsible for initiating the diffusing computation to assign initial TDMA slots to all sensors. Furthermore, the base station could be co-located with another sensor in the grid, or it could be assigned a separate grid position. For simplicity, we assume that the base station is located at the left-top position [at location $\langle 0, 0 \rangle$]. However, it can be placed in any grid position. The extension for this case is straightforward and, hence, is omitted.

Topology Initially, we assume that sensors are arranged in a grid where each sensor knows its location in the network (geometric position). Each message sent by a sensor includes this geometric position. Thus, a sensor can determine the position, direction, and distance (with respect to itself) of the sensors that send messages to it. Initially, for simplicity, we assume that the sensor network has a perfect grid topology and no sensors have failed or are in a sleeping state. By making these assumptions, we can design algorithms for perfect grid-based sensor networks. Then, we extend the algorithms to deal with the case where sensors (other than the base station) have failed. (Note that the assumption about nonfailure of base station is acceptable; base station is responsible for exfiltrating data from the network. Hence, if it fails, another base station must be available to utilize the sensor network.)

Initially, we assume a rectangular grid topology. Later, we extend the algorithms for hexagonal and triangular grids. We will also show that our algorithm works for the case where sensors are randomly deployed in a geometric distribution. Finally, we also show that our algorithm works even if the sensor nodes are mobile. The last two extensions require localization service [10, 11] so that the sensors can identify their location in the field.

Communication and Interference Ranges We assume that each sensor has a communication range and an interference range. Communication range is the distance up to which a sensor can communicate with certainty/high probability. Interference range is the distance up to which a sensor can communicate, although the probability of such a communication may be low. This assumption is based on the ability of sensors to communicate with each other in a geometric topology or the existence of some sensors that have larger communication range due to higher power levels or elevation in the sensor plane. Hence, to introduce uniformity in the communication capabilities of the sensors, we consider interference range. In such a scenario, one possible way to estimate the interference range is to take the maximum value of the communication ranges of all sensors. (Note that the communication range that is used in this case should be the minimum of the communication ranges of all sensors.)

Now, based on these assumptions, given two sensors, j and k, if k is in the interference range of j but k is not in the communication range of j, then k receives messages sent by j with a low probability. However, if k receives another message while j is sending a message, it is possible that collision between these two messages can prevent k from receiving either of those messages. Based on the definition of the interference range, it follows that it is at least equal to the communication range. Also, we assume that the sensors are aware of their communication range and interference range. Initially, we assume communication range $= 1$, that is, a sensor can only talk to its neighbors in the grid. Then, we extend the algorithms to deal with other communication ranges.

Time Division Multiple Access
Assign time slots to each sensor such that,
If two sensors j and k transmit at the same time, then
($j \notin$ collision group of sensor k).

Figure 4.1 Problem statement of TDMA.

Remark We only consider collisions where at least one of the messages was expected to be received at the destination. In other words, when j and k send a message that collide at l, then that collision is counted only if l is in the communication range of either j or k. If l is outside the communication range of both j and k, l was not expected to receive either of the two messages. Hence, such a collision is not counted.

4.2.3 TDMA Service for Sensor Networks

In this section, we present time division multiple access (TDMA) algorithms for sensor networks. Time division multiplexing is the problem of assigning time slots to each sensor. Two sensors j and k can transmit in the same time slot if j does not interfere with the communication of k and k does not interfere with the communication of j. In this context, we define the notion of a collision group. The *collision group* of sensor j includes the sensors that are in the communication range of j and the sensors that interfere with the sensors in the communication range of j. Hence, if two sensors j and k are alloted the same time slot, then j should not be present in the collision group of k and k should not be present in the collision group of j. Thus, the problem of time division multiple access is defined in Figure 4.1.

Now, we present the algorithm for allotting time slots to the sensors. We also present the TDMA algorithm for broadcast, the TDMA algorithm for convergecast, and the algorithm for local gossip.

TDMA Service for Broadcast In this section, we present our TDMA algorithm for broadcast. Consider a grid network where a sensor can communicate with sensors that are distance away and interfere with sensors that are distance y, $y \geq 1$ away (cf. Fig. 4.2, where $y = 2$). To present the TDMA algorithm, we first present an algorithm for initial slot assignment. Then, we present an algorithm for subsequent slot assignments.

Initial Slot Assignment Let us assume that the base station starts a diffusing computation to assign initial slots at time slot 0. From Figure 4.2, we observe that sensors $\langle 1, 0 \rangle$ and $\langle 0, 1 \rangle$ will receive the diffusion message. Now, both sensors $\langle 1, 0 \rangle$ and $\langle 0, 1 \rangle$ should not forward

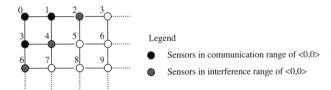

Figure 4.2 Initial slot assignment for broadcast where communication range = 1 and interference range = 2. The number associated with each sensor denotes the slot at which it should forward the diffusion message.

---

const $P_b = (y + 1)^2 + 1;$
// Initial slot assignment for broadcast
When sensor j receives a diffusion message from k
  if (k is west neighbor at distance 1)
    transmit after 1 slot.
  else if (k is north neighbor at distance 1)
    transmit after $(y + 1)$ slots.
  else // duplicate message
    ignore

// TDMA algorithm for broadcast
If sensor j transmits a diffusion message at time slot t_j,
  j can transmit at time slots, $\forall c : c \geq 0 : t_j + cP_b$.

Figure 4.3   TDMA algorithm for broadcast.

the diffusion message at the same time, as their message would collide at sensor $\langle 1, 1 \rangle$. Hence, sensor $\langle 1, 0 \rangle$ is allowed to forward the message at slot 1. Sensor $\langle 2, 0 \rangle$ is allowed to forward the message at slot 2. We note that sensors $\langle 2, 0 \rangle$ and $\langle 0, 1 \rangle$ should not transmit at the same time as their messages will collide at $\langle 1, 1 \rangle$ and $\langle 2, 1 \rangle$. Hence, sensor $\langle 0, 1 \rangle$ is allowed to forward the message at slot $(y + 1)$ (i.e., slot 3 in Fig. 4.2). In general, if a sensor receives a diffusion message from its west neighbor, it forwards after 1 slot, and if it receives the message from its north neighbor, it forwards after $(y + 1)$ slots. In short, if the base station initiates the diffusion at slot 0, sensor $\langle i, j \rangle$ will forward the message at slot $i + (y + 1)j$.

TDMA Slots   Once the initial slot (the slot in which a sensor forwards the diffusion message) is determined, a sensor determines its future slots using the TDMA period. Let j and k be two sensors such that j is in the collision group of k. Let t_j (respectively, t_k) be the initial slot of j (respectively, k). We propose an algorithm where the slots assigned for j are $t_j + c \times$ MCG where $c \geq 0$ and MCG captures information about the maximum collision group in the system.

From the initial slot assignment algorithm, we know that $t_j \neq t_k$. Now, future messages sent by j and k can collide if $t_j + c_1 \times$ MCG $= t_k + c_2 \times$ MCG, where $c_1, c_2 > 0$. In other words, future messages from j and k can collide iff $|t_j - t_k|$ is a multiple of MCG. More specifically, to ensure collision freedom, it suffices that for any two sensors j and k such that j is in the collision group of k, MCG does not divide $|t_j - t_k|$. We can achieve this by choosing MCG to be $\max(|t_j - t_k| : j$ is in the collision group of $k) + 1$.

If j is in the collision group of k, then $|t_j - t_k|$ is at most $(y + 1)^2$; such a situation occurs if j is at distance of $y + 1$ in north/south of k. Hence, the TDMA period is $(y + 1)^2 + 1$. The algorithm for assigning TDMA slots is shown in Figure 4.3.

The algorithm assigns time slots for each sensor based on the time at which it transmits the diffusion message. Thus, a sensor (say, j) can transmit in slots: $t_j, t_j + ((y + 1)^2 + 1)$, $t_j + 2((y + 1)^2 + 1), \ldots$, etc. Figure 4.4 shows a sample allocation of slots to the sensors.

Theorem 4.2.1   The initial slot assignment for broadcast in the above algorithm is collision-free.

PROOF   Let us assume that the source sensor $\langle 0, 0 \rangle$ (i.e., base station) starts transmitting at slot 0. By induction, we observe that sensor $\langle i, j \rangle$ will transmit at time slot $t = i + (y + 1)j$.

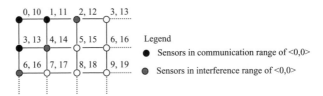

Figure 4.4   TDMA slot assignment for broadcast where communication range $= 1$ and interference range $= 2$. The numbers associated with each sensor denote the slots at which it can send a message.

Now, we show that collisions will not occur in this algorithm. Consider two sensors $\langle i_1, j_1 \rangle$ and $\langle i_2, j_2 \rangle$. Sensor $\langle i_1, j_1 \rangle$ will transmit at time slot $t_1 = i_1 + (y + 1)j_1$ and $\langle i_2, j_2 \rangle$ will transmit at time slot $t_2 = i_2 + (y + 1)j_2$. Collision is possible only if the following conditions hold:

- $t_1 = t_2$, that is, $(i_1 - i_2) + (y + 1)(j_1 - j_2) = 0$.
- The Manhattan distance between $\langle i_1, j_1 \rangle$ and $\langle i_2, j_2 \rangle$ is less than or equal to $(y + 1)$, that is, $|i_1 - i_2| + |j_1 - j_2| \leq y + 1$.
- $\langle i_1, j_1 \rangle$ and $\langle i_2, j_2 \rangle$ are distinct, that is, $|i_1 - i_2| + |j_1 - j_2| \geq 1$.

From the first condition, we conclude that $(i_1 - i_2)$ is a multiple of $(y + 1)$. Combining this with the second condition, we have $|i_1 - i_2| = 0$ or $|j_1 - j_2| = 0$. However, if $|i_1 - i_2| = 0$ (respectively, $|j_1 - j_2| = 0$), then from the first condition $(j_1 - j_2)$ [respectively, $(i_1 - i_2)$] must be zero. If both $(i_1 - i_2)$ and $(j_1 - j_2)$ are zero, then the third condition is violated. Thus, collision cannot occur in this algorithm.   ∎

Theorem 4.2.2   The above algorithm satisfies the problem specification of TDMA.

PROOF   Consider two distinct sensors j and k such that j is in the collision group of k. The time slots assigned to j and k are $t_j + cP_b$ and $t_k + cP_b$, respectively, where c is an integer and $P_b = (y + 1)^2 + 1$ is the period between successive slots. Suppose a collision occurs when j and k transmit a message at slot $t_j + c_1P_b$ and $t_k + c_2P_b$, respectively, where $c_1, c_2 > 0$. In other words, $t_j + c_1P_b = t_k + c_2P_b$. From Theorem 4.2.1, we know that $t_j \neq t_k$. Therefore, collision will occur iff $|t_j - t_k|$ is a multiple of P_b. However, since j is in the collision group of k, $|t_j - t_k|$ is at most $(y + 1)^2$ (less than P_b). In other words, $|t_j - t_k| \leq (y + 1)^2 < P_b$. Hence, if j and k transmit at the same time, then they are not present in the collision group of each other. This is a contradiction. Thus, collisions cannot occur in this algorithm.   ∎

Remark   In our algorithms, we have used Manhattan distance in the calculation of interference range. If we consider geometric distance instead, our algorithms can be extended appropriately by using a larger interference range that accommodates all sensors where interference may occur.

TDMA Service for Convergecast   Suppose sensor $\langle 1, 1 \rangle$ sends a message to the base station at slot 14 (cf. Fig. 4.4). Sensors $\langle 0, 1 \rangle$ and $\langle 1, 0 \rangle$ receive the message. However, these sensors have just missed their slots (11 and 13, respectively) and hence, need to wait until

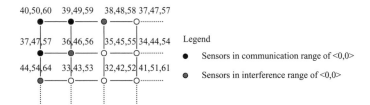

Figure 4.5   TDMA slot assignment for convergecast where communication range = 1 and interference range = 2. The numbers associated with each sensor denote the slots at which it can send a message. Some initial slots are not shown.

the next slot before forwarding the message to the base station. Therefore, the algorithm in Section 4.2.3 introduces a significant delay for convergecast, where a group of sensors send data (e.g., information about the activities of an intruder in the field [1]) to the base station. Hence, we customize the TDMA algorithm in Figure 4.3 for convergecast.

To reduce the delay for convergecast, we change the slot assignment as follows: If j receives a message from its left neighbor, then it chooses to transmit the diffusion in (-1)th slot (in circular sense). In other words, j transmits in the $(P-1)$th slot, where P $(=(y+1)^2+1)$ is the interval between slots assigned to a sensor and y is the interference range of the sensors. If j receives a message from its top neighbor, then it transmits in the $(-(y+1))$th slot. (For example, see Figure 4.5 for slot assignment for the case where $y=2$.) After the first slot is determined, the sensors can then transmit once in every P slots.

As we can see from Figure 4.5, with the above slot assignment, delay for convergecast is reduced. Specifically, when a sensor transmits a message that is to be relayed by sensors closer to the base station (left-top sensor), such a relay introduces only a small (respectively, no) delay. Thus, the TDMA algorithm customized for convergecast is shown in Figure 4.6.

Theorem 4.2.3   The above algorithm satisfies the problem specification of TDMA.

PROOF   The proof is similar to Theorem 4.2.2.   ∎

```
const P_c = (y + 1)² + 1;
// Initial slot assignment for convergecast
When sensor j receives a diffusion message from k
    if (k is west neighbor at distance 1)
        transmit in the P_c + (−1)th  slot.
    else if (k is north neighbor at distance 1)
        transmit in the P_c + (−(y + 1))th slot.
    else // duplicate message
        ignore


// TDMA algorithm for convergecast
If sensor j transmits a diffusion message at time slot t_j,
    j can transmit at time slots, ∀ c : c ≥ 0 : t_j + cP_c.
```

Figure 4.6   TDMA algorithm for convergecast.

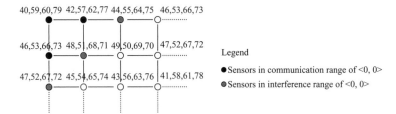

Figure 4.7   TDMA slot assignment for gossip where communication range = 1 and interference range = 2. The numbers associated with each sensor denote the slots at which it can send a message. Some initial slots are not shown.

TDMA Service for Local Gossip   For local gossip, the communication is in all directions. Hence, we need an approach that combines the slot assignment for broadcast and convergecast. We proceed as follows: We increase the value of the period (P) to $2((y + 1)^2 + 1)$, twice the previous value. With this increased value, each sensor gets two slots (even and odd) in this period. Let the slots assigned to the base station be 0 and $P - 1$. To simplify the presentation, let us assume that the base station starts a diffusion in its even or the 0th slot. When j receives the diffusion from its left neighbor, it chooses the slot that is 2 higher than that used by the left neighbor. Likewise, when j receives the diffusion from its top neighbor, it chooses the slot that is $2(y + 1)$ higher than that used by the top neighbor. (For example, see Fig. 4.7 for slot assignment for the case where $y = 2$.) Note that the diffusion messages are forwarded in the even slots. In our solution for gossip, whenever sensor k transmits in the even slot, say t_k, it can also transmit in $((P - 1) - t_k)$ mod P, the odd slot. Thus, the TDMA algorithm customized for local gossip is shown in Figure 4.8.

Theorem 4.2.4   The above algorithm satisfies the problem specification of TDMA.

PROOF   The proof is similar to Theorems 4.2.2 and 4.2.3. Note that in the gossip algorithm, even slots behave like the broadcast algorithm and odd slots behave like the convergecast algorithm.   ∎

---

const $P_g = 2((y + 1)^2 + 1)$;
// Initial slot assignment for local gossip
When sensor j receives a diffusion message from k
  if (k is west neighbor at distance 1)
    transmit after 2 slots.
  else if (k is north neighbor at distance 1)
    transmit after $2(y + 1)$ slots.
  else // duplicate message
    ignore

// TDMA algorithm for local gossip
If sensor j transmits a diffusion message at time slot t_j,
  j can transmit at time slots,
    $\forall c : c \geq 0 : t_j + cP_g$,
      $(((P_g - 1) - t_j) \bmod P) + cP_g$.

Figure 4.8   TDMA algorithm for local gossip.

Based on Figure 4.7, in the case where TDMA is customized for local gossip, the interval between two successive slots of a sensor can be twice as much as in the case where TDMA is customized for broadcast/convergecast. Thus, if a sensor needs to transmit a message, then the worst-case delay is larger when the TDMA service is customized for local gossip. In spite of this deficiency, the TDMA service provides substantial benefits for broadcast and convergecast even if it is customized for local gossip. To see this, observe that once the base station sends the broadcast message in its even slot, any sensor receiving it can forward it with a small delay (cf. Fig. 4.7). Likewise, if a sensor transmits a convergecast message in the odd slot, any sensor receiving it can forward it with a small delay. In fact, as seen from Figure 4.7, in the TDMA service customized for local gossip, if any sensor wants to transmit a message in any given direction (east, west, north, south, southeast, southwest, northeast, or northwest), then any sensor that receives that message can forward it with a small delay.

Based on the above discussion, if the most common communication pattern is known to be broadcast or convergecast, we can customize the TDMA service accordingly. Even if the communication pattern is unknown or varies with time, customizing the TDMA service for local gossip provides a significant benefit for other communication patterns.

4.2.4 SS-TDMA: Properties

In this section, we present some of the properties of our algorithms. First, we show how stabilization can be added to the algorithms in Section 4.2.3. Then, we show that under certain conditions, the delay in delivering a broadcast message using the algorithm in Section 4.2.3 is optimal. Finally, we show that the algorithms proposed in Section 4.2.3 can be used in the context of power management.

Stabilization and Reliability We now add stabilization to the TDMA algorithms discussed in Section 4.2.3, that is, if the network is initialized with arbitrary clock values (including the case where there is a phase offset among clocks), we ensure that it recovers to states from where collision-free communication is achieved. The TDMA algorithm in Section 4.2.3 relies on the initial slot assignment algorithm. We modify that algorithm to obtain self-stabilizing TDMA (SS-TDMA). Specifically, in SS-TDMA, the base station periodically sends a diffusion message in a slot it believes to be its TDMA slot (according to the algorithm in Section 4.2.3). Whenever a sensor receives the diffusion message, it recalculates its TDMA slot based on the appropriate algorithm in Section 4.2.3. Then, it forwards the diffusion message to that slot.

If the clock values are corrupted, then some sensors may not receive the diffusion message. To deal with this case, in SS-TDMA, whenever a sensor does not get the diffusion message for certain consecutive number of times, the sensor shuts down, that is, it will not transmit any message until it receives a diffusion message from a sensor closer to the base station. The network will eventually reach a state where the diffusion message can be received by all sensors. From then on, the sensors can use the TDMA algorithm in Section 4.2.3 to transmit messages. Based on the above description, we observe that if there are no faults in the network and the links are reliable, then no sensor will ever shut down. Moreover, if faults perturb clock values, then eventually they will be restored so that subsequent communication is collision-free. Thus, we have the following theorem.

Theorem 4.2.5   Starting from arbitrary initial states, SS-TDMA recovers to states from where collision-free communication among sensors is restored.

Remark   We do not specify the parameters such as the period between successive diffusing computations, the number of diffusions a sensor waits before shutting down, and so forth. This choice depends on how frequently we want to perform validation of slots to account for clock drifts, acceptable overhead when no clock drifts occur, and acceptable time for recovery. Since we use a time synchronization service [12] along with SS-TDMA, the clock drift among sensors is very small. Further, the overhead incurred by the time synchronization service is very low (one beacon every 15 s). Hence, the period between successive diffusing computations to revalidate the time slots could be higher. Thus, the frequency of diffusing computations is expected to be very low. However, we do not consider the issue of optimizing these values based on the requirements of the application. This issue is orthogonal to the service proposed in this section.

Dealing with Unreliable Links   So far, we have assumed that if a sensor sends a message, then in the absence of collisions it would be correctly received. However, in a sensor network, the message could be lost due to other environmental factors. In SS-TDMA, such failures are already tolerated. However, in such cases, some sensors may shut down incorrectly. The probability of such shut downs can be reduced. Toward this end, let p be the probability that a sensor receives a message from its neighbor. Also, let n be the number of diffusion periods a sensor waits before shutting down. Now, consider a sensor j that receives a diffusion message after l intermediate transmissions. The probability that this sensor does not receive the diffusion message is $1 - p^l$, and the probability that this sensor shuts down in the absence of faults is $(1 - p^l)^n$. Note that this is an overestimate since a sensor receives the diffusion message from more than one sensor. If we consider $p = 0.90, l = 10$, and $n = 10$, the probability that a sensor that is 10 hops away from the base station will incorrectly shut down is 0.0137. Thus, we have the following theorem.

Theorem 4.2.6   If p is the probability of a successful communication over a link, n is the number of diffusion periods a sensor waits before shutting down, then the probability that a sensor l hops away from the base station shuts down incorrectly is at most $(1 - p^l)^n$.

Corollary 4.2.1   If there are no faults in the network and the links are reliable, then no sensor will ever shut down.

PROOF   If the links are reliable, then $p = 1$. Hence, from Theorem 4.2.6, this theorem follows.   ■

Delay Optimality   In this section, we prove that, under certain assumptions, the broadcast algorithm is optimal and our algorithm reduces the delay in delivering a message to its intended receivers. Toward this end, we prove Theorems 4.2.7 and 4.2.8 next.

Theorem 4.2.7   A broadcast where (1) communication range and interference range of the sensors is 1, (2) every sensor must transmit at least once, (3) if $i_1 \leq i_2$ and $j_1 \leq j_2$ then the slot used by sensor $\langle i_1, j_1 \rangle$ to transmit the broadcast message should be less that or equal to the slot used by sensor $\langle i_2, j_2 \rangle$, and (4) no collisions should occur, will take at least $3(n - 1) + 1$ slots in a $n \times n$ grid where the initiator of broadcast is at one corner.

Figure 4.9 Broadcast in a 2 × 2 grid.

PROOF   We prove this by induction. For the base case, consider a 2 × 2 grid as shown in Figure 4.9. Suppose sensor $\langle 0, 0 \rangle$ starts the broadcast at slot 0. Sensors $\langle 1, 0 \rangle$ and $\langle 0, 1 \rangle$ receive the message. Now, both these sensors cannot transmit in the next slot since there will be a collision at $\langle 1, 1 \rangle$. Hence, without loss of generality, let sensor $\langle 1, 0 \rangle$ transmit the broadcast at slot 1. Now, either sensors $\langle 0, 1 \rangle$ or $\langle 1, 1 \rangle$ can transmit next. However, sensor $\langle 0, 1 \rangle$ should transmit the message before sensor $\langle 1, 1 \rangle$. Hence, sensor $\langle 0, 1 \rangle$ transmits at slot 2 and sensor $\langle 1, 1 \rangle$ transmit at slot 3. Hence, the broadcast takes 4 slots. Thus, the theorem holds for 2 × 2 grid.

For the inductive case, let us assume that the theorem holds for $n \times n$ grid, that is, the broadcast takes $3(n - 1) + 1$ slots. Now, we prove that in a $(n + 1) \times (n + 1)$ grid, the broadcast takes $3n + 1$ slots.

From the induction hypothesis, we know that the last (bottom-right) sensor (i.e., $\langle n - 1, n - 1 \rangle$) in the $n \times n$ grid transmits the broadcast message at slot $3(n - 1)$. Hence, sensor $\langle n, n - 1 \rangle$ to the right in $(n + 1) \times (n + 1)$ grid, transmits the message at slot $3(n - 1) + 1$ (cf. Fig. 4.10). Similar to the argument in 2 × 2 grid, sensor $\langle n - 1, n \rangle$ should transmit the message before sensor $\langle n, n \rangle$. Therefore, sensor $\langle n - 1, n \rangle$ transmits at slot $3(n - 1) + 2$ and sensor $\langle n, n \rangle$ transmits at slot $3(n - 1) + 3 = 3n$. Hence, the broadcast takes at least $3n + 1$ slots. Thus, the theorem holds for $(n + 1) \times (n + 1)$ grid.   ■

Theorem 4.2.8   For communication range and interference range of 1, SS-TDMA requires $3(n - 1) + 1$ slots for broadcast in a $n \times n$ grid.

Thus, under the assumptions of Theorem 4.2.7, SS-TDMA customized for broadcast pattern is delay-optimal. Further, in the broadcast algorithm presented in Section 4.2.3, whenever a sensor sends a message, the sensors farther away from the base station receive the message just before their allotted time slots. Hence, they can transmit the message with a small delay. Similarly, for other communication patterns and interference ranges, a sensor can forward the message to others with a small delay.

Energy Efficiency   In this section, we discuss the energy efficiency of SS-TDMA. Energy-efficient algorithms are important in sensor networks due to the inherent power constraints of the sensors. Energy efficiency can be achieved in SS-TDMA as follows. A

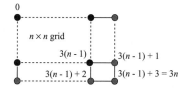

Figure 4.10   Broadcast in a $(n + 1) \times (n + 1)$ grid.

sensor remains in active mode only in its alloted time slots (if it needs to send any data) and in the alloted time slots of the sensors within its communication range. In the remaining slots, the sensor can save energy by turning off its radio and remaining in idle mode. Suppose the communication range of a sensor is 1, then a sensor will have at most 4 neighbors. Let P be the period between successive time slots alloted to a sensor. A sensor will have to be in active mode in its alloted time slot and in the alloted time slot of its 4 neighbors, during every period. In other words, a sensor will have to be in active mode in 5 slots each period. For interference range, $y = 2$, period $P = (y + 1)^2 + 1 = 10$, a sensor will have to be active in 5 slots in every 10 slots, that is, 50% of the time. In general, if the communication range is 1, then a sensor needs to be awake for at most 5 slots in every $(y + 1)^2 + 1$ slots.

We note that more optimizations are possible in the above scheme. In SS-TDMA, whenever a sensor has some message to send, it will send the message at the start of its alloted time slot. Hence, neighboring sensors can decide whether they should continue listening in that time slot. If a sensor does not receive any message in the first part of the time slot, it can turn its radio off. Further, depending on the communication pattern, a sensor can choose to listen only to a subset of its neighbors. For example, in broadcast, a sensor always gets a message from the sensors that are closer to the base station. Hence, sensors can choose to listen only to the slots assigned to their neighbors that are closer to the base station. Thus, energy efficiency can be achieved in SS-TDMA.

4.2.5 SS-TDMA: Simulation and Implementation Results

We have implemented SS-TDMA for different communication patterns on MICA motes [2, 3]. Further, we have simulated our algorithms in Prowler [13], which allows one to simulate arbitrarily large number of sensors (especially MICA motes). Here, we present the middleware architecture of SS-TDMA for MICA motes, the simulation model, and the simulation results.

SS-TDMA: Middleware Architecture   SS-TDMA service includes APIs for initialization, send, and receive. We discuss each of these APIs and their internal details, next.

Initialization   As discussed in Section 4.2.3, one of the parameters to the service is the interference range used by sensors. We assume that the interference range of all sensors is identical. For initialization, SS-TDMA assumes that once the sensor network is deployed, there is a delay before the application begins. This delay is used to perform a diffusing computation and to assign initial slots. Additionally, the diffusion is performed periodically to (re)validate the slots and to deal with clock drift among sensors. Thus, one of the parameters to the service is the period after which diffusion is used to (re)validate the slots that sensors need to use for TDMA.

Yet another parameter for SS-TDMA is the time slot (in physical time) that should be assigned for sending a message. We choose the slot time so that it is larger than the time required to send a message of maximum length (including preamble, Cyclic Redundancy Check (CRC), etc.). SS-TDMA also takes the parameter that identifies the communication pattern for which the service should be customized. The application can use this parameter to customize the communication that occurs most frequently. As discussed in Section 4.2.3, if the commonly occurring communication pattern is not known, then customizing SS-TDMA for local gossip is beneficial.

Send   Although SS-TDMA ensures that when two sensors, say j and k, transmit simultaneously, neighbors of j (respectively, k) receive messages from j (respectively, k) without collision, we still use CSMA. Thus, if j is about to transmit in its TDMA slot and it observes that the channel is busy, then j backs off until the next TDMA slot. Although in our simulations and in experiments with a small number of motes such a back off never occurred, it is expected that it may occur in a larger experimental setup. We expect that using CSMA in addition to TDMA will reduce the collisions that may occur due to unsynchronized clocks, larger interference range than that used by SS-TDMA, or interference range that varies with time or other environment factors.

The send is nonblocking. Hence, if SS-TDMA receives more than two messages and the sum of their lengths is less than the maximum message length, we combine those messages and send them in the next time slot.

Receive   There are no special tasks performed when SS-TDMA receives a message. All received messages are forwarded to upper layer. Additionally, if the received message includes multiple embedded messages, then the receive action separates them.

Simulation Model   In this section, we discuss the simulation model of the experiments. We use a probabilistic wireless network simulator, prowler [13], that is a simulation environment for embedded systems especially for MICA motes [2, 3]. The simulator has a modular design. Each layer of the system architecture is designed as a separate module.

Using prowler, one can prototype different sensor network applications, communication models, propagation models, and topology. For our TDMA simulations, we use the radio/communication model that is based on the algorithms in Section 4.2.3. Using this model, we implement the notion of communication range and interference range. To compare our algorithms with the existing implementation, we use the default radio models provided by prowler. These models include CSMA and a primitive model that uses no access protocol. Finally, we use the rectangular grid as the underlying topology since it reflects the topology used in LITeS [1]. Specifically, in LITeS, the sensors are arranged in a rectangular grid and the base station is placed at one corner in this grid.

Now, we discuss the simulations we performed in the context of these communication patterns based on the experiences with LITeS. Then, we discuss the simulations we performed to study the effect of location errors. SS-TDMA groups up to four messages in the queue into a single message before transmitting.

Broadcast   The base station (sensor at left-top corner) initiates a broadcast. It sends the broadcast message to its neighbors in the communication range. Whenever a sensor receives the broadcast message for the first time, it relays it (for sensors farther from the base station). We conduct the broadcast simulations for different network sizes. In these simulations, we consider the following metrics: maximum delay incurred in receiving the broadcast message, number of sensors that receive the broadcast message, and number of collisions. Since CSMA (respectively, no MAC layer) does not guarantee reception by all sensors, we also consider the delay when a certain percentage of sensors receive the broadcast message. Regarding collisions, we compute the ratio of the number of collisions to the number of messages.

Convergecast   For convergecast, a set of sensors send a message to the base station (approximately) at the same time. In our experiments, we keep the network size fixed at

10×10. We choose a subgrid of varying size; sensors in this subgrid transmit the data to the base station. We assume that the subgrid that sends the data to the base station is in the opposite corner from the base station, that is, the subgrid is farthest from the base station. For these simulations, we compute maximum delay incurred for receiving messages at the base station, the percentage of sensors whose messages are received by the base station and the number of collisions. Once again, as in broadcast simulations, we compute the delay for the case where a certain percentage of messages are received by the base station.

Local Gossip   In local gossip, a subgrid of sensors send the data. The goal is to transmit the data from these sensors to the sensors in the subgrid and the neighbors of the sensors in the subgrid. Thus, local gossip is applicable in *locally* determining the set of sensors that observed a particular event. In our experiments, we keep the network size fixed at 10×10. We choose different sizes of subgrids. For these simulations, we compute the average delay incurred for receiving messages at the sensors that are expected to receive the local gossip and number of collisions.

Location Errors   An important concern for a communication protocol is the errors in sensor location. Errors are introduced in sensor location due to misplacement of sensors or external factors such as wind, vehicle movement, and the like. Communication protocols that depend on sensor location should be able to tolerate this kind of error.

   To model location errors, we randomly perturbed the sensors. In our simulation, the error in sensor location is determined using a normal distribution, $N(\mu, \sigma)$, where μ is the mean error distance and σ is the standard deviation of the error. The direction of perturbation was randomly selected from $0°$ to $360°$. To ensure that the grid remains connected in spite of perturbations, during these simulations we increased the communication range. We note that this is a reasonable assumption in that if we need to tolerate location errors, then the ideal distance between two neighboring sensors should be smaller than the communication range.

Simulation Results   In this section, we present our simulation results that compare the use of SS-TDMA with the case where CSMA is used and with the case where no MAC layer is used. The results presented in this section are the mean of three experiments. Also, we have shown the variance in the graphs, whenever it is more than 5% of the mean.

Broadcast   In Figure 4.11, we present our simulation results for broadcast for the case where communication range and interference range is 1. Figure 4.11*a* identifies the number of collisions that occur in different algorithms. As expected, SS-TDMA is collision-free for all network sizes. By contrast, in CSMA, about 10% of messages suffer from collisions.

   Figure 4.2.11*b* identifies the maximum delay incurred in receiving broadcast messages. Since all sensors may not receive the broadcast message when CSMA is used, we consider the delay when a certain percentage, 80 to 100%, of sensors receive the broadcast message. As we can see, the delay in SS-TDMA is only slightly higher.

   Figure 4.2.11*c* identifies the number of sensors that receive the broadcast message. We find that with SS-TDMA/CSMA, all sensors receive the message. However, without the MAC layer, the number of sensors that receive the message is less than 50%.

   In Figure 4.12, we present the simulation results for broadcast for the case where communication range is 1 and interference range is 2. As we can see, these results are similar to those in Figure 4.11.

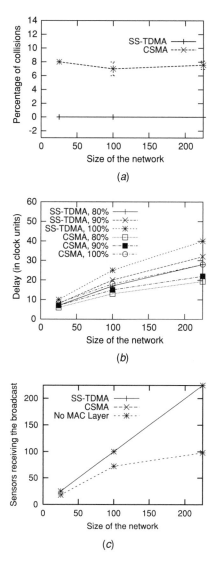

Figure 4.11 Results for broadcast with communication range = 1 and interference range = 1. With SS-TDMA/CSMA, all sensors receive the broadcast, and hence the graphs for them are identical.

Convergecast In Figure 4.13, we present our simulation results for convergecast for the case where communication range and interference range are 1. Figure 4.13a identifies the number of collisions that occur in different algorithms. As we can see from Figure 4.13a, although SS-TDMA is collision-free, there are a significant number of collisions with CSMA. Regarding delay, as we can see from Figures 4.13b and 4.13c, the delay incurred by SS-TDMA is reasonable and the base station receives all the messages sent by the sensors. By contrast, with CSMA, approximately 50% of the messages are received when the number of sensors sending the data to the base station increases. The delay incurred for 75% of the messages to reach the base station is infinity when the field size is greater than 4. This is

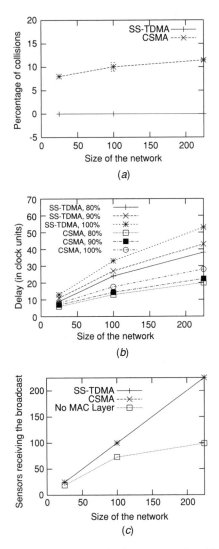

Figure 4.12 Results for broadcast with communication range = 1 and interference range = 2. With SS-TDMA/CSMA, all sensors receive the broadcast, and hence, the graphs for them are identical.

represented in the graph (cf. Fig. 4.13*b*) by an arrow that goes vertically outside the graph. And, without the MAC layer, no data reaches the base station (cf. Fig. 4.13*c*).

In Figure 4.14, we present the simulation results for convergecast for the case where communication range is 1 and interference range is 2. As we can see, these results are similar to those in Figure 4.13.

Local Gossip In Figures 4.15*a* and 4.15*b*, we present our simulation results for local gossip for the case where communication range is 1 and interference range is 1. Figure 4.15*a* identifies the number of collisions as the size of the group performing local gossip increases. As we can see, CSMA-based solutions suffer significant collisions whereas SS-TDMA is collision-free. Also, as seen from Fig. 4.15*b*, the delay in SS-TDMA is somewhat more

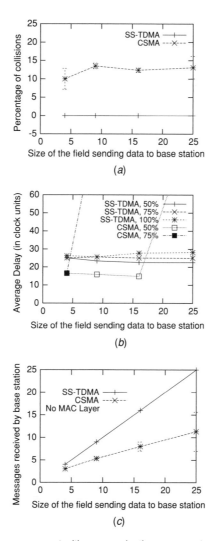

Figure 4.13   Results for convergecast with communication range $= 1$ and interference range $= 1$.

than that in CSMA. However, unlike SS-TDMA where all sensors receive the necessary messages, in CSMA, the sensors receive approximately 50% of messages. Once again, the results are similar for the case where interference range is increased to 2 (cf. Fig. 4.15c and 4.15d).

Effect of Location Errors   In our location error experiments, even if the sensors are perturbed from their ideal position, as long as the perturbation is small and the communication range is increased so that the network remains connected, the results are close to those presented earlier. We introduce location errors in the sensors as follows. Let $\langle a, b \rangle$ be the ideal location of a sensor. Let e_d be the distance a sensor is perturbed from its ideal location, and θ_d be the angle of perturbation. The error distance e_d is determined using the normal distribution $N(\mu, \sigma)$, where μ is the mean error distance and σ is the

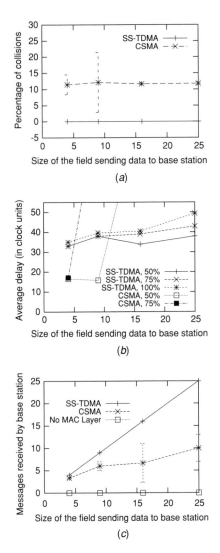

Figure 4.14   Results for convergecast with communication range = 1 and interference range = 2.

standard deviation of e_d. Thus, the error in location on 95% of the sensors is in the range $(-\mu - 2\sigma, \mu + 2\sigma)$. Hence, to determine the topology, we increase the *physical communication range* by $\mu + 2\sigma$. However, the algorithm uses 1 for both the communication and interference ranges. For small perturbations (i.e., $\mu \leq 0.2$), increasing the physical communication range is sufficient to ensure that the network is connected. However, for larger perturbations (i.e, $\mu > 0.2$), if the communication and interference ranges used by the algorithm are 1, the number of collisions increase significantly. Hence, we need to increase the interference range that the algorithm uses, say, to 2. Additionally, if the predicted μ is less than the actual mean error, the algorithm can increase its interference range when it observes significantly higher number of collisions using the approach to change the collision-group size (cf. Section 4.2.6).

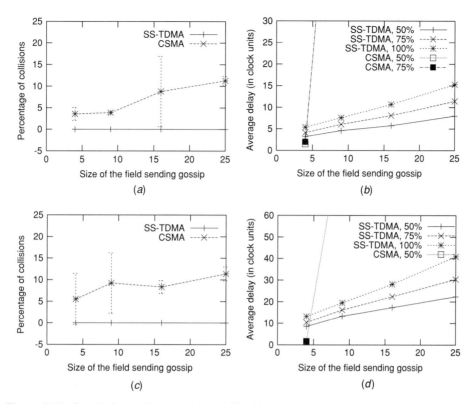

Figure 4.15 Results for local gossip: (a) and (b) with communication range = 1 and interference range = 1, (c) and (d) with communication range = 1 and interference range = 2.

In our simulations, μ takes the following values: 0.0 to 0.4; and σ takes the following values: 0.0 to 0.2. And θ_d is determined using the uniform distribution $U(0, 2\pi)$. Thus, the actual location of the sensor is $\langle a + e_d \cos(\theta_d), b + e_d \sin(\theta_d) \rangle$.

BROADCAST   In Figure 4.16, we present the simulations results for broadcast with location errors. Figure 4.16a identifies the percentage of collisions during broadcast. As we can see, when μ increases, the number of collisions increases. However, the collisions are within 2%. Figure 4.16b identifies the maximum delay involved in delivering the broadcast message to all the sensors. We can note that the delay is within 5% when compared to the case where no location errors are introduced. Finally, Figure 4.16c identifies the number of sensors receiving the broadcast message. As we can observe, all the sensors receive the message except for the case where $\mu = 0.2$ and $\sigma = 0.1$. Even in this case, more than 98% of the sensors receive the broadcast message. Table 4.1 shows the percentage of collision for the case where $\mu = 0.4$ and interference range = 2. As we can observe, the percentage of collisions is small with increased interference range.

CONVERGECAST   In Figures 4.17a and 4.17b, we present the simulation results for convergecast with location errors. Figure 4.17a identifies the number of collisions during the message communication. We note that, as the error in sensor location increases, collisions increase. Further, as observed earlier, the collisions are within acceptable limits, that is,

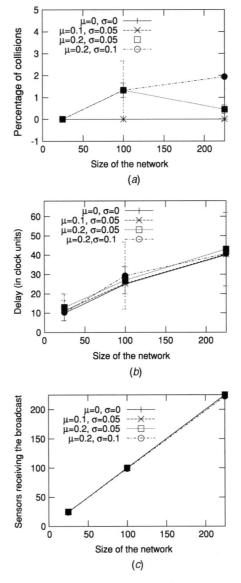

Figure 4.16   Results for broadcast with location errors.

Table 4.1   Percentage of Collisions for $\mu = 0.4$, $\sigma = 0.2$, and interference range $= 2$

	Broadcast			Convergecast		Local Gossip	
Network Size	Mean	Variance	Field Size	Mean	Variance	Mean	Variance
			4	8.46	2.26	5.15	0.21
25	0	0	9	7.64	1.97	5.35	2.35
100	5	4	16	9.55	18.36	7.82	1.67
225	6.31	2.81	25	10.30	3.17	10.07	14.18

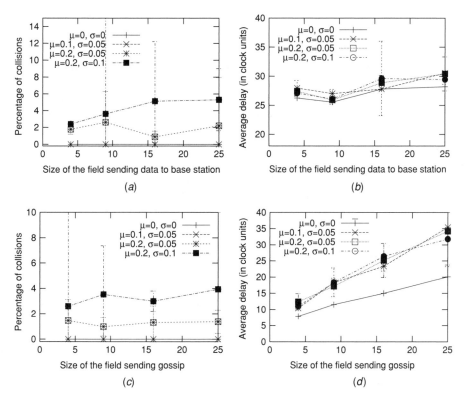

Figure 4.17 Results for convergecast and local gossip with location errors.

within 6%. Figure 4.17*b* identifies the average delay involved in delivering the convergecast messages to the base station; the average delay is within 5% when compared to the case where no location errors are introduced. Further, similar to the case where no location errors are present, the base station receives all the convergecast messages. Moreover, if the mean error distance increases, we can keep the percentage of collisions small by increasing the interference range (cf. Table 4.1).

LOCAL GOSSIP In Figures 4.17*c* and 4.17*d*, we present the simulation results for local gossip with location errors. Similar to the observations made earlier in this section, from Figure 4.17*c*, we observe that the number of collisions during message communication is small (i.e., within 4%). Further, the delay involved in delivering the local gossip messages is within 15% when compared to the case where no location errors are introduced. Finally, all the local gossip messages are delivered to the group that expects such messages. Moreover, if the mean error distance increases, we can keep the percentage of collisions small by increasing the interference range (cf. Table 4.1). From these simulations, we conclude that the location errors do not significantly affect the performance of our TDMA service.

Effect of Grouping In the proposed SS-TDMA service for sensor networks, if the service receives two or more messages and the sum of the message lengths is less than the maximum message length of a TDMA message, SS-TDMA combines these messages into a single TDMA message. In this section, we study the effect of grouping. Specifically, we study the effect of varying the number of messages grouped into a single TDMA message for convergecast and local gossip. Note that in the simulations for broadcast only one message is transmitted, and, hence, we do not consider the issue of grouping for broadcast. Further, in this section, we present results for the delay in delivering the messages. In our simulations, the base station (respectively, the group expecting the gossip messages) receives all the convergecast (respectively, gossip) messages. Also, the percentage of collisions is zero.

In Figures 4.18a and 4.18b, we present the simulation results for convergecast and local gossip where the communication range is 1 and interference range is 1. We consider the following values for grouping constant (GP): 1, 2, and 4. In Figure 4.18a, we can observe that the delay increases when the number of messages grouped into single TDMA message decreases. With GP = 1, only one message is sent in a TDMA slot. With GP = 2, one or two messages are sent in a slot. In this case, SS-TDMA service will group messages in its queue depending on the total message size. We note that the TDMA message queue will also contain messages generated by other sensors. This is due to the fact that a sensor needs to forward a message generated by other sensors in convergecast. Thus, the grouped messages may contain messages from different sensors and, hence, the delay will be different with different group sizes.

In Figure 4.18b, we present the results for local gossip. We observe that the results are similar to convergecast. Specifically, as GP decreases, delay in delivering the local gossip messages increases.

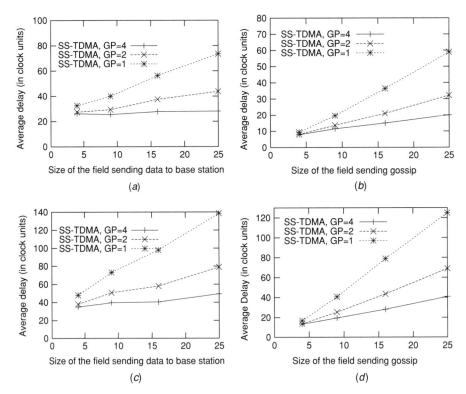

Figure 4.18   Effect of grouping constant; with communication range = 1 and interference range = 1, for (a) convergecast and (b) local gossip, and with communication range = 1 and interference range = 2, for (c) convergecast and (d) local gossip.

Once again, the results for the case where the communication range is 1 and interference range is 2 are similar (cf. Fig. 4.18c and 4.18d).

4.2.6 SS-TDMA: Extensions

So far, we had assumed that sensors are arranged in a rectangular grid as this is the topology used in LITeS [1]. In this section, we discuss some of the extensions to SS-TDMA. Specifically, we show how SS-TDMA can be extended to support larger communication ranges; we present SS-TDMA algorithms for other grid-based topologies as well as for arbitrary geometric topologies; also, we show how this approach can be used to support mobile sensor nodes; and, we show how SS-TDMA deals with failed sensors. Finally, we show how SS-TDMA can be applied to network programming of sensors [5].

SS-TDMA: Larger Communication Ranges In this section, we present the TDMA algorithm for communication range $= 2$ customized for broadcast pattern. We note that similar extensions are possible for even larger communication ranges and other communication patterns (i.e., convergecast and local gossip).

Consider a rectangular grid where a sensor can communicate with its distance 2 neighbors, that is,

$$\text{Communication range} = \text{Interference range} = 2$$

$$y = \left\lceil \frac{\text{interference range}}{\text{communication range}} \right\rceil = 1$$

(cf. Fig. 4.19). The base station is located at the left-top position ($\langle 0, 0 \rangle$) in the network. Given a sensor location $\langle a, b \rangle$, depending upon whether a, b are even or odd, we can split the network into four subgrids: even–even, even–odd, odd–even, and odd–odd. Now, the SS-TDMA algorithm from Section 4.2.3 can be used in each of these subgrids. In these subgrids, communication range = interference range = 1. Let the base station (location $\langle 0, 0 \rangle$), located in the even–even subgrid, start the diffusion to assign initial slots for broadcast pattern at slot 0. Sensors $\langle 1, 0 \rangle$, $\langle 2, 0 \rangle$, $\langle 0, 1 \rangle$, $\langle 1, 1 \rangle$, and $\langle 0, 2 \rangle$ will receive the message. Figure 4.20a shows when these sensors can forward the diffusion message.

From Figure 4.20a, we observe that sensors $\langle 2, 0 \rangle$ and $\langle 0, 2 \rangle$ are allowed to forward the diffusion message after 1 and 2 slot(s), respectively. This is similar to the algorithm in Section 4.2.3 where communication range = interference range = 1. Other sensors transmit in such a way that they do not interfere with the even–even subgrid. Specifically, sensor $\langle 1, 0 \rangle$ forwards the diffusion after $P = (y + 1)^2 + 1$ slots, that is, after 5 slots, sensor $\langle 0, 1 \rangle$ forwards the diffusion message after $2P$ slots, and sensor $\langle 1, 1 \rangle$ forwards the message after

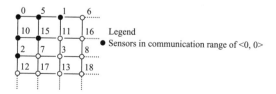

Figure 4.19 TDMA slot assignment in a network with communication range $= 2$.

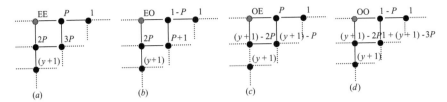

Figure 4.20   Initial slot assignment when the sender is in (a) even–even subgrid, (b) even-odd subgrid, (c) odd–even subgrid, and (d) odd–odd subgrid. The sensors shaded in gray are the senders and the sensors shaded in black are in the communication range of the respective senders.

$3P$ slots (cf. Fig. 4.19). If the sender is in a even–odd subgrid (respectively, odd–even and odd–odd subgrid), then the initial slots are assigned using the slot assignment specified in Figure 4.20b (respectively, Fig. 4.20c and 4.20d).

Once the initial slots are assigned, sensors can determine the future slots using the period between successive slots. The TDMA period for communication range $= 2$ in a rectangular grid is $4P$.

Theorem 4.2.9   The above algorithm satisfies the problem specification of TDMA.

SS-TDMA: Hexagonal Grids   Consider a hexagonal grid network where a sensor can communicate with its distance 1 neighbors and interfere with its distance 2 neighbors (cf. Fig. 4.21). Note that by simple geometry, we can show that the distance between opposite corners of a hexagon is twice as long as an edge of the hexagon. We assume that the base station is located at the leftmost corner on the left-top hexagon in the network (cf. Fig. 4.21).

From Figure 4.21, we observe that whenever the base station transmits, sensors located at the top (say, j) and bottom (say, k) of the base station at geometric distance 1 from the base station can transmit next. However, if both these sensors transmit simultaneously, then collision occurs at the base station. Hence, we proceed as follows: Whenever j receives the diffusion message from the base station, it retransmits the message after 1 slot. Likewise, whenever k receives the diffusion message from the base station, it retransmits the message after $2y$ slots, where y is the interference range of the sensors. Further, whenever a sensor receives a message from its neighbor on the straight edge (cf. Fig. 4.21), it forwards the message after 1 slot.

Once the initial slots are assigned, each sensor can determine future slots based on the time it forwards the diffusion message and the period between successive slots. For a hexagonal

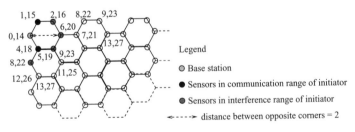

Figure 4.21   TDMA slot assignment in hexagonal-grid network where communication range $= 1$ and interference range $= 2$. The numbers associated with each sensor denote the slots at which it can send a message. Slots for some sensors are not shown.

---

const $P_{hex} = 2y(y + 1) + \lfloor \frac{y}{z} \rfloor + 1;$
// Initial slot assignment in hexagonal grids
   when sensor j receives a diffusion message from k
     if [k is at distance 1 in the same level
       (i.e., $j - k$ is a straight edge)]
       transmit after 1 slot.
     else if (k is at distance 1 in the lower level)
       transmit after 1 slot.
     else if (k is at distance 1 in the upper level)
       transmit after $2y$ slots.
     else // duplicate message
       ignore

// TDMA algorithm for broadcast in hexagonal grids
If sensor j transmits a diffusion message at time slot t_j,
   j can transmit at time slots, $\forall c : c \geq 0 : t_j + c P_{hex}.$

Figure 4.22   TDMA algorithm for hexagonal grid.

grid, the period between successive slots, $P_{hex} = 2y(y + 1) + \lfloor y/2 \rfloor + 1$ suffices. Thus, the SS-TDMA algorithm for hexagonal grids is shown in Figure 4.22.

Theorem 4.2.10   The above algorithm satisfies the problem specification of TDMA.

We note that the above algorithm is customized for broadcast. We can also customize SS-TDMA on a hexagonal grid for convergecast and local gossip. Toward this end, we need to change the initial slot assignments and the TDMA period P_{hex} similar to the modifications discussed for rectangular grids in Section 4.2.3. Specifically, for convergecast, whenever sensor j receives the diffusion message, it forwards the message in its negative slot [i.e., it forwards the message in ($P_{hex} - t_j$)th slot, where t_j is the slot in which it is supposed to forward according to the above algorithm]. Likewise, for local gossip, similar modifications can be applied.

Remark   We observe that it is possible to convert a triangular grid into a hexagonal grid. Once the hexagonal grid is obtained, SS-TDMA for the hexagonal network can be applied to allot time slots to different sensors. However, in this algorithm, the sensors within the hexagon will not get time slots. We can allow the sensors in the boundary of the hexagon (called *boundary* sensors) to share their time slots with the intermediate sensors. In order to allow boundary sensors to share their TDMA slots with the intermediate sensors, we need to increase the collision group size. In this case, the collision group of a sensor includes the sensors that are within distance $y + 2$. For more details of this extension, we refer the reader to [14].

Two-Dimensional Clustering   In this section, we discuss how SS-TDMA can be implemented in other geometric 2D deployments. This approach is based on the virtual grid idea in geographical-adaptive fidelity (GAF) algorithm from [15]. The idea behind this approach is to embed a rectangular grid on the field where the sensors are deployed.

With the help of localization service [10, 11], a sensor can determine its x, y coordinates in the field. Based on these values, the sensor can determine the square in which it is present in the grid embedded on the field. Once the sensor knows its location in the grid, it can determine its communication slots using the algorithm presented in Section 4.2.3. To ensure that the sensors in neighboring squares in the grid can communicate with each other, the distance between the sensors in the neighboring squares should be less than or equal to the communication range of the sensors. If two sensors fall on the longest diagonal between the neighboring squares, then the distance between these two sensors is $\sqrt{5}r$, where r is the length of the square. Hence, we need to ensure that the length of the square, $r \leq$ communication range$/\sqrt{5}$ in order to allow these two sensors to communicate. With this embedding, we now obtain a 2D grid where some sensors have failed; such a scenario will occur if there is no sensor that falls in the given $r \times r$ square. To deal with this problem we can use the extension identified later in this section that handles sensor failures.

In this scheme, however, two or more sensors may fall into the same square. Hence, we need other mechanisms that allows these sensors to collaborate among themselves on deciding how to share the communication slots assigned to that square. Since the number of sensors that fall in the same square are expected to be small, a simple protocol can be easily designed to decide how these sensors will share their TDMA slots. However, this issue is outside the scope of this section.

Remark   In most applications, if more than one sensor is present in a square, these sensors provide redundant information. Hence, it is preferable that only one sensor is active at any instant of time. Periodically, the active sensor can delegate its role to other sensors in the square. Remaining sensors can thus turn off their radio and conserve energy.

Supporting Mobile Sensor Nodes   SS-TDMA can be applied to mobile sensor networks provided localization service [10, 11] is available. In the presence of mobile sensor nodes, time slots used by a sensor keeps on changing due to its motion. Hence, a sensor needs to know its current position. In other words, SS-TDMA can be used in mobile sensor networks if localization service is available.

Dealing with Failed Sensors   In this section, we focus on providing TDMA service in the presence of failed/sleeping sensors. We assume that the base station does not fail and that the network remains connected.

In SS-TDMA, a sensor normally receives the diffusion message for the first time from a sensor that is closer to the base station. In the presence of failed/sleeping sensors, a sensor may receive the diffusion message for the first time from the sensor that is (physically) farther away from the base station. SS-TDMA ignores such messages as duplicates. However, in the presence of failed/sleeping sensors, a sensor should forward such a diffusion message. This ensures that the diffusion message reaches all the active sensors. The algorithm for the initial slot assignment is given in Figure 4.23.

This modification, however, also assigns slots to failed/sleeping sensors. Hence, if the number of such failed/sleeping sensors is large, bandwidth is wasted. To improve the bandwidth utilization, we consider the problem of reducing the collision group size to deal with the sensors that are sleeping or have failed. Our solution involves three tasks: (1) allowing each sensor to determine its collision group, (2) computing the maximum collision group size (MCG) in the network, and (3) communicating the MCG to all sensors.

> **when** sensor j receives a diffusion message for the first time from sensor k
> update local clock;
> determine the ideal diffusion slot;
> find the TDMA slots using the appropriate algorithm from Section 4.2.3;
> transmit in ideal diffusion slot or next TDMA slot, whichever is earlier.

Figure 4.23   Initial slot assignment while dealing with failed sensors.

Determining Collision Group of Each Sensor   Regarding the first part, if a sensor, say j, plans to be inactive for a long time, it should inform the sensors in its collision group before it becomes inactive. This can be achieved as follows: When j wants to become inactive, it informs its neighbors (in a slot assigned by SS-TDMA). These neighbors, in turn, inform their neighbors until the information reaches all sensors in the collision group of j. Alternatively, if j fails (or becomes inactive without informing its neighbors), its neighbors can detect this fact by observing that no communication was received in the slot allotted to j. This information, in turn, will be communicated to the sensors in the collision group of j. A sensor, say k, updates its collision group to max (collision group of $k, \forall i : |t_i - t_k|$) where sensor i is in the collision group of k and t_i is the time slot at which sensor i transmits a diffusion message.

Computing the MCG   Regarding the second part, we use the initial slot assignment algorithm for broadcast (cf. Section 4.2.3) where the communication range is 1 and the interference range can be greater than 1. Once again, for simplicity, we assume that left-top sensor ($\langle 0, 0 \rangle$) communicates the size of the collision group when it initiates the diffusion. Whenever a sensor, say j, propagates the diffusion, it sets the collision group to max (collision group included in a message that was received by j, collision group of j). It follows that the sensor in the right-bottom corner will be able to obtain the MCG in the network. This MCG can then be communicated to the left-top sensor using the current collision-free SS-TDMA algorithm. If the right-bottom sensor has failed, this responsibility can be delegated to other sensors.

Communicating the MCG   Finally, regarding the third part, once the left-top sensor learns the new collision group, it can include this when it initiates the next diffusion. This diffusion will allow the sensors to learn the size of the new collision group that will then be used by SS-TDMA.

We would like to note that the above description is intended to show that it is possible to change the size of the collision group to ensure optimal bandwidth utilization. The parameters involved in changing the collision group are the frequency with which sensors update their collision group and the frequency with which the sensor(s) at the right-bottom communicate the group change information. Also, it is possible to accelerate the change using the distributed reset [16, 17]. However, this issue is outside the scope of this section.

Improving Bandwidth Utilization Further   Bandwidth utilization in SS-TDMA can be improved further by the following techniques. First, the time slot interval used in SS-TDMA can be increased. Now, time slot is divided into two phases; listen phase and send phase. Whenever a sensor wants to send data in its assigned time slot, it will start transmitting at the beginning of the listen phase. With this modification, a sensor that requires more bandwidth will listen in the slots assigned to other sensors in its collision group. If j notices that medium is free in the listen phase of a time slot (assigned to a sensor, say, k), j can use the send phase of this slot to send its data. However, collisions may occur with this scheme, since two or more sensors in the collision group of k can try to access the medium in the send phase simultaneously. Note that the send phase is long enough to transmit a message, and the listen phase is small enough so that other sensors can detect whether the medium is free or not.

Second, some of the slots can be left unassigned. With this modification, whenever a sensor (say, j) requires more bandwidth, it can choose to send the data in the slots not used by any sensor in its collision group. Again, no guarantees can be made about this communication. Another approach is that sensors can collaborate among themselves to dynamically allocate the unassigned slots. With this modification, unassigned slots can be requested by a sensor that requires more bandwidth. Based on a collaborative decision-making algorithm, future slots can be assigned to a sensor. This guarantees that the communication is collision-free.

SS-TDMA: Application to Network Programming   In this section, we discuss how SS-TDMA can be applied to network programming of sensors. As mentioned earlier, sensor network applications often include thousands of sensors that will be deployed in a very large hostile field. Programming/upgrading the software in these sensors is difficult. To deal with this problem, wireless programming [5] is used. However, the solution in [5] is applicable only to single-hop networks. In other words, this solution is useful only if the sensors are one hop away from the base station. Hence, we need an approach for multihop wireless programming of sensors, where intermediate sensors also forward the program. Some of the existing multihop network programming solutions include multihop over-the-air programming (MOAP) [18], Trickle [19], and multihop network reprogramming (MNP) [6]. These approaches are based on CSMA and, hence, rely on advertisement/request and back-off mechanisms for propagating the code in the network. However, this approach can delay programming some sensors considerably.

SS-TDMA can be effectively used to perform network programming. Specifically, we can extend the single-hop programming solution in [5] using SS-TDMA to reprogram a multihop network. Toward this end, we make the following enhancements to [5]. Whenever a sensor receives a program capsule (code segment containing 16 bytes of instructions), it stores the program in its secondary storage or EEPROM in the appropriate address. Additionally, it buffers the capsule in SS-TDMA's internal queue. SS-TDMA will forward the capsule in the sensor's alloted time slot. Hence, in this approach, a sensor need not wait to get all/part of the program before forwarding. Therefore, program capsules are sent in a pipeline. Moreover, all the sensors will be reprogrammed approximately at the same time. Theoretically, to reprogram a $n \times n$ network with a program containing x capsules, SS-TDMA takes $(x - 1)P_b + 2(n - 1)P_b$ amount of time, where P_b is the period between successive slots in broadcast. As a result of pipelining, the last $x - 1$ capsules can be forwarded within $(x - 1)P_b$, since the base station will send one capsule per period. The first capsule takes $2(n - 1)P_b$ to reach the last or the bottom-right sensor in the grid. For time slot = 30 ms, interference range = 2, $P_b = 10 \times$ time slot = 300 ms, a 1000- capsule program can be forwarded in a 10×10 network within 305.1 s or 5.1 min.

Currently, we have implemented network programming using SS-TDMA in MICA motes [2, 3]. Other than the issue of collision considered in this section, SS-TDMA deals with other causes of message loss such as environmental factors. SS-TDMA provides an ability to obtain implicit ACKs; whenever a sensor forwards a message, it acts as an implicit ACK for the predecessor. The network programming service utilizes these implicit ACKs to ensure that no packets are lost and to perform retransmissions only when necessary. The details of this service, however, are outside the scope of this section.

4.2.7 Related Work

In this section, we discuss the related work on MAC protocols for sensor networks. To deal with the problem of message collision, approaches such as collision avoidance and collision freedom based MAC protocols are proposed.

Collision Avoidance Protocols   Collision avoidance protocols like carrier sense multiple access (CSMA) [20,21] try to avoid collisions by sensing the medium before transmitting a message. If the medium is busy, then the protocol retries after a random exponential back-off period. Another example of collision avoidance protocol is carrier sense multiple access and collision detection (CSMA/CD). CSMA/CD [21] is difficult to use in the context of sensor networks as the collisions are often detected at some receivers, whereas other receivers, and sender(s) may not detect the collision.

Ye, Heidemann, and Estrin [22] propose an energy-efficient sensor-MAC (S-MAC) protocol. The authors identify the main sources of energy waste, namely, collisions, overhearing, idle listening, and control packet overhead. To reduce the energy consumption, S-MAC uses periodic listen and sleep cycles and IEEE 802.11-style RTS/CTS/Data/ACK sequence for communication. S-MAC minimizes the number of data collisions by using RTS/CTS control signals before transmitting the data. It reduces the energy spent on overhearing by scheduling network allocation vector (NAV) whenever a sensor overhears RTS/CTS not associated with itself. Until NAV fires, S-MAC allows the overhearing sensor to sleep. Further, S-MAC reduces idle listening by restricting communication to occur only in the receiver's scheduled listen interval. Finally, it reduces the control overhead by transmitting small packets such as RTS, CTS, or ACK. Timeout-MAC (T-MAC) [23] is an extension to S-MAC that allows adaptive duty cycle to handle load variations in the network unlike the fixed duty cycle operation in S-MAC.

S-MAC differs from SS-TDMA in a number of ways. First, SS-TDMA does not use IEEE 802.11-style control for communication. Overhearing and idle listening are not a problem in SS-TDMA since a sensor will listen only in the time slots alloted to its neighbors. And, protocol control overhead is very minimum in SS-TDMA. SS-TDMA just requires the base station to send periodic diffusing computations to validate the slots assigned to the sensors.

However, the frequency of such diffusions can be low since time synchronization service takes care of most clock drifts in the sensors. We emphasize that while time synchronization service is necessary for both S-MAC and SS-TDMA, SS-TDMA requires coarse-grained synchronization. S-MAC requires fine-grained synchronization since whenever a sensor wants to send a message, it needs to know when to send RTS and when to expect the corresponding CTS. Toward this end, the listen interval in S-MAC is split into three parts; first part for SYNC exchanges, which allows a sensor to discover the existence of other sensors and to know their listen/sleep schedules; the second part for RTS exchanges; and the third part for CTS exchanges. Since SYNC/RTS/CTS packets are small, the interval for such control packet transmissions is small. Hence, sensor needs to have highly precise timing information.

Collision-Free Protocols Collision-freedom protocols like frequency division multiple access (FDMA), code division multiple access (CDMA), and time division multiple access (TDMA) ensure that collisions do not occur while the sensors communicate. FDMA [21] ensures collision freedom by allotting different frequencies for the sensors. FDMA is not applicable in the context of sensor networks since the sensors (e.g., MICA motes [2, 3]) are often restricted to transmit only on one frequency. CDMA [24] requires that the codes used to encode the message be orthogonal to each other so that the destination can separate different messages. Thus, CDMA requires expensive operations for encoding/decoding a message. Therefore, CDMA is not preferred for sensor networks that lack the special hardware and that have limited computing power.

TDMA ensures collision freedom by allotting time slots for communication. TDMA-based protocols can be classified as either randomized or deterministic protocols, based on the way time slots are alloted to different sensors or how the startup algorithm works. Randomized TDMA protocols are included in [25–27]. And, deterministic TDMA protocols are included in [28, 29].

Randomized Startup In [25], Claesson, Lönn, and Suri propose a randomized startup algorithm for TDMA. Whenever a collision occurs during startup (synchronization phase), exponential back-off is used for determining the time to transmit next. In contrast, SS-TDMA uses a self-stabilizing deterministic algorithms for assigning the initial slots. Further, the complexity of the algorithm in [25] is $O(N)$, where N is the number of system nodes, whereas the complexity of the initial slot assignment algorithm in SS-TDMA is $O(D)$, where D is the diameter of the network. Moreover, an important assumption in [25] is that each node has a unique message length.

In [26, 27], initially, nodes are in random-access mode, and TDMA slots are assigned to the nodes during the process of network organization. Specifically, in low-energy adaptive clustering hierarchy (LEACH) [27], clusters are formed and each cluster elects a cluster head. All the nodes in the network are assumed to have enough radio power to communicate with the base station. However, only the cluster heads are allowed to communicate with the base station directly (single hop). Other nodes should communicate with their cluster head in order to forward a message to the base station. To achieve this hierarchy, non-cluster-head nodes reduce their radio power in such a way that they communicate only with their cluster heads. Intercluster interference is avoided by using spread spectrum or CDMA.

LEACH differs from SS-TDMA in that the slot assignment in LEACH is randomized. Further, LEACH requires CDMA to prevent intercluster interference. By contrast, SS-TDMA does not need CDMA. Further, SS-TDMA does not assume that all sensors can communicate with the base station, which is reasonable in a large sensor field.

Deterministic Startup In [28], Arisha, Youssef, and Younis propose a clustering scheme to allot time slots to different sensors. Each cluster has a gateway node. The gateway node informs each sensor in its cluster about the time slots in which the sensors can transmit messages and also the time slots in which the sensors should listen. In this algorithm, slot assignment is performed by the gateway and communicated to different sensors. This approach assumes that a node can act as a sensor and/or a gateway. If a node acts as a sensor, it gets one slot in every period, where period depends on the number of the nodes in the network. And, if a node acts as a gateway, it gets multiple slots in every period depending on the number of its children. The approach in [28] differs from SS-TDMA in that in

SS-TDMA slot assignment is uniform. The period in SS-TDMA depends on the collision group size (or the interference range) unlike the approach in [28].

4.2.8 Conclusion and Future Work

We presented SS-TDMA and showed that it can be customized for broadcast, convergecast, and local gossip. While SS-TDMA is designed for grid-based topologies, we showed how we can extend it to deal with nongrid topologies, mobile sensor nodes, and failed sensors. Thus, SS-TDMA can deal with commonly occurring difficulties, for example, failed sensors, sleeping sensors, unidirectional links, and unreliable links, in sensor networks.

As discussed in Section 4.2.3, we recommend that if the application requirements are unknown, then the TDMA service for the local gossip be used. Toward this end, we observe that the period used for local gossip is twice that for the case of broadcast/convergecast. Hence, it is possible that initiator(s) of broadcast/convergecast suffer extra delay when the local gossip solution is used. However, once the initiator sends its message, subsequent relaying occurs quickly. This is due to the fact that the solution for local gossip also ensures that the communication patterns such as broadcast and convergecast incur small delays at intermediate sensors. Thus, the solution for local gossip provides substantial benefit to broadcast and convergecast. In fact, SS-TDMA optimized for local gossip also enables a sensor to send data in any given direction in such a way that the delay incurred by the data at intermediate sensors is small.

SS-TDMA is energy-efficient and it also allows sensors to save more power by turning off the radio completely as long as the remaining sensors remain connected. These sleeping sensors can periodically wake up, wait for one diffusion message from one of its neighbors and return to a sleeping state. This will allow the sensors to save power as well as keep the clock synchronized with their neighbors. Since SS-TDMA is stabilizing fault-tolerant, if all sensors are deactivated for a long time causing arbitrary clock drift, SS-TDMA ensures that eventually the diffusion will complete successfully and collision-free communication will be restored.

One of the important issues in SS-TDMA is the *exposed terminal* problem. For example, consider four sensors A, B, C, and D arranged in a line. If B wants to send a message to A and C wants to send a message to D, either B or C will have to suppress its communication in SS-TDMA, since D (respectively, A) belongs to the collision group of B (respectively, C). One way to overcome this problem is as follows. Unlike vertex coloring in SS-TDMA (where slots are assigned to the individual sensors), slots are assigned to individual edges in the network (edge coloring). Now, B and C can simultaneously send a message to A and D, respectively. However, edge coloring is expensive in the sense that if a sensor wants to send a broadcast message, it has to send up to d messages, where d is the number of its neighbors.

Another important issue in SS-TDMA is to determine the communication and interference range of a sensor. There are several ways to achieve this. Initially, we overestimate the interference range by considering the manufacturer specification about the ability of sensors to communicate with each other. Subsequently, we can use the biconnectivity experiments [30] to determine appropriate communication range and appropriate interference range. Given any two sensors, j and k, these results allow these sensors to determine the probability that j can communicate with k and the probability that k can communicate with j. Using these results, we can update the communication range and the interference range: j is in the communication range of k iff min(probability with which j can communicate with k, probability with which k can communicate with j) exceeds a certain threshold. And, j is in the interference range of k iff max(probability with which j can communicate with k, probability with which k can communicate with j) exceeds another threshold. Using the approach in Section 4.2.6 for dealing with failed sensors, we can communicate the communication range and interference range of all sensors to the base station. Base station can then change the communication range and interference range appropriately.

We have combined SS-TDMA with previous algorithms on time synchronization (e.g., [12]). SS-TDMA and time synchronization complement each other. SS-TDMA is useful in ensuring that messages sent for time synchronization do not collide. And, time synchronization helps to reduce the clock drift and to ensure that the drift does not cause TDMA slots of nearby sensors to overlap.

We have combined SS-TDMA with previous work on *implicit* acknowledgments [1]. We expect that for known communication patterns such as broadcast, convergecast, and local gossip, combining SS-TDMA with implicit acknowledgments will be especially useful. In these communication patterns, when sensor j transmits a message to k, k is expected to retransmit it to its successor (unless k is the last sensor to receive that communication). Since message sent by k is broadcast to all its neighbors, j can also hear that message. Thus, the retransmission by k acts as an implicit acknowledgment for j. With SS-TDMA, j can wait until the TDMA slot is assigned to k; if k does not transmit in that slot, j can conclude that k did not receive its message. Thus, j can reduce the power spent in waiting for the implicit acknowledgment by listening to the radio only in the TDMA slot for k.

There are several questions raised by this work: First, an interesting question is how to determine the initial sensor that is responsible for initiating the diffusion. In some heterogeneous networks where some sensors are more powerful and more reliable, these powerful/reliable sensors can be chosen as the base station. Alternatively, during deployment of sensors (e.g., by dropping them from a plane), we can keep several potential base stations that communicate with each other directly and use the approach in [31, 32] so that one of them is chosen to be the initiator. Another interesting issue is regarding the convergecast communication pattern. As mentioned in Section 4.2.3, during convergecast, a bottleneck is created near the base station. An extension to SS-TDMA then is to account for this bottleneck by allotting more slots to the sensors near the base station compared to the rest of the network.

4.3 COMPREHENSIVE PERFORMANCE STUDY OF IEEE 802.15.4

Jianling Zheng and Myung J. Lee

IEEE 802.15.4 is a new standard uniquely designed for low rate wireless personal area networks (LR-WPANs). It targets low data rate, low power consumption, and low cost wireless networking and offers device-level wireless connectivity. We develop an NS-2 simulator for IEEE 802.15.4 and conduct several sets of experiments to study its various features, including: (1) beacon-enabled mode and non-beacon-enabled mode; (2) association, tree formation, and network autoconfiguration; (3) orphaning and coordinator relocation; (4) carrier sense multiple access with collision avoidance (CSMA-CA), both unslotted and slotted; and (5) direct, indirect, and guaranteed time slot (GTS) data transmissions. In non-beacon-enabled mode and under moderate data rate, the new IEEE 802.15.4 standard, compared with IEEE 802.11, is more efficient in terms of overhead and resource consumption. It also enjoys a low hop delay (normalized by channel capacity) on average. In beacon-enabled mode, an LR-WPAN can be flexibly configured to meet different needs, such as link failure self-recovery and low duty cycle. In both beacon-enabled mode and non-beacon-enabled mode, association and tree formation proceed smoothly, and the network can shape up efficiently by itself. We also discuss some issues that could degrade the network performance if not handled properly.

4.3.1 Background Information

Compared with wired networks, wireless networks provide advantages in deployment, cost, size, and distributed intelligence. Wireless technology not only enables users to set up a network quickly, but also enables them to set up a network where it is inconvenient or impossible to wire cables. The "carefree" feature and convenience of deployment make a wireless network more cost-efficient than a wired network in general.

The release of IEEE 802.15.4 (referred to as 802.15.4 here), "Wireless Medium Access Control (MAC) and Physical Layer (PHY) Specifications for Low Rate Wireless Personal Area Networks (LR-WPANs)" [33],[2] represents a milestone in wireless personal area networks and wireless sensor networks. 802.15.4 is a new standard uniquely designed for low rate wireless personal area networks. It targets low data rate, low power consumption, and low cost wireless networking and offers device-level wireless connectivity. A host of new applications can benefit from the new standard, such as those using sensors that control lights or alarms, wall switches that can be moved at will, wireless computer peripherals, controllers for interactive toys, smart tags and badges, tire pressure monitors in cars, and inventory tracking devices.

IEEE 802.15.4 distinguishes itself from other wireless standards such as IEEE 802.11 (referred to as 802.11 here) [34] and Bluetooth [35] by some unique features (see Section 4.3.2). However, there are no simulations or implementations available so far to test these new features. We develop an NS-2 simulator for 802.15.4 and carry out several sets of experiments to evaluate its performances, in hopes of helping IEEE to verify and/or improve the design, and facilitating researchers and manufacturers to develop products based upon this new standard. 802.15.4 has been designed as a flexible protocol in which a set of parameters can be configured to meet different requirements. As such, we also try to find out how users can tailor the protocol to their needs and where the trade-off is for some applications.

4.3.2 Brief Description of IEEE 802.15.4

The new IEEE standard, 802.15.4, defines the physical layer (PHY) and medium access control (MAC) sublayer specifications for low data rate wireless connectivity among relatively simple devices that consume minimal power and typically operate in the personal operating space (POS) of 10 m or less. An 802.15.4 network can simply be a one-hop star, or, when lines of communication exceed 10 m, a self-configuring, multihop network. A device in an 802.15.4 network can use either a 64-bit IEEE address or a 16-bit short address assigned during the association procedure, and a single 802.15.4 network can accommodate up to

---

[2] All results in this section apply to the IEEE 802.15.4 draft D18 [33].

64k (2^{16}) devices. Wireless links under 802.15.4 can operate in three license-free indus-trial scientific medical (ISM) frequency bands. These accommodate over air data rates of 250 kb/s (or expressed in symbols, 62.5 ksym/s) in the 2.4-GHz band, 40 kb/s (40 ksym/s) in the 915-MHz band, and 20 kb/s (20 ksym/s) in the 868-MHz band. A total of 27 chan-nels are allocated in 802.15.4, with 16 channels in the 2.4-GHz band, 10 channels in the 915-MHz band, and 1 channel in the 868-MHz band.

Wireless communications are inherently susceptible to interception and interference. Some security research has been done for WLANs and wireless sensor networks [36–41], but pursuing security in wireless networks remains a challenging task. 802.15.4 employs a fully handshaked protocol for data transfer reliability and embeds the advanced encryption standard (AES) [42] for secure data transfer. We give a brief overview of the PHY layer, MAC sublayer, and some general functions of 802.15.4. Detailed information can be found in [33].

The PHY Layer The PHY layer provides an interface between the MAC sublayer and the physical radio channel. It provides two services, accessed through two service access points (SAPs). These are the PHY data service and the PHY management service. The PHY layer is responsible for the following tasks:

- *Activation and Deactivation of the Radio Transceiver* Turn the radio transceiver into one of the three states, that is, transmitting, receiving, or off (sleeping) according to the request from the MAC sublayer. The turnaround time from transmitting to receiving, or vice versa, should be no more than 12 symbol periods.

- *Energy Detection (ED) Within the Current Channel* It is an estimate of the received signal power within the bandwidth of an IEEE 802.15.4 channel. No attempt is made to identify or decode signals on the channel in this procedure. The energy detection time shall be equal to 8 symbol periods. The result from energy detection can be used by a network layer as part of a channel selection algorithm or for the purpose of clear channel assessment (CCA) (alone or combined with carrier sense).

- *Link Quality Indication (LQI) for Received Packets* Link quality indication measure-ment is performed for each received packet. The PHY layer uses receiver energy detection (ED), a signal-to-noise ratio, or a combination of these to measure the strength and/or quality of a link from which a packet is received. However, the use of LQI result by the network or application layers is not specified in the standard.

- *Clear Channel Assessment (CCA) for Carrier Sense Multiple Access with Collision Avoidance (CSMA-CA)* The PHY layer is required to perform CCA using energy de-tection, carrier sense, or a combination of these two. In energy detection mode, the medium is considered busy if any energy above a predefined energy threshold is detected. In carrier sense mode, the medium is considered busy if a signal with the modulation and spreading characteristics of IEEE 802.15.4 is detected. And in the combined mode, both conditions aforementioned need to be met in order to conclude that the medium is busy.

- *Channel Frequency Selection* Wireless links under 802.15.4 can operate in 27 different channels (but a specific network can choose to support part of the channels). Hence the PHY layer should be able to tune its transceiver into a certain channel upon receiving the request from the MAC sublayer.

- *Data Transmission and Reception* This is the essential task of the PHY layer. Modulation and spreading techniques are used in this part. The 2.4-GHz PHY layer employs a 16-ary quasi-orthogonal modulation technique in which each four information bits are mapped into a 32-chip pseudo-random noise (PN) sequence. The PN sequences for successive data symbols are then concatenated and modulated onto the carrier using offset quadrature phase shift keying (O-QPSK). The 868/915 MHz PHY employs direct sequence spread spectrum (DSSS) with binary phase shift keying (BPSK) used for chip modulation and differential encoding used for data symbol encoding. Each data symbol is mapped into a 15-chip PN sequence, and the concatenated PN sequences, are then modulated onto the carrier using BPSK with raised cosine pulse shaping.

The MAC Sublayer The MAC sublayer provides an interface between the service-specific convergence sublayer (SSCS) and the PHY layer. Like the PHY layer, the MAC sublayer also provides two services, namely, the MAC data service and the MAC management service. The MAC sublayer is responsible for the following tasks:

- *Generating Network Beacons If the Device Is a Coordinator* A coordinator can determine whether to work in a beacon-enabled mode in which a superframe structure is used. The superframe is bounded by network beacons and divided into *aNumSuperframeSlots* (default value 16) equally sized slots. A coordinator sends out beacons periodically to synchronize the attached devices and for other purposes (described later).
- *Synchronizing to the Beacons* A device attached to a coordinator operating in a beacon-enabled mode can track the beacons to synchronize with the coordinator. This synchronization is important for data polling, energy saving, and detection of orphanings.
- *Supporting Personal Area Network (PAN) Association and Disassociation* To support self-configuration, 802.15.4 embeds association and disassociation functions in its MAC sublayer. This not only enables a star to be setup automatically but also allows for the creation of a self-configuring, peer-to-peer network.
- *Employing the Carrier Sense Multiple Access with Collision Avoidance (CSMA-CA) Mechanism for Channel Access* Like most other protocols designed for wireless networks, 802.15.4 uses CSMA-CA mechanism for channel access. However, the new standard does not include the request-to-send (RTS) and clear-to-send (CTS) mechanism, in consideration of the low data rate used in LR-WPANs.
- *Handling and Maintaining the Guaranteed Time Slot (GTS) Mechanism* When working in a beacon-enabled mode, a coordinator can allocate portions of the active superframe to a device. These portions are called GTSs and comprise the contention free period (CFP) of the superframe.
- *Providing a Reliable Link Between Two Peer MAC Entities* The MAC sublayer employs various mechanisms to enhance the reliability of the link between two peers, among them are the frame acknowledgment and retransmission, data verification by using a 16-bit CRC as well as CSMA-CA.

General Functions The standard gives detailed specifications of the following items: type of device, frame structure, superframe structure, data transfer model, robustness, power

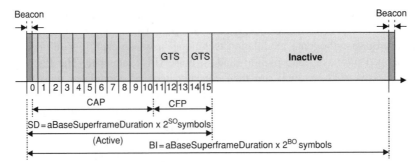

Figure 4.24 Example of the superframe structure.

consumption considerations, and security. Here, we give a short description of those items closely related to our performance study, including type of device, superframe structure, data transfer model, and power consumption considerations.

Two different types of devices are defined in an 802.15.4 network, a full function device (FFD) and a reduced function device (RFD). An FFD can talk to RFDs and other FFDs and operate in three modes serving either as a PAN coordinator, a coordinator, or a device. An RFD can only talk to an FFD and is intended for extremely simple applications.

The standard allows the optional use of a superframe structure. The format of the superframe is defined by the coordinator. From Figure 4.24, we can see the superframe comprises an active part and an optional inactive part and is bounded by network beacons. The length of the superframe (a.k.a. beacon interval, BI) and the length of its active part (a.k.a. superframe duration, SD) are defined as follows:

$$BI = aBaseSuperframeDuration *2^{BO}$$
$$SD = aBaseSuperframeDuration *2^{SO}$$

where, aBaseSuperframeDuration = 960 symbols

$$BO = \text{beacon order}$$
$$SO = \text{superframe order}$$

The values of BO and SO are determined by the coordinator. The active part of the superframe is divided into *aNumSuperframeSlots* (default value 16) equally sized slots, and the beacon frame is transmitted in the first slot of each superframe. The active part can be further broken down into two periods, a contention access period (CAP) and an optional contention free period (CFP). The optional CFP may accommodate up to seven so-called guaranteed time slots (GTSs), and a GTS may occupy more than one slot period.

However, a sufficient portion of the CAP shall remain for contention-based access of other networked devices or new devices wishing to join the network. A slotted CSMA-CA mechanism is used for channel access during the CAP. All contention-based transactions shall be complete before the CFP begins. Also all transactions using GTSs shall be done before the time of the next GTS or the end of the CFP.

Data transfer can happen in three different ways: (1) from a device to a coordinator, (2) from a coordinator to a device, and (3) from one peer to another in a peer-to-peer multihop network. Nevertheless, for our performance study, we classify the data transfer into the following three types:

- *Direct Data Transmission*   This applies to all data transfers, either from a device to a coordinator, from a coordinator to a device, or between two peers. Unslotted CSMA-CA or slotted CSMA-CA is used for data transmission, depending on whether the non-beacon-enabled mode or the beacon-enabled mode is used.
- *Indirect Data Transmission*   This only applies to data transfer from a coordinator to its devices. In this mode, a data frame is kept in a transaction list by the coordinator, waiting for extraction by the corresponding device. A device can find out if it has a packet pending in the transaction list by checking the beacon frames received from its coordinator. Occasionally, indirect data transmission can also happen in non-beacon-enabled mode. For example, during an association procedure, the coordinator keeps the association response frame in its transaction list and the device polls and extracts the association response frame. Unslotted CSMA-CA or slotted CSMA-CA is used in the data extraction procedure.
- *GTS Data Transmission*   This only applies to data transfer between a device and its coordinator, either from the device to the coordinator or from the coordinator to the device. No CSMA-CA is needed in GTS data transmission.

Power conservation has been a research focus for wireless networks [43–49] since most devices in wireless networks are battery powered. The standard was developed with the limited power supply availability in mind and favors battery-powered devices. The superframe structure, the indirect data transmission, and the *BatteryLifeExtension* option are all examples. If the *BatteryLifeExtension* is set to TRUE, all contention-based transactions are required to begin within *macBattLifeExtPeriods* (default value 6) full backoff periods after the interframe space (IFS) period of the beacon frame.

4.3.3 NS2 Simulator

The 802.15.4 NS2 [50] simulator developed at the Joint Lab of Samsung and the City University of New York conforms to IEEE P802.15.4/D18 Draft. Figure 4.25 outlines the

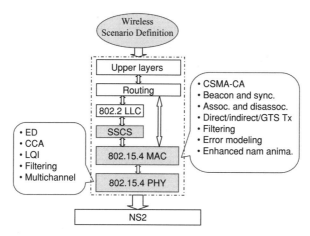

Figure 4.25   NS2 simulator for IEEE 802.15.4.

function modules in the simulator, and a brief description is given below for each of the modules.

- *Wireless Scenario Definition* It selects the routing protocol, defines the network topology, and schedules events such as initializations of PAN coordinator, coordinators and devices, and starting (stopping) applications. It defines the radio propagation model, antenna model, interface queue, traffic pattern, link error model, link and node failures, superframe structure in beacon enabled mode, radio transmission range, and animation configuration.
- *Service-Specific Convergence Sublayer (SSCS)* This is the interface between 802.15.4 MAC and upper layers. It provides a way to access all the MAC primitives, but it can also serve as a wrapper of those primitives for convenient operations. It is an implementation-specific module and its function should be tailored to the requirements of specific applications.
- *802.15.4 PHY* It implements all 14 PHY primitives.
- *802.15.4 MAC* This is the main module. It implements all the 35 MAC sublayer primitives.

4.3.4 Performance Metrics and Experimental Setup

Performance Metrics We define the following metrics for studying the performance of 802.15.4. All metrics are defined with respect to MAC sublayer and PHY layer in order to isolate the effects of MAC and PHY from those of upper layers.

- *Packet Delivery Ratio* The ratio of packets successfully received to packets sent in the MAC sublayer. This metric does not differentiate transmissions and retransmissions and therefore does not reflect what percentage of upper layer payload is successfully delivered, although they are related.
- *Hop Delay* The transaction time of passing a packet to a one-hop neighbor, including time of all necessary processing, backoff as well as transmission, and averaged over all successful end-to-end transmissions within a simulation run. It is not only used for measuring packet delivery latency but also used as a negative indicator of the MAC sublayer capacity. The MAC sublayer has to handle the packets one by one and therefore a long delay means a small capacity.
- *RTS/CTS Overhead* The ratio of request-to-send (RTS) packets plus clear-to-send (CTS) packets sent to all the other packets sent in 802.11. This metric is not applicable to 802.15.4 in which RTS/CTS mechanism is not used. We compare the performances of 802.11 and 802.15.4 to justify the dropping of RTS/CTS mechanism in 802.15.4.
- *Successful Association Rate* The ratio of devices successfully associated with a coordinator to the total devices trying to associate with a coordinator. In our experiments, a device will retry in one second if it fails to associate with a coordinator in the previous attempt. The association is considered successful if a device is able to associate with a coordinator during a simulation run, even if multiple association attempts have been made.
- *Association Efficiency* The average number of attempts per successful association.

- *Orphaning Rate* A device is considered orphaned if it misses *aMaxLostBeacons* (default value 4) beacons from its coordinator in a row. The orphaning rate is defined as the ratio of devices orphaned at least once to the total devices that are in beacon-enabled mode and keep tracking beacons. This metric is not applicable to devices in non-beacon-enabled mode or devices in beacon-enabled mode but not tracking beacons. In our experiments, all devices in beacon-enabled mode track beacons.

- *Orphaning Recovery Rate* Two different versions are defined for this metric. One is the ratio of orphaned devices that have successfully relocated their coordinators, that is, have recovered from orphaning, to the total orphaned devices. The other is the ratio of recovered orphanings to the total orphanings, in which multiple orphanings of a device are counted. No further attempt is made if the orphaning recovery procedure fails.

- *Collision Rate* The total collisions during a simulation run.

- *Collision Rate Between Hidden Terminals* The total collisions that occur between hidden terminals during a simulation run. Hidden terminals prevent carrier sense from working effectively, and therefore transmissions from them are likely to collide at a third node [51]. In 802.11, the RTS and CTS mechanisms are used to tackle this problem [34].

- *Repeated Collision Rate* The total collisions that happen more than once between the same pair of packets during a simulation run.

- *Collision Distribution* The time distribution, within a superframe, of collisions. This metric is only used in beacon-enabled mode.

- *Duty Cycle* The ratio of the active duration, including transmission, reception, and carrier sense time, of a transceiver to the whole session duration.

Experimental Setup Five sets of experiments are designed to evaluate the various performance behaviors of 802.15.4, including those applicable to all wireless networks (such as packet delivery ratio, packet delivery latency, control overhead, and transmission collision) as well as other behaviors specific to LR-WPANs (such as association, orphaning, and different transmission methods). The first set is for non-beacon-enabled mode; the second and third sets are for mixed mode, that is, a combination of beacon-enabled mode and non-beacon-enabled mode; and the fourth and fifth sets are for beacon-enabled mode. The first three sets run in a multihop environment (Fig. 4.26*a*), and the other two sets run in a one-hop star environment (Fig. 4.26*b*). Although a specific network can take a quite different topology, the two topologies used in our experiments represent the topologies currently supported by 802.15.4 and are enough for performance study purposes.

General Parameters Assuming a 10^{-6} to 10^{-5} link bit error rate (BER), we apply a 0.2% statistical packet error rate (PER) to all our experiments. The simulation duration is 1000 s, and the application traffic runs from 20 to 900 s, leaving enough time for the experiment to shut down gracefully. Since the popular constant bit rate (CBR) traffic used in most simulations is too deterministic for nonmobile wireless networks, Poisson traffic is used for all application sessions in our experiments. The application packet size is 90 bytes. Except the fifth set of experiments, all the other experiments use direct data transmission.

The radio propagation model adopted in all our experiments is two-ray ground reflection. Beacon order (BO) and superframe order (SO) take the same value in all beacon-enabled modes, that is, the optional inactive part is not included in superframes. Most experiments

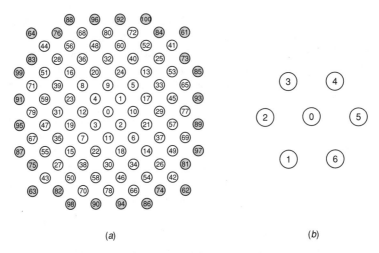

(a) (b)

Figure 4.26 Experiment scenarios.

run 10 times with random seeds, but those with a traffic load of 0.2 packet per second (pps) and those with a traffic load of 0.1 pps run 20 times and 40 times, respectively. Other experiment-specific configuration information is given in the following paragraphs corresponding to each set of experiments.

Experiment Set 1: Comparing 802.15.4 with 802.11   The first set of experiments are used to compare the performances of 802.15.4 and 802.11. Although 802.15.4 and Bluetooth bear more similarities from the application point of view, 802.15.4 and 802.11 are more comparable as far as our performance study is concerned. Both 802.15.4 and 802.11 support multihop network topology and peer-to-peer communications, which are used in our first set of experiments. The dominant topology in Bluetooth, on the other hand, is one-hop star or so-called piconet, which consists of one coordinator and up to seven devices. In a piconet, a device only communicates with its coordinator. Although scatternets can be used to extend the coverage and the number of devices of a Bluetooth network, our research work showed that there are scalability problems in scatternets [52]. Furthermore, all the devices in either 802.15.4 or 802.11 share a single chip code for spread spectrum, while different devices in Bluetooth are assigned different chip codes. Based on the above facts, we select 802.11 instead of Bluetooth for comparison. The performance is evaluated with respect to the following parameters as well as those listed in the previous paragraph:

- 101 nodes evenly distributed in an 80×80 m^2 area (Fig. 4.26a).
- 9-m transmission range, which only covers the neighbors along diagonal direction.
- 802.15.4 operates at an over air data rate of 250 kbps (in the 2.4-GHz ISM band) and in non-beacon-enabled mode, and 802.11 operates at a data rate of 2 Mbps.
- Poisson traffic with the following average packet rates: 0.1 packet per second (pps), 0.2 pps, 1 pps, 5 pps, and 10 pps.
- We apply two types of application traffic: (1) peer-to-peer application traffic, which consists of 6 application sessions between the following nodes: $64 \rightarrow 62$, $63 \rightarrow 61$, $99 \rightarrow 85$, $87 \rightarrow 97$, $88 \rightarrow 98$, and $100 \rightarrow 86$, and (2) multiple-to-one application traffic,

which consists of 12 application sessions from nodes 64, 62, 63, 61, 99, 85, 87, 97, 88, 98, 100, and 86 to node 0. The first type of application traffic is used to study the general peer-to-peer behavior of 802.15.4 and, for comparison, it is applied to both 802.15.4 and 802.11. The second type of application traffic targets the important application of 802.15.4, wireless sensor networks, where traffic is typically between multiple source nodes and a sink. It is only applied to 802.15.4. Although the second type of application traffic is not used for comparing 802.15.4 with 802.11, we include it here to facilitate the comparison of 802.15.4 behaviors under different application traffic. We refer to the second type of application traffic as sink-type application traffic hereinafter.

Experiment Set 2: Association Efficiency   The second set of experiments are designed to evaluate the association efficiency under different number of beaconing coordinators and different beacon orders. The same network topology, transmission range, frequency band, data rate, and peer-to-peer application sessions are used as in the first set of experiments. Except node 0, which is the PAN coordinator, and the leaf nodes, which are pure devices, all the other nodes serve as both a coordinator (to its children) and a device (to its parent). So we have 73 coordinators and 100 devices. This set of experiments run in a mixed mode, with different percentage of coordinators beaconing (0, 25, 50, 75, and 100%). The beacon order varies and takes the values of 0, 1, 2, 3, 4, 5, 6, and 10. The application traffic is fixed at 1 pps.

Experiment Set 3: Orphaning   The third set of experiments are used to study the device orphaning behavior, namely, how often orphanings happen and what percentage of orphanings, in terms of number of orphaned devices or number of orphanings, can be recovered. The experimental setup is the same as that of the second set of experiments.

Experiment Set 4: Collision   The fourth set of experiments target the collision behavior of 802.15.4. The experiments run in a beacon-enabled star environment. Nevertheless, except for some beacon-specific metrics, most of the metrics extracted from this set of experiments are general and can serve for both beacon- and non-beacon-enabled modes. Besides the general parameters given above, the following parameters are used in the experiments:

- Seven nodes form a star with a radius of 10 m, with one coordinator at the center and six devices evenly distributed around it (Fig. 4.26b).
- 15-m transmission range, which enables the coordinator to reach all the devices. However, a device can only reach the coordinator and two devices adjacent to it. In other words, devices are hidden from each other unless they are adjacent to each other.
- Operates at an over air data rate of 250 kbps (in the 2.4-GHz ISM band).
- Poisson traffic with the average packet rate of 1 pps.
- Six application sessions, one for each device, are setup from the devices to the coordinator.
- The beacon order changes from 0 to 8.

Experiment Set 5: Direct, Indirect, and GTS Data Transmissions   The last set of experiments are used to investigate the different features of the three data transmission methods in 802.15.4. We compare the packet delivery ratio, hop delay, and duty cycle of the three different methods. All the parameters are the same as those in the fourth set of

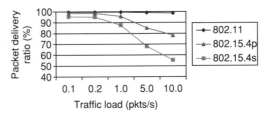

Figure 4.27 Comparing 802.15.4 with 802.11: Packet delivery ratio.

experiments, except that only two application sessions originating from adjacent devices are used, and that three different data transmission methods are used.

4.3.5 Experimental Results

Comparing IEEE 802.15.4 with IEEE 802.11   To distinguish experiment results for 802.15.4 with different application traffic, we use 802.15.4p and 802.15.4s to denote the data series corresponding to peer-to-peer application traffic and sink-type application traffic, respectively (see Figs. 4.27 and 4.28). However, when experiment results are not specific to a certain application traffic (e.g., the data series 802.15.4 in Fig. 4.29) or only one application traffic is applied (e.g., for 802.11), the protocol name is used only to denote the corresponding data series.

For peer-to-peer application traffic, as shown in Figure 4.27, the packet delivery ratio of 802.11 decreases slowly from 99.53 to 98.65% when the traffic load changes from 0.1 to 10 pps. On the other hand, the packet delivery ratio of 802.15.4 drops from 98.51 to 78.26% for the same traffic load change (data series 802.15.4p in Fig. 4.27). For sink-type application traffic, the packet delivery ratio of 802.15.4 drops more sharply from 95.40 to 55.26% when the traffic load changes from 0.1 to 10 pps (data series 802.15.4s in Fig. 4.27). In general, 802.15.4 maintains a high packet delivery ratio for application traffic up to 1 pps (95.70% for 802.15.4p and 87.58% for 802.15.4s), but the value decreases quickly as traffic pps load increases.

The difference of packet delivery ratio between 802.15.4 and 802.11 comes from the fact that the former does not use RTS/CTS mechanism while the latter does. This RTS/CTS overhead proves to be useful when traffic load is high, but obviously too expensive for low data rate applications as of the case of LR-WPANs for which 802.15.4 is designed. From Figure 4.29, we can see the ratio of (RTS + CTS) packets to Poisson data packets is within the scope [2.02, 2.78], which cannot be justified in 802.15.4, considering the less than 4% increase of packet delivery ratio for application traffic up to 1 pps. Note that, even under

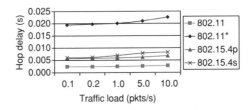

Figure 4.28 Comparing 802.15.4 and 802.11: Hop delay.

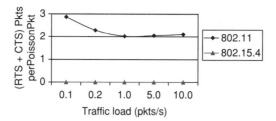

Figure 4.29   Comparing 802.15.4 with 802.11: RTS /CTS overhead.

collision-free condition, the ratio of (RTS + CTS) packets to Poisson data packets is larger than 2.0, because RTS/CTS packets are also used for transmissions of other control packets such as AODV packets. It is clear that the high ratio of (RTS + CTS) packets to Poisson data packets for 0.1 pps must come from the high ratio of other control packets to Poisson data packets, since collisions are ignorable under such low traffic load.

The RTS/CTS mechanism also affects the network latency. We measure the average hop delay for both protocols in comparison, and the results are depicted in Figure 4.28. The initial results show that 802.11 enjoys a lower delay than 802.15.4 (data series 802.11 and 802.15.4p in Fig. 4.28). Nevertheless, this comparison is unfair to 802.15.4, since it operates at a data rate of 250 kbps while 802.11 operates at 2 Mbps in our experiments. Taking this into account, we normalize the hop delay according to the media data rate, which gives us a different view that the hop delay of 802.11 is around 3.3 times that of 802.15.4 (data series 802.11* and 802.15.4p in Fig. 4.28). The hop delay for sink-type application traffic is from 6.3% (for 0.1 pps) to 20.9% (for 10 pps) higher than that for peer-to-peer application traffic (data series 802.15.4s and 802.15.4p in Fig. 4.28). The increment of delay is expected since all the traffic flows now need to converge on the sink node.

Association Efficiency   The typical scenario of an LR-WPAN is a densely distributed unattended wireless sensor network. Self-configuration in deployment and autorecovery from failures are highly desirable features in such a network [53]. For this purpose, 802.15.4 includes an association and disassociation mechanism together with an orphaning and coordinator relocation mechanism in its design. We give out the experimental results of association in this subsection, while the experimental results of orphaning will be given in the next subsection.

To associate with a coordinator a device will perform an active channel scan in which a beacon request frame is sent or a passive channel scan in which no beacon request frame is sent to locate a suitable coordinator. Active channel scan is used in our experiments since a device needs to explicitly request for beacons in non-beacon-enabled environment. When a coordinator receives the beacon request frame, it handles it differently depending on whether it is in beacon-enabled mode or non-beacon-enabled mode. If the coordinator is in beacon-enabled-mode, it discards the frame silently since beacons will be bent periodically anyway. Otherwise, the coordinator needs to unicast a beacon to the device soliciting beacons. In our experiments, we vary the percentage of beaconing coordinators to see the different effects of beaconing coordinators and nonbeaconing coordinators.

In general, the successful association rate is very high (more than 99%) for different combinations of beaconing coordinators and nonbeaconing coordinators, as illustrated in Table 4.2. From Figure 4.30, we can see that a device gets an almost equal chance to

Table 4.2   Successful Association Rate vs. Beaconing Coordinator Ratio

Beacon coordinator ratio (%)	0	25	50	75	100
Successful association rate (%)	100	100	100	99	100

associate with a beaconing coordinator or a nonbeaconing coordinator. However, this result is obtained for beacon order 3, and it may be different for other beacon orders. Normally, a beaconing coordinator with a larger beacon order (i.e., longer superframe) reacts slowly to a beacon request, which means it will not get the same chance to serve as a coordinator for a certain device, when competing with other nonbeaconing coordinators or beaconing coordinators with smaller beacon orders.

The association efficiency shown in Figure 4.31, in terms of attempts per successful association, is high. The association procedure is a multistep procedure as briefly described by the following pseudocode (for device part only):

```
1: channel scan
2: if coordinators not found
3:    association fail
4: elseif no coordinators permit association
5:    association fail
6: else
7:    select a proper coordinator
8:    send association request to the coord.
9:    wait for ACK
10:   if ACK not received
11:   association fail
12: else
13:     send data request to the coord.
14:     wait for ACK
15:     if ACK not received
16:         association fail
17:     else
18:         wait for association response
19:         if asso. response not received
20:             association fail
21:         elseif association not granted
22:             association fail
23:         else
24:             association succeed
```

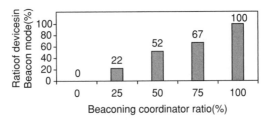

Figure 4.30   Devices associated with beaconing coordinators.

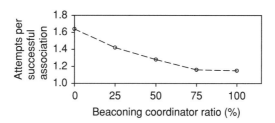

Figure 4.31 Association efficiency vs. beaconing coordinator ratio.

If there are multiple nonbeaconing coordinators around, they all will try to unicast a beacon, using unslotted CSMA-CA, to the device asking for beacons. These beacons are likely to collide at the device due to the hidden terminal problems as a fact of lacking RTS/CTS, that is, even the first step of the association may fail. The situation is better if there are multiple beaconing coordinators around since they will continue beaconing as usual even if a beacon request is received. Of course, if beacons are sent with high frequency (low beacon order), then the collisions will increase, which will bring down the association efficiency. In summary, nonbeaconing coordinators are likely to affect the first step of the association procedure, while the beaconing coordinators can affect all the steps. As revealed by our experimental results, beaconing coordinator as a whole is a better choice regarding association efficiency, provided the beacon order is not too small. See Figure 4.32.

Table 4.3 gives out the distribution of association attempts, which shows that most of the devices succeed in their first association attempt, a small part of the devices try twice or three times, and three devices try four times.

Association is the basis of tree formation in a peer-to-peer multihop network. The efficiency of tree formation is directly related to association efficiency. A tree is a useful structure and can be used by network layer, especially for routing purpose. In this set of experiments, a tree is quickly formed thanks to the high association efficiency. Various configurations are also done during this procedure, such as select a channel and an identifier (ID) for the PAN, determine whether beacon-enabled mode or non-beacon-enabled mode to be used, choose the beacon order and superframe order in beacon-enabled mode, assign a 16-bit short address for a device, and set the *BatteryLifeExtension* option and many other options in the MAC layer PAN information base (MPIB). The smooth procedure of association and tree formation indicates that an 802.15.4 network has a feature of self-configuration and can shape up efficiently.

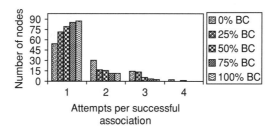

Figure 4.32 Attempts per successful association vs. beaconing coordinator (BC) ratio.

Table 4.3   Distribution of Association Attempts (expressed in number of devices)

	1 attempt	2 attempts	3 attempts	4 attempts
0% beaconing coordinators	54	30	14	2
25% beaconing coordinators	71	16	13	–
50% beaconing coordinators	79	15	5	1
75% beaconing coordinators	85	11	3	–
100% beaconing coordinators	87	11	2	–

Orphaning   The orphaning study is conducted in an environment with all coordinators beaconing. Specifically, we examine the orphaning behavior for different beacon orders. Orphaning mechanism, work only if a device is successfully associated with a beaconing coordinator, and the device keeps tracking the beacons from the coordinator. Since orphaning is related to association, here we also give out the association results. Table 4.4 and Figure 4.33 suggest that the performance of beacon-enabled modes with small beacon orders is not so good as that with large beacon orders. For example, the attempts per successful association for beacon order 0 is "outstanding" among its peers. And the successful association rate for beacon order 1 and beacon order 2 is also slightly lower than others.

Unsurprisingly, orphaning is also more serious in those beacon-enabled modes with smaller beacon orders (Fig. 4.34). The percentage of devices orphaned in beacon order 0 or beacon order 1 is about the same (around 58%), and is 29 times that in beacon order 2. There is no orphaning in beacon order 3 or up. In an environment with a high rate of orphaning, the chance an orphaned device successfully recovers from all orphanings is very low (2% for beacon order 0 and 4% for beacon order 1 as shown by data series "Devices Recovered"), but the recovery rate of orphaning itself is not that bad (from 30 to 89% as shown by data series "Orphanings Recovered"). One point worth mentioning is that, a device that failed to recover from all orphanings still benefits from the recovery mechanism since its association with the coordinator is prolonged, though not to the end of the session.

Collision   It is clearly shown in Figure 4.35 that more collisions happen in low beacon orders than in high beacon orders. And the network virtually loses its control in beacon order 0, due to large number of collisions. This type of "beacon storm" problem is alleviated in high-order beacons. Due to the broadcast nature of wireless networks, broadcast-based storm is not a rare phenomenon [54]. It necessitates careful handling.

As expected, the majority of collisions happen between hidden terminals (Fig. 4.36), that is, between any two devices not adjacent to each other in our experiments (see Section 4.3.4). However, probability of collisions between nonhidden terminals in low beacon orders is not trivial either. This means the slotted CSMA-CA can no longer work effectively if the beacon order is very small, and the chance that two nonhidden terminals jump to the channel simultaneously is significantly increased.

Table 4.4   Successful Association Rate vs. Beacon Order

Beacon order	0	1	2	3	4	5	6	10
Successful association rate (%)	99	96	95	100	99	100	100	99

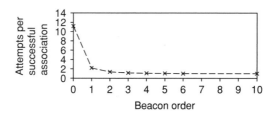

Figure 4.33 Association attempts vs. beacon order.

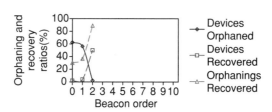

Figure 4.34 Orphaning and recovery.

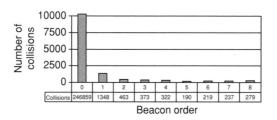

Figure 4.35 Collisions vs. beacon order.

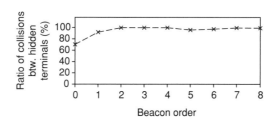

Figure 4.36 Ratio of collisions between hidden terminals.

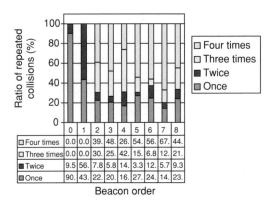

	0	1	2	3	4	5	6	7	8
□ Four times	0.0	0.0	39.	48.	26.	54.	56.	67.	44.
□ Three times	0.0	0.0	30.	25.	42.	15.	6.8	12.	21.
■ Twice	9.5	56.	7.8	5.8	14.	3.3	12.	5.7	9.3
■ Once	90.	43.	22.	20.	16.	27.	24.	14.	23.

Beacon order

Figure 4.37 Ratio of repeated collisions.

Unexpectedly, the ratio of repeated collisions is very high, as manifested in Figure 4.37. By tracking these collisions, we find the reason is that the suggested backoff length in 802.15.4 is too short, especially for long frames (physical protocol data unit larger than 100 bytes). This short backoff length results from the consideration of energy conservation, but a too short backoff length will cause repeated collisions and defeat the initial design goal. The fact that no collisions repeated more than twice in beacon order 0 and beacon order 1 is somewhat misleading. It is not because the collisions can be resolved within the first two backoffs but that the enormous number of collisions make it impossible in effect for a packet to collide with another packet more than twice before it reaches its retransmission threshold.

The last metric we extract from this set of experiments is the time distribution of collisions within a superframe. In beacon-enabled mode, a transaction (transmission of a frame as well as reception of an acknowledgment frame if required) using slotted CSMA-CA is required to be completed before the end of the contention access period (CAP). Otherwise, the transaction should be delayed until the beginning of next superframe. In such a design, more collisions are expected at the beginning of a superframe, especially a short superframe (low beacon order) in which more transactions are likely to be delayed until the beginning of the next the frame. This is confirmed by our experimental results shown in Figure 4.38. For beacon order 0, for example, about 75% of collisions happen within the first millisecond of a superframe (but 1 ms is only about 6.5% of a superframe of beacon order 0).

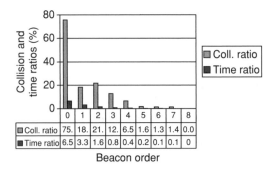

	0	1	2	3	4	5	6	7	8
□ Coll. ratio	75.	18.	21.	12.	6.5	1.6	1.3	1.4	0.0
■ Time ratio	6.5	3.3	1.6	0.8	0.4	0.2	0.1	0.1	0

Beacon order

Figure 4.38 Ratio of collisions within the first millisecond of a superframe.

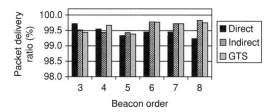

Figure 4.39 Different data transmission methods: packet delivery ratio.

Direct, Indirect, and GTS Data Transmissions   In this set of experiments, we compare three different data transmission methods, that is, direct, indirect, and guaranteed time slot (GTS) data transmissions (DIG). The focus is latency (Fig. 4.40) and duty cycle (Fig. 4.41), but packet delivery ratio is also given (Fig. 4.39) for the sake of completion. Small beacon orders 0, 1, and 2 are not shown in the above figures, since, in GTS data transmission, we only allocate one slot for each device and the slot is too short for holding a data frame.

No significant difference has been observed in the packet delivery ratio among the three data transmission methods. Nevertheless, the hop delay varies, which will definitely affect the packet delivery ratio in upper layers. The hop delay in direct data transmission is much shorter than those in indirect and GTS data transmissions.

One fundamental aspect of 802.15.4 is low power consumption, which is very desirable in a wireless sensor network, as the replacement of batteries is very cumbersome due to the large number of sensors. Most power-saving mechanisms in 802.15.4 are based on beacon-enabled mode. In direct data transmission, if the *BatteryLifeExtension* option is set to TRUE, the receiver of the beaconing coordinator is disabled after *macBattLifeExtPeriods* (default value 6) backoff periods following the interframe space (IFS) period of the beacon frame. Using default configuration, this means that the transceiver of a coordinator or a device is required to be turned on for only about 1/64 of the duration of a superframe, if no data to be exchanged. If the value of *BatteryLifeExtension* is FALSE, the receiver of the beaconing coordinator remains enabled for the entire CAP. In indirect data transmission, a device can enter a low power state, like sleeping state, if it finds there are no pending packets by checking the beacon received from its coordinator.

As shown in Figure 4.41, the duty cycle is around 2% in indirect data transmission, and about 1% in GTS data transmission. However, there are two slots or 12.5% of a superframe allocated for GTS data transmission in our experiments, which means that

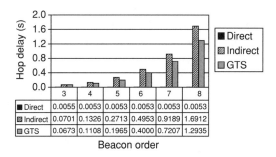

	3	4	5	6	7	8
■ Direct	0.0055	0.0053	0.0053	0.0053	0.0053	0.0053
▨ Indirect	0.0701	0.1326	0.2713	0.4953	0.9189	1.6912
▣ GTS	0.0673	0.1108	0.1965	0.4000	0.7207	1.2935

Beacon order

Figure 4.40 Different data transmission methods: Hop delay.

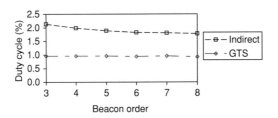

Figure 4.41 Different data transmission methods: Duty cycle.

$(12.5 - 1)/12.5 = 92\%$ of the allocated GTS slots are wasted. This result shows that GTS is too expensive for low data rate applications.

The above duty cycle measurement is based on the traffic load of one packet per second, and it shall vary when traffic load changes. Perfect synchronization among devices is also assumed in the measurement, which is generally not true in practice. Some margin should be provided for the nonperfect synchronization, which means an increment in duty cycle. One more point about power conservation is that, it is acquired at the cost of delay, as clearly shown in Figure 4.40. The power consumption mechanisms employed in 802.15.4 are based on the assumption of low data rate and should be used properly.

4.3.6 Conclusions

At its heart, the new IEEE 802.15.4 standard, which is designed for low rate wireless personal area networks (LR-WPANs), is an enabling standard. It brings to light a host of new applications as well as changes many other existing applications. It is the first standard to allow simple sensors and actuators to share a single standardized wireless platform.

To evaluate the general performance of this new standard, we develop an NS2 simulator, which covers all the 802.15.4 PHY and MAC primitives, and carry out five sets of experiments, that is, experiments of: (1) comparing the performance between 802.15.4 and 802.11, (2) association and tree formation study, (3) orphaning and coordinator relocation investigation, (4) examination of unslotted CSMA-CA and slotted CSMA-CA behaviors, and (5) comparing three different data transmissions, namely, direct, indirect, and guaranteed time slot (GTS) data transmissions. Detailed experimental results are presented, and analyses and discussions are given.

In non-beacon-enabled mode and for low rate applications (traffic load $\leq$ one packet per second), the packet delivery ratio of 802.15.4 is similar to that of 802.11. However, 802.15.4 shows clear advantage over 802.11 regarding control overhead and transaction latency. The experimental results endorse the non-RTS/CTS CSMA-CA approach.

Association and tree formation in 802.15.4 proceed smoothly in both beacon-enabled mode and non-beacon-enabled mode, which implies 802.15.4 possesses a good self-configuration feature and is able to shape up efficiently without human intervention. The orphaning and coordinator relocation (recovery from orphaning) mechanism provides for a device a chance of self-healing from disruptions. The orphaning recovery probability is about 30% for the worst case and about 89% for the best case in our experiments. Notwithstanding, the chance that an orphaned device is completely recovered, that is, it recovers each time it is orphaned, is very low.

For the lack of RTS/CTS, 802.15.4 is expected to suffer from hidden terminal problems. Our experiment results match this expectation. But for low data rates up to one packet per second, the performance degradation is minor. The default CSMA-CA backoff period in 802.15.4 is too short, which leads to frequent repeated collisions. Superframes with low beacon orders can also lower the slotted CSMA-CA backoff efficiency and lead to high collision probability at the beginnings of superframes.

Our study shows that 802.15.4 is an energy-efficient standard favoring low data rate and low power consumption applications. GTS data transmission is an expensive approach for low data rate applications, as can be seen from our experimental results.

4.4 PROVIDING ENERGY EFFICIENCY FOR WIRELESS SENSOR NETWORKS THROUGH LINK ADAPTATION TECHNIQUES

Song Ci, Hamid Sharif, Krishna Nuli, and Prasa Raviraj

Technological advances in low-power digital signal processors, radio frequency (RF) circuits, and micromechanical systems (MEMS) have led to the emergence of wirelessly interconnected sensor nodes. A large number of tiny intelligent wireless sensor nodes when combined offer radically new technological possibilities. The sensor nodes are typically battery operated and therefore energy constrained. Hence, energy conservation is one of the foremost priorities when designing protocols for wireless sensor networks (WSNs). Limited power resources and the bursty nature of the wireless channel are the biggest challenges in WSNs. Link adaptation techniques improve the link quality by adjusting Medium access control (MAC) parameters such as frame size, data rate, and sleep thereby improving energy efficiency. Our study emphasizes optimizing WSNs by building a reliable and adaptive MAC without compromising fairness and performance. In this section, we present link adaptation techniques at the MAC layer to enhance energy efficiency of the sensor nodes. The proposed MAC uses a variable frame size instead of a fixed frame size for transmitting data. To reduce the com, we used two new approaches called extended Kalman filter (EKF) and unscented Kalman filter (UKF) to predict the optimal frame size for improving energy efficiency and goodput, while minimizing the sensor memory requirement. We designed and verified different network models to evaluate and analyze the proposed link adaptation schemes. The correctness of the proposed theoretical models have been verified by conducting extensive simulations. The simulation results show that the proposed algorithms improve the energy efficiency by 15%.

4.4.1 Background Information

The integration of sensing, computation, and communication into a single tiny device made wireless sensor networks a reality. With the advent of sophisticated mesh networking protocols and with the possibility of deploying hundreds of devices, sensor networks offer radical new technological possibilities. With hundreds of powerful sensors configuring and communicating through sophisticated protocols, we have a very powerful platform with a huge potential to change the remote monitoring applications. Sensing applications represent a new paradigm for network operation; one that is different from more traditional wireless data networks. The power of WSNs lies in their ability to monitor the physical environment through ad hoc deployment of numerous self-configuring sensor nodes.

Wireless sensor networks are useful in a wide spectrum of applications ranging from residential, industrial, and environmental to military and space. Sensor networks are being increasingly deployed in outdoor environments, where they can monitor natural habitats, ecosystems, disaster sites, weather conditions, nuclear accidents, and battle fields. WSNs, being robust and distributed, are suitable for deployment in inaccessible environments and under inhospitable weather conditions where maintenance is inconvenient or impossible [55–58].

The most straightforward application of a wireless sensor network is to monitor remote environments. For example, a remote forest area can be monitored by deploying hundreds of sensors that configure themselves to form a network and immediately report upon detection of any event such as fire. Unlike traditional wired networks, the deployment of WSNs is relatively simple and inexpensive. Moreover, such networks can be easily extended by simply adding more devices without any rework or complex reconfiguration. The sensor nodes can ideally run for over a year on a single set of batteries. Given the cost of these sensor nodes, it is not feasible to discard dead sensor nodes, and it is also not possible to replace the batteries on these sensor nodes. Hence, there is a great need for energy-efficient protocols that can greatly reduce power consumption and increase the lifetime of wireless sensor nodes.

The third-generation wireless microsensor node, such as MICA2DOT [59], is approximately the size of a quarter. It can be seen that the biggest challenge in wireless sensor networks is to cope with the harsh resource constraints posed by their limited size. The most difficult resource constraint to meet is the power consumption. Being tiny, they prohibit use of large and long-lasting energy sources. The technology is driving these devices to become smaller by the day, thereby putting more constraints on the resources onboard a sensor node. Hence, a wireless sensor network platform must provide support for power-aware and energy-efficient protocols that can drastically reduce power consumption.

Such architectures increase the operating life of a sensor network by minimizing the amount of energy spent on its operations. In general, energy efficiency of any system can be defined as the ratio between the amount of data transmitted and the energy consumed for that operation. Hence, an energy-efficient system is one that minimizes the total amount of energy spent on its operations. Looking at the general operation of a sensor node, we can say that there are ways to improve the energy efficiency of WSNs. In a sensing application, the observer is interested in monitoring the behavior of the phenomenon under some specified performance requirements. In a sensor network environment, the individual sensors sample local values and disseminate information as needed to other sensors and eventually to the observer (controller node). The integrated information at the controller node is used to derive appropriate inferences about the environment. The radio interface of a sensor node hence consumes a major part of the energy. Table 4.5 lists the power consumption measurements of a typical sensor node for CPU and the radio for active, sleep, transmitting and receiving modes. The energy consumption can be minimized by reducing the retransmissions and the average number of transmissions. The wireless channel, being time-varying and bursty, dictates the packet losses in the network by causing frame errors. These erroneous frames are generally discarded at the destination resulting in retransmissions resulting in a waste of valuable energy. There is a need to identify the primary reasons for these frame errors and minimize them to improve the energy efficiency of WSNs. The data transmitted

Table 4.5   Experimental Node Current Consumption [58]

CPU	Radio (OOK Modulation)
2.9 mA (Active)	5.2 mA (Tx-19.2 Kbps)
1.9 mA (Sleep)	3.1 mA (Tx-2.4 Kbps)
1.0 μA (Off)	4.1 mA (Rx)
	5.0 μA (Sleep)

over a wireless channel accumulates errors due to three kinds of impairments: interference, slow fading, and fast fading.

Interference is either caused by white noise or by other users using the same frequency band. For example, Bluetooth, microwave ovens, IEEE 802.11b, or cordless phones degrade the performance of the wireless sensor networks. Slow fading is usually composed of path losses and shadowing. Fast fading is generally the result of delay spreading or Doppler frequency shifting, or both. The problems posed by the wireless channel, if addressed, can greatly improve the energy efficiency of the wireless sensor nodes. As mentioned above, retransmissions are the primary cause for the wastage of energy resources.

This section presents link adaptation mechanisms that enhance energy efficiency of WSNs. Our work focuses on improving the MAC layer since MAC is responsible for the successful operation of a network by providing access to the shared medium and transmitting data over the physical channel. The emphasis is on reducing the energy consumption without compromising MAC fairness and performance. The proposed algorithms reduce retransmissions that are primarily caused by frame errors and collisions. Since the rate of frame errors is sensitive to the frame size, we adjust the amount of data transmitted at any instant depending on the channel quality. We use a variable frame size approach instead of a fixed frame size, which is generally used in WSNs. We optimize the frame size with respect to the channel quality by using an optimal frame size predictor that improves the energy efficiency. The other advantage of having a variable frame size is improved throughput; sensor nodes transmit large data frames when the channel is clear. Thus, a variable frame size not only reduces energy consumption but also ensures faster delivery of data.

MAC design has been a broad research area, and much research has been conducted in the area of wireless sensor networks to improve energy efficiency. In [60] an energy-efficient MAC protocol called sensor-MAC (s-MAC) was proposed. Data link layer design issues for WSNs is discussed in [61]. In [56], a new scheme for energy efficiency in WSNs was proposed. Packet optimization was explored in WSNs, but a fixed frame size was adopted [55,62]. An adaptive data transmission rate control mechanism was proposed for WSNs [63]. An adaptive radio is proposed in [64]. Extensive work has been done in developing adaptive frame size algorithms [65] for wireless networks, but to the best of our knowledge no research has been conducted in the area of WSNs. Optimal frame size predictions using the traditional Kalman filter approach were proposed for wireless networks in [56, 66]. Work has been done in developing efficient hardware and software platforms for WSNs [15]. Recently, link adaptation has also been receiving extensive research effort [56, 63, 65–67]. In [56, 65–67] the frame length was adaptive, in [63] the data transmission rate was adaptive, in [62] the radio was adaptive, and in [60] the sleeping time was adaptive. The motivating factor behind these research efforts is the need to adapt the network with respect to the time-varying and bursty wireless channel quality. However, it is difficult to know how to track the changes of channel quality over a period of time and use this information to update a set of parameters in a systematic way. There are some research efforts trying to answer this problem [56, 65–67], but they are focusing on a general case and consider the system as a whole. They are not specific for WSNs, where energy efficiency is the primary concern and, quite often, there is no TCP/IP stack. Thus, in this section, we propose a link adaptation algorithm by adapting the frame size for energy efficiency in a WSN environment. We introduce Extended Kalman Filter (EKF) and Unscented Kalman Filter (UKF) approaches in this section to track the wireless channel quality and predict optimal frame size for data transmissions. The advantages of EKF and UKF—their simplicity and their ability to give accurate estimation and prediction results—will allow us to update the system settings accurately and systematically in an energy-efficient fashion as well as minimize the usage of the limited amount of sensor memory. Our work is different from the previous efforts in two important aspects:

1. A variable frame size was used instead of a fixed frame size.
2. EKF and UKF were used for predicting optimal frame size depending on the channel quality.

Simulation results showed that the proposed approach reduces the energy consumption in comparison to the fixed frame size and traditional Kalman filter approaches.

4.4.2 Proposed Approaches

Link adaptation techniques have received much research effort in the area of wireless data networks. Efforts have been made to develop algorithms that adapt MAC parameters such as frame size and data transmission rate to link quality. By adjusting the system parameters, we improve the link quality. For example, in an adaptive transmission rate system, depending on the channel quality, different modulation schemes are used to obtain desired data rates. The primary objective of these ideas is to adapt to a time-varying bursty wireless channel and to meet a certain level of QoS requirements imposed by certain applications. But it is difficult to track the changes of channel quality during a period of time and use this information to update a set of system parameters in a systematic way. Thus, we propose link adaptation algorithms to adapt the frame size with respect to the changing wireless channel quality.

The proposed algorithms predict an optimal frame size whenever a sensor has data to transmit based on the network parameters such as channel quality, frame length, protocol overheads, and collisions. A prediction scheme is necessary because of the time-varying and bursty nature of the wireless channel and also because the system performance is sensitive to frame size. When large frames are transmitted under a noisy channel, they accumulate bit errors, which triggers retransmissions. These retransmissions consume valuable energy and degrade the energy efficiency of the entire system.

An optimal frame size is predicted using a Kalman filter approach. The Kalman filter has been widely used as a predictor and corrector technique in linear discrete-data systems. The filter keeps track of the history of the system being modeled. Based on this history, the filter predicts the future state and then corrects the predicted value to a more accurate value using the current measurements of the system state. The MAC packages the data in a frame and transmits them over the wireless channel. For WSNs a fixed frame size is used, which if not properly set to an optimal value will result in frame errors and retransmissions. Studies have been carried out to come up with an optimal frame size depending on the application and frequency of data transmissions, but they ignored the most important factor that should influence the frame size, the channel quality. We use filtering techniques to keep track of the channel history and predict the optimal frame size given the present channel quality. The focus of this work is to improve the energy efficiency of wireless sensor networks at the MAC layer without affecting the performance. Our approach reduced energy consumption as well as yielded better throughput.

Extended Kalman filters and unscented Kalman filters are used for optimal frame size predictions. The advantages of using an optimal frame size are:

1. When the channel is noisy, the frame size is decreased, which reduces the chances of frame errors minimizing the retransmissions.
2. When the channel is free of noise, the frame size is increased, which improves the throughput of the sensor node.

We introduce the Kalman filters, the extended Kalman filters, and the unscented Kalman filter in next and also discuss the optimal frame size predictor.

Kalman Filter Link adaptation techniques adjust the system parameters in order to improve the overall network performance. If MAC parameters such as frame size and data rate are optimized depending on the link quality, it will definitely improve the energy efficiency of WSNs. However, it is not easy to predict optimal values for MAC parameters by tracking the channel characteristics. It becomes necessary to develop a prediction mechanism that can predict optimal values for different system-level parameters by keeping track of the network characteristics and the wireless channel quality. The problem of estimating the current state of a discrete-data system based on the past history has been of paramount importance in many fields of science, engineering, and finance. Estimation techniques called filters have been proposed to model the system under observation using a set of mathematical equations. Filtering can be defined as a technique that estimates the state of a system as a set of observations become available. The Kalman filter, the Weiner filter, and particle filters are well-known filtering techniques [68]. Due to the complexity and implementation overheads of both Weiner and particle filters, the Kalman filter is more widely used. The Kalman filter was proposed in 1960 as a solution to discrete-data linear filtering problems. Since then it has been the subject of extensive research and application. The filter has the unique capability of estimating the past, present, and future states of a system even without the precise knowledge of the modeled system [55, 68].

Before getting into the details of the Kalman filter, let us see how a general state-space model can be modeled using the Kalman filter, approach. The general state-space model is generally broken down into a state process and state measurement model:

$$p(x_{k+1}|x_k) \tag{4.1}$$

$$p(z_{k+1}/x_{k+1}) \tag{4.2}$$

where $x_{k+1} \in R_x^n$ denotes the states of the system at time $k + 1$ and $z_{k+1} \in R_z^n$ denotes the observations. The states follow a first-order Markov process, and the observations are assumed to be independent given the states. For example, we are interested in a nonlinear, non-Gaussian system; the model can be expressed as follows:

$$x_{k+1} = f(x_k, w_{k+1}) \tag{4.3}$$

$$y_{k+1} = h(u_{k+1}, x_k, v_{k+1}) \tag{4.4}$$

where $z_{k+1} \in R_z^n$ denotes the output observations, $u_{k+1} \in R_u^n$ denotes the input observations, $x_k \in R_x^n$ denotes the state of the system, $w_{k+1} \in R_w^n$ denotes the process noise, and $v_{k+1} \in R_v^n$ denotes the measurement noise. The mappings $f : R_x^n X R_v^n \longmapsto R_x^n$ and $h : R_x^n X R_v^n \longmapsto R_z^n$ represent the deterministic process and measurement models. To complete the specification of the model, the prior distribution (at $k = 0$) is denoted by $p(x_0)$. The posterior density $p(x_{0:k}|y_{1:k})$, where $x_{0:k} = x_0, x_1, \ldots, x_k$ and $y_{1:k} = y_1, y_2, \ldots, y_k$ constitutes the complete solution to the sequential estimation problem. In many applications, such as tracking, it is of interest to estimate one of its marginal probability density functions, namely the filtering density $p(x_k|y_{1:k})$. By computing the filtering density recursively, we do not need to keep track of the complete history of the states. Thus, from a storage point of view, the filtering density is more parsimonious than the full posterior density function. If we know the filtering density, we can easily derive various estimates of the system's states including means, modes, medians, and confidence intervals.

To estimate any system one starts by modeling the evolution of the system and also the noise in measurements. The resulting models are called process model and measurement model, and the equations that predict and correct the estimates are called time update and measurement update equations, respectively. The time and measurement update equations are used to estimate the past, present, and future states of the system being modeled. Consider a general prediction problem with noisy data. Suppose we are trying to estimate the state variable $x \in R^n$ of a discrete-time controlled process. The current estimate of the state variable x_{k+1} is given by a linear equation of the form

$$x_{k+1} = Ax_k + Bu_{k+1} + w_{k+1} \tag{4.5}$$

and a measurement $z \in R^m$ given by

$$z_{k+1} = H_{k+1}x_k + v_{k+1} \tag{4.6}$$

In these equations x_k is the system state at time k, x_{k+1} and is the system state at time $k + 1$, and u_{k+1} is the system control input at time $k + 1$. The random variables w_{k+1} and v_{k+1} represent the process and measurement noise. These are independent of each other and have normal probability distributions. The term A relates the state at time step k to the state at step $k + 1$, B relates the control input u to the current state, and H relates the current state of the system to the measurement z_{k+1}.

The Kalman filter estimates a process by using a form feedback control: The filter estimates the process state at any instant and obtains the feedback in the form of measurements. The equations used by a Kalman filter are categorized into two types: *time update equations* and *measurement update equations*. The time update equations are responsible for projecting the current state and error covariance estimates forward to obtain the a priori estimates for the next time step. The measurement update equations are responsible for the feedback, that is, incorporating a new measurement into the a priori estimate to obtain an improved a posteriori estimate. The time update equations can also be considered as predictor equations, while the measurement update equations can be considered as corrector equations. Hence, the final Kalman filter algorithm resembles a predictor-corrector algorithm. The equations for the time and measurement updates Follow:

$$\hat{x}'_{k+1} = A\hat{x}'_k + Bu_{k+1}$$

$$P'_{k+1} = AP_k A^T + Q_{k+1} \tag{4.7}$$

$$K_{k+1} = P'_{k+1} H^T_{k+1} \left(H_{k+1} P'_{k+1} H^T_{k+1} + R_{k+1} \right)^{-1}$$

$$\hat{x}_{k+1} = \hat{x}'_{k+1} + K_{k+1} \left(z_{k+1} - H_{k+1}\hat{x}'_{k+1} \right)$$

$$P_{k+1} = (I - K_{k+1} H_{k+1}) P'_{k+1} \tag{4.8}$$

where K_{k+1} is known as the Kalman gain, Q_{k+1} is the variance of the process noise, and R_{k+1} is the variance of the measurement noise.

The Kalman filter, even though one of the widely used mechanisms for estimating linear systems, has limitations in modeling nonlinear and non-Gaussian systems. New filtering techniques such as the extended Kalman filter (EKF) and the unscented Kalman filter were proposed to overcome the limitations of the Kalman filter. These schemes are efficient estimating schemes for nonlinear and non-Gaussian systems [56]. In this section we propose a novel approach for optimal frame size predictions based on the extended Kalman filter and the unscented Kalman filter under the time-varying, bursty wireless channel. Earlier efforts [56, 66] in this area used a Kalman filter approach for wireless data networks. In this work we utilize both EKF and UKF approaches, which are proven to be more effective estimation techniques than the traditional Kalman filter [69].

Extended Kalman Filter   The EKF is a minimum mean-square error (MMSE) estimator that allows incorporating the latest observations into a prior updating routine. The filter is based upon the principle of linearizing the measurements and evolution models. The capability of linearizing the models to incorporate nonlinear and non-Gaussian systems make EKF different from the traditional Kalman filter. EKF is based on the Taylor series expansion of the current estimate and measurement nonlinear functions f and h. If the process has a state vector $x \in R^n_x$, the current estimate x_{k+1} is given by the process model f and a measurement z_{k+1}, where $z \in R^m_z$ is given by the observation model h as shown below:

$$x_{k+1} = f(x_k, u_{k+1}, w_{k+1})$$

$$z_{k+1} = h(x_k, v_{k+1}) \tag{4.9}$$

where w_k and v_k are process noise and measurement noise, respectively, and u_k is the control vector for the random variable x. Using only the linear expression terms, it is easy to derive the following update equations for the mean $\hat{x}_{k+1}$ and covariance P_{k+1} of the Gaussian approximation to the posterior distribution of the states $x \in R^n_x$. The time update and measurement update equations are as shown below:

$$\hat{x}'_{k+1} = f\left(\hat{x}'_k, 0\right)$$

$$P'_{k+1} = F_{k+1} P'_k F^T_{k+1} + G_{k+1} Q_{k+1} G^T_{k+1}$$

$$K_{k+1} = P'_{k+1} H^T_{k+1} \left[\left(U_{k+1} R_{k+1} U^T_{k+1} + H_{k+1} P_{k+1} H^T_{k+1} \right) \right]^{-1} \tag{4.10}$$

$$\hat{x}_{k+1} = \hat{x}'_{k+1} + K_{k+1}\big[z_{k+1} - h\big(\hat{x}'_{k+1}, 0\big)\big]$$
$$P_{k+1} = P'_{k+1} - K_{k+1}H_{k+1}P'_{k+1} \tag{4.11}$$

where K_{k+1} is the Kalman gain, Q_{k+1} is the variance of the process noise, R_{k+1} is the variance of the measurement noise, and F_{k+1} and G_{k+1} are the Jacobians of the process model and H_{k+1} and U_{k+1} of the measurement model, respectively:

$$F_{k+1} = \frac{\partial f(x_{k+1})}{\partial x_{k+1}}\bigg|(x_{k+1} = \hat{x}_{k+1})$$

$$G_{k+1} = \frac{\partial f(w_{k+1})}{\partial w_{k+1}}\bigg|(w_{k+1} = \overline{w}) \tag{4.12}$$

$$H_{k+1} = \frac{\partial h(x_{k+1})}{\partial x_{k+1}}\bigg|(x_{k+1} = \hat{x}_{k+1})$$

$$U_{k+1} = \frac{\partial h(v_{k+1})}{\partial v_{k+1}}\bigg|(v_{k+1} = \overline{v}) \tag{4.13}$$

EKF calculates the posterior mean and covariance accurately to the first order with all higher order moments truncated.

Unscented Kalman Filter   The UKF is a recursive minimum mean-square error (RMMSE) estimator. The UKF also incorporates the latest observations into a prior updating routine. In addition, the UKF generates proposal distributions that match the true posterior more closely and has the capability of generating heavier tailed distributions than the well-known Kalman filter. The UKF addresses some of the approximation issues of the Kalman filter. Unlike the Kalman filter, UKF does not approximate the process and observation models. It uses the true nonlinear models and rather approximates the distribution of the state random variable. In the UKF, the state distribution is represented by a Gaussian random variable (GRV) and is specified using a minimal set of deterministically chosen sample points. These sample points completely capture the true mean and covariance of the GRV. When propagated through the true nonlinear system, they capture the posterior mean and covariance accurately to the second order for any nonlinearity, with errors only introduced in the third and higher orders. The process described above is called unscented transformation (UT), which is a method for calculating the statistics of a random variable that undergoes a nonlinear transformation and builds on the principle that it is easier to approximate a probability distribution than an arbitrary nonlinear function.

In UKF the state random variable (RV) is redefined as the concatenation of original state and noise variables: $x_{k+1}^a = [x_{k+1}^T w_{k+1}^T v_{k+1}^T]^T$. The scaled unscented transformation σ point scheme is applied to this new augmented state RV to calculate the corresponding σ matrix, x_{k+1}^a. The complete UKF algorithm that updates the mean $\hat{x}$ and covariance P of the Gaussian approximation to the posterior distribution of the states is given by the following set of equations. After concatenating the original state and noise variables into x_{k+1}^a, we calculate the σ points at time step $k + 1$ using the following equation:

$$X_{k+1}^a = \Big[\hat{x}_k^a \, \hat{x}_k^a \pm \sqrt{(v_a + \lambda)P_k^a}\Big] \tag{4.14}$$

Using the σ points, the process, and the measurement models, the following time update equations have been developed:

$$X_{k+1|k}^x = f(X_k^x, X_k^w)$$

$$\hat{x}_{k+1|k} = \sum_{i=0}^{2n_a} W_i^{(m)} X_i^x, k+1|k$$

$$P_{k+1|k} = \sum_{i=0}^{2n_a} W_i^{(c)} [X_i^x, k+1|k - \hat{x}_{k+1|k}][X_i^x, k+1|k - \hat{x}_{k+1|k}]^T$$

$$Y_{k+1|k} = h(X_{k+1|k}^x X_k^w)$$

$$\hat{y}_{k+1|k} = \sum_{i=0}^{2n_a} W_i^{(m)} Y_i, k+1|k \tag{4.15}$$

These estimates of the mean and covariance are accurate to the second order of the Taylor series expansion for any nonlinear function. Errors are introduced in the third and higher order moments but are scaled by the choice of the system parameters. In comparison, EKF only calculates the posterior mean and covariance accurately to the first order with all higher order moments truncated. Based on these estimated values, Kalman gain K_{k+1} is computed, and then the mean $\hat{x}_{k+1}$ and covariance P_{k+1} are updated using the following measurement update equations.

$$P_{\hat{y}_{k+1}\hat{y}_{k+1}} = \sum_{i=0}^{2n_a} W_i^{(c)} [Y_i, k+1|k - \hat{y}_{k+1|k}]$$

$$|X|[Y_i, k+1|k - \hat{y}_{k+1|k}]^T$$

$$P_{x_{k+1}y_{k+1}} = \sum_{i=0}^{2n_a} W_i^{(c)} [X_i^x, k+1|k - \hat{x}_{k+1|k}]$$

$$|X|[Y_i, k+1|k - \hat{y}_{k+1|k}]^T$$

$$K_{k+1} = P_{x_{k+1}y_{k+1}} P_{\hat{y}_{k+1}\hat{y}_{k+1}}^{-1}$$

$$\hat{x}_{k+1} = \hat{x}_{k+1|k} + K_{k+1}(y_{k+1} - \hat{y}_{k+1|k})$$

$$P_{k+1} = P_{k+1|k} - K_{k+1} P_{\hat{y}_{k+1}\hat{y}_{k+1}} K_{k+1}^T \tag{4.16}$$

where $x^a = [x^T w^T v^T]^T$, $X^a = [X^{xT} X^{wT} X^{vT}]^T$, λ is the composite scaling parameter, $n_a = n_x + n_w + n_v$, Q is the process noise covariance, R is the measurement noise covariance, K_{k+1} is the Kalman gain, and W_i are the weights of the σ points. Note that no explicit calculation of Jacobians or Hessians are necessary to implement this algorithm. UKF outperforms EKF in accuracy and robustness and it does so at no extra computational cost. The superior performance of the UKF over EKF is reported in [69].

4.4.3 Optimal Frame Size Predictor

The wireless channel, due to its time-varying nature, is highly inefficient when a fixed frame size scheme is used. As discussed in the previous sections, a variable frame size minimizes the rate of frame errors caused by transmitting large packets through a noisy channel, which in turn reduces the number of retransmissions. When the channel quality is good, large frames are transmitted, which increases the throughput of the system. By using an optimal frame size, we not only save energy. but also increase the system throughput.

EKF and UKF were used to predict the frame size for any transmission based on the history of the system, which includes the channel quality, frame size, protocol overheads, and collisions. Also note that due to changes in a network environment such as the number of users, as well as different channel qualities sensed at transmitter and receiver, the frame size is a local optimal. Every sensor node uses the frame predictor before it transmits data. In order to use EKF or UKF to predict optimal frame size, we need to develop process model and observation model that will fit into the traditional EKF and UKF models. As our focus is to optimize the throughput of the sensor nodes at any instant, the channel throughput is considered for developing process and observation models. The channel throughput depends on the frame size, protocol delays, channel quality, collision rate, and data rate; the following equation gives the relationship between throughput and the frame size:

$$\rho = \frac{LR}{\begin{aligned}&(L + H)[(1 - P_b)^{-(L+H)} + N]\\&+(B + D + h + o)[(1 - P_b)^{-(L+H)} + N]R\\&+ACK(1 - Pb)^{-\text{ACK}}\\&+O[(1 - P_b)^{-\text{ACK}}R + N*C]\end{aligned}} \tag{4.17}$$

where

L	=	payload size of a frame
R	=	transmission data rate
H	=	MAC protocol header of a frame
N	=	average number of collisions occurred between two renewal points
B	=	average number of random backoff time slots under a certain network load and channel quality. The backoff procedure follows the algorithm specified by the CSMA/CA standard
D	=	DIFS specified by the standard
h	=	preamble overhead introduced by the PHY layer
P_b	=	bit error rate under a known channel quality
ACK	=	frame length of an ACK frame
o	=	protocol overhead such as MAC and PHY processing delays, Tx/Rx switching time durations
O	=	overhead of ACK
C	=	average length of collisions

Equations (1.20) and (1.21) give the frame lengths at time k and $k + 1$, respectively

$$\rho_k = \frac{L_k R_k}{(L_k + H)[(1 - P_{b_k})^{-(L_k+H)} + N_k]}$$
$$+ (B + D + h + o)[(1 - P_{b_k})^{-(L_k+H)} + N_k]R_k$$
$$+ ACK(1 - P_{b_k})^{-ACK}$$
$$+ O[(1 - P_{b_k})^{-ACK}R_k + N_k C_k] \tag{4.18}$$

$$\rho_{k+1} = \frac{L_{k+1} R_{k+1}}{(L_{k+1} + H)[(1 - P_{b_{k+1}})^{-(L_{k+1}+H)}}$$
$$+ N_{k+1}] + (B + D + h + o)[(1 - P_{b_{k+1}})^{-(L_{k+1}+H)}$$
$$+ N_{k+1}]R_{k+1} + ACK(1 - Pb_{k+1})^{-ACK}$$
$$+ O[(1 - P_{b_{k+1}})^{-ACK}R_{k+1} + N_{k+1}C_{k+1}] \tag{4.19}$$

Our goal is to maximize the channel throughput ρ, at every transmission by predicting an optimal frame size, with considerations of collisions and frame errors. Hence, the following equations can be derived by combining the above two equations. The optimal frame size at time $k + 1$ is obtained by solving the following equation for L_{k+1}:

$$L_k P_{b_{k+1}} L_{k+1}^2 + u L_{k+1} + v = 0$$
$$L_{k+1} = \frac{-u \pm \sqrt{u^2 - 4L_k Pb_{k+1}v}}{2L_k Pb_{k+1}} \tag{4.20}$$

where u and v are given by

$$u = 2L_k P_{b_{k+1}} H + L_k S R_k P_{b_{k+1}}$$
$$- P_{b_k} L_k^2 - 2L_k P_{b_k} H - H - P_{b_k} H^2 - H N_k$$
$$- SR_k - SR_k P_{b_k} L_k - SR_k P_{b_k} H - SR_k N_k$$
$$- ACK(1 - P_{b_k})^{-ACK} - O R_k(1 - P_{b_k})^{-ACK}$$
$$- N_k C_k \tag{4.21}$$

$$v = L_k H + L_k P_{b_{k+1}} H^2 + L_k H N_k + L_k S R_k$$
$$+ L_k SR_k P_{b_{k+1}} H + L + k SR_k N_k$$
$$+ L_k ACK(1 - P_{b_{k+1}})^{-ACK}$$
$$+ L_k O R_k(1 - P_{b_{k+1}})^{-ACK}$$
$$+ L_k N_k C_k \tag{4.22}$$

The frame size prediction model is therefore defined. In general, most communication network systems have restrictions on frame sizes such as $L_{\max}$ and $L_{\min}$. The following equations are then developed as the state transition model of the frame size predictor. The process model is

$$
\begin{aligned}
L_{k+1} &= L_{\max} && L_{k+1} > L_{\max} \\
&= L_{\mathrm{opt}} && L_{\min} < L_{k+1} < L_{\max} \\
&= L_{\min} && L_{k+1} < L_{\min}
\end{aligned}
\tag{4.23}
$$

where

$$
L_{\mathrm{opt}} = \frac{-u \pm \sqrt{u^2 - 4L_k P_{bk+1} v}}{2 L_k P_{bk+1}}
\tag{4.24}
$$

And the observation model is

$$
\rho_{k+1} = Q(L_{k+1}, Pb_{k+1})
\tag{4.25}
$$

where ρ_{k+1} is the observation at time $k + 1$ and Q is the observation function. In [56, 66, 67] it is shown that the Kalman filter when used for estimation reduces the root-mean-square (rms) errors considerably. Additionally, in [68] it is stated that the UKF has lower rms errors while estimating a nonlinear and non-Gaussian system. UKF is more accurate than a traditional Kalman filter and EKF in its predictions for nonlinear systems.

4.4.4 Simulation Environment

We designed and verified various network scenarios to conduct the simulations. The parameters used for the simulator were chosen from widely used sensor network devices so as to make the simulations close to real networks. The primary objective for running the simulations was to measure the energy consumed by the radio interfaces onboard the sensor nodes. Figure 4.42 depicts a sample topology consisting of five nodes, where nodes 1 through 4 are sensor nodes that send the sensed information to node 5. This topology was chosen because information from the sensor nodes is usually collected at a central node for processing, typically a controller node. The controller node collects the information from these sensor nodes, processes it, and decides what action(s) to take.

Depending on the location of the controller node, the sensor node either directly sends the data, to the controller node, or by routing the data through its neighboring nodes. Apart from the sensed data, the information that is exchanged by the sensor nodes is the routing information. We computed the energy consumed at the sensor nodes by keeping track of all the transmissions. All simulations were run using Qualnet [70] by plugging the extended Kalman filter and the unscented Kalman filter modules into the simulator. These modules were implemented in C and have been tested extensively to make sure it complies with the model in terms of results obtained. In [69] the authors have provided sample data and results of the experiments conducted with different filtering techniques including the EKF and UKF. The code was tested by running these experiments, and the results obtained were compared with the sample data to validate the code's accuracy.

MAC Layer Settings The MAC layer in the simulator was modeled to suit the wireless sensor networks. For the adaptive approach, lower and upper bounds were used (20 and 250 bytes, respectively) to keep the predicted frame size within a desired range. For the fixed

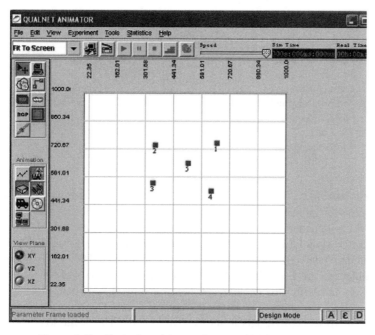

Figure 4.42 Illustration of the topology used by the simulations.

frame approach, a frame size of 150 bytes was used. The fixed frame size was set to 150 after running simulations with different sizes in the range of 50 to 500 bytes. Since WSNs are not restricted just to applications such as weather monitoring where the data is in the order of few bytes, we chose a larger message size of 1 byte. There have been applications where the sensors were mounted on a robot and transmitted video to the controller node. In applications such as seismic and robotic, the data can get as big as hundreds of kbytes depending on the level of activity. Also when there is a high activity around a sensor, huge amounts of data are generated. This requires larger packet sizes and efficient schemes to conserve energy. Unlike MAC for wireless networks, we considered fragmentation for packets larger than the frame size used for transmitting data (either fixed or predicted). The MAC parameters used for the simulations are listed in Table 4.6.

Table 4.6 MAC Layer Parameters Used for the Simulation

MAC parameters	
SIFS	770 μs
DIFS	1540 μs
Time slot	55 μs
Data rate	19.2 kbps
MAC header length	64 bits
ACK length	80 bits
Maximum frame length	250 bytes
Minimum frame length	20 bytes
CWmin	32
CWmax	256

Table 4.7 Physical Channel Parameters Used for the Simulation

PHY Parameters	
Operating frequency	916.30– 916.70 MHz
Modulation type	OOK
Demodulation type	Non-coherent
Transmitting current	12 mA
Receiving current	4.5 mA
Data rate	19.2 kbps
FEC scheme	None

PHY Layer Settings   The physical (PHY) layer in the simulator was also modeled to suit the wireless sensor networks. To make the simulations more close to the real system, we chose the PHY parameters from TR1000, one of the most widely used wireless transceivers developed by RF Monolithic Inc. [71]. The power settings, operating frequency, modulation scheme, and data rate were set from the TR1000's data sheet. The PHY parameters used for the simulations are listed in Table 4.7. The location of the sensors was considered to be relatively fixed after the deployment; hence a slow fading channel was considered. In this simulation, we chose 1 kbyte as the message size. Upon detecting an event the sensor node sends the sensed data to the controller node. The traffic load was varied by changing the interarrival times between the packets. After the simulation environment was set up as per the above mentioned parameters, we ran extensive simulations to verify the proposed model. A simple energy consumption model was used to make the computations. We measured the time spent by each node in transmitting data, receiving data, and listening (idle). The duration of each operation was multiplied by the corresponding power setting used in the simulator. The simulations were run for both approaches with similar traffic load and channel conditions. The results suggest that our approach reduces energy consumption. In Section 4, We further discuss the performance of the proposed scheme in comparison to the fixed frame approach.

4.4.5   Results

The simulation parameters were set to simulate close to real-time sensor network scenarios. We have run extensive simulations with EKF and UKF implementations of the optimal frame size predictor. The algorithms were tested for performance evaluation under different channel quality conditions by modifying the PHY layer parameters. The results obtained are presented and discussed in detail in the following sections. The EKF and UKF implementations of the optimal frame size predictor were run by plugging the algorithms into the MAC implementation of the simulator. We have considered four network scenarios with 2, 5, 10, and 20 nodes, respectively. All the network scenarios have sensor nodes and controller nodes, where the sensor nodes send the sensed information periodically to the controller node. The controller nodes are usually plugged into a continuous power supply source and the sensor nodes are battery operated. Hence, the power consumed by each of these sensor nodes is computed and recorded toward the end of the simulation. Simulations were conducted for all the network scenarios and the average power consumed by each sensor node is computed.

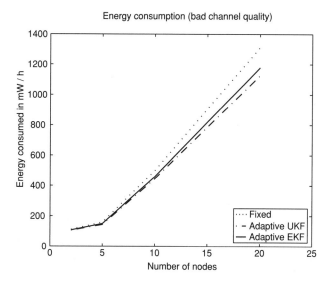

Figure 4.43 Comparison of energy efficiency using EKF and UKF approaches under a noisy channel.

Similar experiments were conducted with the fixed frame size approach. The average power consumed by each sensor node is computed for all the different network scenarios considered. The energy efficiency of the EKF and UKF approaches, in comparison to the fixed frame approach, can be seen in Figures 4.43 and 4.44. The solid lines denote the energy consumption by EKF approach, the dashed lines represent the energy consumption by UKF approach, and the dotted lines represent the energy consumed by sensors using fixed frame size. The simulations were run for two different channel conditions; Figure 4.43

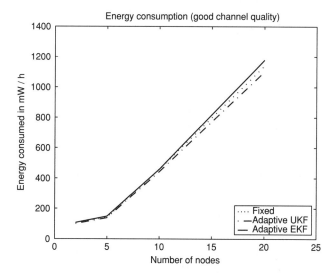

Figure 4.44 Comparison of energy efficiency using EKF and UKF approaches under a clear channel.

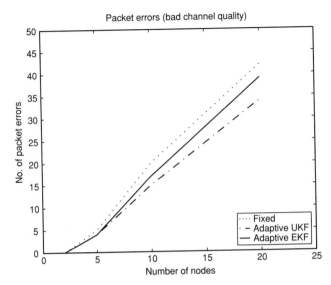

Figure 4.45   Comparison of packet errors using EKF and UKF approaches under a noisy channel.

shows the energy consumed by the sensor nodes under a noisy channel (a noisy channel is simulated by raising the noise floor of the receiver). Similarly, Figure 4.44 shows the energy consumed under a clear channel. In both cases we can see improved energy efficiency, but it can be seen that the proposed algorithms performed better under noisy channel conditions than in a clear channel. The reason for this behavior is that under noisy channel the probability of frame errors is high; by adapting to the bad channel we achieve better energy efficiency. As the objective of this research work is to reduce retransmissions by minimizing frame errors, we decided to monitor frame errors. The frame errors indicate the number of frames that are discarded and retransmitted. Figures 4.45 and 4.46 show the number of frame errors occurring for different network scenarios. The results were observed for both adaptive frame size approaches (EKF and UKF) and the fixed frame size approach. The experiments were again conducted for different channel quality conditions. The results shown in Figures 4.45 and 4.46 indicate that the frame errors are reduced considerably by the adaptive algorithms, which in turn reduced the energy consumption. Our results analysis showed an improvement of energy efficiency with the enhanced MAC. The simulations were run for different channel conditions. A noisy channel was simulated by increasing the noise floor of the receiver. It can be seen that energy consumption is reduced as a result of using an optimal frame size for every transmission. It can also be observed that our approach yielded better results under noisy channel conditions. By using the EKF, we achieved a 10% reduction in energy consumption for 20 nodes under a noisy channel, a 4% reduction under a clear channel. With UKF we achieved a 14% reduction in energy consumption for 20 nodes under a noisy channel, a 5% reduction under a clear channel. It can be seen that the proposed algorithms resulted in lower frame errors than the fixed frame size approach. When the channel was noisy the frame size was adjusted to a lower value, reducing the chances of packets accumulating errors. The throughput of the system is also improved by transmitting larger frames when the channel is free of noise. These results highlight the advantages of the link adaptation techniques.

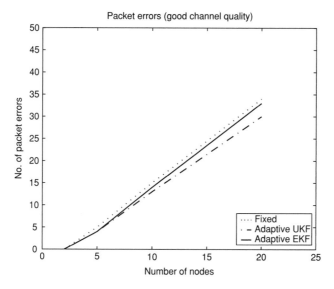

Figure 4.46 Comparison of packet errors using EKF and UKF approaches under a noise-free channel.

The algorithms were also tested for performance evaluation with mobile sensor nodes. We have considered similar network scenarios with 2, 5, 10, and 20 nodes moving randomly in the terrain. All the network scenarios have sensor nodes and controller nodes, where the sensor nodes are mobile and the controller node is stationary. The power consumed by each of these sensor nodes is computed and recorded. The energy efficiency of the EKF and UKF approaches, in comparison to the fixed frame approach, can be seen in Figures 4.47 and 4.48.

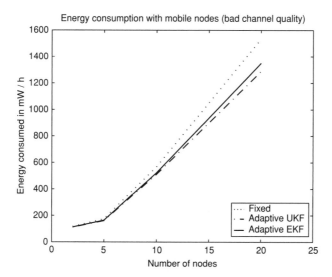

Figure 4.47 Comparison of energy efficiency using EKF and UKF approaches under a noisy channel with mobile nodes.

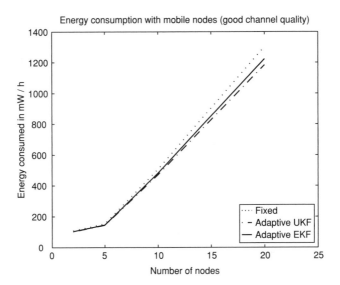

Figure 4.48   Comparison of energy efficiency using EKF and UKF approaches under a clear channel with mobile nodes.

The solid lines denote the energy consumption by EKF approach, the dashed lines represent the energy consumption by UKF approach, and the dotted lines represent the energy consumed by sensors using fixed frame size. It can be seen that the proposed algorithms have similar performance even with mobile nodes. There is an increase in energy consumption in both the adaptive and fixed approach due to an increase in channel interference. As the nodes move, we have a fast fading effect that degrades the channel further; that is the reason we see higher energy consumption. There is also a slight increase in performance; we see 11% reduction in energy consumption by using the EKF for 20 nodes under a noisy channel, a 5% reduction under a clear channel. With UKF we achieved a 16% reduction in energy consumption for 20 nodes under a noisy channel, a 6% reduction under a clear channel.

The algorithms were also tested for performance evaluation with large sensor nodes with a longer message size of 10 kbytes . We have considered network scenarios with 2, 5, 10, 20, 50, and 100 nodes. The energy efficiency of the EKF and UKF approaches, in comparison to the fixed frame approach, can be seen in Figures 4.49 and 4.50. The solid lines denote the energy consumption by the EKF approach, the dashed lines represent the energy consumption by the UKF approach, and the dotted lines represent the energy consumed by sensors using fixed frame size. It can be seen that the proposed algorithms have similar performance even with large networks and larger message sizes.

The algorithms were then tested for throughput performance evaluation with a message size of 10 kbytes. We have considered network scenarios with 2, 5, 10, 20, 50, and 100 nodes. The energy efficiency of the EKF approach, in comparison to the fixed frame approach, can be seen in Figure 4.51. The solid lines denote the energy consumption by the EKF approach, and the dotted lines represent the energy consumed by sensors using fixed frame size. It can be seen that the proposed algorithms have better throughput performance than the fixed frame size approach sizes.

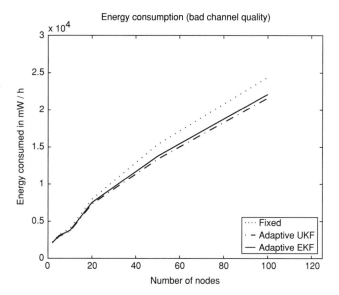

Figure 4.49 Comparison of energy efficiency using EKF and UKF approaches under a noisy channel with longer messages.

The results suggested that UKF performed better than the EKF. Since EKF only uses the first-order terms of the Taylor series expansion of the nonlinear functions, it often introduces large errors in the estimated statistics of the posterior distributions of the states. In contrast, UKF does not approximate the nonlinear process and observation models; it uses the true nonlinear models and rather approximates the distribution of the state random variable. The

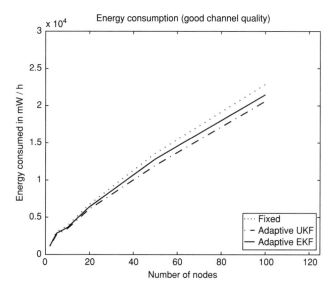

Figure 4.50 Comparison of energy efficiency using EKF and UKF approaches under a clear channel with longer messages.

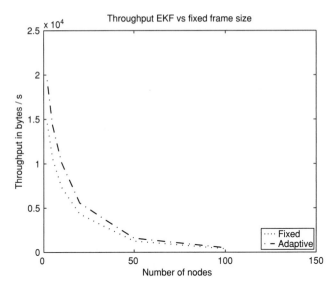

Figure 4.51   Comparison of throughput using EKF approach and fixed frame size approach.

results of the UKF approach in comparison to the EKF approach showed more accuracy in predictions and more efficiency in energy conservation [72].

4.4.6   Conclusion

We proposed energy-efficient link adaptation algorithms for wireless sensor networks. The primary objective of these enhancements at the medium access control layer was to reduce power consumption. We investigated the strict power requirements for WSNs and also the effect of link quality on power consumption. We studied various link adaptation schemes that address the adversity of the wireless channel and be energy efficient at the same time. The proposed link adaptation algorithms adjust the frame size depending on the wireless channel quality. By utilizing an optimal frame size predictor, the proposed architecture is able to support flexible communications protocols without sacrificing fairness and efficiency.

We employed the most widely used predictor-corrector technique (Kalman filter) for developing the optimal frame size predictor by which we can keep track of variations in channel quality accurately. The two most efficient filtering schemes that are based on the traditional Kalman filter for modeling nonlinear and non-Gaussian systems are the extended Kalman filter and unscented Kalman filter. The EKF and UKF were found to derive accurate estimates of states of the system being modeled. The algorithms have been validated through the development and evaluation of software modules for the simulation package. The correctness of our model is verified by both simulative and analytical results. The new MAC is found to deal with the power constraints and time-varying wireless channel quality. The simulation results suggest that the algorithms resulted in improved energy efficiency without compromising the network performance. This study has a major impact on supporting link adaptation in wireless sensor networks because of its detailed analysis of optimal frame size predictors.

WSNs is a new but fast growing area; there are many unresolved problems that are to be explored. In this section, the work has been carried out within the context of the MAC protocol. Keeping in mind that improving energy efficiency is a system-wide task, we should also consider the performance of upper layer protocols over the wireless medium and how their performance could be improved by using link adaptation algorithms. Many wireless sensor network standards have been proposed and implemented by different groups of researchers. The development of a link adaptation algorithm with considerations of compatibility among different vendors is very important in building a seamless wireless sensor network platform in the future. We have also identified that the platform must be flexible to meet the wide range of application requirements posed by the sensor networks. The application scenarios for the sensor networks range from simple data collection applications to more complex environment monitoring applications. As each scenario has substantially different communication patterns, the algorithms developed must suit this kind of variety.

Our future work will focus on designing a full-fledged MAC. Along with the current enhancements and the enhancements discussed in this work, it also includes priority-based queuing, locality of sensor nodes, clustering of nodes, and data aggregation techniques. These improvements will further enhance the energy efficiency of wireless sensor networks.

REFERENCES

1. A. Arora, P. Dutta, S. Bapat, V. Kulathumani, H. Zhang, V. Naik, V. Mittal, H. Cao, M. Demirbas, M. Gouda, Y.-R. Choi, T. Herman, S. S. Kulkarni, U. Arumugam, M. Nesterenko, A. Vora, and M. Miyashita, "A line in the sand: A wireless sensor network for target detection, classification, and tracking," in *Computer Networks*, (Elsevier), 2004, to appear.

2. J. Hill, R. Szewczyk, A. Woo, S. Hollar, D. E. Culler, and K. Pister, "System architecture directions for network sensors," in *Proceedings of the International Conference on Architectural Support for Programming Languages and Operating Systems (ASPLOS)*, Cambridge, MA, Nov. 2000.

3. D. E. Culler, J. Hill, P. Buonadonna, R. Szewczyk, and A. Woo, "A network-centric approach to embedded software for tiny devices," in *EMSOFT*, Series Lecture Notes in Computer Science, vol. 2211, Springer, 2001, pp. 97–113.

4. A. Mairwaring, J. Polastre, R. Szewczyk, D. Culler, and J. Anderson, "Wireless sensor networks for habitat monitoring," in *Proceedings of the ACM International Workshop on Wireless Sensor Networks and Applications (WSNA)*, Atlanta, 2002.

5. "TinyOS: A component-based OS for the networked sensor regime," http://www.tinyos.net. Latest source available at http://sourceforge.net/cvs/?group "id=28656, June 2005.

6. S. S. Kulkarni and L. Wang, "MNP: Multihop network reprogramming service for sensor networks," Technical Report MSU-CSE-04-19, Department of Computer Science, Michigan State University, May 2004.

7. S. S. Kulkarni and M. Arumugam, "TDMA service for sensor networks," in *Proceedings of the Third International Workshop on Assurance in Distributed Systems and Networks*, Tokyo, Mar. 2004.

8. E. W. Dijkstra, "Self-stabilizing systems in spite of distributed control," *Communications of the ACM*, vol. 17, no. 11, pp. 643–644, 1974.

9. S. Dolev, *Self-Stabilization*, Cambridge, MA: MIT Press, 2000.

10. G. Agha, W. Kim, Y. Kwon, K. Mechitov, and S. Sundresh, "Evaluation of localization services (preliminary report)," DARPA NEST Program, http://osl.cs.uiuc.edu/docs/nest-localization-report-2003/, 2003.

11. T. He, C. Huang, B. Blum, J. Stankovic, and T. Abdelzaher, "Range-free localization schemes for large scale sensor networks," in *Proceedings of the Ninth Annual International Conference on Mobile Computing and Networking (MOBICOM)*, San Diego, 2003, pp. 81–95.

12. T. Herman, "NestArch: Prototype time synchronization service," http://www.ai.mit.edu/people/sombrero/nestwiki/index/ComponentTimeSync, Jan. 2003.

13. G. Simon, P. Volgyesi, M. Maroti, and A. Ledeczi, "Simulation-based optimization of communication protocols for large-scale wireless sensors networks," in *Proceedings of the IEEE Aerospace Conference*, Big Sky, MI, Mar. 2003.

14. U. Arumugam, "Collision-free communication in sensor networks," Master's thesis, Computer Science and Engineering, Michigan State University, Sep. 2003. Available http://www.cse.msu.edu/~arumugam/research/MastersThesis/main.ps.

15. Y. Xu, J. Heidemann and D. Estrin, "Geography-informed energy conservation for ad hoc routing," in *Proceedings of the ACM/IEEE International Conference on Mobile Computing and Networking (MOBICOM)*, Rome, July 2001, pp. 70–84.

16. A. Arora and M. Gouda, "Distributed reset," *IEEE Transactions on Computers*, vol. 43, no. 9, pp. 1026–1038, 1994.

17. S. S. Kulkarni and A. Arora, "Multitolerance in distributed reset," *Chicago Journal of Theoretical Computer Science*, 1998. Available: http://citeseer.nj.nec.com/kulkarni98multitolerance.html.

18. T. Stathopoulos, J. Heidemann, and D. Estrin, "A remote code update mechanism for wireless sensor networks," Technical Report CENS-TR-30, University of California, Los Angeles, Center for Embedded Networked Computing, Nov. 2003.

19. P. Levis, N. Patel, S. Shenker, and D. Culler, "Trickle: A self-regulating algorithm for code propagation and maintenance in wireless sensor network," in *Proceedings of the First USENIX/ACM Symposium on Networked Systems Design and Implementation (NSDI)*, San Francisco, 2004.

20. A. Woo and D. Culler, "A transmission control scheme for media access in sensor networks," in *Proceedings of the Seventh Annual International Conference on Mobile Computing and Networking*, Rome, 2001, pp. 221–235.

21. R. Rom and M. Sidi, *Multiple Access Protocols: Performance and Analysis*, Springer-Verlag, 1989. Also available http://www.comnet.technion.ac.il/rom/PDF/MAP.pdf.

22. W. Ye, J. Heidemann and D. Estrin, "An energy-efficient mac for wireless sensor networks," in *Proceedings of the Twenty-First International Annual Joint Conference of the IEEE Computer and Communications Societies (INFOCOM)*, New York, June 2002, pp. 1567–1576.

23. T. van Dam and K. Langendoen, "An adaptive energy-efficient protocol for wireless sensor networks," in *Proceedings of the First International Conference on Embedded Networked Sensor Systems (SenSys)*, Los Angeles, Nov. 2003, pp. 171–180.

24. A. J. Viterbi, *CDMA: Principles of Spread Spectrum Communication*, Redwood City, CA: Addison Wesley Longman, 1995.

25. V. Claesson, H. Lönn, and N. Suri, "Efficient TDMA synchronization for distributed embedded systems," in *Proceedings of the Twentieth IEEE Symposium on Reliable Distributed Systems (SRDS)*, New Orleans, Oct, 2001, pp. 198–201.

26. K. Sohrabi and G. J. Pottie, "Performance of a novel self-organization protocol for wireless ad-hoc sensor networks," in *Proceedings of the IEEE Vehicular Technology Conference*, Houston, 1999, pp. 1222–1226.

27. W. B. Heinzelman, A. P. Chandrakasan, and H. Balakrishnan, "An application-specific protocol architecture for wireless microsensor networks," *IEEE Transactions on Wireless Communications*, vol. 1, no. 4, pp. 660–670, Oct. 2002.

28. K. Arisha, M. Youssef, and M. Younis, "Energy-aware TDMA-based MAC for sensor networks," in *Proceedings of the IEEE Workshop on Integrated Management of Power Aware Communications, Computing and Networking (IMPACCT)*, New York, May 2002.

29. S. S. Kulkarni and U. Arumugam, "Collision-free communication in sensor networks," in *Proceedings of the Sixth Symposium on Self-stabilizing Systems (SSS)*, Lecture Notes in Computer Science, vol.2704, June 2003, pp. 17–31.

30. Y.-R. Choi, M. Gouda, M. C. Kim, and A. Arora, "The mote connectivity protocol," in *Proceedings of the Twelfth International Conference on Computer Communications and Networks*, Dallas, Oct. 2003, pp. 533–538.

31. L. Gasieniec, A. Pelc, and D. Peleg, "The wakeup problem in synchronous broadcast systems," *SIAM Journal of Discrete Mathematics*, vol. 14, no. 2, pp. 207–222, 2001.

32. B. S. Chlebus, L. Gasieniec, A. Gibbons, A. Pelc, and W. Rytter, "Deterministic broadcasting in ad hoc radio networks," *Distributed Computing*, vol. 15, no. 1, pp. 27–38, 2002.

33. "Low rate wireless personal area networks," IEEE P802.15.4/D18, Draft Standard, Feb. 2003.

34. IEEE, "Wireless LAN medium access control (MAC) and physical layer (PHY) specifications," IEEE 802.11, Part 11, Aug. 1999.

35. *Bluetooth SIG, Bluetooth Specifications, Version* 1.0, July 1999.

36. C. Karlof and D. Wagner, "Secure routing in wireless sensor networks: Attacks and countermeasures," paper presented at the *First IEEE International Workshop on Sensor Network Protocols and Applications*, Anchorage, AK, 2003.

37. A. Perrig, R. Canetti, D. Song, and D. Tygar, "The tesla broadcast authentication protocol," in *RSA Cryptobytes*, 2002.

38. Y. Hu, D. B. Johnson, and A. Perrig, "Sead: Secure efficient distance vector routing for mobile wireless ad hoc networks," in *Proceedings of the Fourth IEEE Workshop on Mobile Computing Systems and Applications*, Calicoon, NY, June 2002, pp. 3–13.

39. L. Eschenauer and V. Gligor, "A key-management scheme for distributed sensor networks," in *Conference on Computer and Communications Security, Proceedings of the 9th ACM conference on Computer and communications security*, Washington DC, 2002.

40. R. D. Pietro, L. V. Mancini, and A. Mei, "Random key assignment for secure wireless sensor networks," in *Proceedings of the 2003 ACM Workshop on Security of Ad Hoc and Sensor Networks*, Fairfax, VA, Oct. 2003.

41. A. D. Wood and J. A. Stankovic, "Denial of service in sensor networks," *IEEE Computer Magazine*, vol. 35, no. 10, pp. 54–62, Oct. 2002.

42. F. I. P. S. P. 197, *Advanced Encryption Standard (AES)*, Publication 197, U.S. Department of Commerce/NIST, Springfield, VA, Nov. 2001.

43. J. Heidemann, W. Ye, and D. Estrin, "An energy-efficient mac protocol for wireless sensor networks," in *Proceedings of the Twenty-First International Annual Joint Conference of the IEEE Computer and Communications Societies*, New York, June 2002.

44. E. Shih et al., "Physical layer driven protocol and algorithm design for energy-efficient wireless sensor networks," in *Proc. MOBICOM*, Rome, 2001.

45. A. Y. Wang, S. Cho, C. G. Sodini, and A. P. Chandrakasan, "Energy efficient modulation and mac for asymmetric rf microsensor systems," in *IEEE International Symposium on Low Power Electronics and Design*, Huntington Beach, CA, 2001.

46. V. Raghunathan, C. Schurgers, S. Park, and M. B. Srivastava, "Energy-aware wireless microsensor networks," *IEEE Signal Processing Magazine*, vol. 19, no. 2, pp. 40–50, Mar. 2002.

47. D. Ganesan et al., "Complex behavior at scale: An experimental study of low-power wireless sensor networks," Technical Report UCLA/CSD-TR 02-0013, UCLA Computer Science Department, 2002.

48. A. Woo and D. Culler, "A transmission control scheme for media access in sensor networks," in *Proceedings of the Seventh Annual International Conference on Mobile Computing and Networking*, New York, 2001, pp. 221–235.

49. M. Zorzi and R. R. Rao, "Multihop performance of energy-efficient forwarding for ad hoc and sensor networks in the presence of fading," in *IEEE/ICC '04*, Paris, France, June 20–24, 2004.

50. *Network Simulator—NS2*, Marina del Rey, CA: Information Sciences Institute.

51. J. H. Schiller, *Mobile Communications*, Reading, MA: Addison-Wesley, 2000.

52. M. L. Liu and T. Saadawi, "A bluetooth scatternet-route structure for multi-hop ad hoc networks," *IEEE Journal on Select Areas in Communications*, vol. 21, no. 2, pp. 229–239, Feb. 2003.

53. A. Cerpa and D. Estrin, "Adaptive self-configuring sensor networks topologies," in *Proceedings of the IEEE INFCOM*, New York, June 2002.

54. S. Ni, Y. Tseng, Y. Chen, and J. Sheu, "The broadcast storm problem in a mobile ad hoc network," in *Proceedings of the Fifth Annual ACM/IEEE International Conference on Mobile Computing and Networking*, Seattle, 1999, pp. 152–162.

55. I. Akeyildiz, Y. Sankarasubramanniam, and S. McLaughlin, "Energy efficiency based packet size optimization in wireless sensor networks," in *Proceedings of the First IEEE International Workshop on Sensor Network Protocols and Applications*, Anchorage, AK, 2003, pp. 1–8.

56. S. Ci, H. Sharif, and D. Peng, "A new scheme for energy efficiency in wireless sensor networks," in *Proceedings of the IEEE ICC 2004*, Paris, 2004.

57. A. Wang, W. Heinzelman, and A. Chandrakasan, "Energy-scalable protocols for battery-operated microsensor networks," in *Proceedings of the SiPS'99*, Taipei, Taiwan, 1999, pp. 483–492.

58. P. Sung, A. Savvides, and M. Srivastava, "Simulating networks of wireless sensors," in *Proceedings of the Simulation Conference' 01*, Arlington, VA, 2001, pp. 1330–1338.

59. "Mica2dot data sheet," C. T. Inc., 2003.

60. W. Ye, J. Heidemann, and D. Estrin, "An energy-efficient mac protocol for wireless sensor networks," in *Proceedings of the IEEE INFOCOM'02*, New York, 2002, pp. 1567–1576.

61. L. Zhong, J. Rabaey, G. Chunlong, and R. Shah, "Data link layer design for wireless sensor networks," in *IEEE MILCOM'01*, Anaheim, CA, 2001, pp. 352–356.

62. E. Modiano, "An adaptive algorithm for optimizing the packet size used in wireless arq protocols," *Wireless Networks*, vol. 5, pp. 279–286, 1999.

63. A. Woo and D. Culler, "A transmission control scheme for media access in sensor networks," in *ACM SIGMOBILE*, Rome, 2001.

64. C. Chien, M. Srivastava, R. Jain, P. Lettieri, V. Aggarwal, and R. Sternowski, "Adaptive radio for multimedia wireless links," *IEEE Journal on Selected Areas in Communications*, vol. 17, pp. 793–813, 1999.

65. P. Letteri and M. Srivastava, "Adaptive frame length control for improving wireless link throughput, range, and energy efficiency," in *Proceedings of the IEEE INFOCOM'98*, San Francisco, CA, 1998, pp. 564–571.

66. S. Ci, H. Sharif, and A. Young, "Frame size adaptation for indoor wireless networks," *IEEE Electronics Letters*, vol. 37, no. 18, pp. 1135–1136, 2001.

67. S. Ci and H. Sharif, "Adaptive approaches to enhance throughput of ieee 802.11 wireless lan with bursty channel," in *Proceedings of the Twenty-Fifth Annual IEEE Conference on Local Computer Networks*, Tampa, 2000.

68. G. Welch and G. Bishop, "An introduction to the kalman filter," Technical Report TR95-041, University of North Carolina, Chapel Hill, 2000.

69. R. Merwe, A. Doucet, N. Freitas, and E. Wan, "The unscented particle filter," Technical Report, 2000.

70. "Qualnet by scalable network technologies," http://www.scalable-networks.com, 2003.

71. "Ash transceiver," R. M. Inc., 2003.

72. K. Nuli, H. Sharif, S. Ci, and L. Cheng, "An adaptive medium access control for energy efficiency in wireless sensor networks," in *IEEE ICNSC'04*, Taipei, Taiwan, 2004.

5

NETWORK ROUTING

5.1 INTRODUCTION

Network routing is responsible for delivering data from a source to a sink across a network. These protocols are divided into two functions. First, path discovery is responsible for finding all, in the best case, but at least one, in the worst case, path through a network between the source and sink. Second, path selection chooses one path (typically) over which the data are sent. There are many options for the path discovery phase; path selection is performed to best meet the requirements of the application. In this chapter we present three contributions that discuss various aspects of routing.

In sensor networks, unlink in the Internet, sinks typically require a type of data from a certain area and hence are not concerned with a specific sensor. The act of receiving data from a sensor, or set of sensors, is often called data dissemination and is closely tied to routing.

Path discovery can take place using one of two main methods. In the first, called proactive path discovery, nodes exchange information to discover paths between all nodes in the network even if no data are being sent between the two nodes. In these networks information is typically exchanged periodically and if the topology of a network changes. These protocols are very efficient for static networks, that is, in networks that have a stable topology. They also reduce latency because when two nodes wish to communicate, a path between them already exists. In the second, called reactive path discovery, paths are discovered between two nodes only when they need to communicate. This method is useful in networks that have unstable topologies because it does not generate routing traffic unnecessarily. However, there is latency in data transfer because when two nodes wish to communicate, it is likely that no path exists between them and one must be established before data transfer can take place.

In practice, proactive routing protocols are used in sensor networks that do not have mobile sensors and do not have a tremendous number of source–sink pairs. Reactive routing

Sensor Network Operations, Edited by Phoha, LaPorta, and Griffin
Copyright © 2006 The Institute of Electrical and Electronics Engineers, Inc.

protocols are used in networks that support mobile nodes because network topology changes frequently in these networks. In fact, these routing protocols are very similar in nature to protocols used in ad hoc mobile networks.

Path selection is performed using a variety of conditions, including application performance requirements, power cost, and security. If an application requires low latency, the path with the lowest latency among those returned by the path discovery algorithm is selected. If power consumption is a concern, a node may instead select a path that uses links that require less transmission power even though latency may be higher. Security is a major concern in routing protocols because a malicious node may advertize false routes, thus disabling a network.

Section 5.2, by Sha, Shi, and Sellumuthu, presents an algorithm that balances the load in a sensor network. By balancing load across nodes, power is expended more equally, and hence network lifetime is increased. To support operation in a large network, a hierarchy of nodes is introduced, and nodes are grouped into clusters. Section 5.3, by Hamdaoui and Ramanathan, presents a routing protocol that is closely linked to a medium access control (MAC) protocol. As discussed in Chapter 4, MAC protocols must resolve contention on a link and have a large impact on power consumption. This routing protocol selected paths to limit the contention experienced by the MAC protocol. In Section 5.4, by Hu, Ramesh, and Siddiqui, a secure routing protocol is presented. The novelty in this protocol is that a security and routing hierarchy is introduced to improve scaling and performance.

5.2 LOAD-BALANCED QUERY PROTOCOLS FOR WIRELESS SENSOR NETWORKS

Kewei Sha, Weisong Shi, and Sivakumar Sellamuthu

Wireless sensor networks (WSNs) have attracted a lot of attention recently because they hold the potential to revolutionize many segments of our economy and life. Energy-efficient network routing is one of the biggest challenges to real sensor network deployment and operations due to physical constraints of sensors. A great deal of research work on energy-efficient routing protocols has been proposed aiming to extend the lifetime of the sensor network; however, the load imbalance problem resulting from both high-level query protocols and low-level routing protocols has been neglected in the past. In this section, we intend to answer two questions: (1) *Can previous proposed algorithms generate a balanced load among all sensors?* (2) *Does the energy-efficient path always result in the maximum lifetime of the whole sensor network system?* We find that few previous work notices the load imbalance problem; thus we propose a simple but efficient query protocol called indirect query (IQ) to avoid the load imbalance problem, which is of great importance to the sensor network operations. To tackle the second question, we first propose a formal definition and model of the lifetime of the whole sensor network, the load imbalance factor, and the effect of the depleted sensors. Based on these models, a detailed analysis is presented. Furthermore, we mathematically prove that IQ can effectively balance the load and extend the lifetime of the whole sensor network. Simulation results validate that the IQ protocol and its variant based on probability (IQBP) indeed balance the load among sensors very well and extend the lifetime of the sensor network to at least 7 and 10 times longer than that of the traditional query model.

5.2.1 Background Information

As new fabrication and integration technologies reduce the cost and size of microsensors and wireless sensors, we will witness another revolution that involves observation and control of our physical world [1–6], as networking technologies have done for the ways individuals and organizations exchange information. The commercialization of microsensors with wireless communication, such as Motes from Crossbow [7] and Intel, enables interesting applications where current wired sensor technology may fail in conditions such as unattended environment monitoring [8], habitat monitoring [9], military surveillance [10], and others.

Although sensor networks are a miniature of traditional distributed systems and mobile ad hoc networks, there are several major differences between them. First, the number of sensor nodes is much more than the nodes in other systems. Second, the sensor nodes are densely distributed, but they fail more easily in the sensor networks. Third, sensors have limited bandwidth and small memory, which probably will disappear gradually with the development of new fabrication technologies. The last but most significant difference is the energy constraint of the sensors. Most sensors are battery-backed; thus each sensor has limited physical capability such as power and calculation ability due to the energy limitation. When the sensor runs out of power, the lifetime of the sensor is ended. Then either we should replace the failed sensors in the sensor networks, which is impossible in some situations, such as when the sensors are deployed to an area where human beings cannot access (e.g., hazardous areas and waste containment cover clay soil [8]), or the whole sensor network will be down because just a few sensors are down. These differences make the previous routing protocols and approaches in wireless ad hoc networks or distributed systems insufficient to be used in wireless sensor networks. Therefore, how to optimize sensor network operations has become a very hot research topic.

Key challenges in wireless sensor network design and operations, including routing protocols, information collection protocols, query forwarding, and in-network processing, are saving energy and extending the overall lifetime of the system, usually by trading increased computation for reduced radio communication, which is relatively expensive in terms of energy cost [11]. For example, in the μAMPS sensor node described elsewhere [12], the energy required for communication is ∼1 nJ/bit, and the processor can compute approximately 150 instructions per bit for communication.

A great deal of energy-efficient network routing and query mechanisms that consider the characteristics of sensor nodes along with application and architecture requirements have been proposed in the literature [13–28]. The characteristics investigated in these studies include maximum available power, minimum energy, minimum hop length, and minimum available power. They do find some good routing protocols to save the energy of sensor nodes by selecting energy-efficient paths; however, the global load imbalance problem has been neglected in most of the previous efforts, which may result in another kind of energy efficiency. For example, if some depleted sensor nodes separate the whole sensor network into several parts, the whole network will become isolated and useless even though a majority of sensor nodes still have a lot of remaining energy.

In this section we want to answer the following two related questions. First, can previous proposed algorithms generate a balanced load among all sensors? We find that few previous efforts have taken this into account during the design of algorithms and protocols; however, we believe that maintaining a relatively balanced load of each sensor node plays a vital role in extending the lifetime of the whole system in sensor network operations. Second, does the energy-efficient path always result in the maximum lifetime of the whole sensor network system? Our conjecture to this question is that the minimum energy path is not necessary to guarantee the maximum lifetime of the whole sensor network.

To address the first problem, we first identify the load imbalance problem in the traditional query model and in routing protocols by using the Greedy Perimeter Stateless Routing (GPSR) routing algorithm [29] as an example. Normally, there are two ways to balance the load of each sensor node. On the one hand, the routing protocol should guarantee that the load of each sensor (i.e., packet forwarding) is balanced, as exploited by Dai and Han in [30], which aims to create a balanced routing tree. We call the load balance schemes proposed at the routing level the *microlevel* approach. On the other hand, load balance is also possible at the high level, such as query level and aggregation level, where the load of each sensor (i.e., query handling) needs to be balanced. For example, rotating the cluster head in clustering algorithms [17,31,32] belongs to this category. We call this type of high-level load balancing the *macrolevel* approach. These two levels complement each other. In this section, we focus on how to balance a load at the macrolevel, which we believe is more effective than the microlevel approach to achieve the goal of maximizing the lifetime of the sensor network. Based on this observation, a novel load-balanced query protocol named IQ is proposed. The basic idea of IQ is randomly (or intelligently) choosing a delegate for each query and allowing the delegate to handle the rest of the query processing.

Our approach to address the second question consists of two parts: *modeling followed by theoretical analysis* and *experimental verification by simulation*. We propose a formal definition and model of the remaining lifetime of each sensor, the remaining lifetime of the whole sensor network, the lifetime of the whole sensor network, and the effect of depleted sensors to the lifetime of the sensor network. Based on these models, a detailed analysis and comparison between traditional approaches and our proposed approach are illustrated in the context of two different routing scenarios: *point to point (unicast)* and *point to area multicast (broadcast)*. The analysis results match very well with the performance evaluation results by simulation.

Both the analysis and experimental results show that the IQ protocol and its variant IQBP provide a perfect load balance among all sensor nodes and are much more energy efficient than the traditional query protocol. For example, after 400 queries, the maximum energy consumptions of all sensor nodes are 0.6 J (IQ) and 0.4 J (IQBP), while that of the traditional query model is 3 J, which equals the total energy of each sensor. In terms of the maximum number of queries that can be processed by the whole network, the lifetimes of IQ and IQBP are at least 7 and 10 times longer than that of the traditional query model.

Our contributions include the following five aspects:

- We identify the importance of the load imbalance problem and its relation with the lifetime of the whole sensor network both at the macrolevel (query protocol level) and the microlevel (routing protocol level).
- We give a formal model for the remaining lifetime of each sensor, the remaining lifetime of the whole sensor network, the lifetime of the whole sensor network, the load imbalance factor of the sensor network, and the effect of the depleted sensor to the operation of the whole sensor network. To our knowledge, this is the first effort to give a formal analysis of the lifetime of sensors and sensor networks.
- We propose a simple but efficient query protocol, IQ, to balance the workload and to extend the lifetime of the sensor network. In addition to the basic model, we also propose

two optimized variants, IQBP and IQ based on history (IQBH), to optimize the trade-off between the query response time and the load balance.

- We mathematically prove that the IQ protocol does balance the load and extends the lifetime of the sensor network.

- We give a comprehensive performance evaluation to answer the questions raised in this section and to compare the difference between the traditional query protocol and the proposed IQ protocols. The results agree with previous analytical results.

5.2.2 Load Imbalance Problem

We illustrate the load imbalance problem in wireless sensor networks, which consists of two types of imbalance: *routing protocol related* and *query protocol related*. The routing protocol related imbalance results from the intrinsic nature of the specific routing protocol, so it is protocol-dependent. On the contrary, the query protocol related imbalance results from the traditional query model, independent of low-level routing protocols. These two types of load imbalance are orthogonal and should be addressed separately in the real deployment. They correspond to the microlevel and macrolevel load balancing schemes defined in Section 5.2.1.

Traditional Query Model A common communication architecture in a wireless sensor network is illustrated in Figure 5.1, where the sensor nodes are scattered in the sensor field. They are responsible for collecting data and transferring data in multihops back to the sink (or gateway), which is usually a powerful machine with more computing power, large memory, and infinite energy. The sink then communicates with the task manager, which is probably far from the sensor field, via the Internet or other commercial networks.

The traditional query process (denoted as "traditional query model" or "traditional" in the rest of the section) in a wireless sensor network is as follows. After a query is generated, it will go through the sink and the sensor nodes near the sink, as indicated by the arrows in Figure 5.1, and finally reach the destination after routing through multiple sensors. The sensor nodes that have the data for the query will generate reply messages and route them back to the sink. The number of replied messages for each query depends on whether the data aggregation/optimization protocol is used and the characteristics of the query itself. The possible query patterns are unicast, area multicast, and broadcast [33].

Query-Related Load Imbalance In the simple query model, we find that if the query message is sent to the sensor nodes near the sink, it will not proceed to the farther sensor nodes; if the query message is sent to sensor nodes far from the sink, it will go through the nearby sensors first and thus the sensors near the sink will have more chance to deplete their energy. More seriously, in a broadcast query, which results in many reply messages, many more messages will go through the sensor nodes near the sink than through the sensor

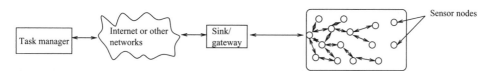

Figure 5.1 Traditional query model in wireless sensor networks.

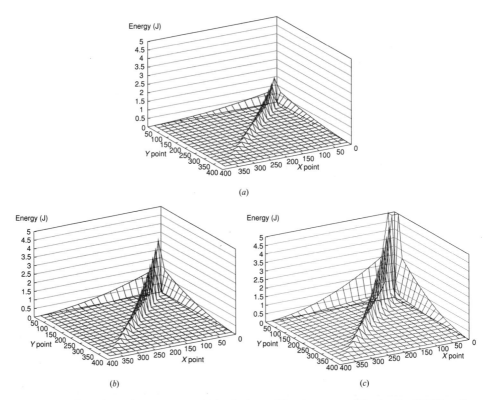

Figure 5.2 Snapshots of energy consumption in the traditional query model using the GPSR routing protocol: (*a*) after 100 queries, (*b*) after 200 queries, and (*c*) after 400 queries.

nodes far from the sink. Unfortunately, a high percentage of queries in the sensor network are actually multicast- or broadcast-based queries because of the attribute-based naming and routing protocols [34]. Thus, the sensor nodes near the sink will fail first because they run out of energy quickly. After certain nodes near the sink fail, the sink cannot gather any information from the other sensor nodes in the sensor network. In this sense, the whole sensor network is considered to be down (see Fig. 5.2*c*).

Figure 5.2 shows the effect of the load imbalance using the traditional query model. The data are collected from our experiment conducted on Capricorn [35], in which 400 sensor nodes are scattered in a 20×20 grid. Each sensor's initial energy is 5 Js. Figure 5.2 shows the snapshot of energy consumption in the traditional query model using the GPSR routing protocol [29], where the x axis and y axis specify the location of the sensor node and the z axis denotes the energy consumption of each sensor. Figure 5.2*a* is a snapshot of energy consumption after 100 queries and shows that the sensor nodes located on the two edges of the sensor field and the diagonal line connected to the sink consume much more energy than others. Figure 5.2*b* reports the snapshot of energy consumed after 200 queries and denotes that the load imbalance is enlarged with more processed queries. Figure 5.2*c* is a snapshot of energy consumption after 400 queries and shows that all the sensor nodes nearest the sink have already run out of their total energy. At that time, even though there is a lot of energy remaining in sensor nodes far from the data sink, the whole sensor network is down.

These three snapshots of energy consumption report that the load is unbalanced in the traditional query model and the lifetime of the sensor network is greatly reduced due to the death of the sensor nodes near the sink. To this end, we believe that it is the unbalanced load that leads to inefficient energy utilization.

Routing Protocol-Related Load Imbalance   Another type of the load imbalance is caused by the routing protocols, which concentrate on the energy-efficient path for each message and neglect the load imbalance in the global view. Next, we report this kind of load imbalance caused by the greedy feature of the GPSR routing protocol [29], which is widely used in mobile ad hoc networks and sensor networks.

In the GPSR routing protocol, the message is routed to the destination greedily based on the distance. Thus, if sensors are deployed dense enough, the sensors near the forwarding sensor will not have a chance to be involved at all. We select the 20 sensor nodes located on the diagonal line in the square sensor field as a sample to investigate the load imbalance in the GPSR routing protocol. Figure 5.3 shows the remaining energy of these sensor nodes. In the figure, the x axis is the sensor node ID, and the y axis is the value of the remaining energy of the corresponding sensor node after 100 queries are processed. The data are collected from our simulation, where 400 sensor nodes are scattered uniformly in the sensor field, and three types of density are used in the experiment, including three nodes in each 1000 m^2, 27 nodes in each 1000 m^2, and 75 nodes in each 1000 m^2.

In Figure 5.3, the line with diamonds reports the remaining lifetime of the diagonal sensors with the sensor density of three nodes in each 1000 m^2. In this case, the nodes near the sink have less remaining lifetime than farther ones. When the density increases, the load imbalance caused by the routing protocol occurs. Only the sensor node that is the closest to the destination within the communication range of the sender (forwarders) consumes a lot of energy, while others consume very little energy; for instance, the 21st and 42nd nodes in the line with triangles and the 21st, 42nd, 63rd, and 84th nodes in the line with stars consume

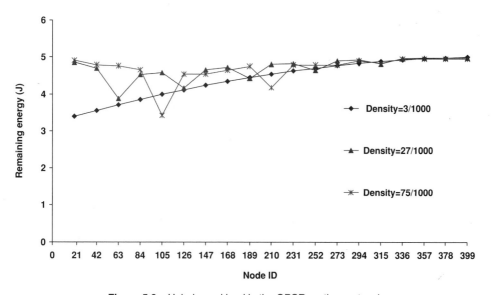

Figure 5.3 Unbalanced load in the GPSR routing protocol.

very little energy. When the density is 27 nodes in each $1000 \, m^2$, the load is concentrated to the 63rd, 126th, and 189th nodes, and when the density increases to 75 nodes in each $1000 \, m^2$, only two nodes, the 105th and 210th, are extensively used, while the rest of the sensors nodes are idle.

To this end, we argue that the load imbalance is a very important reason for the reduction in lifetime of the sensor network. From the above analysis, we find that the short lifetime of the whole sensor network is caused by the unbalanced load to the different sensor node. Although some nodes still have a lot of energy remaining, the sensor networks cannot operate correctly, that is, a lot of energy is wasted. Thus, we should consider maximizing the lifetime of the whole sensor network rather than just the lifetime of some sensor nodes. In fact, energy-efficient path routing can only save energy and enlarge the lifetime of the individual sensor; the load balance at the routing level can prolong the lifetime of the whole path, while the query level load balance can extend the lifetime of the whole sensor network. Thus we consider that the query level load balance is more significant than the routing level load balance from the perspective of lifetime [36], although it complements these lower level protocols very well. The effective way to extend the lifetime of sensor networks is to evenly balance the load to all the sensor nodes. This is the goal of the new proposed query protocol, which will be described in the following section.

5.2.3 IQ: Indirect Query Protocol

From the above analysis, we find that the short lifetime of the sensor network is caused by the two types of load imbalance in the sensor networks. So, if we can balance the load to all the sensor nodes, the lifetime of the whole sensor network can be prolonged. Here we will focus only on the query level load balance (i.e., macrolevel), which is more significant to the lifetime of sensor network than the routing level (i.e., microlevel) load balance. Moreover, we believe our approach will automatically benefit from the routing level load balancing schemes. To achieve this goal, we propose a simple but efficient query protocol, called indirect query, as shown in Figure 5.4. The basic idea of IQ is forwarding the query from the sink to another randomly chosen sensor (called the *delegate*) first and allowing the chosen sensor to take care of data collection for that query and then send the query results back to the data sink. The rationale behind this algorithm is the observation that the lifetime of each sensor has a different effect on the lifetime of the whole sensor network; that is, the nodes close to the data sink are more important than the nodes far from the data sink. We estimate that the proposed IQ protocol can balance the load (message transmission and computation overhead) of the whole system so as to extend the lifetime of the whole sensor network.

The pseudocode of the IQ protocol is shown in Figure 5.5*b*, which consists of three steps. First, the data sink randomly selects a sensor as the query delegate and forwards the query to the delegate (*query forwarding* in the figure) when the query is a broadcast query. Second,

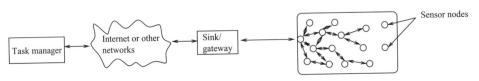

Figure 5.4 Example of IQ protocol.

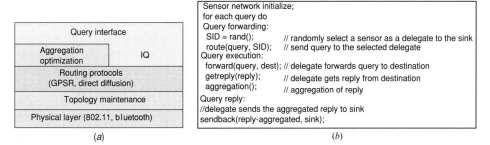

Figure 5.5 (*a*) General view and (*b*) the pseudocode of the IQ protocol.

the delegate gets the query and conducts the query processing on behalf of the data sink and then aggregates the replies (*query execution* in the figure). Third, the delegate sends the aggregated reply back to the data sink (*query reply phase* in the figure). Comparing with the traditional query model, two extra steps, *query forwarding* and *query replying*, are added in the IQ protocol.

Figure 5.5*a* shows the general view of the IQ protocol, which is a high-level protocol. In fact, as we described before, the load balance problem can be addressed across all layers, including the load-balanced routing protocol [30, 37], topology maintenance [31, 32, 38, 39], and MAC-aware routing [15]. The IQ protocol will benefit from the in-network processing and aggregation protocol, such as TAG [22] and direct diffusion [18].

Comparing with the traditional query model, we list the advantages and disadvantages of the proposed IQ protocol:

- Indirect query can balance the load by moving some workload to other nodes.
- Statistically, the load of the sensors located in the middle of the sensor network will have more load than those at the perimeter.
- For a point-to-point communication pattern (such as unicast), the performance of IQ is the same as that of the traditional model by choosing the sink as the delegate, but for point-to-area multicast (and broadcast), such as direct diffusion and flooding (broadcast) based approaches, it will be very helpful.
- The load balance is achieved by trading off the query response time with the load balance. Indirect query has two more steps than the traditional query model. We hope this is deserved by extending the lifetime of the whole network.
- We expect that an optimized delegate selection algorithm can amortize this extended latency to some extent.

Two Optimizations: IQBP and IQBH The key to the IQ protocol is how to choose the delegate. In the basic IQ protocol, the delegate of each query is randomly chosen, which risks extending the query response time if the delegate is far from the replied sensors. Thus, we propose two optimizations to the basic model by intelligently choosing delegates to amortize the extended query response time and the load imbalance.

The load imbalance in the basic IQ protocol is caused by the fact that the sensors in the central have more load than others if each sensor has the same probability to be chosen as a delegate. Intuitively, if we shift some of the load from load-intensive nodes to lightly loaded nodes, the load will be balanced. The first optimization we made is called indirect query based on probability. As the name suggests, in IQBP we will always choose the sensor nodes that have high probability to be lightly loaded as delegates, which always consume less energy than others. We know that the sensors located in the area far from the sink consume less energy than the sensors in other areas, that is, these sensors have higher probability to be lightly loaded than the nodes nearer the data sink. In the implementation, the probability of each node to be chosen as a delegate is proportional to its distance to the data sink.

While the basic IQ and the IQBP balance the load of sensors in most cases, we should note that the load balance is achieved by trading off the (possible) latency stretch that resulted from the two extra steps introduced by the IQ. To address this problem, we propose another optimization to reduce the latency and to maintain the load balance at the same time. The rationale behind this optimization is exploiting the historical information of queries and replies.

In our simple IQ protocol, we assume that the probability that each sensor node generates a reply message is the same; however, this is not always true. For example, if the query is "the sensors with temperature higher than 85," there may be some regions in which the temperature will never be higher than 85, so the sensors in that region will never generate any reply message. Therefore, the probability of the sensors generating reply messages will always vary from one to another. Given that sensors have varied reply possibility, we can select a delegate from those sensors with higher probability of generating a reply than others. We can define the probability of each sensor node generating a reply according to the type of query and the distribution characteristic of the query. To define the probability, we need to record the reply history of all sensor nodes. Thus, this query protocol is called indirect query based on history and can effectively short the path that most messages go through and thus extend the lifetime of the sensor network.

It is worth noting that there is a trade-off between balancing the load and reducing the query response time in IQBH. For instance, if the areas interested by queries (application scientists) are distributed evenly, the algorithm works perfectly; however, if there is a hot-spot area, we need to migrate the load by intentionally choosing delegates far from the hot-spot area to achieve a balanced load. Thus the benefit of this protocol depends on the distribution of high-level queries. In our current study, since we have not collected enough queries from real applications, we decide to detain the performance evaluation of this protocol to future work, after we get some experience with the waste containment application [8].

By using these three IQ protocols, we can extend the lifespan of the whole sensor network. Both the theoretical analysis (Section 5.2.4) and simulation results (Section 5.2.5) show that the IQ protocol can definitely balance the workload in the sensor network so as to prolong the lifetime of the sensor network.

Analysis and Discussion   In this section we analyze the IQ protocol in more detail and compare it with other work closely related to IQ. For a more general comparison, see Section 5.2.7.

Analysis   As mentioned in the previous section, the beauty of IQ is balancing the load among sensors by migrating the load of the sensors near the base station to other sensors; however, whether this type of migration is useful depends significantly on the potential of aggregation of query results. Thus, it is necessary to examine the IQ protocol in more detail. Generally, a query consists of two stages: *query propagation* and *query reply collection*. Let M represent the number of sensors toward which a query is routed and N represent the number of sensors that send back replies. Here, each sensor is assumed to send back at most one reply for each query. Next, we will analyze the behavior of IQ in three scenarios: $1:1$, $M:1$, and $M:N$, where the two digits represent M and N, respectively.

In the $1:1$ case ($M = 1$, $N = 1$), there is one destination and one reply message. If the sink is chosen as the delegate, IQ will perform exactly at the traditional query protocol.

Otherwise, IQ will introduce two extra communication overheads. Thus, the sink is always chosen as the delegate in IQ for this scenario.

In the $M:1$ case ($M = M$, $N = 1$), the query is sent to multiple sensors, and only one reply is needed, as in the *area-anycast* scenario. Since there is no aggregation needed in this case, IQ will ask the reply sensor to send the reply back to the sink directly, instead of to the delegate first. Note that the IQ protocol is still useful at the query propagation stage by shifting the load to the delegate and reducing the load of the sensors near the sink; however, the gain at this stage depends on the specific implementation of broadcast (or multicast) algorithms, such as flooding.

Finally, in the case of $M:N$ ($M = M$, $N = N$), where N is less than or equal to M but is far larger than 1, the query is sent to multiple sensors, and multiple reply messages are expected to be generated from these sensors. Indirect query works perfectly at both stages when the reply messages can be aggregated at the delegate, which is the ideal case for the proposed IQ protocol. However, when the reply messages could not be aggregated at the delegate, IQ is not suitable to be used at the query reply collection stage because of the one extra step introduced by the protocol, while it works fine at the query propagation stage, as in the case of $M:1$.

Discussion As we mentioned earlier, IQ is an application level query protocol that seeks for a balanced load for each sensor. Therefore, it complements other energy-efficient routing protocols very well. For example, the IQ protocol will benefit greatly from the MAC-aware routing protocol proposed by Hamdaoui and Ramanathan [15], which takes the MAC contention into consideration.

The idea of using the delegate to forward queries is similar to clustering approaches, such as Low Energy Adaptive Clustering Hierarchy (LEACH) [17], that assume each cluster head will communicate with the sink directly. However, there are two differences between these approaches. First, the delegates in the IQ protocol are not necessary to have the capability to communicate with the sink directly, which fits very well with wireless sensor networks that prefer multihop routing to direct transmission, as suggested elsewhere [6]. Second, it is possible to use a clustering hierarchy to collect data in the last step of the IQ protocol. However, we doubt the applicability of clustering-based protocols in a real sensor network because of the large maintenance overhead, (e.g., cluster head selection). This has been verified by the recent University of California–Berkeley TinyOS project [40], which found that few projects use clustering-based protocols because of the complexity of protocols. Thus, we choose to let all sensors send their data back to delegate directly in our protocol.

It is worth noting that the IQ protocol works at a layer higher than the data aggregation layer, as shown in Figure 5.5a. So our query protocol can automatically benefit from the data aggregation protocol to further extend the lifetime of the sensor network. However, the data aggregation is an application-specific task that we will evaluate in future work.

5.2.4 Model of Lifetime

As specified in the previous sections, our goal is to balance the load and extend the lifetime of the whole sensor network. The *lifetime of a wireless sensor network* is an application-specific, flexible concept. However, we can first abstract the *remaining lifetime of the wireless sensor network*, which is defined as the weighted sum of the lifetimes of all the sensors in the network. Given that, we can define the lifetime of the whole sensor network for three major application categories: active query, event-driven, and passive monitoring.

In an active query application, the lifetime of the whole sensor network can be defined as the maximum number of queries the sensor network can handle before the sensor network terminates. For an event-driven application, the lifetime of the whole sensor network can be defined as total number of events the sensor network can process before terminating. For passive monitoring, the lifetime of the whole sensor network can be defined as the total amount of time before the sensor network terminates. The termination of the sensor network is defined as the time when the remaining lifetime of the wireless sensor network starts to stabilize, which implies that the sensor network loses connectivity or the number of sensors with zero remaining lifetime exceeds a threshold and the sensor network becomes useless. Here, we assume the energy consumption of regular maintenance overhead is negligible and will be considered later.

Because the remaining lifetime of the whole sensor network is defined based on the remaining lifetime of sensors in the sensor network, in this section, we define the remaining lifetime of the single sensor first, then model the remaining lifetime of the whole sensor network, and finally define the lifetime of the whole sensor network.

Assumptions and Definitions of Parameters Several assumptions made by our model are listed here (the symbols used in this analysis are listed in Table 5.1):

- All the sensors are homogeneous, that is, the physical capacity of each sensor is the same.
- The location is available, either by physical devices such as the Global Positioning System (GPS) or by topology discovery algorithms [38,41].
- The sensors in the sensor network are evenly distributed and dense enough to cover the whole area.
- The location of each sensor is stationary.
- The sensor's power is limited so that it can only communicate with its neighbors within its communication range. Multihop is required to communicate with others outside the communication range.
- The data sink is fixed, which is usually true in the real deployment.

Definition of Remaining Lifetime of Sensors The *remaining lifetime of the individual sensor* is defined as the remaining normalized energy of the sensor at some moment, N_m. Here we normalize the remaining initial energy of all the sensors in the sensor network as 1. In the time during which a query is executed, the energy is consumed when the sensor receives or sends a query message and a reply message. So the remaining lifetime of the sensor is the total energy of each sensor minus the energy consumed when the messages go

Table 5.1   Variables Used in This Section

Variables	Description
ϵ_{jiq}	In the ith query the amount of energy consumed when one query message goes through the jth sensor, specifically $\epsilon_{jiq,\mathrm{rcv}}$ for receiving and $\epsilon_{jiq,\mathrm{snd}}$ for sending. When we assume all the query messages are the same, ϵ_{jiq} can be reduced to ϵ_q.
ϵ_{jir}	In the ith query the amount of energy consumed when one reply message goes through the jth sensor, specifically $\epsilon_{jir,\mathrm{rcv}}$ for receiving and $\epsilon_{jir,\mathrm{snd}}$ for sending. When we assume all the reply messages are the same, ϵ_{jir} can be reduced to ϵ_r.
E_j	The total initial energy of the jth sensor. When we assume all the sensors are homogeneous, $E_j = E$.
$1/f$	An application-specific parameter to determine the possibility of a sensor generating a reply to a query
N_{jiq}	Number of query messages that pass the jth sensor during the ith query
N_{jir}	Number of reply messages that pass the jth sensor during the ith query
r	Communication range of each sensor
S_q, S_r	Sizes of query message and reply message, respectively
P_{jiq}	Probability that the ith query message go through the jth sensor node
P_{jir}	Probability that the reply messages for ith query goes through the jth sensor
P_{ji}	Probability that the query message or reply message will go through the jth sensor node during the ith query
$P(B)_{ij}$	Probability that the sensor farther than the jth sensor to the sink is selected as the destination in the ith query
$P(A)_{ij}$	Probability that the message goes through the jth sensor in the ith query when that message goes through the circular area where the jth sensor node is located
N_{far}	Number of sensors that are farther than the jth sensor node to the sink
N_{total}, N_n	Number of total sensor nodes in the sensor network
$N_{nbr\,s}$	Number of sensors in the communication range
N_{fail}	Number of sensors that are out of power
θ	Maximum number of depleted sensors when the sensor network functions incorrectly
ρ	Density of the sensor nodes in the sensor network
d_{jis}	Distance from the jth sensor to the sink (or delegate) during the ith query
N_m	Sequence number of queries processed in active query, sequence number of event-driven, events and amount of time used in passive monitoring
w_j	Weight(importance) of the jth sensor node in the sensor network
d_{max}	Ratio of the maximum distance between every two sensor nodes in the sensor network to the communication range r
$\mathcal{L}$	Remaining lifetime of the whole sensor network
$L(j)$	Remaining lifetime of the jth sensor
LFT	Lifetime of the whole sensor network

through the sensor:

$$L(j) = 1 - \sum_{i=1}^{N_m} \frac{\epsilon_{jiq} N_{jiq} + \epsilon_{jir} N_{jir}}{E_j}$$

We borrow the energy model and symbols used elsewhere [17] to calculate the energy consumption of each message transmission. The energy consumed when the sensor receives a message of size k is

$$\epsilon_{\text{rcv}} = \epsilon_{\text{elec}} k$$

and the energy consumed on sending a message of size k is

$$\epsilon_{\text{snd}} = \epsilon_{\text{elec}} k + \epsilon_{\text{amp}} r^2 k$$

So, we have

$$\epsilon_{jiq} = \epsilon_{jiq,\text{rcv}} + \epsilon_{jiq,\text{snd}}$$

and

$$\epsilon_{jir} = \epsilon_{jir,\text{rcv}} + \epsilon_{jir,\text{snd}}$$

Probability of ith Message Through jth Sensor   To calculate $L(j)$, we should calculate N_{jiq} and N_{jir} first, which are related to the probability of the message going through the jth sensor in the ith query. Thus we first need to define the probability of the ith message going through the jth sensor.

As we observe, query messages directed to sensors far from the sink and reply messages from sensors far from the sink will both go through some sensors nearer the sink than these sensors, while messages from or to sensors near the sink will not go through the farther ones. Thus if all sensors have the same probability to be the query destination, the probability that the message goes through sensors near the sink is larger than the probability of going through sensors far from the sink. Figure 5.6 shows the message propagation in the sensor network from a macro view. A message is routed from the sink to the jth sensor step by step, and at each step the message must be in one circle, as shown in the figure. If there are n sensors in one circle, the probability that the message is held by one specified sensor in that circle is $1/n$. In point-to-point routing, we can assume that the probability for the ith query message to go through the jth sensor is the same as the probability for the reply messages for that query to go through the jth sensor, that is, $P_{jiq} = P_{jir}$. Here we use P_{ji} as the probability that the message will go through the jth sensor in the ith query. In the broadcast, $P_{jiq} = 1$, as every sensor will get the query. Let A denote the event that the message goes through the jth sensor when it goes through the circle the node locates and B be the event that the

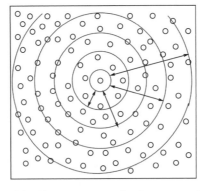

Figure 5.6   Message propagation in sensor networks.

destination of the message i is farther from the sink than the jth node, where A and B are two independent events. Let $P(B)_{ij}$ be the probability of B and $P(A)_{ij}$ be the probability of A. Then the probability that the ith message will go through the jth node is

$$P_{ij} = P(A|B)_{ij} = P(A)_{ij} P(B)_{ij}$$

Here,

$$P(B)_{ij} = \frac{N_{\text{far}}}{N_{\text{total}}} = \frac{N_n - \pi d_{jis}^2 \rho}{N_n}$$

and

$$P(A)_{ij} = \frac{1}{\pi \left\{ \left[(\lfloor d_{jis}/r \rfloor + 1)r \right]^2 - (\lfloor d_{jis}/r \rfloor r)^2 \right\} \rho}$$

$$= \frac{1}{\pi (2\lfloor d_{jis}/r \rfloor + 1)r^2 \rho}$$

Thus

$$P_{ij} = \frac{1}{\pi (2\lfloor d_{jis}/r \rfloor + 1)r^2 \rho} \frac{N_n - \pi d_{jis}^2 \rho}{N_n}$$

Now we can calculate the remaining lifetime of the sensor in a unicast Traditional query and both a broadcast traditional query and broadcast IQ. In the above formula, the difference between a traditional query and IQ lies in d_{jis}, which is constant in a traditional query for each sensor because the sink is fixed all the time and is changing in IQ because of the shifting delegate.

Remaining Lifetime of Sensors in Unicast Traditional Query In the case of unicast, the message will go through only the sensors on the path between the sink and the destination once for each query, and so does the reply message for that query. Thus the remaining lifetime of each sensor in a unicast traditional query can be defined as

$$L(j) = 1 - \sum_{i=1}^{N_m} \frac{\epsilon_{jiq} N_{jiq} + \epsilon_{jir} N_{jir}}{E_j}$$

$$= 1 - \sum_{i=1}^{N_m} \frac{\epsilon_{jiq} P_{ji} 1 + \epsilon_{jir} P_{ji} 1}{E_j}$$

Because we assume the sensors are homogeneous and the same type messages have the same size, $E_j = E$, $\epsilon_{jiq} = \epsilon_q$, and $\epsilon_{jir} = \epsilon_r$, the remaining lifetime of each sensor in the point-to-point routing is

$$L(j) = 1 - \frac{1}{E} \frac{(\epsilon_q + \epsilon_r) N_m}{\pi (2\lfloor d_{jis}/r \rfloor + 1)r^2 \rho} \frac{N_n - \pi d_{jis}^2 \rho}{N_n}$$

Remaining Lifetime of Sensors in Broadcast Traditional Query   Next, we analyze the remaining lifetime of each sensor when the query message is broadcast to all sensors from the sink. In this case, the query message floods to all the sensors while the reply message will go through the sensors on its path to the sink. For each query message, if we assume the probability that one sensor will generate a reply message is $1/f$, the remaining lifetime of each sensor in a broadcast traditional query is

$$L(j) = 1 - \sum_{i=1}^{N_m} \frac{\epsilon_{jiq} N_{jiq} + \epsilon_{jir} N_{jir}}{E_j}$$

$$= 1 - \sum_{i=1}^{N_m} \frac{\epsilon_{jiq} N_{nbrs} + \epsilon_{jir} P_{jir} N_n (1/f)}{E_j}$$

Because the query is routed by flood, that is, each sensor will get a query from its neighbors and send a query to its neighbors, the remaining lifetime of the sensor in the broadcast traditional query is

$$L(j) = 1 - \frac{N_m \epsilon_q N_{nbrs}}{E} - \frac{\epsilon_r N_m}{E f \pi r^2 \rho} \frac{N_n - \pi d_{jis}^2 \rho}{2\lfloor d_{jis}/r \rfloor + 1}$$

Remaining Lifetime of Sensors in IQ   In IQ, a query is directed to a randomly selected delegate, then the delegate acts as the sink to take care of query forwarding, data collection, and data transmitting back to the sink. If the probability of each sensor to be a delegate is the same, when total of N_m queries have been processed and there are N_n sensors in the sensor network, for each sensor the possible time it is selected as a delegate is N_m/N_n. The number of times that the sensor is located in the area with $d_{jis} = kr$ to the delegate is the number of sensors located in the circle between kr and $kr + 1$ to that sensor. The number of sensors located in the circle between kr and $kr + 1$ is $\pi(2k + 1)r^2\rho$. Thus the remaining lifetime of each sensor in IQ can be defined as

$$L(j) = 1 - \sum_{i=1}^{N_m} \frac{\epsilon_{jiq} N_{jiq} + \epsilon_{jir} N_{jir}}{E_j}$$

$$= 1 - \frac{\epsilon_q N_m N_{nbrs}}{E} - \frac{1}{E} \frac{N_n}{f} \sum_{i=1}^{N_m} \epsilon_r N_{jir} - \frac{N_m}{N_n} \frac{N_n}{f E} \epsilon_{r,rcv}$$

$$= 1 - \frac{\epsilon_q N_m N_{nbrs}}{E} - \frac{1}{E} \frac{N_n}{f} \sum_{k=1}^{d_{max}} \left(\epsilon_r \frac{N_m}{N_n} P_{jis} \left(\pi\rho \left\{ [(k+1)r]^2 - (kr)^2 \right\} \right) \right) - \frac{N_m}{f E} \epsilon_{r,rcv}$$

$$= 1 - \frac{\epsilon_q N_m N_{nbrs}}{E} - \frac{\epsilon_r N_m d_{max}}{E f} + \frac{\epsilon_r N_m}{E f} \frac{\pi\rho r^2}{2 N_n} d_{max}(d_{max} + 1) - \frac{N_m}{f E} \epsilon_{r,rcv}$$

Based on the remaining lifetime of sensors, we define the remaining lifetime of the whole sensor network in the following section. First we define the importance of sensors.

Importance of Different Sensor Nodes   The failure, where we consider only the failure resulting from the depletion of energy, of sensors will cause the sensor network to

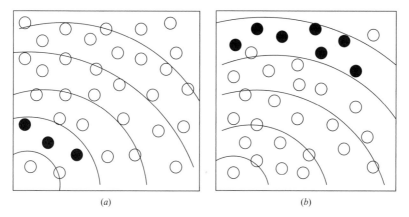

(a) (b)

Figure 5.7 Example of the importance of different sensors, assuming the data sink is located at the lower left corner: (a) Sensors near the sink are dead. (b) Sensors far from the sink are dead.

act improperly, but the level of the damage it causes is different, which is why the previous definition of lifetime as the time before first sensor failure or first message failure without considering the location of the sensor is unsatisfactory. For the same number of failure sensors, sometimes the damage may be very slight and the sensor network still perform almost normally, while at other times it may be very serious and cause the sensor network to lose most of its functionality. Two cases are described in the Figures 5.7a and b. In the figure, the black nodes represent the sensors that have run out of energy and the white ones denote the ones that are still alive. In Figures 5.7a and b the sensor networks cannot act as they should since in both cases the sensor network cannot get data from some sensors, while in Figure 5.7a, although there are only three failed sensors, the sink cannot get data from most of the sensors. And in Figure 5.7b, there are seven dead nodes but the sink can still get data from most of the sensors in the sensor network. So the damage to the sensor network by the failure sensors is related not only to the number but also to the location of the failed sensors. To this end, different sensors in a network are important for different reasons and we use the *weight* of a sensor to determine the importance of that sensor. Based on the above analysis, the nearer a sensor is to the sink, the more important it is. We define the weight of each sensor as:

$$w_j = c \frac{1}{d_{jis}^2}$$

where c is a constant.

Remaining Lifetime of the Whole Sensor Network   In previous sections we defined the remaining lifetime for sensors. Now we are in a position to examine the remaining lifetime of the whole sensor network. We consider the *remaining lifetime of the whole sensor network* as the sum of the weighted remaining lifetimes of all sensors in the sensor network:

$$\mathcal{L} = \sum_{j=1}^{N_n} w_j L(j)$$

From this definition, we can easily get the remaining lifetime of the whole sensor network in a unicast and broadcast traditional query and IQ. Specifically, in a unicast query, the remaining lifetime of the whole sensor network is defined as

$$
\mathcal{L} = \sum_{j=1}^{N_n} c \frac{1}{d_{jis}^2} \left(1 - \frac{1}{E} \frac{(\epsilon_q + \epsilon_r) N_m}{\pi(2\lfloor d_{jis}/r \rfloor + 1) r^2 \rho} \frac{N_n - \pi d_{jis}^2 \rho}{N_n} \right)
$$

$$
\approx c\pi\rho \left(2\ln d_{\max} + \frac{\pi^2}{6} \right) - \frac{c(\epsilon_q + \epsilon_r) N_m \pi^2}{6Er^2} + \frac{c(\epsilon_q + \epsilon_r) N_m}{Er^2 d_{\max}}
$$

In a broadcast traditional query, the remaining lifetime of the whole sensor network is defined as

$$
\mathcal{L} = \sum_{j=1}^{N_n} c \frac{1}{d_{jis}^2} \left(1 - \frac{N_m \epsilon_q N_{nbrs}}{E} - \frac{\epsilon_r N_m}{E f \pi r^2 \rho} \frac{N_n - \pi d_{jis}^2 \rho}{2\lfloor d_{jis}/r \rfloor + 1} \right)
$$

$$
\approx c\pi\rho \left(2\ln d_{\max} + \frac{\pi^2}{6} \right) \left(1 - \frac{\epsilon_q N_m N_{nbrs}}{E} \right) - \frac{c\pi^2 \epsilon_r N_m N_n}{6Efr^2} + \frac{c\epsilon_r N_m \pi \rho d_{\max}}{Ef}
$$

Similarly in IQ, the remaining lifetime of the whole sensor network is defined as

$$
\mathcal{L} = \sum_{j=1}^{N_n} c \frac{1}{d_{jis}^2} \left(1 - \frac{\epsilon_q N_m N_{nbrs}}{E} - \frac{\epsilon_r N_m d_{\max}}{Ef} + \frac{\epsilon_r N_n}{Ef} \frac{\pi \rho r^2}{2N_n} d_{\max}(d_{\max} + 1) - \frac{N_m}{Ef} \epsilon_{r,rcv} \right)
$$

$$
\approx c\pi\rho \left(1 - \frac{\epsilon_q N_m N_{nbrs}}{E} - \frac{\epsilon_r N_m d_{\max}}{Ef} + \frac{\epsilon_r N_m}{Ef} \frac{\pi \rho r^2}{2N_n} d_{\max}(d_{\max} + 1) - \frac{N_m}{fE} \epsilon_{r,rcv} \right)
$$

$$
\times \left(2\ln d_{\max} + \frac{\pi^2}{6} \right)
$$

Based on the remaining lifetime of the whole sensor network, the lifetime of the sensor network can be formally defined as

$$
\text{LFT} = \{ N_m \mid \mathcal{L}(N_m - 1) < \mathcal{L}(N_m) \text{ and } \mathcal{L}(N_m + 1) = \mathcal{L}(N_m) \text{ or } N_{\text{fail}} \geq \theta \}
$$

where θ is a predefined threshold of the maximum number of failure sensors in the sensor network and N_{fail} is the number of failure sensors.

Based on these models, we depict the detailed analysis of different query protocols in the following section.

5.2.5 Analytical Comparison: Traditional Query Versus IQ

One of the goals of modeling is to evaluate the performance of different protocols. Now we are in the position to compare a traditional query with IQ in terms of the remaining lifetime of sensors and the remaining lifetime of the whole sensor network.

To quantitatively compare these two query protocols, we adopt the practical values of sensor parameters obtained from Berkeley motes [11], including the initial energy and the energy consumption rate. In [11] two 1.5-V batteries rated at 575 mAh are used for each

Table 5.2   Comparison of Remaining Lifetimes of Different Nodes in Different Locations

Distance to Sink	Traditional Query		
	Unicast	Broadcast	IQ
$0 < d < r$	$1 - \dfrac{158N_m}{10^6}$	$1 - \dfrac{8280N_m}{10^6}$	$1 - \dfrac{2149N_m}{10^6}$
$d = 7r$	$1 - \dfrac{8N_m}{10^6}$	$1 - \dfrac{2000N_m}{10^6}$	$1 - \dfrac{2149N_m}{10^6}$
$d = 14r$	1	$1 - \dfrac{1670N_m}{10^6}$	$1 - \dfrac{2149N_m}{10^6}$

sensor, so the initial total energy of each sensor is 1.725 J. The energies needed to transmit and receive a single bit are 1 and 0.5 µJ, respectively. We assume the sizes of the query message and reply message to be 240 and 1200 bits, respectively. Thus it takes 240 µJ to transmit a query message and 120 µJ to receive a query message, and it takes 1200 and 600 µJ to transmit and receive a reply message. If we assume the total number of sensors in the sensor network is 1500 and the density of the sensor network is 1 per 1000 m², the maximum distance between every two sensors is $14r$, where r is the communication range equal to 50 m. We also assume that the probability of one sensor generating a corresponding reply message is $\frac{1}{30}$; thus $f = 30$.

First we compare the sensors located at different areas in the sensor network based on the remaining lifetimes of the sensors. Table 5.2 depicts the remaining lifetimes of the sensors located at different regions in the sensor field in the context of different query protocols. We select sensors with distance to the sink within one communication range, with 7 times the communication range, and with 14 times the communication range. The last range is the largest distance to the sink. From the values in the table, we find that the remaining lifetime of the sensor increases with the increase in the distance between the sensor to the sink in a traditional query, but it increases quicker in the broadcast communication than in unicast communication, because the load imbalance is accumulated faster in the broadcast case. Thus the sensors near the sink will consume a lot of energy and fail very quickly, which results in earlier termination of the whole sensor network. From this observation, we find that the unbalanced load results in a short lifetimes of the sensor network. On the contrary, as we expected, the remaining lifetimes of sensors located in different regions using the IQ protocol are almost the same, which denotes that IQ does a good job in balancing the load among all sensors.

We also compare the remaining lifetime of the sensor network here. The results of the comparison of three types of query protocols are listed in Table 5.3. From these deduced results, we find that the sensor network using the IQ protocol has a longer remaining

Table 5.3   Comparison of Remaining Lifetime of Whole Sensor Network

	Traditional Query		
	Unicast	Broadcast	IQ
Remaining lifetime	$\dfrac{21,733 - 0.8N_m}{10^6}$	$\dfrac{21,733 - 69N_m}{10^6}$	$\dfrac{21,733 - 46N_m}{10^6}$

lifetime than that using the traditional query by providing a global optimization to balance the load to the whole sensor network. Furthermore, considering Tables 5.2 and 5.3, in a broadcast traditional query, when $N_m = 121$, although there is still a large amount of energy ($13,384/10^6$ from Table 5.3) remaining in the sensor network, it will never be used because the sensor network is down when all sensors within the sink's communication range are down (see in Table 5.2 when $N_m = 121$). However, in IQ, because the load is balanced, no sensor will run out of energy much earlier than others. So most of the energy of each sensor will be effectively used in IQ. To this end, we think that IQ is more energy efficient than a traditional query, that is, the energy utilization is much higher in IQ than in a traditional protocal. From the above analysis, we conclude that IQ extends the lifetime of the whole sensor network because it balances the load to all the sensors in the sensor network, which again validates our argument that load balance plays a very important role in the lifetime of the whole sensor network.

Load Imbalance Factor To capture the load balance feature, we propose a new performance metric called a load imbalance factor (LIF). Here we define the LIF of the sensor network as the total variation of the remaining lifetime of the sensor nodes. The larger the variation, the more imbalance there is in the workload. The LIF is defined as

$$B = \sum_{j=0}^{N_n} \frac{(L_j - \bar{L})^2}{N_n}$$

Here $\bar{L}$ is the average lifetime of all the sensors in the sensor network, that is,

$$\bar{L} = \frac{\sum_{i=0}^{N_n} L_i}{N_n}$$

The LIF reflects the extent of the load imbalance problem in the system and will be used as a performance metric to compare different query protocols in the performance evaluation section.

5.2.6 Performance Evaluation

Currently, two popular approaches are widely used to evaluate the behavior of wireless sensor networks in the community: ns-2 with wireless extension [42] and sensor network emulators such as Tossim [43] and EmStar [44]. The former is good at simulating the details of packet routing and network protocols but is limited by its scalability. However, ns-2 with wireless extension will run very slowly when more than 200 nodes are simulated. The latter is limited to supporting only TinyOS, so it is not good at general simulation, which is what we want in this section. Thus, a large-scale discrete-event-driven sensor network simulator called Capricorn [35] is used to conduct the simulation.

We first describe our simulation configuration, then compare the IQ protocol with the traditional query protocol in terms of *the remaining lifetime of the sensor network, the lifetime of the sensor network, the energy consumption in three approaches, the load imbalance factor*, and *the effect of depleted sensors on the normal operation of the sensor network*. Furthermore, a detailed comparison between the IQ and the IQBP protocol is presented to illustrate the benefit of the proposed optimization.

Table 5.4   Simulation Parameters

Variables	Values
Communication range	30 m
Number of nodes	400
Total energy of each sensor	3 J
Packet size	240, 1200 bits
Energy dissipated for receiving	50 nJ/bit
Energy dissipated for transmission	50 nJ/bit
Energy dissipated for transmit amplifier	100 pJ/bit/m^2
Bandwidth	40 kbps

Simulation Configuration   Capricorn has three components: a *topology generator,* a *discrete-event simulator*, and a *log analysis module*. The topology generator is used to generate the topology of the sensor field, including both regular and irregular fields. The simulator takes input as queries in the Structured Query Language (SQL) which are processed by the *query agent* to generate appropriate messages to be delivered by the underlying sensor network. A snapshot is generated for each tick of the simulator. The collected logs will be analyzed by the log analysis module. In our simulation, 400 nodes are scattered to a 600×600-m^2 field. We choose the GPSR routing protocol in the simulator to deliver messages. The values of parameters used in the simulation are listed in Table 5.4.

Remaining Lifetime of Sensor Networks   The goal of the proposed IQ protocol is to extend the lifetime of the whole sensor network, where the lifetime of the whole sensor network is defined based on the lifetime of the sensors. Thus the remaining lifetime of the whole network is a very important metric to measure the performance of IQ. In this experiment, we compare different protocols in terms of the remaining lifetime of the whole sensor network, which is defined in Section 5.2.4. In the simulation, the initial energy of each sensor is set to 3 J. According to the default simulation parameter values, the initial remaining lifetime of the sensor network is 21, which is calculated from the formula in Section 5.2.4. Figure 5.8 shows the simulation results, where the x axis is the number of queries and the y axis represents the remaining lifetime of the whole network.

From Figure 5.8 it can be seen that the remaining lifetime of the whole sensor network decreases with the increase in the number of processed queries. In the traditional query model, the remaining lifetime drops very quickly from 21 to less than 10 after 300 queries have been processed. After 300 queries have been processed, the remaining lifetime of the whole network using the traditional query model remains stable. This is because the sensor network is already dead after 300 queries. In other words, no more query messages can be sent from the sink to other live sensors; thus there is no more energy consumption. This does not mean that the remaining energy in the live sensors is saved. On the contrary, this energy is wasted and can never be used in the future. In IQ and IQBP, the remaining lifetime of the whole sensor network drops very slowly and smoothly. After 400 queries have been processed, the remaining lifetime of the sensor network is still good using both IQ and IQBP. There is still a lot of energy saved that can be used for future queries. From this point of view, we argue that IQ and IQBP are more energy efficient than the traditional query model. In terms of the remaining lifetime, IQ and IQBP are comparable. From Figure 5.8, it is easy to see that the IQBP protocol is a little better than IQ because of the good

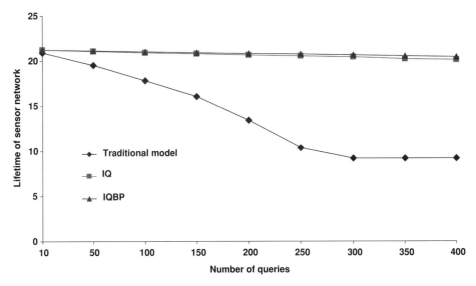

Figure 5.8 Lifetimes of traditional model, IQ, and IQBP.

choice of delegates. More detailed comparison between these two protocols is provided in Section 5.2.6.

Lifetime of Sensor Networks We compare the lifetimes of a sensor network using a traditional query and IQ, which is more interesting to application scientists and system designers. We set the value of θ (the threshold to determine the energy of the sensor network) to 10%. The comparison between traditional and IQ protocols is reported in Figure 5.9, where the x axis is the initial energy of each sensor and the y axis is the lifetime of the sensor network. From the figure, it can be easily seen that the lifetime increases almost linearly with the increase of initial energy in both a traditional query and IQ. However, the lifetime of the whole sensor network increases much faster in IQ than in a traditional query, where the lifetime in the traditional model is about one-seventh of that in IQ. Additionally, if we increase the value of θ, the gap between the traditional query protocol and the proposed IQ protocol will become much larger. Thus we conclude that IQ indeed extends the lifetime of the whole sensor network several times compared to a traditional protocol.

Energy Consumption We are also interested in how much energy is consumed in the traditional query model, IQ, and IQBP. In this simulation, we compare the energy consumption of these three approaches. Figure 5.10 reports the energy consumption of these three approaches after 400 queries have been processed, where the x and y axes provide the location of each sensor node and the z axis depicts the value of energy consumption.

Figure 5.10*a* shows that some sensor nodes in the traditional query model consume a lot of energy, especially the sensor nodes located along the two edges and the diagonal line of the sensor field to which the data sink belongs. So these sensor nodes are energy hungry, consuming all 3 J, while sensors located outside this region consume as little as 0.1 J after 400 queries. Obviously, the energy consumption in the traditional query model is very unbalanced. On the contrary, the load in IQ and IQBP is balanced very well, as shown

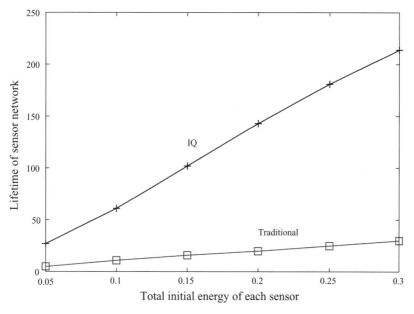

Figure 5.9 Comparison of the lifetime of a sensor network using a traditional query and IQ.

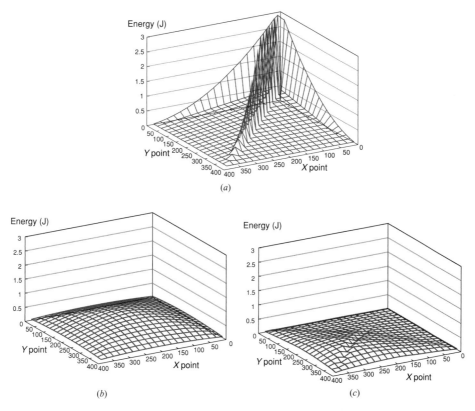

Figure 5.10 Comparison of energy consumption of the traditional query model, IQ, and IQBP using GPSR: (*a*) energy consumption of sensor nodes using the traditional query model, (*b*) energy consumption of sensor nodes using IQ, and (*c*) energy consumption of sensor nodes using IQBP.

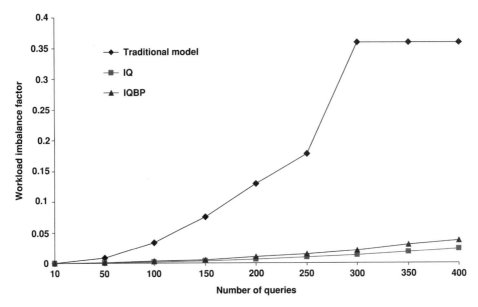

Figure 5.11 Comparison of the LIF.

in Figures 5.10*b* and *c*, where there are no energy-intensive nodes. In IQ the maximum energy consumption is 0.6 J and the minimum energy consumption is 0.04 J; in IQBP, the maximum energy consumption is 0.4 J and the minimum energy consumption is 0.01 J. In other words, in the IQ protocol, by running out of the total 3 J, the sensor network can process at least 2000 queries, and in IQBP, it can at least execute 3000 queries. Thus the lifetimes of the sensor networks using the IQ and IQBP protocols are at least 7 and 10 times longer than that of the traditional query model, and IQBP is slightly better than the basic IQ protocol.

Load Imbalance Factor As defined in the Section 5.2.5, the LIF is a very useful metric to characterize the load balance. Here we compare the LIFs of three query protocols: the traditional query model, the IQ, and the IQBP.

Figure 5.11 shows the values of the LIF, where the *x* axis depicts the number of queries processed and the *y* axis the value of the LIF. As shown in the figure, the LIF in the traditional query model increases from 0 to approximately 0.35 when the number of queries processed increases from 0 to 300. After 300 queries have been processed, the LIF in the traditional query model remains stable because the sensor network is already dead. The fact that the LIF increases with the number of queries shows that the heavily loaded sensors keep busy while the lightly loaded sensors keep idle. Thus, the more queries are processed, the larger the LIF will be. This unbalanced behavior will lead to the case that some sensors run out of energy very quickly while others still have most energy remaining.

It can be seen from Figure 5.11 that the LIFs using IQ and IQBP are much smaller than that of the traditional query model. The LIF using the traditional query model is 0.35, which is seven times larger than those using IQ and IQBP. The rate of increase in the LIF using the traditional query model is much faster than that using IQ and IQBP. The small LIF value and slow increase in the ratio indicate that the proposed IQ protocol makes the load almost balance to the whole sensor network.

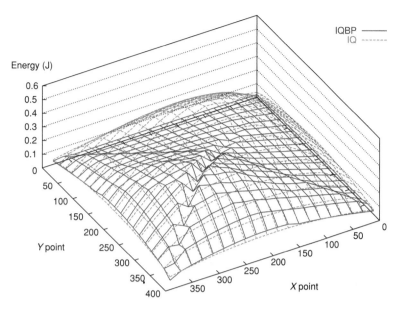

Figure 5.12 Detailed comparison of energy consumption between IQ and IQBP after 400 queries.

From Figure 5.11 we can see that IQ and IQBP are comparable in terms of LIF. However, the LIF for IQ is a bit smaller than that of IQBP. This is because in IQ the chance for each sensor to be selected as a delegate is the same, while in IQBP only sensors located in some specific areas are chosen as delegates, which makes the sensors in that area a bit more heavily loaded.

Effect of Optimized Delegate Selection Now we are in a position to see the effect of the optimized delegation selection. Currently, we have implemented the IQBP protocol. As mentioned in the previous section, both IQ and IQBP are good load balance query protocols to achieve the goal of extending the lifetime of sensor networks. Here we want to examine the difference between them in detail. The comparison of energy consumption between IQ and IQBP is restated in Figure 5.12. This figure shows that the heavily loaded sensors have been moved from the central part of the sensor field to a region far from the sink in the IQBP protocol, and are selected to be the delegates; at the same time the figure shows IQBP reduces the maximum energy consumption in the sensor.

To explore the benefit we can get from the optimization of IQBP, we inspect the detailed energy consumption in both the IQ and IQBP protocols. Figure 5.13 shows the cumulative distribution function of the energy consumption of all 400 nodes. The x axis is the energy consumption and the y axis is the percentage of sensors whose energy consumption is equal to or less than the corresponding energy consumption denoted at the x axis.

From Figure 5.13 we can see that the maximum energy consumption using IQ (i.e., 0.6 J) is larger than that using IQBP (i.e., 0.4 J). When the IQ protocol is used, about 30% of the nodes consume more than 0.4 J, yet none of the sensors consume more than 0.4 J when the IQBP protocol is used. About 70% of nodes consume less than 0.2 J using the IQBP protocol, while only less than 40% of nodes using IQ consume energy less than 0.2 J. Thus, using IQBP the wireless sensor network will have more remaining energy than using IQ.

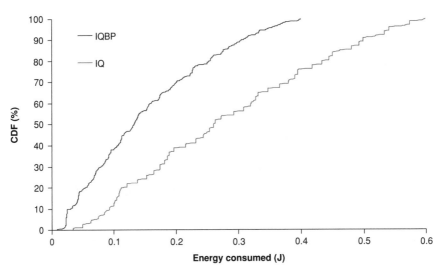

Figure 5.13 The cumulative distribution function of energy consumption in IQ and IQBP protocols after 400 queries.

Effect of Depleted Sensors Based on the above observations, we can deduce that the sensor nodes near the sink are more important than those further away. To evaluate the effect caused by the failure of sensor nodes on the whole sensor network, we propose a new metric called the effect of depleted sensors factor (EDF), defined as the total importance of the failed (depleted) sensor nodes:

$$ \text{EDF} = \sum_{i \in S} w_i $$

where S is the set of failed sensor nodes defined as

$$ S = \left\{ S_j \left\| \left\lfloor \frac{E - L(j)}{E} \right\rfloor = 1, j = 1, 2, 3, \ldots, N_n \right\} $$

So

$$ \text{EDF} = \sum_{i \in S} w_i = \sum_{j=0}^{N_m} w_j \left\lfloor \frac{E - L(j)}{E} \right\rfloor $$

A comparison of EDFs for the three query protocols is given in Table 5.5.

From Table 5.5, we find that the EDFs associated with IQ and IQBP are always 0, because all sensors in the network still contain energy even after processing 400 queries. We estimate that the EDFs for IQ and IQBP will start increasing after processing approximately 2000 and 3000 queries, respectively. However, the EDF increases to 2.86 just after processing 200 queries using the traditional query model, which implies that some sensors already start failing. Ultimately after the EDF reaches 3.06 it becomes a constant when the traditional query model is used, which means no new sensor nodes die but the whole sensor network

Table 5.5 Comparison of EDFs of Three Protocols

Query Protocols	Number of Queries			
	100	200	300	400
Traditional query model	0	2.86	3.06	3.06
IQ	0	0	0	0
IQBP	0	0	0	0

is down. Based on these analyses, we argue that the lifetime of the whole network using the proposed IQ query model (and its variant) is indeed extended about 7 or 10 times compared with the traditional query model.

Discussion From the above analysis, we conclude that the routing level optimization does not necessarily extend the lifetime of the whole sensor network if query is processed using the traditional query model. However, the proposed IQ and its variants are perfect query protocols which approximately balance the load to the whole sensor network, extend the lifetime of the whole sensor network to at least 7 times (using IQ) and 10 times (using IQBP) more than that of the traditional query model. In general, IQ and IQBP are comparable, but the maximum value of energy consumption using IQBP is smaller than that using IQ, which implies that the basic IQ protocol can still be improved by intelligently choosing the delegates. This will be our future work.

5.2.7 Related Work and Discussion

Although energy-efficient routing protocols and optimizations to maximize the lifetime of sensor networks have been widely studied in the literature [45], few of the previous efforts address the load balancing problem. Additionally, few of the previous efforts have been done to formally model the lifetime of the sensor network. Our work is the first step in this direction. In other work, the lifetime of the sensor network has been defined as the time for the first node to run out of power such as in [14, 17, 19, 46, 47] or a certain percentage of network nodes to run out of power as in [32, 48]. We think that these definitions of the lifetime of the sensor network are not satisfactory. The former is too pessimistic, as when only one node fails the rest of the nodes can still provide appropriate functionality to the whole sensor network, while the latter does not consider the importance of the different sensors in a sensor network.

Zhang and Hou have investigated the theoretical upper bound of the lifetime of a wireless sensor network based on the network coverage [49]. Their model does not consider the practical usage of wireless sensor networks and assumes a perfect topology control. Their definition of the lifetime of the whole sensor network is similar to ours; however, we use the predetermined threshold for the termination of the sensor network lifetime. On the other hand, our definition is based on the real communication and energy of sensors, a definition that is more applicable than theirs. Their work defines the upper bound of the lifetime of a wireless sensor network and thus complements our work.

The work done by Chang and Tassiulas [14] is the closest one to ours. They propose a routing approach based on a network flow approach to get maximum lifetime routing. Kalpakis et al. [19] have modeled the data routes set up in sensor networks as the maximum lifetime data-gathering problem and present a polynomial time algorithm, but their work is still based on the traditional query model. Moreover, they define the lifetime of a sensor network as the time from when the sensor network starts to run to the time the first sensor node fails. This definition is not accurate because even if some far-away sensor nodes are dead, the sensor network can still perform the correct function.

Energy-aware routing [27] is proposed by Shah et al. to use a set of suboptimal paths to increase the lifetime of the network. This approach always uses the minimum-energy path. It uses one of multiple paths with a certain probability to increase the lifetime of the whole network. This approach is somehow similar to the approach in [30]. They balance the load of each data path, but the sensors near the sink are still the bottleneck of the sensor network. In our work the load is evenly distributed to delegates which can balance the load among all sensor nodes.

Servatto and Barrenechea propose the constrained random-walk routing protocol to achieve a certain load balancing property by routing messages along all possible routes between source and destination with a certain probability [37]. This routing protocol balances the load to different paths. However, the nodes near the injecting and exiting points have much more load than others. Unlike Servatto and Barrenechea's effort to develop load balance at the routing level, we try to address this problem from a different higher level angle. Therefore, their work complements our work very well.

Hamdaoui and Ramanathan [15] propose a MAC-aware routing for data aggregation. They take MAC contention constraints into consideration and deal with two objectives: maximizing the network lifetime and minimizing the total consumed energy. Their approach can achieve longer network lifetime, but the proposed IQ protocol is at a higher level than their approach and can definitely benefit from their results.

Dai and Han in [30] have constructed a load-balanced tree in sensor networks to balance the load in different branches of the routing tree. They have made the whole load almost evenly distributed to nodes that are the children of the root; however, these nodes will become the bottleneck of the whole sensor network because all the data from the nodes lying at a lower layer must go through these nodes. Thus these nodes will run out of energy very quickly, while the nodes lying at the leaves of the tree will still have a lot of energy remaining. Thus, the lifetime of the whole sensor network cannot be extended too much in their approach.

At the high level, several researchers propose to balance the specific load of sensors by rotating their functionality, including coordinators in topology management [31], grid zone headers in Geographic Adaptive Fidelity (GAF) routing [32], cluster headers in hierarchical protocols [17], and multipath data routing [50] for fault tolerance. However, these approaches are different from the proposed IQ protocol in the scope of load balance. Most of them guarantee that the load of a localized area is balanced but cannot control the load of the whole network, which is addressed by the IQ protocol.

The random choice of a delegation for query is inspired by the recent proposed data-centric storage [51], which leverages the nature of the load balance of a distributed hash table [52–55] technique to build an indirect infrastructure for queries. However, the advantage of this approach is limited to the distribution of predefined events. To our knowledge, there are no performance results available so far. In contrast, IQ is totally independent of the distribution of events and queries.

5.2.8 Conclusions

We first identify the load imbalance problem that resulted from both the traditional query protocol and the GPSR routing protocol. Based on this observation, we propose a novel query protocol which migrates the load of sensors close to the data sink evenly to other sensors by randomly choosing a delegate for each query. Two optimized protocols based on the basic IQ protocol are also proposed to exploit the fact that sensors vary in their importance based on their distance to the data sink and to exploit the application-specific information.

We also propose a formal definition and model of the remaining lifetime of each sensor, the remaining lifetime of the whole sensor network, the lifetime of the whole sensor network, and the effect of depleted sensors on the lifetime of the sensor network. Based on these models, a detailed analysis and comparison between traditional approaches and our proposed approach is depicted. We mathematically prove that the IQ protocol can balance the load of the whole sensor network, so as to extend the lifetime of sensor network. These results are validated by the evaluation results conducted on Capricorn [35], a large-scale sensor network simulator. The simulation shows that the IQ protocol and its variant IQBP balance the load among sensors and so extend the life of the sensor network to at least 7 and 10 times longer than that of the traditional query model.

5.3 ENERGY-EFFICIENT AND MAC-AWARE ROUTING FOR DATA AGGREGATION IN SENSOR NETWORKS

Bechir Hamdaoui and Parmeswaran Ramanathan

Wireless sensor networks consist of many self-organizing and self-coordinating nodes that cooperatively maintain connectivity without any need for wired infrastructure. Because nodes are battery-powered, efficient use of their available energy resources is important. As a result, recent reported routing schemes for wireless networks mainly focus on conserving energy. Even though these schemes generally save power, they do not account for the MAC contention constraints associated with the shared medium. As a result, routing solutions may not be feasible in the sense that the medium will likely not be able to support the data rates.

In this section, we propose an energy-efficient and MAC-aware routing scheme for data aggregation in sensor networks that (i) accounts for MAC contention constraints and (ii) systematically deals with both objectives of maximizing the network lifetime and minimizing the total consumed energy. Through simulations, we demonstrate that the proposed scheme is more likely to result in physically feasible solutions for networks that are infeasible under the reported schemes. We also infer that the rate solutions obtained under the proposed routing scheme are always feasible whereas those obtained under any of the reported schemes may not be feasible. Further, we show that the proposed approach achieves longer network lifetime than the alternative approaches.

5.3.1 Background Information

Wireless sensor networks are being considered for many military and civilian applications such as target tracking, environmental monitoring, disease transmissions in animals, and smart kindergarten. They are comprised of a large number of low-cost nodes that collaborate to carry out a certain signal processing task. This collaboration typically occurs in the form of information fusion from nodes in the same geographic vicinity to arrive at a consensus decision related to the signal processing task. A local coordinator node in a geographic region gathers the data from the nearby nodes, fuses the information, and arrives at the necessary consensus decisions.

If the local coordinator node is within the transmission range of another node, then the information is directly forwarded to the coordinator. Otherwise, the nodes rely on other intermediate nodes to forward the information to the coordinator. Since the nodes are often battery-powered, energy conservation during information exchange is critical for increasing the lifetime of the network. As a result, energy conservation has been a focus of several recent papers [56–66].

In [56–60], reduction in energy consumption is achieved through in-network processing and data aggregation at intermediate nodes. For instance, in [56], a method called diffusion routing is proposed for collecting information from a set of nodes. Diffusion routing supports in-network processing to reduce the amount of information exchanged, thereby saving energy. Similar advantages of in-network processing are shown in [59]. Reduction in energy is achieved through careful selection of nodes for data aggregation in [57,60]. An energy-aware medium access protocol for sensor networks is proposed in [67]. A technique called braided multipath routing is proposed in [68] to increase the robustness of information exchange to route failures while conserving energy. By routing through a series of geographic regions instead of a series of nodes, the scheme in [69] allows many nodes to sleep and conserve energy without disrupting the ongoing communications.

Routing schemes have also been developed to suit the energy-constrained wireless ad hoc networks [46, 61–66]. One class of routing schemes is based on finding routes that minimize the total consumed energy [61, 63, 65]. Another class of schemes focuses on maximizing the lifetime of the network by avoiding routes with nodes having the least amount of available energy [66]. Some schemes formulate the routing problem as a linear programming optimization where the objective function is to optimize energy consumption either by minimizing the total used energy or maximizing the time until the first node runs out of energy resources [46, 62]. Even though these reported routing approaches reduce energy consumption, they all ignore the MAC contention constraints associated with the shared wireless medium. As a result, the number of flows routed through nodes in the same neighborhood may be such that the shared medium may not be able to provide the net data rate required to support these flows. If this happens, the data rate requirements of the traffic flows cannot be satisfied by the network. A primary reason for this discrepancy is that most of the approaches reported in the literature essentially perform network layer optimization without considering the effects of the underlying MAC layer.

In this section, we propose an energy-efficient and MAC-aware routing approach for data aggregation in sensor networks that (i) accounts for MAC contention constraints and (ii) systematically deals with both objectives of maximizing the minimum available energy—increasing the lifetime of the network—and minimizing the total energy consumption. Some of the existing solutions, however, deal with both objectives in an ad hoc fashion (e.g., [66]) based on a parameter that the designer must choose in advance. Through simulation results, we illustrate the importance of considering MAC contention constraints into routing formulations and their effects on the feasibility of the routing solutions. We show that the proposed scheme achieves longer network lifetimes than most of the reported approaches. We also show that the proposed routing scheme can be extended to support extra routing features.

5.3.2 Related Work

Routing protocols for wireless networks have been the subject of extensive research [70–76]. Generally speaking, these studies focus on the problem of finding and maintaining routes between a pair of nodes despite topology changes due to node mobility. The schemes are evaluated in terms of performance metrics such as routing overhead, latency for route discovery, and packet drop rate. The use of energy consumption as an optimization measure in determining the routes is more recent [46, 61–66, 69, 77–81]. One class of energy-aware routing techniques is based on grouping nodes into clusters where each cluster is represented by a clusterhead node [78–80]. Nodes in this class communicate their data to their clusterhead instead of the processing center, and the clusterheads communicate the aggregated data to the processing center. These clustering techniques reduce the overall power consumption since nodes do not need to communicate their data all the way to the processing center, which is typically several hops further away from the nodes than the clusterheads. In [78], the authors propose a distributed algorithm to organize nodes of wireless sensor networks into clusters. Another class of energy-aware routing algorithms is based on conserving energy by turning off as many nodes as possible while assuring network connectivity for the active nodes [69, 77]. In these techniques, nodes operate in one of the three distinct modes: active, idle, or sleep. Routing algorithms of this class typically require coordination beyond the local neighborhood to decide which nodes should remain in the active mode at a given time. Such coordination is difficult to achieve without either tight synchronization or considerable message overhead.

A complementary approach to the above two classes is to make efficient use of energy resources available at the nodes in determining the routes. Some schemes minimize the total energy consumed by the nodes in a route for each traffic flow with an overall goal of minimizing the total energy consumption in the network [61, 63, 65]. If the nodes do not adjust their transmit power based on the distance to the intended receiver, then this is akin to minimizing the number of hops. On the other hand, if the nodes adjust their transmit power based on the distance to the receiver, then this approach tends to choose paths with large number of hops [65]. In [61], the energy spent in packet retransmission is also considered in finding routes that minimize the total energy consumed by the nodes. Unfortunately, these approaches do not generally maximize the network lifetime.

Recent papers have focused on finding routes that account for the energy resources available at the nodes [46, 62, 64, 66, 81]. Some of these techniques [64] maximize the lifetime by selecting routes that maximize the total of the remaining energy of nodes forming the routes. Others (e.g., [65, 66]) maximize the lifetime by finding routes that maximize the minimum remaining energy of all nodes among all paths. In [66], the routing scheme switches its objective depending on the amount of energy available at the nodes. If among all possible routes there exists a nonempty set of routes for which the minimum remaining energy of all nodes in the set is above a prespecified threshold, then a path minimizing the total consumed energy is chosen among the set. If, on the other hand, there is no route for which the minimum remaining energy of all nodes of the route is above the threshold, then the route that maximizes the minimum remaining energy of all nodes forming the path is selected. In [46, 62], a distributed heuristic is proposed for finding routes with an objective of maximizing the time until at least one node exhausts its energy resources. The performance of the heuristic is compared to an "optimal solution" obtained by solving a linear programming problem. The comparison shows that the heuristic often performs well relative to the optimal solution.

In this section, we will compare the proposed energy-efficient and MAC-aware routing scheme to the following reported routing schemes. We only present a brief description of the schemes. Refer to the original papers for more details.

1. *Minimum Total Energy [65, 63]*   This scheme finds routes that minimize the total energy consumed by all nodes. We assume that nodes do not adjust their transmission power, that is, they all use the same amount of energy per transmitted bit. Thus, this scheme is equivalent to finding the shortest path.

2. *Max–Min [65, 64]*   This approach finds routes that maximize the minimum remaining energy of all nodes constituting the path.

3. *Conditional Max–Min [66]*   The conditional max–min scheme switches between the minimum total and the max–min approaches based on the energy level of the nodes' batteries. If among all possible routes there exists a set of routes for which the minimum remaining energy of all nodes is above a prefixed threshold, then the scheme uses the minimum total energy scheme to select a route from that set. Otherwise, it uses the max–min scheme to find a route among all possible routes. We use a prefixed threshold equal to 50% of the maximal energy capacity of the nodes' batteries.

4. *Redirect Flow [46, 62]*   To find a route from node n to node m, this scheme first uses the max–min scheme to find all routes from each of n's neighbor to m. Then, it balances the rate at which n is communicating with m among all those routes so that all the routes have the same lifetime.

5.3.3 Routing Model

We model the wireless sensor network as a directed graph $G = (\mathcal{N}, \mathcal{F})$ of a finite nonempty set $\mathcal{N}$ of nodes and a set $\mathcal{F}$ of flows. Each flow f corresponds to an ordered pair of distinct nodes (n, m) such that m is within n's transmission range—m is a neighbor of n—and n needs to transmit to m. The set $\mathcal{N}$ consists of one coordinator node (CN) and many sensor nodes (SNs). We consider the multiple-sources-to-single-destination routing model in which data traffic is only generated by SNs and all destined to the CN and no traffic is generated by the CN. If a SN is not within direct communication range of the CN, then it relies on other SNs to send its relevant information to the CN. We further assume that G is connected; that is, for each node n there exists at least one path (a set of nodes) through which n can communicate with the CN.

5.3.4 Routing Optimization

Consider the wireless network $G = (\mathcal{N}, \mathcal{F})$ defined in Section 5.3.3 where each node $i \in \mathcal{N}$ generates data traffic destined to the CN at a rate of R_i bits per second. Let $B_i(t)$ denote the energy resources available at node i for network communications at a given time instant t. Also, let ϵ_{ij} denote the energy required to transmit a bit from node i to node j. Let x_{ij} denote the number of bits per second forwarded by node i to a neighboring node j. Given the required rate vector $R = [R_i]_{1 \leq i \leq |\mathcal{N}|}$, we aim at finding a routing solution that minimizes energy consumption. In the remainder of this section, we describe the proposed routing approach.

Routing Constraints   Independently of the routing objectives, given the required rate vector $R = [R_i]_{1 \leq i \leq |\mathcal{N}|}$, the following set of constraints must be satisfied:

- Flow Balance Constraints   At each sensor node, the total outgoing traffic rate must equal the sum of the incoming traffic rate and the traffic generated at the SN. For the CN, the total incoming traffic rate must equal the total traffic generated by all SNs. That is, for each node $i \in \mathcal{N}$,

$$\sum_{j \in \mathcal{N}} x_{ji} + R_i = \sum_{j \in \mathcal{N}} x_{ij} \qquad i \neq \text{CN} \tag{5.1}$$

$$\sum_{j \in \mathcal{N}} x_{ji} = \sum_{j \in \mathcal{N}} R_j \qquad i = \text{CN} \tag{5.2}$$

$$x_{ij} \geq 0 \qquad j \in \mathcal{N} \tag{5.3}$$

The above constraints are needed to assure conservation of the rate flows at each node.

- Energy Consumption Constraints   If node i has $B_i(t_0)$ amount of energy at a particular time t_0, then the remaining energy at any future time $t \geq t_0$ must be greater than or equal to zero. That is, for each sensor node i,

$$B_i(t_0) - (t - t_0) \sum_{j \in \mathcal{N}} \epsilon_{ij} x_{ij} \geq 0 \tag{5.4}$$

We assume that the center node has an infinite amount of energy. These constraints are imposed to assure that only nodes having enough energy resources can forward and/or send traffic.

- MAC Contention Constraints   The contention constraints on sharing the medium depend on the MAC protocol. Without imposing medium contention constraints on the routing formulation, routing solutions may not be feasible in the sense that the MAC may not be able to assure these rate solutions. For instance, in an IEEE 802.11 MAC protocol [82] network, if node i is in communication with node j, then all nodes within the same transmission range of i or j cannot communicate. Therefore, if one does not impose certain constraints that prevent nodes within each other's transmission ranges from transmitting at the same time, resulting rate solutions may not be feasible. Let Ψ_{ij} denote the set of all ordered pairs of nodes that cannot communicate with each other as long as node i is transmitting to node j. Then, the rate vector $[x_{ij}]_{1 \leq i,j \leq \mathcal{N}}$ is feasible—that is, it satisfies the medium access constraints—if for each ordered pair $(i, j) \in \mathcal{F}$ of nodes the following MAC contention constraints hold [83]:

$$x_{ij} + \sum_{(p,q) \in \Psi_{ij}} x_{pq} \leq C \tag{5.5}$$

where C is the maximum rate supported by the wireless medium. Without loss of generality, in the remainder of this section we will assume that C equals 1.

Routing Formulation   Let t_0 be the initial time and T be a time horizon over which the energy-aware optimization is performed. The routing problem is to determine the values of x_{ij} subject to constraints (5.1) to (5.5) with the following conflicting objectives: (i) maximize the minimum available energy among all nodes at time $t_0 + T$ and (ii) minimize the total consumed energy by all the nodes over the period T. The first objective tends to increase the lifetime of the network[1] since the objective avoids finding routes with nodes having the least amount of available energy. The second objective, on the other hand, tends to conserve the overall energy.

We propose to divide the multiple objective optimization problem into two linear programming problems (LPPs). In the first LPP (LPP-I), the objective function is to maximize the minimum available energy among all nodes subject to constraints (5.1) to (5.5). In the second LPP (LPP-II), the objective function is to minimize the total consumed energy by all the nodes subject not only to constraints (5.1) to (5.5) but also to maintaining the maximum value of the objective function achieved by solving LPP-I. Thus, the optimal solution of LPP-II is also an optimal solution of LPP-I. The idea here is that there are often multiple optimal solutions to LPP-I each corresponding to a data rate vector. If this is the case, LPP-II then chooses from among these multiple rate vectors the one that optimizes the objective function of LPP-II. This is illustrated in Example 5.3.1.

We introduce a fictitious variable, denoted λ, and add a new set of constraints that ensure that the energy available at each node after the network is operational for T seconds is at least λ. We call these constraints *lifetime constraints*. Formally, the *lifetime constraints* can be written as follows. For every sensor node i,

$$\lambda \leq B_i(t_0) - T \sum_{j \in \mathcal{N}} \epsilon_{ij} x_{ij} \tag{5.6}$$

Maximizing λ subject to constraints (5.1) to (5.6) can be shown [84] to be equivalent to maximizing the minimum available energy among all nodes at time $t = t_0 + T$ subject to constraints (5.1) to (5.5).

Therefore, LPP-I can be expressed as

> **Maximize** λ
> **Subject to:**
>   Flow balance constraints
>   Energy consumption constraints
>   MAC contention constraints
>   Lifetime constraints

Let λ^* denote the optimal value of λ obtained from solving LPP-I. Then, LPP-II can be stated as

> **Maximize** $\sum_{j \in \mathcal{N}, j \neq CN} B_j(t_0 + T)$
> **Subject to:**
>   Flow balance constraints
>   Energy consumption constraints
>   MAC contention constraints
> $\lambda^* \leq B_i(t_0) - T \sum_{j \in \mathcal{N}} \epsilon_{ij} x_{ij} \quad i \neq CN$

The fourth set of constraints in LPP-II enforces that the minimum available energy at all nodes at time $t_0 + T$ is above the optimal energy value λ^* determined by LPP-I. In the

---

[1] As defined in [46,66], the lifetime of the network under rate vector x is the amount of time that runs out the first battery.

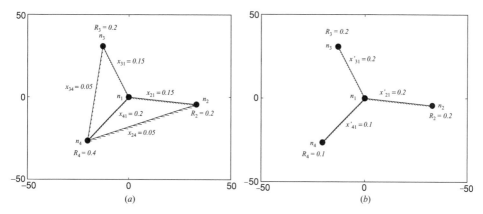

Figure 5.14 Rate solutions for Example 5.3.1: (*a*) solution to LPH-I and (*b*) solution to LPP-II.

following example, we illustrate the two-level energy optimization feature of the proposed routing approach.

Example 5.3.1   Consider a network of one center node n_1 and three sensor nodes n_2, n_3, and n_4. Assume that all the nodes are neighbors of each other. Suppose that $R_2 = R_3 = 0.2$ and $R_4 = 0.1$ bits per second and all batteries are initially full and all have 1 J worth of energy ($B_2 = B_3 = B_4 = 1$). Also, assume that $\epsilon_{ij} = 0.01$ J per bit for all i and j and the optimization horizon T is 1 s. The rate vector x, shown in Figure 5.14a, where $x_{21} = x_{31} = 0.15$, $x_{41} = 0.2$, $x_{24} = x_{34} = 0.05$ bits per second and $x_{ij} = 0$ otherwise, is a solution obtained by solving LPP-I. The rate solution x' obtained by solving LPP-II is such that $x'_{21} = x'_{31} = 0.2$, $x'_{41} = 0.1$, and $x'_{ij} = 0$ otherwise (see Fig. 5.14b). Note that both x and x' are optimal solutions to LPP-I since they both achieve the same maximal value of λ^* and both result in the maximal network lifetime of 500 s. (The value of λ^* depends on the initial energy level of the batteries; e.g., at first when all batteries are full, $\lambda^* = 1 - 1 \times 0.01 \times 0.2 = 0.998$ J) However, observe that x' consumes less total energy [$(0.2 + 0.2 + 0.1) \times 500 \times 0.01 = 2 + \frac{1}{2}$ J] than that consumed by x[$(0.2 + 0.2 + 0.2) \times 500 \times 0.01 = 3$ J]. In fact, if we route the traffic using the rate vector x, all nodes simultaneously die after 500 s. Alternatively, routing with x' also takes the same amount of time (500 s) for the first node to die (nodes n_2 and n_3 die together), but node n_4 will be left with a battery of $\frac{1}{2}$ J worth of energy, which keeps it alive for another 500 s. Thus, the proposed two-level linear programming formulation yields to an optimal solution that balances between a longer lifetime and less overall consumed power; that is, the formulation minimizes the total consumed energy while maximizing the lifetime. The gain of the two-level energy optimization is studied in greater detail in Section 5.3.5

As mentioned in earlier sections, the proposed routing formulation incorporates MAC contention constraints whereas those reported in the literature (e.g., [46,62]) do not consider them. If one does not account for the multiple access nature of the wireless medium, the resulting solution may not be feasible [83]. That is, the medium will likely not be able to support the net data rate needed to deal with all the traffic flows routed through some spatial neighborhoods. When this happens, the packet queues at the corresponding nodes will grow

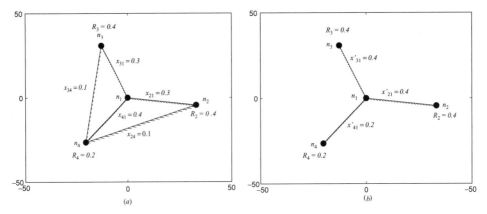

Figure 5.15   Rate solutions for Example 5.3.2: (*a*) without MAC contention constraints and (*b*) with MAC contention constraints.

and there will be either larger delays or packet losses, both of which result in violation of quality-of-service (QoS) requirements. We illustrate the importance of including the MAC contention constraints to routing formulations in Example 5.3.2.

Example 5.3.2   Let us again consider the same network defined in Example 5.3.1 where the network consists of one center node n_1 and three sensor nodes n_2, n_3, and n_4. Now, suppose that $R_2 = R_3 = 0.4$ and $R_4 = 0.2$ bits per second and all batteries are again initially full ($B_2 = B_3 = B_4 = 1$). Also, assume that $\epsilon_{ij} = 0.01$ J per bit for all i and j. In this example, we solve the routing scheme defined in [46] twice: with and without the MAC contention constraints. Figure 5.15*a* shows a solution x obtained by solving the scheme exactly as defined in [46]—that is, without MAC contention constraints. The rate solution x', shown in Figure 5.15*b*, is obtained when the MAC contention constraints are included. First, observe that both solutions result in the same maximal lifetime of 250 s; thus they both are optimal solutions to the problem formulated in [46]. However, note that the solution obtained without imposing the MAC contention constraints (Fig. 5.15*a*) is not physically feasible if IEEE 802.11 [82] is the medium access scheme. This is because the summation of all the rates x_{ij} is equal to 1.2, which exceeds the wireless medium capacity $C(C = 1)$. Alternatively, when the MAC contention constraints are included, solving the same problem results in a feasible rate solution ($\sum_{1 \leq i,j \leq 4} x'_{ij} = 1$). This example shows the importance of including the physical contention constraints to routing formulations. In Section 5.3.5, a more detailed illustration is provided.

Routing Implementation   We assume that the CN has an unlimited amount of energy, which is reasonable since typically the CN is connected to an accessible power-supplied infrastructure. Provided that the CN has a reliable power supply, it is responsible for most of the intensive computational processing such as solving the two LPPs.

The CN first solves the two linear programming problems LPP-I and LPP-II. The optimal rate solution x obtained by solving LPP-II is then sent to the sensor nodes. Once a node i receives the rates from the CN, it forwards every packet to its neighbor j with a *packet forwarding probability*, p_{ij}, computed as

$$p_{ij} = \frac{x_{ij}}{\sum_{k \in \mathcal{A}_i} x_{ik}}$$

where $\mathcal{A}_i$ is the set of i's neighbors. This routing process is repeated every T seconds; that is, periodically, the CN solves the routing problem and sends the rates to all the nodes. Nodes use the rate solution to forward packets for the next T seconds. At the end of each optimization horizon T, each node sends its battery level information and its neighbor list to the CN, which uses them to determine the optimal rates for the next horizon. It is worth noting that since initially the CN would not have the battery information of the SNs or the topography of the network, it is not possible for it to compute the routing solution (e.g., x_{ij}) for the first horizon. The SNs, on the other hand, would not be able to send their battery information to the CN because of the lack of routing paths. One way of solving this initial phase is to use a flooding approach; that is, at start time, the SNs use flooding techniques to pass their battery information and neighbor lists to the CN. Once the CN receives such information, it then solves the routing problem and sends the rates to all SNs. During later horizons, the SNs would then use these routes to send their state information. The value of the optimization horizon T is a design parameter which we will discuss in Section 5.3.5.

There are two points worth noting. First, our routing approach assures that the rates provided by the optimal solution are met on the average. The instantaneous rates, however, may deviate considerably from the optimal rates. If T is sufficiently large, then the rates averaged over this horizon are likely to be close to the optimal rates. Second, since all the traffic is going to the CN, packets are forwarded to a given neighbor based only on the forwarding probability and not on the origin of the packets. This works perfectly so long as there is no packet looping during the routing process of the traffic. Packet looping occurs if packets could return to where they came from during the routing process and results in packets never reaching the CN. The proposition below shows that the solution resulting out of LPP-II is cycle-free,[2] and thus the above forwarding scheme ensures that all the packets will reach the CN.

Proposition 5.3.1   If x is an optimal solution to LPP-II, then x is cycle-free.

PROOF   We will prove that if x has a cycle, then it is not an optimal solution. Let x be a solution that contains at least one cycle. Without loss of generality, let $\mathcal{C} = (1, 2, \ldots, k)$ denote the sequence of nodes forming the cycle. That is, $x_{k1} > 0$ and $x_{ii+1} > 0$ for $1 \le i < k$. Let $\xi = \min\{x_{k1}, \min\{x_{ii+1} : i \in \{1, 2, \ldots, k-1\}\}\}$. Consider the vector $[\hat{x}_{ij}]_{1 \le i, j \le N}$ such that $\hat{x}_{k1} = x_{k1} - \xi$, $\hat{x}_{ii+1} = x_{ii+1} - \xi$ for $1 \le i \le k - 1$ and $\hat{x}_{ij} = x_{ij}$ otherwise. Clearly, the objective function value obtained by $\hat{x}$ is bigger than that obtained by x since $\hat{x}_{ij} \le x_{ij}$. Thus, by showing that $\hat{x}$ satisfies all the constraints stated by LPP-II, we will prove that x is not an optimal solution to LPP-II. Note that each node in $\mathcal{C}$ has both its incoming rate and its outgoing rate reduced by exactly ξ. All other nodes which do not belong to $\mathcal{C}$ have their rates remain the same. Hence, under $\hat{x}$, the flow balance constraints are met. Moreover, since, for all i, j, $0 < \hat{x}_{ij} \le x_{ij}$, then $B_i(t_0) \ge (t - t_0) \sum_{j \in \mathcal{N}} \epsilon_{ij} x_{ij} \ge (t - t_0) \sum_{j \in \mathcal{N}} \epsilon_{ij} \hat{x}_{ij}$,

---

[2] We use the traditional definition of cycles; i.e., x has a cycle if there exists a sequence $(n_1, n_2, \ldots, n_k)$ of nodes in $\mathcal{N}$ such that $x_{k1} > 0$ and $x_{ii+1} > 0$ for $1 \le i < k$.

and hence, both the energy consumption constraints and the lifetime constraints are also met. Finally, because $x_{ij} \geq \hat{x}_{ij}$ for all $i, j \in \mathcal{N}$ and x satisfies the MAC contention Constraints, $\hat{x}$ also satisfies these constraints. ∎

5.3.5 Performance Evaluation

We evaluate the performance of the proposed routing scheme. The evaluation consists of two simulation studies. In the first study, we illustrate the effect of MAC contention constraints on the feasibility of the rate solutions, whereas, in the second study, we evaluate the performance of the scheme in terms of network lifetime and energy consumption.

Effect of MAC Contention Constraints To further demonstrate the significance of including the MAC contention constraints to routing formulations, we simulate and compare the proposed scheme and all the schemes described in Section 5.3.2.

Simulation Method and Metrics One hundred random wireless sensor networks are generated, each with 50 SNs and 1 CN. The SNs are randomly and uniformly distributed in a cell of size 100×100 m^2. The CN is placed in the center of the cell. Each SN is assumed to have data traffic with a rate requirement of $0.01C$, where C is the maximum data rate of the wireless medium. Since there are 50 SNs, this corresponds to an incoming traffic load of 50% at the CN. We consider three values of the transmission range: 21, 25, and 35 m. For each transmission range, we compute the *average node degree* (δ), which is the average of each node's number of neighbors in all the 100 graphs. These average node degrees are 4.98, 7.53, and 14.65 respectively for the ranges of transmission 21, 25, and 35. All the 100 graphs are connected—that is, each SN is able to communicate with the CN either directly or through a set of nodes.

For each of the three average node degrees δ, we simulate the 100 graphs for the four power-aware routing algorithms described in Section 5.3.2. We also simulate these graphs using the proposed routing approach but without including the MAC contention constraints—that is, LPP-I and LPP-II are not subject to Eq. (5.5). Figure 5.16 illustrates the effect of not including the MAC contention constraints on the physical feasibility of the solutions obtained by solving the power-aware routing schemes. Each bar of the figure corresponds to a combination of a routing scheme and an average node degree and represents the number of feasible graphs out of the 100 simulated graphs. A simulated graph is considered feasible if the solution obtained by solving the corresponding routing algorithm satisfies the MAC contention constraints [83] given by Eq. (5.5). Since the aim here is to study the feasibility of the solutions based only on whether they meet the MAC contention constraints, we set a small value for the optimization horizon T so that the energy consumption constraints are met and all schemes have rate solutions. We assume that the energy per bit ϵ_{ij} is constant for all nodes and does not depend on the distance between neighbor nodes.

Result Analysis Figure 5.16 shows that for each of the five schemes the higher the average node degree δ, the higher the percentage of feasible graphs. Since, on average, nodes in graphs with higher δ have more neighbors, they have more alternatives for routing. If the average node degree is low, then the likelihood that different nodes are forced to route through the same nodes is higher than if these nodes have more neighbors. Thus, graphs with lower average node degrees result in more physical infeasibility of rate solutions. A

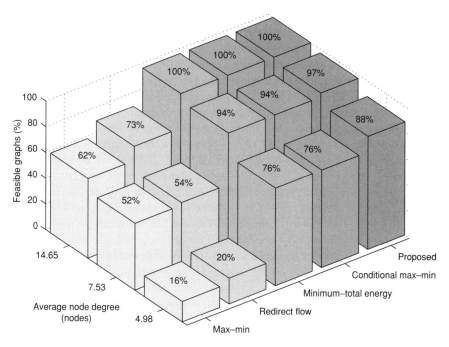

Figure 5.16 Graph feasibility: Number of SNs = 50, cell width = 100 m, network traffic load = 50 %.

second observation we deduce from the figure is that the percentages of feasible solutions obtained by solving the minimum total energy and the conditional max–min schemes are the same. Since the conditional max–min scheme is identical to the minimum total energy scheme when battery energy levels are above a prefixed threshold, then for small optimization horizons T, both algorithms result in the same solutions for the first period. In our simulations, the prefixed threshold is set to be 50% of the initial battery energy level. Note that the max–min approach has the least number of feasible solutions. This is because the scheme tends to choose nodes with maximum remaining energy, and hence nodes with higher energy levels are more likely to be part of many routes. This in turn increases the chances that some neighborhoods will have higher contention, resulting in physical infeasibility of the solution. The redirect flow algorithm is similar to the max–min algorithm except that senders rely on more than one neighbor to forward their traffic, and thus the resulting routes are balanced over more nodes. However, this still results in a higher number of infeasible solutions since the distribution is local (at the first hop only), and further nodes in the route will still cause higher contention by being involved in many routes. The minimum total energy scheme does better than the max–min and the redirect flow schemes due to the following reason. Since ϵ_{ij} is constant for all nodes, finding routes that minimize the total power is the same as finding the shortest paths—that is, routes with the least number of hops. For graphs with uniformly distributed nodes, a shortest path algorithm finds routes such that nodes within the same hop distance from the CN are part of the same number of routes. Thus, the minimum total energy scheme results in routing solutions that are more distributed over all nodes than solutions found by the max–min and redirect flow algorithms. This balances the contention over all neighborhoods, resulting in an increase of the number of feasible graphs. Finally, we observe that the proposed routing approach performs the best in terms of solution feasibility. In all cases, the proposed scheme achieves a slightly greater number of feasible graphs than the minimum total energy and conditional max–min schemes and a substantially greater number than the max–min and redirect flow schemes. The key here is that the proposed formulation finds rates which result in optimal distribution over all the nodes. As in the case of the minimum total energy scheme, an evenly distributed rate solution does better in balancing the physical contention over all neighborhoods. The proposed approach does better by achieving an optimally balanced contention. However, observe that there are a few graphs for which the solutions obtained by the proposed scheme are not physically feasible. Out of the 100 graphs, this happens only for 3 and 12 graphs when the average node degrees are respectively 4.98 and 7.53. When nodes do not have many neighbors, including the CN, then the likelihood that nodes must forward their traffic through the same node (usually nodes which are direct neighbors of the CN) is very high. This in turn increases the contention of neighborhoods around the CN. In these cases, the proposed two linear programming formulation does not have a feasible solution.

Given the above, we can make two inferences. First, the proposed scheme is more likely to result in feasible rate solutions for networks that are infeasible under the reported schemes. This is due to the fact that the rate solutions obtained by solving the proposed scheme achieve more balanced contention over all neighborhoods. Second, because the proposed routing formulation accounts for the MAC contention constraints, the rate solutions obtained under it are always feasible whereas those obtained under any of the reported schemes may not be feasible.

Network Lifetime and Energy Consumption   We evaluate the lifetime performance of the proposed routing scheme by comparing it to the four reported schemes described in Section 5.3.2.

Simulation Method and Metrics   Each scheme is simulated for the 16, 52, and 62 graphs out of the 100 graphs that are found feasible in Section 5.3.5 respectively for the range of transmission values of 21, 25, and 35 m. Let d denote the transmission range. The average node degrees δ corresponding to these transmission ranges of the 16, 52, and 62 feasible graphs are 5.23, 7.64, and 14.70, respectively. In each simulation run, each sensor node is assumed to have infinite data traffic of $0.01C$ bits per second to send to the CN where the medium capacity C is assumed to be 1 bit per second. Initially, all nodes' batteries are full and have 1 J worth of energy $[B_i(t_0) = 1 \ \forall i]$. The amount of energy consumed due to the communication of one bit ϵ_{ij} is assumed to be $10^{-8} \times d^2$ for any pair (i, j) of neighbor nodes. Note that, once the transmission range d is chosen, the amount of energy per transmitted bit is the same for all nodes and does not depend on the distance between the sender node and its receiving neighbor.

Each simulation consists of (i) solving the scheme, that is, determining the routes to the CN; (ii) routing the network traffic using the obtained solution for T seconds; and (iii) updating the battery levels and the lifetimes of all active nodes. These three steps are repeated until every sensor node is either depleted of its energy resources or disconnected from the coordinator node. For each of the three studied average node degrees, the measured lifetime is averaged over all the simulated feasible graphs. Unlike in Section 5.3.5, here the proposed routing scheme considers the MAC contention constraints.

Let *average maximal lifetime* be the average over all nodes of the amount of time that runs out each node's full battery provided that each node adjusts its transmit power so that it communicates directly with the CN without relying on other nodes to forward its

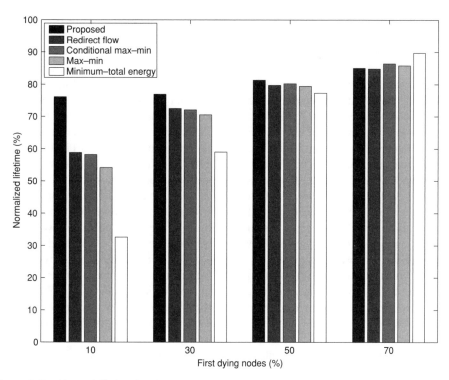

Figure 5.17 Network lifetime for $d = 35$ and $F = 100$—averaged over the 62 feasible graphs whose $\delta = 14.70$.

traffic. We define the *normalized lifetime* of a node as the ratio of its absolute lifetime achieved under a given routing scheme to the average maximal lifetime. We also define the *computational frequency, F*, as the average maximal lifetime over the optimization horizon T. The parameter F represents the number of T periods the CN computes and forwards the routes until approximately half of the nodes die. For example, if T is 5 s and the average maximal lifetime is 50 s—meaning that, on average, every SN would be alive for 50 s if SNs are not cooperative and thus SNs would send directly their signals to the CN—then the CN would have to compute and forward routing solutions $\frac{50}{5} = 10$ times before most nodes die.

Result Analysis   Figures 5.17 to 5.19 show the normalized lifetime as a function of the percentage of dying nodes respectively for d values of 35, 25 and 21. (A bar in a figure corresponding to an x-axis value of $x\%$ and a y-axis value of $y\%$ can be interpreted as $x\%$ of the nodes live only for $y\%$ of their maximal lifetime.) Figure 5.17 illustrates the normalized lifetime of the nodes averaged over 62 feasible graphs whose $\delta = 14.70$ with a computational frequency $F = 100$. The figure shows that when the proposed routing scheme is used, the first 10% dying nodes live for approximately 78% of their maximal lifetimes. However, 10% of the nodes live only for approximately 59, 58, 55, and 32% of their maximal lifetimes when the redirect flow, conditional max–min, max–min, and minimum total energy schemes are used, respectively. Note that even up to the first 50% dying nodes, the proposed scheme provides a longer lifetime than all of the other four schemes. Since the proposed approach finds routes that optimize energy consumption with respect to all the nodes as opposed to only a set of nodes, our scheme results in longer lifetimes for the first dying nodes. The minimum total scheme, on the other hand, provides the shortest lifetime for the first 50% dying nodes among all the studied schemes because it always chooses the paths with the least number of hops. This results in the shortest lifetime of the first dying nodes since nodes that happen to be in the shortest path are always chosen independently of their available energy. However, since the minimum total energy approach minimizes the total consumed amount of energy, the last dying nodes of the network tend to live for a longer time than those under the other schemes, as evidenced in Figure 5.17. In fact, note that under the minimum total energy scheme, the last 50% dying nodes live longer than those that live under any of the other four schemes. The other three schemes, max–min, redirect flow, and conditional max–min, yield lifetimes between the proposed and minimum total energy schemes. They all find routes that somehow maximize the lifetime of the network, which explains why they do better than the minimum total energy scheme. However, their optimization is not global, which explains why they do worse than the proposed scheme.

In brief, the proposed routing technique tends to maintain the life of as many nodes as possible, which increases the operational time of the network, in detriment of decreasing the lifetimes of the last dying nodes—that is, all nodes tend to die simultaneously. Conversely, the other schemes achieve shorter lifetimes for nodes that die first while achieving longer lifetimes for those that die last. We believe that the lifetimes of the first dying nodes (e.g., 10 to 30%) are more important and crucial to the network mission than those of the last dying nodes since losing a large number of nodes results in less likely useful network.

Effect of δ on the Lifetime   To study the effect of the average node degree on the network lifetime, we also collect the normalized lifetimes of the nodes for $\delta = 7.64$ and $\delta = 5.23$, as shown in Figures 5.18 and 5.19. Only the feasible graphs are simulated—that is, the 52 graphs for $\delta = 7.64$ and the 16 graphs for $\delta = 5.23$. The computational frequency again equals 100.

Note that the smaller the average node degree, the narrower the gap between the lifetimes achieved by the different schemes (see Figs. 5.17, 5.18, and 5.19). Figure 5.19 ($\delta = 5.23$) shows that up to the first 50% of dying nodes all routing schemes result in almost the same lifetime, except for the minimum total energy scheme. The gap, however, gets wider as δ gets larger. This is simply because when the average node degree is small—that is, nodes have few neighbors—the likelihood that nodes have very few (e.g., one or two) possible routes is high. In other words, the fewer the average number of neighbors, the fewer the number of possible routes. Therefore, all routing schemes end up selecting almost the same routes. This results in achieving almost the same lifetimes under all the routing schemes.

The minimum total energy scheme, however, maintains a relatively large lifetime difference from all the other four schemes. Unlike the other schemes, the minimum total energy scheme chooses the paths that minimize the total consumed energy as opposed to increasing the lifetime, which again explains why the minimum total energy scheme always has the shortest lifetimes for the first dying nodes while the longest lifetimes for the last dying nodes.

Effect of F on the Lifetime   So far in this section we have used a computational frequency equal to 100, which means that approximately half of the nodes stay alive during 100 optimization horizons. In this section, we study and discuss the effect of the computational frequency on the performance of the network. To do so, we simulate the 52 feasible graphs whose $\delta = 7.64$ for two more frequencies: 200 and 300. For ease of comparison, we only

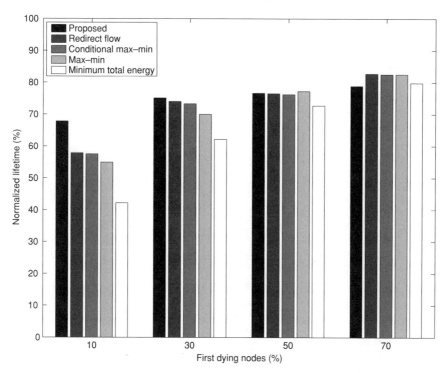

Figure 5.18 Network lifetime for $d = 25$ and $F = 100$—averaged over the 52 feasible graphs whose $\delta = 7.64$.

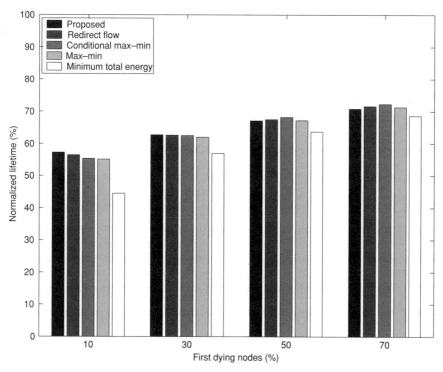

Figure 5.19 Network lifetime for $d = 21$ and $F = 100$—averaged over the 16 feasible graphs whose $\delta = 5.23$.

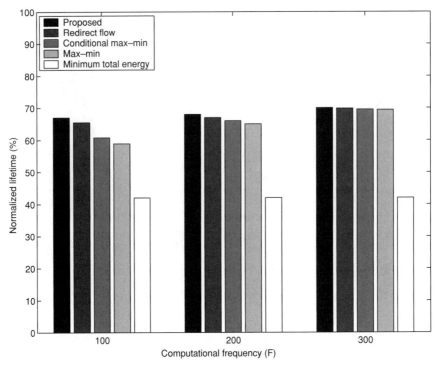

Figure 5.20 Network lifetime of the first 10% dying nodes for $d = 25$—averaged over the 52 feasible graphs whose $\delta = 7.64$.

consider the lifetimes of the first 10% dying nodes. The collected results with F equal to 100, 200, and 300 are shown in Figure 5.20. Observe that as the frequency increases (i.e., period decreases), the lifetimes achieved under the reported schemes (except the minimum total energy scheme) all tend to increase toward the lifetime achieved under the proposed scheme, which is almost independent of the value of T. As mentioned earlier, one major feature of the proposed scheme is that it optimizes the lifetime in a global manner—it determines a node's route by taking into account the routes of all other nodes. In contrast, max–min, redirect flow, and conditional max–min schemes do not consider the number of paths of which a particular node is already a part. In other words, a particular node whose energy level is relatively high may be selected for a path regardless of the total number of routes of which it is a part. Therefore, if the period T during which nodes route their traffic is large, then these high-energy-level nodes will be exhausted rapidly under any of the reported routing schemes, thereby resulting in shorter lifetimes. This explains the relatively large difference between the achieved lifetimes under the different schemes when $F = 100$. However, if T is small (e.g., $F = 300$), then the nodes with high energy levels will not be depleted of their energy rapidly. First, since T is small, these nodes will not lose a substantial amount of their energy during a given horizon T. Second, since these nodes have their energy levels lowered, they are less likely to be selected again when the paths are recomputed for the next horizon T. Consequently, nodes tend to live for a longer time under smaller optimization horizons. Note that the lifetimes achieved by the minimum total energy scheme do not depend on how often the routes are recomputed. Again, this is because the shortest paths remain the same independent of the period length.

In studying the effect of the optimization horizon, we notice that one can increase the lifetimes achieved under the reported schemes by using small values of T and therefore may achieve similar performance to that of the proposed scheme. However, this typically results in greater computation and higher control message overhead, which in turn results in more energy consumption. Therefore, when taking both the energy consumption and the lifetime into consideration, the proposed scheme performs better.

5.3.6 Extension to the Active–Idle–Sleep Mode

The packet forwarding scheme described in Section 5.3.4 makes the proposed scheme suitable and more practical for networks where nodes must transition between active, idle, and sleep modes to save energy. It is shown [69, 77] that idle nodes (nodes which neither transmit nor receive) consume a significant amount of energy; in fact, the amount of consumed energy is almost as high as that consumed by nodes in the receive (active) mode. As a result, to conserve energy, it is important to put nodes in the sleep mode rather than in the idle mode whenever they are neither transmitting nor receiving. Therefore, power-aware routing schemes should be suitable and efficient for such active–idle–sleep networks. If nodes are allowed to sleep as in these networks, then the reported routing schemes require a coordination mechanism by which they are able to discriminate between nodes that are in the sleep mode and those that are not. Such a mechanism adds complexity and restriction to the routing algorithm. For example, if a node of a given path transitions to the sleep mode, most of the reported routing approaches (e.g., [66]) need to rerun their algorithms to find new paths. Our approach does not require such extra recomputation and is not concerned with the extra complexity added by the sleep awareness mechanism. To satisfy the rate solution, nodes often need to forward their traffic to more than one neighbor with different packet forwarding probabilities. Therefore, nodes can manage locally to send more data (higher rate) to one neighbor while the other neighbor is sleeping. When the sleeping node transitions back to the active mode, then the sending node can transmit at a higher rate to that recently awakened node. This is managed so that the overall average rates are equal to the optimal rates. Hence, the proposed scheme requires only local coordination between nodes and their direct neighbors. Note that, as we show in Proposition 5.3.1, the proposed energy-aware routing solution is cycle-free. This guarantees that all traffic eventually will reach the access point without looping. Our proposed solution does not require any cooperation between the sleep-aware mechanism and the routing algorithm. In addition, there is no need to recompute the routing solution (e.g., paths) when one or more nodes transition to the sleep mode. Therefore, the proposed routing scheme is more suitable and easier to implement in networks where the active–idle–sleep mode is required.

5.3.7 Conclusion

We propose an energy-efficient and MAC-aware routing scheme for data aggregation in sensor networks that accounts for MAC contention constraints and systematically deals with both objectives of maximizing the network lifetime and minimizing the total consumed energy. We demonstrate the importance of coupling between the medium access and network layer solutions by studying the effect of not including the medium access constraints on the physical feasibility of the routing solutions. Through simulations, we show that the proposed scheme is more likely to result in feasible solutions for networks that are infeasible under the reported schemes. Further, we infer that rate solutions obtained under the proposed routing scheme are always feasible whereas those obtained under any of the reported schemes may not be feasible.

In addition, we show that the proposed scheme achieves longer network lifetime than those of the reported schemes. The reported schemes achieve shorter lifetimes for the first dying nodes while achieving longer lifetimes for those that die last. However, since the network becomes less useful after losing a large number of its nodes, it is more important to optimize the lifetimes of the first dying nodes than the last dying ones. Because the proposed scheme achieves longer lifetimes for the first dying nodes in detriment of decreasing the lifetimes of the last dying ones, it yields viable and more practical networks when used.

5.4 LESS: LOW-ENERGY SECURITY SOLUTION FOR LARGE-SCALE SENSOR NETWORKS BASED ON TREE-RIPPLE-ZONE ROUTING SCHEME

Fei Hu, Vaithiyam Ramesh, and Waqaas Siddiqui

Wireless sensor networks can be applied in many exciting scenarios such as terrorist detection in crucial areas, disaster recovery in highly dangerous domains, and battlefield parameter collections. Data transmission should be confidential and authenticated in many cases to prevent active network attacks. Security in WSNs is a largely unexplored research area compared to security research in other types of networks such as wired Internet and wireless LAN (local area network). This research provides a WSN security solution that can achieve the following goals: (1) low complexity through small calculation and communication overhead, (2) low energy through *symmetric* cryptography, (3) scalability through our hierarchical *tree-ripple-zone* (TRZ) routing architecture, and (4) robustness through the tolerance of wireless transmission errors and rekeying packet losses. Our *LESS* (*low-energy security solution*) is based on a promising two-level routing scheme, which is different from other security schemes that usually assume common routing architectures based on cluster or flat topology. Through the management of multiple keys such as ripple keys and zone keys, LESS can secure in-networking processing with little energy and in a distributed way. In addition, LESS can adapt to a *dynamic* WSN topology that may be caused by sensor mobility, malfunction, and addition. Our simulation and analytical results show that LESS has low energy consumption and good scalability.

5.4.1 Background Information

Progress in wireless networking and *microelectromechanical systems* (MEMS) is contributing to the formation of a new computing domain—WSNs [85]. There are some solutions for securing *general ad hoc* networks in the literature [86–88]. However, those solutions cannot be applied efficiently in sensor networks where there can be thousands of energy-limited nodes. We have comprehensively discussed the security differences between WSNs and general ad hoc networks in [89].

Other authors have proposed some novel security schemes such as key establishment protocol by Tassos et al., authentication protocols by Andre et al., a group security scheme by TianFu et al., a cooperative security protocol by Roberto, and security energy analysis by Alireza and Ingrid. In this section, we will show that we can seamlessly integrate security with a hierarchical routing scheme in a large-scale WSN to achieve robust, low-energy rekeying and broadcast authentication without the need of time synchronization.

We further summarize the current WSN security works as follows:

1. *Broadcast Authentication* The pioneer work on securing sensor networks is SPINS [90, 91], which proposes µTESLA, an important innovation for achieving broadcast authentication of any messages sent from the base station. An improved multilevel µTESLA key-chain mechanism was proposed in [92, 93].

2. *Key Predistribution* How could we make sure that any pair of sensors could find a shared key to encrypt/decrypt their messages, that is, they have a pairwise key? Note that we cannot use a *public key (asymmetric) scheme* to secure data transmission since the limited memory of a sensor may not be large enough to hold a typical public key of a few thousand bytes [90]. We also cannot use a *trusted-server scheme* since in WSNs we generally cannot assume any trusted infrastructure [95]. Thus, we have only the third choice: to bootstrap secure communications among sensors using key predistribution to preload some *symmetric* keys in sensors. A *key-pool* scheme was suggested in [96] to guarantee that any two nodes share at least *one* pairwise key with a certain probability. Multiple pairwise keys may be found between nodes through the schemes in [96]. A Blom-based key predistribution scheme was discussed in [94]. Key predistribution schemes utilizing location information were described in [98–100].

3. *Intrusion Detection* In WSNs, a challenging problem is to detect compromised sensors and then to tolerate intrusions by bypassing the malicious sensors [101]. A basic requirement is to limit the impact of compromised sensors into a limited local area instead of allowing an intrusion to affect the majority of the network [102, 103]. The base station may be responsible for most of the intrusion detection since the sensors are too limited in memory and energy [104].

4. *Other Security Issues* There also exist other research on typical security issues in WSNs, including *denial-of-service* (DOS) attacks [105], routing security [106], group communication security [107, 108], multiple-key management [109, 110], and security performance analysis [111–113].

Recently a security scheme called LiSP was proposed in [141]. It has some ideas close to our work. However, it has the following drawbacks:

1. LiSP cannot scale to very large WSNs because in its two-level topology the low-level nodes use just a flat routing architecture, which can bring complex routing/security maintenance when there are thousands of sensors in a typical WSN. We propose a RTZs-based self-organization algorithm to assign low-level sensors to different "zones" and "ripples." Thus our security scheme can better employ localized routing/data aggregation operations to achieve global tasks.

2. LiSP does not consider the integration of security with routing schemes. However, different routing strategies can bring significantly variable security overhead. We argue that any WSN security should consider particular network routing architecture.

3. Our scheme uses a multikey-based approach to meeting different security requirements in low-energy WSN applications. For instance, we define "ripple key" to achieve asynchronous broadcast authentication.

4. We propose a "ripple-to-ripple"-based failure recovery scheme to recover lost keys, which can adapt to the unreliable wireless communication nature.

More advantages of our scheme will be discussed in detail later.

The existing research work on WSN security has generated some consistent conclusions as follows:

1. *Use symmetric-key-based schemes:* Memory use is the major concern in WSNs and prohibits us from using memory-intensive *asymmetric* keying schemes.

2. *Low transmission energy:* Key generation and distribution protocols with too many hash computations and global key delivery rounds can quickly drain the limited power of a node.

3. *Scalable implementation:* Since WSNs usually include hundreds, if not thousands, of nodes and network topology/density can change as time progresses due to sensor deaths and new sensor additions, a centralized or flooding-based security scheme cannot scale well.

Thus distributed algorithms and localized coordination to achieve global convergence are preferred.

In this work, we manage to solve some *new* issues that have not been considered or not efficiently addressed by previous research on WSN security. Specifically, we will propose a LESS that makes the following two *contributions*:

1. *Seamless Integration of Security with Scalable Routing Protocols* We have found that most of the existing WSN security strategies focus only on key management/security algorithms *themselves* while ignoring the important idea that *any security scheme should be seamlessly integrated with the special characteristics of WSN architecture, especially routing protocols*; otherwise, the security scheme may NOT be practical or energy efficient from the network protocol point of view. For example, all existing key predistribution schemes try to establish pairwise keys between each pair of nodes. However, most sensors do not need to establish a *direct* point-to-point secure channel with sensors multiple hops away since WSNs use hop-to-hop communication techniques to achieve long-distance transmission. One of the most famous schemes, SPINS [91], simply assumes a flooding-based spanning-tree architecture with the base station as the tree root. However, the establishment and maintenance of a global spanning tree in a large-scale WSN with a large footprint may not only bring unacceptable communication overhead (and thus increased energy consumption[3]) but also cause a large transmission delay and make the assumption of time synchronization in μTESLA (a *broadcast authentication* protocol [91]) impractical. We should keep in mind that sensor networks have particular application scenarios, such as telemedicine and battle-field monitoring, and *an energy-efficient, scalable, low-cost routing architecture should be established first before the adoption of any security protocols*.

Our proposed LESS is based on a *sink–supernode–sensor (3S)* network model, which is a popular WSN self-organization architecture that can be applied in many situations. We then propose an integrated scheme that closely combines our proposed TRZ-based routing architecture with a multiple-key-based distributed security protocol. LESS can save a significant amount of communication energy through a two-level (i.e., sink-to-supernode and

---

[3] Most of a sensor's power is consumed by communication, not instruction processing [1]. The energy used to communicate one bit of data can be used to execute over 1000 local instructions [6]. High-energy consumption can drain sensors quickly and thus shorten the network lifetime.

supernode-to-sensor) hierarchical self-organization algorithm and through in-networking processing such as data aggregation in *supernodes* and *zone masters* (details are given later).

Another important feature of LESS is that it has a *robust* hop-to-hop transmission scheme and can recover from multiple key losses.

2. *Robust Rekeying for Data Aggregation Purposes*   A dominant feature of the WSN is its *dynamic* network topology. Sensors may run out of battery power and new sensors may be added to the network. More importantly, opponents might compromise sensors and all security information in those sensors may be obtained. Therefore after key predistribution and sensor deployment, a *rekeying* scheme should be used to update some types of keys such as group keys and session keys. To the best of our knowledge, current work has not efficiently established a rekeying scheme that has the following three features simultaneously: (a) low complexity, (b) scalability, and (c) robustness. To minimize the complexity, LESS uses a one-way hash function to generate symmetric keys based on streaming ciphers for all confidentiality and authentication cases and thus avoids high-cost public-key-based algorithms that require large keys and complex calculations. To achieve scalability, LESS utilizes a TRZ-based routing protocol to deliver new keys to only the supernodes or zone masters instead of to each sensor directly. Our TRZ routing scheme can converge to an optimal status within a reasonable duration even in a high-density network (such as 10 sensors per square meters) and thus has good scalability. Because keys can be lost due to unreliable wireless transmissions, LESS achieves robust security through the following two schemes (details later):

- Repairing a broken one-way hash chain to deduce new keys that are lost
- Conducting ripple-to-ripple key loss recovery

To the best of our knowledge, no energy-efficient rekeying scheme has been proposed to secure *data aggregation,* which is an important operation to save WSN communication energy [114]. In our 3S model, data aggregation happens in each zone master and supernode. To secure data aggregation, we argue that *multiple keys* (such as a global key, a zone key, and a pairwise key) should be generated and refreshed to encrypt/decrypt sensing data from *multiple sensors* instead of using a *single key*. The simple flooding scheme in [115] only updates a single global key, which makes the whole network insecure once a single sensor is captured.

5.4.2 LESS—TRZ-Based Routing Architecture

TRZ-Based Routing Architecture   We argue that a low-complexity security scheme should be based on a hierarchical network architecture because a flat topology[4] can cause large key generation/distribution/maintenance overhead. Most energy-efficient sensor network protocols assume a multilevel network architecture in order to limit the sensor communication range to a local area and to perform data aggregation in some high-level sensors [120].

In LESS, we assume a general 3S hierarchical network model. In the 3S model, the majority of sensors (usually $>90\%$) have little memory ($<10\,$k bytes of RAM [85]), little

---

[4] In a one-level flat topology, all sensors play the same roles. Each sensor sends data to the sink *individually*. The system needs to maintain a routing architecture in the *whole* network. Key sharing schemes should be global. That is, a sensor may need to establish a pairwise key with each of the other sensors even if communication may never happen between them. Key distribution is performed in the whole network and thus causes much delay.

energy, and low cost. A much smaller number of supernodes (usually <10%), with more memory, more powerful microprocessors, and a longer wireless transmission range due to their greater energy capacity, are deployed randomly throughout the sensor network. Our 3S model fits many sensor network applications very well. The following are a few typical application scenarios that can apply the 3S model:

Scenario 1: Building Monitoring.   Assume hundreds of sensors are deployed randomly in a large building that also has a *wireless local area network* (WLAN), where the WLAN consists of distributed *access points* (APs) that collect sensing data from nearby sensors that can be used to detect humidity, temperature, and chemical materials. Here we can regard the AP as the supernode. An Internet server that connects a few APs in order to *indirectly* collect data from the sensors can be regarded as a sink.

Scenario 2: Telemedicine.   The PI has built a telemedicine prototype [117, 118] that consists of tiny medical sensors attached to each patient. Medical phones are possessed by serious patients and the elderly. The doctors' remote servers send query commands to the patients' sensors and collect medical data from them. Again, we can apply the 3S model to this network architecture: sensors (medical sensors)–supernode (medical phones)—sink (doctors' servers).

Scenario 3: Battlefield Monitoring.   In the battlefield WSN, the soldiers carry tiny sensors while the tanks (i.e., supernodes) communicate with the soldiers nearby. A command center can communicate with tanks and soldiers to monitor the battlefield environment.

Scenario 4: Industrial Control.   In typical industrial applications, we can deploy tiny sensors in the production assembly lines to report abnormal data. Based on the received sensing parameters, we may issue control instructions to some high-power actuators (which correspond to the supernodes in our 3S model) to change the producing behavior.

The utilization of the small number of nodes to perform more load-intensive operations was also used in ROP (resource-oriented protocol) [140] to achieve low-delay communication among heterogeneous sensors. However, we will further propose the concepts of ripple, concentric-circle, and MST (minimum spanning tree) in each zone in order to implement our low-energy security schemes.

Based on the 3S model, LESS uses a *two-level* key management scheme to perform scalable and low-energy WSN security. First, among supernodes (level 1) we use a robust MST as the wireless backbone routing algorithm [117]. Based on the MST-based routing architecture, we apply a *buffer-based key-chain* security scheme that can tolerate multiple key transmission losses (details in Section 5.4.3). Second, around each supernode (level 2), to reduce the energy used for key distribution and data aggregation among neighboring sensors, we first use a low-cost probing protocol to select some sensors as masters. These masters then use *zone keys* to securely collect data from their neighboring sensors and then generate *ripple keys* among masters with the same number of hops from a supernode. The concepts of masters, *zone keys*, and *ripple keys* will be further explained in Section 5.4.3.

Figure 5.21 shows our proposed 3S model and two-level zone–ripple routing architecture.

The left part in Figure 5.21 shows level 1 nodes, that is, supernodes. In the right part, around each supernode, we first choose some sensors as masters. Each master can aggregate data from its local zone that usually consists of one-hop neighboring sensors. A ripple consists of the masters with the same distance from the supernode.

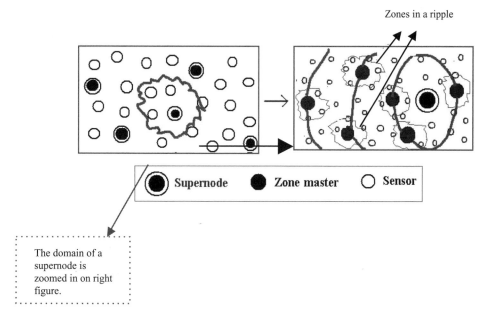

Zones in a ripple

| ⊙ **Supernode** | ● **Zone master** | ○ **Sensor** |

The domain of a supernode is zoomed in on right figure.

Figure 5.21 3S network model.

To reduce the number of pairwise keys and the energy consumed for routing management, our MST algorithm reduces the full-connection topology (used in [91]) to a sparse tree in which we can still always find a path between any two supernodes. Figure 5.22 shows the efficiency of our MST routing algorithm. Because the number of supernodes is not large and supernodes are assumed to have more power and calculation capability than sensors, our MST-based key management scheme forms a robust wireless "backbone" consisting of supernodes that serve the heavy-load data aggregation and control the rekeying among the sensors belonging to the *domain* of a supernode. An efficient *domain determination* algorithm was described in [117].

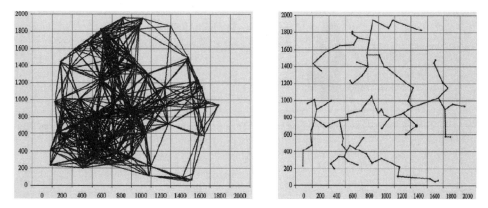

Figure 5.22 MST topology-forming algorithm among supernodes. Note: X and Y axes represent a physical area of sensors (in meters).

Robustness to Dynamic Topology   Some WSN applications have mobile sensors. For instance, in battlefield networks, both soldiers (with sensors) and tanks (with supernodes) are moving; in telemedicine networks also, patients with sensors are likely mobile. In addition, sensor density can change from time to time due to sensor malfunction and new sensor addition. LESS adapts to the dynamic topology in the following two ways:

1. Rerouting Protocols   Periodically the network runs a self-organization protocol to establish a new MST and new zones. The updating period is determined by mobility rate and other factors. We provided the rationale for selecting the update period in [117].

2. Rekeying Scheme (details later)   In the following two sections, we will discuss the *rekeying* procedures in both levels: In level 1 (i.e., supernodes) rekeying is based on a *one-way hash function* and a buffer-based key chain, while in level 2 (i.e., all sensors belong to the domain of a supernode) rekeying regenerates multiple symmetric keys such as zone keys and group keys.

5.4.3 LESS—Level 1 Security

In the level 1 MST-based backbone architecture, we define two types of keys. (1) A *session key* (SK) is used for the encryption/decryption of *data packets*. (2) A *backbone key* (BK) is used to secure *control packets* that include SK rekeying information. Figure 5.23 shows the relationship between these two keys. Note that SKs need to be rekeyed *periodically* to defeat active attacks. However, the BK is refreshed in an *event-triggered* way. Typical *events* include new supernode insertion, death, or compromise. The sink can use any well-known

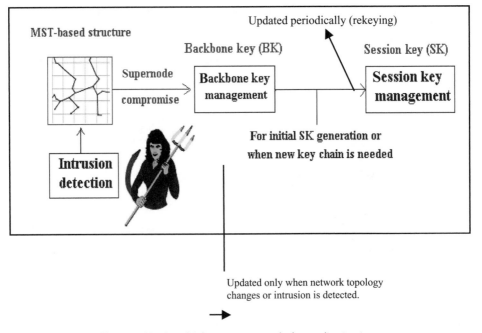

Figure 5.23   Level 1 (among supernodes) security structure.

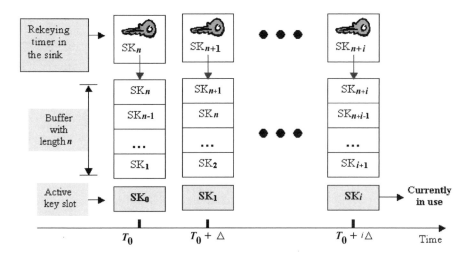

Figure 5.24   Key-chain scheme in WSN backbone.

group communication protocol [124–126] to update the BK, that is, BK rekeying. We will focus on the rekeying of SKs since frequent SK renewal during *data packet* transmission is crucial to defend against *keystream reuse attacks*.[5] For BK management, refer to [40–42].

To make our discussion more straightforward, we list the main elements of our level 1 security design principle as follows:

1. Use One-Way Hash Function [90] to Reduce Security Complexity   To ensure secure SK rekeying, the sink initially uses BK management schemes [124–126] to encrypt, authenticate, and transmit a control packet, which includes a SK to be used in the current sensing data transmission. Note that control packets are different from data packets. The former is secured by BK management schemes while the latter should be secured by SKs. Over a certain interval, a new SK will be distributed to all supernodes. All new SKs are derived from a one-way hash function H as in [91]. We do not use a *message-authenticated code* for SK authentication since receivers can use the *one-way* property of the SK sequence to check whether the received SK belongs to the same sequence or not. We call such a SK authentication concept *implicit authentication,* which works based on the following key-chain scheme (Fig. 5.24), which is also used in the LiSP scheme [141]:

The sink first precomputes a long one-way sequence of keys: $\{SK_M, SK_{M-1}, \ldots$ $SK_n, SK_{n-1}, \ldots, SK_0\}$(size $M \gg n$), where $SK_i = H(SK_{i+1})$. Initially only SK_n (instead of the whole M-size key sequence) is distributed to each supernode. Then a supernode can utilize H to figure out $SK_{n-1}, \ldots, SK_0$. The n keys $\{SK_n, SK_{n-1}, \ldots SK_1\}$ are stored in a local *key buffer* of length of n. However, SK_0 is not in the buffer because it is used

---

[5] Keystream reuse attack: To save energy, a WSN protocol should minimize the amount of data transmitted. Thus the symmetric stream cipher is a good choice for WSN security because the size of the ciphertext is the same as that of the plaintext [7]. A keystream is generated as a function of the message key and the initialization vector and is XORed with the plaintext to produce the ciphertext. Stream ciphers usually encrypt packets with a per-packet initialization vector (IV), but due to the limited IV space (only 24 bits in IEEE 802.11 WEP [48]), it is vulnerable to practical attacks, as reported in [49].

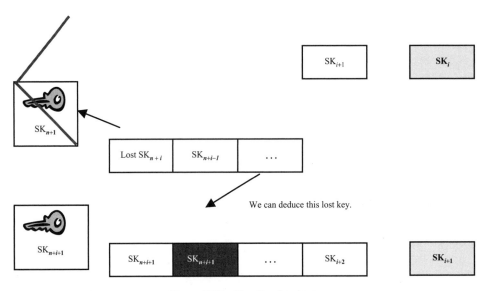

Figure 5.25 Handling key loss.

for the current data packet encryption/decryption. After the initial SK_n delivery, the sink periodically sends $SK_{n+1}, SK_{n+2}, \ldots, SK_M$ (one key distribution in each period) to all supernodes. To achieve communication confidentiality, each new key is encapsulated into a rekeying control packet that is encrypted with the BK or the currently active SK. After receiving a new SK, the supernode keeps applying H to it for some time, trying to find a key match in its key buffer. For instance, assume that a supernode receives a new key SK_j. Also assume that its key buffer already holds n SKs as follows: $\{SK_i, SK_{i-1}, \ldots SK_{i-n+1}\}$. *If the supernode finds out that $H(H(H \ldots (H(SK_j))) \notin \{SK_i, SK_{i,\notin}, \ldots, SK_{i\in+1}\}$ that is, implicit authentication fails, SK_j will be discarded. Otherwise, if implicit authentication is successful, the key buffer is shifted one position.* The SK shifted out of the buffer is pushed into the "active key slot" to be used as the current SK (see Fig. 5.24), and the empty position is filled out with a new key SK', where SK' is derived from the received SK_j through H and meets the following two conditions: $SK' = H(H(H(\ldots H(SK_j)))$ *and* $H(SK') = SK_i$.

2. Robustness to Multiple SK Losses   The reason for using a key buffer is to tolerate multiple key losses. As shown in Figure 5.25, even if up to $n - 1$ SKs are lost due to unreliable wireless channels (such as radio shading, fading, and communication noise), as long as the nth SK received is authenticated, we can still restore the lost SKs in the key buffer by applying the one-way hash function. By using a key buffer, the lost SK will not influence the current security session until a maximum delay of $n - 1$ rekeying periods. This feature also makes SK management robust to clock skews.

3. Assumption of Weak Time Synchronization in Supernodes   Because level 1 supernodes typically have higher transmission power and are fewer than ordinary sensors and also because our MST algorithm can establish a low-maintenance, low-delay routing architecture among them, the assumption of weak synchronization among supernodes is practical [137]. This assumption is also made in other WSN security schemes such as [91]. However, the assumption of synchronization may not be practical for level 2 sensors due to their large

number and limited transmission power. Thus, we adopt the ripple key approach to address the level 2 broadcast authentication problem (discussed in the next section).

Note that here we use a one-way key sequence that is different from the SPINS scheme [91] where the buffer holds all unauthenticated data packets until the node receives the correct key from the sink. If a few key disclosures are missed, SPINS suffers from high latency and is critically limited by the buffer size. In contrast, LESS holds only n small SKs and does not disrupt the ongoing data security since it delivers each SK well before its use.

As discussed above, SK rekeying is conducted *periodically* to overcome active attacks (see Fig. 5.24). However, the SK should be reset immediately under any of the following conditions:

- One of the supernodes is compromised by attackers.
- A supernode runs out of energy and thus stops operations.
- A new supernode is added.

cannot deduce any lost keys.

ls to build a new key sequence.

ue to supernode addition/death) through topol-
or detects node compromise through *intrusion*
nmediately renews the backbone key to BK_{new}
124–126] and generates a new key sequence
$(M \gg n')$, $SK_i' = H(SK_{i+1}')$. Where n' is the
sink then sends out a new group of parameters

$$E_{BKnew}\left(\Delta'\,|\,n'\,|\,SK_{n'}'\right) \tag{5.7}$$

ernode will perform the following tasks: (1)
2) Rebuild a new buffer with length of n',
..., SK_1', SK_0'} based on $SK_{n'}'$ and the hash
e session key for data encryption. (5) Invoke
timer expires, the sink sends out a new key

$$\left(_{n'+i+1}'\right) \quad \text{at time } t = i\,\Delta' \tag{5.8}$$

uence (after $n \times \times'$), it will also send out a
pernode misses n keys, the sink must use a
ıd a message with $E_{BK}\left(\Delta'|n'|SK_C'\right)$ to that
y.

... using a shorter rekeying period Δ' (i.e., more frequent key updating). However, this will lead to greater communication overhead and increased

---

[6] Each time we require the supernodes to create buffers with a new size n' to make attacks more difficult.

hash calculation. There also exist other trade-offs. For instance, the larger the buffer size n', the more tolerant to key loss a supernode is, because it can recover more missed keys based on the one-way hash function. However, a large buffer requires more memory and calculation overhead.

The software implementation of level 1 security in the sink and the supernodes is shown in Figure 5.26. This algorithm periodically recognizes supernode compromises, additions, and deaths. The sink and all supernodes use MST algorithms to generate the level 1 network topology.

5.4.4 LESS—Level 2 Security

Design Principle   LESS handles the secure transmission problem *around each supernode* based on a ripple–zone routing architecture. It first uses the *member recognition protocol* (*MRP*) to tell each sensor its security domain in terms of a supernode ID. The basic idea of MRP is as follows. Each supernode broadcasts a *hello* packet to the whole network. When the hello packet arrives at a sensor, the hop number field in the packet header is increased by 1. Redundant packets (with the same supernode ID and the same hop-number field) are discarded to avoid loops. Each sensor only keeps the hello packet with the minimum hop number and discards others, thus automatically choosing the closest supernode. Each sensor determines its parent's supernode ID and sends a *confirmation* packet to its parent supernode along the reverse path through which it received the hello packet. Once the supernodes receive these confirmations, they know their domain members and can aggregate data from those sensors using the *cluster keys*, *ripple keys*, and *pairwise keys* (discussed later) and then propagate aggregated data to other supernodes using the level 1 security protocol until data arrive securely at the sink.

Next, LESS needs to achieve hierarchical key management by forming all members of each supernode into *zones*. Each zone has a data aggregator called a master. All the masters of a supernode domain are selected using a low-complexity *master selection* algorithm whose brief description is given in Figure 5.27. There is a maximum of one hop between a sensor and its master for the convenience of utilizing pairwise keys between neighboring sensors.

On the fairness of the master selection algorithm, to avoid the overexhaustion of power storage in the masters, we adopt the role rotation scheme used in LEACH [122] to switch the role of the master among the sensors in a zone. After the execution of our algorithm (Fig. 5.27), all sensors will know to which zone they belong. For every certain period Γ' (its value is chosen based on the power exhaustion speed of average sensors), a master will choose a member in its zone that has the least time of being a master to become the new master.

We define a ripple that consists of all masters with the same transmission cost (LESS uses number of hops to the supernode as the cost metric). Communications happen only between two masters in *neighboring* ripples. The importance of ripples lies in the use of ripple keys (RKs) to achieve broadcast authentication without the need for time synchronization (discussed later).

A unique issue in WSN security is that the selection of key sharing schemes should consider the impact on *in-networking* processing [109]. For example, data aggregation is necessary for reducing communication overhead from redundant sensed data. If we adopt one type of key, that is, a pairwise key that is shared between only two nodes [94], memory limitations will prohibit a master from maintaining all the keys necessary to aggregate data

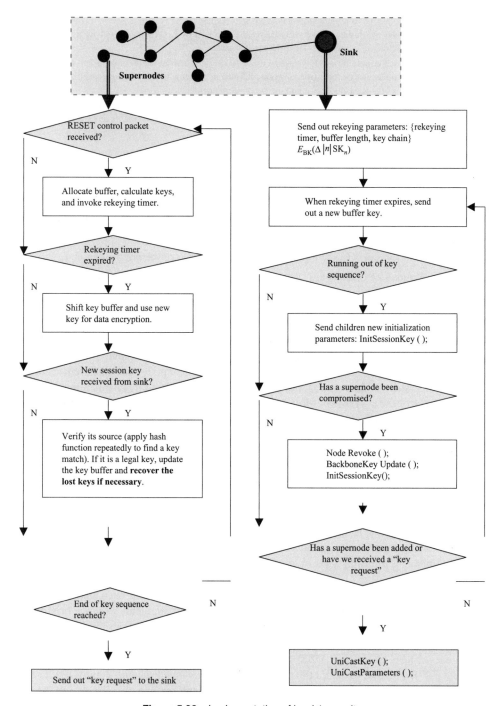

Figure 5.26 Implementation of level 1 security.

Rule 1: The supernode sends a probing packet with a ripple-ID field (initially 0) to its sensor members. When a member receives the probing packet, it increases the ripple ID by 1. Only members with ripple IDs of the form $3M$ ($M = 1,2,\ldots$) may elect themselves as masters.

Rule 2: A member with ripple ID of $3M$ invokes a timer T_1 after it sends out the probing packet. If the timer expires and the node does not hear from any other masters in the same ripple, it chooses itself as a master and immediately broadcasts its decision to the neighboring members.

Rule 3: Suppose member B with ripple ID of $3M + 1$ receives the probing packet from member A with ripple ID $3M$. If B finds out that A is a master, B sets its master to A; otherwise B starts its timer T_2. If B hears from another master C before its timer expires, it chooses C as its master; otherwise, B sends a message to A to solicit it to become a master.

Rule 4: A member D with ripple ID of $3M + 2$ starts its master election timer T_3 after it forwards the probing packet to a member which has a ripple ID of $3M + 3$. If it hears from a master E at ripple $3M + 3$ before T_3 expires, it sets its master to E. Otherwise, D arbitrarily chooses any of the members at ripple $3M + 3$ to be its master and informs the member of its decision.

Rule 5: Any member at ripple $3M$ that has not selected itself as a master but later receives an explicitly master solicitation message from another member will choose itself as a master and inform others.

Other rules: There are other details such as choosing of T_1, T_2, T_3 and loop avoidance problem.

Figure 5.27 Master selection algorithm within the domain of a supernode.

from its member nodes. We cannot simply build an *end-to-end* secure channel between each sensor and the sink, such as in [91], because *intermediate* sensors (such as masters) may need to decrypt and authenticate the data collected from multiple sensors.

Since different types of messages exchanged among sensor nodes have different security requirements (such as data aggregation security), a *single* keying mechanism may not be suitable for all cases. Thus, *multiple* keys need to be introduced in level 2. Again, we integrate the *key management* with our routing architecture consisting of zones and ripples. To save security overhead, LESS generates a new key based on a family of *pseudorandom functions* (PRFs) $\{f\}$ as follows:

$$K'_{K,x} = f_K(x) \qquad \text{where } K \text{ is the last key and } x \text{ is a random number} \tag{5.9}$$

Function f has the following characteristics [127]: (1) Given f and x, it is computationally infeasible to calculate K' without knowing K. (2) Given $K'_{K,x1}, K'_{K,x2}, \ldots, K'_{K,xi}$, it is computationally infeasible to calculate $K'_{K,xj}(x_1 \neq x_2 \neq \cdots \neq x_i \neq x_j)$.

LESS manages multiple types of keys for different security purposes as follows:

1. *Master-to-Supernode Key (MSK):* An MSK is shared between each master and its supernode. When a supernode needs to forward a query command (which is originally from the sink) to a specific zone area, an MSK is used to establish a secure end-to-end channel between the supernode and the relevant master. An MSK is generated based on a level 1 SK as follows: MSK = f_{SK} (Master-ID). In our master selection algorithm (Fig. 5.27), note that we adopt the *reverse path establishment* scheme as in AODV [128] so that each supernode may be informed of all of its children's master IDs. It only needs to store the MSKs for the masters instead of for all sensors, which saves much memory space in the supernode.

NOTE: Each time a supernode receives a new SK through the level 1 security protocol, it will forward a control packet (which includes the new SK and is encrypted with the last SK) in a ripple-to-ripple way to all masters within its domain so that eventually all masters know the new MSK. In this way, we achieve level 2 rekeying through level 1 SKs.

2. *Inter–Master Pairwise Key (MPK):* Occasionally secure channels need to be established between two masters that belong to two supernode domains. The inter–master pairwise key can be established with the help of the MSK using a similar scheme as in SPINS [91], where a node-to-node pairwise key can be determined via two node-to-sink negotiations.

3. *Sensor-to-Master Pairwise Key (SPK):* A sensor-to-master pairwise key is shared between a master and each of the sensors in its zone. Different sensor-to-master pairs have different SPKs. An SPK is derived from a PRF, *f*, and the current SK as follows:

$$SPK = f_{K_0}(\text{Sensor-ID}) \qquad \text{where } K_0 = f_{\text{SK}}(\text{Master-ID}) \qquad (5.10)$$

4. *Zone Key (ZK):* Zone keys are used for data aggregation and for the propagation of a query message to the whole zone. Each ZK is shared among all sensors in one zone. A master transmits a new ZK to each of its zone sensors through different SPKs.

5. *Ripple Key (RK):* A RK is used for broadcast authentication in a supernode domain. LESS does not use μTESLA [91] for two reasons: (a) μTESLA needs loose time synchronization that is not practical among a large number of low-cost sensors. (b) The delayed release of the authentication key needs a long-size data packet buffer in each sensor, which is a high requirement due to the very limited memory of a tiny sensor. LESS overcomes these shortcomings by using a RK that is shared by all masters belonging to the same ripple. The RK is determined by the supernode, which sends different RKs for different ripples through control packets encrypted by the MSK. A supernode will send out a broadcast message that needs to be authenticated for multiple times. Each time the supernode uses a different RK to encrypt it. Thus only the masters in the corresponding ripple can decrypt it.

Discussions

Robustness to Key Losses   In level 1, we use the one-way property of hash functions to recover the lost keys between two rekeying events. In level 2, since we do not use a one way key chain (because it is not suitable for sensors that have less memory and computational capability than supernodes), we recover lost control packets that include key information through ripple-to-ripple retransmission scheme and a *local path repair* routing protocol [131]. A simple example demonstrating our key loss recovery scheme is shown in Figure 5.28; after master B sends out a control packet to A (the closest master to B), B sets up a timer T_1. If T_1 expires and B has still not received a positive acknowledgment packet from A, B will retransmit the packet to A. If B attempts more than three retransmissions without success, it regards A as unreachable (possibly due to an unreliable wireless link), and LESS then finds another good path as follows: B checks the reachability of A's neighboring masters H and E. If either of them is reachable, B will transmit the packet to one of them. If both H and E are unreachable, LESS goes back one ripple, that is, to master K. Then K discovers B's neighboring masters and tries to deliver the packet. Eventually a new ripple-to-ripple good path will be discovered. We will verify its efficiency through simulation experiments in the next section.

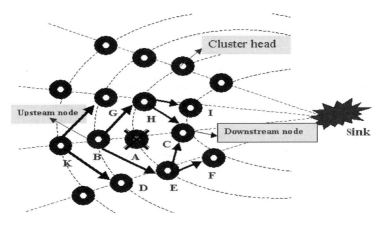

Figure 5.28 Ripple-to-ripple key loss recovery.

Relationship between SPKs and ZKs   Zone beys are very important for securing data aggregation in a local sensing area since they enable the whole zone to encrypt and decrypt data packets without the use of multiple keys. Note that ZK rekeying occurs after SPK rekeying. The basic procedure is as follows: Whenever a master M regenerates a new random key $\text{ZK}' = f_{\text{ZK}}(x)$ (where ZK is the last zone key and x is a random number), it sends ZK' to each of its zone members using the corresponding SPKs as follows:

$$M \rightarrow \text{supernode } \Theta_i : \quad \text{rekeying\_\#}|\text{Master\_ID}|E_{\text{SPK}(i)}(\text{ZK}'|\text{Rekeying\_\#}|\text{Master\_ID})$$

Each zone member analyzes and decrypts this message, then stores the new ZK' in its buffer. The rekeying of ZK is performed periodically (the control timer is the same as level 1 Δ).

Improvement on Our RK Scheme   The aforementioned RK scheme can overcome the *time synchronization* problem [91] when used for broadcast authentication; that is, each zone ensures that a broadcasting message does come from its own supernode versus other sources. To disseminate a message, a supernode broadcasts a separate message (called Command_MSG) for different ripples as follows:

$$Supernode \rightarrow Ripple\ [i]:$$
$$Message|Super\text{-}ID|Tran\text{-}ID|MAC_{RK(i)}(Message|Super\text{-}ID|Tran\text{-}ID|)$$

where Super-ID helps each master distinguish its own supernode from others, and MAC is message-authenticated code Tran-ID helps avoid "routing loops" as follows: a master identifies different commands through the transaction ID (i.e., Tran-ID) and discards duplicate ones. The message-authenticated code is derived from the RK of ripple level i. Intermediate masters that do not belong to ripple i cannot tamper with the above message without detection since they do not know RK i. After each master receives an authenticated message, it will use the ZK to broadcast to all sensors in its zone.

If a message arrives at the masters in the same ripple with variable delays, there could exist the following attack: A master A in ripple i that receives a message MSG earlier than other masters in the same ripple can fake a new message MSG′ and generate a new MAC using its RK i, and then send it to another master B in ripple i. If B has not received MSG yet, it will be unable to identify the forged message MSG′. To protect the system from this attack, we use our aforementioned inter-MPK as follows:

1. We require that each master in ripple i maintain a few MPKs in its cache (called ripple MPKs) with its close *upstream* and *downstream* masters that are located in ripple $i − 1$ (upstream) and $i + 1$ (downstream), respectively. We use the *received signal strength* (*RSS*) to determine the geographically close masters according to the following relationship between the RSS and the distance x from the transmitter [136]:

$$\text{RSS}_{\text{dB}} = 10\gamma \log(x) \tag{5.11}$$

where γ is the propagation path loss coefficient [136].

2. Each time a master in ripple $i − 1$ receives a message and finds out that it is supposed to be received by its next ripple, it will encrypt it by the corresponding downstream ripple MPKs and relay the following message to the downstream masters (in ripple i):

$$\text{To downstream} : \ i\,|\,E \ (\text{Ripple-MPK, Command-MSG}) \tag{5.12}$$

3. When a master receives the above message, it will try each of its upstream ripple MPKs to recover the contents of the original message, that is, Command-MSG.

   The above scheme can prevent *message spoofing* from masters in the same ripple since a master only receives and decrypts messages from its upstream masters.

Since the messages encrypted by RKs are transmitted in a hop-by-hop way, nodes on the smallest circle (i.e., closest to the supernode) will take a big relaying load and die more quickly than others. This is a common problem in the uplink (sensor-to-supernode/sink) transmission of any WSN routing protocols. We will put off this issue as part of our future work. However, in the downlink (supernodes to sensors), we can simply use the broadcast scheme since a supernode has enough antenna power to cover the entire domain.

5.4.5 Performance Analysis

Simulation Analysis for Our TRZ Routing Architecture   The ns-2 simulator with a modified routing/transport layer is used for all evaluations. We mark 8% of all sensors as supernodes. The network size is varied between 100 and 1000 nodes. Sensors are assumed to move according to a *random-point model* [123]. Initially they are randomly deployed in a $1000 \times 1000\,\text{m}^2$ area. The sink is located at the center of one of the edges of the square. The transmission range of each sensor is assumed to be 20 m while a supernode is assumed to have a communication radius of 60 m. Each data packet is 20 bytes and each control packet (i.e., key management packet) is 5 bytes. All results are calculated to 95% confidence intervals. The *packet error rate* in the *data link layer* due to unreliable wireless transmission is assumed to be 10%. The data packet generation rate is assumed to be 10 packets per second.

Energy Efficiency of LESS Because energy consumption is one of the biggest concerns in WSN protocols, we investigate the energy efficiency of LESS compared to two other *non-TRZ-based* security schemes: a flat-topology-based scheme such as in [94] and a simple one-level cluster-based scheme such as in [115]. The choice of a wireless energy model is important because different assumptions regarding the radio features (e.g., energy dissipation in transmit and receive modes) vary the performance analysis results notably. In this project, we adopt a popular sensor energy model from [122] to calculate the total *average* energy dissipation for key delivery and general data transmission in each sensor. To transmit/receive a k-bit packet over a distance d, the radio expends the following energy based on the theories of radio propagation fading loss [122]:

$$
E_{\text{transmit}}(k, d) = \begin{cases} E_{\text{elec}} + 10kd^2 & d < 87m \\ E_{\text{elec}} + 0.0013kd^4 & d \geq 87m \end{cases}
$$
$$
E_{\text{receive}}(k, d) = E_{\text{elec}}k \tag{5.13}
$$

where E_{elec} is the electronics energy as a function of local modulation, filtering, and digital coding. We adopt the value given in [122] as $E_{\text{elec}} = 50$ nJ/bit.

We first use simulations to collect data and calculate the average energy consumption in our TRZ-based routing scheme for transmitting and receiving control packets *(which have keying information)* and data packets in a sensor network consisting of 500 sensors. We then change our TRZ routing protocol to a flat topology [91] and then to one-level cluster-based topology [115]. We observe and compare the results. Figure 5.29 clearly shows the energy efficiency of our LESS scheme based on the TRZ routing protocol versus the flat topology or the one-level cluster-based routing.

Validity of "Master" Selection Algorithm The low-complexity of LESS depends largely on the efficiency of the ripple–zone organization in level 2. We attempt to use a small number of zones to cover an *entire* supernode domain in order to more effectively perform data aggregation and reduce the number of keys maintained. We cannot choose too many sensors as masters since too many zones bring intolerant level 2 key maintenance overhead.

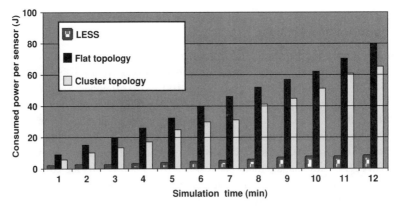

Figure 5.29 Energy consumption for control packet and data packet transmissions.

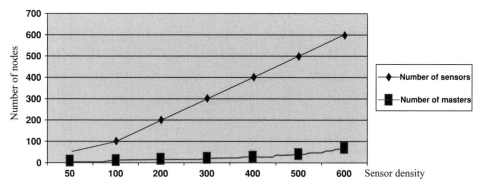

Figure 5.30   Sensor density versus zones per supernode domain.

As shown in Figure 5.30, our master selection algorithm can choose a nonlinear number of masters with the linear increase of sensors.

Simulation Analysis for Security Schemes

Robustness to Key Loss   Based on our integrated routing and security management strategy, even though radio transmissions in WSNs have a much higher packet error/loss rate than *wired* networks, LESS can minimize the effective control packet *loss rate* by using a one-way hash function in level 1 and ripple-to-ripple loss recovery in level 2. As shown in Figure 5.31, other data dissemination protocols, such as ACQUIRE [134], which simply uses cluster-to-cluster forwarding, and TAG, which is based on a simple spanning tree WSN architecture [129], have a much higher control packet loss rate (i.e., key loss rate) than our LESS, which is based on a TRZ architecture.

Low Complexity of the Security Protocols   A crucial metric for the *complexity* of WSN security protocols is the number of control packets transmitted between individual nodes. We compare our LESS key distribution scheme that is based on a TRZ routing architecture to three WSN security schemes that are based on other ad hoc routing protocols, including

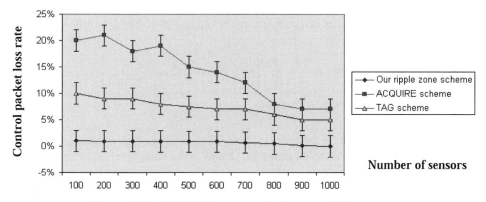

Figure 5.31   LESS: Robustness to wireless transmission errors.

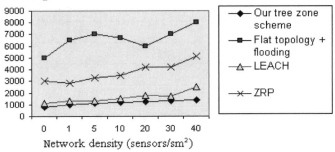

Packets transmitted in
level 1 routing

Figure 5.32 Control packet communication overhead [117].

flat topology [91], LEACH [122], and ZRP [119], and calculate the number of control packets transmitted in the level 1 routing architecture.

As shown in Figure 5.32, both LEACH [122] and our LESS have a satisfactory control packets communication overhead compared to others. LEACH [122] adopts a cluster-based routing scheme, which has a certain similarity to our level 2 routing management. However, it asks each cluster head to directly communicate to the sink, which may not be practical since long-distance wireless communication consumes too much energy [85]. Our two-level TRZ routing scheme only requires that each sensor communicate with its local master, which further transmits data to the *next-ripple master*, which limits communication in a small area and thus saves much energy.

Analytical Model on Security Overhead of LESS We used a first-order *Markov chain* model to analyze the calculation and communication overhead when incorporating our security features into level 1 supernode communications. In local sensor processing, calculations involving the one-way hash function consume the most energy [90]. We therefore focus on the cost of computing hash functions during each rekeying session. A supernode may fail to receive a new session key, or it may receive an incorrect session key that cannot be authenticated by using the hash function. Incorrect session keys may come from opponents attempting denial-of-service attacks. If the buffer length is n, the probability of *key loss* is P_{loss} and the probability of *key corruption* is $P_{\text{corruption}}$. We derive the expected times for hash function calculations in a rekeying cycle, E_{rekeying} [#-of-hash], as follows [135]:

$$E_{\text{rekeying}} [\text{\#-of-hash}] = \frac{2.5 - P_0}{2 - P_0^n} \left\{ n P_0^{n-1} + (1 - P_{\text{failure}}) \sum_{i=1}^{n} i P_0^i \right\} \qquad (5.14)$$

where $P_0 = P_{\text{loss}} + P_{\text{corruption}}$

Assuming $P_{\text{corruption}} = 0.25$, we vary P_{loss} from 0.0 to 0.5 and compare the simulation and analytical results. Figure 5.33 clearly shows the validity of our model.

Security Analysis Now we analyze the protection of a WSN through our LESS from various attacks.

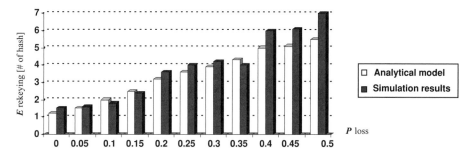

Figure 5.33 Analytical and simulation results.

1. *BK Attacks* Because the distribution of new SKs depends on control packets encrypted by BK that is managed by group security schemes [124–126], it is possible for an attacker to compromise the current BK and thus can attack any future SK disclosures. LESS can minimize the impact of this attack through our *buffered key-chain* scheme. Due to the SK buffer, there is a delay between receiving a new SK and actual applying this new SK. Suppose the distribution interval is Δ' (i.e., the rekeying period) and n is the buffer length. Then $n \times \Delta$' later we may use this new SK. As long as we can detect the BK compromise within $n \times \Delta$' and renew the BK and SKs through a new level 1 *initialization procedure* that can help each supernode rebuild a new SK buffer chain, the data packets will maintain normal security performance.

2. *SK Attacks* The attacker may modify the transmitting SK, inject phony SK, or use wireless channel interference to damage control packets. These attacks may also result in data replay and DOS attacks. LESS can easily defeat these attacks through implicit authentication. Due to the one-way characteristics of the hash function keys, any false SKs cannot pass the authentication test, that is, after L times ($L \leq n$) of using the hash function, if we still *cannot* satisfy the following formula, we will regard that it is a false SK (in this formula, SK_{FAKE} is a false SK and SK_{NOW} is the SK currently used):

$$\underbrace{H(H(\cdots(H(SK_{FAKE})\cdots)))}_{L} = SK_{NOW} \tag{5.15}$$

We have also included a rule into our level 1 security protocol: Whenever a sensor continuously fails the authentication test three times, it will send back an *attack detection* message to the sink to trigger a new level 1 initialization procedure.

3. *Relay Attacks* Cannot succeed because a sensor ensures that a new received SK will not only pass the authentication test but also be unique.

4. *Periodical Attacks* Possibly an attacker can issue an attack at a fixed time if he or she knows the value of Δ' (i.e., the rekeying period). We address this problem through the addition of a random time interval $\varepsilon(t)$ to Δ'; that is, the new rekeying period is $\Delta' + \varepsilon(t)$, where $\varepsilon(t) \leq \Delta'/8$.

5. *Main-in-the-Middle Attacks* LESS can also defeat main-in-the-middle attacks (where an attacker fools the level 1 nodes as if he or she were a legal BK supernode) through a transmission of the math-authentical code in the level 1 initialization procedure as follows:

$$\text{Sink} \rightarrow \text{supernode} : E_{BKnew}\left(\Delta'|n|SK_0|MAC|(\Delta'|n|SK_0)\right) \tag{5.16}$$

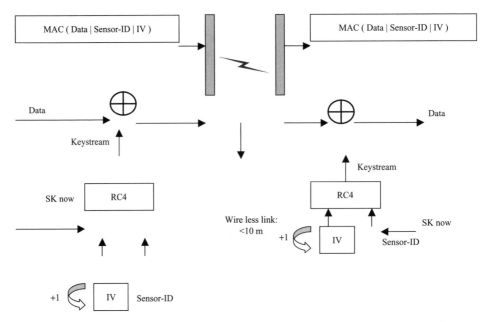

Figure 5.34   Cryptographic procedure of LESS (MAC = message-authenticated code).

6. *Data Level Attacks* LESS defeats it through SK rekeying every Δ' and inclusion of Sensor-ID and a per-packet *initialization vector* (IV) (which will also be updated from packet to packet) in the generation of keystreams to counter the *keystream reuse* problem.

Cryptographic Implementation LESS uses a *stream cipher* RC4 [139] to implement encryption/decryption algorithms because the stream cipher has a lower complexity of security algorithm compared to a *block cipher*. To address the keystream reuse problem, a sender includes its own sensor ID[7] into the generated keystream. In addition, for each message sent, the sender increments its own per-packet IV by 1. Thus, we can ensure that each keystream is unique. Figure 5.34 shows the cryptographic procedure of LESS. Notice that we include message-authenticated code.

In addition, to generate multiple secret keys, we adopt the *pseudorandom functions* (PRFs) $\{f\}$ to derive new secret keys based on the current session key SK_{now} and a random number x as follows:

$$KEY_{NEW} = f(SK_{now}, x) \tag{5.17}$$

The generation of a random number x is based on the counter approach as in [91].

5.4.6 Conclusions and Future Work

This work addressed some challenging security issues in an important information infrastructure—large-scale and low-energy WSNs. The salient advantages of this work compared

[7] Sensor-ID: We can ensure that no two sensors have the same sensor ID using a random number function before we deploy the sensors in an area. In fact, we can generate unique sensor IDs through any schemes used to generate a nonce [54] (which is a random number so long that exhaustive search of all possible nonces is not feasible; no two nonces are the sample in a generation procedure).

to other related ones are as follows:

1. Instead of purely focusing on security research as in most of the current literature, we argued that WSNs have specific network constraints and data transmission requirements compared to general ad hoc networks and other wireless/wired networks. We proposed to seamlessly integrate WSN security with a promising routing architecture that proves to be scalable and energy efficient.

2. To protect from active attacks in mobile sensor networks, we proposed two-level rekeying/rerouting schemes that can not only adapt to a dynamic network topology but also securely update keys for each data transmission session.

3. Due to the importance of secure in-networking processing such as data aggregation in WSNs, LESS defined a multiple-key management scheme closely related to the proposed TRZ routing architecture.

It will be interesting to investigate tighter integration of security with routing in WSNs in our future work. If there are no predetermined supernodes, how can we use wireless backbone construction algorithms to select supernodes that are evenly distributed in a WSN in a way that guarantees maximum connection with neighboring sensors? We will investigate the dominating set algorithms [130] for finding those supernodes. Directed diffusion [116] is also an interesting protocol for establishing paths and may be used in our level 2 routing organization.

REFERENCES

1. I. F. Akyildiz, W. Su, Y. Sankarasubramaniam, and E. Cayirci, "A survey on sensor networks," *IEEE Communications Magazine*, vol. 40, no. 8, pp. 102–114, Aug. 2002.

2. D. Estrin, D. Culler, K. Pister, and G. Sukhatme, "Connecting the physical world with pervasive networks," *IEEE Pervasive Computing*, vol. 1, no. 1, pp. 59–69, 2002.

3. D. Estrin, R. Govindan, J. Heidemann, and S. Kumar, "Next century challenges: Scalable coordination in sensor networks," in *Proceedings of the Fifth Annual ACM/IEEE International Conference on Mobile Computing and Networking (MobiCom'99)*, Aug. 1999.

4. D. Estrin, W. Michener, and G. Bonito, "Environmental cyberinfrastructure needs for distributed sensor networks," paper presented at the National Science Foundation Sponsored Worksop, Aug. 2003.

5. R. Min, M. Bhardwaj, S. Cho, E. Shih, A. Sinha, A. Wang, and A. Chandrakasan, "Low Power Wireless Sensor Networks," in *Proc. 14th International Conference on VLSI Design*, pp. 205–210, Bangalore, India, Jan. 2001.

6. G. Pottie and W. Kaiser, "Wireless integrated network sensors," *Communications of the ACM*, vol. 43, no. 5, pp. 51–58, May 2000.

7. "Crossbow technology inc." Available: http://www.xbow.com, June 2005.

8. W. Shi and C. Miller, "Waste containment system monitoring using wireless sensor networks," Technical Report MIST-TR-2004-009, Wayne State University, Mar. 2004.

9. H. Wang, D. Estrin, and L. Girod, "Preprocessing in a tiered sensor network for habitat monitoring," *EURASIP Journal on Applied Signal Processing*, vol. 4, pp. 392–401, 2003.

10. H. O. Marcy et al., "Wireless sensor networks for area monitoring and integrated vehicle health management applications," in *AIAA Guidance, Navigation, and Control Conference and Exhibit*, Aug. 1999.

11. J. Hill, R. Szewczyk, A. Woo, S. Hollar, D. Culler, and K. Pister, "System architecture directions for networked sensors," in *Proceedings of the Ninth ASPLOS'00*, Cambridge, MA, Nov. 2000, pp. 93–104.

12. A. Wang and A. Chandrakasan, "Energy-efficient dsps for wireless sensor networks," *IEEE Signal Processing Magazine*, vol. 19, no. 4, pp. 68–78, July 2002.

13. D. Braginsky and D. Estrin, "Rumor routing algorithm for sensor networks," in *Proceedings of the First Workshop on Sensor Networks and Applications (WSNA'02)*, Atlanta, Oct. 2002.

14. J. H. Chang and L. Tassiulas, "Maximum lifetime routing in wireless sensor networks," in *Proceedings of Advanced Telecommunications and Information Distribution Research Program (ARIRP'00)*, Mar. 2000.

15. B. Hamdaoui and P. Ramanathan, *Energy-Efficient and MAC-Aware Routing for Data Aggregation in Sensor Networks*, New York: IEEE Press, 2004.

16. W. Heinzelman, J. Kulik, and H. Balakrishnan, "Adaptive protocols for information dissemination in wireless sensor networks," in *Proceedings of the Fifth Annual ACM/IEEE International Conference on Mobile Computing and Networking (MOBICOM'99)*, Seattle, Aug. 1999, pp. 174–185.

17. W. R. Heinzelman, A. Chandrakasan, and H. Balakrishnan, "Energy-efficient communication protocol for wireless microsensor networks," in *Proceedings of HICSS'00*, Maui, Hawaii, Jan. 2000.

18. C. Intanagonwiwat, R. Govindan, and D. Estrin, "Directed diffusion: A scalable and robust communication paradigm for sensor networks," in *Proceedings of the Sixth Annual ACM/IEEE International Conference on Mobile Computing and Networking (MOBICOM'00)*, Seattle, Aug. 2000.

19. K. Kalpakis, K. Dasgupta, and P. Namjoshi, "Maximum lifetime data gathering and aggregation in wireless sensor networks," in *Proceedings of IEEE Networks'02 Conference (NETWORKS02)*, Aug. 2002.

20. S. Lindsey, C. Raghavendra, and K. M. Sivalingam, "Data gathering algorithms in sensor networks using energy metrics," *IEEE Transactions on Parallel and Distributed Systems*, vol. 13, no. 8, pp. 924–936, Sept. 2002.

21. S. Lindsey, C. S. Raghavendra, and K. Sivallingam, "Data gathering in sensor networks using the energy*delay metric," in *Proceedings of the IPDPS Workshop on Issues in Wireless Networks and Mobile Computing*, Fort Lauderdale, FL, Apr. 2001.

22. S. Madden, M. J. Franklin, J. Hellerstein, and W. Hong, "Tag: A tiny aggregation service for ad-hoc sensor network," in *Proceedings of the Fifth USENIX Symposium on Operating Systems Design and Implementation*, San Francisco, Dec. 2002.

23. A. Manjeshwar and D. P. Agrawal, "TEEN: A protocol for enhanced efficiency in wireless sensor networks," in *Proceedings of the First International Workshop on Parallel and Distributed Computing Issues in Wireless Networks and Mobile Computing*, San Francisco, Apr. 2001.

24. A. Manjeshwar and D. P. Agrawal, "Apteen: A hybrid protocol for efficient routing and comprehensive information retrieval in wireless sensor networks," in *Proceedings of the Second International Workshop on Parallel and Distributed Computing Issues in Wireless Networks and Mobile Computing*, Fort Lauderdale, FL, Apr. 2002.

25. N. Sadagopan, B. Krishnamachari, and A. Helmy "The acquire mechanism for efficient querying in sensor networks," in *Proceedings of the First International Workshop on Sensor Network Protocol and Applications*, Anchorage, AK, May 2003.

26. C. Schurgers and M. Srivastava, "Energy efficient routing in wireless sensor networks," in *MILCOM Proceedings on Communications for Network-Centric Operations: Creating the Information Force*, Vienna, VA, 2001.

27. R. Shah and J. Rabaey, "Energy aware routing for low energy ad hoc sensor networks," in *Proceedings of the IEEE Wireless Communications and Networking Conference (WCNC'02)*, Orlando, FL, Mar. 2002.

28. A. Wand and A. Chandrakasan, "The cougar approach to in-network query processing in sensor networks," *ACM SIGMOD Record*, vol. 31, no. 3, pp. 9–18, Madison, WI, Sept. 2002.

29. B. Karp and H. T. Kung, "Gpsr: Greedy perimeter stateless routing for wireless networks," in *Proceedings of the Sixth Annual ACM/IEEE International Conference on Mobile Computing and Networking (MOBICOM'00)*, Seattle, Aug. 2000.

30. H. Dai and R. Han, "A node-centric load balancing algorithm for wireless sensor networks," in *Proceedings of IEEE GLOBECOM'03*, San Francisco, Dec. 2003.

31. B. Chen, K. Jamieson, H. Balakrishnan, and R. Morris, "SPAN: An energy-efficient coordination algorithm for topology maintenance in ad-hoc wireless networks," in *Proceedings of the Seventh Annual ACM/IEEE International Conference on Mobile Computing and Networking (MobiCom'01)*, Rome, July 2001.

32. Y. Xu, J. Heidemann, and D. Estrin, "Geography-informed energy conservation for ad hoc routing," in *Proceedings of the Seventh Annual ACM/IEEE International Conference on Mobile Computing and Networking (MOBICOM'01)*, Rome, July 2001.

33. T. He, J. A. Stankovic, C. Lu, and T. Abdelzaher, "Speed: A stateless protocol for real-time communication in sensor networks," in *Proceedings of IEEE ICDCS'03*, Providence, Rhode Island, May 2003.

34. J. Heidemann, F. Silva, and D. Estrin, "Matching data dissemination algorithms to application requirements," in *Proceedings of the First ACM SenSys'03*, Los Angeles, Nov. 2003.

35. K. Sha, Z. Zhu, S. Sellamuthu, and W. Shi, "Capricorn: A large scale wireless sensor network simulator," Technical Report MIST-TR-2004-001, Wayne State University, Detroit, Jan. 2004.

36. K. Sha and W. Shi, "Revisiting the lifetime of wireless sensor networks," in *Proceedings of ACM SenSys'04*, Baltimore, Nov. 2004.

37. S. D. Servetto and G. Barrenechea, "Constrained random walks on random graphs: Routing algorithms for large scale wireless sensor networks," in *Proceedings of the First ACM Workshop on Wireless Sensor Networks and Applications (WSNA'02)*, Atlanta, Sept. 2002.

38. B. Deb, S. Bhatnagar, and B. Nath, "A topology discovery algorithm for sensor networks with applications to network management," in *IEEE CAS Workshop (Short Paper)*, Pasadena, CA, Sept. 2002.

39. C. Schurgers et al., "Topology management for sensor networks: Exploiting latency and density," in *Proceedings of the MobiHoc'02*, Lausanne, Switzerland, June 2002.

40. P. Levis et al., "The emergence of networking abstractions and techniques in tinyos," in *Proceedings of the First USENIX/ACM Networked System Design and Implementation*, San Francisco, Mar. 2004.

41. Z. Zhu and W. Shi, "HALF: A highly accurate landmark-free localization algorithm in wireless sensor networks," Technical Report MIST-TR-2004-008, Wayne State University, Detroit, Mar. 2004.

42. "The network simulator - ns-2," http://www.isi.edu/nsnam/ns/.

43. P. Levis, N. Lee, M. Welsh, and D. Culler, "Tossim: Accurate and scalable simulation of entire tinyos applications," in *Proceedings of the First ACM SenSys'03*, Los Angeles, Nov. 2003.

44. L. Girod et al., "Emstar: A software environment for developing and deploying wireless sensor networks," in *Proceedings of the 2004 USENIX Annual Technical Conference*, Boston, June 2004.

45. K. Akkaya and M. Younis, "An energy-aware qos routing protocol for wireless sensor networks," in *Proceedings of the IEEE Workshop on Mobile and Wireless Networks (MWN'03)*, May 2003.

46. J. Chang and L. Tassiulas, "Routing for maximum system lifetime in wireless ad-hoc networks," in *Proceedings of the Thirty-Seventh Annual Allerton Conference on Communication, Control and Computing*, Monticello, IL, Sept. 1999.

47. I. Kang and R. Poovendran, "Maximizing static network lifetime of wireless broadcast adhoc networks," in *IEEE 2003 International Conference on Communications*, Anchorage, AK, May 2003.

48. E. Duarte-Melo and M. Liu, "Analysis of energy consumption and lifetime of heterogeneous wireless sensor networks," in *Proceedings of IEEE Globecom*, San Francisco, Nov. 2003.

49. H. Zhang and J. Hou, "On deriving the upper bound of α-lifetime for large sensor networks," in *Proceedings of the Fifth ACM International Symposium on Mobile Ad Hoc Networking and Computing*, May 2004.

50. D. Ganesan et al., "Highly resilient, energy efficient multipath routing in wireless sensor networks," *Mobile Computing and Communications Review (MC2R)*, vol. 1, no. 2, 2002.

51. S. Shenker, S. Ratnasamy, B. Karp, D. Estrin, and R. Govindan, "Data-centric storage in sensornets," in *Proceedings of the First ACM Workshop on Hot Topics in Networks (HotNets-I)*, Princeton, Oct. 2002.

52. S. Ratnasamy, P. Francis, M. Handley, R. Karp, and S. Schenker, "A scalable content addressable network," in *Proceedings of ACM SIGCOMM'01*, San Diego, 2001.

53. I. Stoica, R. Morris, D. Karger, M. F. Kaashoek, and H. Balakrishnan, "Chord: A scalable peer-to-peer lookup service for internet applications," in *ACM SIGCOMM'2001*, San Diego, 2001.

54. B. Zhao, J. Kubiatowicz, and A. Joseph, "Tapestry: An infrastructure for fault-tolerant wide-area location and routing," Technical Report UCB/CSD-01-1141, Computer Science Division, University of California, Berkeley, Apr. 2001.

55. A. Rowstron and P. Druschel, "Pastry: Scalable, distributed object location and routing for large scale peer-to-peer systems," in *IFIP/ACM Middleware 2001*, Heidelberg, 2001.

56. C. Intanagonwiwat, R. Govindan, D. Estrin, J. Heidemann, and F. Silva, "Directed diffusion for wireless sensor networks," *IEEE/ACM Transactions on Networking*, vol. 11, no. 1, pp. 2–16, Feb. 2003.

57. M. Ahmed, S. Krishnamurthy, S. Dao, and R. Katz, "Optimal selection of nodes to perform data fusion in wireless sensor networks," in *Proceedings of the International Society for Optical Engineering*, vol. 4396, Bellingham, WA, 2001, pp. 53–64.

58. B. Krishnamachri, D. Estrin, and S. Wicker, "The impact of data aggregation in wireless sensor networks," in *Proceedings of the Twenty-Second Inter Conference on Distributed Computing Systems Workshops*, Vienna, July 2002, pp. 575–578.

59. D. Perovic, R. Shah, K. Ramchandran, and J. Rabaey, "Data funneling: Routing with aggregation and compression for wireless sensor networks," in *Proceedings of the First IEEE International Workshop on Sensor Network Protocols and Applications*, May 2003, pp. 156–162.

60. K. Dasgupta, K. Kalpakis, and P. Namjoshi, "An efficient clustering-based heuristic for data gathering and aggregation in sensor networks," in *Proceedings of IEEE Wireless Communications and Networking Conference*, vol. 3, 2003, New Orleans, pp. 1948–1953.

61. S. Banerjee and A. Misra, "Minimum energy paths for reliable communication in multi-hop wireless networks," in *Proceedings of ACM MOBIHOC*, Lausanne, Switzerland, 2002, pp. 146–156.

62. J.-H. Chang and L. Tassiulas, "Energy conserving routing in wireless ad-hoc networks," in *Proceedings of IEEE INFOCOM*, Tel Aviv, 2000, pp. 22–31.

63. L. Li and J. Halpern, "Minimum-energy mobile wireless networks revisited," in *Proceedings of ICC*, Helsinki, June 2001, pp. 287–283.

64. X. Li, P. Wan, Y. Wan, and O. Frieder, "Constrained shortest paths in wireless networks," in *Proceedings of IEEE MILCOM*, Mclean, VA, 2001, pp. 884–893.

65. S. Singh, M. Woo, and C. Raghavendra, "Power-aware routing in mobile ad hoc networks," in *Proceeding of ACM MOBICOM*, Dallas, 1998, pp. 181–190.

66. C.-K. Toh, H. Cobb, and D. Scott, "Performance evaluation of battery-life-aware routing schemes for wireless ad hoc networks," in *Proceedings of ICC*, Helsinki, June 2001, pp. 2824–2829.

67. W. Ye, J. Heidemann, and D. Estrin, "An energy efficient MAC protocol for wireless sensor networks," in *Proceedings of IEEE INFOCOM*, New York, vol. 3, June 2002, pp. 1567–1576.

68. D. Ganesan, R. Govindan, S. Shenker, and D. Estrin, "Highly resilient, energy efficient multipath routing in wireless sensor networks," *Mobile Computing and Communication Review*, vol. 1, no. 2, June 2002.

69. Y. Xu, J. Heidemann, and D. Estrin, "Geography-informed energy conservation for ad hoc routing," in *Proceedings of ACM MOBICOM*, Rome, July 2001, pp. 70–84.

70. S. Basagni, I. Chlamtac, and V. Syrotiuk, "A distance routing effect algorithm for mobility," in *Proceedings of ACM MOBICOM*, Dallas, 1998, pp. 76–84.

71. T. Camp, J. Boleng, B. Williams, L. Wilcox, and W. Navidi, "Performance comparison of two location based routing protocols for ad hoc networks," in *Proceedings of IEEE INFOCOM*, New York, 2002, pp. 1678–1687.

72. Z. Hass, J. Halpern, and L. Li, "Gossip-based ad hoc routing," in *Proceedings of IEEE INFO-COM*, New York, vol. 3, 2002, pp. 707–717.

73. D. Johnson, D. Maltz, Y. Hu, and J. Jetcheva, "The dynamic source routing protocol for mobile ad hoc networks (DSR)," Internet-draft-ietf-manet-dsr-07.txt, 2002.

74. S. Chul and M. Woo, "Scalable routing protocol for ad hoc networks," *Wireless Networks*, vol. 7, pp. 513–529, 2001.

75. Y. Ko and N. Vaidya, "Location-aided routing (LAR) in mobile ad-hoc networks," *Wireless Networks*, pp. 307–321, 2000.

76. C. Perkins, E. Belding-Royer, and S. Das, "Ad hoc on demand distance vector (AODV) routing," Internet-draft-ietf-manet-aodv-11.txt, 2002.

77. V. Raghunathan, C. Schurgers, S. Park, and M. Srivastava, "Energy-aware wireless microsensro networks," *IEEE Signal Processing Magazine*, pp. 40–50, Mar. 2002.

78. S. Bandyopadhyay and E. Coyle, "An energy efficient hierarchical clustering algorithm for wireless sensor networks," in *Proceedings of IEEE INFOCOM*, San Francisco, Mar. 2003.

79. A. Amis, R. Prakash, T. Vuong, and D. Huynh, "Max-min D-cluster formation in wireless ad hoc networks," in *Proceedings of IEEE INFOCOM*, New York, Mar. 2000.

80. C. Srisathapornphat and C. Shen, "Coordinated power conservation for ad hoc networks," in *Proceedings of ICC*, 2002, pp. 3330–3335.

81. K. Woo, C. Yu, H. Youn, and B. Lee, "Non-blocking, localized routing algorithm for balanced energy consumption in mobile ad hoc networks," in *Proceedings of the International Workshop on Modeling, Analysis and Simulation of Computer and Telecommunication Systems*, Cincinnati, 2001, pp. 117–124.

82. "Wireless LAN medium access control (MAC) and physical layer (PHY) specification: High speed physical layer in the 5 GHz band," IEEE Standard 802.11a, Sept. 1999.

83. B. Hamdaoui and P. Ramanathan, "Rate feasiblity under medium access contention constraints," in *Proceedings of IEEE GLOBECOM*, San Francisco, vol. 6, Dec. 2003, pp. 3020–3024.

84. R. J. Vanderbei, *Linear Programming: Foundations and Extensions*. Kluwer Academic, 1997.

85. D. W. Carman, P. S. Kruus, and B. J. Matt, "Constraints and approaches for distributed sensor network security," Technical Report 00-010, NAI Labs, 2000.

86. S. Carter and A. Yasinsac, "Secure position aided ad hoc routing protocol," in *Proceedings of the IASTED International Conference on Communications and Computer Networks*, Cambridge, MA, Nov. 3–4, 2002.

87. B. Dahill, B. N. Levine, E. Royer, and C. Shields, "A secure routing protocol for ad hoc networks," in *Proceedings of the Tenth Conference on Network Protocols (ICNP)*, Paris, Nov. 2002.

88. Y. Hu, D. B. Johnson, and A. Perrig, "Sead: Secure efficient distance vector routing for mobile wireless ad hoc networks," in *Proceedings of the Fourth IEEE Workshop on Mobile Computing Systems and Applications (WMCSA 2002)*, Calicoon, NY, June 2002, pp. 3–13.

89. F. Hu and N. K. Sharma, "Security considerations in wireless sensor networks," *Ad Hoc Networks Journal*, vol. 2, no. 4, 2004.

90. A. Perrig, R. Szewczyk, V. Wen, and A. Woo, "Security for smartdust sensor network," http://www.cs.berkeley.edu/ vwen/classes/f2000/cs261/project/sensor_security.html.

91. A. Perrig, R. Szewczyk, V. Wen, D. Culler, and J. D. Tygar, "Spins: Security protocols for sensor networks," in *Proceedings of Seventh Annual International Conference on Mobile Computing and Networks MOBICOM*, Rome, July 2001.

92. D. Liu and P. Ning, "Efficient distribution of key chain commitments for broadcast authentication in distributed sensor networks," in *Proceedings of the Tenth Annual Network and Distributed System Security Symposium*, San Diego, CA, Feb. 2003, pp. 263–276.

93. D. Liu and P. Ning "Multi-level u-tesla, a broadcast authentication system for distributed sensor networks," Technical Report TR-2003-08, North Carolina State University, Department of Computer Science, 2003.

94. W. Du, J. Deng, Y. S. Han, and P. Varshney, "A pairwise key pre-distribution scheme for wireless sensor networks," in *Proceedings of the Tenth ACM Conference on Computer and Communications Security (CCS)*, Washington, DC, Oct. 27–31, 2003.

95. H. Chan, A. Perrig, and D. Song, "Random key predistribution schemes for sensor networks," in *IEEE Symposium on Research in Security and Privacy*, Berkeley, 2003.

96. R. D. Pietro, L. V. Mancini, and A. Mei, "Random key assignment for secure wireless sensor networks," in *2003 ACM Workshop on Security of Ad Hoc and Sensor Networks*, George W. Johnson Center at George Mason University, Fairfax, VA, 2003.

97. L. Eschenauer and V. D. Gligor, "A key-management scheme for distributed sensor networks," in *Proceedings of the Ninth ACM conference on Computer and Communications Security*, Washington, DC, 2002.

98. D. Liu and P. Ning, "Establishing pairwise keys in distributed sensor networks," in *Tenth ACM Conference on Computer and Communications Security*, Washington DC, Oct. 2003.

99. D. Liu and P. Ning, "Location-based pairwise key establishments for relatively static sensor networks," in *2003 ACM Workshop on Security of Ad Hoc and Sensor Networks*, George W. Johnson Center at George Mason University, Fairfax, VA, Oct. 31, 2003.

100. W. Du, J. Deng, Y. Han, S. Chen, and P. Varshney, "A key management scheme for wireless sensor networks using deployment knowledge," in *IEEE INFOCOM '04*, Hong Kong, Mar. 7–11, 2004.

101. J. Deng, R. Han, and S. Mishra, "Insens: Intrusion-tolerant routing in wireless sensor networks," Technical Report CU-CS-939-02, Department of Computer Science, University of Colorado, 2002.

102. S. Doumit and D. P. Agrawal, "Self-organized criticality and stochastic learning-based intrusion detection system for wireless sensor networks," in *MILCOM 2003*, Monterey, CA, 2003.

103. J. Deng, R. Han, and S. Mishra, "A performance evaluation of intrusion-tolerant routing in wireless sensor networks," in *Second International Workshop on Information Processing in Sensor Networks*, Palo Alto, CA, Apr. 2003.

104. J. Deng, R. Han, and S. Mishra, "Security support for in-network processing in wireless sensor networks," in *ACM Workshop on Security of Ad Hoc and Sensor Networks*, George W. Johnson Center at George Mason University, Fairfax, VA, Oct. 2003.

105. A. D. Wood and J. A. Stankovic, "Denial of service in sensor networks," *IEEE Computer Magazine*, vol. 35, no. 10, pp. 54–62, 2002.

106. C. Karlof and D. Wagner, "Secure routing in wireless sensor networks: Attacks and countermeasures," in *First IEEE International Workshop on Sensor Network Protocols and Applications*, Anchorage, AK, May 2003.

107. G. Wang, W. Zhang, G. Cao, and T. L. Porta, "On supporting distributed collaboration in sensor networks," in *MILCOM 2003*, Monterey, CA, Oct. 2003.

108. J. Zachary, "A decentralized approach to secure group membership testing in distributed sensor networks," in *MILCOM 2003*, Monterey, CA, Oct. 2003.

109. S. Zhu, S. Setia, and S. Jajodia, "Leap: Efficient security mechanisms for large-scale distributed sensor networks," in *Tenth ACM Conference on Computer and Communications Security*, Washington DC, 2003.

110. Y. W. Law, S. Etalle, and P. H. Hartel, "Key management with group-wise pre-deployed keying and secret sharing pre-deployed keying," EYES project (Europe), available through GOOGLE search engine, 2004.

111. M. P. S. Slijepcevic, V. Tsiatsis, S. Zimbeck, and M. B. Srivastava, "On communication security in wireless ad-hoc sensor network," in *Eleventh IEEE International Workshops on Enabling Technologies: Infrastructure for Collaborative Enterprises*, Pittsburgh, PA, June 10–12, 2002.

112. M. Chen, W. Cui, V. Wen, and A. Woo, "Security and deployment issues in a sensor network," http://citeseer.nj.nec.com/chen00security.html, 2004.

113. Y. Law, S. Dulman, S. Etalle, and P. Havinga, "Assessing security-critical energy-efficient sensor networks," Technical Report TR-CTIT-02-18, Department of Computer Science, University of Twente, 2002.

114. B. Krishnamachari, D. Estrin, and S. B. Wicker, "The impact of data aggregation in wireless sensor networks," in *International Workshop on Distributed Event-Based Systems*, Vienna, Austria, July 2002.

115. S. Basagni, K. Herrin, D. Bruschi, and E. Rosti, "Secure pebblenets," in *Proceedings of the 2001 ACM International Symposium on Mobile Ad Hoc Networking and Computing*, Long Beach, CA, Oct. 2001, pp. 156–163.

116. C. Intanagonwiwat, R. Govindan, and D. Estrin, "Directed diffusion: A scalable and robust communication paradigm for sensor networks," in *Proceedings of the Sixth Annual International Conference on Mobile Computing and Networking*, Boston, MA, Aug. 2000.

117. F. Hu and S. Kumar, "Wireless sensor networks for mobile telemedicine: Qos support," in *World Wireless Congress*, San Francisco, CA, May 25–28, 2004.

118. F. Hu and S. Kumar, "Multimedia query with qos considerations for wireless sensor networks in telemedicine," in *Proceedings of International Conference on Internet Multimedia Management Systems*, Orlando, FL, Sept. 7–11, 2003.

119. Z. Haas and M. Pearlman, "The zone routing protocol (zrp) for ad hoc networks," IETF Internet draft for the Manet Group, 1999.

120. W. Heinzelman, A. Chandrakasan, and H. Balakrishnan, "Energy-efficient communication protocol for wireless microsensor networks," in *Proceedings of the Hawaii International Conference on System Sciences*, Maui, Jan. 2000, pp. 1–10.

121. K. Sohrabi, J. Gao, V. Ailawadhi, and G. Pottie, "Protocols for self-organization of a wireless sensor network," *IEEE Personal Communications*, pp. 16–27, Oct. 2000.

122. W. Heinzelman, "Application-specific protocol architectures for wireless networks," Ph.D. dissertation, Massachusetts Institute of Technology, Cambridge, MA, June 2000.

123. A. Helmy, "Mobility-assisted resolution of queries in large-scale mobile sensor networks (marq)," *Computer Networks Journal*, Aug. 2003.

124. H. Harney and C. Muchenhirn, "Group key management protocol (gkmp) architecture," RFC 2094, 1997.

125. X. S. Li, Y. R. Yang, M. G. Gouda, and S. S. Lam, "Batch rekeying for secure group communications," in *Proceedings of Tenth International Word Wide Web Conference*, Hong Kong, 2001.

126. C. K. Wong, M. G. Gouda, and S. S. Lam, "Secure group communications using key graphs," in *Proceedings of ACM SIGCOMM'98*, Vancouver, Sept. 1998.

127. O. Goldreich, S. Goldwasser, and S. Micali, "How to construct random functions," *Journal of the ACM*, vol. 33, no. 4, pp. 210–217, 1986.

128. "Ad hoc on-demand distance vector routing," http://moment.cs.ucsb.edu/AODV/aodv.html, home page, University of California, Santa Barbara.

129. S. Madden, M. J. Franklin, J. M. Hellerstein, and W. Hong, "Tag: A tiny aggregation service for ad-hoc sensor networks," in *Proceedings of the Fifth Annual Symposium on Operating Systems Design and Implementation*, Dec. 2002.

130. I. Stojmenovic, M. Seddigh, and J. Zunic, "Dominating sets and neighbor elimination-based broadcasting algorithms in wireless networks," *IEEE Transactions on Parallel and Distributed Systems*, vol. 12, no. 12, Dec. 2001.

131. D. Tian and N. Georganas, "Energy efficient routing with guaranteed delivery in wireless sensor networks," in *Proceedings of IEEE Wireless Communications and Networking Conference*, New Orleans, LA, Mar. 2003.

132. "Wireless lan medium access control (mac) and physical layer (phy) specifications," IEEE Standard 802.11-1997, Part ii, http://www.drizzle.com/ aboba.

133. J. R. Walker, "Unsafe at any key size; an analysis of the wep encapsulation," IETF, 2000.

134. N. Sadagopan, B. Krishnamachari, and A. Helmy, "The ACQUIRE mechanism for efficient querying in sensor networks," in *First IEEE International Workshop on Sensor Network Protocols and Applications*, Anchorage, AK, May 2003.

135. W. Siddiqui, "Two-level-based sensor network security and performance model," Master's thesis, Computer Engineering Department, Rochester Institute of Technology, 2004.

136. H. G. Ebersman and O. K. Tonguz, "Handoff ordering using signal prediction priority queuing in personal communication systems," *IEEE Transactions on Vehicular Technology*, vol. 48, no. 1, pp. 20–35, Jan. 1999.

137. J. Elson and D. Estrin, "Time synchronization for wireless sensor networks," in *Proceedings of IPDPS Workshop on Parallel and Distributed Computing Issues in Wireless Networks and Mobile Computing*, San Francisco, April 2001.

138. W. Stallings, *Cryptography and Network Security: Principles and Practices*, 3rd ed., Upper Saddle River, NJ: Prentice-Hall, 2003.

139. I. Kumar, *Cryptology*, Laguna Hills, CA: Aegean Park, 1997.

140. Y. Ma, S. Dalal, M. Alwan, and J. Aylor, "ROP: A resource oriented protocol for heterogeneous sensor networks," in *Virginia Tech Symposium on Wireless Personal Communications*, 2003.

141. T. Park and K. G. Shin, "Lisp: A lightweight security protocol for wireless sensor networks," *ACM Transactions on Embedded Computing Systems*, vol. 3, no. 3, pp. 634–660, 2004.

6

POWER MANAGEMENT

6.1 INTRODUCTION

Power management is a critical issue in sensor networks because these networks are typically characterized by severely power-limited nodes. The nodes are power-limited in two dimensions: First, even when operating at full power, the nodes have small batteries and therefore cannot expend a great deal of power to complete tasks; and, second, their power supplies cannot typically be replenished; so once exhausted, a node is useless. Because of this second characteristic, intelligent and efficient use of power directly impacts the length of time each node can operate within a network. As nodes dissipate their power, the sensor network eventually loses its fidelity, and perhaps its coverage, until it is no longer operable.

Power management is required at all layers and components of the sensor network and impacts the design of all algorithms. Each node must be designed to minimize its power consumption while still being able to complete its task. Each network must have algorithms designed to maximize its utility, that is, keep it alive long enough to complete its mission or, in long-term monitoring applications, maximize the lifetime of the network. Several discussions already presented in this book have addressed different aspects of power management. This included discussions on sensor placement, medium access control (MAC) layer protocols, moving nodes to reduce overall power consumption, intelligent routing, and data dissemination. Many other techniques exist as well.

For example, in many sensor networks redundant nodes are often deployed for a variety of reasons, such as to ensure with a certain probability that at least one sensor is present in every critical area if nodes are deployed randomly. In these networks, distributed algorithms are designed so that only the minimum number of nodes required to complete a mission are active and all others are idle so that they may conserve their energy. This type of algorithm may run on every node, or may run among nodes with special functions, for example, nodes responsible for providing localization beacons. In other networks, nodes

Sensor Network Operations, Edited by Phoha, LaPorta, and Griffin
Copyright © 2006 The Institute of Electrical and Electronics Engineers, Inc.

self-organize into a hierarchy of clusters to make communication, data dissemination, and sensing more efficient. Clearly, cluster leadership must rotate to balance the processing and communication so that the energy in cluster heads is not quickly depleted.

Other solutions include designing algorithms and protocols that trade-off processing power and communication power to maximize the lifetime of the network. Still other solutions manage the power consumption of each individual hardware component of a sensor node. This solution may even trade-off precision in data for power by, for example, reducing the number of bits/sample transmitted.

In addition to algorithms that conserve power, it is critical to be able to characterize power consumption accurately so that network lifetime can be predicted, and new power management algorithms may be developed. In this chapter we present four discussions on power management that present both algorithms and models for characterizing power consumption.

In Section 6.2, by Hoang and Motani, the operation of each component in a node is carefully controlled to save power. Trade-offs are considered in sensing and reporting data; that is, nodes attempt to eliminate the transmission of redundant data. In Section 6.3, by Jain and Liang, sensor network lifetime is carefully analyzed for two regular deployment scenarios, a square grid and a hex grid. In Section 6.4, by Dasika, Vrudhula, and Chopra, sensors compute sets of nodes that provide all possible covers in the network. The use of these sets is then rotated so that the load is balanced across nodes. Both a centralized and distributed algorithm is presented. In Section 6.5, by Jurdak, Lopes, and Baldi, a model is developed to characterize lifetime in underwater networks. The impact on power consumption of four dimensions of operation is determined: intermediate node distance, transmission frequency, data frequency, and number of nodes.

6.2   ADAPTIVE SENSING AND REPORTING IN ENERGY-CONSTRAINED SENSOR NETWORKS

Anh Tuan Hoang and Mehul Motani

Wireless sensor networks may have to operate without battery recharges for long periods of time while still providing acceptable quality of service (QoS). We focus on strategies to increase the lifetime of such networks by translating a system-level QoS specification to the node level. This allows us to focus on the problem of controlling a single sensor node in which energy is consumed in the sensing, processing, and transceiving components. Our objective is to jointly control the operating modes of these components so that the node lifetime is increased while guaranteeing the required event missed rate. Increasing the lifetime of each node directly translates into an increase in the sensor network lifetime. We apply this approach to a cluster-based wireless sensor network that consists to two types of nodes, that is, sensing nodes and data-gathering/relaying nodes. The advantage of controlling individual nodes is scalability, which is important in a network with potentially thousands of nodes. Furthermore, when our adaptive schemes are applied to individual nodes in a network, the energy consumption in all the nodes is statistically the same. This leads to uniform energy consumption and graceful network degradation, which are desirable for mission-critical sensor networks. Finally, we present numerical results that demonstrate the energy savings of the proposed adaptive control policies.

6.2.1 Background Information

A sensor network consists of a large number of sensor nodes that are responsible for collecting specific information about the environment in which they are deployed. Examples of such networks can be found in health, military, and home applications [1,2]. Figure 6.1 shows a generic model for a cluster in a hierarchical cluster-based wireless sensor network. Nodes in each cluster monitor a particular geographical area and report their measurements to the cluster head, who then may act on it or may aggregate and forward the data to the decision center. If these nodes are deployed in remote areas, they may have to operate without battery recharges or replacements for long periods of time while still providing acceptable QoS. This points to one of the most challenging issues in designing wireless sensor networks, that is, to do the job well for as long as possible. This is precisely the problem we consider in this section—how to provide acceptable QoS while maximizing the system lifetime.

As system-level performance is the primary concern, much of the existing literature in wireless sensor networks concentrates on system-level design and analysis. Examples include energy-conserving routing techniques [3,4], data gathering and aggregation [5–7], and identifying bounds for network lifetime [8]. Our approach is somewhat different. In this section, which is the extension of our work in [9], we focus on controlling a single node in a sensor network. Even though the final goal is to improve the system-level performance, many system-level performance parameters can be filtered down to the node level. For example, increasing individual node lifetime is directly related to prolonging network lifetime. Furthermore, in dense sensor networks, most communication occurs between nodes within close range. In this situation, the energy consumed in the signal processing and electronics of the sensor node is comparable to the energy consumed by the transceiver. Therefore, finding an appropriate trade-off between energy used for communication and that for signal processing and electronics in the node is important. Finally, as sensor networks will likely contain thousands of nodes, joint control of sensors may not be feasible and node-level control policies will ensure scalability.

A nice property of our node-level control approach is that it results in control schemes that, when applied to all nodes in the sensor network, make the energy consumption in all the nodes statistically the same. In other words, the energy consumption is uniform across the network and all nodes in the network will tend to die together. This is a desirable property for those sensor networks in which the system performance degrades significantly when a certain number of nodes cease to operate, regardless of whether the rest of the network is still functioning. Examples include mission-critical intruder detection and tracking applications, where holes in the network caused by dead nodes can have serious undesirable consequences.

In our problem, events of interest happen within the sensing region of a sensor node according to some stochastic law. Each event lasts for a random period of time during which it can be captured if the sensing unit of the node is active. Events are considered missed if they disappear before being captured. If an event is captured, it is processed into data packets that are buffered in the memory unit of the sensor node. As the sensor node is only equipped with limited memory, if all storage space is used up, data packets generated from sensed events are lost and the corresponding events need to be recaptured. The transceiver unit of the sensor node transmits data packets, that is, reports collected information, to the cluster head at a fixed rate over a time-varying wireless channel. Since the channel fluctuates in time, the transmit power required to ensure reliable communication also varies with time.

When controlling a sensor node, we face the following trade-off. On one hand, the sensing and transceiving units of the node should be active as much as possible so that the event missed rate is minimized. On the other hand, as events occur randomly, there are periods when there are no events to sense and no data packets to transmit. In those periods, the sensing and transceiving units should be turned off (or idle) to conserve energy. For the same reason, when the wireless channel is in a deep fade, the transceiving unit should be turned off (or idle). So the objective is to jointly control the operating modes of the sensing and transceiving components in the sensor node so that energy is conserved while guaranteeing the required event-missed rate. While doing so, we assume that the event arrival statistics, the buffer occupancy, and the channel quality are available for us on which to base the control decisions.

We note that in this section, the performance metric of the sensor network is in terms of the event-missed rate and energy consumption. In many application, apart from the event-missed rate, data reporting latency is also an important factor. Therefore, taking both event detecting probability and data reporting latency into account will be considered in our future work.

Works related to ours include [10–17]. In [10, 11], the problem of adapting the transmission to the data arrival statistics and buffer and channel conditions has been considered. We extend [10, 11] by incorporating the event-sensing/data-gathering process into the problem and considering adapting the operations of both sensing and transmitting units. In [12], the authors study a power-aware methodology that reduces sensor node energy consumption by heuristically exploiting both sleep state and active power management. Our work has a similar objective, however, we structure the problem as a Markov decision process (MDP) so that optimal control policies can be obtained. Our problem is also related to techniques of adaptive power management, which are considered in [14–16]. However, our formulation is more specific for cases of wireless sensor nodes, with joint control of the sensing and transceiving units.

The problem we are looking at is related to the CEO problem [18], in which the center controller (or CEO) receives noisy observations of a random source from a set of L observers over a rate-constrained link. The objective is to minimize the distortion of the CEO's estimate of the source subject to the rate constraints of the observers. In [19], an achievable rate distortion region is computed for the CEO problem in which the source is an independent identically distributed (i.i.d.) Gaussian sequence, and the observations are corrupted by i.i.d. Gaussian noise. In [20], the results of [19] are extended by deriving an achievable rate distortion region for a CEO problem with an unreliable sensor network in which all sensor nodes are always functional but the links to the CEO are subject to erasure. Unlike [20], in our model, all links are assumed to be reliable as long as enough transmit power is used. However, since nodes can be turned off (or idle) and miss an event, we have an equivalent notion of erasure. Using the results in [20], we can determine how many sensors should capture an event so that the distortion in the cluster head is below some threshold and then translate this into an allowable event-missed rate requirement for each sensor.

The main contributions of our section are as follows.

- We translate the sensor network system design problem into a sensor node design problem and then formulate an optimization in which the sensing and transmitting components of the sensor node are jointly controlled to increase its lifetime.

- We structure the above optimization problem as a tractable Markov decision process and use dynamic programming techniques to obtain the solution.
- We present numerical results to illustrate the energy savings of our jointly adaptive sensing and transmitting policies.

6.2.2 Cluster-Based Wireless Sensor Network

Network Architecture   We consider a cluster-based wireless sensor network, which is depicted in Figure 6.1. In a cluster-based network, sensor nodes are organized into clusters and each cluster is usually responsible for monitoring one specific geographical area. We adopt the heterogeneous model in which there are two different types of nodes. Type I nodes are normal sensors whose main responsibility is to sense the surrounding environment to

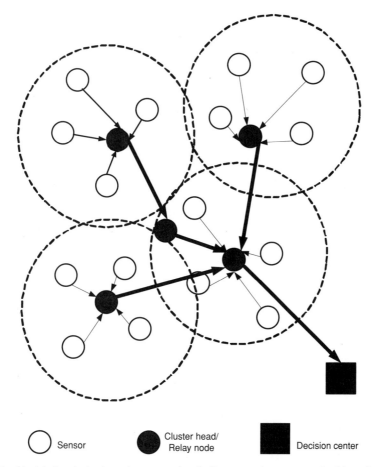

Figure 6.1   Model of a cluster-based sensor network. Sensor nodes are organized into clusters, and each cluster is responsible for monitoring one specific geographical area (marked by a circle). There are two types of nodes, i.e., sensing nodes (type I) and data-gathering/relaying nodes (type II). Sensor nodes (type I) transmit collected data directly to the corresponding cluster heads (type II) that then route data toward a decision center.

capture events of interest (more details will be given shortly). These type I nodes are usually equipped with very limited energy source. Type I nodes make up each cluster, and they transmit collected data directly to cluster heads that are type II nodes. The main functions of type II nodes are data gathering and/or relaying. They aggregate the data collected in their corresponding clusters and relay them toward a decision center, where the data is interpreted and decisions are made. We assume that type II nodes are less energy-constrained than type I nodes are. In fact, the number of type II nodes is usually much less than that of type I nodes so that we can afford the cost of providing them with more energy storage.

There are several advantages of adopting a cluster-based architecture for wireless sensor networks. First, a wireless sensor network usually consists of a very large number of nodes, and organizing these nodes into clusters makes the control and management more scalable [21]. Second, a cluster-based architecture makes it easier to carry out data compression and/or aggregation. Data collected by nodes in the same cluster are usually strongly correlated. Therefore, it is natural for nodes in the same cluster to compress their data based on one another, and for cluster heads to carry out data fusion/aggregation [21]. As the redundancy in collected data is removed before data leaves the cluster, it helps reduce network congestion and saves energy spent in relaying information. Finally, by organizing nodes into clusters and allowing direct communication between sensor nodes and cluster heads, the total number of hops (on average) to the decision center is reduced, hence reducing the average data-gathering latency.

Even though in this section we consider a cluster-based network that consists of two types of nodes, that is, sensing nodes and cluster heads, the approach developed can be applied to cluster-based networks of homogeneous sensors by electing one node to be the cluster head. Since cluster heads consume more energy than sensing nodes, one way to balance the energy consumption among all nodes is to randomly rotate the role of the cluster head [22].

Event Sensing and Data Routing   We focus on event detecting applications in which the objective is to capture random events that happen within the coverage area of a sensor network. Time is divided into slots of duration τ and time slot i denotes the time interval $[\tau i, \tau(i + 1))$. Without loss of generality, let $\tau = 1$. We assume that events happen in each time slot according to a Bernoulli distribution with probability λ. Suppose an event happens at time t within time slot i, that is, $i \leq t < i + 1$, the quantity $t - i$ is assumed to be uniformly distributed in $(0, 1)$. We also assume that the event exists in the sensing region for a period that is uniformly distributed in $(0, D]$, where D is the maximum event lifetime. The tasks of event sensing and data gathering are carried out using the following mechanism.

- Within each cluster, after detecting events and processing them into data packets, sensors send the data packets directly to the cluster head. We assume that there is a certain MAC scheme that allows reliable packet transmission (TDMA, CDMA).
- We assume that there is no intercluster interference. For example, this can be achieved by assigning nonoverlapping frequency bands to adjacent clusters, similar to what has been done in GSM cellular systems.
- Upon receiving data collected in their clusters, cluster heads carry out necessary data fusion/aggregation tasks and route the aggregated data to the decision center.

We note that in this section, the performance metric of the sensor network is in terms of the event-missed rate and energy consumption. In many application, apart from the

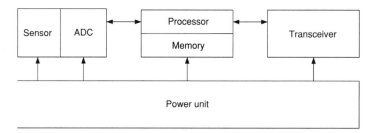

Figure 6.2 Components of a typical wireless sensor node. Each sensor node comprises of four main units, i.e., the sensing and analog-to-digital converter (ADC) unit, the processing and storage unit, the transceiver unit, and finally the power supply unit.

event-missed rate, data reporting latency is also an important performance factor. Taking both event detecting probability and data reporting latency into account will be considered in our future work.

Model of Sensor Nodes   A generic model for a wireless sensor node is depicted in Figure 6.2. In particular, each sensor consists of four main components: the sensing and analog-to-digital conversion (sensing/ADC) unit, the processing/storage unit, the transceiver, and the power supply. Each operation of sensing, processing, buffering, and transmitting costs a certain amount of energy. In this section, we focus on controlling the sensing/ADC and transceiving units and assume that the operation of the processor/storage unit is determined by the corresponding modes of the sensing/ADC and transceiving units.

The sensing/ADC unit and the transceiving unit can support different operating modes, which correspond to different amounts of power consumed and work done. In particular, we assume that the sensing/ADC unit can operate in *active* and *off* modes while the transceiving unit can operate in *active, idle*, and *off* modes.

Operating Modes of Sensing/ADC Unit

Active Mode of Sensor   When the sensing unit is active, the power consumed per time slot is P_{sa}. In this mode, the sensing unit is able to capture all events happening within its sensing region. If the sensing unit is turned off for more than one time slot, then when it wakes up there may be several events within its sensing region that have not been captured. In that case, we assume that the sensing/ADC unit can process multiple events simultaneously. If at least one event is captured within a time slot, then A packets will be generated and added to the buffer. Moreover, we assume that in the case when multiple events are processed within one time slot, they can be represented using A packets. This is based on the observation that the properties of events happening closely to one another in time and space are likely to be highly correlated. In that case, the amount of data required to represent multiple correlated events is not much larger than that required to represent each individual event.We also assume that packets generated from processing all events that happen prior to or within time slot i are only added to the buffer at the end of time slot i. The buffer can stored up to B packets, and when there is no space left, packets generated from sensed events are dropped and the corresponding events need to be recaptured.

Off Mode of Sensor When the sensing/ADC unit is off in a time slot, it consumes no power, that is, $P_{so} = 0$. We can exploit this to conserve energy. For example, it may happen that the event lifetime is long enough so that the sensing unit can be turned off and woken up some time later before the event disappears. Another example is when the buffer is highly loaded, in which case, even if the sensing/ADC unit captures an event, the resulting packet will likely be dropped due to buffer overflow. When the sensing component is in off mode, no sensing is carried out. Therefore, any event that happens and disappears during this period is missed.

Mode Transition Latency for Sensor Switching the sensing/ADC unit between active and off modes takes time. It is usually the case that the time needed to turn off some component is much less than what is required to turn it back on. We assume that the time to turn the sensing/ADC unit off is equal to one time slot while the time needed to switch this component back on is $L_{so} = k\tau$ for some integer k. During the transient period between active and off modes, we assume that the power consumed per time slot of the sensing/ADC unit is $(P_{sa} + P_{so})/2$. This assumption is also made in [23], where the authors model the increase or decrease in the power consumption during state transient period as linear functions of time.

Operating Modes of Transceiving Unit When considering the operation of the transceiver, we mainly focus on controlling the transmission of data packets out of the buffer. We assume there is a separate channel to exchange control and channel information. The transceiver circuit can be in one of the three possible modes, that is, active, idle, and off.

Active Mode of Transceiver When the transmitter is in active mode, packets are transmitted out of the buffer at the maximum rate of r packets per time slot. The power cost when transmitting at rate r depends on the condition of the wireless channel. The channel is represented by a stationary and ergodic K-state Markov chain. We assume that the channel remains in the same state during each time slot. Let G and Γ denote the instantaneous channel state and fading gain, respectively. Here G is an integer with $1 \leq G \leq K$ and Γ is a positive real number. When the channel is in state $G = k, 0 \leq k < K$, the fading gain is $\Gamma = \gamma_k$, $\gamma_k > 0$. The K-state Markov channel model is completely described by its stationary distribution of each channel state k, denoted by $p_G(k)$, and the probability of transiting from state k into state j after each time slot, that is, $P_G(k, j), 0 \leq k, j < K$.

Let P_{ta} be the power consumed in the active mode when the channel gain is unity. When the channel gain is γ, the actual power needed to transmit at rate r is P_{ta}/γ. When the transmitter is active, apart from the transmission power, there is also power consumed in the electronics, denoted by P_{tc}.

Idle and Off Modes of Transceiver The transceiver can also operate in idle and off modes. In both of these modes, no data are transmitted out of the buffer. The main purpose of operating in these modes is to save power when there is no urgent need to transmit data, that is, when the buffer is lightly loaded, or when the channel conditions are not good, that is, the channel is in a deep fade. The power consumed by the transceiver in its idle and off modes are P_{ti} and P_{to}, respectively. We assume $P_{ti} = P_{tc}$ and $P_{to} = 0$.

Mode Transition Latency for Transceiver We assume that, for the transceiver unit, mode transitions must either originate from or terminate at the active mode. In other words, transitions are not allowed between idle and off modes. Similar to the case of the sensing/ADC unti, we assume that the powers consumed when transiting between active and idle modes is $(P_{tc} + P_{ti})/2$ and between active and off modes is $(P_{tc} + P_{to})/2$.

Similar to the case of sensing/ADC unit, we assume that the time delay for switching the transceiving unit from active mode to either idle or off mode is equal to one time slot. The latency for switching the transceiving unit from idle and off modes back to active mode are denoted by L_{ti} and L_{to}, respectively.

Discrete-Time Adaptive Power Management At the beginning of each time slot, given that the knowledge of the event process, the buffer occupancy and the channel conditions are available, we have to make a control decision on the operating modes of the sensing/ADC unit and the transmitting unit. While doing so, we face the trade-off between conserving energy, so that the system lifetime can be increased, and providing good quality of service, in terms of event-missed rate. Here, increasing lifetime is equivalent to reducing the time-averaged power consumed. In order to take both of the factors into account, we formulate the following optimization problem: *At the beginning of each time slot, select the operating modes of the sensing/ADC and transmitting units so that the weighted sum of the average power consumed and event-missed rate is minimized.*

6.2.3 Markov Decision Process

We structure the above optimization problem as a Markov decision process. This involves defining the system states, the set of valid control actions for each system state, the cost associated with taking a particular control action in a given system state, the dynamics of the system state given a control action, and finally the objective function over which the optimization will be carried out.

System States The system state at time slot i consists of the following components:

- G_i is the channel state during time slot i, $0 \le G_i < K$.
- B_i is the number of data packets in the buffer at the beginning of time slot i, $0 \le B_i \le B$.
- S_i is the state of the sensing/ADC unit at the beginning of time slot i. Apart from two states *active* and *off*, the sensing component can also be in one of the transitional states $\text{TR}(k), k = 1, 2, \ldots (L_{so} - 1)$. The sensing unit is in state $\text{TR}(k)$ when it is in the process of transitioning from the off state to the active state, and there are k time slots remaining before the active state is assumed.
- T_i is the state of the transceiving unit at the beginning of time slot i. Similar to the case of the sensing unit, the possible states of the transceiving unit are *active*, *idle*, *off*, and $\text{TR}(k), k = 1, 2, \ldots (L_{to} - 1)$.
- C_i is the number of consecutive time slots (until the end of time slot $i - 1$) during which the sensing/ADC unit has been *effectively off*. By saying the sensing/ADC unit is effectively off during a particular time slot, we mean that no event is successfully captured during that time slot. The first scenario is when the sensing/ADC unit is actually turned off and no processing is done. The second scenario is when the sensing/ADC unit is active, however, the buffer is full so that processed data are lost and corresponding

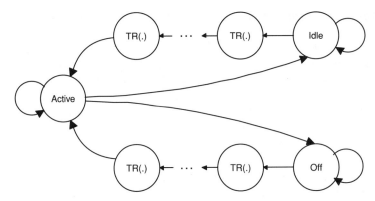

Figure 6.3 State diagram of the operation of the sensing/ADC and transceiving units. Note that the number of transitions from off to active and idle to active is variable and can take different values for both the sensing/ADC and transceiving units.

events need to be recaptured. C_i is important in estimating how many events currently exist in the system and require processing. As the maximum lifetime of an event is D time slots, we only need to consider $0 \leq C_i \leq D$.

The system state at time slot i is $\mathbf{X}_i = (G_i, B_i, S_i, T_i, C_i)$.

Control Actions   At the beginning of time slot i, given knowledge of the event process and the system state $\mathbf{X}_i$, we need to decide whether to leave each component of the system in the same operating mode or to transition to another mode. Note that the set of possible control actions depends on the current system state. Let $\mathbf{U}(\mathbf{x})$ denote the set of all possible control actions when the system is in state $\mathbf{x}$. Figure 6.3 depicts the state diagram of the operation of the sensing/ADC and the transceiving units. A complete state transition diagram for the system state $\mathbf{X}_i$ is too complex to be shown, so we omit it from the section.

System Dynamics   The channel dynamics do not depend on control actions. Due to limited space, we do not discuss the dynamics of the sensing/ADC and transmitter states here. Given a particular control action, it is straightforward to determine the dynamics of these variables. Instead, we concentrate on characterizing B_{i+1}, C_{i+1}, assuming that B_i, S_i, T_i, C_i and S_{i+1}, T_{i+1} are already known. Let

$$S = \mathbb{1}(S_i = active \text{ and } S_{i+1} = active) \tag{6.1}$$

and

$$T = \mathbb{1}(T_i = active \text{ and } T_{i+1} = active) \tag{6.2}$$

where $\mathbb{1}(e)$ is the indicator function for the event e. $S = 1$ means that the sensing/ADC unit is active *during* time slot i while $S = 0$ means that the sensing/ADC unit is effectively off during this time slot. Similarly, variable T indicates whether or not the transceiving unit is active during time slot i. Now let

$$L = \begin{cases} B - \max(0, B_i - r) & \text{if } T = 1 \\ B - B_i & \text{otherwise} \end{cases} \tag{6.3}$$

then L is the space left in the buffer during time slot i. During time slot i the sensing/ADC unit will successfully process incoming events if it is in the active mode and there is enough space left in the buffer to store data resulting from event processing. Therefore,

$$C_{i+1} = \mathbb{1}(S = 0 \text{ or } L < A) \times \min(C_i + 1, D) \tag{6.4}$$

At the beginning of time slot i, the sensing/ADC unit has been effectively off for C_i time slots. Given that events happen in each time slot according to an i.i.d. Bernoulli process with probability λ, we can write down the probability of having at least one event to process in time slot i as

$$W(\lambda, C_i) = 1 - \prod_{j=0}^{C_i} (\lambda P_m(C_i - j) + 1 - \lambda) \tag{6.5}$$

where $P_m(t)$ is the probability that an event happening in time slot a disappearing before the beginning of time slot $a + t$. We have

$$P_m(t) = \begin{cases} 1 & \text{if } t \geq D + 1 \\ \dfrac{t - 0.5}{D} & \text{if } 1 \leq t \leq D \\ 0 & \text{if } t \leq 0 \end{cases} \tag{6.6}$$

Now the buffer occupancy at the beginning of time slot $i + 1$ can be written as

$$B_{i+1} = \begin{cases} B - L & \text{if } C > 1 \\ B - L + E & \text{otherwise} \end{cases} \tag{6.7}$$

where $E = A$ with probability $W(\lambda, C_i)$ and $E = 0$ with probability $1 - W(\lambda, C_i)$.

Cost of Actions In a particular time slot i, given the system state $\mathbf{X}_i = (G_i, B_i, S_i, T_i, C_i)$, we define the cost function of a control action U_i as

$$C(\mathbf{X}_i, U_i) = \beta C_m(\mathbf{X}_i, U_i) + C_p(\mathbf{X}_i, U_i) \tag{6.8}$$

Here $C_m(\mathbf{X}_i, U_i)$ is the expected number of events that are missed by the sensing unit during time slot i, and $C_p(\mathbf{X}_i, U_i)$ is the total power consumed by the sensing and transceiving units. β is a positive weighting factor that gives the priority of reducing number of missing events over conserving power. In particular, by increasing β, we tend to keep the system component in the active modes more frequently to lower the event loss rate at the expense of using more power. On the other hand, for smaller values of β, the average consumed power will be reduced at the cost of missing more events.

In (6.8), $C_p(\mathbf{X}_i, U_i)$ can be readily calculated given the power consumed in each mode and in the mode-transitioning process of each unit (see Sections 6.2.2 and 6.2.2). The expected number of events missed during time slot i can be calculated as follows:

$$C_m(\mathbf{X}_i, U_i) = \mathbb{1}(C_{i+1} > 0) \times \sum_{j=0}^{C_i} \lambda \left(P_m(C_i + 1 - j) - P_m(C_i - j) \right)$$

$$= \mathbb{1}(C_{i+1} > 0) \times \lambda P_m(C_i + 1) \tag{6.9}$$

Note that C_{i+1} and $P_m(C_i + 1)$ are calculated using (6.4) and (6.6).

Problem Objective   Let $\pi = \{\mu_0, \mu_1, \mu_2, \ldots\}$ be a policy that maps system states into transmission rates for each time slot i, that is, $U_i = \mu_i(X_i)$. We then have the optimization problem

$$\pi^* = \arg\min_{\pi} J_{\text{avr}}(\pi)$$

$$= \arg\min_{\pi} \left\{ \limsup_{T \to \infty} \frac{1}{T} \mathbb{E} \left\{ \sum_{i=0}^{T-1} C(X_i, U_i) \right\} \right\} \tag{6.10}$$

This problem can be solved efficiently using dynamic programming techniques [24]. In particular, for our system in which all the states are connected, there exists a stationary policy π, that is, $\mu_i \equiv \mu$ for all i, which is the solution of (6.10).

6.2.4 Numerical Results and Discussions

In this section, we study the performance of the adaptive control policies obtained by solving the MDP formulated in Section 6.2.3. In particular, we look at the following adaptive policies:

- *Fully Adaptive*: The operating modes of both sensing and transmitting units are adapted to the channel and buffer conditions. The sensing/ADC unit supports active and off modes while the transmitting unit can operate in active, idle, and off modes. It is expected that this scheme will perform best over all adaptive policies being considered.
- *No Off Mode for Sensor*: The transmitting unit is adaptive to the buffer and channel states. The sensing unit is always active. Our purpose of looking at this scheme is to determine whether it is essential to have a sensing component that can adapt to the system conditions.
- *No Off Modes for Sensor and Transmitter*: The sensing and transmitting units only support two modes, that is, active and idle. By comparing the performance of this scheme to that of the fully adaptive scheme, we can determine the importance of allowing the sensing and transmitting components to turn off when the system work load is low.

The policies described above are obtained by solving the optimization problem formulated in the previous section. Different adaptive schemes are obtained by defining the cost function in different ways. For example, to prohibit the off mode, we simply set the cost of going from active to off mode to be very large.

The channel is modeled as a two-state Markov channel (sometimes called the Gilbert–Elliot channel) shown in Figure 6.4. The channel can be either in a *good* or *bad* state and

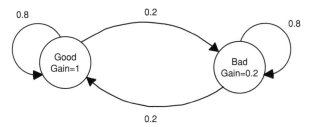

Figure 6.4 State diagram of two-state Markov channel.

will transition between the states with the probabilities shown in the state diagram. The channel gain in the bad state is $\gamma_0 = 0.2$ and in the good state is $\gamma_1 = 1$. The other system parameters are as follows.

Event Process

- Events happen in each time slot according to a Bernoulli distribution with parameter $\lambda = 0.2$.
- Each event lasts for a random period that is uniformly distributed within $(0, D]$, with $D = 2$ and 4 time slots.

Sensing/ADC Unit

- Power consumed in off and active modes are $P_{so} = 0$ and $P_{sa} = 10$, respectively.
- Transition latency from off to active state is $L_{so} = 1$ time slot.

Storage Unit

- Each event requires $A = 2$ storage units (or packets) in the buffer.
- The buffer length is $B = 4$ packets.

Transceiving Unit

- Power consumed in active modes are $P_{ta} = 6$ and $P_{tc} = 4$. Power consumed in idle mode is $P_{ti} = 4$.
- Transition latency from off to active state is $L_{to} = 3$ time slots and from idle to active state is $L_{ti} = 1$ time slot.
- When the transceiving unit is active, data is transmitted out of the buffer at the rate of $r = 1$ packets per time slot. Note that the transceiver can operate in the active mode when there is at least one packet in the buffer.

The system parameters are summarized in Table 6.1. In Figures 6.5 and 6.6, the performance of the those schemes described above are given. Note that the values of latencies are in time slots and the values of power levels are relative. For example, the ratios of power consumed in active, idle, and off states of the transceiving unit is $10 : 4 : 0$. How large one unit of power is depends on the actual system. For example, if in a certain sensor, the power consumed by the transceiving unit in the active mode is 100 mW, then each unit of power

Table 6.1   System Parameters for Numerical Results

λ	D	A	B	γ_0	γ_1	r	P_{ta}
0.2	2, 4	2	4	0.2	1	1	6
P	P	P	L	L	P	P	L
4	4	0	3	1	10	0	1

here corresponds to 10 mW. So the x axis of Figures 6.5 and 6.6 should be interpreted accordingly.

It can be observed from Figures 6.5 and 6.6 that by adapting the operating modes of the sensing/ADC and transceiving units to the system condition, that is, the event arrival process, buffer occupancy, and channel fading, the performance of the sensor node is improved. As it is expected, the fully adaptive scheme always offers the best performance in terms of event-missed rate versus average power consumed. It is also evident that the relative performance gain is higher when the average consumed power decreases (and the corresponding event-missed rate increases). When plenty of power is available, it is not essential to adapt to the system conditions. As can be seen, when the average power used is large, the performance of the three schemes tested are relatively close.

6.2.5 Sensor Networks

From the previous sections, we know how to optimally control the sensing and transmitting units of a single sensor. We now consider how to relate the individual control policy to

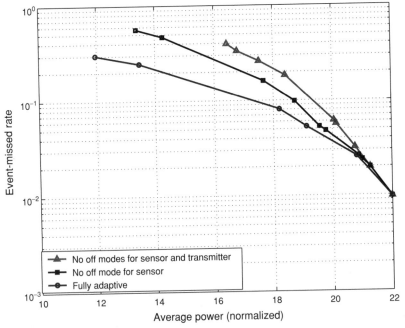

Figure 6.5 Event-missed rate versus (normalized) average power for single sensor. System parameters are as in Table 6.1, channel in Figure 6.4, and maximum event lifetime of $D = 2$ time slots.

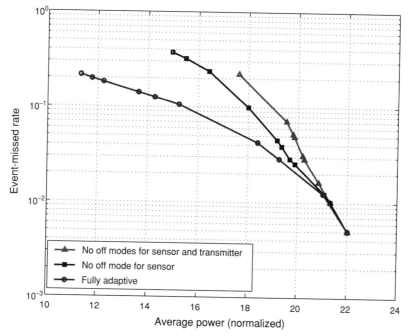

Figure 6.6 Event-missed rate versus (normalized) average power for single sensor. System parameters are as in Table 6.1, channel in Figure 6.4, and maximum event lifetime of $D = 4$ time slots.

the sensor network system shown in Figure 6.1. The sensor network system has N sensors that monitor the environment and report information back to a cluster head. To ensure low distortion, it may be that an event needs to be captured by more than one sensor. We assume that an event is successfully detected if at least M out of the N sensors capture it.

The parameter M can be determined from the system-level performance requirements, for example, probability of missed detection. We link the system-level performance requirements, in terms of the event-missed rate, to a node-level performance requirement. Then, the optimization problem we have formulated so far can be applied to control each individual sensor, so that each sensor guarantees its individual node-level event-missed rate and contributes to guarantee an overall system-level event-missed rate.

Let P_n and P_s be the node-level and system-level event-missed rates, respectively. Let $\mathbf{V}$ be the set that contains all subsets of N sensors. Define the function $P : \mathbf{V} \longrightarrow [0, 1]$ as

$$P(\mathbf{v}) = \Pr\{\text{all sensors in } \mathbf{v} \text{ miss an event}\}, \mathbf{v} \in \mathbf{V} \qquad (6.11)$$

There are $Q = \binom{N}{N-M+1}$ distinct subsets in $\mathbf{V}$ of cardinality $(N - M + 1)$, and we denote these Q subsets as $\mathbf{v_1}, \mathbf{v_2}, \ldots \mathbf{v_Q}$. We also denote by $\mathbf{v_k}$ the event that all sensors in the set $\mathbf{v_k}$ miss detection.

Recall that an event is considered missed if there are less than M out of N sensors catching it. This means that

$$P_s = \Pr\{\text{at least } N - M + 1 \text{ sensors miss the event}\}$$

$$= \Pr\{\mathbf{v_1} \cup \mathbf{v_2} \cup \cdots \cup \mathbf{v_Q}\} \qquad (6.12)$$

By the inclusion–exclusion principle, P_s can be written as

$$P_s = \sum_{i=1}^{Q} P(\mathbf{v_i}) - \sum_{i_1=1}^{Q-1} \sum_{i_2=i_1+1}^{Q} P(\mathbf{v_{i_1}} \cap \mathbf{v_{i_2}})$$

$$+ \sum_{i_1=1}^{Q-2} \sum_{i_2=i_1+1}^{Q-1} \sum_{i_3=i_2+1}^{Q} P(\mathbf{v_{i_1}} \cap \mathbf{v_{i_2}} \cap \mathbf{v_{i_3}}) - \cdots + (-1)^{Q-1} P(\mathbf{v_1} \cap \mathbf{v_2} \cap \cdots \mathbf{v_Q}), \quad (6.13)$$

where $\cap$ denotes set intersections. By the union bound, P_s can be upper bounded as

$$P_s \leq \sum_{i=1}^{Q} P(\mathbf{v_i}) \tag{6.14}$$

Since all sensors in the system have identical buffer lengths, i.i.d. channel processes, and adaptive policies, their event-missed processes are identically distributed. Therefore, (6.14) can be simplified as

$$P_s \leq Q \times P(\mathbf{v_1}) \leq Q \times P_n \tag{6.15}$$

where the second inequality follows from Bayes theorem. So (6.15) gives us an upper bound for the system event-missed rate P_s in terms of the node event-missed rate P_n. This is in fact a loose bound but can be tighten if we can characterize the correlation among the system states of different sensors.

Before moving on, let us discuss in more detail the performance requirements in the sensor network. The performance requirements can be specified in terms of probability of missed detection at the decision center, which determines the tolerable distortion between the reconstructed data at the decision center and the original source. In [20], the authors consider the CEO problem with unreliable links (subject to erasure) between the rate-constrained sensors and the decision center. For this situation, they find an achievable rate distortion region, when K out of N sensors are able to relay their data to the center. So given N rate-constrained sensors and a maximum tolerable distortion, we can determine the minimum number of sensors M which need to relay their measurement to the center. Furthermore, [20] proves that if out of N sensors, M of them can relay their observations to the central node, then the achievable distortion is the same as if there are only M sensors in the system and their links to the central node are always reliable. This interesting observation justifies turning each sensor on and off independently, without taking other sensors' actions into account.

We now take a look at the numerical results for the performance of a system with four sensors. Figure 6.7 plots the performance in terms of the system event-missed rate versus average power consumed for different requirements on the number of sensors needed to successfully catch an event. All four sensors have the same operating parameters given in Table 6.1. The channels seen by the four sensors are assumed to be i.i.d. and represented in Figure 6.4. The simulations were run for 10 million time slots, which corresponds to about 2 million events being generated. We also plot the corresponding upper bound for the system event-missed rate for each of the cases. As expected, when the number of sensors required to catch each event increases, the system event-missed rate increases. We would like to briefly comment on the nonmonotonic behavior seen in the performance. We believe that

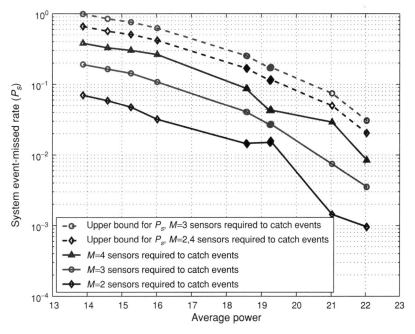

Figure 6.7 Event-missed rate versus average power. System consists of $N = 4$ sensors, with parameters given in Table 6.1, channel in Figure 6.4, and maximum event lifetime of $D = 4$ time slots.

this is due to the fact that the missed event processes across sensors are correlated and the individual control policy does not account for this phenomenon. An interesting avenue of further research is to characterize this correlation and use it in designing the control policy.

The fact that a system-level requirement can be translated into a node-level requirement points to a design methodology for sensor networks. The first step is to build a database that maps the system-level event-missed rate to the node-level event-missed rate. This database can be obtained either through analysis (in certain instances) or extensive simulation. In the second step, given a system-level requirement, we determine the corresponding node-level requirement with the help of the database. Finally, the approach developed in the previous sections can be applied to optimally control each sensor node so that the node-level requirement is met. As an example, in Figure 6.8, we plot the system-level event-missed rate as a function of the node-level event-missed rate for the system of four sensors described above. Now suppose we are given a system specification requiring that at least two out of the four sensors must catch the event and that, at most, 3 in 100 events can be missed ($P_s = 3 \times 10^{-3}$). Then from Figure 6.8, we see that a node capable of catching at least 1 in 10 events ($P_n = 10^{-2}$) is sufficient and the optimal policy in (6.10) guarantees that each node can meet that requirement for the maximum time possible.

6.2.6 Conclusion

In this section, we looked at strategies to jointly control the modes of operation of the sensing and transceiving units of individual nodes in wireless sensor networks. The strategies take advantage of the event and channel statistics to conserve energy while providing a certain

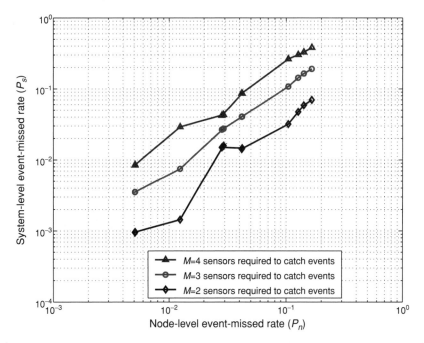

Figure 6.8 System-level event-missed rate versus node-level event-missed rate. System consists of $N = 4$ sensors, with parameters given in Table 6.1, channel in Figure 6.4, and maximum event lifetime of $D = 4$ time slots.

QoS. We argued that increasing the lifetime of individual nodes will increase the lifetime of the sensor network. In that light, we showed that a system-level performance requirement can be filtered down to the node level. Our simulation results show the importance of exploiting the idle and off modes in sensor nodes to conserve energy when the workload is low. One advantage of controlling nodes individually is scalability, which is important in a network with potentially thousands of nodes.

It is noted that our adaptive control schemes operate the sensor nodes in statistical idling manners, resulting in uniform energy consumption over all the nodes. This leads to graceful network energy degradation, which is a desirable property for many mission-critical sensor networks.

It may be of interest to consider approaches in which multiple sensor nodes interact and cooperate (in a distributed manner) in carrying out the sensing tasks. While this might improve system performance, it increases the complexity of the computation and message passing needed for cooperation. The challenge then is to design efficient, scalable, cooperative algorithms.

6.3 SENSOR PLACEMENT AND LIFETIME OF WIRELESS SENSOR NETWORKS: THEORY AND PERFORMANCE ANALYSIS

Ekta Jain and Qilian Liang

Increasing the lifetime of energy-constrained wireless sensor networks is one of the key challenges in sensor network research. In this section, we present a *bottom-up* approach to evaluating the lifetime performance of networks employing two basic sensor placement schemes: square grid and hex grid. Lifetimes of individual sensor nodes as well as lifetimes of networks are modeled as random variables, and their probability density functions (pdf) are obtained theoretically. Reliability theory is used for the network lifetime analysis and provides a methodology for similar studies. Simulation results show that the actual pdf's are very close to those obtained theoretically. The theoretical results provided in this section will serve as a basis for other related research such as analysis of other sensor placement schemes, lifetime and sensor density trade-off study, and performance of energy-efficiency-related algorithms.

6.3.1 Background Information

Sensor networks represent a significant advancement over traditional invasive methods of monitoring and are more economical for long-term data collection and monitoring when compared to the traditional personnel-rich methods. Although most military applications require random deployment of sensor nodes, a number of nonmilitary applications allow the explicit placement of sensors at specific locations. Such placement-friendly sensor networks are widely used for infrastructure security, environment and habitat monitoring, traffic control, and the like [25]. An in-depth study of a real-world 32-node habitat monitoring sensor network system presented in [26] is deployed on a small island off the coast of Maine to study nesting patterns of petrels. Another such application [27] involves the use of sensors in buildings for environmental monitoring, which may include chemical sensing and detection of moisture problems. Structural monitoring [28] and inventory control are some other applications of such networks.

Wireless sensor networks comprise of small, energy-constrained sensor nodes. Each node basically consists of a single or multiple sensor module, a wireless transmitter–receiver module, a computational module, and a power supply module. These networks are usually employed to collect data from environments where human intervention after deployment, to recharge or replace node batteries, may not be feasible, resulting in limited network lifetime. Most applications have prespecified lifetime requirements, for instance, the application in [26] has a lifetime requirement of at least 9 months. Thus estimation of lifetime of such networks prior to deployment becomes a necessity. Prior works on evaluation of lifetime have considered networks where sensor nodes are randomly deployed. Bhardwaj and co-workers) [29] give the upper bound on lifetime that any network with the specified number of randomly deployed nodes, source behavior, and energy can reach while [30] discusses the upper bounds on lifetime of networks with cooperative cell-based strategies.

We are concerned with networks that allow the placement of sensor nodes in specific positions. These positions depend on a number of factors that include, but are not limited to, sensing and communication capabilities of the sensor nodes, sensor field area, and cost of deployment, which is connected to the density of the network.

In this section, we deal with the issues of sensor placement and lifetime estimation and present a bottom-up approach to lifetime evaluation of a network. As a first step in this direction we investigate the lifetime behavior of a single sensor node and then apply this knowledge to arrive at the network lifetime behavior for two basic placement schemes. We believe that these simple schemes can serve as a basis for the evaluation of more complex schemes for their lifetime performance prior to deployment and help justify their deployment costs. We remark that our investigation can be applied to a network employing any routing protocol and energy consumption scheme. Although in a less direct way, our analysis can also be used to assess various energy-efficiency-related algorithms for their lifetime performance. Also, since our analysis incorporates lifetime evaluation for minimum as well as high-density placement, the trade-off between cost of deployment and lifetime can be studied because higher density implies higher cost. Our analytical results are based on the application of reliability theory. These results are supported by extensive simulations, which validate our analysis.

6.3.2 Preliminaries

Basic Model   Consider identical wireless sensor nodes placed in a square sensor field of area A. All nodes are deployed with equal energy. Each sensor is capable of sensing events up to a radius r_s, the sensing range. We also define a communication range r_c, beyond which the transmitted signal is received with signal-to-noise ratio (SNR) below the acceptable threshold level. We assume the communication range r_c to be equal to the sensing range r_s. The communication range depends on the transmission power level, gains of the transmitting and receiving antennae, interference between neighboring nodes, path loss, shadowing, multipath effects, and the like. Direct communication between two sensor nodes is possible only if their distance of separation r is such that $r \leq r_c$. We call such nodes *neighbors*. Communication between a sensor node and its nonneighboring node is achieved via peer-to-peer communication. Thus the maximum allowable distance between two nodes who wish to communicate directly is $r_{\max} = r_c = r_s$. A network is said to be deployed with minimum density when the distance between its neighboring nodes is $r = r_{\max}$.

Placement Schemes   The simplest placement schemes involve regular placement of nodes such that each node in the network has the same number of neighbors. We arrive at two basic placement schemes by considering cases where each sensor nodes has four and three neighbors. This leads us to the *square-grid* and *hex-grid* placement schemes shown in Figures 6.9a and 6.9b, respectively. A sensor node placement scheme that uses two neighbors per sensor node has been described in [31]. We believe that these three elementary placement schemes can serve as basis for other placement schemes because a placement scheme of any complexity can be decomposed into two-neighbor, three-neighbor, and four-neighbor groups. Both grids shown have the same number of nodes,[1] and nodes in both grids are equidistant from their respective neighbors (with distance of separation r).

Coverage and Connectivity   Coverage and connectivity are two important performance metrics of networks and hence a discussion on them becomes imperative before the lifetime of the network can be defined. Coverage scales the adequacy with which the network covers the sensor field. A sensor with sensing range r_s is said to cover or sense a circular region of radius r_s around it. If every point in the sensor field is within distance r_s from at least one sensor node, then the network is said to provide complete or 100% coverage.

Various levels of coverage are acceptable depending on the application. In critical applications, complete coverage is required at all times. Any loss of coverage leads to a sensing gap in the field. Such gaps cause breach of security in case of surveillance applications. Also, in applications that require data with high precision, a sensing gap leads to inaccuracies.

---

[1] The hex grid has lower density than the square grid. With 36 nodes deployed, the network with square grid covers an area of $25r^2$, and the hex grid covers an area of $48r^2$, almost double that of the square grid.

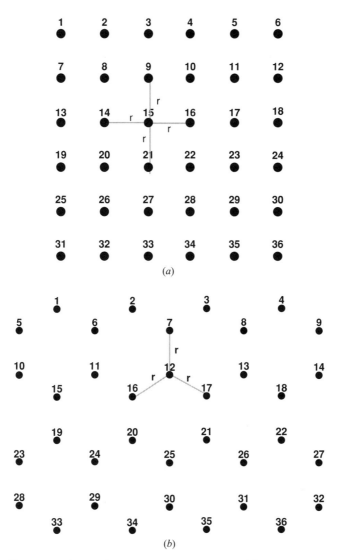

Figure 6.9 Placement schemes for a 36-node sensor network: (*a*) square grid and (*b*) hex grid.

For such networks any loss of coverage renders the network nonfunctional. In some other applications a small loss of coverage may be acceptable.

Connectivity scales the adequacy with which nodes are able to communicate with their peers. One of the strengths of sensor networks arises from their ability to aggregate data collected from different sensor nodes. This requires adequate communication between sensor nodes. Any node should be able to communicate with any other node for proper functioning of the network. If a single node gets isolated due to failure of all its neighbors, it will be unable to communicate with the rest of the network. If a large number of nodes fail due to lack of energy, a part of the network may get completely disconnected from the rest. In our analysis we assume that only 100% connectivity is acceptable and the network fails with any loss of connectivity.

An example of a sensor placement scheme that concentrates mainly on coverage as its parameter of interest can be found in [32], where a sensor placement algorithm for grid coverage has been proposed. In our analysis we require the network to provide complete coverage and connectivity. We give equal importance to both parameters and declare the network nonfunctional if either of them falls below their desired levels.

Lifetime   The basic definition of lifetime, or more precisely the postdeployment active lifetime of a network, is the time measured from deployment until network failure. Based on the levels of coverage and connectivity required to deem a network functional, network failure can be interpreted in different ways. Since only complete coverage and connectivity are acceptable to us, network failure corresponds to the first loss of coverage or connectivity.

In this section, we concentrate on finding the minimum lifetime of a network, the worst-case scenario. To be able to evaluate this minimum lifetime, we need to know the lifetime of a single sensor node, the minimum number of node failures that cause network failure, and the positional relationship[2] between the failed nodes.

Minimum Density Networks   Consider the square grid and the hex grid deployed with minimum density. Both schemes survive the failure of a single node without loss of either connectivity or coverage, implying that the minimum number of node failures that can lead to network failure is greater than one. Failure of any two neighboring nodes causes loss of coverage and hence network failure as indicated in Figures 6.10a and 6.10b.

Thus the minimum number of node failures that cause network failure is two, and these two nodes must be adjacent to each other (neighbors). A network may undergo multiple node failures and still be connected and covered if any of the failed nodes are not neighbors. But the absolute minimum number of node failures that can cause network failure is two.

Network Lifetime for Dense Placement   Let N_{min} be the number of sensor nodes deployed in a sensor field of area A with minimum density. In a minimum density network, the separation between neighbors is r_{max}, the maximum range of the sensor nodes. A network deployed with higher density would require a larger number of nodes N, $N > N_{min}$ and smaller separation between neighboring nodes r, $r < r_{max}$. Since the range of a sensor node r_{max} is greater than the distance to its immediately adjacent nodes, it is now possible to communicate not only with the nodes immediately adjacent to it but also with other nodes that fall in its range. In other words, each node now has more numbers of neighbors that any node in a minimum density network.

We discussed earlier that the minimum lifetime of a square-grid or hex-grid network with $r = r_{max}$ (minimum density) is the time to failure of two neighboring nodes. High-density networks can lose $(N - N_{min})$ nodes and still be deployed with minimum density and fail only after the further loss of a minimum of two neighboring nodes. Since the network after $(N - N_{min})$ node failures is deployed with minimum density, the time taken for the failure of any two neighboring nodes would be equal to the minimum lifetime of a minimum density network (T_{md}), which was discussed in the preceding section. If t_{dense}, the time taken for the failure of $(N - N_{min})$ nodes can be found, we define the

---

[2] Positional relationship between two nodes can be that the two nodes are diagonal, adjacent, or completely unrelated.

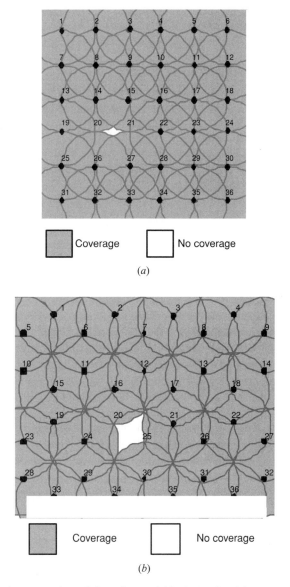

Figure 6.10 Loss of coverage due to failure of two neighboring nodes: (*a*) square grid: Failure of nodes 20 and 21 causes loss of coverage. (*b*) Hex grid: Failure of nodes 20 and 25 causes loss of coverage.

lifetime of a high-density network (T_{hd}) as

$$T_{hd} = t_{dense} + T_{md} \qquad (6.16)$$

6.3.3 Node Lifetime Evaluation

We begin our analysis with the estimation of the lifetime of a single sensor node. A node is said to have m possible modes of operation, and at any given time the node is in one of these m modes. Let w_i be the fraction of time that a node spends in the ith mode, where

w_i's are such that

$$\sum_i w_i = 1 \qquad i = 1, 2, \ldots, m \tag{6.17}$$

The w_i's do not remain constant in general and vary for different networks scenario's. Hence they are modeled as random variables that take values from 0 to 1 according to specific probability distribution functions (pdf)'s that we attempt to find.

We assume that the total node energy at the time of deployment is E_{total} and that the node spends a power of P_i watts/unit time, while in the ith mode. If T_{node} is the lifetime of the node, then the node spends $w_i T_{node}$ time units in mode i. We define the lifetime of a node as the time from deployment to when its energy drops below a threshold energy value E_{th}. The following equation follows from this definition and from the fact that Eth is negligible compared to E_{total}:

$$E_{total} - \sum_i w_i P_i T_{node} \geq E_{th}$$

$$T_{node} \geq \frac{E_{total}}{\sum_i w_i P_i} \tag{6.18}$$

We observe from this equation that T_{node} is a function of the random variables w_i and hence is a random variable itself. This is an important observation as it implies that since the modes of the various nodes cannot be known a priori, the lifetime of a single node can be best represented as a random variable that takes different values with certain probabilities defined by its probability density function $f_t(t)$. In order to obtain the pdf of T_{node}, we first find the pdf of $w_i \, \forall i$.

To this purpose we assume that at any given time a node can be in one of its two modes of operation, *active* or *idle*. A node is active when it is either transmitting or receiving and is idle otherwise. If p is the probability of node being active, $1 - p$ is the probability of the node being idle at any given time.

$$\text{Pr}\{\text{node is active}\} = p \tag{6.19}$$

$$\text{Pr}\{\text{node is idle}\} = 1 - p \tag{6.20}$$

Note that the actual values of p will depend on the energy-efficiency-related algorithms used such as energy-efficient routing protocols and energy consumption schemes. Let w_1 and w_2 be the fraction of time the node is in active and idle mode, respectively. Since we assume only two modes of operation, $w_2 = 1 - w_1$. The random variable w_1 is clearly binomial in nature, with probability of success, that is, probability of node in active mode being p. We observe the node over T time units. Since w_1 has a binomial distribution, the probability of the node being active for x time units of a total of T time units is give by

$$P\{w_1 = x\} = C_T^x p^x (1 - p)^{T-x} \tag{6.21}$$

As T becomes large, the binomial distribution can be approximated to a normal distribution with mean $\mu = Tp$ and variance $\sigma^2 = np(1 - p)$. Since $w_2 = 1 - w_1$, w_2 also follows normal distribution with mean $\mu = T(1 - p)$ and variance $\sigma^2 = np(1 - p)$. Hence we conclude that the fraction of time that the node spends in any mode follows the normal distribution.

We observe from Eq. (6.18) that the denominator of the expression is a normal random variable because it is the sum of m normal random variables. This helps us draw an important conclusion that the reciprocal of the lifetime of a node follows the normal distribution.

6.3.4 Network Lifetime Analysis Using Reliability Theory

Since the lifetimes of an individual node is not constant but a random variable, it follows that the network lifetime is also a random variable. We apply reliability theory to find the distribution of the network lifetime, given the node lifetime distribution. The following section treats the basics of reliability theory before going on to its application.

Reliability Theory   Reliability theory is concerned with the duration of the useful life of components and systems of components [33–35]. We model the lifetime (of individual nodes and the network) as a continuous nonnegative random variable T. Below we describe three of the many ways of defining this nonnegative random variable T.

Probability Density Function   The pdf gives the probability of the random variable taking a certain value [36]. It is represented as $f(t)$ and has a probabilistic interpretation

$$f(t)\, \Delta t = P\,[t \leq T \leq t + \Delta t] \tag{6.22}$$

for small values of Δt. All pdf's must satisfy two conditions, $\int_0^\infty f(t)\, dt = 1$ and $f(t) \geq 0, \forall t$.

The lifetime of any item can be thought of as the time until which the item survives, or the time at which the item fails, that is, time to failure. The pdf represents the relative frequency of failure times as a function of time. We have the distribution of the lifetime of a node in the form of its pdf, in that we know that the reciprocal of the node lifetime has a normal pdf.

Cumulative Distribution Function (cdf)   The cdf corresponding to the pdf $f(t)$ is denoted by $F(t)$ and is very useful as it gives the probability that a randomly selected component or system will fail by time t. It can be expressed in three different ways:

- $F(t)$ = the area under the pdf $f(t)$ to the left of t.
- $F(t)$ = the probability that a single randomly chosen new component will fail by time t.
- $F(t)$ = the proportion of the entire population that fails by time t.

The relationship between the cdf and pdf of a random variable is given below:

$$F(t) = \int_0^t f(s)\, ds \tag{6.23}$$

The pdf and the cdf are the two most important statistical functions in reliability theory. They are closely related as shown by Eq. (6.23). They give a complete description of the probability distribution of the random variable. The knowledge of these two functions enables us to find any other reliability measure.

From the relationship between the cdf and the pdf defined in Eq. (6.23), and the interpretation of the pdf as the relative frequency of failure times, we interpret the cdf of the lifetime T as the probability that the item in question will fail before the associated time value, t. As a result the cdf is sometimes called the *unreliability* function.

Survivor Function   The *survivor function* $S(t)$, also known as the *reliability function* $R(t)$ can be derived using the interpretation of the cdf as the unreliability function. It is defined as the probability that a unit is functioning at any time t:

$$S(t) = P[T \geq t] \quad t \geq 0 \tag{6.24}$$

Since a unit either fails or survives, and one of these two mutually exclusive alternatives must occur, we have

$$S(t) = 1 - F(t) = 1 - \int_0^t f(s)\,ds \tag{6.25}$$

All survivor functions must satisfy three conditions: $S(0) = 1$, $\lim_{t \to \infty} S(t) = 0$, and $S(t)$ is nonincreasing.

There are two interpretations of the survival function. First, $S(t)$ is the probability that an individual item is functioning as time t. This interpretation will be used later in finding the lifetime distribution of the network from the lifetime distribution of its constituent nodes. Second, if there is a large population of items with identical distributed lifetimes, $S(t)$ is the expected fraction of the population that is functioning at time t. This interpretation will be used in finding the time taken for $(N - N_{min})$ node failures for estimation of the minimum lifetime of dense networks.

We obtain the pdf of the lifetime of a single sensor node from Section 6.3.3. Using Eqs. (6.23) and (6.25) we obtain the survivor function $S(t)$ of a single sensor node. Note that all sensor nodes are assumed to be identical with survivor functions $S(t)$.

System Reliability   The main objective of system reliability is the construction of a distribution that represents the lifetime of a system based on the lifetime distributions of the components from which it is composed. To accomplish this, we consider the relationship between components. This approach to finding system lifetime has the inherent advantage that it is often easier and cost effective to extensively test a single component or subsystem rather the whole system.

Reliability Block Diagram   Reliability block diagram (RBD) is a graphical representation of the components of the system and provides a visual representation of the way components are reliability-wise connected. Thus the effect of the success or failure of a component on the system performance can be evaluated.

Consider a system with two components. If this system is such that a single component failure can render the system nonfunctional, then we say that the components are reliability-wise connected in series. If the system fails only when both its components fail, then we say that the components are reliability-wise connected in parallel. Note that the physical connection between the components may or may not be different from their reliability-wise

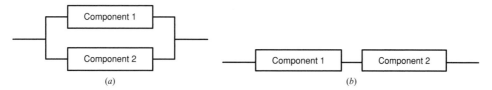

Figure 6.11 Reliability block diagrams (RBD) for a system of two components. (*a*) RBD with series-connected components and (*b*) RBD with parallel connected components.

connection. The RBDs for both cases are given in Figure 6.11. If $S_1(t)$ and $S_2(t)$ are the survival functions of the two components, the system survival functions $S(t)_{\text{series}}$ and $S(t)_{\text{parallel}}$, for the series and parallel cases are given in Eqs. (6.26) and (6.27), respectively:

$$S_{\text{series}}(t) = S_1(t)S_2(t) \tag{6.26}$$

$$S_{\text{parallel}}(t) = 1 - [(1 - S_1(t))(1 - S_2(t))] \tag{6.27}$$

Any complex system can be realized in the form of a combination of blocks connected in series and parallel, and the system survival function can be obtained by using Eqs. (6.26) and (6.27). In our analysis, the network is the system under consideration and the sensor nodes are the components of the system. All sensor nodes are assumed to have identical survival functions $S(t)$, and their failures are supposed to be independent on one another.

Lifetime of Minimum Density Networks

Square Grid   As defined in Section 6.3.2, the minimum network lifetime is the time to failure of any two neighboring nodes. We know that the failure of any single node does not cause network failure. The failure of any node coupled with the failure of any of its neighbors causes network failure. Using this definition we build the RBD for the square grid as shown in Figure 6.12.

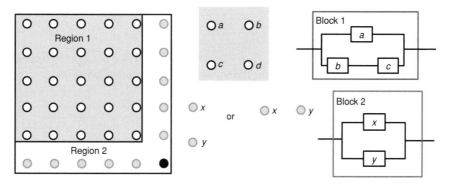

Figure 6.12 RBD of a single node in a square grid. Nodes belonging to region 1 are modeled as block 1 and nodes belonging to region 2 are modeled as block 2. The network RBD consists of $(\sqrt{N_{\text{min}}} - 1)^2$ block 1's in series with $2(\sqrt{N_{\text{min}}} - 1)$ block 2's.

Figure 6.12 shows the RBD block for a single node in the network. A node can be modeled in two ways depending on its position in the sensor field. This distinction, based on its position, is made due to a simple observation that nodes at the right edge of the sensor field (region 2) do not have any right neighbor (node b) as opposed to nodes in region 1. Also, nodes at the bottom edge of the sensor field (region 2) do not have a bottom neighbor (node c) as opposed to the nodes in region 1. From simple calculations we see that an N node network has $(\sqrt{N_{\min}} - 1)^2$ sensor nodes in region 1 and $2(\sqrt{N_{\min}} - 1)$ nodes in region 2. Hence, the network RBD consists of $(\sqrt{N_{\min}} - 1)^2$ block 1's in series with $2(\sqrt{N_{\min}} - 1)$ block 2's. Note that as every node in a square grid, node a has four neighbors, but its relationship with only two neighbors is modeled in its RBD block. This is because the relationship with the other two neighbors will be modeled when their RBD blocks are constructed. If this is not followed, then the relationship between every node–neighbor pair will be modeled twice.

If $S_a(t)$, $S_b(t)$, and $S_c(t)$ are the survival functions of nodes a, b, and c, respectively, then the survival function of block 1, s_{block1} is[3]

$$s_{\text{block1}} = 1 - (1 - s_a)(1 - s_b s_c) \tag{6.28}$$

Since all nodes are identical, they have identical survivor functions s. Hence (6.28) is simplified to

$$s_{\text{block1}} = 1 - (1 - s)(1 - s^2)$$
$$= s + s^2 - s^3 \tag{6.29}$$

A similar analysis is carried out for nodes belonging to region 2. If s_x and s_y are the survivor functions of nodes x and y, respectively, then the survivor function of block 2, s_{block2} is

$$s_{\text{block2}} = 1 - (1 - s_x)(1 - s_y) \tag{6.30}$$

Since all nodes are identical, they have identical survivor functions s. Hence (6.30) is simplified to

$$s_{\text{block2}} = 1 - (1 - s)(1 - s)$$
$$= 2s - s^2 \tag{6.31}$$

Since the network RBD consists of $(\sqrt{N_{\min}} - 1)^2$ block 1's and $2(\sqrt{N_{\min}} - 1)$ block 2's in series, the network survivor function for the square-grid placement scheme is

$$s_{\text{network}} = (s_{\text{block}-1})^{(\sqrt{N_{\min}}-1)^2}(s_{\text{block}-2})^{2(\sqrt{N_{\min}}-1)} \tag{6.32}$$

The required cdf and pdf of the network lifetime can be obtained from this survival function using (6.25) and (6.23).

---

[3] For notational convenience we use s_i to represent the survivor function of any node i, instead of $S_i(t)$, without any loss in generality.

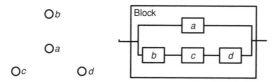

Figure 6.13 RBD block for a single node in the hex grid: The network RBD consists of $N/2$ such blocks in series.

Hex Grid The analysis for the hex grid is carried out on the same lines as that of the square grid. Figure 6.10b shows that as in the case of a square grid, two neighboring node failures cause network failure. The RBD block of a single node is shown in Figure 6.13. Since the relation between a node and all of its neighbors is modeled by its corresponding RBD block, the RBD block for the neighbors is not constructed as this causes the relationship between the nodes to be considered twice. Hence the network RBD consists of only $N_{min}/2$ RBD blocks in series. If nodes a, b, c, and d have identical survival functions, as in the square grid, then the survival function of the block shown in the figure is given by

$$s_{block} = 1 - (1 - s)(1 - s^3) \tag{6.33}$$

Since $N_{min}/2$ such blocks connected in series represent the network, the network survival function is given by:

$$s_{network} = (s_{block})^{N/2} \tag{6.34}$$

Once again after survival function of the network is obtained, the cdf and pdf are obtained from it using (6.25) and (6.23)

Lifetime of High-Density Networks The network lifetime for dense placement of sensor nodes is given by Eq. (6.16) and was defined previously as the time to failure of any two neighboring nodes after $(N - N_{min})$ nodes have failed, where N is the number of nodes deployed and N_{min} is the number of sensor in a minimum-density deployment. We now concentrate on finding the time taken for $N - N_{min}$ node failure, t_{dense}. This situation can be modeled as a k-out-of-$n : F$ system. The k-out-of-n system is a special case of parallel redundancy. An n component system that fails if and only if at least k of the n components fail is called a k-out-of-$n : F$ system.

Recall that the survivor function $S(t)$ of an item was interpreted as the expected fraction of the population that functions at time t. Since t_{dense} is the time taken for $N - N_{min}$ node failures, the fraction of nodes that are functioning at time t_{dense} is N_{min}/N. The time t_{dense} is such that

$$S(t_{dense}) = \frac{N_{min}}{N} \tag{6.35}$$

where $S(t)$ was already defined as the survivor function of any sensor node. Since $S(t)$ can be found with the knowledge of the pdf of the node lifetime, t_{dense} can be evaluated from Eq. (6.35) and the minimum lifetime of a high-density network can be evaluated using (6.16)

6.3.5 Simulation

We ran extensive simulations to:

- Evaluate the pdf of lifetime of a single node.
- Validate the theoretical analysis for the pdf of lifetime of a single node.
- Evaluate the pdf of network lifetime for a square grid.
- Validate the theoretical analysis for the pdf of the network lifetime for a square grid.
- Evaluate the pdf of network lifetime for a hex grid.
- Validate the theoretical analysis for the pdf of the network lifetime for a hex grid.

Node Lifetime Distribution   The first set of simulations was aimed at estimating the pdf of the lifetime of a node, theoretically as described in Section 6.3.3 and through simulations. The node, at any time, can be one of its two modes of operation, active or idle. The probabilities with which a node remains in each of these modes were defined by the network protocol used. Simulation results, which are reported in Figure 6.14, show that the theoretical pdf matches very closely with the actual pdf.

Network Lifetime Distributions   The second set of simulations were aimed at estimating the pdf of the lifetime of the network, when it uses the square-grid placement and the hex-grid placement, using the theoretical analysis described in Section 6.3.4. In both cases 36 nodes were deployed, and the distance between neighboring nodes was assumed to be the same. Equations (6.32) for the square grid and (6.34) for the hex grid were used with $N = 36$. Figures 6.15a and 6.15b show the theoretical and actual pdf obtained for the square grid. Figures 6.16a and 6.16b show the theoretical and actual pdf of a 36-node hex grid. Figures 6.15 and 6.16 indicate that the theoretical results agree closely with the actual results.[4] Also, for networks that fail when the first node dies [31], the network lifetime analysis will be very similar to the work in this section. The network can then be modeled as a simple series-connected block diagram, and the survival function of the network will simply be the product of the survival function of the N nodes that constitute the network.

6.3.6 Conclusion

One of the key challenges in networks of energy-constrained wireless nodes is maximization of the network lifetime. If the application allows placement of sensor nodes, our goal of maximizing the lifetime can be aided by choosing a suitable placement pattern. In this section we evaluated the lifetime of a network employing two simple placement patterns. In evaluating the lifetime, we came up not with any particular value but a probability density function pdf for minimum network lifetime. We followed a *bottom-up* approach by first evaluating the node lifetime pdf and then going on to find the network lifetime pdf. Theoretical results as well as our methodology will enable analysis of other sensor placement schemes, study of lifetime cost trade-offs, and performance analysis of energy-efficiency-related algorithms.

---

[4] Note that edge effects are neglected in the theoretical analysis described in Section 6.3.4.

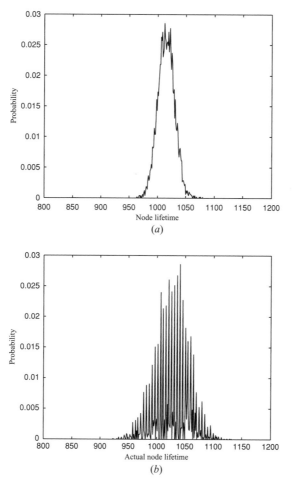

Figure 6.14 Probability density function of the lifetime of a node: (*a*) theoretical pdf, and (*b*) actual pdf.

6.4 ALGORITHMS FOR MAXIMIZING LIFETIME OF BATTERY-POWERED WIRELESS SENSOR NODES

Sridhar Dasika, Sarma Vrudhula, and Kaviraj Chopra

In this section we present a new approach for energy management in a wireless sensor network. We assume that the network consists of a large and dense collection of sensors that are distributed to *cover* a geographical area. Each sensor is powered by a battery and hence has a limited energy capacity. We present two novel approaches to maintain the maximum possible coverage for the longest duration. The algorithms determine a schedule for transitioning sets of sensors between active and inactive states while satisfying user specified performance constraints. We use an efficient and compact representation of *all possible covers* and develop algorithms for maximizing the lifetime of the network by switching between sensor covers. The algorithms take into account the transmission/reception costs of sensors, a user-specified quality constraint, and a novel model that accounts for the

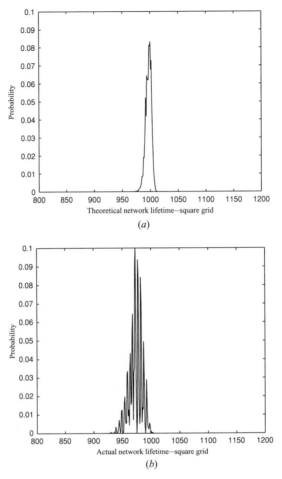

Figure 6.15 Probability density function of the network lifetime employing the square-grid placement scheme: (*a*) theoretical pdf and (*b*) actual pdf.

rate-dependent capacity effect and *charge recovery* effect of batteries during idle periods. Two different scenarios are examined: (1) a centralized scheme, referred to as central sleep state control, in which a base station decides on the covers to be activated and deactivated, and (2) a distributed scheme called distributed sleep state control, in which the decision to be active or inactive is made by each sensor based on local neighborhood information. The objective is to maximize the network lifetime while maintaining the maximum possible coverage. We include experimental results that demonstrate (1) the efficiency of the representation for large networks, (2) the improvement in network lifetime due to battery charge recovery, and (3) the improvement in network lifetime when the proposed sensor management schemes are incorporated into existing routing protocols.

6.4.1 Background and Motivation

Advances in processor, memory, and wireless communication technology have made it possible to manufacture small and inexpensive units capable of sensing, data processing,

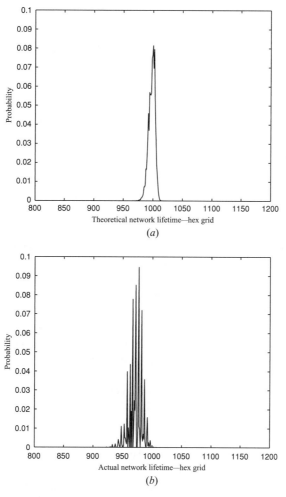

Figure 6.16 Probability density function of the network lifetime employing the square-grid placement scheme: (*a*) theoretical pdf and (*b*) actual pdf.

and wireless communication. A large network of such units, linked by radio frequency (RF), can be deployed over a geographical area to sense, process, and report some phenomena. Wireless sensor networks (SN) are envisioned to be used for a wide range of applications such as battlefield surveillance, environment and health monitoring, failure detection and diagnosis, inventory tracking, homeland security, and many others [37, 38].

Each individual sensor has one or more sensing devices, a processor with memory, a transceiver, a power unit, and is capable of covering only a small portion of the complete sensing region R. The sensing unit converts the analog-sensed data into digital signals. The processing unit executes algorithms that are used for the sensors to cooperate and perform the sensing tasks. The transceiver unit connects a sensor node to the network. The power unit consists of a battery and a direct current–direct current DC–DC converter. In a large network, consisting of hundreds or thousands of densely deployed sensors, each sensor would have to be inexpensive [37] and for ease of deployment and have a very small form factor. This severely limits the energy capacity of the sensor's battery, which in turn limits its lifetime. Because replacement of batteries in such a network is generally not feasible, a key challenge is to efficiently utilize the finite energy resources so as to achieve the maximum possible coverage for the longest duration.

The transceiver is often the most power hungry component of a sensor. For example, for the Rockwell WINS sensor nodes [39], the sensor consumes 64 mW, whereas with the processor on, the communication unit consumes 1016, 687, 663, and 352 mW in the transmit, receive, idle, and sleep modes, respectively. For this reason, the techniques for achieving greater energy efficiency is to reduce the energy cost due to communication. This can be done at all levels of the network hierarchy—at the physical layer as in [40], the MAC layer [41,42], in the network layer [43–48], and at the application layer as in [49–51]. These methods attempt to maximize lifetime by reducing the communication overhead on each sensor node and do not explicitly consider *coverage* as a criterion.

Sensor management techniques that do consider coverage exploit the fact that the networks are redundant (density of 20 nodes/m^3 [40]). In such a scenario, there often exist many subsets of sensors (covers) that can cover the given region R. Substantial energy savings can be achieved by keeping only one cover active at anytime and turning off the communication subsystem of the remaining sensors. This leads to the problem of identifying the covers or subsets and scheduling their activity periods to maximize the operational lifetime of the network. The method described in [52] *randomly* activates and deactivates sensors. Thus, it is difficult to guarantee a given level of coverage even if it were possible. Moreover, energy may be wasted since the number of active sensors turned on may be more than the required minimum number of sensors. In [53] subsets of sensors, referred to as *feasible sets*, that can cover a given region are identified, and such sets are scheduled for activation and deactivation. Determining the set of all feasible sets is of exponential complexity and, hence, only a small fraction of such sets are considered. What is needed is a systematic approach to identify covers and methods to transition between them based on a well-defined criterion.

The existing techniques for energy management in sensor networks do not *explicitly* consider the characteristics of the battery when making decisions. By minimizing the energy consumption of the sensors, these techniques reduce the *load* on the battery, leading to longer battery life. However, energy minimization is not sufficient for maximizing the battery lifetime [54]. This is because batteries exhibit several unique characteristics that affect their capacity. For instance, when a battery is discharged for some time and then put to rest (i.e., load disconnected to simply reduced), the terminal voltage at the end of the rest period can be higher than the terminal voltage just prior to the start of the rest period—a phenomena referred to as *charge recovery* [54]. The amount of charge recovery depends on the history of the discharge profile. Another important characteristic of the battery is that the capacity of a battery decreases as the discharge rate increases. Thus, if a battery is discharged by applying a sequence of constant loads, $i_1, i_2, \ldots, i_n$, each of some duration, then the available capacity at the end of the last load i_n will depend on the order in which these loads were applied, while the total energy consumption will be the same. Thus, a more appropriate criterion to maximize is the *residual charge* in the battery. Scheduling covers using this criterion will result in better utilization of the sensor network.

Contributions of This Section   In this section, we present novel algorithms to maximize the lifetime of a sensor network by exploiting the redundancy in the sensor covers. The approach uses a *reduced ordered binary decision diagram* (ROBDD) [55] to compute all the covers of a given sensor network. This representation is extremely efficient and can accommodate very large numbers of covers (e.g., $> 10^{16}$ covers with 100 sensors). Using ROBDDs, two different approaches to sensor management are presented. In the *centralized* approach, the base station computes a global optimal cover and then passes the active/inactive control decisions to the individual sensors. We refer to this as the *centralized sleep state control* or CSSC. The max–min minimal cover (M3C) algorithm is developed to compute the optimal cover for maximizing the lifetime of the network, while satisfying user-specified performance constraints (e.g., 100% coverage, or data quality, etc.). We show that one can further improve the network lifetime by switching between optimal covers, due to the battery charge recovery effect. Although the representation of all covers is very efficient (number of covers represented is many orders of magnitude greater than any previously considered [53]), the computation time of the centralized approach ceases to be practical for very large networks. We eliminate this problem with a *distributed* approach. This is referred to as *distributed sleep state control* (DSSC). In DSSC, the decision to transition between the active and inactive state is locally determined at each node based on the battery charge of the node and its neighbors. Due to the autonomous decision making, the DSSC approach can be applied to arbitrarily large networks and it guarantees maximum possible coverage. In both schemes, the cost function accounts for the transmission and reception energy, the network topology, and the battery characteristics such as the *rate-dependent capacity effect* and *charge recovery effect*. The proposed schemes are evaluated for two existing communication protocols, namely the direct transmission and LEACH [46]. Experimental results demonstrate that by leveraging the network redundancy and battery recovery effect, significant improvement in the network lifetime can be achieved.

6.4.2 Notation and Terminology

Covering

- $\prod, \sum$ denote Boolean conjunction and disjunction.
- R is the set of *points* to be covered.
- $S = \{s_1, s_2, \ldots, s_n\}$ is the set of sensors that are deployed.
- When a sensor is deployed, it covers a subset of the R. Hence, a sensor s can simply be viewed as a subset of R, the subset being the set of points of R *covered* by s.
- *Note*: We associate a Boolean variable with each element $s_i \in S$ using the *same symbol*.
- S_p is the set of all sensors that cover a given point p.
- The covering constraint for a point p denoted by λ_p is given by

$$\lambda_p = \sum_{i \in S_p} s_i \tag{6.36}$$

where $\lambda_p = 1$, if there is at least one sensor that covers point p.

- Let $P \subseteq R$, and $S' = \{s_{i_1}, s_{i_2}, \ldots, s_{i_k}\} \subseteq S$. Then the *(sensor) cover function* of P with respect to S' is a characteristic function denoted by $\chi_P(s_{i_1}, s_{i_2}, \ldots, s_{i_k})$, and is defined as follows:

$$\chi_P(s_{i_1}, s_{i_2}, \ldots, s_{i_k}) = \prod_{p \in P} \lambda_p \tag{6.37}$$

Note: If $S' = S$, we omit S', that is, $\chi_P = \chi_P(S)$.

- A sensor s' is a *neighbor* of s if $s \cap s' \neq \phi$.
- $N(s) \subseteq S - s$ is a set of neighbors of s. *Note:* The definition of *neighbor* is intentionally abstract as it can be specified using any one of several criteria.
- $NC(s) = \{s' \in N(s) \| \cup_{s'} = s\}$ is a (neighbor) cover of s, formed by its neighbors.
- A sensor s is *essential* if $\chi_s(N(s)) = 0$.
- ρ is the minimum number of sensors that are required to cover each point in R. We use ρ as a *data quality* constraint.

Sensor States and Energy Level

- $w_s(t)$ is a weight assigned to sensor s at time t. It will be used to represent the residual battery life of sensor s at time t.
  Note: The battery and energy models are used to determine the weight function $w_s(t)$ associated with each sensor. A detailed description of the model and how $w_s(t)$ is computed appears in Section 6.4.7.
- $\text{weak}_s(t) = \{s' \in N(s) \mid w_{s'}(t) < w_s(t)\}$. $\text{weak}_s(t)$ is the set of neighbors of s, whose weight is less than that of s.
- The state of a sensor s is denoted by q_s, where $q_s \in \{0, 1, X\}$, representing the sensor being off, on, and undecided, respectively.
- $p_s \in \{0, 1\}$ denotes the present state of sensor s.
- $\mu_s(t)$ is a tuple representing the states of all sensors in $\text{weak}_s(t)$.
- τ is the minimum duration that a cover needs to be *active*.
  Note: τ is a user-specified parameter and depends on the type of application. For example, the minimum value of τ is the transmission time for a single packet. The information content required by a specific application and the data rate of the system determine τ.
- *Dead Sensor*: At any time t, a sensor s is considered *dead*, if its weight $w_s(t) = 0$.

6.4.3 Representation of Covers

In an SN, sensors are deployed in high density making the network redundant. This means that there exist many subsets (covers) of sensors that cover the complete region. The SN management schemes described in this section are aimed at maximizing the lifetime of the network while maintaining maximum possible coverage. This requires activating and deactivating covers. For this to be possible, it is essential that all the possible covers be represented. Since a cover is a subset of sensors and the number of covers grows exponentially with the number of sensors, a compact representation of sets of subsets is needed.

Given a sensing region R, n sensors $s_1, s_2, \ldots, s_n$ and their respective sensing radii $r_1, r_2, \ldots, r_n$, the determination of all covers is equivalent to finding all n-tuples $< s_1, s_2, \ldots, s_n >$ such that χ_R is satisfied. This is identical to the classical unate covering problem [56]. Figure 6.17 shows R as a 4×4 grid with 5 sensors $S = \{s_0, s_1, s_2, s_3, s_4\}$. Figure 6.18 shows the corresponding *covering matrix*. The rows correspond to the points in R, and the columns correspond to the sensors. An entry of 1 in a (row, col) = $((x, y), s)$ means that sensor s can cover grid point (x, y). Thus the sum (disjunction) of all the variables in row (x, y) represents the conditions for covering point (x, y), that is, λ_p. The conjunction of the row sums represents the conditions for covering every point, that is, χ_R. For the given example, we obtain (after simplification) $\chi_R = s_1 s_2 s_3 s_4 + s_0 s_1 s_2 s_3$.

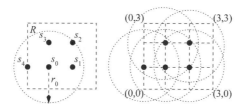

Figure 6.17 Sensor covering.

The sensor cover function, being a Boolean function, allows incorporation of various types of constraints. For instance, a common way to improve the quality of data in noisy environments is to require that each point in the region R be monitored by at least ρ sensors. Let $\chi_p^\rho(\mathcal{S}')$ denote the sum of all products of ρ sensors taken from $\mathcal{S}'$ that contain point p. Then the sensor cover function that prescribes that at least ρ sensors cover each point in R is given by

$$\chi_R(\mathcal{S}') = \prod_{p \in R} \chi_p^\rho(\mathcal{S}') \tag{6.38}$$

For example, the point $(0, 1)$ in the covering matrix shown in Figure 6.18 can be covered by sensors s_0, s_3, and s_4. If $\rho = 2$, then $\chi_{(0,1)}^\rho = s_0 s_3 + s_0 s_4 + s_3 s_4$.

Heterogenous Sensor Nodes In a general scenario, different practical limitations such as sensor node calibration, location, terrain, and design may result in an irregular shape of the sensing region. Thus, all the sensor nodes deployed in the geographical region can have different shapes for their respective sensing regions. The covering paradigm presented can be easily extended to such scenarios. The algorithms presented abstract the shape of the sensing region to a set of points by approximating the complete sensing space using a

(x,y)	s_0	s_1	s_2	s_3	s_4	$\lambda_p(S)$
(0,0)	1	0	0	0	1	$s_0 + s_4$
(0,1)	1	0	0	1	1	$s_0 + s_3 + s_4$
(0,2)	1	0	0	1	1	$s_0 + s_3 + s_4$
(0,3)	0	0	0	1	0	s_3
(1,0)	1	1	0	0	1	$s_0 + s_1 + s_4$
(1,1)	1	1	1	1	1	$s_0 + s_1 + s_2 + s_3 + s_4$
(1,2)	1	1	1	1	1	$s_0 + s_1 + s_2 + s_3 + s_4$
(1,3)	0	0	1	1	0	$s_3 + s_2$
(2,0)	1	1	0	0	0	$s_0 + s_1$
(2,1)	1	1	1	1	1	$s_0 + s_1 + s_2 + s_3 + s_4$
(2,2)	1	1	1	1	1	$s_0 + s_1 + s_2 + s_3 + s_4$
(2,3)	0	0	1	1	0	$s_2 + s_3$
(3,0)	0	1	0	0	0	s_1
(3,1)	0	1	1	0	0	$s_1 + s_2$
(3,2)	0	0	1	0	0	s_2
(3,3)	0	0	1	0	0	s_2

Figure 6.18 Covering matrix for *R*.

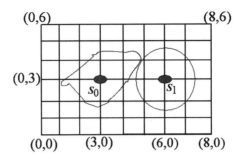

Figure 6.19 Heterogenous sensor nodes.

two-dimensional grid. Hence, every sensor represents a set of points it covers, independent of the shape of its sensing region.

Consider the example shown in Figure 6.19 with sensors s_0, s_1. Sensor s_1 has a circular sensing region and s_0 has a irregular sensing region. For the example, shown in Figure 6.19, s_0 covers the points $\{(3, 2), (3, 3), (3, 4), (4, 4), (4, 3), (2, 3)\}$, and s_1 covers the points $\{(5, 2), (5, 3), (5, 4), (6, 2), (6, 3), (6, 4), (7, 2), (7, 3), (7, 4)\}$. Thus, depending upon the design and calibration of a sensor node, the set of points covered by the sensors change, affecting the computation of the sensor cover function.

For the case of simplicity, we assume a homogenous sensing radii for all the sensor nodes in our experiments. However, the presented approach as described above can be easily extended to sensors with any irregular sensing region.

Binary Decision Diagrams At the heart of many recent advances in logic synthesis, testing, and verification is the invention of reduced ordered binary decision diagrams (ROBDD) [55, 57] for representing Boolean functions. ROBDDs have two very useful properties. First, for a given ordering of variables, they are *canonical*. This means that there is a unique ROBDD structure for a given Boolean function. Second, they are extremely compact when representing large combinatorial sets. It is this property that we exploit, since the set of all covers is simply a very large set of subsets, described by χ_R. The literature on BDDs and their many variants is enormous. Here we simply provide a definition and focus on the most relevant properties and operations. The reader is referred to [58] for a detailed description on BDDs.

Definition 6.4.1 A ROBDD [55] with respect to a given ordering of Boolean variables X is a rooted, directed, acyclic graph representing a Boolean function of the variables. It has two terminal nodes, labeled by **0** and **1**, that denote the two constant functions. Each internal node u is labeled by a variable $l(u) \in X$, and has two children, $T(u)$ and $E(u)$. The function denoted by node u is $l(u)f_T + l(u)'f_E$, where f_T and f_E are the functions denoted by nodes $T(u)$ and $E(u)$. The order in which the variables appear along a path is consistent with the given ordering of variables. The functions denoted by $T(u)$ and $E(u)$ are distinct and subgraphs rooted at any two nodes u and v are not isomorphic.

The use of ROBDDs for solving covering problems has been well documented [56,59]. Following the method described in [59], we use ROBDDs to represent the sensor cover function χ_R. Figure 6.20 shows the ROBDD of $\chi_R = s_1s_2s_3s_4 + s_0s_1s_2s_3$. The one edge and zero edge are shown by solid and dotted lines. The efficiency of this representation for sensor covering is demonstrated in Section 6.4.8.

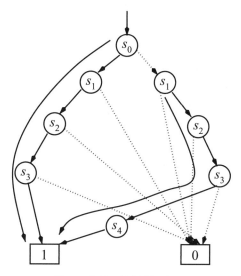

Figure 6.20 ROBDD of χ_R.

There are several important characteristics of ROBDDs [60] that should be noted for better understanding of the sensor management algorithms.

- The nodes in the ROBBD of χ_R represent sensors and the ROBDD represents the set of all covers.
- A path to the terminal node **1** (a *1 path*) represents a *feasible* solution of the sensor cover function (i.e., a cover). If the path contains the 1 edge $(u, T(u))$ from a node u, then the corresponding variable appears in the cover. For example, the ROBDD of Figure 6.20, shows 2 covers.
- ROBDDs are well behaved for many functions, although for certain functions their size grows exponentially in terms of the number of variables, regardless of the ordering [60].
- The time and space complexity of logic operations is polynomial in the number of variables.
- Covering problems can be solved in time linear to the size of the ROBDD representing the constraints.

A key operation that will be performed repeatedly in the sensor management algorithms is the deactivation of a sensor. This is achieved by Boolean operation called *universal quantification* [60], which can be efficiently implemented using ROBDDs.

When a sensor s is deactivated, the ROBDD must be updated so that it still represents the set of all covers. Any point that is covered by s and by at least one other sensor must remain in the cover when s is deactivated. When a node s is dead, we must update the ROBDD so that it still represents the set of all covers. Any point that is covered by s and by at least one other node must remain covered when s is dead. Therefore, the sensor cover function when s is dead is obtained by evaluating χ_s at $s = 0$.

Consider the example shown in Figure 6.21 with neighbor set $N(s_0) = \{1, 2, 3, 4\}$. The ROBDD representing the (neighbor) covers of s_0 is shown in Figure 6.22a. We now show

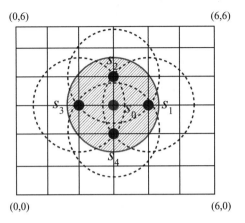

Figure 6.21 Placed sensors.

an example of a computation that a sensor performs using ROBDDs. It shows how a sensor determines whether its points can be covered by its neighbors given the state of some or all of them. This is a key operation that is performed in DSSC.

Consider evaluating χ_{s_0} given the state of two of its neighbors, s_2 and s_4. Recall, s denotes a set of points in R and χ_s is the sensor cover function of those points. Figure 6.22b shows that the ROBDD obtained after χ_{s_0} is evaluated at $\{s_2 = 0, s_4 = 0\}$. The ROBDD represents the remaining covers for s_0, that is, $\{s_1, s_3\}$. The fact there still exists a path to 1 means that after deactivating s_2 and s_4, it is still possible to cover the points of s_0. The ROBDD obtained after χ_{s_0} is evaluated at $\{s_2 = 1, s_4 = 1\}$ is shown in Figure 6.22c. The ROBDD evaluates to a 1. This means that by turning on sensors s_2, s_4, sensor s_0 is covered. Figure 6.22(d) shows the ROBDD obtained when χ_{s_0} is evaluated $\{s_2 = 0, s_3 = 0\}$. The ROBDD represents the fact that no cover exists for s_0 if s_2 and s_3 are turned off. Thus, by evaluation of χ_s using the state of its neighbors, a sensor s can decide whether or not to be on or off.

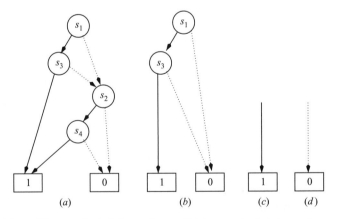

Figure 6.22 (a) χ_{s_0}, (b) χ_{s_0} evaluated at $s_2 = 0$, $s_4 = 0$, (c) χ_{s_0} evaluated at $s_2 = 1$, $s_4 = 1$, and (d) χ_{s_0} evaluated at $s_2 = 0$, $s_3 = 0$.

Constructing the Sensor Cover Function As described in the previous section, the sensor cover function denotes the set of covers for a specific set of points. In the centralized sleep state control, the sensor cover function is computed for the entire sensing region while in the distributed sleep state control it is for the set of points covered by a sensor.

Given the network parameters such as the number of sensor nodes, sensing radius, grid size (number of points) a network is generated by randomly distributing the sensors in the grid. For each point in R, the set of sensors covering it is computed and a BDD representing the covering constraint is constructed. [Recall that the covering constraint is the disjunction (OR) of all the sensors.] In CSSC, the conjunction (AND) of all the BDDs representing the covering constraints at every point in the geographical region R is computed resulting in χ_R. In DSSC, χ_s is computed by taking the conjunction of all the BDDs representing the covering constraints of points covered by sensor s.

6.4.4 Centralized Sleep State Control

We present the sensor management algorithms for a centralized sleep state control (CSSC). Recall that the set of all covers of R is represented by χ_R. With CSSC, the base station selects covers (e.g., 1 paths in χ_R) based on the weights of the sensors. The SN is considered *alive* if and only if there exists at least one valid cover. The earliest time at which no cover exists is defined as the *lifetime* of the network. Let χ_{on} denote the characteristic function (representing the set of sensors) in the current active cover. The method to compute the weight of the sensor s is described in Section 6.4.7. In the following section, we describe how covers are selected to maximize the lifetime of the network.

Minimal Max–Min Cover (M3C) When a sensor dies, all covers containing it become invalid. Thus the weakest sensor in a cover determines the lifetime of a cover. Hence, by maximizing the lifetime of the weakest sensor in a cover, we delay the death of that cover. Moreover, a sensor often belongs to many covers. An optimal cover needs to be turned on among all the covers in such a way that the lifetime is maximized. This is accomplished, by selecting a cover whose minimum weighted node (min) is maximum (max) over all existing covers. By keeping max–min cover on, we avoid discharging all other covers whose weakest sensor node weights are smaller than the current weakest node, thus increasing the lifetime of the network. Such a cover is referred to as the *max–min* cover. Activating the max–min cover ensures that all covers with weaker sensors (sensors whose weight is smaller than the current weakest sensor) are not expending energy.

Definition 6.4.2 Let $S_1 = (s_{1,1}, s_{1,2}, \ldots, s_{1,n_1})$ and $S_2 = (s_{2,1}, s_{2,2}, \ldots, s_{2,n_2})$ be two covers, arranged in increasing order of their weights. Let j be the smallest index such that $s_{1,j} \neq s_{2,j}$. Then the max–min cover of S_1 and S_2 is equal to S_1 if $w_{s_{1,j}} > w_{s_{2,j}}$ and is equal to S_2 if $w_{s_{2,j}} > w_{s_{1,j}}$.

Definition 6.4.3 A cover, which is both minimal (i.e., no proper subset is also a cover), and max–min is called the minimal max–min cover (M3C). The characteristic function of a M3C is denoted by χ_{M3C}.

The pseudocode to compute χ_{M3C} is presented in Figure 6.23. Its arguments are the characteristic function χ_R, which represents the region R, the characteristic function χ_{on} of the previous on cover, and the set of sensors S. It returns the χ_{M3C}. After initialization, sensors are sorted in ascending order by their weights (step 2). Initially, when χ_{on} is empty, the index of the weakest sensor is assigned 0 and no sensor is deleted from χ_{M3C} (step 7). If

M3C $\{\chi_R, \chi_{on}, \mathcal{S}\}$
(1) $\chi_{M3C} \leftarrow \chi_R$;
(2) $\mathcal{S} \leftarrow$ sort_by_weight$\{\mathcal{S}\}$;
(3) if χ_{on}=NULL
(4) $\quad$ i$\leftarrow$ 0;
(5) else
(6) $\quad$ i $\leftarrow$ Min_Weight_Sensor_Index$\{\chi_{on}\}$;
(7) $\chi_{M3C} \leftarrow \chi_{M3C} (s_1 = 0 \ldots s_i = 0)$
(8) if χ_{M3C} is ϕ
(9) $\quad$ return χ_{on};
(10) for each j $\leftarrow$ (i+1 to length of array $\mathcal{S}$)
(11) $\quad$ result $\leftarrow \chi_{M3C}(s_j = 0)$
(12) if result $\notin \phi$
(13) $\quad \chi_{M3C} \leftarrow$ result;
(14) return χ_{M3C};

Figure 6.23 Pseudo code to compute M3C.

χ_{on} is nonempty, then the index of the weakest sensor in the on cover is found (step 6). In step 7, all sensors weaker than the weakest sensor of χ_{on} are removed by evaluating χ_{M3C} with respect to all weaker sensor nodes set to 0. If χ_{M3C} is empty then χ_{on} is returned (step 9). Otherwise a minimal cover is obtained by removing all sensors that are not necessary to retain a cover (steps 10 through 13). The result is χ_{M3C}.

Complexity Analysis Evaluating χ_{M3C} with respect to a variable s set to 0 is $O(B)$ in complexity [60], where B is the size of the ROBDD. Let m be the total number of sensors in the network. For computing M3C (see Fig. 6.23), the evaluation is performed with respect to i sensors in step 7. This is to obtain the set of max–min covers. In step 10, the evaluation operation is repeated for the remaining $m - i$ sensors to obtain a minimal cover from the max–min covers. Thus, χ_{M3C} is evaluated for m times. Hence, the worst-case complexity is $O(mB)$. However, the size of the BDD is changed due to the evalaution operation. Let B_{max} be the maximum size of the BDD formed over all m quantification operations. Thus, the worst-case complexity for computing a M3C cover is $O(mB_{max})$.

Sensor Scheduling

Sequential Schedule The baseline scheduling algorithm is a simple sequential scheduling scheme. A cover is on until the first sensor dies, at which point another cover is activated. Note that the batteries in the sensors that are in the first cover and not in the next *will recover some charge*. This is accounted for in the battery model.

The procedure of computing the sequential schedule is shown in Figure 6.24. The inputs are χ_R and $\mathcal{S}$. Using the energy and battery models (see Section 6,4.7), the algorithm iteratively generates the load profile (current drawn from the battery as a function of time) for each sensor and computes the battery lifetime T. At the start of each iteration χ_{on} is computed by making an initial call to M3C (step 3). Because in this scheme the cover is kept on until it ceases to be a valid, for each iteration the duration of the on time Δ_k is limited by the lifetime of the weakest sensor in the cover χ_{on}. This lifetime is computed using the battery model by applying a constant transmission load (each sensor transmits to the base station, which is at a fixed distance away).

M3C_Sequential_Schedule$\{\chi_R, \mathcal{S}\}$
(1) $T \leftarrow 0; \chi_{on} \leftarrow$ NULL;
(2) while $\chi_R \notin \phi$
{
(3) $\chi_{on} \leftarrow$ M3C $\{\chi_R, \chi_{on}, \mathcal{S}\}$;
(4) $\Delta_k \leftarrow$ Weakest_Sensor_Lifetime$\{\chi_{on}\}$
(5) $T \leftarrow T + \Delta_k$;
(6)   Update_Weights $\{ \mathcal{S}, \chi_{on}, \Delta_k, T \}$;
(7)   for each $s \in \mathcal{S}$
(8)   if s is dead
(9) $\chi_R \leftarrow \chi_R(s = 0)$;
}
(10) return T;

Figure 6.24 Pseudocode for sequential schedule.

In step 5 the time T is advanced by Δ_k. In step 6, weights of all sensors are updated. This uses the battery model to compute the available charge at T. The set of dead sensors are removed from χ_R in steps 7, 8, and 9, thereby ensuring that the set of invalid covers are pruned out from the search space. The procedure terminates when no covers exists in the network (i.e., χ_R is empty).

Switching Schedule   We now examine a different scheduling scheme that explicitly puts sensors to sleep to exploit the charge recovery characteristics of batteries. The lifetime of the weakest sensor in a cover determines the lifetime of the cover. The set of weakest sensors in all the covers can be treated as a set of multiple batteries. The reason for doing so is that recent research on scheduling multiple batteries [61] has shown that the lifetime of a multiple battery system can be improved by appropriately switching between them. Based on this argument, we present a switching scheme that aims at maximizing the recovery effect at each sensor using the M3C. After every τ units of time (recall τ is the minimum amount of time a cover needs to be on), we switch between different M3Cs, thus ensuring that sensors with the highest battery reserves bear the transmission load, giving greater opportunity for other sensors in the network to recover.

The basic steps in the switching schedule are shown in Figure 6.25. Given χ_R and S as inputs, the algorithm computes the lifetime of the network. The iterative procedure starts

M3C_Switching_Schedule$\{\chi_R, \mathcal{S}\}$
(1) $T \leftarrow 0; \chi_{on} \leftarrow$ NULL;
(2) while $\chi_R \notin \phi$
{
(3) $\chi_{on} \leftarrow$ M3C $\{\chi_R, \chi_{on}, \mathcal{S}\}$;
(4) $\Delta_k \leftarrow \tau$
(5) $T \leftarrow T + \Delta_k$;
(6)   Update_Weights $\{ \mathcal{S}, \chi_{on}, \Delta_k, T \}$;
(7)   for each $s \in \mathcal{S}$
(8)   if s is dead
(9) $\chi_R \leftarrow \chi_R(s = 0)$;
}
(10) return T;

Figure 6.25 Pseudocode for switching schedule.

by computing the M3C (step 3). The selected M3C is turned on for τ units of time and the network on time is incremented in step 5. In step 6, the weights of all sensors are updated. Steps 7, 8, and 9 describe a method to identify the dead sensors in the network. These sensors are removed from the χ_R. The steps from 2 to 10 are repeated until no more valid covers exist, finally returning the value of the lifetime. Procedure UPDATE_WEIGHTS first updates the *load profile* of each sensor over the current time interval (Δ_k) and then updates the weight (available battery charge) of the sensors in the current cover.

6.4.5 Distributed Sleep State Control

The previous network control method was centralized. All the scheduling is done at the base station and this can be precomputed for the entire lifetime of the network. It requires no significant computational resources on the part of a sensor. However, the algorithms do not scale well as the size of the network increase. We now present a distributed control strategy called distributed sleep state controls, or DSSC. Key features of DSSC are as follows: (1) It uses self-configuration capabilities of sensors; (2) it is a distributed or localized algorithm and scales linearly with the size of the network; (3) it operates above the network layer but is independent of any application; (4) it guarantees maximum possible coverage, while allowing a large fraction of sensors to be inactive; and (5) it can be used with any protocol to improve the lifetime of the network.

When the sensor network is deployed, a *neighbor discovery phase* is performed. We assume that the sensors are deployed in a two-dimensional region, and each sensor knows its location using different location finding systems [62–64] and has a maximum sensing *radius* of r. Each sensor transmits hello messages with the sensor ID, its location for a distance of $2r$. A sensor also receives hello messages from all other sensors in its neighborhood. Based on the number of such messages received, a sensor can determine its neighbors. We assume the sensors transmit and receive hello messages using CSMA-MAC protocol with the minimum amount of power for the required distance. Once a sensor receives all information about its neighbors, it computes (constructs the ROBDD) χ_s. At the end of this phase, each sensor has complete knowledge of its set of covers. It should be noted that the neighbor discovery process needs to be performed only once, immediately after the network deployment.

Following the initial neighborhood discovery phase, the operation of the sensor network is divided into cycles (see Fig. 6.26). The duration of the cycle is a user-specified parameter and depends on the specific application. Each cycle consists of an *arbitration* phase and a *sensing* phase. At the beginning of the arbitration phase, the sensors first exchange energy information with their neighbors using CSMA-MAC protocol. This is followed by each sensor making a decision to turn itself off or on. The logic for determining a sensor's state is described below. Once a sensor decides its state, it remains in that state for the rest of the

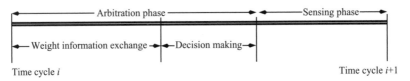

Figure 6.26   Sequence of steps in a time cycle.

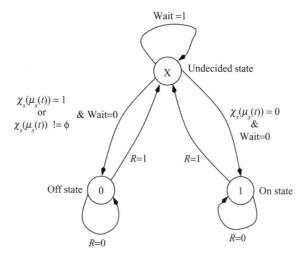

Figure 6.27 State transitions diagram.

cycle. Following the arbitration phase, active sensors sense for a specified time interval and transmit the information to their destination sensor. It should be noted that the destination sensor depends on the underlying communication protocol.

On/Off Decisions We now describe the logic employed by each sensor to decide its state. Figure 6.27 shows the state transition diagram of a sensor. This strategy ensures the maximum possible coverage, while keeping as many sensors as possible in the sleep state. In the arbitration phase, once the sensors have exchanged their weight status information, each sensor knows the subset of its neighbors that are weaker than itself, that is, the tuple $\mu_s(t)$. A sensor s decides its state as follows:

1. It remains in the *undecided* state until all its weaker neighbors decide their state. When a sensor decides its state, it transmits that information to its neighbors.
2. Once the states of each of its weaker neighbors has been decided, it computes $\chi_s(\mu_s(t))$. That is, it evaluates the ROBDD representing all the covers of its points with the known values of the sensors in $\mu_s(t)$. There are three possible outcomes of this computation:
   a. $\chi_s(\mu_s(t)) = 1$. This means that the state of its weaker neighbors provides a cover for s. Hence s puts itself in the off state.
   b. $\chi_s(\mu_s(t)) = 0$. This means that the state of its weaker neighbors does not provide a cover of s. In this case, s puts itself in the on state.
   c. If $\chi_s(\mu_s(t)) \neq 0, \neq 1$. This means the current state of its weaker sensors does not provide a cover of s, and some points still remain to be covered. However, the existence of a 1 path in the ROBDD means that the points not covered by the weaker neighbors can be covered by some subset of sensors stronger than s. The coverage of those points of s is guaranteed (see Theorem 6.4.1). Hence, in this case s puts itself in the off state.

Theorem 6.4.1 If the weaker neighbors of a sensor s do not cover s, then either s will be on or s will be covered by some subset of its stronger neighbors.

PROOF   Let $p \in s$ be a point not covered by any weaker neighbor of s. Then if $\chi_s(\mu_s(t)) = 0$ (i.e., no weaker neighbor covers p), s will turn itself on (condition 2b). If $\chi_s(\mu_s(t)) \neq 0$, then $\chi_s(\mu_s(t)) \neq 1$. Then there exists a stronger neighbor of s, say s' such that $p \in s'$. Now we have the same situation with s' replacing s. Repeating the above argument, we reach the situation where s is the strongest sensor. If none of the weaker neighbors of s cover p, s will be essential and therefore will be on, covering p.   ∎

The boundary conditions concern the weakest and strongest sensors. The weakest sensor puts itself in the on state if it is essential (condition 2b), otherwise it puts itself in the off state (condition 2c). The strongest sensor turns itself off if its weaker neighbors provide a cover (condition 2a), otherwise it turns itself on (condition 2b). The *domino* style of decision making ensures that no deadlock situation occurs. The control algorithm also ensures that coverage is not lost when possible.

The procedure executed by each sensor is shown in Figure 6.28. Given χ_s, $N(s)$, t, reset, and p_s, the algorithm computes the next state q_s of a sensor. At the beginning of every cycle, as soon as a sensor receives a reset signal (this signal can be issued by the cluster head or the time a sensor has to reset itself can be preprogrammed), it transmits its energy information to all its neighbors. Consequently, a sensor receives the energy status of all its neighbors. Following exchange of *weight* information (Section 6.4.7), a sensor pushes itself to an *undecided* state (steps 2, 3, and 4). In steps 5 through 8, a sensor deletes all its dead

```
State_Decide { χₛ, N(s), t, reset, pₛ }
(1) if reset=1
    {
(2)    Transmit wₛ(t) to s' ∈ N(s);
(3)    Receive wₛ'(t) from s' ∈ N(s);
(4)    qₛ ← x;
(5)    for each s' ∈ N(s)   do
       {
(6)       if s'  is  dead
          { (7)      χₛ ← χₛ(s' = 0);
(8)       N(s) = N(s)/s';
       }
(9)       Sort (All s' ∈ N(s)) w.r.t wₛ'(t);
(10)      Compute weakₛ(t);
(11)      Wait ← 0;
(12)      While ∃ s' ∈ weakₛ(t) s.t. qₛ' = x
          { (13)      Wait ← 1;
          } (14)      if Wait=1
          { (15)         qₛ ← x;
          } (16)      else
             { (17)      if χₛ{μₛ(t)}=0
                { (18)         qₛ ← 1;
                } (19)      else
                   { (20)         qₛ ← 0;
(21)          Broadcast qₛ to s' ∈ N(s);
(22)          pₛ ← qₛ;
       }
(23)else
    { (24)   qₛ ← pₛ;
    }
```

Figure 6.28   Pseudocode of STATE_DECIDE.

neighbors from its neighbor sets and recomputes its sensor cover function, by quantifying out the dead neighbors. In steps 9 and 10, weak neighbors (w.r.t the given sensor) are identified by sorting the weight information of all the sensors in $N(s)$.

In step 12, a sensor checks if all its weaker neighbors have decided their status. If all weaker neighbors have decided their respective states, a sensor proceeds to decide its state, otherwise it remains in an undecided state. To decide its state, a sensor evaluates χ_s using the known states of all sensors in $weak_s(t)$. When the sensor cover function evaluates to an empty set, a sensor turns itself on; otherwise it turns itself off (steps 14 through 22). After a decision is made, a sensor broadcasts its decision to all its neighbors. In the case when no reset signal is given, a sensor remains in its present state (step 24).

Consider the example shown in Figure 6.29 with five sensors, $s_0, \ldots, s_4$, randomly distributed. The weight of the sensors is shown beside each sensor. Based on increasing

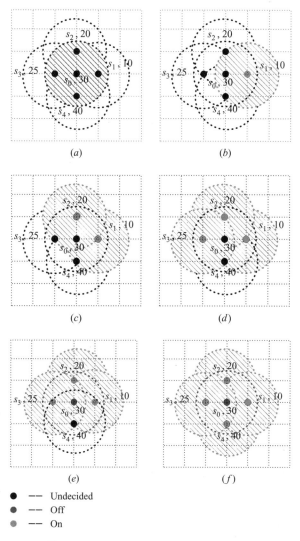

(a) (b)

(c) (d)

(e) (f)

● —— Undecided
● —— Off
● —— On

Figure 6.29 Illustration of decision making.

weight, the sensors can be ordered as $\{s_1, s_2, s_3, s_0, s_4\}$, s_1 being the weakest and s_4 being the strongest sensor node. At the beginning of the arbitration phase, sensor nodes exchange their weight information (for identifying weaker neighbors) and push themselves to an *undecided* state as shown in Figure 6.28a. Each sensor node waits for all its weak neighbors to decide their state. Following this condition, sensor s_1 (being the weakest) decides to turn itself on, as it is an essential sensor (shown in Fig. 6.28b). As soon as s_1 decides its status, sensor s_2 is allowed to decide its state, as all its weaker neighbors have decided their state. Since s_2 is essential, it turns on (shown in Fig. 6.28c). Similarly, s_3 is turned on. When sensors s_1, s_2, s_3 have decided their state, s_0 is allowed to take its decision. After evaluation of its sensor cover function, s_0 turns itself off as it is covered by its neighbors (shown in Fig. 6.28e). Once s_0 decides its state, s_4 takes a decision to turn on (see Fig. 6.28) as it is essential.

The principal characteristics of the above algorithm are that it uniformly distributes the load on all the sensors in the network and attempts to maximize the lifetime of the network. Using residual charge information facilitates weaker sensors in the neighborhood to make a decision before the stronger ones. The responsibility of turning on and sensing the region is pushed onto the stronger sensors. A sensor is put to sleep more often if there are larger numbers of covers (in the network). However, a cover ceases to be valid if at least one sensor is *dead*. Therefore the lifetime of a cover is limited by the lifetime of the weakest sensor in the cover. Hence maximizing the lifetime of the weakest sensors in a cover delays the death of a cover. We show that the cover obtained at the end of the arbitration phase is an M3C.

Corollary 6.4.1 If sensors decide their state based on conditions 2a, 2b, and 2c, then:

1. maximum possible coverage is achieved, and
2. the resulting cover is an M3C.

PROOF1. Let point $p \in s$. If p is not covered by the weaker neighbors of s, then by Theorem 6.4.1, p is either covered by s or it is covered by the stronger neighbors of s. Hence, no point p is left uncovered, resulting in complete coverage. ∎

PROOF2. Let C be the resulting cover in the network. If C is not minimal, then there exists a sensor s that is on even though it is covered by a subset of its neighbors. In such a case, either condition 2a or 2c is violated at sensor s. Hence C is minimal.

Let s be the weakest sensor in C. Let C' be a max–min cover and suppose C is not. For C' to be the max–min cover, there must exist a sensor s' that covers the points of s in C' such that $w_{s'} > w_s$. This implies condition 2c would be violated at s, since s would not turn itself on, if there exists a stronger neighbor to cover it. Hence C is a max–min cover. ∎

Complexity Analysis Let B_s be the size of the BDD representing χ_s and $|N(s)|$ is the total number of neighbors for s. The evaluation operation (equivalent to finding the zero cofactor of χ_s w.r.t a given variable) requires to be performed w.r.t all the dead neighbors. Hence, the worst-case complexity of the algorithm shown in Figure 6.29 is $O(|N(s)|B_s)$. The size of BDD changes as a result of the quantification operation. Let $B_{s_{max}}$ be the maximum size of the BDD over all quantification operations. Hence, the worst-case complexity is $O(|N(s)|B_{s_{max}})$.

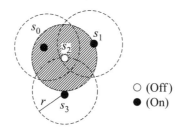

Figure 6.30   Redundant transmissions.

6.4.6 Energy Efficiency of the Algorithms

The proposed algorithms were developed with key focus on generic applications of sensor networks including environmental monitoring, data acquisition, and the like where a sensor node requires to sense and transmit data continuously. In such a scenario, reducing the redundancy in sensing costs implies a decrease in the transmission costs as well. To see this, consider the example shown in Figure 6.30.

If all four sensors s_0, s_1, s_2, s_3 are turned on, redundant information is transmitted about the sensing region of s_2. By identifying that s_2 is redundant (already covered by its neighbors), we reduce the number of transmissions, while ensuring coverage is not affected. Thus, the algorithms attempt to maximize the lifetime by reducing the number of redundant transmissions, while ensuring coverage. This is the key distinction between our approach and those presented in the other existing schemes such as SPAN [44], ASCENT [49], and LEACH [46].

The proposed approach being orthogonal to the existing topology control algorithms can be used in conjunction with them to further improve the lifetimes achieved by either of them individually. Since the source code for LEACH is publically available, we demonstrate this on LEACH protocol. However, in principle similar improvements can also be achieved using SPAN [44], ASCENT [49], or any other topology control methods. Additionally, the presented algorithms can be directly implemented in case of networks with active sensors such as microwave and infrared sensors, where sensing cost is not negligible.

6.4.7 Computation of Sensor Weight $w(t)$

The weight $w(t)$ represents the amount of *residual charge* in the battery. The residual battery charge at time t depends on the *load profile* $i(t)$ as seen by the battery. We first describe the model for computing $i(t)$. We then present a brief description of the battery model and the computation of the residual charge based on the load profile.

Load Profile Computation   A node draws varying amount of current from its battery depending on the tasks it performs, for example, transmission, reception, processing, and so forth. We use the energy model described in [46] to construct the model for the load profile. Let $E_T(k, d)$ and $E_R(k)$ denote the total energy dissipated to transmit k bits, over distance d and energy dissipated to receive k bits, respectively. Let E_{amp} denote the transmitter amplification factor and E_{elec} denote energy consumed by the transmitter or receiver electronics per bit. Then from [46], we have

$$E_{T_x}(k, d) = E_{\text{elec}}k + E_{\text{amp}}kd^2 \tag{6.39}$$

$$E_{R_x}(k) = E_{\text{elec}}k \tag{6.40}$$

The energy consumed increases quadratically with the transmission distance from the base station and linearly with the number of bits to be transmitted.

During a time cycle of duration T, a sensor performs its sensing task, processes the (fixed amount) of data, and transmits a packet. Depending on the protocol, it may also receive some data. Let $I_{T_x}(k, d)$, $I_{R_x}(k, d)$ represent the total current for transmission and reception of k bits over a distance d, respectively, and $I_p(k)$ represent the current drawn by the data processing unit. Let I_s denote the total load current on the battery of sensor s.

Assuming that the supply voltage of the sensor is fixed, expressions for these current components are obtained directly from Eqs. 6.39 and 6.40.

$$I_{T_x}(k, d) = I_{\text{elec}}k + I_{\text{amp}}kd^2 \tag{6.41}$$

$$I_{R_x}(k) = I_{\text{elec}}k \tag{6.42}$$

$$I_p(k) = I_{\text{proc}}k \tag{6.43}$$

where I_{elec} is the current constant for the transceiver electronics, I_{amp} is the amplification constant of the transmitter, and I_{proc} is the average current per bit associated with the data processing unit.

Direct transmission Protocol   In a direct transmission protocol (DTP), all nodes in the network transmit information directly to the base station. Hence, the load current on the battery of sensor s, denoted by I_s, is given by

$$I_s = I_{T_x}(k, d_s) \tag{6.44}$$

where d_s represents the distance of node s to the base station. I_s is drawn by the node s throughout the time cycle. In other words, for the DTP, the load on a battery of a sensor is constant and depends only on the distance of the sensor to the base unit. Moreover, we assume that transmission current dominates the current for sensing and data processing at each sensor node. If this is not the case, then the other components can simply be added.

LEACH Protocol   When using DSSC with the LEACH protocol (see Section 6.4.8), certain sensors, called cluster head, act as *leaders*. Cluster heads act as local base stations to which the other sensors transmit data. The cluster heads are responsible for transmitting to the central base station. A sensor then can be in one of the following three states:

- On and is a cluster head.
- On and is not a cluster head.
- Off and is not a cluster head.

If a sensor s is a cluster head with η_s number of other sensors in its cluster, then its load current denoted by $I_{s,c}$ is given by

$$I_{s,c} = I_{T_x}(k, d_s) + \eta_s I_{elec}k + \eta_s k I_{proc} \qquad (6.45)$$

In Eq. 6.45, $I_{T_x}(k, d_s)$ is the current for transmission to the base station, $\eta_s I_{elec}k$ is the current drawn for reception, and $\eta_s k I_{proc}$ is the current required for processing the data. Unlike the individual sensors, for a cluster head it is reasonable to include the cost of data processing. This value of current $I_{s,c}$ is drawn by the node for the complete time cycle.

If a node s is not a cluster head, and is turned on, then its load current denoted by $I_{s,on}$ is given by

$$I_{s,on} = I_{T_x}(k, d_{s,c}) + I_{elec}k + I_{elec}kN(s) + I_{T_x}(k, 2r) \qquad (6.46)$$

where $d_{s,c}$ represents the distance from the node s to the cluster head. In Eq. (6.46), $I_{T_x}(k, d_{s,c})$ is the current for transmission to the cluster head, $I_{elec}k$ is the current drawn for the reception process from the cluster head, $I_{T_x}(k, 2r)$ and $I_{elec}kN(s)$ represents the overhead for transmitting and receiving the weight information to/from all alive neighbors. The current $I_{s,on}$ is drawn by the node all through the time cycle.

If a node s is not a cluster head and is turned off, then its load current $I_{s,off}$ for the time cycle is given by

$$I_{s,off} = I_{elec}k + I_{elec}kN(s) + I_{T_x}(k, 2r) \qquad (6.47)$$

This current value $I_{s,off}$ is drawn for a very small fraction of the time cycle. The node draws no current for the rest of the time cycle, as it goes to the sleep state. The load on the battery of sensor s is given by one of Eqs. (6.44), (6.45), (6.46), or (6.47), depending on the state of sensor s and the protocol.

Once the entire load profile up to time t is known, we can compute the residual charge in a battery. This is described in the following section.

Battery Model   We employ a recently developed [65] analytical model of battery that captures, to a first order, two important characteristics: the *rate-dependent capacity effect* and the *charge recovery* effect. The rate-dependent capacity effect refers to the fact that the *usable* or *available* capacity of the battery at time t is a function of the *discharge rate*. The charge recovery effect refers to the fact that when a battery is put to rest, its capacity increases beyond what it was just before it was put to rest.

The model is based on the diffusion process inside a *symmetric* cell. It is characterized by two parameters, denoted by α and β (see Fig. 6.31). The α parameter models the battery's theoretical capacity (expressed in Ah), while β is a measure of the rate at which the active charge carriers are replenished at the electrode surface. The *lifetime L* of the battery is defined as the time when the carrier concentration at the electrode surface falls below a threshold, and the battery cuts off. A "large" value of β means that the diffusion process can "keep up" with the rate of charge withdrawl at the electrode. In this case the second term becomes negligible, and the battery acts like an ideal source. Figure 6.31 shows a useful visualization of the model, which explains how it captures the charge recovery effect.

At any time t, the total battery capacity, which is *constant*, is a sum of three quantities— the charge $l(t)$ that is *lost* to the load by t, the charge $a(t)$ that *available* at t, and their

$$\alpha = \int_0^L i(\tau)\,d\tau + \sum_{m=1}^{\infty} \int_0^L i(\tau)e^{-\beta^2 m^2 (L-\tau)}\,d\tau$$

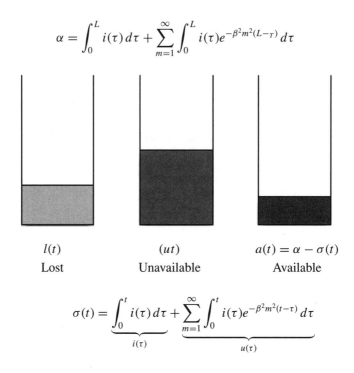

$$l(t) \qquad\qquad (ut) \qquad\qquad a(t) = \alpha - \sigma(t)$$

Lost   Unavailable   Available

$$\sigma(t) = \underbrace{\int_0^t i(\tau)\,d\tau}_{i(\tau)} + \underbrace{\sum_{m=1}^{\infty} \int_0^t i(\tau)e^{-\beta^2 m^2 (t-\tau)}\,d\tau}_{u(\tau)}$$

Figure 6.31 Battery model.

difference $u(t)$, which is the charge *trapped* or *unavailable* at t. When a load draws current from the battery, the charge is delivered from the container $a(t)$. The *apparent charge lost* is $\sigma(t) = l(t) + u(t)$. When $a(t)$ reaches zero [i.e., $\sigma(t) = \alpha$], the battery is cutoff. At this point, the load is zero. Keeping the load at zero, as t increases, $u(t)$ decreases, as does $\sigma(t)$. Therefore $a(t)$ increases. This means that the unavailable charge is being made available, representing charge recovery. For a given battery, a computationally efficient method for estimating the parameters α and β based on a sequence of constant load tests is given [65].

For a sequence of constant current loads, the equations shown in Figure 6.31 can be further simplified. Consider a load profile given as a sequence of N constant current values $I_0, I_1, \ldots, I_{N-1}$, applied to the battery until the battery is fully discharged (up to time $t = L$). The load I_k starts at time t_k and is applied for a duration $\Delta_k = t_{k+1} - t_k$. Then the equations for α and $\sigma(t)$ in Figure 6.31 are replaced by

$$\alpha = \sum_{k=0}^{N-1} I_k \Delta_k + 2 \sum_{k=1}^{N-1} I_k \sum_{m=1}^{\infty} \frac{e^{-\beta^2 m^2 (L-t_k-\Delta_k)} - e^{-\beta^2 m^2 (L-t_k)}}{\beta^2 m^2} \tag{6.48}$$

$$\sigma(t) = \underbrace{\sum_{k=0}^{n-1} I_k \Delta_k}_{l(t)} + \underbrace{2 \sum_{k=1}^{n-1} I_k \sum_{m=1}^{\infty} \frac{e^{-\beta^2 m^2 (t-t_k-\Delta_k)} - e^{-\beta^2 m^2 (t-t_k)}}{\beta^2 m^2}}_{u(t)} \tag{6.49}$$

Once the load profile of a sensor is estimated, the battery model is used to compute the weight of a sensor $w_s(t) = a_s(t)$, where $a_s(t) = \alpha - \sigma(t)$ is the charge available at time t at sensor s.

6.4.8 Experimental Results

Efficiency of Representation On a grid of size 50×50, n sensors were randomly distributed. The results reported in all simulations were averaged over 100 randomly generated networks for each value of n. The programs were implemented using CUDD decision diagram package [66] and were executed on a 1.8-GHz Pentium 4, with 512 MB of memory. It should be noted that the Y axis represents numbers in the log scale for Figures 6.32 to 6.35.

To examine the efficiency of the representation, we varied the network density by increasing the number of sensors from 50 to 100, each with a sensing radius of $r = 10$. Figure 6.32 shows that while the total number of covers increases significantly with respect to n, the size of the ROBDD representation increases at a much smaller rate. In fact, in Figure 6.33, the ratio of the ROBDD size to the number of sensing covers approaches zero as the number of covers increase.

Next, we varied the sensing radii of every sensor from 8 to 20, while keeping the $n = 75$. Figure 6.34 shows that the number of covers increases as expected with respect to the sensing radii. However, the size of ROBDD for χ_R initially increases for smaller radii and thereafter decreases for larger radii. This is due to the fact that for smaller radii, the cover matrix is sparsely filled. As a result, the ROBDD representations are more compact. Similarly for larger sensing radii, the cover matrix is very dense, and this again results in a compact representation. However for a moderate radius, when the matrix is neither very sparse or dense, the ROBDD representation is larger.

Figure 6.35 shows how the number of covers varies with ρ. Recall, that ρ is a user-specified *data quality* constraint that is expressed as the minimum number of sensors required to cover each point in R. ρ was varied from 1 to 4, with r and n being 10 and 90, respectively. As expected the number of covers and ROBDD size decreases as ρ increases. This is again due to sparsity of the covering matrix.

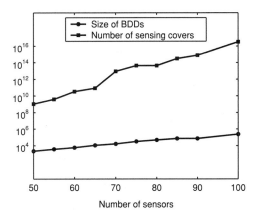

Figure 6.32 Impact of network density.

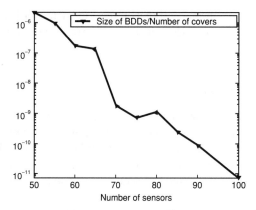

Figure 6.33 Ratio of size of ROBDDs to number of sensing covers.

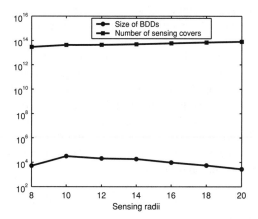

Figure 6.34 Increasing sensing radii.

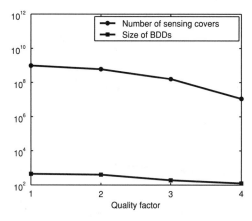

Figure 6.35 Increasing ρ.

Note that even for a very sparse network with 50 sensors and sensing radii of 10, the number of covers is several orders of magnitude greater than those considered by any of the previous approaches [53]. The run times to compute all covers varied from 2 to 426 s with increase in network density from 50 to 100 sensors. Thus, the presented approach provides an ideal platform for developing more efficient sensor management techniques.

Centralized Sleep State Control   We present the results of experiments to verify the approach and the improvement in the network lifetime due to battery recovery effect. The battery parameters α and β were estimated by direct measurements of a 3-V lithium ion battery [54]. For the current set of simulation results, the parameters $\alpha = 40,000$, $\beta = 0.2$, $\tau = 10$ min and current constants $I_{elec} = 1\,\mu A/bit$, $I_{amp} = 2\,nA/bit/m^2$ per packet. The packet size was kept at 4000 bits. These current parameters are derived from the energy constants specified in LEACH [46]. In our energy model, we do not consider the energy spent in the idle mode. The proposed algorithms will result in significant improvements over the baseline scheme by considering the idle mode energy consumption. The reason is explained as follows. In our baseline scheme, with which we compare our results, radios of all sensors are turned on throughout the time cycle. Hence, all sensors drain their battery resources even when they are in the idle state for the entire time cycle. On the other hand, the proposed algorithm turns off all the suboptimal redundant nodes in the network after the initial arbitration phase and only a subset of nodes keep their radios on for the complete time cycle. Hence, greater energy savings can be achieved by additionally considering the idle mode energy consumption. On a 50×50 grid, we randomly distributed n sensors where n varies from 50 to 100 sensors.

We studied lifetimes in two different scenarios: direct transmission with and without CSSC. With CSSC, the improvement in lifetime was as high as $3.7X$ (see Fig. 6.36) when using M3C sequential schedule. An improvement in lifetime of $3.9X$ (see Fig. 6.37) is observed when the M3C switching schedule is used. The percentage improvements of M3C switching schedule normalized with respect to the M3C sequential schedule are shown in Figure 6.38. Results shown in Figure 6.38 were performed by varying n and keeping $\rho = 1$ and $r = 10$. The results indicate up to 5% improvement in the sensor network lifetime is possible for the same M3C sensor management scheme by introducing switching of covers (Fig 6.38). The improvement of switching over the sequential discharge mechanism is a function of the location of the base station. The farther the base station the greater is the

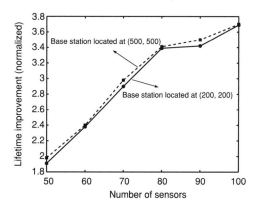

Figure 6.36   Lifetime improvement with M3C sequential scheme.

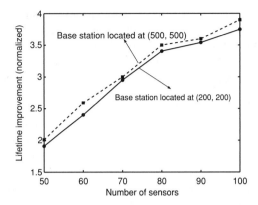

Figure 6.37   Lifetime improvement with M3C switching scheme.

current discharge at each sensor and the higher is the charge recovery effect. Hence, the M3C switching schedule performs much better as the distance to the base station increases. In the sequential scheme, when a sensor is exhausted, it is removed from further consideration. Thus no battery recovery effect is present. In the switched scheme, each sensor is discharged for a fixed amount of time, and it can become part of a cover in a subsequent iteration. Consequently, during its idle period, some of its unavailable charge will become available. To demonstrate this, the percentage improvement in the network lifetime was computed as a function of the battery parameter β. Recall that the large value of β indicates that the diffusion process can *track* the discharge and, consequently, $\sigma(t)$ represents the actual charge consumed, and there is no unavailable charge to recover. Figure 6.39 clearly shows how the switching scheme exploits the charge recovery capability. For $n = 100$, the maximum improvement shown in Figure 6.38 is approximately 5%. All of this improvement is due to a charge recovery as shown in Figure 6.39. As β increases, the improvement of the switching scheme over the sequential scheme decreases.

Distributed Sleep State Control   We implemented DSSC with different protocols to verify the approach and examine the improvement in the network lifetime. This was done for both direct transmission and LEACH protocols. Since the source code for LEACH is

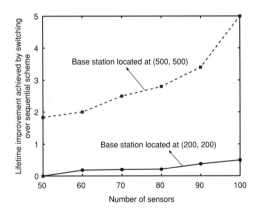

Figure 6.38   Lifetime improvement achived by switching over sequential scheme .

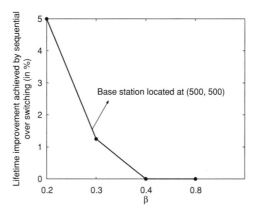

Figure 6.39 Effect of β on lifetime for $n = 100$, $r = 10$.

publically available, we demonstrate this on LEACH protocol. However, in principle lifetime improvements can also be achieved by implementing our algorithms with SPAN [44], ASCENT [49], or any other protocols.

For DSSC, much larger grids and networks can be supported. For this reason, a grid of 100×100 was used, and n sensors were randomly distributed. Again, results reported in all simulations were averaged over 100 randomly generated networks for each value of n. We used a packet size of 4000 bits and assumed that the base station is located at (600,600).

Each sensor is *on* for a time cycle. A time cycle is the minimum time a sensor needs to be turned on, and it is a user-specified information. The information content required by a specific application and the data rate of the system determine the value of the time cycle. The overhead in implementing DSSC involves transmission and reception of the weight information messages after every time cycle. The sizes of the weight information messages is assumed to be small, as they carry only sensor ID and energy information. We assume the size of these messages is 100 bits in our simulation results.

Direct Transmission Protocol   Direct transmission protocol (DTP) is a simple protocol in which each sensor transmits directly to the base station. DTP was implemented with and wihtout DSSC. The lifetime improvement achieved by DSSC is shown in Figure 6.40.

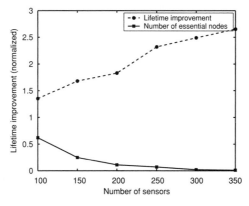

Figure 6.40 Lifetime improvement DSSC over direct transmission ($r = 8$).

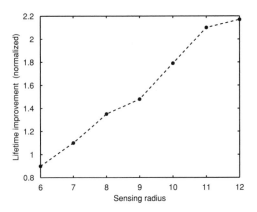

Figure 6.41   Effect of r on DSSC with direct transmission, $n = 100$.

Results were obtained for network sizes varying from 100 to 350, keeping $r = 8$. DSSC exploits the redundancy in the network to enhance the lifetime of the network by $2.5X$. DSSC allows sensors to go to sleep as often as possible, enabling sensors to recover some amount of unavailable charge. As the number of essential sensors in the network decreases (i.e., as redundancy increases), the improvement in the lifetime also increases. Lifetime improvement is highly correlated with the amount of redundancy in the network. The effect of increasing the sensing radius on the lifetime improvement is shown in Figure 6.41. As the sensing radius was increased from 6 to 12 for a 100-sensor network, the lifetime is improved by about $2.2X$. As the sensing radius is increased, the number of covers increases, facilitating the sensors to go to sleep more often.

LEACH Protocol   LEACH [46] is one of the more well-known protocols in wireless sensor networks. In LEACH, a limited percentage of sensors elect themselves as cluster heads and the rest of the sensors associate themselves to a cluster head. Each sensor transmits its messages to the local cluster head, and the local cluster head fuses the data from all the individual sensors in its cluster and finally transmits the fused data packet to the base station. The responsibility of being a cluster head changes after each cycle.

The lifetime improvement for LEACH due to DSSC is shown in Figures 6.42, 6.43, and 6.44 as the number of sensors n vary from 100 to 350 and $r = 8$. An improvement of about 18% due to DSSC is observed for a network with 350 sensors. We studied the lifetime improvement for values of the cycle time equal to 10, 20, and 30 min and $\beta = \{0.2, 0.4, 0.8\}$. Figures 6.42, 6.43, and 6.44 show the impact of increasing cycle time for different values of β. As the cycle time increases, the improvement in lifetime due to DSSC increases. A large cycle time allows larger number of sensors to sleep, thereby allowing them to recover the trapped charge in their batteries resulting in the lifetime improvement.

Additionally, we observe that the lifetime improvement of DSSC is strongly dependent on the redundancy in the network. The improvement in lifetime decreases as the redundancy in the network decreases. As the redundancy decreases in the network, the number of covers for each sensor decrease, forcing a sensor to turn itself *on* more often to provide adequate coverage.

Figure 6.45 shows the times at which 5, 10, 25, 50, 75, and 100% of the sensors die using LEACH with and without DSSC. The percentage improvement in lifetime at which the sensors die over LEACH is shown in Figures 6.46 and 6.47. Implementing DSSC with

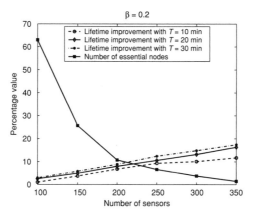

Figure 6.42 Lifetime improvement with DSSC over LEACH for varying cycle durations, $\beta = 0.2$.

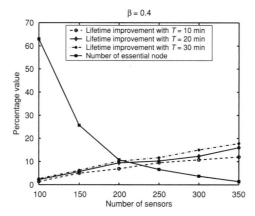

Figure 6.43 Lifetime improvement with DSSC over LEACH for varying cycle durations, $\beta = 0.4$.

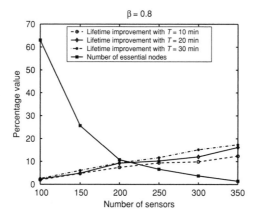

Figure 6.44 Lifetime improvement with DSSC over LEACH for varying cycle durations, $\beta = 0.8$.

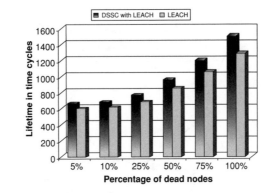

Figure 6.45 Performance results for $n = 350$, $r = 8$, $T = 30$ min and $\beta = 0.8$.

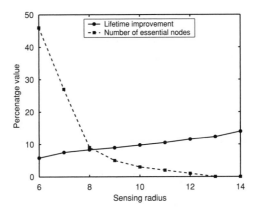

Figure 6.46 Effect of r on DSSC with LEACH for $n = 200$, $\beta = 0.8$.

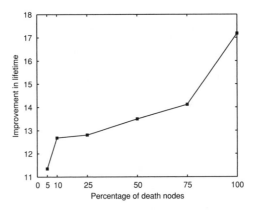

Figure 6.47 Improvement in lifetime for varying percentages of sensor deaths in the network.

LEACH shows that lifetimes can be increased. The key observation is that DSSC allows a uniform distribution of energy in the network, thereby keeping alive as many sensors as possible. From Figure 6.45, we observe that when using DSSC, sensors die gradually, thereby providing maximum possible coverage for a longer time.

6.4.9 Conclusions

Sensor networks are designed to be redundant, and the presented approach exploits redundancy in the network to maximize the lifetime of the network. The method provides precise formalism for sensor covering and includes a novel battery model to maximize the available charge in each sensor. Also as demonstrated the formulation naturally extends to computing covers with specified quality factor ρ. In the centralized approach, the simulation results show that by switching between covers, the battery recovery effect can be exploited, and network lifetime can be further improved. We extended the centralized covering algorithm to a distributed one and developed a novel localized coordination strategy using the residual battery charge at each node. The coordination strategy puts a large number of nodes to sleep, without comprising on the coverage. The distributed nature of the algorithm, allows it to scale linearly as the size of the network increases. Simulation results show that implementing DSSC, with existing communication protocols can substantially enhance the lifetime of the network.

6.5 BATTERY LIFETIME ESTIMATION AND OPTIMIZATION FOR UNDERWATER SENSOR NETWORKS

Raja Jurdak, Cristina Videira Lopes, and Pierre Baldi

Acoustic technology has been established as the exclusive technology that provides robust underwater communications for military and civilian applications. One particular civilian application of interest is the deployment of underwater acoustic sensor networks. The main challenges of deploying such a network are the cost and the limited battery resources of individual sensor nodes. Here, we provide a method that addresses these challenges by estimating the battery lifetime and power cost of shallow water networks, in terms of four independent parameters: (1) internode distance, (2) transmission frequency, (3) frequency of data updates, and (4) number of nodes per cluster. Because transmission loss in water is dependent on both frequency and distance, we extend the general method to exploit topology-dependent distance and frequency assignments. We use the method to estimate the battery life for tree, chain, and grid topologies for various combinations of internode distance, frequency, and cluster size in a shallow water setting. The estimation results reveal that topology-dependent assignments prolong battery life of the tier-independent method by a factor of 1.05 to 131 for large networks. In the case of a linear network deployed along a coastline with a target battery life of 100 days, topology-dependent assignments could increase the network range and aggregated sensor data of the topology-independent method by a factor of 3.5.

6.5.1 Background Information

Underwater acoustic communication has been used for a long time in military applications. Compared to radio waves, sound has superior propagation characteristics in water, making it the preferred technology for underwater communications. The military experience with this technology has led to increased interest for civilian applications, including the development of underwater networks. The main motivation for underwater acoustic networks is their relative ease of deployment since they eliminate the need for cables and they do not interfere with shipping activity. These networks are useful for effectively monitoring the underwater medium for military, commercial, or environmental applications. Environmental applications include monitoring of physical indicators [67] (such as salinity, pressure, and temperature) and chemical/biological indicators (such as bacteria levels, contaminant levels, and dangerous chemical or biological agent levels in reservoirs and aqueducts).

The work presented in this section is part of an interdisciplinary effort at University of California, Irvine to develop a shallow-water underwater sensor network for real-time monitoring of environmental indicators, similar to current air quality monitoring systems. One of the major considerations for the development of such a network is the power consumption at individual nodes. This work is motivated by the practical need to estimate the battery life of sensor nodes, which has implications on the usefulness, topology, and range of the network. Estimating the battery life of sensor networks prior to design and deployment of the actual network requires an analytical method that coarsely captures the behavior of a shallow-water sensor network. On the theoretical level, this work is driven by the need to develop a generic method for battery lifetime estimation that combines both the networking and medium-specific aspects in sensor networks.

Most of the existing work has focused on modeling the battery lifetime of sensor networks in air [68, 69]. The goals of this work are:

- To provide an estimation method for network battery lifetime specific to the conditions of underwater acoustic sensor networks
- To propose topology-dependent optimizations for power consumption
- To use the estimation method to evaluate the benefits of the optimizations for a typical shallow-water sensor network

6.5.2 Related Work

Interest in underwater acoustics dates back to the early twentieth century when sonar waves were used to detect icebergs [70]. Later, the military started using underwater acoustics for detecting submarines [70] and mines [71, 72]. Underwater acoustic applications further extended to seafloor imaging [73], object localization and tracking [74], and data communication [70] for ocean exploration and management of coastal areas. The previous experiences with underwater acoustics have led to the design of underwater sensor networks that include a large number of sensors and perform long-term monitoring of the underwater environment [75]. In underwater sensor networks, the issue of limited battery resources at the sensors is particularly important because of the difficulty and cost of recharging sensor batteries once the network is deployed.

In the recent literature, several approaches address estimation and optimization of the lifetime of energy-constrained networks. In the context of underwater networks, Fruehauf and Rice [76] propose the use of steerable directional acoustic transducers for signal transmission and reception in underwater nodes to reduce the energy consumption and thus prolong the lifetime of a node. Among other approaches that apply to more general energy-constrained networks, Tilaky et al. [77] assess the trade-offs involved in the design and topology of sensor networks. Marsan et al. [78] consider techniques to maximize the lifetime of a Bluetooth network by optimizing network topology and argue that their optimization techniques are also applicable to general ad hoc networks. Several routing [79–82] and MAC [83] algorithms have been developed for energy-efficient behavior in sensor network in order to maximize network lifetime. For example, Misra and Banerjee [79] present a routing algorithm to maximize network lifetime by choosing routes that pass through nodes with currently high capacity. The capacity of a node according to [79] is a combined measure of the remaining battery energy and the estimated energy spent in reliably forwarding data of other nodes in the network. Panigrahi et al. [84] derive stochastic models for battery behavior to represent realistic battery behavior in mobile embedded systems. In our work, we model battery behavior as a function of the acoustic transmit and receive power, which are the dominant sources of power consumption in underwater transceivers [85]. Some models [68,69] attempt to derive an upper bound on the lifetime of a sensor network, in terms of a generic set of parameters. Some of the parameters in our method, such as the internode distance and the number of nodes that relay data to the sink, are also considered in [68] and [69]. However, both of the previous models assume a path loss inversely proportional to d^n, where d is the distance between a sender and receiver. This assumption applies to most aerial wireless networks but does not capture the specific conditions of underwater networks in which the path loss depends on frequency as well as distance [see Eqs.(6.52) and (6.53) below]. Furthermore, delay and multipath propagation effects in underwater networks are certainly different from aerial networks. The case of relatively infrequent data updates is addressed in [86], which focuses on radio frequency sensor networks where nodes periodically send data updates toward the central node. In our method, we also consider the case of infrequent data updates toward a central node in underwater acoustic networks, and as in [86], we attempt to derive algorithms for data gathering and aggregation that maximize the lifetime of the network.

6.5.3 Network Battery Life Estimation Method

The challenges of designing shallow-water acoustic networks include the following:

1. *Spectrum Allocation:* The limited available acoustic spectrum [87] in underwater environments makes this issue particularly challenging.
2. *Topology:* Internode distances and number of forwarding nodes are factors that impact the overall performance of the network [68,78,86].
3. *Shallow-Water Environment:* This environment tends to have distinct multipath characteristics [87,88], for instance, due to surface reflection of the signal. Shallow-water noise also follows distinct patterns because of various noise sources [89], such as winds and shipping activity.

Design choices that address these challenges affect the battery lifetime of the network, which is our main metric of interest. The network battery life must be sufficiently long to avoid recharging or replacing node batteries frequently. A related metric that can be formulated is the power consumption to throughput ratio (PCTR), indicating the power cost of transmitting bits in the network.

Maximizing battery lifetime while minimizing PCTR typically requires networks to have less frequent data updates, lower spatial density, or shorter range [77]. All of these characteristics yield lower accuracy in the sensed data. Thus, there is a trade-off involved between prolonging network lifetime and maximizing the accuracy of sensed data.

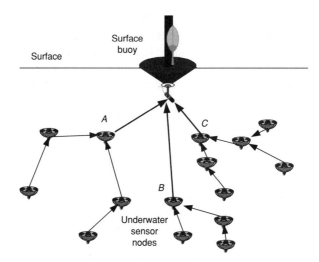

Figure 6.48   Example underwater sensor network.

Consequently, the first step in our network battery life estimation method is to identify the design parameters that impact battery lifetime and power consumption, which are highly dependent on the network scenario. Next, the method investigates the signal propagation characteristics in the deployment region of interest as a function of the independent variables to derive the required transmission power for successful data reception. Third, we exploit the fact that data dissemination in our network is periodic, and we compute the power cost of data delivery during one update period. Finally, the method uses the data delivery power cost during an update period to estimate the battery lifetime and power cost of the network. Each of the remaining subsections in this section focuses on one of these steps.

Network Design Parameters   Figure 6.48 illustrates our generalized network topology to analyze the tradeoffs of accurate underwater environmental indicator monitoring and power efficiency. The network in Figure 6.48 has a multihop centralized topology in which several trees are rooted at the base station, and data flow is always toward the base station. The convergence of data at the base station is appropriate for underwater sensor networks because sensor data in these networks is typically sent to shore for collection and analysis.

In the topology of Figure 6.48, nodes monitor their surrounding environmental conditions and periodically send the collected information toward a central shore or surface station, which subsequently collects and processes the data. We consider the transmit and receive power to be the main sources of power consumption at each node [68, 86], and we assume that the sensing and processing powers are negligible.

Channel allocation is trivial for sparse networks since the data updates can be scheduled so that all nodes can use the same frequency channel at different times. However, as the network density increases, nodes must tightly synchronize their transmissions to avoid collisions on the common channel. Requiring tight synchronization among sensors adds implementation and communication cost to the network. Thus in the case of fairly dense networks, the first challenge is to provide a multiple access technique that does not rely on node synchronization and that allows simultaneous transmissions by several nodes. We consider frequency division multiplexing as a multiple access technique for our method.

Because the transmissions of nodes are separated through distinct frequency channels, a node A that uses a channel with a higher frequency consumes more power than a node B using a lower frequency channel because underwater signal propagation depends on both frequency and distance. As a result, the battery resources at A run out earlier than the resources at B. Thus, the maximum frequency (f) in any spectrum allocation scheme determines the worst case for battery lifetime and power consumption of the network.

Another factor that impacts network battery lifetime and power consumption is the frequency of data updates from sensors. One reasonable technique to prolong battery life is to increase the update period (R), which yields a lower power consumption rate. Significant variations in underwater medium conditions occur on the scale of a few minutes to the scale of decades [90, 91]. For example, managing a recreational beach area requires measuring danger from currents and wave sizes every several minutes. In contrast, coastal zone pollution management requires measurements in the time scale of years. Thus, an update period in the order of 20 min is sufficient to capture the environmental variations that occur in the shorter time scale.

To avoid consuming power for sending signals over long distances, we consider a multihop topology in which nodes that are closer to the base station[5] forward the signals of nodes further away from the base station (see Fig. 6.48).

As such, nodes that are further away from the base station need only consume transmit power to get the signal to the next hop. Thus, the internode distance (d) (or the length of one hop) has significant impact on power considerations of a multihop network. A multihop topology extends the range of operation of the network, but it raises the issue of increased power overhead at intermediate nodes, which forward to the data of nodes further away. For example, if traffic routing is based solely on distance, then the nodes closest to the base station must forward the data of all the other nodes in the network. As such, it is important that the power costs of forwarding do not overburden the forwarding nodes.

To address this issue, nodes are divided into clusters that are defined by proximity. Within each cluster, nodes are segmented into tiers. Figure 6.49 shows a network topology with four clusters and three tiers per cluster. The nodes at the lowest tier (tier 3 in Fig. 6.49) are the furthest away from the base station and transmit messages to other nodes in the same cluster at the next higher tier (tier 2); tier 1 nodes, which are closest to the base station, finally transmit the accumulated data to the base station. Therefore, tier 1 nodes represent the bottlenecks in terms of battery lifetime because they carry the burden of transmitting the messages of all other nodes in their respective clusters. Thus, the number of nodes in a cluster (M) is an important design choice of the network. The choice of M depends on the data sampling granularity that the application requires. M also establishes a trade-off between the power consumption for transmissions over large distances and the power overhead of forwarding data. Note that forwarding nodes could aggregate or fuse their own data [92] with data arriving from more distant nodes in order to compress the overall amount of data to be transmitted, and, ultimately, to save on transmission power. Our method does not consider aggregation and thus presents a conservative estimate of the power consumption at the forwarding nodes.

In sum, we identified four important network design parameters that impact the battery lifetime and power consumption of an underwater sensor network: (1) the transmission frequency f, (2) the update period R, (3) the average signal transmission distance d, and (4) the number of nodes in a cluster M.

---

[5] A base station could be mounted on a surface buoy or on a nearby location on shore.

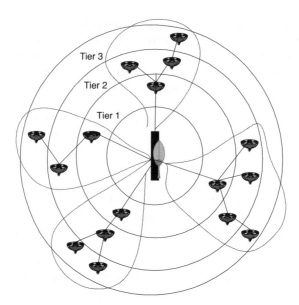

Figure 6.49 A Network with four clusters and three tiers.

Underwater Acoustics Fundamentals

Passive Sonar Equation The passive sonar equation [70] characterizes the signal-to-noise ratio (SNR) of an emitted underwater signal at the receiver:

$$SNR = SL - TL - NL + DI \tag{6.50}$$

where SL is the source level, TL is the transmission loss, NL is the noise level, and DI is the directivity index. All the quantities in Eq. (6.50) are in dB re μPa, where the reference value of 1 μPa amounts to 0.67×10^{-22} W/cm^2 [70]. In the rest of the section, we use the shorthand notation of dB to signify dB re μPa.

Factors contributing to the noise level NL in shallow-water networks include waves, shipping traffic, wind level, biological noise, seaquakes, and volcanic activity, and the impact of each of these factors on NL depends on the particular setting. For instance, shipping activity may dominate noise figures in bays or ports, while water currents are the primary noise source in rivers. For the purpose of this analysis, we examined several studies of shallow-water noise measurements under different conditions [70,89]. As a result, we consider an average value for the ambient noise level NL to be 70 dB as a representative shallow-water case. We also consider a target SNR of 15 dB [70] at the receiver.

The directivity index DI for our network is zero because we assume omnidirectional hydrophones. Note that this is another conservative assumption, since using a directive hydrophone as described in [76] reduces power consumption.

Through the above assumptions, we can express the source level SL intensity as a function of TL only:

$$SL = TL + 85 \tag{6.51}$$

in dB.

Transmission Loss   The transmitted signal pattern has been modeled in various ways, ranging from a cylindrical pattern to a spherical one. Acoustic signals in shallow waters propagate within a cylinder bounded by the water surface and the seafloor, so cylindrical spreading applies for shallow waters. Urick [70] provides the following equation to approximate the transmission loss for cylindrically spread signals:

$$TL = 10 \log d + \alpha d \times 10^{-3} \tag{6.52}$$

where d is the distance between source and receiver in meters, α is the frequency-dependent medium absorption coefficient, and TL is in dB.

Equation (6.52) indicates that the transmitted acoustic signal loses energy as it travels through the underwater medium, mainly due to distance-dependent attenuation and frequency-dependent medium absorbtion. Fisher and Simmons [93] conducted measurements of medium absorbtion in shallow seawater at temperatures of 4 and 20°C.

We derive the average of the two measurements in Eq. (6.53), which expresses the average medium absorption at temperatures between 4 and 20°C:

$$\alpha = \begin{cases} 0.0601 \times f^{0.8552} & 1 \le f \le 6 \\ 9.7888 \times f^{1.7885} \times 10^{-3} & 7 \le f \le 20 \\ 0.3026 \times f - 3.7933 & 20 \le f \le 35 \\ 0.504 \times f - 11.2 & 35 \le f \le 50 \end{cases} \tag{6.53}$$

where f is in kHz, and α is in dB/km.

Through Eq. (6.53), we can compute medium absorbtion for any frequency range of interest. We use this value for determining the transmission loss at various internode distances through Eq. (6.52), which enables us to compute the source level in Eq. (6.51) and subsequently to compute the power needed at the transmitter.

Transmission Power   We have shown how the source level SL relates to internode distance and frequency through Eqs. (6.51), (6.52), and (6.53). SL also relates to the transmitted signal intensity at 1 m from the source according to the following expression:

$$SL = 10 \log \frac{I_t}{1 \, \mu Pa} \tag{6.54}$$

where I_t is in μPa. Solving for I_t yields

$$I_t = 10^{SL/10} \times 0.67 \times 10^{-18} \tag{6.55}$$

in W/m², where the constant converts μPa into W/m².

Finally, the transmitter power P_t needed to achieve an intensity I_t at a distance of 1 m from the source in the direction of the receiver is expressed as [70]

$$P_t = 2\pi \times 1 \, m \times H \times I_t \tag{6.56}$$

in watts, where H is the water depth in meters.

In short, we have presented a method to obtain the required transmitter power for signal transmissions at a given distance d and frequency f. First, we can compute the transmission loss TL in terms of f and d, and we subsequently compute the source level SL, which yields the source intensity I_t. Finally, we can compute the corresponding transmit power P_t needed to achieve a source intensity of I_t.

Data Delivery   We now present the tier-independent method for the estimation of battery lifetime and power consumption. In Section 6.5.4 we consider more sophisticated tier-dependent frequency and distance assignments that build on the tier-independent method. Without loss of generality, we assume that the size of data packets is 1 kbit, which is enough to report sixteen 8-byte measurements, such as temperature, pressure, and salinity at every node in a 20-min interval. We also assume that the bandwidth of each acoustic channel is 1 kHz. Thus, the available bit rate for each node is 1 kbit/s, which is well within the bit rates of current hydrophones [85], and the packet transmission time is 1 s. P_t is thus the power needed to transmit one packet in a contention-less environment. Note that a bandwidth of 1 kHz could be achieved through a combination of spread spectrum and frequency division multiplexing to achieve a higher number of coexisting nodes. Even if these multiple access techniques are used, packet collisions and corruptions remain possible. Furthermore, in each update period, a node not only sends its own data but also the data of other nodes that are further away from a data sink.

We consider a generic medium access control (MAC) protocol where a node accesses the channel, sends a data packet, and awaits an acknowledgment, which has a size of 200 bits. In case the acknowledgment times out, the node retransmits the data packet. Assuming a 0.1 packet loss rate, then each data packet and each acknowledgment is correctly received with a probability of 0.9. Consequently, the probability that both a packet and its corresponding acknowledgment are correctly received is 0.81, implying that each packet must be sent $1/0.81 = 1.23$ times on average. The node consumes power for sending and receiving data packets, as well as sending and receiving acknowledgments. The receive power of each message is typically around one fifth of the transmit power in commercially available hydrophones [85]. Thus, the average power in watts consumed by a node during each update period (frame) is

$$P_{\text{frame}} = 1.23 P_t \times N \left(1 + \tfrac{1}{5} + \tfrac{1}{5} + \tfrac{1}{25}\right) \tag{6.57}$$

where N is the number of data packets that the node forwards during an update period. The first two terms in Eq. (6.57) account for sending and receiving data packets, while the last two terms account for sending and receiving acknowledgments.

This section considers two specific cases of cluster organizations: a linear chain, which represents the worst-case scenario for network lifetime and applies to environmental monitoring along coastlines, rivers, or aqueducts, and a grid topology, which applies to other practical environmental monitoring applications such as in a lake or bay. In the rest of this section, the discussion focuses on the chain topology, and in Section 6.5.5, we apply the method to sensors placed in a grid topology. In the chain architecture, the average number of packets N forwarded by a node is equal to $M/2$. [6]

As mentioned earlier, tier 1 nodes represent the bottleneck for network battery life since they have the highest forwarding burden of all nodes. Thus, we express the maximum

---

[6] This is a conservative estimate.

amount of power consumed during one frame at a tier 1 node as:

$$P_{max} = 1.23 P_t \times N_{max}\left(1 + \tfrac{1}{5} + \tfrac{1}{5} + \tfrac{1}{25}\right) \qquad (6.58)$$

in watts, where N_{max} is the maximum number of packets forwarded by a tier 1 node. In the chain architecture, tier 1 nodes send their own data packet and forward the packets of all other nodes in the cluster during each update period, so N_{max} for this architecture is equal to M.

Network Lifetime and Power Consumption A good measure of overall network power consumption is the ratio of overall power consumption to throughput. During each update period, each node in a cluster of M nodes sends its own data packet and forwards any pending data packets of its neighbors, yielding an average PCTR of:

$$\text{PCTR} = \frac{M \times P_{frame}}{M \times 1000 \text{ bits}} = \frac{P_{frame}}{1000} \qquad (6.59)$$

in watts/bit. Next we want to determine the limit on the battery lifetime of a network, which depends mainly on tier 1 nodes. The time that a node's transceiver is active during one update period is important for battery life considerations. Each node uses a store and forward mechanism to forward a sequence of packets as it receives them in order to minimize the active time of its transceiver. Taking into account collisions and retransmissions, the total active time for a tier 1 transceiver in one update period is

$$T_{total} = 1.23 \left(N_{max} + \frac{N_{max}}{5} \right) \qquad (6.60)$$

in seconds.

The next step is selecting a power source. We consider that we have 3 off-the-shelf 9-V, 1.2-A-h batteries at each node. The total energy available at each node is

$$E_t = 3 \times 9 \times 1.2 = 32.4 \qquad (6.61)$$

in Volts per ampere–hour. The total active time of a transceiver is therefore the ratio of the total energy to the power consumed in one frame:

$$T_{active} = \frac{E_t}{P_{frame}} = \frac{32.4}{P_{frame}} \qquad (6.62)$$

in hours. A node's transceiver is only active for a fraction of the time in each update period of R seconds. Therefore, the battery lifetime of a node is expressed by:

$$T_{lifetime} = \frac{T_{active}}{T_{total}} \times \frac{R}{24} \qquad (6.63)$$

in days, where R is in seconds.

6.5.4 Topology-Dependent Optimizations

The tier-independent battery life and power consumption estimation method in Section 6.5.3 treats all network nodes equally, by assuming all internode distances are the same and by assigning frequency values randomly. However, the tier-independent method disregards the fact that tier 1 nodes carry a heavier power burden than other nodes. Consequently, applying measures that favor tier 1 nodes can yield improvements in battery life and power consumption. For this purpose, we propose two enhancements to the tier-independent battery life and power consumption estimation method: tier-dependent frequency assignment and tier-dependent distance assignment.

Tier-Dependent Frequency Assignment   Equations (6.52) and (6.53) indicate that the transmission loss increases at higher frequencies, which implies that nodes using high frequencies must transmit acoustic signals at higher power. Thus, we assign tier 1 nodes the lowest frequency band, and we assign each subsequent tier the next higher frequency band, until nodes at the lowest tier are assigned the highest frequency band. This assignment allows nodes with higher forwarding load to use lower frequencies and thus save power.

Tier-Dependent Distance Assignment   Equation (6.52) also shows that distance is the other independent variable that impacts transmission loss. Therefore, it would be beneficial to assign distances in a way that reduces the power load on nodes at lower tiers. Thus, we place tier 1 nodes at the shortest internode distance from the base station, and we increase internode distance for each subsequent tier.

Required Modifications   One goal of tier-dependent assignments is to reduce the overall power consumption per frame in the network. Thus, tier-dependent assignments require modifications to Equations (6.57), (6.58), and (6.60) in the general method, where N becomes

$$N = M - i + 1 \tag{6.64}$$

for each tier i. As a result, P_{frame}, P_{max}, and T_{total} should be computed for each tier individually. We also modify the expression for PCTR to reflect the distinction among tiers:

$$\text{PCTR} = \frac{\sum_{i=1}^{M} P_{\text{frame}}^i}{M \times 1000} \tag{6.65}$$

in watts/bit, where P_{frame}^i is the power that a node at tier i consumes during one update period.

   The other goal of tier-dependent assignments is to move the bottleneck tier away from the base station. Thus, Eqs. (6.62) and (6.63) use the individual tier values for P_{frame} and T_{total} to compute the battery lifetime of each tier. This modification shifts the dependence of the network battery lifetime from tier 1 to the bottleneck tier i.

6.5.5 Case Study

The requirements of our underwater environmental sensor network effort provided concrete values for some of the parameters discussed above. The deployment region of the

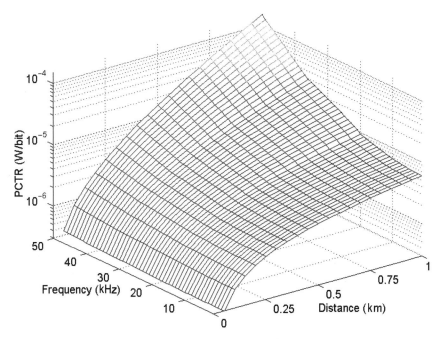

Figure 6.50 PCTR versus distance and frequency, for a cluster size of 500 nodes.

network has a maximum depth of 10 m. To effectively monitor environmental indicators in the water, the recommended internode distances are in the range of 50 m to 1 km. The update period R is 20 min. Furthermore, maintenance work (such as cleaning) must be performed on the sensors themselves every 100 days or so, suggesting a target battery life of 100 days.

In the tier-independent method, we establish bounds for other parameters and analyze the results within those bounds. The maximum frequency varies from 1 to 50 kHz, in steps of 1 kHz. [7] The maximum separation distance, which was established to be between 50 m and 1 km, is increased in steps of 50 m. Finally, we consider that a set of M nodes are communicating within a cluster, where M varies from 1 to 500 with a step of 1.

The rest of this section is as follows. We first derive the PCTR and battery lifetime of the chain topology for each combination of distance, frequency, and cluster size using the tier-independent method. Then, we derive results for the tier-dependent assignment methods, and we compare them to the tier-independent method. Finally, we estimate and compare the battery life and power consumption for a grid topology using the tier-independent and frequency-dependent methods.

Tier-Independent Method   Figure 6.50 shows the power consumption to throughput ratio (PCTR) plotted in terms of the maximum frequency and internode distance for a cluster size of 500 nodes. The PCTR increases with higher transmission frequencies at internode distances above 250 m, whereas frequency has little effect on PCTR at distances below 250 m. The maximal impact of frequency on PCTR can be seen at an internode distance of

---

[7] This is in line with the capabilities of existing hardware.

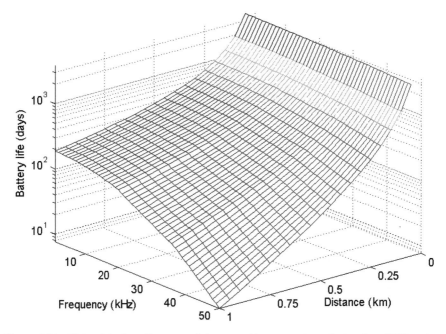

Figure 6.51 Network battery life versus distance and frequency for a cluster size of 500 nodes.

1 km, where transmission frequencies of 1 and 50 kHz exhibit PCTR values of 5.7 and 148 µW/bit, respectively. In contrast, varying internode distances from 50 m to 1 km does cause PCTR to increase for both low and high frequencies, with the sharpest increase of PCTR with distance occurring at 50 kHz.

Figure 6.51 illustrates the variation of the network battery lifetime according to the internode distance and the maximum frequency. The network battery life decreases sharply with increasing distance. When internode distances are small and the nodes transmit at low frequencies, the impact of medium absorption is negligible and most of the consumed power is due to signal attenuation [Eq. (6.52)]. Medium absorbtion plays a larger role as the transmission frequency increases above 10 kHz resulting in shorter battery life. Transmitting at high frequencies over large distances shortens the battery life even further.

Tier-Dependent Assignments Now we derive results for the tier-dependent assignment methods in order to compare them with the tier-independent method. Within the tier-dependent frequency assignment, we consider two subcases:

1. Constant frequency band (CFB): We assign tier i nodes a frequency of i kHz, as long as i is less than 50. For values of i greater than 50, all tiers use a frequency of 50 kHz.
2. Variable frequency bands (VFB): Frequency assignments for VFB are the same as CFB for cluster sizes within 50 nodes. For cluster sizes above 50, we divide up the spectrum into bands of $50/M$, and we assign the lowest frequency band to tier 1 nodes. Each subsequent tier uses the next higher frequency band.

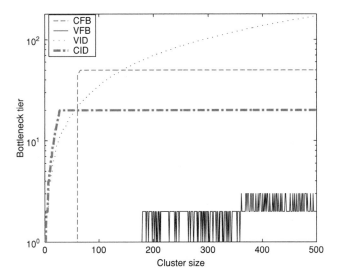

Figure 6.52 Bottleneck tier versus cluster size: The plots for the distance dependent cases are for a frequency of 50 kHz, and the plots for frequency-dependent cases are for a distance of 1 km.

Similarly, tier-dependent distance assignment has two subcases:

1. Constant internode distance (CID): The internode distance of tier i is $50i$ meters for i less than 20, and 1 km for the remaining tiers.
2. Variable internode distances (VID): Internode distances in VID for cluster sizes below 20 are the same as for CID. For cases in VID where the cluster size is greater than 20, the increase in internode distance as we move up one tier is $1/M$ kilometer.

Figure 6.52 provides insight into the impact of tier-dependent assignments on the tier with the shortest battery lifetime (bottleneck tier). The bottleneck tier in the constant frequency band method remains at tier 1 for cluster sizes below 60 nodes. For higher cluster sizes, tier 50 becomes the bottleneck tier since nodes at tier 50 are both using the 50-kHz band (which has the highest power cost) and forwarding the data packets of other nodes. In the variable frequency band method, the bottleneck tier remains at 1 for small cluster sizes, fluctuates between tiers 1 and 2 for moderate cluster sizes, and between tiers 2 and 3 for larger cluster sizes. The bottleneck tier remains close to the base station since only nodes furthest away from the base station are using the highest frequency bands. The bottleneck tier for constant internode distances exhibits a similar behavior to CFB. The bottleneck tier shifts from tier 1 to tier 20 and remains there once the cluster sizes starts to grow. In the case of variable internode distances, the bottleneck tier continues moving away from the base station as M increases to 500, and for a cluster size of 500 nodes, tier 227 is the bottleneck tier.

Figure 6.53 shows the variations of the PCTR for the tier-independent, CFB, VFB, CID, and VID cases as a function of M. The PCTR in the tier-independent case increases linearly with M as a direct consequence of Eqs. (6.57) and (6.59). For the constant frequency band case, PCTR increases at a lower rate for small cluster sizes, where the maximum frequency in the network is less than 50 kHz. At cluster sizes above 50 nodes, PCTR for the constant

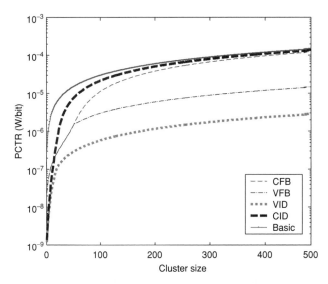

Figure 6.53 PCTR versus cluster size: The plot for the tier-independent method shows PCTR for a distance of 1 km and a frequency of 50 kHz. The plots for the frequency-dependent assignments show PCTR for an internode distance of 1 km, and the plots for the distance-dependent assignments show the PCTR for a frequency of 50 kHz.

frequency band case increases linearly at the same rate as the tier-independent case, since each additional tier uses the frequency of 50 kHz and thus contributes a constant portion of additional power. The two plots converge for large cluster sizes. In the case of variable frequency bands, the PCTR is the same as CFB for cluster sizes below 50 nodes. However, the PCTR for variable frequency bands increases at a lower rate for cluster sizes larger than 50 nodes because VFB uses smaller frequency bands to accommodate additional tiers.

The average power consumption for the constant internode distance method is lower than the frequency-dependent cases only for cluster sizes below 14 nodes. For larger cluster sizes, CID achieves less power savings than the frequency-dependent methods but still improves on the tier-independent case. PCTR in the CID case increases linearly at about the same rate as constant frequency band and the tier-independent case, since each additional tier has an internode distance of 1 km and thus contributes a constant portion of additional power. As a result, the PCTR of the constant internode distances method converges with that of CFB and the tier-independent method for large clusters. Finally, the plot for the variable internode distance case exhibits the lowest PCTR of all cases. It follows the same behavior as CID for cluster sizes within 20, and then it increases slowly toward 3 μW/bit for 500-node clusters. As in the variable frequency band case, the slower rate of increase in PCTR for the variable internode distance case stems from its use of smaller distance increments as the cluster size increases.

Figure 6.54 shows the variation of the network battery life as a function of cluster size using each of the five methods. The results in Figure 6.54 are a natural extension of the results in Figure 6.53. Both CID and CFB yield a longer battery life than the tier-independent case for smaller cluster sizes. The battery life for CID drops more steeply than the battery life for CFB for smaller clusters, but the two plots converge together with the plot of the tier-independent method for high cluster sizes. The improvements in battery life for VFB and VID are more significant. For a cluster size of 500 nodes, variable frequency bands

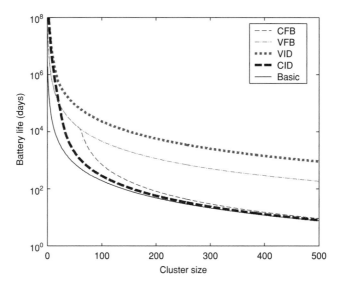

Figure 6.54 Network battery life versus cluster size: The plot for the tier-independent method shows PCTR for a distance of 1 km and a frequency of 50 kHz. The plots for the frequency-dependent assignments show PCTR for an internode distance of 1 km, and the plots for the distance-dependent assignments show the PCTR for a frequency of 50 kHz.

yield a 24-fold improvement in network battery life, whereas variable internode distances prolong the battery life by 131 times compared to the tier-independent method. The ratio of battery life for VID and VFB remains around 5 for medium and large cluster sizes.

Grid Topology The estimation method uses the same equations for the grid topology as the ones for the chain topology, except for the values of $N_{\max}$ and N. In an $S \times S$ grid, $N_{\max}$ takes the value of S and N takes the value of $(S + 1)/2$.

Figure 6.55 illustrates a typical grid topology of nine nodes. The node indices indicate the order in which nodes are placed in the grid coverage area. Once nodes form a perfect square, we begin adding sensors on tier 1 in a new column, then at tier 2, and so on, until we reach the highest tier. In Figure 6.55, once the first four nodes are in place, nodes 5 and 6 are added at tiers 1 and 2 in column 3. Once all existing tiers have a sensor in the

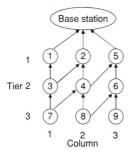

Figure 6.55 Grid topology network with nine nodes: The indices of nodes indicate the order in which the nodes are added to expand the network. The arrows indicate the possible forwarding paths for each node.

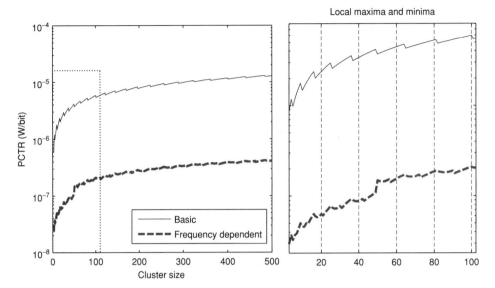

Figure 6.56 PCTR versus cluster size for the grid topology: The plot for the tier-independent method shows PCTR for a distance of 1 km and a frequency of 50 kHz. The plot for the frequency-dependent assignments show PCTR for an internode distance of 1 km.

new column, any additional sensors are placed in a new tier from left to right, until we get another perfect square topology.

Within the grid topology, nodes self-organize into a triangular lattice, as shown in Figure 6.55. This architecture allows two nodes with the same child to share the load of forwarding that child's data. Load sharing is beneficial when one of the two parent nodes has fewer children than the other since the parent nodes can take turns in forwarding the common child's data packets.

We estimate and compare the battery life and power consumption of the grid topology network for the tier-independent and the tier-dependent frequency assignment methods. Because the main application of a grid topology is environmental monitoring at uniform distances, we do not consider tier-dependent distance assignments for this topology.

Figure 6.56 shows the average power consumption in the network as the cluster size grows. An interesting observation of Figure 6.56 is the local maxima at perfect square cluster sizes. For those cases, the forwarding load is evenly split among the nodes of each tier, so load sharing does not yield any benefits. Adding an extra node to a perfect square network at tier 1 enables load sharing among the nodes of tier 1, which yields lower overall average power consumption. There are also local maxima in the plot of the frequency-dependent method at cluster sizes that correspond to a rectangular grid of size $k \times (k + 1)$ for any k. To explain these local maxima, consider again Figure 6.55 for $k = 2$. There are 6 nodes in the network, with 3 in each tier. This symmetry among nodes of the same tier reduces the benefits of load sharing as in the perfect square case. The ratio of battery life of the tier-dependent frequency method to the tier-independent method remains constant with a 30-fold improvement for cluster sizes larger than 50. The power savings that the tier-dependent frequency method achieves over the tier-independent method grow from 0.58 μW/bit for small clusters to 12.5 μW/bit for 500-node clusters.

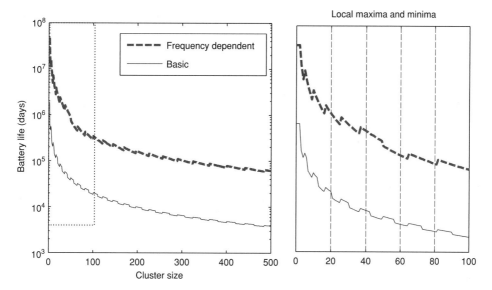

Figure 6.57 Battery life versus cluster size for the grid topology: The plot for the tier-independent method shows PCTR for a distance of 1 km and a frequency of 50 kHz. The plot for the frequency-dependent assignments show PCTR for an internode distance of 1 km.

Figure 6.57 shows the network battery life for the tier-independent and tier-dependent frequency methods as the cluster size grows. The local minima in the plots correspond to the perfect square cluster sizes, where the power consumption peaks (Fig. 6.56). In the tier-independent method, battery lifetime also drops steeply whenever adding a node corresponds to creating a new tier. In contrast, the tier-dependent frequency method does not have sharp drops for creating new tiers, primarily because tiers with high forwarding load use lower frequency bands, so the impact of nodes at a new tier is minimal. The tier-dependent frequency assignment method prolongs the battery life of the tier-independent method by a factor of 15. Even for large cluster sizes of 500 nodes in a 22×22 km^2 area, the battery life for both the tier-independent and tier-dependent methods is in the order of years, which is a significant improvement over the chain topology. This effect stems from the fact that in the grid topology, a fewer number of packets need to be forwarded by low tier nodes and neighboring nodes at the same tier can benefit from load sharing.

6.5.6 Discussion and Conclusion

Maximum Range Alternatives   One of the requirements of our particular shallow-water network is that the sensor nodes should be retrieved and cleaned every 100 days or so. This requirement implies that the network battery lifetime must be at least 100 days. We can derive the options for achieving the target battery life for the chain topology from Figure 6.54.

The options that achieve the target battery life of 100 days are shown in Figure 6.58. The right side of Figure 6.58 shows a magnified view of the overlapping plots in the left side. Using the tier-independent method limits M to 138 nodes per cluster, which provides a network range of 138 km with a density of 1 node/km. The constant internode distance

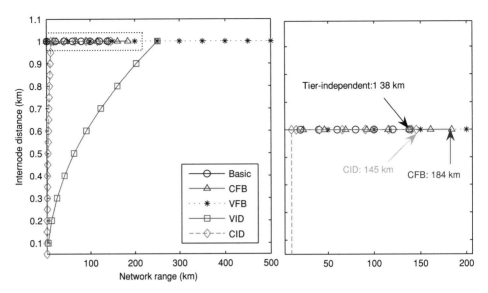

Figure 6.58 Internode distance versus network range for a battery lifetime of 100 days.

method achieves a slightly higher network range of 145 km, with a cluster size of 155 nodes. The node density for CID decreases steadily from 20 nodes/km to 1 node/km for the first 20 tiers, and it remains at 1 node/km for the remaining tiers. The constant frequency band method supports 184 nodes per cluster for a battery life of 100 days, and as a result it further extends the network range to 184 km with a density of 1 node/km. For variable internode distances, the node density decreases steadily from 500 nodes/km at tier 1 to 1 node/km at tier 500, achieving a network range of 250.5 km. The variable frequency band method achieves the highest network range of 500 km, with a cluster size of 500 nodes and a density of 1 node/km. Compared to the tier-independent method, VFB increases the cluster size, network range, and aggregated sensor data by a factor of 3.5. If we prolong the maintenance cycle to 1 year instead of 100 days, the cluster sizes of CFB, CID, VID, VFB, and the tier-independent method drop to 120, 89, 500, 358, and 72, respectively.

In the grid topology, both the tier-independent and the tier-dependent frequency method achieve a battery life of more than a year for 500-node cluster sizes, with a density of 1 node/km and a coverage area of 22×22 km^2.

Method Comparison As the results in Figure 6.58 indicate, tier-dependent distance assignments provide fine-grained sampling of areas that are closer to the base station and less granular data in areas further away. For example, these methods are suitable for networks that require granular coastal data and coarser data from waters beyond coastal areas. In theory, variable internode distance appears to provide for the longest battery life among the five methods considered. However, if nodes cannot be easily anchored at the seafloor at specific distances, then waves may move the sensors, and, as a result, the sensors would have to continuously discover distances from neighbors in order to adjust the transmit power accordingly. Furthermore, as cluster size increases, it becomes more difficult and expensive to realize the shorter internode distances and larger number of sensors that VID requires. The constant internode distance method improves on the tier-independent method, but it has a shorter battery life and a shorter range than VID, VFB, and CFB. However, CID has looser requirements on node placement than VID, which makes it more practical. Since only the first 20 nodes in CID are placed at progressively increasing distances, it is easier and cheaper to place these 20 nodes at the specified distances and subsequently place all other nodes at large approximate distances.

Tier-dependent frequency assignments have looser sensor placement requirements and provide data with uniform granularity. Thus, both frequency assignment methods are suitable for many environmental applications that require sampling of underwater data at regular distance intervals or for applications that tolerate approximate sensor placement. Frequency-dependent assignments are also suitable for self-organizing sensor networks in which the sensors must discover the topology themselves and choose frequency bands according to their position in the topology. Constant frequency bands add only minimal complexity to the tier-independent scheme by requiring that nodes are aware of their position in the topology in order to choose an appropriate frequency. The variable frequency band method, which achieves the longest network range, adds more signal processing complexity, since it requires the same channel rate using a smaller frequency bandwidth.

Grid Topology   Applying the estimation methods to a grid topology with uniformly placed nodes yielded longer network lifetime than all cases of the converging chain network, which is to be expected since the chain topology represents the lower bound on network lifetime. As mentioned earlier, networks with a grid topology are useful for environmental monitoring of lakes or bays. The estimation results that we derived cover for a maximum area of 22×22 km^2. To apply the results to larger areas, a relay station at the edge of each cluster can collect the data and forward to the base station. Alternatively, the network can still use a single base station and simply expand cluster sizes to cover the larger area.

Self-Recharging Sensors   Battery lifetime in sensor networks becomes less of an issue if there is some way of recharging battery resources at individual nodes without human intervention. In an underwater sensor network, nodes can derive mechanical, chemical, or solar energy from their surrounding environment. For example, nodes could absorb and store mechanical energy from water flows through small windmill-like devices. Whether the benefits of such devices overweigh the cost of building them into sensor nodes remains an open issue.

Method Applicability   Although we applied our method to a shallow seawater network, the method also applies to networks at any depth and any fluid. In deeper waters, the impact of both distance and frequency on transmission loss changes. One obvious distinction is that the signal undergoes spherical spreading for deeper waters, as opposed to cylindrical spreading in shallow water. Medium absorption is also depth dependent, and several studies [94] have explored this dependence through measurements. Other factors, such as the noise level, should also be modified to represent deep-water environments. Applying the method to other fluids also requires similar changes to the path loss and noise models. Finally, the network deployment setting may require other changes to the method. For instance, there is no signal spreading in pipes and the transmission loss beyond a certain range is independent of distance.

Conclusion   In sum, we derived a method to estimate the battery life and power cost for underwater sensor networks. Our method first identifies the main independent variables

(f, d, M, R) that impact network battery life and power consumption. Next, the method investigates the signal propagation characteristics in the deployment region of interest as a function of the independent variables (f and d in this case) to derive the required transmission power for successful data reception. Third, the transmission power estimate is combined with the relevant independent variables (M and R in this case) to compute the power cost of data delivery during one update period. Finally, the method uses the data delivery power cost during an update period to estimate the average node battery life and average network power cost.

We applied this estimation method and its tier-dependent variants to a set of shallow-water network scenarios that are representative of our underwater sensor network effort. We found that, for the chain topology, the variable internode distance method achieves the longest battery life compared to the tier-independent and frequency assignment methods, and it provides better fine-grained sampling compared to the other methods for the same target battery life. On the other hand, the variable frequency band method maximizes network range for a given cluster size, provides data samples at uniform granularity, and still achieves a comparatively long battery life.

We also applied the method to a grid topology with uniformly placed sensors to estimate the network battery life and power consumption. The battery life was expectedly longer in the grid topology than the chain topology, and the tier-dependent frequency assignments prolonged battery life nearly by a factor of 15 over the tier-independent method. Because our method is applicable to any topology or fluid medium, researchers can adapt the method to estimate power consumption and network battery life in the initial design and planning stages of fluid sensor networks.

REFERENCES

1. I. Akyildiz, W. Su, Y. Sankarasubramaniam, and E. Cayirci, "A survey on sensor networks," *IEEE Communications Magazine*, pp. 102–114, Aug. 2002.

2. C. Chong and S. P. Kumar, "Sensor networks: Evolution, opportunities, and challenges," *Proceedings of the IEEE*, vol. 91, no. 8, pp. 1247–1256, Aug. 2003.

3. J. Chang and L. Tassiulas, "Routing for maximum system lifetime in wireless ad-hoc networks," in *Proceedings of Thirty-Seventh Annual Allerton Conference on Communication, Control, and Computing*, Monticello, IL, 1999.

4. R. Shah and J. Rabaey, "Energy aware routing for low energy ad hoc sensor networks," in *Proceedings of IEEE Wireless Communications and Networking Conference (WCNC)*, Orlando, FL, 2002.

5. C. Intanagonwiwat, R. Govindan, and D. Estrin, "Directed diffusion: A scalable and robust communication paradigm for sensor networks," in *Proceedings of ACM MOBICOM'00*, Boston, 2000.

6. K. Kalpakis, K. Dasgupta, and P. Namjoshi, "Maximum lifetime data gathering and aggregation in wireless sensor networks," in *Proceedings of IEEE Networks'02 Conference*, Atlanta, 2002.

7. B. Krishnamachari, D. Estrin, and S. Wicker, "Impact of data aggregation in wireless sensor networks," in *Proceedings of DEBS'02*, Vienna, 2002.

8. M. Bhardwaj, A. Chandrakasan, and T. Garnett, "Upper bounds on the lifetime of sensor networks," in *Proceedings of IEEE International Conference on Communications*, Helsinki, 2001.

9. A. T. Hoang and M. Motani, "Adaptive sensing and transmitting for increasing lifetime of energy constrained sensor networks," in *Proceedings of Thirty-Eighth Conference on Information Sciences and Systems, CISS'2004*, Princeton, NJ, 2004.

10. A. T. Hoang and M. Motani, "Buffer and channel adaptive modulation for transmission over fading channels," in *Proceedings ICC'03*, pp. 2748–2752, July 2003.

11. R. A. Berry and R. G. Gallager, "Communication over fading channels with delay constraints," *IEEE Transactions on Information Theory*, pp. 1135–1149, May 2002.

12. A. Sinha and A. Chandrakasan, "Dynamic power management in wireless sensor networks," *IEEE Design and Test of Computer*, pp. 62–74, Mar. 2001.

13. E. Shih, S. Cho, N. Ickes, R. Min, A. Sinha, A. Wang, and A. Chandarakasan, "Physical layer driven protocol & algorithm design for energy-efficient wireless sensor networks," in *Proceedings of ACM SigMobile*, Rome, 2001.

14. L. Benini, A. Bogliolo, and G. Paleologo, "Policy optimization for dynamic power management," *Transactions on CAD*, 1999.

15. E. Chung, L. Benini, and G. D. Micheli, "Dynamic power management for non-stationary service requests," in *Proceedings of DATE*, Munich, 1999.

16. Q. Qiu, Q. Wu, and M. Pedram, "Stochastic modeling of a power-managed system: Construction and optimization," in *Proceedings of ISLPED*, San Diego, 1999.

17. J.-F. Chamberland and V. Veeravalli, "The art of sleeping in wireless sensing networks," paper presented at the IEEE Workshop on Stat. Sig. Proc., St. Louis, Sept. 2003.

18. T. Berger, Z. Zhang, and H. Viswanathan, "The CEO problem," *IEEE Transactions on Information Theory*, pp. 887–902, May 1996.

19. Y. Oohama, "The rate-distortion function for the quadratic gaussian CEO problem," *IEEE Transactions on Information Theory*, pp. 1057–1070, May 1998.

20. P. Ishwar, R. Puri, S. S. Pradhan, and K. Ramchandran, "On rate-constrained estimation in unreliable sensor networks," in *Proceedings of Information Processing in Sensor Networks, Second International Workshop, IPSN 2003*, Palo Alto, CA, Apr. 22–23, 2003.

21. D. Melo and M. Liu, "Analysis of energy consumption and lifetime of heterogeneous wireless sensor networks," in *Proceedings of IEEE Globecom 2002*, Taipei, 2002.

22. W. Heinzelman, A. Chandrakasan, and H. Balakrishnan, "Energy-efficient communication protocol for wireless sensor networks," in *Proceedings of the Hawaii International Conference on System Sciences*, Maui, 2000.

23. E. Shih, S. Cho, N. Ickes, R. Min, A. Sinha, A. Wang, and A. Chandrakasan, "Physical layer driven protocol and algorithm design for energy-efficient wireless sensor networks," in *Proceedings of ACM MOBICOM*, Rome, 2001.

24. P. R. Kumar and P. Varaiya, *Stochastic Systems: Estimation, Identification, and Adaptive Control*, Englewood Cliffs, NJ: Prentice-Hall, 1986.

25. C. Y. Chong and S. P. Kumar, "Sensor networks: Evolution, opportunities, and challenges," *Proceedings of the IEEE*, vol. 91, no. 8, pp. 1247–1256, Aug. 2003.

26. A. Mainwaring, J. Polastre, R. Szewczyk, D. Culler, and J. Anderson, "Wireless sensor networks for habitat monitoring," in *Proceedings of WSNA'02*, Atlanta, Sept. 28, 2002.

27. B. Healy, "The use of wireless sensor networks for mapping environmental conditions in buildings," in *ASHRAE Seminar*, July 2, 2003. Available: http://www.nist.gov/tc75/ASHRAE Summer2003SeminarHealy.pdf.

28. V. A. Kottapalli, A. S. Kiremidjian, J. P. Lynch, E. Carryer, and T. W. Kenny, "Two-tierd wireless sensor network architecture for structural health monitoring," in *Proceedings of SPIE*, San Diego, Mar. 2003.

29. M. Bhardwaj, T. Garnett, and A. Chandrakasan, "Upper bounds on the lifetime of sensor networks," in *Proceedings of IEEE International Conference on Communications*, Helsinki, 2001, pp. 785–790.

30. D. M. Blough and P. Santi, "Investigating upper bounds on network lifetime extension for cell-based energy conservation techniques in stationary ad hod networks," in *Proceedings of MOBICOM'02*, Atlanta, Sept. 2002.

31. K. Kar and S. Banerjee, "Node placement for connected coverage in sensor networks," in *Proceedings WiOpt*, Sophia-Antipolis, France, Mar. 2003.

32. S. S. Dhillon, K. Chakrabarty, and S. S. Iyengar, "Sensor placement for grid coverage under imprecise detections," in *Proceedings of FUSION*, 2002.

33. "Life data analysis reference," http://www.weibull.com/lifedatawebcontents.htm.

34. L. M. Leemis, *Reliability: Probabilistic Models and Statistical Methods*, Englewood Cliffs, NJ: Prentice-Hall, 1995.

35. D. Kececioglu, *Reliability Engineering Handbook*, vols. 1 and 2, Englewood Cliffs, NJ: Prentice-Hall, 1991.

36. A. Papoulis and S. U. Pillai, *Probability, Random Varibales and Stochastic Processes*, 4th ed., New York: McGraw-Hill, 2002.

37. I. F. Akyildiz, W. Su, Y. Sankarasubramaniam, and E. Cayirci, "Wireless sensor networks: A survey," *Computer Networks*, vol. 38, no. 4, pp. 393–422, Mar. 2002.

38. A. Cerpa, J. Elson, D. Estrin, L. Girod, M. Hamilton, and J. Zhao, "Habitat monitoring: Application driver for wireless communications technology," *Mobile Computing and Networking*, pp. 174–185, 2001.

39. V. Raghunathan, C. Schurgers, S. Park, and M.B.Srivastava, "Energy aware wireless microsensor networks," *IEEE Signal Processing Magazine*, vol. 19, no. 2, pp. 40–50, 2002.

40. E. Shih, S. H. Cho, N. Ickes, R. Min, A. Sinha, A. Wang, and A. Chandrakasan, "Physical layer driven protocol and algorithm design for energy-efficient wireless sensor networks," *Mobile Computing and Networking*, pp. 272–287, 2001.

41. A. Woo and D. E. Culler, "A transmission control scheme for media access in sensor networks," *Mobile Computing and Networking*, pp. 221–235, 2001.

42. W. Ye, J. Heidemann, and D. Estrin, "An energy-efficient mac protocol for wireless sensor networks," in *Annual Joint Conference of the IEEE Computer and Communications Societies (INFOCOM)*, vol. 3, New York, 2002, pp. 1567–1576.

43. J. H. Chang and L.Tassiulas, "Energy conserving routing in wireless ad-hoc networks," in *Annual Joint Conference of the IEEE Computer and Communications Societies*, New York, 2000, pp. 22–31.

44. B. Chen, K. Jamieson, H. Balakrishnan, and R. Morris, "Span: An energy-efficient coordination algorithm for topology maintenance in ad hoc wireless networks," *Mobile Computing and Networking*, pp. 85–96, 2001.

45. W. Heinzelman, J. Kulik, and H. Balakrishnan, "Adaptive protocols for information dissemination in wireless sensor networks," *Mobile Computing and Networking*, pp. 174–185, 1999.

46. W. Heinzelman, "Application-specific protocol architectures for wireless networks."

47. C. Intanagonwiwat, R. Govindan, and D. Estrin, "Directed diffusion: a scalable and robust communication paradigm for sensor networks," *Mobile Computing and Networking*, pp. 56–67, 2000.

48. S. Lindsey and C. S. Raghavendra, "Pegasis: Power-effcient gathering in sensor information systems," in *IEEE Aerospace Conference*, Big Sky, Montana, 2002.

49. A. Cerpa and D. Estrin, "Ascent: Adaptive self-configuring sensor networks topologies," in *Annual Joint Conference of the IEEE Computer and Communications Societies*, vol. 3, 2002, pp. 1278–1287.

50. S. Goel and T. Imielinski, "Prediction-based monitoring in sensor networks: Taking lessons from mpeg," *ACM Computer Communications*, Charlottesoille, VA, Oct. 2001.

51. Y. Xu, J. Heidemann, and D. Estrin, "Geography-informed energy conservation for ad hoc routing," *Mobile Computing and Networking*, pp. 70–84, 2001.

52. D. Tian and N. Georganas, "A coverage-preserving node scheduling scheme for large wireless sensor networks," in *ACM Workshop on Wireless Sensor Networks and Applications*, Atlanta, 2002, pp. 32–41.

53. M. Perillo and W. Heinzelmann, "Optimal sensor management under energy and reliability constraints," in *IEEE Wireless Communications and Networking Conference*, vol. 3, New Orleans, 2003, pp. 1621–1626.

54. D. Rakhmatov and S. Vrudhula, "Energy management for battery-powered embedded systems," *ACM Transactions on Embedded Computing Systems*, vol. 2, no. 3, pp. 277–324, Aug. 2003.

55. R. E. Bryant, "Graph-based algorithms for boolean function manipulation," *IEEE Transactions on Computers*, vol. 35, no. 8, pp. 677–691, 1986.

56. O. Coudert, "On solving covering problems," in *Design Automation Conference*, Geneva, 1996, pp. 197–202.

57. K. S. Brace, R. L. Rudell, and R. E. Bryant, "Efficient implementation of a bdd package," in *Design Automation Conference*, 1990, pp. 40–45.

58. *International Journal of Software Tools for Technology Transfer*, Special issue, vol. 3, no. 2, pp. 171–181, 2001.

59. T. Villa, T. Kam, R. K. Brayton, and A. S. Vincentelli, "Explicit and implicit algorithms for binate covering problems," *IEEE Transactions on Computer Aided Design*, vol. 16, no. 7, pp. 677–691, 1997.

60. G. D. Hactel and F. Somenzi, *Logic Synthesis and Verification Algorithms*, Kluwer Academic, 1996.

61. R. Rao, S. Vrudhula, and D. Rakhmatov, "Analysis of discharge techniques for multiple battery systems," in *International Symposium on Low Power Design*, 2003, pp. 44–47.

62. J. Hightower and G. Borriello, "Location systems for ubiquitous computing," *IEEE Computers*, vol. 34, no. 8, pp. 57–66, 2001.

63. N. Bulusu, J. Heidmann, and D. Estrin, "Gps-less low-cost outdoor localization for very small devices," *IEEE Personal Communications*, pp. 28–34, Oct. 2000.

64. N. Priyantha, A. Chravarthy, and H. Balakrishnan, "The cricket location-support system," *Mobile Computing and Networking*, pp. 32–43, 2000.

65. D. Rakhmatov and S. Vrudhula, "A model for battery lifetime analysis for organising applications on a pocket computer," *IEEE Transations on Very Large Scale Integration*, vol. 11, no. 6, pp. 1019–1030, 2003.

66. F. Somenzi, "Cudd decision diagram package, release 2.3.1," ECE Department, University of Colorado, 1999.

67. "National oceanic and atmospheric administration," http://www.csc.noaa.gov/coos/hawaii.html, 2003.

68. M. Bhardwaj, T. Garnett, and A. Chandrakasan, "Upper bounds on the lifetime of sensor networks," in *International Conference on Communications*, vol. 3, Helsinki, 2001, pp. 785–790.

69. J. Zhu and S. Papavassiliou, "On the energy-efficient organization and the lifetime of multi-hop sensor networks," *IEEE Communications Letters*, vol. 7, no. 11, pp. 537–539, 2003.

70. R. J. Urick, *Principles of Underwater Sound*, New York: McGraw-Hill, 1983.

71. J. Groen, J. Sabel, and A. Hètet, "Synthetic aperture processing techniques applied to rail experiments with a mine hunting sonar," in *Proc. UDT Europe*, 2001.

72. P. Gough and D. Hawkins, "A short history of synthetic aperture sonar," in *Proceedings of IEEE International Geoscience and Remote Sensing Symposium*, Seattle, 1998.

73. P. Chapman, D. Wills, G. Brookes, and P. Stevens, "Visualizing underwater environments using multi-frequency sonar," *IEEE Computer Graphics and Applications*, 1999.

74. G. Trimble, "Underwater object recognition and automatic positioning to support dynamic positioning," in *Proceedings of the Seventh International Symposium on Unmanned Untethered Submersile Technology*, Durham, NH, 1991, pp. 273–279.

75. X. Y. et al., "Design of a wireless sensor network for longterm, in-situ monitoring of an aqueous environment," *Sensors*, vol. 2, pp. 455–472, 2002.

76. N. Fruehauf and J. Rice, "System design aspects of a steerable directional acoustic communications transducer for autonomous undersea systems," in *OCEANS*, vol. 1, 2000, pp. 565–573.

77. S. Tilaky, N. B. Abu-Ghazalehy, and W. Heinzelman, "Infrastructure tradeoffs for sensor networks," in *WSNA*, Atlanta, 2002.

78. M. Marsan, C. Chiasserini, A. Nucci, G. Carello, and L. D. Giovanni, "Optimizing the topology of bluetooth wireless personal area networks," in *INFOCOM*, vol. 2, 2002, pp. 572–579.

79. A. Misra and S. Banerjee, "Mrpc: Maximizing network lifetime for reliable routing in wireless environments," in *Wireless Communications and Networking Conference*, vol. 2, 2002, pp. 800–806.

80. J. H. Chang and L.Tassiulas, "Energy conserving routing in wireless ad-hoc networks," in *Proceedings of the Annual Joint Conference of the IEEE Computer and Communications Societies*, Tel Aviv, 2000, pp. 22–31.

81. Y. Xu, J. Heidemann, and D. Estrin, "Geography-informed energy conservation for ad hoc routing," *Mobile Computing and Networking*, pp. 70–81, 2001.

82. S. Lindsey and C. S. Raghavendra, "Pegasis: Power-effcient gathering in sensor information systems," in *Proceedings of IEEE Aerospace Conference*, Big Sky, MI, 2002.

83. W. Ye, J. Heidemann, and D. Estrin, "An energy-efficient mac protocol for wireless sensor networks," in *Proceedings of IEEE INFOCOM' 02*, vol. 3, 2002.

84. D. Panigrahi, C. Chiasserini, S. Dey, R. Rao, A. Raghunathan, and K. Lahiri, "Battery life estimation of mobile embedded systems," in *Fourteenth International Conference on VLSI Design*, Banglore, India, 2001, pp. 57–63.

85. "Underwater acoustic modem," www.link-quest.com.

86. K. Kalpakis, K. Dasgupta, and P. Namjoshi, "Maximum lifetime data gathering and aggregation in wireless sensor networks," in *International Conference on Networking*, 2002, pp. 685–696.

87. J. G. Proakis, J. Rice, and M. Stojanovic, "Shallow water acoustic networks," *IEEE Communications Magazine*, vol. 11, 2001.

88. M. Stojanovic, "Recent advances in high speed underwater acoustic communications," *Oceanic Engineering*, vol. 21, no. 4, pp. 125–136, 1996.

89. S. A. L. Glegg, R. Pirie, and A. LaVigne, "A study of ambient noise in shallow water," http://www.oe.fau.edu/ acoustics/.

90. A. B. Boehm, S. B. Grant, J. H. Kim, S. L. Mowbray, C. D. McGee, C. D. Clark, D. M. Foley, and D. E. Wellman, "Decadal and shorter period variability of surf zone water quality at Huntington Beach, California," *Environmental Science and Technology*, 2000.

91. R. Holman, J. Stanley, and T. Ozkan-Haller, "Applying video sensor networks to nearshore environment monitoring," *Pervasive Computing*, pp. 14–21, 2003.

92. Z. Xinhua, "An infromation model and method of feature fusion," in *ICSP*, 1998.

93. F. H. Fisher and V. P. Simmons, "Sound absorption in sea water," *Journal of Acoustical Society of America*, vol. 62, no. 558, 1977.

94. F. H. Fisher, "Effect of high pressure on sound absorption and chemical equilibrium," *Journal of Acoustical Society of America*, vol. 30, no. 442, 1958.

7

DISTRIBUTED SENSING AND DATA GATHERING

7.1 INTRODUCTION

In sensor networks, there are many techniques for gathering and disseminating data. Even though sensors are typically limited in their processing capabilities, they often posses enough intelligence to perform some rudimentary processing of data as it is received. In addition, the sensor nodes may process data as they are relaying it on behalf of other nodes. Finally, nodes may exchange information with each other so that their sensing mission is more efficiently and effectively accomplished.

There are two motivations for this cooperation among sensor nodes. First, multiple sensors may "see the same phenomena, albeit from different views. If this data can be reconciled into a more meaningful form as it is passes through the network, it becomes more useful to an application. Second, as discussed in previous chapters, in many applications far more sensors are deployed than are actually needed to accurately sense the target phenomena. There are several reasons for the overdeployment: to ensure that even if nodes are randomly distributed there is, with a high probability, a node present in a desired location; to provide backup nodes in case of failure or battery depletion; or to have extra nodes available if more thorough data gathering is required. In this case, it makes sense to limit the number of sensors that are active at any time to extend network lifetime.

Intelligently gathering and disseminating data in a cooperative fashion has several benefits. First, if data is processed as it is passed through the network, it may be compressed thus occupying less bandwidth. This also reduces the amount of transmission power expended by nodes on the forwarding path from the source to the sink. In some cases, sensors may even be placed into an idle mode if their data is redundant, allowing them to save battery power for later operations. Finally, the information is more useful to the sink receiving it.

There are many challenges to achieving efficient sensing and data gathering. First, nodes must determine if the data they are sensing is redundant. This is difficult to do in isolation, so some communication is required for nodes to coordinate. Second, in the case when nodes are redundant, they must coordinate so that the correct nodes are put in an idle mode. Third, in cases of sensors cooperating to provide multiple views of data, the sensors must send their data on a common path so that it may be properly merged. An additional complication arises if data is being sent securely. In this case, cooperating nodes must be able to process each others data, and therefore have established trust relationships, or data must be coded in a way that it can be merged without being decoded.

In this chapter we present four studies on distributed sensing and data gathering. Section 7.2, by Çam, Ozdemir, Sanli, and Nair, proposes a cluster architecture in which the head places sensors with redundant data in to sleep mode. It also proposes a coding scheme so that data may be sent securely and still combined. Section 7.3, by Choi and Das, proposes a method in which only the minimum number of sensors required for an application report their data. They developed four algorithms for selecting the active sensors. Section 7.4, by Zhu, Yang, and Papavassiliou, incorporates quality of service (QoS) into data aggregation. Depending on the latency requirements of data, sensors may hold data so that it may be aggregated before forwarding. In Section 7.5, by Wang, Qi, and Beck, a progressive approach to gathering data is proposed. In this technique, sensors make decisions based on local observations and partial data received from other nodes.

7.2 SECURE DIFFERENTIAL DATA AGGREGATION FOR WIRELESS SENSOR NETWORKS

Hasan Çam, Suat Ozdemir, H. Ozgur Sanli, and Prashant Nair

Data aggregation protocols aim at eliminating redundant data transmissions and, therefore, are essential to improve the lifetime of energy-constrained wireless sensor networks. Conventional data aggregation protocols are designed to eliminate redundant data at cluster heads or data aggregators only. Since transmission of raw redundant data from sensor nodes to the cluster heads still causes significant energy wastage, this section presents a secure differential data aggregation (SDDA) protocol for wireless cluster-based sensor networks to reduce the amount of redundant data transmitted from sensor nodes to cluster heads, in addition to providing security in communication. SDDA prevents redundant data transmission from sensor nodes by implementing the following schemes: (1) SDDA transmits differential data rather than raw data, (2) SDDA performs data aggregation on pattern codes representing the main characteristics of sensed data, and (3) SDDA employs a sleep protocol to coordinate the activation of sensing units of sensor nodes in such a way that only one of those sensor nodes sensing the same data is activated. In the SDDA data transmission scheme, the raw data from sensor nodes are compared with reference data, and then only the difference data are transmitted, where reference data is obtained by taking the average of previously transmitted data. SDDA establishes secure connectivity among sensor nodes with a key predistribution scheme by taking advantage of deployment estimation of sensor nodes. The fact that SDDA does not require cluster heads to examine the sensed data for data aggregation also helps implementing secure data transmission because the encrypted data of sensor nodes need not be decrypted by cluster heads along the transmission path. Performance of SDDA is evaluated in terms of bandwidth efficiency and energy consumption. Simulation results show that SDDA is more energy and bandwidth efficient than conventional data aggregation techniques, and the efficiency of SDDA increases as the data redundancy increases.

7.2.1 Background Information

Wireless sensor networks for monitoring environments have recently been the focus of many research efforts and emerged as an important new area in wireless technology. In a wireless sensor network, a large number of tiny, sensor-equipped and battery-powered nodes communicate with each other using low data-rate radio [1–3]. The advent of low power, low-cost digital signal processors and radio frequency (RF) circuits [20] accelerate the feasibility of the inexpensive sensor networks applications from military networks (surveillance, intelligence) to environmental monitoring networks (climate and habitat monitoring). The key attributes of wireless sensor networks are the storage, power, and processing limitations. These limitations and their specific architecture motivate our work for the design of energy-efficient and secure communication protocol. This section presents an SDDA protocol that incorporates both data aggregation and security concepts together for cluster-based wireless sensor networks.

Sensor networks are densely deployed with high node redundancy to deal with node failure based on connectivity and coverage problems since sensor nodes are severely energy constrained, and it is not feasible to replace the batteries of thousands of sensor nodes. However, dense deployment also causes neighboring sensor nodes to have highly overlapping sensing regions. This results in multiple nodes to generate a packet whenever an event occurs in the overlap of their sensing regions. This is due to the fact that each node observes the physical region of overlap independent of its neighbors. It is important to emphasize that redundant packets may not be eliminated immediately on their way to the base station with data aggregation since packets can follow different paths in the clustering hierarchy. Therefore, it is a promising idea to schedule the coverage of neighboring sensor nodes to enable each point in the target region to be covered by a minimum number of sensor nodes at any time. To accomplish this, this section introduces an algorithm to coordinate the sleep and active modes when sensor nodes have overlapping sensing ranges.

An important problem in sensor networks is to maximize the lifetime of sensor nodes under the limited energy constraint. Communication constitutes an important share, 70%, of the total energy consumption of the sensor network [2], therefore area coverage and data aggregation techniques [4–7] can greatly help conserve the scarce energy resources by eliminating redundant data transmissions and reducing the number of unnecessary broadcasts and collisions. In [7], a cluster head collects the representative data items showing how sensor readings change over a predefined time interval from each sensor node and sends only the ones matching critical events, such as sudden temperature drop forecasting a storm, to the base station. SPIN (sensor protocols for information via negotiation) [18] eliminates redundant data using the meta-data negotiations by nodes. In directed diffusion [6], gradients are set up to collect data, and data aggregation makes use of positive and negative reinforcement of paths. In addition to data aggregation protocols, clustering and load balancing protocols such as the low-energy adaptive clustering hierarchy (LEACH) protocol [10] extend network lifetime by evenly distributing the energy load among the sensors by utilizing randomized rotation of cluster heads.

In the proposed SDDA protocol, only one of those sensor nodes sensing the same data is activated and the others are put into the sleep mode in order to prevent as much as possible the redundant data from being sensed. Instead of transmitting any raw sensed data, sensor nodes with active sensing units first generate and transmit the pattern codes that represent the main characteristics of the actual data. The pattern codes are basically representative data items that are extracted from the actual data in such a way that every pattern code has certain characteristics of the corresponding actual data. The extraction process may vary depending on the application type of the actual data such as selecting certain pixels from the actual image data. Cluster heads identify the distinct pattern codes and then request only one sensor node for each distinct pattern code to send the actual data to the base station. SDDA uses a differential data transmission scheme for sending patterns and sensor readings to improve energy savings achieved with the use of pattern codes with small size. In this transmission scheme, the raw data from the sensor nodes are compared with reference data and then only the difference data are transmitted.

Since security in data communication is a very important issue in wireless sensor networks, data aggregation protocols should work with the data communication security protocols, as any conflict between these protocols may create loopholes in network security. In wireless sensor networks, the first step to establishing security is to efficiently distribute shared secrets between sensors and the base station. Therefore, an efficient key distribution scheme with low storage overhead is a prerequisite for the design of a secure data aggregation protocol. SDDA establishes secure connectivity among sensor nodes with a key predistribution scheme by taking advantage of deployment estimation of sensor nodes. SDDA establishes secure links among sensor nodes using a key predistribution scheme with low memory overhead. Security of the communication between the base station and the sensor nodes is further strengthened with the use of pattern codes based on the fact that SDDA does not require cluster heads to decrypt the actual sensed data for data aggregation. The preliminary versions of this security protocol appeared in [8, 9].

7.2.2 Secure Differential Data Aggregation

Data aggregation is essential in wireless sensor networks due to the fact that the same areas are covered by possibly more than one sensor node. This section first explains why the proposed data aggregation protocol is differential. Then, the two protocols of SDDA, namely, differential data aggregation and security protocol, are described. Finally, to increase the gain of bandwidth utilization and energy efficiency as a result of differential data aggregation, a coverage scheme for reducing the number of active sensor nodes is presented.

Communication dominates the energy consumption of a sensor node [10]. It is therefore critical to reduce the amount of data to be transmitted. In conventional data aggregation algorithms, sensors transmit their raw data to cluster heads. This may cause wastage of energy and bandwidth if the successive transmissions have common raw data. However, the proposed protocol transmits the differential data rather than the actual data, where the *differential data* refers to the difference between the reference data and the actual data. Assuming that the data of sensor readings are mapped to numbers, the *reference data* in this section refers to the average of all the sensor readings in the most recent data transmission session. Each sensor node computes its reference data and sends it to the cluster head. As an example, let $102°F$ denote the current temperature reading of a sensor node. If $100°F$ is the reference temperature, the sensor node can send only the difference (i.e., $2°F$) as the current sensor reading. Hence, differential data aggregation has great potential to reduce the amount of data to be transmitted from sensor nodes to cluster heads. The basic motivation behind differential data aggregation is that significant changes in sensor readings can occur only when a *critical* event (e.g., a fire event for sensor network monitoring temperature) happens in the environment. In general, these so-called critical events occur much less frequently than ordinary events in sensor networks. To take advantage of differential data aggregation in every data transmission from sensor nodes to cluster heads, the proposed SDDA protocol makes use of the differential data aggregation while transmitting pattern codes and actual data as explained below.

Not only sensor nodes but also cluster heads benefit from differential data aggregation because the latter receive and process less amount of data. The efficiency of this technique increases as the volume of data gets larger than that of the differential data because the reference data is transmitted once only.

To be able to use differential data aggregation, SDDA initially determines the reference values for both pattern nodes and actual data based on the past sensor readings. Then, cluster heads and base station store these reference values in order to recover the actual raw pattern codes and data when they receive differential pattern codes and data. SDDA first generates pattern codes by finding out the most important features of the sensor readings and then determines the differential pattern code by obtaining the difference between the current pattern code and the reference pattern code. Subsequently, the differential pattern code is transmitted to the cluster head. When the cluster head requests actual data, sensor node transmits the differential actual data.

Differential Data Aggregation   The differential data aggregation protocol is described next in three phases.

Phase 1   Reference values for pattern codes are sent to the cluster head only, whereas reference values for actual data are sent to the base station via cluster heads at the beginning of each data transmission session. After sensing new data, sensor node generates actual pattern codes by finding out main characteristics of the current sensed data. Then, sensor nodes transmit differential pattern codes to cluster heads.

Phase 2   After receiving all differential pattern codes, cluster heads recover the actual pattern codes by adding reference pattern codes. Then, cluster heads determine distinct pattern codes and request only one sensor node to transmit data for each distinct pattern code.

Phase 3   Each of those sensor nodes that receive a data transmission request from the cluster head first generates the differential data by finding the difference between the current sensed data and its reference data. Then, the differential data is encrypted and sent to the cluster head, which forwards it to the base station. Differential data is used to obtain the actual data at the base station using reference data of sensor node.

In what follows, these three phases are explained in detail, along with the pseudocode implementation of the algorithms.

Phase 1: Pattern Code Generation and Transmission   Differential data aggregation protocol represents the characteristics of the sensor data by *parameters*. A pattern generation (PG) algorithm first maps the sensor data to a set of numbers. Then, based on the user requirements and *precision* defined for the environment in which the network is deployed, this set of numbers is divided into *intervals* such that the boundaries and width of intervals are determined by the predefined *threshold* values. PG algorithm then computes the *critical values* for each interval using the *pattern seed* and generates the *interval* and *critical value* lookup tables. The interval lookup table defines the range of each interval and the critical value lookup table maps each interval to a critical value. For example, the critical values may be assigned as 1 for the first interval and vary through 9 being the last interval. These critical values form the base for the generation of pattern codes.

Upon sensing data from the environment, the sensor node compares the characteristics of data with the intervals defined in the lookup table of the PG algorithm. Then, a corresponding critical value is assigned to each parameter of the data; concatenation of these critical values forms the pattern code of that particular data. Before a pattern code is transmitted to the cluster head, the time stamp and the sender sensor ID are appended to end of the pattern code. The cluster head runs the pattern comparison algorithm to eliminate the redundant pattern codes resulting in prevention of redundant data transmission. Cluster heads choose a sensor node for each distinct pattern code to send corresponding data of that pattern code; then chosen sensor nodes send the data in encrypted form to the base station over the cluster head.

The PG algorithm uses a secret random number called *pattern seed* to improve the confidentiality of the pattern codes. The pattern seeds are generated and broadcasted in encrypted format by cluster heads. We will explain how to broadcast pattern seeds secretly. Pattern seeds are changed at regular time intervals; therefore, the PG algorithm generates a different pattern code for the same data at different time intervals.

Algorithm 7.2.1   Pattern Generation (PG)

Input: Sensor reading D.
        Data parameters being sensed.
        Threshold[ ] : Array of threshold levels of data intervals
        for each data type.
        Data precision requirements of the network for each data
        parameter.
        S(critical[ ], seed): Function to shuffle the mapping of
        critical values to data intervals.

Output: Pattern code (PC)

Begin

    1. **Variable** PC = [];   // Initialize the pattern code
    2. **for** each data parameter
    3. **Declare** n;   // Number of data intervals for
                                            this data type

    4. **Declare** *interval*[n]*criticalvalue*[n]; // Lookup tables
    5. Extract the data from D for the corresponding data
        parameter.
    6. Round off data for the precision required by the
        corresponding data parameter.
    7. **if** (new pattern seed sent by cluster head) **then**
    8.   // Refresh the data interval widths and mapping of
            critical values to data intervals
    9.   **for** i = 1 to n
   10.   *interval*[i] = threshold[i - 1] - threshold[i] ;
   11.   *criticalvalue*[i] = *S(criticalvalue*[i], seed) ;
   12.   **endfor**
   13. **endif**
   14. Find the respective critical value for each current data
        sensed using *interval* and *criticalvalue* lookup tables.
   15. PC = PC + [critical value] ;   // Concatenate critical
                                                value to pattern code
   16. **endfor**;
   17. PC = PC + [Timestamp] + [Sensor ID] ; // Append timestamp
                                                    and sensor id

End

Table 7.1   Lookup Tables for Data Intervals and Critical Values

Threshold values	30	50	70	80	90	95	100
Interval values	0–30	31–50	51–70	71–80	81–90	91–95	96–100
Critical values	**5**	**3**	**7**	**8**	**1**	**4**	**6**

Pattern Generation Example   This example explains how PG algorithm generates a pattern code. Let $D(d1, d2, d3)$ denote the sensed data with three parameters $d1, d2$, and $d3$ representing temperature, pressure, and humidity, respectively, in a given environment. Each parameter sensed is assumed to have threshold values between the ranges 0 to 100 as shown in Table 7.1. The pattern generation algorithm performs the following steps:

- Pattern code to be generated is initialized to empty pattern code (line 1).
- The algorithm iterates over sensor reading values for parameters of data that are being sensed. In this case, it first considers temperature (line 2).
- Temperature parameter is extracted from sensor reading D (line 5).
- For the temperature parameter, the algorithm first checks whether a new pattern seed is received from the cluster head (line 7). Arrival of a seed refreshes the mapping of critical values to data intervals. As an example, the configuration in Table 7.1 can be reshuffled as $\{8, 3, 5, 1, 7, 6, 4\}$, changing the critical value for the first interval ([0–30]) from 5 to 8 while keeping the second interval ([31–50])'s critical value as 3, same with the previous case.
- The data interval that contains the sensed temperature is found from the interval table. Then, from the interval value, corresponding critical value is determined from critical value table (line 14). Table 7.2 shows the critical values for different sensor readings if the same lookup tables of Table 7.1 are used for temperature, pressure, and humidity.
- PC is set to the new critical value found (line 15). For the pressure and humidity, corresponding critical values are appended to the end of partially formed PC.
- Previous steps are applied for the pressure and humidity readings
- When full pattern code is generated, time stamp and sensor identifier is sent with the pattern code to the cluster head (line 17).

Pattern codes with the same value are referred as a *redundant set*. In this example, data sensed by sensor 1 and sensor 3 are the same with each other as determined from the comparison of their pattern code values (pattern code value 747) and those for the redundant set 1. Similarly, data sensed by sensor 2, sensor 4, and sensor 5 are the same (pattern code value 755), redundant set 2. The cluster head selects only sensors from each redundant set (sensor 1 and sensor 4 in this example) to transmit the actual data of that redundant set based on the time stamps according to the pattern comparison algorithm described below.

In a pattern comparison algorithm, upon receiving all of the pattern codes from sensor nodes in a period of T, a cluster head classifies all the codes based on redundancy (Table 7.2). The period T varies based on the environment and the application type of the sensor network. Unique patterns are then moved to the "selected set" of codes. The sensor nodes that correspond to the unique pattern set (selected set) are then requested to transmit the actual data. ACK signals may be broadcast to other sensors ("deselected set") to discard their (redundant) data.

Table 7.2 Pattern Codes Generation Table

	Sensor 1	Sensor 2	Sensor 3	Sensor 4	Sensor 5
Data	D(56, 92, 70)	D(70, 25, 25)	D(58, 93, 69)	D(68, 28, 30)	D(63, 24, 26)
Critical value for d1	7	7	7	7	7
Critical value for d2	4	5	4	5	5
Critical value for d3	7	5	7	5	5
Pattern code	747	755	747	755	755

LEGEND

Redundant Set #1 (Sensor 1, 3)

Redundant Set #2 (Sensor 2,4,5)

Selected unique Set #1 (Sensor 1,4)

Algorithm 7.2.2 Pattern Comparison

```
Input: Pattern codes
Output: Request sensor nodes in the selected set to send actual
        encrypted data.
Begin
    1. Broadcast "current seed" to all sensor nodes
    2. while(current seed is not expired)
    3.     time counter = 0
    4.     while(time counter < T)
    5.         get pattern code, sensor ID, time stamp
    6. end while
    7. Compare and classify pattern codes based on redundancy
       to form "classified set".
    8. selected set={one pattern code from each classified set}
    9. deselected set = classified set - selected set
   10. if (sensor node is in selected set)
   11.        Request sensor node to send actual data
   12. endif
   13. endwhile
End
```

While pattern-based data aggregation ensures security, it has limited precision as specified by user requirements since a number of values in the same interval are mapped to a critical value, which forms part of a pattern code. One way of enhancing precision is to increase the number of intervals while keeping the interval sizes small. This increases the number of critical values correspondingly and the number of bits to represent the pattern codes. This increased pattern code size problem is solved with the help of reference and differential pattern codes. While this increases the computation overhead at the sending and receiving nodes, due to the significant energy consumption difference between the computation and communication [2, 10], the overall gains achieved by transmitting a smaller number of pattern code bits overcomes the computational energy required at either ends.

Once pattern codes are generated, they are transmitted to the cluster head using the following algorithm SDT (session data transmission) that can also be used for transmitting actual data values in *Phase 3*. SDT is implemented in every session of data transmission, where session refers to the time interval from the moment the communication is established between a sensor node and the cluster head until the communication terminates. Each session is expected to have a large number of packets. In the beginning of each session, the cluster head receives the reference data along with the first packet and stores it until the end of the session. After a session is over, the cluster head can remove its referenced data.

Algorithm 7.2.3   Session Data Transmission

```
Begin
  For each session, do
   While(sensor node has pattern codes or data packets for
   transmission)
    1.if(first pattern codes or data packet of session)
    2.  send the pattern codes or data packet along with the
        reference data
    3.else
    4.  send the differential pattern codes or data packet
    5.endif
   EndWhile
End
```

Phase 2: Choosing Sensor Nodes for Data Transmission by Cluster Heads   The technique of using lookup tables and pattern seed ensures that the sensed data cannot be regenerated from the pattern codes, which in turn help the sensor nodes to send pattern codes without encryption. Only sensor nodes within the cluster know the pattern seed, which ensures the security of the sensed data during the data aggregation. The security of actual data transmission is provided by our security protocol as described in phase 3.

Phase 3: Differential Data Transmission from Sensor Nodes to Cluster Head   After the cluster head identifies which sensor nodes should send their data to the base station, those nodes can send their differential data to the base station. The differential data is securely sent to the base station using the security protocol described in this section. The base station converts the differential data to the raw data using reference data.

Security Protocol   This section presents the key distribution scheme and security algorithm for having secure data transmission during the phase 3 of the SDDA protocol presented above.

Key Distribution Scheme   Traditional network or public key exchange and key distribution protocols are impractical for large-scale sensor networks because of their limitations on communication range, operational power, computation power, and network characteristics. Eschenauer and Gligor [11] present a random key predistribution scheme for wireless sensor networks where each sensor node receives a random set of keys from a key pool, and each pair of nodes can communicate with each other only if they share a common key. This scheme is improved in [12–14]. In [12] authors use the estimated location information of sensor nodes to reduce memory space and computational overhead due to key distribution. Although their scheme is similar to the key distribution scheme proposed in this section, our scheme differs from their technique by employing different key pools for local and global connectivity of the network, which further improves the memory and power utilization. The key distribution scheme in [13] is very similar to Eschenauer and Gligor's [11] scheme except that their proposed scheme requires that any pair of nodes should have q common key within their key set. Liu and Ning [14] present a key distribution scheme based on polynomial-based key predistribution that reduces the computational needs of sensor nodes.

Eschenauer and Gligor's [11] key distribution scheme randomly distributes the keys to all sensors in the network. This approach may result in secure peer-to-peer communication links between sensor node pairs that are far away from each other. However, because of the sensor nodes' low transmission range, in sensor networks secure peer-to-peer communication links are usually desired between sensors that are neighbors or very close to each other. Our scheme is based on [11] and takes advantage of the estimated location information of sensors that can be predicted with some probability from the way sensors are deployed. Estimated location information does not allow long secure peer-to-peer communication links to happen in sensor network. Moreover, estimated location information helps our scheme to reduce the memory and computational consumption of sensor nodes due to key distribution.

To obtain the estimated location information of sensors, deployment of the sensor nodes over the network area must be done by following a procedure. Assuming the sensor network area is a *square*, the following is the deployment procedure of our key distribution scheme.

- Sensor nodes are divided equally into *n deployment groups*.
- Sensor deployment area is divided equally into *n deployment grids*.
- The center points of these *n*grids are determined as *deployment points*.
- A group of sensors is dropped from each deployment point.

Figure 7.1 presents the sensor network area with *deployment grids* and *deployment points*, each deployment point is represented by a small circle while a gray square represents a deployment grid. Deployment grids are denoted by notation $n(i, j)$. Figure 7.1 also has *neighboring grids* that are represented by white squares ($S_{i,j}$). These neighboring grids overlap with the common neighboring areas of gray deployment grids.

KEY DISTRIBUTION   Before the deployment, sensors are divided into n groups of size k and n random key pools of size s are generated. Each key pool S_i consists of s random keys and their corresponding key IDs. Each S_i is assigned to a sensor group (SG_i), and then each sensor node in group SG_i randomly selects m keys out of S_i without the replacement of selected keys. These selected keys and their key IDs form the sensor node's *local key ring*. Sensor nodes use their local key rings to establish secure communication links within the group members and provide *local connectivity* in the group. In [11], connectivity is defined as the probability that any two neighboring nodes share one key. However, *global connectivity* is also required within the sensor network. To ensure the global connectivity, each sensor needs a *global key ring* that provides an interconnection between neighboring deployment grids. In Figure 7.1, each *neighboring grid* covers the common neighboring area of four or less deployment grids. The global connectivity of the sensor network is ensured, if all of the deployment grids that are covered by a neighboring grid are connected. Therefore, a key pool ($S_{i,j}$) of size t is generated for each neighboring grid and each sensor $n_{i,j}$ selects n keys from the key pool $K = U\{S_{i,j}\}$. The set $\{S_{i,j}\}$ denotes the set of key pools that are overlapping with the deployment grid of $n_{i,j}$. The selected keys and their

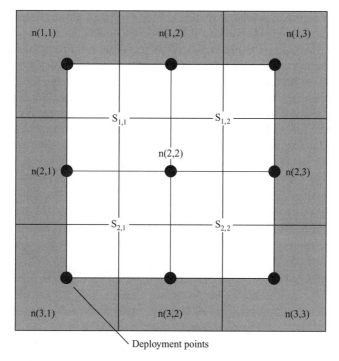

Figure 7.1 Sensor network area with subsquares and deployment points.

respective key IDs form the sensor node's *global key ring*. Thus, before the deployment each sensor has $(n + m)$ keys in its local and global key rings.

SHARED KEY DISCOVERY   Once the network is deployed, sensor nodes start the *shared key discovery* phase. This phase is implemented in two steps:

- Shared key discovery within the group members.
- Shared key discovery between neighboring grids.

In step 1, sensors use their local key rings. First, each node broadcasts its node ID and key IDs in its local key ring to find out the neighboring nodes that share a common key. Once every sensor node discovers its neighbors that it shares a key with, then a challenge–response protocol is performed between neighbor nodes who wish to mutually verify that they really have knowledge of the key. Neighboring sensor nodes who do not share a key can also establish secure peer-to-peer communication links using already established secure communication links. In step 2, neighboring sensors from different deployment grids find out if they share a common key from their global key rings. Each node broadcasts its node ID and key IDs in its global key ring and discovers its neighboring nodes with which it shares a key. These neighboring nodes are more likely to be from neighboring deployment grids since global key rings are selected from key pools assigned to neighboring grids.

COMPARISON WITH ESCHENAUER AND GLIGOR'S SCHEME   Since the number of sensor nodes in a deployment group is much smaller than the number of sensor nodes in the whole network, our key distribution scheme requires smaller key pool and key ring sizes compared to Eschenauer and Gligor's [11] scheme. Moreover, local and global connectivity is higher than Eschenauer and Gligor's scheme since the sensor nodes that are sharing a key are more likely to be in each other's transmission range. Let us consider the same example used in [11], with a sensor network of 10,000 nodes and one key pool of 100,000 keys. In this example, Eschenauer and Gligor show that if each node has 200 keys in their key ring the sensor network would have 0.33 global connectivity degrees. Let us assume the same network size for our key distribution scheme and further assume that network area is divided into 100 deployment grids. For each deployment grid our scheme requires a key pool of size 1000. The local connectivity degree of our key distribution scheme is presented in Figure 7.2. Also, Figure 7.3 shows the global connectivity rate of different key pool sizes and key ring sizes. In [11], connectivity is defined as the probability that any two neighboring nodes share one key. The graphs are obtained by using the connectivity formula defined by Eschenauer and Gligor in [11].

$$p = 1 - \frac{((S-k)!)^2}{(S-2k)!S!} \qquad (p: \text{ connectivity degree } S: \text{ key pool size } k: \text{ key ring size})$$

Let us assume that a key pool size of 1000 for the deployment grids and 3500 for the neighboring grids are used. Figures 7.2 and 7.3 indicate that, if all the sensors have a local

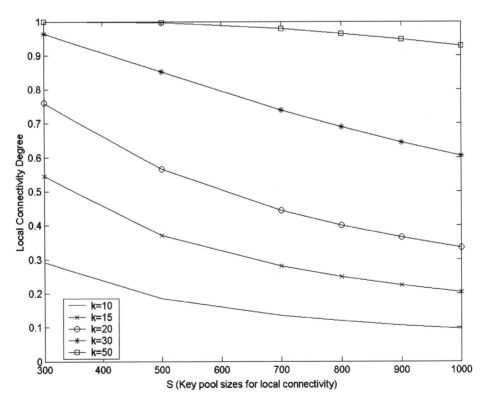

Figure 7.2 Local connectivity degree for various S and k values.

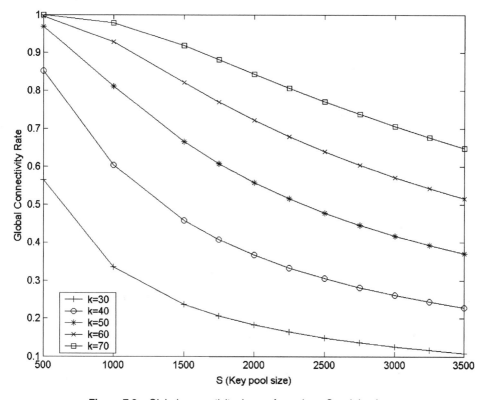

Figure 7.3 Global connectivity degree for various S and k values.

key ring of size 30 and global key ring of size 70, then our key distribution scheme provides 0.61 local connectivity and 0.65 global connectivity degree. Thus, our scheme provides a minimum 0.61 connectivity degree with 100 keys in each sensor. This means that our key distribution scheme reduces memory usage due to key storage by 50% and doubles the network connectivity compared to [11]. Moreover, since the key pools are distinct, the total number of keys in all of our key pools are much bigger than the key pool size in [11], which increases the resilience to node capture.

Security Algorithm   Our aim is to provide an energy-efficient security algorithm to mitigate the problems of denial of service (DoS) attacks [17] and maintain the confidentiality of data within the cluster. While there are similar attempts in the literature [2], they include counter synchronizations, keys, and nonce transmissions before the data transmission is initiated. Due to these synchronizations and nonce transmissions, the algorithm in [2] is energy expensive. Some of the key features of the proposed security protocol are as follows: (1) mitigation of attacks on the network, (2) data freshness, authentication, and confidentiality, and (3) more energy and time efficiency compared to previous work [2].

The proposed security algorithm concentrates on three main aspects of security, namely authentication, integrity, and confidentiality in a time and energy-efficient manner. The inputs to the algorithm are the data, sensor node ID#, the seed key K_B, and the randomly selected key (k) from the node's key ring K. Based on the shared key discovery phase of the key distribution, the sensor nodes form a secure communication path between themselves and the cluster head. The sensor nodes use this secure communication path to receive the "*seed*" key from the cluster head at sensing the very first data. The seed key is used in the generation of the security codes using a cryptographic pseudorandom generator [16]. Depending on the number of bits (b) of the seed key, the pseudorandom generator generates 2^b codes within each sensor node sequentially. Once the sensor node has utilized all the generated codes, it asks for a new seed key from the cluster head for the next data sensed. The usage of sensor node ID# makes the security scheme and intermediate keys used in the encryption procedure specific and *unique* to the respective sensor node. This security scheme is carried out by algorithms 7.2.4 and 7.2.5 implemented at sensor nodes and base stations, respectively.

Algorithm 7.2.4   Sensor Node and Cluster Head Functions

Step 1: If sensor node i wants to send data to its cluster head, go to next step, otherwise exit the algorithm.

Step 2: Sensor node i selects one of the keys (k) randomly from its key ring K.

Step 3: Based on the seed key, sensor node generates security code $C_{\{m\}}$ for $0 < m < 2^b$.

Step 4: Sensor node i XORs the security code $C_{\{m\}}$, with k and its ID# to compute the encryption key K_C. This K_C changes whenever C and k change.

Step 5: Sensor node i encrypts the data with K_C and appends its ID# and the key ID# to the encrypted data along with the time stamp.

Step 6: Cluster head receives the data, appends the hash of the seed key, and then sends them to the base station. Go to step 1.

In Algorithm 7.2.4, the cluster head appends it own ID# to the data in order to help the base station to locate the originating sensor node. The cluster head appends the hash of the seed key only for the first data to be transmitted. It need not send the seed key to sensor nodes unless the latter have utilized all the security codes generated within. In the case of a node capture, this prevents disclosure of common keys used for encryption at the neighbor nodes.

Algorithm 7.2.5   Base Station (BS) Functions

Step 1: For any incoming data compute the encryption key, K_C, using the key ID#, sequential security code and the sensor node ID#. BS computes the sequential security code C by using the seed key transmitted in the first session by the cluster head.

Step 2: Check if the current encryption key K_C has decrypted the data perfectly. This leads to check the creditability of the key ID# and the ID#.

Step 3: Process the decrypted data and obtain the message sent by the sensor nodes.

The base station maintains a log file of the data transmissions using the sensor node ID. Using this log file and the time stamp provided along with the data, base station keeps track of the security codes used in the encryption.

Reducing the Number of Active Sensor Nodes   SDDA eliminates redundant packets at the cluster head level, but it is still possible to conserve energy by analyzing the data-driven communication nature of the wireless sensor nodes and using peer-to-peer local communication of sensor nodes. The technique is based on the fact that if a node's sensing region is covered by the union of its neighbors' sensing regions, pattern codes transmitted from this node to its cluster head will be redundant since its neighbors will also observe the same events. Identifying the nodes that have overlapping sensing ranges and turning off the sensing units of some of those nodes for a bounded amount of time will reduce the redundant data to be transmitted to the cluster head. The gains achieved by the proposed approach become more significant if packet transmission paths of neighboring sensor nodes overlap at the higher levels of the clustering hierarchy only. It is possible to argue that an event-driven sleep coordination method may not achieve significant energy savings in comparison with the differential data transmission and use of pattern codes since most energy models in the literature emphasize the communication aspect of the wireless nodes. However, it is important to point out that energy savings due to the use of the sleep protocol are based on the prevention of redundant data transmission from sensor nodes to cluster heads rather than the circuit level energy savings achieved by deactivation of the sensing units.

The sleep protocol is a distributed protocol in which sensor nodes cooperate with their neighbors to identify the overlapping coverage regions. The term *sleep* refers to the deactivation of the sensing units of nodes rather than their radio. The sleep protocol identifies redundant nodes in terms of coverage by capturing the dynamic properties of the observed phenomena rather than following a geometry-based static circle coverage approach. In the protocol, the total lifetime of the network is divided into fixed length slots of duration T as shown in Figure 7.4. Each slot consists of *observation*, *learning*, and *decision* phases. Each node that is awake updates its local buffer based on the events in its observation phase. In the learning phase, nodes exchange *summary* of their buffers contents, such as a set of hash values of events observed, with their neighbors. In the decision phase, each node evaluates its eligibility to keep their sensing unit on/off for a duration Z' (multiple of T) and broadcasts its decision to its neighbors. The order of broadcasts for nodes is arranged as $(Z/Z_{max})B$, where B is a constant smaller than T, Z (multiple of T) is the previous sleep duration of the node, and Z_{max} (multiple of T) is the maximum duration that a node can sleep, respectively. This serializes the broadcasts of nodes in their local neighborhood and gives priority to redundant nodes that stayed active in the previous decision phases to turn off their sensing units first. When its timer expires, each node performs the following algorithm.

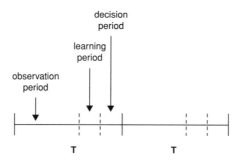

Figure 7.4 Representation of a time slot T consisting of observation, learning, and decision phases.

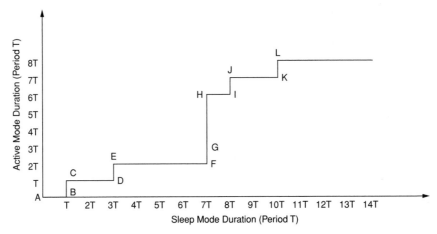

Figure 7.5 Representation of the sleep protocol: Node *X* is initially put to sleep mode at point *A*, and it sleeps for a period of time (*T*) until point *B*. During the period between *B* and *C*, node *X* observes the environment and inserts the pattern codes into its buffer; then it exchanges its buffer with active neighboring sensors. At point *C*, there is a match between buffers, so node *X* again turns on its sensing capability and goes to sleep for 2*T* until point *D*. When buffers do not match, like in point *G*, then node *X* has to be active until one of the neighbors starts exchanging buffers and node *X* has the right to go to sleep mode (point *H*).

Algorithm 7.2.6   OnBroadcastTimerExpire( )
Begin

```
      1. if (events in buffer observed by neighbors which
              has not broadcast their decision)
      2.       Z' = min (2Z, Zmax)
      3.       Broadcast sleep decision to neighbors
      4.       Turn off sensing unit for duration Z
      5. else
      6.       Z' = 0.5T
      7.       Stay awake for next slot
      8. endif
      9. Flush event buffer
End
```

The arrangement of sleep times is similar to the binary exponential back off algorithm. At the end of its sleep time, each node stays awake for slot duration to base its decision criteria on the recent observations, changes in the network, and the environment. The performance analysis results for the sleep protocol are given in Section 7.2.3 and see Figure 7.5.

7.2.3 Performance Evaluation

In this section, we give the performance evaluation of the protocols presented. The performance metrics considered for the evaluation are bandwidth utilization, energy efficiency, and processor execution time. Bandwidth utilization and energy efficiency are used to measure the impact of SDDA and sleep protocol. Processor execution time is used to measure the computational complexity of encryption algorithms.

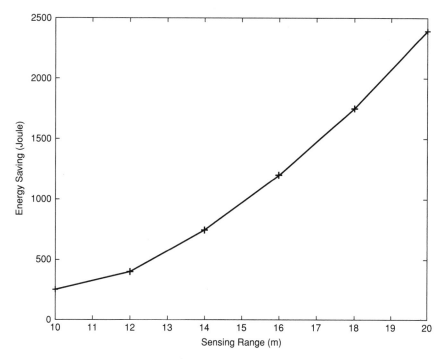

Figure 7.6 Energy efficiency due to sleep protocol.

Evaluation of the Sleep Protocol   In this section, we evaluate the performance of the sleep-active mode protocol. The target area to be monitored is a $100\,\text{m} \times 100\,\text{m}$ square region. The base station is placed at coordinate $(0, 0)$. The results for each variable are averaged over 10 iterations for a specific value for the variable. Each instance of the network is connected and provides full coverage initially. Event buffer size of the nodes is 5, while maximum windows size for sleep duration of the nodes is set to 8. The values for number of nodes, transmission range, and number of events per second inserted to the network are 200, 20 m, and 10, respectively. The time values are represented in terms of slot time T. At each slot, 10 events are inserted at random locations in the network. The resolution of the grid for placement of the nodes and events is 1 m. Figures 7.6 and 7.7 illustrate the effect of increasing sensing range of the nodes on energy consumption of the network and number of undetected events due to the blind spots that may occur due to the sleeping protocol with 1-byte patterns. As can be observed from Figure 7.6 increasing the sensing range of the nodes enables nodes to sleep longer, yielding higher energy savings. Similarly, Figure 7.7 shows that increasing the sensing range enhances the coverage of the network, yielding less percentage of the events to go undetected when the sleep protocol is used.

The result illustrated in Figure 7.6 is a lower bound on the actual savings of the protocol since only redundant transmissions eliminated from sensor nodes are considered. Remaining nodes also have higher coverage due to their increased sensing range, which reduces percentage of undetected events.

Energy Efficiency and Bandwidth Utilization of SDDA   In conventional data aggregation, the cluster head eliminates the redundancy after obtaining the entire actual data

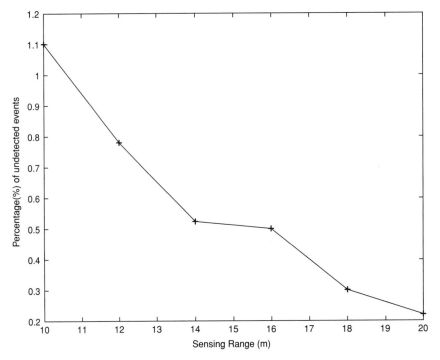

Figure 7.7 Blind spot effect of the protocol on coverage of the network.

from different sensor nodes. However, in SDDA, data aggregation is performed by examining *differential pattern codes* in cluster heads, without sending the actual data. In what follows, we show that SDDA is more energy efficient than the conventional and pattern-based data aggregation techniques because of the following factors:

- The amount of data transmitted in SDDA is much less than the conventional data aggregation.
- The pattern code and data sizes of SDDA are smaller than the pattern code and data sizes of pattern-based data aggregation.

Let T denote the total number of packets that sensor nodes transmit in a session and R denote the number of distinct packets, where R is less than or equal to T. In conventional data aggregation algorithms since the cluster head receives all data packets prior to eliminating redundant data, the total number of packets transmitted from sensor nodes to cluster head would be T. After eliminating redundancy, the cluster head sends R packets to base station. Therefore, the total number of packets transmitted from sensor nodes to base station is $(T + R)$.

In SDDA the cluster head receives T differential pattern codes from all sensor nodes. After eliminating redundancy based on differential pattern codes, cluster requests selected sensor nodes to transmit their data. Since selected nodes are the nodes that have distinct packets, the total number of packets transmitted from sensor nodes to cluster head would be R, which are later forwarded to base station. Therefore, the total number of packets transmitted from sensor nodes to base station is $(2R)$. In wireless sensor networks, often various sensor nodes detect common data and hence R is usually much less than T and $(T + R > R)$. Therefore, SDDA is energy efficient when compared to the conventional data aggregation algorithm.

To assess the energy efficiency of SDDA, we wrote a simulator in C and used GloMoSim [9,15]. Our simulator is used in differential pattern generation and differential pattern comparison aspects of SDDA. GloMoSim is used to simulate the transmission of data and differential pattern codes from sensor nodes to cluster head. In order to compare with SDDA, we also simulated the conventional and pattern-based data aggregation schemes. Simulation results show that SDDA improves energy efficiency significantly by reducing the pattern code size and number of packets transmitted in data communication. The efficiency of SDDA can be further improved if sensor nodes transmit their *differential data* instead of raw data during the actual data transmission phase. The simulation results are presented in Figure 7.8.

In our simulations, we define the occupied bandwidth rate as the ratio of actual bandwidth usage and bandwidth usage when all generated data is sent from sensor nodes to base station. When compared to conventional data aggregation algorithms, as the redundancy increases, the bandwidth usage of SDDA decreases (Fig. 7.8). At 100% redundancy, the bandwidth usage of SDDA is close to zero since SDDA eliminates redundancy before sensor nodes

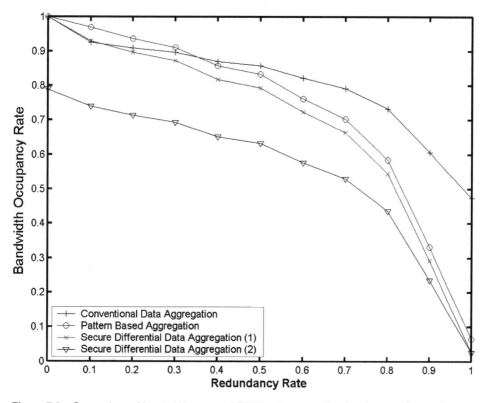

Figure 7.8 Comparison of bandwidth usage of SDDA with conventional and pattern-based data aggregation algorithms. SDDA (1) represents the bandwidth usage when only differential pattern codes are used, and SDDA (2) represents the bandwidth usage when differential pattern codes and differential data are used together.

transmit the actual data packets. However, in conventional data aggregation bandwidth usage is more than 50% of the total bandwidth since all sensor nodes transmits the actual data to be aggregated at the cluster head.

Evaluation of Security Protocol This section presents the performance evaluation of the security algorithm introduced in phase 3 of SDDA using simulators written in C language. The performance metrics are energy efficiency and data confidentiality.

Energy Efficiency of Security Protocol In this subsection we evaluate the energy efficiency of proposed security algorithm in comparison with SPIN introduced in [2]. We first explain the drawbacks of SPIN and then compare it with the proposed security algorithm.

SPIN focuses on providing data freshness and data confidentiality in wireless sensor networks. Some of the drawbacks of this protocol are listed below:

1. High communication overhead due to nonce and counter transmissions.
2. Nonce and counter transmissions are broadcast without encryption.
3. In order to strengthen SPINS long counters and nonce are required.
4. SPIN does not consider information leakage through covert channels and compromised sensor nodes.
5. This security algorithm uses the RC5 cryptographic algorithm, which requires more clock cycles compared to Blowfish [21], which is used in our proposed security algorithm.

The proposed security algorithm maintains the data confidentiality and integrity of the data without any synchronization or nonce transmissions. Hence, it reduces the communication and computation overhead between the cluster head and the sensor nodes, thereby saving energy. The security protocol also considers a key distribution technique, which provides the data confidentiality in any peer-to-peer communication within the cluster. This key distribution technique also mitigates the problems due to sensor node tampering and existence of any covert channels. In addition, the security algorithm uses Blowfish for encryption that reduces the energy consumption as compared to SPIN using RC5.

Knowing that 70% of the energy is utilized in data transmissions; the proposed encryption algorithm saves energy compared to SPIN. Figure 7.9, shows the energy consumption and execution time of SPIN and the proposed security algorithm using the same amount of data.

Cryptographic Strength of the Security Protocol In this subsection we will explain the strength of the proposed security algorithm in terms of brute force attacks. The use of security code C generated by the cryptographic pseudorandom generator [16] strengthens the data confidentiality in the proposed security algorithm. In addition, the brute force attacks are made difficult by the random key distribution technique and the unique key ID representation, as explained in the Section 7.2.3.

In this encryption algorithm, we also consider that the security code outputs are uniformly distributed, that is, the probability that $K_C(M) = K_C(M')$ is 2^{-N} where N is the number of security code bits used in encryption. This design of the security code does not allow the attacker to complete a brute force attack on the message. In addition, there is no correlation between the key used in encryption, security code, and the key ID#. Thus, the proposed encryption algorithm provides authenticity and confidentiality by using different security codes and key (k) each session.

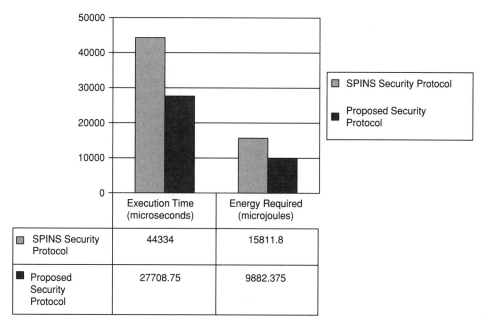

	Execution Time (microseconds)	Energy Required (microjoules)
▣ SPINS Security Protocol	44334	15811.8
■ Proposed Security Protocol	27708.75	9882.375

Figure 7.9 Comparison of encryption algorithms using Strong-Arm SA-1100 profiling [19].

The downside of the proposed security algorithm is that the base station would spend more time and energy for the decryption of the data. But considering the fact that the base station has no energy or computation power constraints, this drawback does not hamper the efficiency of the proposed security algorithm.

7.2.4 Conclusion

This section has introduced a bandwidth and energy-efficient secure data aggregation protocol called SDDA. In contrast with conventional data aggregation protocols, SDDA avoids the transmission of redundant data from sensor nodes to cluster heads. In SDDA, cluster heads are not required to examine the data received from sensor nodes. Instead, cluster heads compare the pattern codes received from sensor nodes and decide on which sensor nodes should be requested to transmit the actual data. This leads to better bandwidth utilization and energy efficiency since no redundant data is transmitted from sensor nodes to cluster heads. This data aggregation does not require the encrypted sensed data to be decrypted by cluster heads since they perform data aggregation on pattern codes. So, only base station decrypts the encrypted data of sensor nodes. Consequently, no encryption/decryption key is broadcast and, therefore, data confidentiality is maintained.

The section also introduces a novel key distribution algorithm implemented prior to the deployment of sensor nodes. Initially, each sensor node is assigned a number of encryption keys by taking advantage of the deployment knowledge of sensor nodes. Whenever two neighboring sensor nodes share the same encryption key, they establish a secure peer-to-peer communication that may lead one of them to switch to idle mode if they sense the same data.

Simulation results show that, as the redundancy rate increases, SDDA bandwidth occupancy decreases, thereby improving its bandwidth efficiency. As the density of sensor nodes increases, the energy saved by using the sleep-active mode protocol also increases, so that it contributes to the enhancement of the overall energy and bandwidth efficiency.

Future work includes developing secure data aggregation protocols for integrated sensor and cellular networks with relays. Also, secure communication and data aggregation algorithms should be developed for those sensor nodes that need to communicate with the neighboring cluster heads due to the congestion in their own cluster.

7.3 ENERGY-CONSERVING DATA GATHERING STRATEGY BASED ON TRADE-OFF BETWEEN COVERAGE AND DATA REPORTING LATENCY IN WIRELESS SENSOR NETWORKS

Wook Choi and Sajal K. Das

A wireless sensor network is a task-specific information gathering platform in which the sensors sense their vicinity and route the sensed information to a data gathering point, thus consuming the limited energy resource that may not often be replenishable. Indeed, an important issue in designing data gathering algorithms for sensor networks is to maximize energy conservation and hence the network's lifetime. Further enhancement to energy conservation rate can be achieved based on such requirements as data delivery latency and desired sensing coverage of the monitored area, as specified by the users or applications. In this section, we propose a novel energy conserving strategy for sensor data gathering that is based on a trade-off between coverage and data reporting latency. The basic idea is to select in each round only a minimum number of sensors as data reporters, which are sufficient for the desired sensing coverage. Sensors that are not responsible for reporting data in the current round cache their sensed data while waiting for the next round and thus save energy. Additionally, selecting the minimum number of reporters reduces the amount of traffic flow to the data gathering point in each round and avoids network congestion as well as channel interference/contention. We make use of four efficient schemes for k-sensor selection. These schemes help us experimentally evaluate such fundamental issues as event detection integrity and data reporting latency, which can be critical in deploying the proposed data gathering strategy. Simulation results demonstrate that the average data reporting latency is hardly affected and the real-time event detection ratio is greater than 80% when the sensing coverage is at least 80%. It is also shown that the sensors can conserve a significant amount of energy with a small trade-off: the higher the network density, the higher is the energy conservation rate without additional computation cost.

7.3.1 Background Information

The rapid advancement of microelectromechanical systems (MEMS) and wireless communications technologies based on short-range radio ushers the advent of highly sophisticated sensor networks in the near future. Such sensors will be equipped with data processing, wireless communication, and sensing units so they are able to cooperate autonomously with each other to form a network. Sensor networks can be characterized by high density and highly limited resources such as bandwidth, energy, computational capability, and storage space. These features distinguish them from the traditional ad hoc networks [22].

Furthermore, sensor networks are usually task-specific such that sensors sense phenomena and transmit their sensed data over the network to the user who, based on the data analysis, may update the sensors' behavior by sending a new task specification. Sensor networks can be deployed both indoors and outdoors, substituting for our sensory organs in inaccessible or inhospitable areas, in a variety of applications such as environment or equipment monitoring [23], smart home/smart space [24], intrusion detection, and surveillance [22, 25], to name a few. In most of the applications, sensors are operated by battery, which is severely limited and often not replenishable. Therefore, an important challenge in designing protocols for sensor networks is to make them more energy efficient so as to maximize their lifetime.

Sensor networks are densely populated by mostly static sensors that sense their vicinity (*sensing coverage*) and deliver sensed data to a gathering point through a multihop routing path, in case the data gathering point is not directly reachable by radio communication range. The data delivery frequency depends on the models that can be classified into continuous, event-driven, on-demand, or hybrid types, based on the applications or the user's interest [26]. The *continuous* model requests all sensors to transmit their sensed data periodically while they are alive. A cluster-based continuous data gathering scheme, called LEACH, is proposed in [10]. Further improvement to energy conservation achieved by the LEACH is shown in [27], which connects all the sensors as a linear chain. Forming a chain requires sensors to have global knowledge, which makes the data gathering scheme unscalable. The clustering scheme requires sensors to consume a certain amount of energy while forming and maintaining clusters. Moreover, the role of cluster heads leads to a relatively large amount of energy consumption as compared to an ordinary cluster member (i.e., noncluster head). Thus rotating the role of cluster heads is necessary to reduce the time variance of sensor failures caused by energy depletion. To this end, we proposed a two-phase clustering (TPC) [28], which reduces the cluster head's workload and thus the cluster head rotation by requiring cluster members to maintain two types of paths to the cluster head: direct control link (one hop) and data relay link (multihop). Unlike the continuous model, sensors in the *event-driven* data gathering model [29] start reporting their sensed data only when a specific event occurs. Whereas in the *on-demand* model [30], they report sensed data only at the users' request.

Due to high density of the network, it is common for multiple sensors to generate and transmit redundant sensed data that results in unnecessary power consumption and hence significantly decreases the network's lifetime. Among the sensors' actions, such as data transmission and target sensing, the energy consumption for wireless data transmission is the most critical. Therefore, minimizing the number of data transmissions between sensors by eliminating redundant data without losing data accuracy, or aggregating multiple sensed data, saves a significant amount of energy. For this purpose, many protocols have been designed for routing and managing topological connectivity, as summarized below.

Data-centric routing [31] attempts to reduce duplicate transmissions by aggregating multiple packets cached for a certain amount of time (i.e., in-network data processing with some data transmission delay), thereby increasing the energy conservation. In [32], a sensing coverage preserving scheme is proposed that turns off the sensors having overlapped coverage area with other sensors. More recently, two elegant algorithms, called connected sensor cover [33] and coverage configuration protocol [34], have been proposed that consider coverage and connectivity problems simultaneously. The former selects a minimum number of sensors to cover a specified area for query execution, thus reducing unnecessary energy consumption from redundant sensing. The latter selects a minimum number of sensors to guarantee that any point within a monitored area is covered by at least k sensors. These protocols find a relatively small set of (connected) sensors by running an algorithm with relatively high computational complexity, exchanging control information with local neighbors to cover the entire monitored area by 100%. The execution and implementation of such algorithms, however, are challenging because the sensors have highly limited resources. In fact, finding the smallest set of connected sensors that completely cover a given monitored area is an NP-hard problem [33].

For more efficient data gathering, sensors can be configured to serve more intelligently (i.e., in making a local decision for data reporting time, data aggregation, or operation mode such as sleep/idle based on the users' quality control requirements such as end-to-end delay, event-missed rate, and desired sensing coverage.

7.3.2 Contributions of This Section

Our motivation stems from a belief that depending on the specific applications, energy conservation can further be enhanced while meeting the user's requirements such as data delivery latency and desired sensing coverage of a monitored area. In this section, we propose a novel energy conserving data gathering strategy for the continuous data gathering model, based on a trade-off between coverage and data reporting latency with an ultimate goal of maximizing the network lifetime. The proposed strategy attempts to select at every data reporting round only a minimum of k sensors as data reporters, which are sufficient to cover as much of the monitored area as the user/application requests. Only these k sensors transmit data to the gathering point, while the others cache their sensed data waiting for the next reporting round, thus saving energy. All the sensors take turns in being selected as a data reporter. The parts of the area not covered by the first set of selected k sensors are covered by a subsequent set of selected k sensors with some delay. The smaller the desired sensing coverage, the longer is the data reporting latency in each sensor; whereas the energy conservation rate is inversely proportional to the coverage. Hence, the proposed energy conserving data gathering strategy is based on a trade-off between sensing coverage and data reporting latency.

Detection of an event may cause congestion if all the involved sensors attempt to transmit their sensed data simultaneously (causing high channel interference if a contention-based channel access scheme is used). In [35], a congestion control mechanism running in a data gathering point is proposed that controls the frequency of the sensors' data transmission upon detecting congestion. Besides the enhanced energy conservation, there is also subsequential benefit such as congestion avoidance and low channel interference/contention, which can be achieved by our proposed strategy using only k sensors in each reporting round. Hence, this contributes to additional energy savings, improving the overall network performance.

The proposed strategy adopts four schemes for k-sensor selection: nonfixed randomized selection (NRS), nonfixed and fixed disjoint randomized selections (N-DRS and F-DRS), and random-walk-based selection (RWS). They differ from one another in terms of data reporting latency, data aggregation, and implementation simplicity. The computational complexity of these four k-sensor selection schemes is constant (i.e., independent of network density and size), thereby providing high scalability. In addition, they do not use (periodic) control information exchange with local neighbors in selecting the k sensors. Thus, the proposed strategy is well suited for sensor networks that run for a long time under highly limited resource constraints. By intensive simulation studies, we also evaluate such fundamental issues as event detection integrity and data reporting latency, which are critical in deploying the proposed strategy. Experimental results demonstrate that our selection schemes can meet the desired sensing coverage by making approximately k sensors report their sensed data in each reporting round. More specifically, in a $100 \times 100\,\mathrm{m}^2$ network field with sensors having 30 m circular sensing range, the real-time event detection ratio is more than 90% using only $k \approx 14$ sensors, which are selected based on an 80% desired coverage of the entire monitored area. It is also shown that the average sensed data reporting latency is hardly affected when the desired coverage is greater than 80%. Furthermore, since the selection of k sensors is not affected by the network density, the energy conservation rate increases without additional computation cost as the network size grows. A preliminary version of this work appeared in [36].

7.3.3 Motivation and Problem Description

A sensor network is deployed to continuously sense the user's interest occurring in a specific application space. The sensor placement to cover the application space can be deterministic or randomized. The deterministic sensor placement is based on a specific topological structure model such as triangular, hexagonal, or mesh topology. On the other hand, the randomized sensor placement is based on a random distribution such as uniform or Poisson. The deterministic sensor placement may not often be practical or feasible due to the cost involved and environmental (geographical) constraints. In this section, we mainly focus on the (uniformly distributed) randomly deployed sensor networks, but the application of the proposed strategy is not limited to such networks only. Our motivation lies in the fact that depending on the type of applications used, the network lifetime can be much more critical than covering the entire monitored area at every data reporting round. The user may desire that only a certain portion of the area be covered at every data reporting round for the extended network lifetime, if the sensed data for the entire monitored area can be acquired within a fixed delay. For example, for a sensor network deployed for statistical study of scientific measurements in a certain area, it may be accurate enough to monitor the status of the area's certain conditions if the network covers approximately 80% of the field on an average in each round. Another example is for a sensor network monitoring slowly-moving objects, it may be acceptable if the network covers only 50% of the area at every round on the condition that the sensed result covering the entire area can be collected within a specified delay.

The problem is how to select a minimum number of sensors, satisfying a desired sensing coverage specified by the user. Figures 7.10a and 7.10b illustrate the problem. The black solid dots within a small solid circle (i.e., $s_1, s_2, s_3, \ldots, s_6$) in both of the figures represent the currently selected sensors, and the hollow dots in Figure 7.10b represent the previously selected sensors. The large solid circle represents the sensing coverage of each sensor. Suppose that the first six selected sensors cover a desired portion of the area but not the entire sensing area (i.e., shaded area is not covered), as shown in Figure 7.10a. The shaded area is being covered by the second set of selected six sensors, as shown in Figure 7.10b. Therefore, the user receives the sensed result for the entire monitored area with a fixed delay (i.e., two consecutive reporting rounds). We thus define the problem under consideration as follows:

Problem Definition   Given a set of $\mathcal{N}$ sensors that are placed over a region $\mathcal{T}$ by a random deployment scheme such that each $i \in \mathcal{N}$ has sensing region SR$_i$. A minimum of

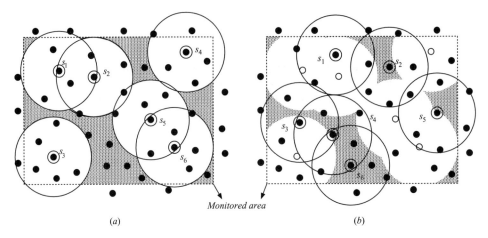

Monitored area

(a)   (b)

Figure 7.10 Illustration of data gathering based on trade-off between coverage and data reporting latency.

k sensors are to be chosen from $\mathcal{N}$ such that

$$\left[\left(\bigcup_{i=1}^{k} SR_i\right) \bigcap \mathcal{T}\right] \geq \text{desired sensing coverage}$$

To solve this problem, we apply a geometric probability theory dealing with random circles on geometrical figures [37], which is the study of probabilistically measuring how much area will be overlapped when a circle is randomly placed over a geometrical figure.

7.3.4 Network Model and Assumptions

A wireless sensor network consists of a large number of sensors, single or multiple base station(s) serving as data gathering point(s) and control center(s), and wireless links representing direct communication between sensors, or between sensors and data gathering point(s), within the radio range. Densely deployed sensors form a static network and learn their local connectivity at the network deployment time. They adapt to topology changes caused by sensor failures, based on the local connectivity learned. We assume that the sensor network is a two-dimensional flat plane. Although there are feasible means to make sensors aware of their location, such as the Global Positioning System (GPS) or directional beaconing [38, 39], we do not assume that sensors are located by any specific coordination system because such localization mechanisms may not be available or practical in building low-cost, low-power sensors with a small form factor.

Formally, we shall define a sensor network as an undirected connected graph $G = (V, E)$, where V and E are the set of nodes (sensors and data gathering points) and the set of edges (bidirectional wireless links), respectively. A sensor node, s_i, generates a fixed-size data packet for a time unit as a sensed result. We call this time unit as a *data reporting round*, and the interval between two consecutive data reporting rounds is denoted by Δt. All the sensors are supposed to forward the generated data packet to the data gathering point using a routing path, making the communication pattern many-to-one. A sequence of edges in

G forms the path, $\mathcal{P} = ((s_1, s_2), (s_2, s_3), \ldots, (s_{i-1}, s_i), (s_i, s_*))$ for $1 \le i \le |V| - 1$, where $s_i \in V$ is a sensor and $s_* \in V$ is a data gathering point. Thus, $\mathcal{P}$ is considered as a multihop routing path and each node $s_i \in \mathcal{P}$ acts as an individual router. Since the proposed strategy is considered as a data gathering protocol running on top of the routing layer without any location information, in this section we assume a nongeographical sensor routing protocol that connects all sensors at the deployment time [40, 41]. A control message from s_* is delivered to the sensor nodes through flooding, as in [29]. Each node $s_i \in V$ has its specific radio and sensing ranges, both of which form a circular area A_{s_i} with radius r. A sensor s_i can directly communicate with any nodes within its radio range area A_{s_i}. Now, let us introduce two definitions:

Definition 7.3.1 *Local neighbors* of sensor s_i are defined as

$$N(s_i) = \{s_j \mid d(s_i, s_j) \le r, \ i \ne j, \} \cup \{s_* \mid d(s_i, s_*) \le r\}$$

such that $|N(s_i)| \le |V| - 1$ where $d(s_i, s_j)$ represents the *Euclidean* distance between two nodes s_i and s_j.

Definition 7.3.2 Let $\hat{d}$ be the expected distance from s_i to $s_j \in N(s_i)$. Since s_j can be anywhere within A_{s_i}, we have

$$\hat{d} = \int_0^{2\pi} \int_0^r \ell^2 f(\ell \cos \theta, \ell \sin \theta) \, \partial \ell \, \partial \theta = \frac{2r}{3}$$

where $f(\ell \cos \theta, \ell \sin \theta) = 1/\pi r^2$ is the probability that s_j is located at a point within A_{s_i}. Note that the *polar coordinate system* is used for this measurement.

7.3.5 Desired Sensing Coverage and Minimum *k* Data Reporters

In this work, the term *desired sensing coverage* (*DSC*) represents a probabilistic percentage for covering any point within the entire monitored area. The user specifies the DSC as the desired "quality of service" to be achieved by sensor data gathering. Thus, we define the desired sensing coverage as a trade-off factor for energy conservation. Only a minimum number of sensors, which is enough to cover the DSC, report their sensed data while the others wait for the next rounds. This relieves the sensors' responsibility of reporting sensed data, and hence saves energy. The DSC is proportional to the amount of sensed data traffic over the network and inversely proportional to both the energy conservation rate and data reporting latency. The question is: In order to meet the DSC specified by the user, how many sensors do we need to select at each data reporting round? To answer this question, we first define the following basic concepts:

Definition 7.3.3 A *monitored area*, denoted by Q, is the actual area to be monitored by sensors. We consider this area as an $\alpha \times \alpha$ square.

Definition 7.3.4 A *sensor-deployed area*, denoted by Λ, is a square area including all sensors that have an effect on covering Q such that the square will have rounded corners with distance less than or equal to r (radius of sensing range) from the boundary of $Q \subset \Lambda$. Thus, the circular sensing range of a sensor residing in $\Lambda - Q$ is not fully overlapped with the area Q.

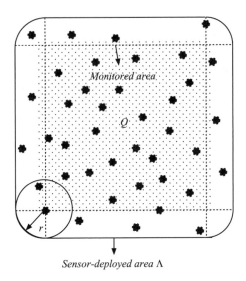

Sensor-deployed area Λ

Figure 7.11   Illustration of monitored and sensor-deployed areas.

Definition 7.3.5   A *probabilistic sensing coverage*, denoted by ψ, is the probability of any point in Q being covered by at least one of the selected k sensors' (residing in Λ) circular sensing range with radius r. This is provided by the users or applications as the desired sensing coverage (DSC).

When sensors are randomly deployed over Q, it is likely that they may be placed slightly beyond the boundary of Q in order to completely cover the monitored area Q or due to the inaccuracy in deployment. Thus, there is a separate definition of the sensor-deployed area, Λ, in measuring the probabilistic sensing coverage. Figure 7.11 illustrates the basic terms defined above. A circle centered at a point at the bottom-left corner in the sensor-deployed area is a sensing coverage of a sensor with radius r.

Let $q \subseteq Q$ be the part covered by the circular sensing ranges of $k \leq |V| - 1$ sensors residing in Λ. Then, the fraction q/Q is the user's DSC at each reporting round. Any point $(x, y) \in Q$ is *covered* if it is inside the circular sensing coverage of a selected sensor in Λ. To measure the probabilistic sensing coverage, we first measure the probability $P_q(x, y)$ that a point $(x, y) \in Q$ will not be covered by a selected sensor, s_i. Let $A(x, y)$ be a circular area centered at (x, y) with radius r. Then, the point will not be covered when $s_i \in \Lambda - A(x, y)$. Therefore, the probability that the point (x, y) is not covered by a randomly selected sensor, is given by

$$P_q(x, y) = \int_{\Lambda - A(x,y)} \int \phi(x, y) \, dx \, dy \tag{7.1}$$

where $\phi(x, y) = 1/\Lambda$ is the probability that s_i is located on a point $(x, y) \in \Lambda$. Equation (7.1) represents the fraction of the sensor-deployed area, Λ, not covered by a sensor's circular sensing range. Thus, the probability that a point is not covered by k randomly selected sensors, is obtained as

$$P_q^k(x, y) = \prod_{i=1}^{k} (P_q(x, y))^i$$

Let $\bar{q}$ be the area of Q not covered. For randomly selected k sensors, the expected value of $\bar{q}$ can be obtained as:

$$E[\bar{q}] = \int_Q \int P_q^k(x, y) \, dx \, dy \qquad (7.2)$$

As mentioned earlier, we consider how much area in Q can be covered by k randomly selected sensors. For this purpose, we first consider the fraction of Q not covered by these k sensors within Q. This can be obtained by dividing $E[\bar{q}]$ by the area α^2 of Q, assuming all (x, y) points are uniformly (randomly) distributed over Q. Applying Eqs. (7.1) and (7.2), the fraction of Q not covered by k selected sensors is given as:

$$E\left[\frac{\bar{q}}{\alpha^2}\right] = \left(\frac{\Lambda - A(x, y)}{\Lambda}\right)^k = \left(\frac{\alpha^2 + 4\alpha r}{\alpha^2 + 4\alpha r + \pi r^2}\right)^k$$

Finally, when k sensors are uniformly (randomly) selected from Q, the probabilistic sensing coverage (ψ) that any point of Q will be covered by at least one of k selected sensors' circular sensing range is given by:

$$\psi = 1 - E\left[\frac{\bar{q}}{\alpha^2}\right] = 1 - \left(\frac{\alpha^2 + 4\alpha r}{\alpha^2 + 4\alpha r + \pi r^2}\right)^k \qquad (7.3)$$

Therefore, the smallest integer k that satisfies the desired sensing coverage, ψ, can be derived as:

$$k = \left\lceil \frac{\log(1 - \psi)}{\log\left(\frac{\alpha^2 + 4\alpha r}{\alpha^2 + 4\alpha r + \pi r^2}\right)} \right\rceil \qquad (7.4)$$

In order to verify the correctness in measuring k, we simulate the analytical model and compare the simulation results with the numerical results obtained from Eq. (7.3). Figures 7.12a and 7.12b show the comparison results in covering a requested portion of

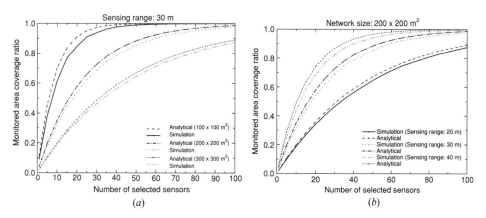

Figure 7.12 Comparison of simulation and analytical results for covering a monitored area.

the monitored area with varying network sizes and sensor's circular sensing ranges. The simulation results shown in each plot correspond to the average of 100 simulation runs. Regardless of the size of the network field and the sensing range, we observe in Figures 7.12a and 7.12b that both the numerical and simulation results match well.

7.3.6   *k*-Sensor Selection Schemes

Based on the DSC, the size of the monitored area, and the sensor's sensing range, Eq. (7.4) gives the required minimum number (k) of sensors that should be selected from all the sensors uniformly distributed over the sensor-deployed area, Λ. Obviously, sensors will experience a certain latency in reporting their sensed data to the data gathering point. This is because they are allowed to report only when they become one of the selected k sensors. We use randomization techniques to select those k reporting sensors in each round. Depending on the randomized selection technique used, the data reporting latency will vary in each sensor. The selection scheme can be centralized or distributed. However, taking into consideration that the wireless sensor network is composed of a large number of sensors, a centralized scheme is not the best choice for scalability. In the following, we propose a family of distributed, k-sensor selection schemes for our data gathering strategy in which sensors individually make a decision about when to report sensed data. These schemes differ from one another in terms of the data reporting latency and implementation simplicity. They are described below.

Nondisjoint Randomized Selection (NRS)   In the NRS scheme, a sensor elects itself as one of the k sensors in each round based on the probability,

$$P_{\mathrm{nrs}} = \frac{k}{|V| - 1}$$

Each sensor draws a random number uniformly distributed in [0, 1]. If the random number is less than or equal to P_{nrs}, the sensor becomes a data reporter. Such a sequence of independent random selection trials to become a reporter can be modeled as a *geometric distribution* having P_{nrs} as the success probability. Thus, the expected number of the trials is $1/P_{\mathrm{nrs}}$. This implies that the data reporting latency (i.e., the elapsed time to be selected as a reporter again) could be very large when the DSC is small. In other words, the NRS scheme cannot guarantee gathering the sensed result for the entire monitored area within a fixed delay. Hence, this scheme is not appropriate for real-time or time-constrained monitoring for the entire area. Furthermore, the NRS scheme does not have any control operation to force already-selected sensors to wait without reelecting themselves as a reporter until all sensors take their turn to report. This implies the set of selected k sensors at the current reporting round may not be disjoint from the set selected in the previous round. However, due to the long-term steady-state behavior of the randomized selection, all the sensors in the deployed area Λ will approximately have the same number of chances to transmit the sensed data. This scheme can be used for applications that require the sensed result for only a specific portion of the monitored area at each data reporting round.

Disjoint Randomized Selection (DRS)   Unlike NRS, the DRS scheme covers the entire monitored area within a fixed delay by allowing all the sensors to have a chance to report their sensed data within a fixed number of reporting rounds. Once the sensors report

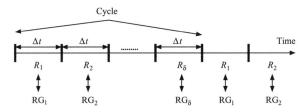

Figure 7.13 Reporting cycle ($\mathcal{C}$), round (R_i), and data reporting group (RG$_i$).

their data, they do not elect themselves as a reporter again until all the sensors have had an equal chance to do so. Before proceeding further, we introduce the following definitions.

Definition 7.3.6 A *data reporting group* (RG) is a set of data reporting sensors selected in a given round. There are

$$\delta = \left\lfloor \frac{|V| - 1}{k} \right\rfloor$$

such disjoint groups, denoted by RG$_i$ where $1 \leq i \leq \delta$. Each sensor belongs to exactly one group such that $\cap_{1 \leq i \leq \delta}\text{RG}_i = \phi$ and $\cup_{1 \leq i \leq \delta}\text{RG}_i = |V| - 1$.

Definition 7.3.7 A *reporting cycle*, denoted by $\mathcal{C}$, is the time periodicity when a sensor s_i reports its sensed data. As shown in Figure 7.13, the reporting cycle contains δ reporting rounds, denoted by R_i for $1 \leq i \leq \delta$ such that R_i corresponds to RG$_i$. Sensors belonging to the group RG$_i$ report their sensed data only in R_i and wait until they complete the cycle. Thus, all the sensors have an equal chance of reporting data within a fixed delay. Note that the total delay in acquiring the sensed result for the entire monitored area depends on the DSC since δ is decided by k measured from Eq. (7.4).

Definition 7.3.8 A *reporting sequence* of a sensor s_i, denoted by RS$_{s_i}$, is a sequence of bits. Each bit maps to each round R_i in $\mathcal{C}$, and hence the number of bits is equal to the number of rounds in $\mathcal{C}$. The sequence is initialized to all zero (i.e., off) and only one bit is flipped (i.e., on) depending on which round within $\mathcal{C}$ is selected. Thereby, RS$_{s_i}$ indicates the round within $\mathcal{C}$ in which the sensor s_i reports its sensed data. For example, RS$_{s_i}$ = "00100" for $k = 4$ and $|V| - 1 = 20$, represents that s_i selects the third round R_3 within $\mathcal{C}$, thus becoming a member of the group RG$_3$.

The users may consider some other constraints along with DSC while receiving data from the sensors. For example, an area is required to be monitored with a uniform fixed delay in each round, or the monitoring pattern should not be learned by an adversary attempting to circumvent the sensing activity (i.e., the uncovered part of the monitored area at every reporting round should be unpredictable). In this regard, we provide two selection schemes to form the reporting groups RG$_i$ for $1 \leq i \leq \delta$: nonfixed disjoint randomized selection (N-DRS) and fixed disjoint randomized selection (F-DRS), described below.

- *Nonfixed disjoint randomized selection* (*N-DRS*): At every cycle $\mathcal{C}$, each sensor elects itself as a reporter by drawing a round randomly from δ reporting rounds within $\mathcal{C}$ so that the set of selected k sensors at each round is memoryless. Therefore, the monitoring pattern is not known beforehand. The reporting latency of each sensor in the N-DRS

Algorithm Construct_RS $(k,|V|)$

1: $\delta \leftarrow \left\lfloor \frac{|V|-1}{k} \right\rfloor$

2: Allocate a bit array $A[\delta]$ and initialize all the entires
    with zero

3: $i \leftarrow \text{RAND}[1,\delta]$

4: $A[i] \leftarrow 1$ /* $1 \leq i \leq \delta$ */

5: **return** $A[\delta]$ /* reporting bit sequence with
    ith bit on */

End_Algorithm

Figure 7.14   Algorithm for constructing reporting sequence (RS$_s$) of sensor s.

scheme ranges from $\delta \times \Delta t$ to $(2\delta - 1) \times \Delta t$ where Δt is the time interval between two consecutive reporting rounds.

- *Fixed disjoint randomized selection (F-DRS)*: Similar to N-DRS, sensors select a reporting round randomly from the δ reporting rounds; however, they do not select at every cycle. Instead, they keep the initially selected reporting round. Therefore, at every cycle C, all the sensors have the same sequence of reporting round and hence a uniform fixed data reporting latency of $\delta \times \Delta t$.

Figure 7.14 presents an algorithm for constructing the reporting sequence RS$_{s_i}$ where $A[\delta]$ is a bit array of length δ and RAND$[1, \delta]$ is a function that returns a uniformly distributed random integer between 1 and δ. Sensors participate in a reporting procedure only when a bit corresponding to the current round in $A[\delta]$ is equal to 1. Thereby, δ disjoint reporting groups $\{RG_i\}_{i=1}^{\delta}$ are formed, and all the sensors transmit their sensed data exactly once during a reporting cycle C. After the cycle C, if sensors use the N-DRS scheme, they acquire a new reporting sequence $A[\delta]$ by running the algorithm in Figure 7.14. Otherwise (i.e., for the F-DRS scheme), they keep the initial reporting sequence until an update request is received. Due to the time-bounded data gathering of the entire monitored area, these two DRS schemes can be used more widely than the NRS scheme.

In NRS and DRS, approximately k sensors report their sensed data to the data gathering point in each round. Other sensors cache their sensed data waiting for their turn to report. If the sensors have data aggregation capability, they can generate a sensed result that is more meaningful than a one-time sensed result at the reporting instant by using sensed data cached during their off-duty time. For example, let a sensor detect an event and report it at round R_1. Then at the next reporting round, the sensor reports "there has been an event since R_1" or "the event disappeared after R_1" depending on the cached sensed data instead of simply reporting "there is an event" or "there is no event."

Note on Selected k Sensors   The NRS and DRS schemes do not guarantee selection of exactly k sensors in each round. The resulting number of selected sensors could be a little more or less than k, due to random selection without any message exchange with neighboring sensors. In fact, as the network size grows keeping the same density, the probability of selecting the exact k sensors also becomes larger. This is because the ratio of k to the total number of sensors becomes smaller (refer to Fig. 7.15), implying that the ratio of the number of sensors that attempt to be one of k sensors becomes larger (i.e., more trials and hence more successes to be one of k sensors). For the two variants of the DRS

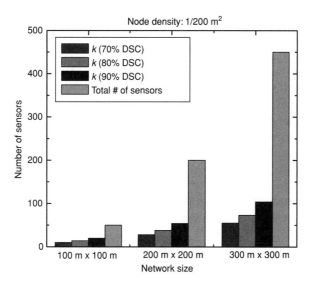

Figure 7.15 Selected k sensors vs. total number of sensors.

scheme, the same effect can be achieved by controlling the number of rounds in the reporting cycle. Considering that the sensor networks are usually large-scale with a high node density, the difference between the number of selected sensors in each round and k will be minimal. This will be validated later while presenting experimental results in Section 7.3.7. Besides the implementation simplicity, these randomized selection schemes have another important advantage: The higher the network density, the higher is the energy conservation rate without any additional computation and communication costs. In the case of the F-DRS scheme having a fixed reporting schedule, we can balance the number of sensors in each round $R_i \in C$ using a minimum of message exchanges. Figure 7.16 illustrates such a balancing procedure where P_{rd} is the probability of being selected as a candidate for drawing a new reporting round. Figure 7.16a shows the initial distribution in which R_1 and R_4 have "2" and "1" overflows, respectively, while R_2 and R_3 have "2" and "1" underflows, respectively. We assume that the data gathering point maintains such an initial distribution in a table format shown in Figure 7.16b and floods it using a control message. Upon receiving the control message, sensors construct a table as shown in Figure 7.16c if the chosen data reporting round is either R_1 or R_4 (i.e., overflowed round). Based on this table, the sensors attempt to

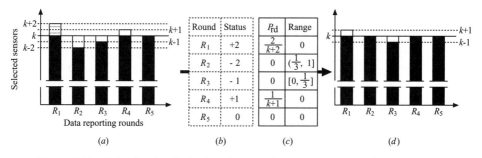

Figure 7.16 Balancing the distribution of sensors in each round $R_i \in C$ ($1 \leq i \leq \delta = 5$).

draw a new data reporting round among the underflowed ones. Suppose a sensor s_i chooses R_1 initially in this example. Then s_i draws a random number within [0,1]. If the random number $\leq 2/k + 2$, then it draws another random number within [0,1]. Finally, s_i chooses either R_2 or R_3 depending on which range the second random number falls in. Figure 7.16d shows the distribution of the number of sensors in each round after the balancing procedure.

Random-Walk-Based Selection (RWS)   We propose another randomization technique based on random walk, to select k sensors at each reporting round. In the RWS scheme, we focus more on the in-network data aggregation than on time-bounded data gathering for further energy conservation. The random walk can be considered as a mobile virtual cluster head that uniformly visits its cluster members to collect what they sensed, thus distributing energy consumption for aggregating and collecting sensed data on-the-fly. There is no need for a cluster head rotation procedure (as in the LEACH protocol [10]) to balance the remaining energy level in each sensor, nor to maintain cluster membership. This concept of a mobile virtual cluster head is different from the passive cluster head concept that waits for all the cluster members to transmit their sensed data. Depending on the number of random walks, only one or more sensors that are currently visited by the random walk will be in a forwarding action, thus experiencing low channel interference/contention between sensors in the transmission mode.

Random Walk Travel   Initially, the random walk is launched by the sensors chosen by the data gathering point. The random walk, denoted by $\mathcal{R}_{s_j}^{s_i}$, travels over the sensor-deployed area, Λ, by (uniformly) randomly choosing the next stopover s_j from the local neighbor set $N(s_i)$ of the current stopover s_i except the previous stopover (Fig. 7.17). Once $\mathcal{R}_{s_j}^{s_i}$ stops over s_j, two sensed data—one in s_j and the other carried by the random walk— are aggregated before moving on to the next stopover. When $\mathcal{R}_{s_j}^{s_i}$ makes the kth stopover, it forwards the aggregated data to the gathering point and then starts making another k stopover. However, Eq. (7.4) is based on uniform random selection of k sensors without

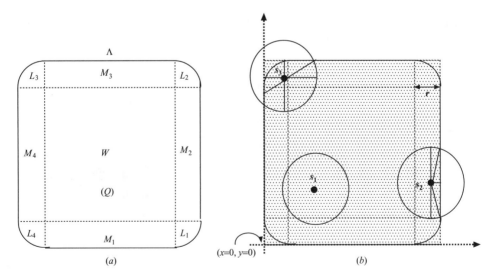

(a)   (b)

Figure 7.17   Valid radio range area depending on sensor's position in sensor deployed area Λ.

Table 7.3 Notations Used in Measuring the Connectivity of Selected k Sensors

Term	Description
$E[A_{s_i}]$	Expected valid radio range (having area A_{s_i}) of sensor s_i within the sensor-deployed area Λ
β	Set of selected and connected k sensors. Initially, β includes only one of the selected k sensors but eventually $|\beta|$ increases up to k
$\bar{\beta}$	Set of selected k sensors. Initially $\bar{\beta}$ includes $k-1$ of the selected k sensors and $|\bar{\beta}|$ decreases to 0.
P_s	Probability that a sensor (randomly chosen from Λ) is within a circular area centering at a specific point within Λ (i.e., sensor's radio range)
Pr_s^l	Probability that a sensor in β finds at least one neighbor in $\bar{\beta}$ in step l
Pr_c^l	Probability that a sensor chosen from $\bar{\beta}$ is connected to at least a sensor in β in step l
Pr_c^k	Probability that all selected k sensors are connected
$\hat{k}$	Selected k sensors plus additional sensors required to ensure that the selected k sensors are connected

considering their connectivity. Thus, when k is small (i.e., the desired sensing coverage is small), the distribution of selected k sensors is relatively sparse, and hence they are disconnected with high probability. This implies that more sensors need to be involved in order to achieve the DSC using the random walk since each random walk's traveling range is limited to the area A_{s_i}. Therefore, we first need to consider the connectivity of k sensors that will help us estimate the additional number of sensors for the random walk travel. Before proceeding further, let us define some notations (summarized in Table 7.3) used in measuring the connectivity of selected k sensors.

Connectivity of k-Sensor Selection We measure the connectivity of the selected k sensors in a probabilistic manner. The objective is to find an additional number of sensors to ensure all the selected k sensors connected, in order to allow the random walk $\mathcal{R}_{s_j}^{s_i}$ to travel while satisfying the desired sensing coverage. Basically, in wireless sensor networks, node connectivity depends on the size of the sensor's radio range A_{s_i}. This is because the larger the radio range, the higher is the probability that a sensor can have neighbors. Thus, based on the probability P_s, we test the connectivity of the k sensors in $k-1$ steps, as shown in Figure 7.18a.

Some part of A_{s_i} that does not overlap with Λ cannot be considered as a valid area in deriving P_s. In addition, the overlapped area of A_{s_i} varies depending on the location of s_i within Λ (refer to Fig. 7.17). Thus, we average out all possible overlapped cases to estimate $E[A_{s_i}]$. We consider a square network field Λ as in Figure 7.17b such that both x and y coordinates are positive. The area difference between the square Λ and a square Λ with rounded corners as in Figure 7.17a is $4r^2 - \pi r^2$. Since the sensor-deployed area Λ is usually much larger than a sensor's radio range, the ratio of the area difference to Λ is so minimal. Thus, the impact on estimating the expected valid radio range of sensors is also minimal when the square Λ is used. In Figure 7.17b, the entire radio range of the sensors (e.g., s_1) in area W is overlapped with Λ. On the other hand, depending on their location, some portion of the radio range of the sensors (e.g., s_2 in M_2 and s_3 in L_3) in regions M_i and L_i ($1 \leq i \leq 4$) is not overlapped with Λ. To compute the average of the overlapped radio range areas of all the sensors in Λ, we first add the overlapped radio range areas of

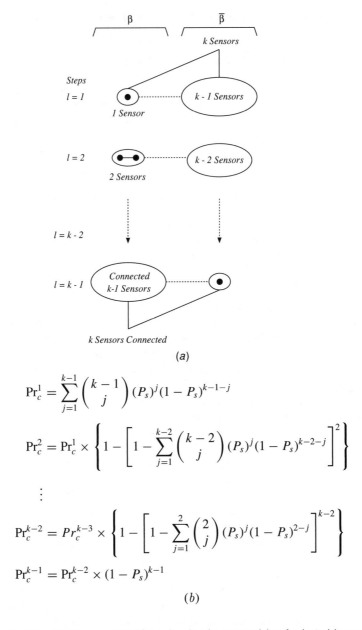

(a)

$$\mathrm{Pr}_c^1 = \sum_{j=1}^{k-1} \binom{k-1}{j} (P_s)^j (1-P_s)^{k-1-j}$$

$$\mathrm{Pr}_c^2 = \mathrm{Pr}_c^1 \times \left\{ 1 - \left[1 - \sum_{j=1}^{k-2} \binom{k-2}{j} (P_s)^j (1-P_s)^{k-2-j} \right]^2 \right\}$$

$$\vdots$$

$$\mathrm{Pr}_c^{k-2} = Pr_c^{k-3} \times \left\{ 1 - \left[1 - \sum_{j=1}^{2} \binom{2}{j} (P_s)^j (1-P_s)^{2-j} \right]^{k-2} \right\}$$

$$\mathrm{Pr}_c^{k-1} = \mathrm{Pr}_c^{k-2} \times (1-P_s)^{k-1}$$

(b)

Figure 7.18   Probabilistic model for estimating the connectivity of selected k sensors.

all sensors in the regions M_i, L_i, and W denoted by $q(M_i)$, $q(L_i)$, and $q(W)$, respectively. The overlapped radio range area of the sensors outside the boundary of the monitored area Q (i.e., sensors in M_i and L_i) can be simply calculated by dividing the overlapped area like the ones of s_2 and s_3 shown in Figure 7.18b.

Now,

$$q(M_i) \approx \sum_{\forall (x,y) \in M_i} \frac{\pi r^2}{2} + y\sqrt{r^2 - y^2} + r^2 \arcsin\left(\frac{y}{r}\right)$$

Similarly,

$$q(L_i) \approx \sum_{\forall (x,y) \in L_i} \frac{\pi r^2}{4} + xy + \frac{y\sqrt{r^2 - y^2}}{2} + \frac{x\sqrt{r^2 - x^2}}{2} + \frac{r^2[\arcsin(\frac{x}{r}) + \arcsin(\frac{y}{r})]}{2}$$

Since the entire radio range area of any sensor within W is overlapped,

$$q(W) = \sum_{\forall (x,y) \in W} \pi r^2$$

Therefore, the total sum of the valid radio range areas of all the sensors in $\Lambda = \alpha^2 + 4\alpha r + 4r^2$ is given by

$$\sum_{i=1}^{4} [q(M_i) + q(L_i)] + q(W)$$

and the expected valid radio range is derived as:

$$E[A_{s_i}] \approx \frac{\sum_{i=1}^{4} [q(M_i) + q(L_i)] + q(W)}{\alpha^2 + 4\alpha r + 4r^2} \tag{7.5}$$

Using $E[A_{s_i}]$, let us now measure the probabilistic connectivity of the selected k sensors.

As illustrated in Figure 7.18a, the connectivity test starts with two sets: β with one sensor and $\bar{\beta}$ with $k - 1$ sensors. Then a sensor from the set $\bar{\beta}$ is chosen at each step l and is considered to be connected to at least one sensor already in the set β with probability Pr_c^l. As a result, $|\bar{\beta}|$ decreases by 1 and $|\beta|$ increases by 1 after each step, such that $|\beta| + |\bar{\beta}| = k$ at every step. Moreover, all sensors in β are considered to be connected at every step. This test procedure is repeated until $\bar{\beta} = \phi$ (empty set).

In step l, a sensor in β finds at least one neighboring sensor from $\bar{\beta}$ with the probability Pr_s^l that is given by:

$$\mathrm{Pr}_s^l = \sum_{j=1}^{k-l} \binom{k-l}{j} (P_s)^j (1 - P_s)^{k-l-j}$$

where $P_s = E[A_{s_i}]/\Lambda$. Since any sensor already in β can be connected to at least one sensor in $\bar{\beta}$ with the probability Pr_s^l, we can choose any sensor from $\bar{\beta}$ and include that sensor in β with the probability Pr_c^l which is given by:

$$\mathrm{Pr}_c^l = \bigcup_{j=1}^{l} \mathrm{Pr}_s^l = 1 - \prod_{j=1}^{l} \overline{\mathrm{Pr}_s^l} \tag{7.6}$$

where $\overline{\mathrm{Pr}_s^l} = 1 - \mathrm{Pr}_s^l$. Finally, the probability that all the selected k sensors are connected is (refer to Fig. 7.18b for the details):

$$\mathrm{Pr}_c^k = \bigcap_{l=1}^{k-1} \mathrm{Pr}_c^l = \prod_{l=1}^{k-1} \mathrm{Pr}_c^l \tag{7.7}$$

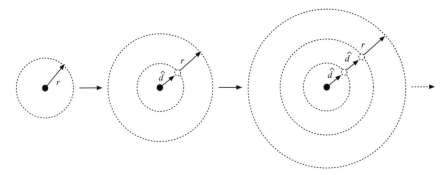

Figure 7.19   Increasing P_s by extending the radio range area.

Probabilistically, $k(1 - \mathrm{Pr}_c^k)$ sensors are disconnected. Now, suppose a disconnected sensor s_i chooses a neighbor s_j from $N(s_i)$, and s_j makes s_i connected or closer to any of the other selected sensors that are already connected. Then this neighbor s_j is counted as an additional sensor, besides k, for the travel of random walk. So, in this connectivity test, the number of additional sensors required is $k - \lfloor Pr_c^k \times k \rfloor$, which is the same as the number of disconnected sensors. For these disconnected sensors, we recalculate P_s by extending the radio range, that is, assuming each of them chose a neighbor (see Figure 7.19). Then using the recalculated P_s, we measure the connectivity of the selected k sensors again from the $\lfloor \mathrm{Pr}_c^k \times k \rfloor$th step (initially, $|\beta| = \lfloor \mathrm{Pr}_c^k \times k \rfloor$ and $|\bar{\beta}| = k - |\beta|$). In this way, we can recursively run the connectivity test with only disconnected sensors from the previous test until all the selected k sensors are considered to be connected and thus obtain the total number of additional sensors required. In every recurrent connectivity test, the radius of the radio range is extended by only $\hat{d}$ (refer to Fig. 7.9). Recall that $\hat{d}$ is the expected distance to a neighboring sensor that can be anywhere in A_{s_i}. Figure 7.20 describes a recursive algorithm that computes the additional number of sensors to probabilistically guarantee the connectivity of the selected k sensors. In this algorithm, $P(E[\pi(\omega + \hat{d})^2]/\Lambda, \gamma)$ represents Pr_c^k remeasured from step $l = \gamma$ with recalculated success probability $P_s = E[\pi(\omega + \hat{d})^2]/\Lambda$, and $E[\pi(\omega + \hat{d})^2]$ is calculated by Eq. (7.5) based on the circular area with radius $\omega + \hat{d}$. Notice that in line 6 of the algorithm (see Fig. 7.20), every recursive call proceeds with only disconnected sensors (i.e., $\hat{k} - \lfloor P \times \hat{k} \rfloor$) with the radio range extended by $\hat{d}$. The algorithm

Alogorithm Find $\hat{k}$ **Begin**
 1. **procedure** FK($P, \hat{k}, \omega, \gamma$) **begin**
 2.   **if** (($1-P$)$\times \hat{k} \leq 1$)
 3.   **then** $\hat{k} \leftarrow \hat{k}+1$;
 4.   **else** $P' \leftarrow P\left(\frac{E[\pi(\omega+\hat{d})^2]}{\Lambda}, \gamma \right)$;
 5. $\gamma \leftarrow \gamma + \lfloor P \times \hat{k} \rfloor$;
 6. $\hat{k} \leftarrow \hat{k} + \mathrm{FK}(P', \hat{k} - \lfloor P \times \hat{k} \rfloor, \omega + \hat{d}, \gamma)$;
 7.   **end_if**
 8.   **return** $\hat{k}$;
 9. **end_procedure**
10. FK($Pr_c^k, k, r, \gamma = 1$) * Main Algorithm */
End_Algorithm

Figure 7.20   Algorithm for finding the total number $\hat{k}$ of sensors for RWS scheme.

Table 7.4 Numerical Results of $\hat{k}$ for k (Sensing Range: 30 m)

ψ	$100 \times 100\,\text{m}^2$		$200 \times 200\,\text{m}^2$		$300 \times 300\,\text{m}^2$	
	k	$\hat{k}$	k	$\hat{k}$	k	$\hat{k}$
0.5	6	11	17	33	32	64
0.6	8	13	22	35	42	71
0.7	10	14	28	40	55	80
0.8	14	17	38	46	73	92
0.9	20	21	54	59	104	115

returns $\hat{k} \geq k$, which is the total number of sensors that have to be traveled by the random walk to achieve the DSC (desired sensing coverage). Table 7.4 shows numerical results of corresponding $\hat{k}$ for k measured from Eq. (7.4).

Constraining Random Walk As in NRS, it is clear that the RWS scheme does not guarantee gathering the sensed result for the entire monitored area within a fixed delay. Furthermore, a set of $\hat{k}$ sensors in each reporting round may not be totally disjointed due to the random selection of next stopovers. In order to reduce the time variance of $\mathcal{R}_{s_j}^{s_i}$'s visiting time in each sensor, we introduce a constrained version of the random walk that moves only to the next hop that is chosen based on the least recently visited (LRV) policy.

Whenever each sensor overhears or receives the random walk $\mathcal{R}_{s_j}^{s_i}$ forwarded by s_j, it marks the current local time as $\mathcal{R}_{s_j}^{s_i}$'s visiting time of s_j if s_j is in its local neighbor set. Thereby, each sensor locally recognizes which neighboring sensor is least recently visited. We run experiments in 200 m × 200 m network field with 200 sensors (1 sensor/200 m^2) for 16,000 s using 5 different sensor deployments. In these experiments event occurrences are uniformly (randomly) distributed over the monitored area with an interval uniformly distributed within [5, 10] seconds, and the random walk moves every 1 s and transmits data after every $\hat{k} = 10$ visits. Figure 7.21 shows the distribution of the number of visits

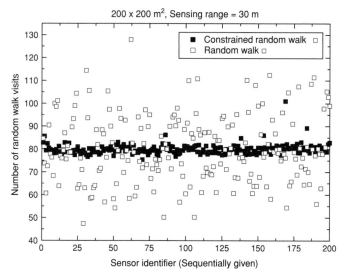

Figure 7.21 Distribution of random walk visits.

Table 7.5 **Random Walk's Visiting Interval and Data Reporting Latency (unit: seconds)**

	Random Walk Type	Average	Min	Max
Reporting latency	Constrained random walk	59.12	1.21	380.42
	Random walk	180.64	1.08	2393.37
Visiting time	Constrained random walk	198.51	1	499.4
	Random walk	196.20	1	5869.4

in both random walk and its constrained version. Clearly, the constrained random walk is able to drastically reduce the time variance of the visiting time, and thus the data reporting latency in each sensor. Table 7.5 summarizes the statistical information gathered from the experiments.

Discussion on Routing   The routing is the most energy consuming operation in a multihop wireless sensor network. Hence, the frequency of the routing service in sensors is a critical factor in determining their remaining lifetime. The amount of total traffic routed to the data gathering point depends on the number of sensors wishing to report their sensed data in each round. Therefore, the routing load (RL) in the network is expressed as:

$$\text{RL} = \sum_{\forall s_i \in F} (1 - \mu)\lambda_{s_i}$$

where λ_{s_i} is the sensed data generation rate in a sensor s_i, F is the set of sensors that report sensed data ($|F| = k$ in our case), and $\mu \leq 1$ is the ratio of data redundancy to the sensed data already reported in s_i, which depends on the sensor density and sensed data reporting interval. Then the total routing load in each round is equivalent to the total number of sensed data packets generated by all $s_i \in F$. Let $\aleph$ be the number of packets a sensor has to forward as a router toward the data gathering point such that the individual routing load of a sensor serving as a router satisfies $\frac{RL}{|F|} \leq \aleph \leq \text{RL}$. That is, the individual routing load depends on the number of routes in which the sensor is involved as a router.

The individual sensor's remaining energy level is more important than the total remaining energy level of the network, since sensor failures may disconnect the network or reduce the quality of monitoring. That is, the variance of $\aleph$ between sensors is a critical factor that should be considered with a high priority for the longevity of the network. In the case that a minimum number of sensors is found to meet the desired sensing coverage, the additional number of sensors involved in routing will be minimum if each of the data reporting sensors chooses the shortest path to the data gathering point and the disjointness of the chosen paths by each reporter becomes minimum (refer to Fig. 7.22). However, this is clearly

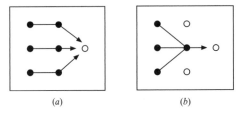

(a)   (b)

Figure 7.22   Route Disjointness: (a) max. disjointness of routes and (b) min. disjointness of routes.

not an optimal routing for the network longevity. The optimal routing should dynamically distribute RL, the routing load, over multiple sensors based on their remaining energy level even though more sensors are involved [42]. Such an optimal routing is needed for any data gathering scheme without regard to the optimality of the number of active sensors in terms of sensing coverage and connectivity. In this section, our main focus has been to find a minimum number of sensors meeting the DSC and a mechanism on how to select them. As investigated in [43], the radio range of currently available sensor hardwares is much larger than the sensing range. This will definitely and positively affect the connectivity of the selected k sensors and hence both the routing and sensor scheduling problems. Currently, we are investigating a data gathering tree that meets the DSC and simultaneously handles the routing as well as sensor state scheduling problems, as a part of our future work.

7.3.7 Performance Evaluation

In this section we evaluate the performance of the proposed energy conserving data gathering strategy using the four k-sensor selection schemes: NRS, N-DRS, F-DRS, and RWS. A discrete-time event simulator, called *simlib* [44], was used to implement our selection schemes and collect statistical information. As mentioned earlier, in this study we focus on the fundamental issues such as event detection integrity and sensed data reporting latency, of the trade-off-based data gathering strategy. We also present the energy conservation capability of the selection schemes.

Methodology The performance of the proposed schemes are measured by evaluating the following three metrics with varying DSC from 0.1 to 0.9 with an interval of 0.1.

- *Real-Time Event Detection Ratio* It is defined as the fraction of the number of event occurrences detected by at least one sensor. Events occur in the monitored area based on uniform distribution. Only k sensors that are selected according to the desired sensing coverage report detected events in each round. Thus, if an event occurs but none of the k sensors detects it, we consider it as a failure of real-time event detection.
- *Data Reporting Latency* It is the time difference between an event detection and its reporting to the data gathering point. All sensors are sensing the monitored area, but only k selected sensors report what they sensed to the data gathering point in each round. The remaining sensors cache what they sensed and wait for the next reporting round, thus introducing latency in reporting the sensed data.
- *Energy Conservation Capability* This is measured by the number of data transmissions by a sensor during a time unit since the energy consumed for data transmission is the most critical in determining the sensor's lifetime.

The performance of metrics measured in the proposed schemes are compared with those of a common data gathering (CDG) scheme in which all sensors transmit their sensed data to the data gathering point at every round. We generate three sensor network fields: $100 \times 100 \, \text{m}^2$, $200 \times 200 \, \text{m}^2$, and $300 \times 300 \, \text{m}^2$. Homogeneous sensors are scattered randomly and uniformly over the network field based on density, 1 sensor/$200 \, \text{m}^2$. Therefore, the total number of sensors for these fields are 50, 200, and 450, respectively, and based on the total number, an identifier is sequentially given to each sensor. We ran 10 experiments, each with different sensor distributions, for different desired sensing coverage in each network

Table 7.6   Simulation Parameter Values

	$100 \times 100\,\mathrm{m}^2$	$200 \times 200\,\mathrm{m}^2$	$300 \times 300\,\mathrm{m}^2$
Network field			
Number of sensors (1 sensor/200 m²)	50	200	450
Sensing range (= transmission range)		30 m and 40 m	
Distribution of event occurrence		Uniform distribution	
Event occurrence interval		Uniformly distributed within [5,10] seconds	
Data reporting interval (Δt)		10 s	
Simulation time		16,000 s	

field. Two different circular sensing ranges are used: $r = 30\,\mathrm{m}$ and 40 m. The transmission radio range is assumed equal to the sensing range. After the deployment, in the case of RWS, sensors broadcast their existence so that they can learn who their neighbors are, and the number of random walks used in these experiments is one. Events occurred in the network space based on uniform distribution, and their occurrence interval is uniformly distributed between 5 and 10 s. Table 7.6 summarizes the parameters used. Note that the implementation does not include any feature of the medium access control (MAC) layer and wireless channel characteristics since our main objective in this simulation study is to measure the real-time event detection capability and data reporting latency, when the desired sensing coverage is specified by the user.

Real-Time Event Detection Ratio   Sensors continuously monitored the environment to detect specific events or collect statistical information for the user. Figures 7.12a and 7.12b showed that only k sensors are sufficient to cover as much of the area as the user requests. To detect specific events occurring irregularly over the entire monitored area, we evaluated the real-time event detection ratio using only the selected k sensors. Intuitively, the event detection ratio is approximately the same as the desired sensing coverage if the selected k sensors satisfy the desired sensing coverage. We measure and compare the real-time event detection ratio using two schemes: the fixed random selection (FRS) scheme, which has exactly k sensors at every data reporting round, and the NRS scheme. The FRS is an example of centralized selection. It was only simulated to show the impact of having approximately k sensors upon achieving the DSC.

We count the number of events detected by k sensors randomly selected based on the FRS and NRS schemes. In all network scenarios, Figures 7.23a and 7.23b show a slightly higher real-time event detection ratio than the desired sensing coverage in both schemes. This is due to the integer function applied to k [see Eq. 7.4]. When k is small or the sensor's sensing range is large, the detection ratio is more sensitively affected by the integer function. This explains the network field of $100 \times 100\,\mathrm{m}^2$ in both the figures. The FRS scheme shows almost the same event detection ratio as NRS for the two other network fields. In the case of $100 \times 100\,\mathrm{m}^2$, a relatively small number of sensors are selected to meet the desired sensing coverage, thus the ratio difference becomes large when NRS does not have exactly k sensors. This also explains that the difference of the event detection ratio between FRS and NRS schemes decreases as the DSC increases (i.e., k becomes large).

Figure 7.23c shows the number of selected sensors to meet the required DSC (ψ). Since the trend in the increase of the number of selected sensors is almost identical in both sensing ranges of 30 and 40 m, we present the result only for the case of 30 m. The number of sensors that has to be selected increases drastically for $\psi \geq 0.9$. For $100 \times 100\,\mathrm{m}^2$, the

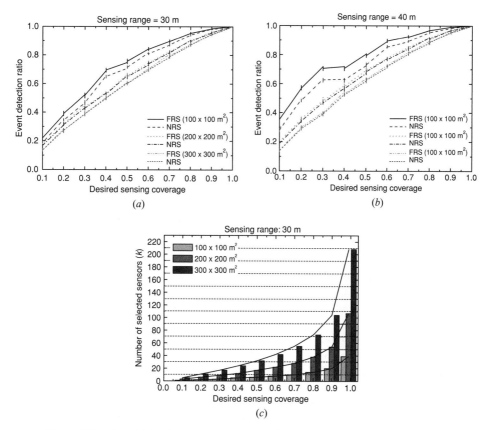

Figure 7.23   Event detection ratio and corresponding number of reporting sensors (k).

event detection ratio is higher than 95% using only 14 selected sensors (i.e., $\psi = 0.8$), which is only 28% of the total number of sensors in this experiment. This implies that the energy conservation can be further increased without overly decreasing the quality of service.

Data Reporting Latency   Since sensors take turns in reporting their sensed data, there is a waiting period before a sensor becomes a data reporter. In addition, as mentioned earlier, the NRS scheme randomly selects k data reporters based on the probability $P_{nrs} = k/|V| - 1$ without having any control operation on the sensors that have been already selected, so that the waiting period is not bounded. On the other hand, the N-DRS, F-DRS, and RWS schemes have control operations to bound the waiting period in each sensor. A sensor keeps sensing and caching until it becomes a reporter. We measure the average and maximum data reporting latency until the sensor acts as one of the selected k reporting sensors and transmits the sensed data to the data gathering point. The number of reporting sensors (k) and the total number of sensors ($|V| - 1$) are the main factors in deciding the latency since they determine δ and P_{nrs}, which affect the waiting interval of a sensor to become a reporter.

Table 7.7 shows k, $\hat{k}$, δ, and the ratio (ρ) of k to the total number of sensors, which are calculated based on the desired sensing coverage, ψ, and the total number of sensors. We can

Table 7.7 Calculated Values for k, $\hat{k}$, δ, and ρ Based on ψ and $|V|$

ψ	$100 \times 100\,\mathrm{m}^2$			$200 \times 200\,\mathrm{m}^2$			$300 \times 300\,\mathrm{m}^2$		
	$\rho\,(\%)$	(k, δ)	$\hat{k}$	$\rho\,(\%)$	(k, δ)	$\hat{k}$	$\rho\,(\%)$	(k, δ)	$\hat{k}$
0.5	12	(6,8.3)	11	8.5	(17, 11.8)	33	7	(32, 14.1)	64
0.6	16	(8,6.25)	13	11	(22, 9.1)	35	9	(42,10.7)	71
0.7	20	(10,5.0)	14	14	(28, 7.1)	39	12	(55, 8.2)	80
0.8	28	(14,3.6)	15	19	(38, 5.3)	46	16	(73,6.16)	92
0.9	40	(20,2.5)	21	27	(54, 3.7)	59	23	(104, 4.3)	115

observe in Table 7.7 that the number, $\hat{k}$, of sensors visited by the random walk increases as the k-sensor selection becomes sparse (i.e., DSC becomes small). We adjust the forwarding interval, defined as $\Delta t / \hat{k}$, of the random walk based on $\hat{k}$ and the data reporting interval Δt, such that the random walk moves fast as the network size increases.

Figures 7.24a, 7.24c, and 7.24e show the average reporting latency of the NRS, N-DRS, F-DRS, and RWS schemes in all the simulated network sizes, and Figures 7.24b, 7.24d, and 7.24f show the corresponding performance for maximum latency. We observe that the F-DRS scheme shows the lowest average data reporting latency with the help of the control operation, which allows all the selected sensors to transmit exactly once within a reporting cycle with a uniform data reporting latency of $\Delta t \times \delta$. On the other hand, NRS has a little larger average reporting latency compared to F-DRS and N-DRS, but it has the highest maximum latency, which implies a large variance of data reporting time among sensors.

This is due to the random selection without any control operation on the selected sensors. In particular, when the desired sensing coverage is low, the maximum latency of NRS becomes very large. This is because a sensor becomes a reporter after several election trials owing to the relatively small success probability, P_{nrs}. In the case of RWS, the maximum latency is not much different from the one in F-DRS and N-DRS, since its travel is constrained to the least recently visited sensor in order to reduce the difference of visiting time between sensors. However, RWS shows the highest average reporting latency among the proposed selection schemes. This is because it took relatively more time for RWS to cover the entire monitored area because the traversal diameter of random work in each round is limited, as well as only one random walk was used in these experiments.

When compared to CDG, Figure 7.24 shows the difference between average and maximum reporting latency decreases as the desired sensing coverage (ψ) increases to 0.9. For $\psi \geq 0.7$, the average reporting latency is less than twice that of CDG in all the cases except for RWS, in which the latency is not a main concern. Particularly, when $\psi = 0.9$ the average latency of the F-DRS, N-DRS, and NRS schemes in all the simulated network fields is almost the same as that in CDG. As shown in Table 7.7, in the case of $\psi = 0.9$, $\rho = 40$, 27, and 23% for $100 \times 100\,\mathrm{m}^2$, $200 \times 200\,\mathrm{m}^2$, and $300 \times 300\,\mathrm{m}^2$, respectively. This implies that the proposed data gathering schemes can significantly save resources with a minimum trade-off between coverage and reporting latency.

Energy Conservation Capability Due to the very limited energy retainment of the sensors, energy conservation is one of the most critical performance metrics in evaluating data gathering schemes. Even though we did not measure the actual amount of energy consumed while the sensors are transmitting their sensed data using the proposed schemes, it is clear that the energy conservation rate is inversely proportional to the desired sensing

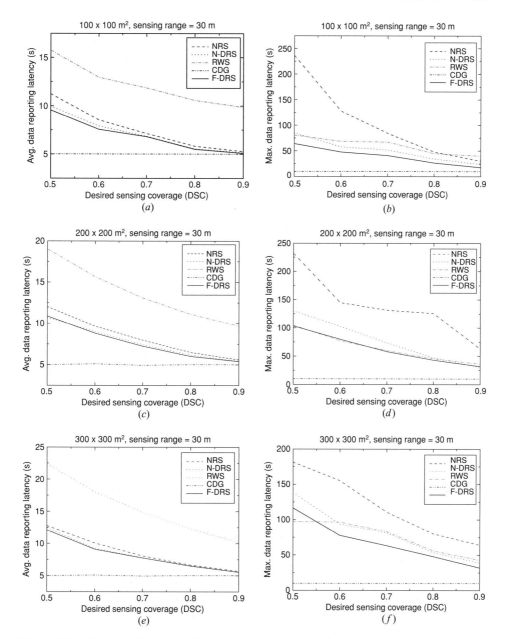

Figure 7.24 Data reporting latency (avg. and max.) of various schemes in each simulated network field.

coverage since approximately k sensors report their sensed data while others delay their reporting duty to conserve their energy. In this section, DRS denotes either of F-DRS and N-DRS schemes since the sensors have the same number of sensed data transmissions in both schemes.

Given a desired sensing coverage, the energy conservation ratio in each sensor depends on k and $|V| - 1$ as in the case of the data reporting latency. This is because k and $|V| - 1$

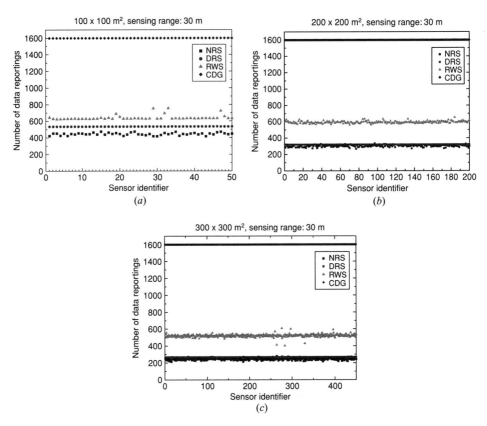

Figure 7.25 Distribution of number of transmissions in sensors (index of energy conservation capability).

determine the parameters δ and P_{nrs}, which affect the frequency of the reporting duty cycle in each sensor. That is, the energy conservation capability is proportional to δ and inversely proportional to P_{nrs}. Figures 7.25a, 7.25b, and 7.25c show the distribution of the number of data reportings (transmissions) generated by all the proposed schemes and CDG in each sensor for $\psi = 0.8$. In the case of CDG, $\delta = 1$ since all the sensors transmit their sensed data at every round. We observe that in all these figures the number of data reportings in each simulated network size is noticeably small as compared to CDG. In the previous subsection we concluded that the data reporting latency is not significantly affected when $\psi \geq 0.7$. This implies that the network lifetime can be significantly increased with a small trade-off. Also, Figures 7.25a, 7.25b, and 7.25c show that the order of the number of data reportings measured in each network size is the same as the descending order of δ values. This demonstrates that the number of data transmissions in each sensor is inversely proportional to δ (see Table 7.7 for the δ values). Thus, the smaller the δ the higher is the energy conservation capability.

Due to the steady-state behavior of random selection and the control operation, the data reporting chances in all the proposed schemes during 16,000-s simulations are well distributed among the sensors, implying a small variance in the amount of energy consumed for data reporting in each sensor. Especially in the case of DRS, all the sensors have the same number of data transmissions as they are designed to do so. As the number of the selected k sensors becomes small, RWS needs a higher number of additional sensors. Thus, the difference in the number of data transmissions increases proportionally to the network size when compared to the ones in DRS and NRS.

7.3.8 Conclusion

In this section, we proposed a novel energy conserving data gathering strategy in wireless sensor networks that is based on a trade-off between coverage and data reporting latency. The ultimate goal of the proposed strategy is to further extend the network lifetime by relieving the quality of service (i.e., sensing coverage of a monitored area) depending on the user's desire. The trade-off represents a negotiation between the user and the network in terms of sensing coverage of a monitored area and data reporting latency, which helps in extending the network lifetime. To achieve this goal, we first presented how to find a minimum of k sensors as data reporters that can meet the desired sensing coverage using geometrical probability. Then we introduced four distributed k-sensor selection schemes with constant computational complexity in order to support the proposed trade-off-based data gathering strategy. The selection schemes are nondisjoint randomized selection (NRS), nonfixed and fixed disjoint randomized selections (N-DRS and F-DRS), and random-walk-based selection (RWS). The NRS (the simplest scheme) senses the monitored area as much as the desired sensing coverage with nondisjoint k sensors in each reporting round so that collecting the sensed results for the entire monitored area within a fixed delay is not guaranteed. The two variants of the DRS scheme have a control operation for the already selected sensors so they cover as much of the monitored area as the desired sensing coverage with a set of disjoint k sensors at every reporting round. Moreover, the sensed data for the entire monitored area are guaranteed to receive with a fixed delay. In RWS, the main design factor is in-network data aggregation rather than time-constrained data gathering. In this scheme, a constrained random walk selects the data reporters by traveling over the monitored area based on the least recently visited policy while choosing the next stopover.

Experimental results show that our proposed strategy can meet the desired sensing coverage specified by the user, using (approximately) k sensors. Moreover, the network lifetime can be significantly increased with a small trade-off: the higher the network density, the higher is the energy conservation rate without any additional computation and communication costs. As part of future work, we plan to integrate the proposed data gathering strategy with routing and MAC layer features and to measure the network lifetime and data reporting latency by running experiments with realistic data gathering scenarios.

7.4 QUALITY-DRIVEN INFORMATION PROCESSING AND AGGREGATION IN DISTRIBUTED SENSOR NETWORKS

Jin Zhu, Jie Yang, and Symeon Papavassiliou

A distributed sensor network is a self-organized system composed of a large number of a low-cost sensor nodes, with limited energy and other resources. In this section a quality-driven information processing and aggregation (Q-IPA) approach is investigated that aggregates data on the fly at intermediate sensor nodes, while satisfying the latency and measurement quality constraints with energy efficiency. One of the main features of the proposed approach is that the task quality of service (QoS) requirements are taken into account to determine when and where to perform the aggregation in a distributed fashion. Furthermore, an analytical statistical model is introduced to represent the data aggregation and report delivery process in sensor networks, with specific delivery quality requirements in terms of the achievable end-to-end delay and the successful report delivery probability. Based on this model we gain some insight about the impact on the achievable system performance, of the various design parameters and the trade-offs involved in the process of data aggregation and the Q-IPA strategy. Finally the performance of the proposed approach is evaluated, through modeling and simulation, in terms of the achievable end-to-end delay and the overall network energy savings, under different data aggregation scenarios and traffic loads.

7.4.1 Background Information

A distributed sensor network is usually a self-organized system composed of a large number of sensor nodes, which are used to measure different parameters that may vary with time and space, and send the corresponding data to a collector center or base station for further processing. There are far-ranging potential applications of sensor networks such as: (1) system and space monitoring, (2) habitat monitoring [45, 46], (3) target detection and tracking [47, 48], and (4) biomedical applications [49–51].

With their focus on applications requiring tight coupling with the physical world, as opposed to the personal communication focus of conventional wireless networks, wireless sensor networks pose significantly different design, implementation, and deployment challenges. For example, the individual sensor nodes usually have several limitations such as limited energy and memory resources, small antenna size, and weak processing capabilities, due to cost-effective considerations and miniature size requirements. However, some time-crucial applications such as those in the battlefield may have very strict performance requirements, and the sensor nodes are required to achieve specific QoS in data collection and transmission, so that the measurement task can be fulfilled within the corresponding latency and resource requirements. Therefore, the sensor nodes in a distributed sensor network have to collaborate with each other, and as a result, effective information gathering and dissemination strategies need to be deployed.

The collaboration between different sensor nodes are mostly realized through multihop network architectures due to their energy efficiency and scalability features [52–55]. Furthermore, since in sensor networks the data in the neighboring nodes are considered highly correlated due to the fact that the observed objects in the physical world are highly correlated, localized data processing and aggregation on the fly may dramatically decrease the amount of information to be transmitted. Therefore, hierarchical infrastructures have been studied to reduce the network traffic, save the energy of sensor nodes, distribute the computation load, and improve the measurement quality in multihop networking environments, where intermediate sensor nodes may be selected to perform data aggregation from the measurement results delivered from different neighboring sensors. For instance, in [56] an information retrieval protocol, APTEEN, for cluster-based sensor networks that can implement data aggregation, has been presented, while the impact of data aggregation on sensor networks is discussed in [57].

It should be noted that several recent efforts have shown the importance of data aggregation in wireless sensor networks and have studied and discussed some of the benefits that can be achieved, by exploiting the features of data correlation [58] and data aggregation [59]. Data aggregation comparison studies have demonstrated the effect of network parameters and the utility of aggregation mechanisms in a wide variety of applications [57,60]. In [61] an SQL-like declarative language for expressing aggregation queries over streaming sensor data is proposed, and it is demonstrated that the intelligent distribution and execution of these aggregation queries in the sensor network can result in significant reductions in communication compared to centralized approaches. Recent work on data aggregation [62] proposed an application-independent data aggregation (AIDA) protocol that resides between the media access control layer and the network layer. The AIDA module combines network units into a single aggregate outgoing AIDA payload to reduce the overhead incurred during channel contention and acknowledgment.

However, although several research works in the literature have discussed the problems of developing efficient routing and data aggregation processes mainly for energy savings/minimization in sensor networks (e.g. [47, 59, 63–66]), several issues associated with the data aggregation process with the specific objective of meeting the task requirements (i.e., QoS-constrained data aggregation) are not yet well addressed. Given the fact that many sensing tasks present some strict reporting quality requirements (e.g., in a time-critical application an obsolete sensor report—individual or aggregated—that may exceed a given time threshold is discarded), development of efficient and feasible strategies that perform data aggregation in a distributed manner and with energy efficiency, in order to meet various quality requirements such as end-to-end latency, is of high research and practical importance.

Objective and Outline   In this section, we study the data gathering and aggregation process in a distributed, multihop sensor network under specific QoS constraints. Specifically, a quality-driven information processing and aggregation (Q-IPA) approach is investigated that aggregates data on the fly at intermediate sensor nodes, while satisfying the latency and measurement quality constraints with energy efficiency. In the proposed approach, each intermediate sensor node determines independently whether or not to perform data aggregation, according to the resource conditions and the specific task requirements when it receives a measurement from other nodes, and needs to deliver it to the data collector center.

One of the main features of the Q-IPA scheme is that the network does not need to be formed into clusters to perform the data aggregation, while the task QoS requirements are taken into account to determine when and where to perform the aggregation in a distributed fashion. Such an approach maintains the flexibility of the network architecture and simplicity in the protocol design, which are essential for the applicability of wireless sensor networks that need to adapt to diverse, unforeseeable, and sometimes hostile environments and situations.

Furthermore, we propose an analytical model to represent the data aggregation and report delivery process in sensor networks, with specific delivery quality requirements in terms of the achievable end-to-end delay and the successful report delivery probability. We apply this model to the Q-IPA algorithm to study the end-to-end latency of a measurement and the corresponding successful report delivery probability. Based on this model we also gain some insight about the impact on the achievable system performance of the various design parameters and the trade-offs involved in the process of data aggregation and the Q-IPA strategy. Moreover, the achievable energy efficiency under the Q-IPA strategy is studied, which further demonstrates the benefits of performing data aggregation in sensor networks. It should be noted that in the literature the study of QoS guarantee in sensor networks is usually focused on the routing protocols tailored to meet the requirements (e.g., [59, 63, 65, 66]), while in this section the emphasis is placed on the study of data aggregation strategies that should be implemented on each individual sensor node in a totally distributed fashion, which is complementary to the applied QoS routing protocols, and can enhance the capability of QoS guarantee in the sensor networks.

7.4.2 Quality-Driven Information Processing and Aggregation (Q-IPA) Approach

Throughout, we consider a multihop communication sensor network as shown in Figure 7.26. We assume that the data collected and/or observed by each sensor node will be transmitted, through this multihop architecture, to a collector center for further processing and decision making. The sensing tasks may be deployed and operated in different ways in terms of the applications: The sensors may be prearranged to collect and send data back periodically; end users may initiate a sensing task; or the sensors may generate and send raw results of a given interest to collector centers, and these centers respond with specific requirements such as report interval, quality and time constraints, and the like.

It should be noted that, as mentioned before, in many applications of sensor networks the data from different neighboring sensors are usually highly correlated, and therefore using collaborative data aggregation processing may improve the overall system performance and reduce the communication load and energy consumption. Therefore, in the following we assume that when a sensor node receives a packet or message from its neighbor, it is able to either perform local processing and aggregation or just forward (relay) it, according to the QoS requirements of the corresponding applications. The procedure that a sensor node locally generates and/or processes a measurement packet in which new data may be aggregated is referred to as reporting, while the corresponding new/updated packet is referred to as a report.

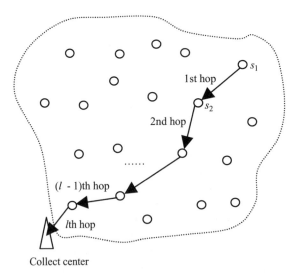

Figure 7.26   Example of multihop sensor network.

Based on the above discussion, the operations of the Q-IPA approach are briefly described as follows:

1. The data collector center broadcasts a measurement task with specific predetermined QoS constraints to the sensor network.
2. Each sensor node decides whether or not to participate in this task according to its conditions and the requirements of the task.
3. If a sensor matches the task type and can fulfill the QoS requirements, the sensor node measures and collects the data for the specific task; otherwise it does not participate in the process.
4. When a sensor node receives a report from its neighbor, it first determines whether or not it would perform data aggregation on the report. The following different cases may occur. (a) If the delay constraint can be satisfied, the sensor node defers the report for a fixed time interval τ with probability γ, during which the node processes and aggregates any reports that arrive and generates a new report before transmitting it to the next hop. With probability of $(1 - \gamma)$ the sensor node will directly try to forward the report without introducing any deferred period. (b) If the delay constraint can be satisfied only if the report is not deferred, the sensor node simply tries to forward this report. (c) If the delay constraint cannot be satisfied in any case, the sensor node will discard the report, to avoid further wasting of any additional resources.

The considered end-to-end QoS constraint is the end-to-end latency requirement D of a report that may aggregate other data or reports along its path from source to the collector center. If the report is delivered to the collector center within the given latency constraint D after its initial generation, it will be considered as a successful delivery. At a sensor node, for a received report, in addition to the possible deferred period τ, there is some additional waiting time caused by the transmission of the previous report at the node. We assume that at each node there can be at most two reports: One is under transmission and the other is waiting to be transmitted. If there are report arrivals while a report is waiting in the sensor node to be transmitted, these newly arrived reports will be aggregated into the waiting report. In the algorithm, γ and τ are configurable system parameters, and their impact on the achievable system performance is analyzed and studied later.

7.4.3 Data Aggregation Modeling

By using proper routing mechanisms, we assume that each report goes through each node only once, and nodes always forward the report to other nodes that are closer to the collector center. Therefore assuming that l nodes are visited from the source node to the collector center, we denote the set of these sensor nodes as $G_l = \{s_1, s_2, \ldots, s_l\}$. Without loss of generality we assume that the distances between the sensor nodes and the collector site are arranged in decreasing order, that is, $d_1 > d_2 > \cdots > d_l$, where d_i is the distance between node s_i and the collector center.

Let us also denote by $t_i^{(R)}$ the reporting time at node s_i, which includes the time period for data aggregation, by $t_i^{(F)}$ the forwarding time at node s_i to next node s_{i+1}, which accounts for the transmission time including the potential retransmission time due to channel contention (this time is related to the report length, the bandwidth, and the communication success probability), and by $t_i^{(P)}$ the propagation time from node s_i to next node s_{i+1}, which depends on the distance between the two nodes. Time periods $t_i^{(R)}$, $t_i^{(F)}$, and $t_i^{(P)}$ are random variables, and in the following their corresponding probability density functions (pdf) are denoted by $f_i^{(R)}(t)$, $f_i^{(F)}(t)$, and $f_i^{(P)}(t)$, respectively. Let us denote by t_i the time interval between the point that node s_i receives a report to the point that this report is delivered to node s_{i+1}. If node s_i does not perform data aggregation, the corresponding time interval is $t_i^{(F)} + t_i^{(P)}$; otherwise, the time interval is $t_i^{(R)} + t_i^{(F)} + t_i^{(P)}$, $i \geq 1$, that is,

$$
t_i = \begin{cases} t_i^{(R)} + t_i^{(F)} + t_i^{(P)} & \text{with reporting} \\ t_i^{(F)} + t_i^{(P)} & \text{without reporting} \end{cases}
$$

and its pdf is denoted by $f_i(t)$. Under the assumption that a sensor node performs reporting with probability γ, we have

$$
f_i(t) = f_i(t|\text{with reporting})\,\gamma
$$
$$
+ f_i(t|\text{without reporting})\,(1 - \gamma)
$$

Let us also assume that the time periods are independent of each other, and denote by $F_i^{(R)}(s)$, $F_i^{(F)}(s)$, and $F_i^{(P)}(s)$ the Laplace transforms of $f_i^{(R)}(t)$, $f_i^{(F)}(t)$, and $f_i^{(P)}(t)$, respectively. Applying the Laplace transform to $f_i(t)$, we have

$$
F_i(s) = E[e^{-st_i}] = \int_0^\infty f_i(t)e^{-st}\,dt
$$
$$
= F_i^{(F)}(s)F_i^{(P)}(s)\left[\gamma F_i^{(R)}(s) + (1 - \gamma)\right] \tag{7.8}
$$

In the following, we first assume that no reports will be discarded due to the delay constraint and obtain the end-to-end delay distribution, which can be used to obtain the probability P_{succ} that the report is delivered to the collector center within the delay constraint D. Then the probability that the report is discarded due to unsatisfactory end-to-end delay performance can be obtained as $(1 - P_{\text{succ}})$. Let us denote the end-to-end delay of a report by $T_L = \sum_{i=1}^L t_i$ and its corresponding pdf by $f_{T_L}(t)$, where the random variable L is the number of hops that are involved in the transmission of a report from the source node to the collector center (including the source node). Thus, the Laplace transform of $f_{T_L}(t)$, denoted by $F_{T_L}(s)$, is given by

$$
F_{T_L}(s) = E[e^{-s(t_1 + t_2 + \dots + t_L)}]
$$
$$
= \sum_{l=1}^N E[e^{-sT_L}|L = l]\Pr[L = l]
$$
$$
= \sum_{l=1}^N p_L(l)\prod_{i=1}^l F_i(s) \tag{7.9}
$$

where $p_L(l)$ is the probability mass function of L, where the random variable L represents the number of hops that are involved in the transmission of the report from the source node to the collector center (including the source node). The pdf of T_L can be obtained by using the inverse Laplace transform of $F_{T_L}(s)$, that is,

$$f_{T_L}(t) = \mathcal{L}^{-1}\left\{F_{T_L}(s)\right\}$$

$$= \sum_{l=1}^{N} p_L(l)\mathcal{L}^{-1}\left\{\prod_{i=1}^{l} F_i(s)\right\} \qquad (7.10)$$

When $f_{T_L}(t)$ is obtained, the successful probability P_{succ} that a report can reach the collector center within the delay constraint D is given by

$$P_{\text{succ}} = P[T_L \leq D] = \int_0^D f_{T_L}(t)\,dt \qquad (7.11)$$

In the Appendix, we derive the pdf of T_L under the assumption that the report arrival process is Poisson, while $t_i^{(F)}$ and $t_i^{(P)}$ are exponentially distributed. However, it is difficult to obtain an analytical expression for P_{succ} in practice, since the distribution of T_L is generally unknown. In the next subsection we use a probabilistic model to lower bound the probability P_{succ}.

7.4.4 Lower Bound on P_{succ}

The end-to-end delay of an independent report that meets the delay constraint and passes through l hops can be represented by

$$T_l = \sum_{i=1}^{l} t_i^{(R)} + \sum_{i=1}^{l} t_i^{(F)} + \sum_{i=1}^{l} t_i^{(P)} \leq D \qquad (7.12)$$

where in general $t_i^{(F)}$ can be upper bounded based on the largest report length and the corresponding data rate of the sensor network, and $t_i^{(P)}$ can be upper bounded by the range of the sensor network and the longest distance between two sensor nodes. Thus, we can assume that D is decomposed as

$$D = D_r(l) + D_f(l) + D_p(l) \qquad (7.13)$$

where $D_r(l)$, $D_f(l)$, and $D_p(l)$ are the upper bounds on the end-to-end reporting time, forwarding time, and propagation time, respectively, when the report needs to be delivered to the collector center using l hops. As a result, in our study the constraint that needs to be satisfied regarding the reporting time can be represented as

$$\sum_{i=1}^{l} t_i^{(R)} \leq D_r(l) \qquad (7.14)$$

Noted that $t_i^{(R)}$ is a function of τ and γ, and (7.14) provides a worst-case bound on the reporting time under the constraint (7.12). Therefore, the lower bound on the probability $p_{\text{succ}}(l)$ that a specific independent report is delivered to the collector center within the end-to-end constraint, when the distance between the source node and the collector center is l hops, is lower bounded by

$$p_{\text{succ}}(l) = P[S_L(t) < D|L = l] \geq P\left[\sum_{i=1}^{l} t_i^{(R)} \leq D_r(l)\right] \tag{7.15}$$

Thus, the probability of success P_{succ} under delay constraint (7.12) is lower bounded by

$$P_{\text{succ}} = \sum_{l=1}^{N} p_L(l)p_{\text{succ}}(l) \geq \sum_{l=1}^{N} p_L(l)P\left[\sum_{i=1}^{l} t_i^{(R)} \leq D_r(l)\right] \tag{7.16}$$

Note that (7.16) provides a lower bound to the probability of a successful report delivery within the QoS constraint for sensor networks with and without data aggregation schemes. When $\gamma = 0$, P_{succ} is reduced to the probability of a successful delivery in a sensor network without any data aggregation scheme, in which each received report will be forwarded as is (without any deferred period). Since the QoS routing algorithm deployed in the sensor network is independent of the proposed Q-IPA approach, we assume that if there is no data aggregation scheme deployed in the sensor network, the report can be delivered to the collector center within its end-to-end delay constraint D, through the use of the deployed routing algorithm. Otherwise the sensor node will not participate in the specific measurement task. That is, we assume $P_{\text{succ,nonaggregation}} = 1$. Furthermore, we assume that under the Q-IPA approach, the generated reports will follow the same path as in the case without data aggregation.

In the Q-IPA approach, if data aggregation is not performed at node i, the reporting time $t_i^{(R)} = 0$, while if data aggregation is performed with probability γ, $t_i^{(R)} = \tau$. It is clear that the longest delay that a report may experience due to data aggregation is $l\tau$, when the number of hops between the source to the collector center is l. If $l\tau \leq D_r(l)$, the end-to-end delay can be guaranteed even if at each node data aggregation is performed, that is, $\gamma = 1$. Thus we can have $p_{\text{succ}}(l) = 1$ when $l\tau \leq D_r(l)$. When $l\tau > D_r(l)$, if all the intermediate nodes perform data aggregation and reporting with a deferred period τ, the end-to-end delay of a report may exceed the delay constraint. The maximum number of data aggregation and reporting that can be performed to guarantee the delay constraint, determined by the upper bound on the reporting time $D_r(l)$, is given by

$$C(l) = \left\lfloor \frac{D_r(l)}{\tau} \right\rfloor \tag{7.17}$$

That is, the lower bound of P_{succ} is equal to the probability that a report experiences at most $C(l)$ times of data aggregations and reporting along its path. Therefore, the probability $p_{\text{succ}}(l)$ is lower bounded by

$$p_{\text{succ}}(l) \geq p_{\text{succ}}^{(\text{LB})}(l) \triangleq \begin{cases} \sum_{k=0}^{C(l)} \binom{l}{k}\gamma^k(1-\gamma)^{l-k} & C(l) < l \\ 1 & C(l) \geq l \end{cases} \tag{7.18}$$

Consequently, the probability of success P_{succ} under delay constraint is lower bounded by

$$P_{\text{succ}} \geq P_{\text{succ}}^{(\text{LB})} \triangleq \sum_{l=1}^{N} p_L(l)p_{\text{succ}}^{(\text{LB})}(l) \tag{7.19}$$

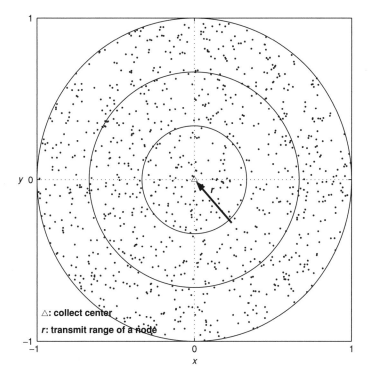

Figure 7.27 Sensor network with uniformly distributed nodes.

Numerical Results and Discussions In the remainder of this section, based on the developed models, we study the impact of parameters γ and τ on the data aggregation model and the performance of Q-IPA algorithm. Among our objectives is to identify the various trade-offs that these parameters present in order to provide guidelines to choose the appropriate values of these design parameters that achieve the desired performance. In the following, let us consider a sensor network where the sensor nodes are uniformly distributed in a disk area with radius R, and each node has a fixed limited transmission range r, as shown in Figure 7.27. We assume that each node always transmits a report as far as possible within its transmitting range, and therefore the maximum number of hops is $M = \lceil R/r \rceil$. If the total number of sensor nodes N is large, the probability $p_L(l)$ can be approximated by

$$p_L(l) = \begin{cases} \dfrac{2l-1}{M^2} & 1 \le l \le M \\ 0 & l > M \end{cases} \tag{7.20}$$

Let us assume the report arrival process at each sensor node follows Poisson distribution with rate λ.

In the first numerical example presented here, our objective is to demonstrate how the lower bound approximation of P_{succ}, given by (7.19), is affected by different values of γ and τ. Specifically, Figure 7.28 shows the lower bound $P_{\text{succ}}^{(\text{LB})}$ for different values of γ and τ for the case with $\lambda = 20$, $M = 20$, and $D_r(l) = 2$ s. It can be seen from this figure that $P_{\text{succ}}^{(\text{LB})}$ decreases with γ and τ since larger values of γ will result in more frequent data

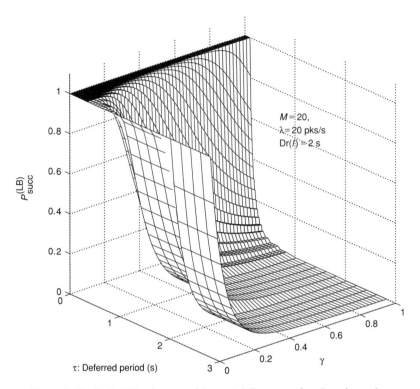

Figure 7.28 Probability of successful report delivery as a function of γ and τ.

aggregation and reporting during the delivery of the report, and larger values of τ will increase the end-to-end delay. As these values both increase, there is higher probability that the end-to-end delay is larger than the constraint, which results in the decrease of P_{succ}.

As shown before, expression (7.19) provides a simple lower bound on the successful report delivery probability P_{succ}. In the following we discuss and evaluate the relation between this lower bound approximation and the actual value of P_{succ}. In Figure 7.29 the curves of the corresponding probabilities are plotted as functions of the deferred period τ, for a sensor network with $M = 10$, $\lambda = 20$, and $\gamma = 0.5$. In this figure the lower bound approximation $P_{succ}^{(LB)}$ is obtained by using (7.18) and (7.19), and the actual P_{succ} is obtained from (7.11) and (7.28) for Poisson report arrivals. In this figure we plot four different curves, which represent the corresponding probabilities for delay constraints $D = 4$ and $D = 2$ s, respectively. Correspondingly in the lower bound calculation, we assume that $D_r(l) = 3.4$ and $D_r(l) = 1.4$.[1] The results in Figure 7.29 demonstrate that, in general, smaller D will result in lower P_{succ}, while the lower bound approximation demonstrates a similar trend with the actual performance of P_{succ}. Based on these results we conclude that $P_{succ}^{(LB)}$ provides an accurate lower bound approximation of the probability of successful report delivery for all values of τ.

---

[1] Since in the calculation of P_{succ} using the Appendix, we use $\lambda_1 = 80$ and $\lambda_2 = 10000$, almost all packets spend less than 0.6 time units for the transmission and propagation through their delivery from the source to the collector center. Thus, in the corresponding lower bound approximation, we assume that $D_r(l) = 3.4$ and $D_r(l) = 1.4$.

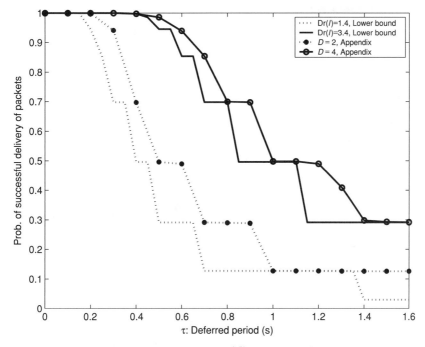

Figure 7.29 Lower bound approximation of P_{succ} ($P_{\text{succ}}^{(\text{LB})}$) and actual P_{succ} for $\gamma = 0.5$ and different delay constraints D, as a function of deferred period τ.

As can be seen from the above discussions, for a given sensor network with specific delay requirement D, P_{succ} depends on parameters τ, γ, and the arrival rate λ. From Figures 7.28 and 7.29 it becomes clear that in order to meet the required quality objectives, there are many different choices for parameters τ and γ, while λ is determined by the nature of the measurement task. Furthermore, it can be noted that if a report is deferred at a node for the time period τ, while no other reports arrive during that period, and as a result no aggregation is actually performed, we do not get any benefits from such an approach, although P_{succ} decreases. In order to enhance the efficiency of the Q-IPA algorithm and maximize its benefits, when determining the optimal τ and γ, P_{succ} can be specified as a QoS requirement of the task or application, together with the delay constraint D. Then the objective function can be to maximize P_{agg}, the probability that a node determines to perform data aggregation and the data aggregation occurs during the deferred period τ. The optimal values of τ and γ can be determined by

$$(\tau_{\text{opt}}, \gamma_{\text{opt}}) = \arg \max_{P_{\text{succ}} \geq P_{\text{req}}} (P_{\text{agg}}) \tag{7.21}$$

where P_{req} is the minimum required probability of successful report delivery to the collector center within the end-to-end delay requirement D. When the report arrival process follows Poisson distribution with rate λ, the probability that there is at least one report arrival during the deferred period τ is $1 - e^{-\lambda \tau}$. In this case, the probability that data aggregation occurs during the deferred period τ is given by

$$P_{\text{agg}} = \gamma \left(1 - e^{-\lambda \tau}\right) \tag{7.22}$$

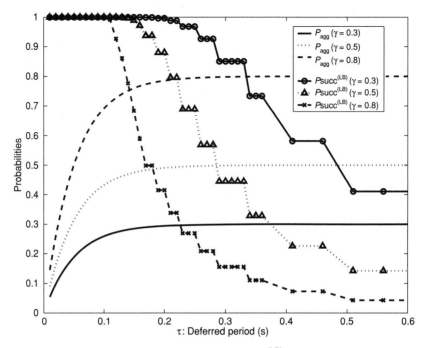

Figure 7.30 Relationship between $P_{\text{succ}}^{(\text{LB})}$ and P_{agg}.

Figure 7.30 shows the lower bound $P_{\text{succ}}^{(\text{LB})}$, and P_{agg}, for different combinations of τ and γ under the assumption of Poisson report arrivals with $\lambda = 20$. From this figure we observe that there is a trade-off between P_{agg} and the probability of successful report delivery, since they follow opposite trends with the change of τ. That is, as τ increases, P_{agg} increases as well—which means that more data will be aggregated in a single report, and, therefore, energy savings will be achieved—while, on the other hand, the probability of successful report delivery decreases. Therefore, if P_{req} is known, we can choose the set of (τ, γ) that can provide the maximum P_{agg}, while the resulting $P_{\text{succ}}^{(\text{LB})} \geq P_{\text{req}}$. It should be noted that since $\gamma \in [0, 1]$, from (7.22), we have $P_{\text{agg}} \leq 1 - e^{-\lambda\tau}$. It can be observed that P_{agg} approaches γ and is insensitive to τ for large values of τ.

Finally, although the objective function considered in this study is the maximization of P_{agg} so that the aggregation efficiency can be maximized, other objective functions can be considered, depending on the metrics of interest. For instance, the objective function could be to maximize the average number of reports aggregated. In this case the corresponding optimal values of τ and γ can be determined by:

$$(\tau_{\text{opt}}, \gamma_{\text{opt}}) = \arg \max_{P_{\text{succ}} \geq P_{\text{req}}} (\lambda\tau)$$

It should be noted that the objective and contribution of the proposed approach and the corresponding models introduced here is twofold. On one hand, for a system with given design parameters, such as the deferred period τ and the aggregation probability γ, based on the models and the strategies developed, we can evaluate various performance metrics, such as the expected successful report delivery probability and the expected end-to-end measurement delay. More importantly, on the other hand, given some specific QoS requirements (such as measurement end-to-end delay constraint and successful report delivery probability requirement) imposed by the task/application under consideration, we can use the proposed approach and accordingly adjust the design parameters τ and γ, in order to fulfill the required QoS and achieve the desired objective (e.g., maximize number of reports aggregated, reduce communications overhead, achieve significant energy savings, extend the sensor network operational lifetime, etc.).

7.4.5 Simulation Results

The achievable performance of the proposed Q-IPA approach, in terms of end-to-end delay (experiment 1) and overall network energy savings (experiment 2) is evaluated, under different data aggregation scenarios and traffic loads, using the OPNET modeling and simulation tool. Specifically, a sensor network consisting of 18 sensor nodes and 1 collector center, distributed in a 100 m × 100 m area, is considered. The transmission range of each node is assumed to be 50 m. When a node begins to transmit, all the neighbors within its transmission range will receive the signal, which is considered as interference for a node if the packet is not destined for it. The media access control (MAC) protocol adapted here is CSMA/CA. Rts/Cts messages are exchanged before a data packet is transmitted if the length of the data packets is more than 64 bytes, otherwise the data packet is transmitted without Rts/Cts exchange. The corresponding power consumption of a node under idle/listen, receiving, and transmitting modes is assumed to be 1, 10, and 36 mW, respectively [67].

Let us also denote by β the aggregation coefficient, which represents the ratio of the new report length after aggregation and reporting, to the total length of all the received packets/reports before aggregation, that is, $\beta = \frac{\text{Length of packet after aggregation}}{\text{Total length of original packets}}$ and $0 < \beta \leq 1$. $\beta = 1$ corresponds to the case that no aggregation is performed, while, in general, the more packets that are correlated and can be aggregated, the smaller β we can achieve.

Furthermore, we assume that the data transmission rate is 1 Mbps, and the packet length is exponentially distributed with mean 100 bytes. It should be noted that next-generation sensor networks are expected to provide data rates even higher than 1 Mbps [62]. Currently, the GNOMES node developed by Texas Instruments and Rice University already provides up to 721 kbps data rate [68]. Several numerical studies reported in the literature have considered transmission rates in the range of 1 to 2 Mbps [59, 69, 70].

Experiment 1: End-to-End Delay   In experiment 1, we compare the end-to-end delay of the sensor network under the Q-IPA approach with the results obtained by a system without any data aggregation. The corresponding results are shown in Figure 7.31. To demonstrate the impact of data aggregation on the sensor network performance in terms of delay, we set $\gamma = 1$, that is, data aggregation is performed at every sensor node. It can be seen that without data aggregation, the delay increases exponentially with the increase of the network load (represented by λ), while under the Q-IPA approach, the delay increases at a much slower rate since performing data aggregation reduces the network traffic load significantly. When the network load is light, the delay induced by the Q-IPA strategy is dominant, due to the fact that the sensor node introduces a deferred period of τ to perform data aggregation, while the corresponding waiting time at each node is negligible. Therefore in this case, the delay in the sensor network under the Q-IPA approach is larger than the one that can be achieved by a system without any data aggregation. However, when the network

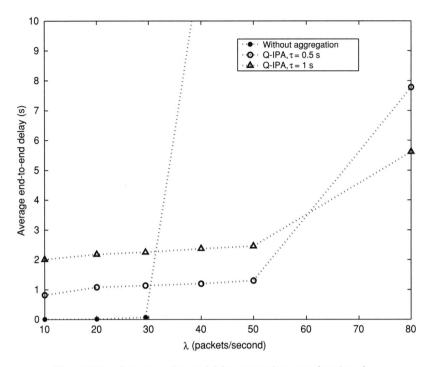

Figure 7.31   Average end-to-end delay comparison as a function of λ.

load is heavy, the waiting time at each node becomes the dominant factor (as compared to τ).

Since the waiting time is significantly affected by the network load, performing data aggregation can reduce the network traffic load and therefore result in the reduction of the end-to-end delay in the sensor network. Therefore, as we can observe from Figure 7.31, for medium-to-heavy traffic load the achievable end-to-end delay under the Q-IPA is significantly lower than the corresponding delay of a system without any data aggregation.

Similarly, for the same reasons, under low-to-medium traffic loads the end-to-end delay for the case with $\tau = 0.5$ is lower than the corresponding delay for $\tau = 1$, while as the traffic load increases the two curves cross each other, since the increased data aggregation results in the reduction of the network traffic load, which in turn results in the reduction of the end-to-end delay under heavy network loads, as explained before.

To evaluate the impact of the delay-related parameters (such as delay constraint and deferred time) on the performance of Q-IPA, in Figure 7.32, we present the probability of successful packet delivery and actual packet dropping probability under the Q-IPA strategy, as a function of the packet origination rate, for different combinations of the delay constraint D and the deferred period τ. For comparison purposes only, we also present the probabilities that the packets arrive at the collector center within a certain end-to-end delay equivalent to the corresponding delay constraints imposed by the Q-IPA, under a strategy that performs data aggregation (similar to Q-IPA) without, however, discarding packets at the intermediate nodes due to any delay constraints (in the graph in Fig. 7.32 we refer to these cases as no-packet-drop). As we expected, the successful packet delivery probability of the system

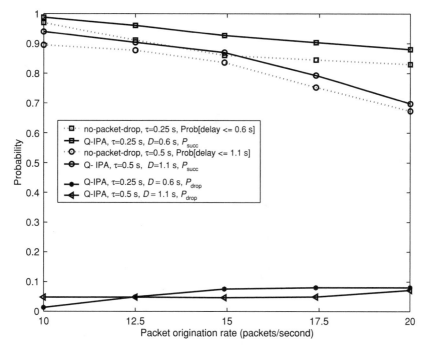

Figure 7.32 Probability of successful packet delivery for different delay constraints.

with delay constraint $D = 0.6$ and $D = 1.1$ s is better than the estimated probability under the strategy that does not discard any packets due to the delay constraints. This happens because the packets that cannot satisfy the imposed delay constraint have been discarded at the intermediate nodes, and therefore the overall traffic has been reduced. Furthermore, as can be observed by this figure, the successful packet delivery probability increases as τ decreases, however, as we see in the next subsection, this happens at the cost of higher energy consumption.

Experiment 2: Energy Efficiency The energy efficiency can be achieved under the operation of Q-IPA strategy at each node. However, due to the data aggregation achieved by the Q-IPA approach, which reduces the overall network traffic load, the energy efficiency can be further enhanced at the network/system level. Throughout this experiment, the energy consumption for the local data processing and aggregation is set to 0.1 nJ/bit. Figure 7.33 depicts the average power consumption in the sensor network, as a function of the traffic load, under three different scenarios. The first one corresponds to the case where no data aggregation is performed in the system (i.e., without aggregation), while the other two scenarios correspond to two different implementations of the Q-IPA approach (i.e., Q-IPA with $\gamma = 1, \tau = 0.5$ s and Q-IPA with $\gamma = 1, \tau = 1$ s, respectively). As can be seen from this figure, the Q-IPA approach outperforms the system without data aggregation, and achieves significant energy savings in the sensor network. For instance, when $\lambda = 50$ packets/second, the Q-IPA system can save 38% of total energy during the same operation period, which means that the network lifetime can be extended up to 1.6 times. Furthermore, it can be also observed from this figure that the energy consumption decreases as

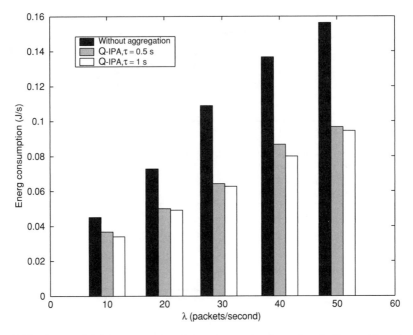

Figure 7.33   Average total energy consumption per second under various traffic conditions ($\gamma = 1$).

τ increases, since when τ increases the average number of packets that can be used for data aggregation increases as well (even for the same traffic load λ).

In Figure 7.34, the trade-off between the energy consumption and average end-to-end delay as a function of parameter γ are shown, for $\lambda = 30$ packets/second, $\tau = 0.5$ s, and $\beta = 0.9$. The corresponding results for $\gamma = 0$ refer to the case that no data aggregation is performed. It can be seen from this figure that as γ increases, the system consumes

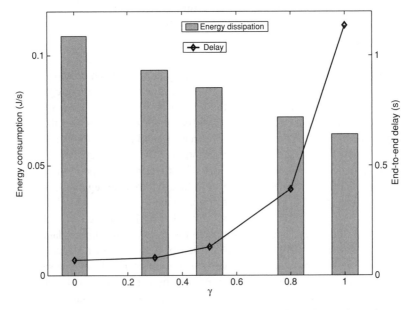

Figure 7.34   Energy consumption and delay comparison for various values of γ.

less energy during the same operation period, while at the same time the average delay increases. Therefore, according to the system design requirements and available resources, the appropriate design and operational parameters can be selected.

7.4.6 Summary

With the development of the information society, sensors are getting more and more new challenges. The detection and monitoring requirements are becoming more complicated and difficult. They trend from single variable to multiple variables, from one sensor to a set of sensors, from simple to complex and cooperative. Networking the sensors to empower them with the ability to coordinate on a larger sensing task will revolutionize information gathering and processing in many situations.

With their focus on applications requiring tight coupling with the physical world, as opposed to the personal communication focus of conventional wireless networks, wireless sensor networks pose significantly different design, implementation, and deployment challenges. In order to meet and fulfill the various task requirements, individual sensor nodes in a distributed sensor network have to collaborate with each other, and, as a result, effective information gathering and dissemination strategies need to be deployed.

To address this problem, first a quality-driven information processing and aggregation (Q-IPA) approach was introduced. The proposed method aggregates data on the fly at intermediate sensor nodes, while satisfying the latency and measurement quality constraints with energy efficiency. One of the main features of the proposed approach is that the task QoS requirements are taken into account to determine when and where to perform the aggregation in a distributed fashion.

Furthermore, an analytical statistical model was introduced, to represent the data aggregation and report delivery process in sensor networks, with specific delivery quality requirements in terms of the achievable end-to-end delay and the successful report delivery probability. Based on this model we gained some insight about the impact on the achievable system performance of the various design parameters and the trade-offs involved in the process of data aggregation and the Q-IPA strategy. Furthermore the simulation results presented demonstrate the effectiveness and efficiency of the proposed Q-IPA approach, in terms of the network energy savings and the achievable end-to-end delay.

Finally it should be noted that the objective of the proposed approach and the corresponding models is twofold. On one hand, for a system with given design parameters, such as the deferred period τ and the aggregation probability γ, we can evaluate the expected successful report delivery probability and the expected end-to-end measurement delay. More importantly, on the other hand, given some specific QoS requirements (such as measurement end-to-end delay constraint and successful report delivery probability requirement) imposed by the task/application under consideration, we can use the Q-IPA approach and accordingly adjust the design parameters τ and γ, in order to fulfill the required QoS, while at the same time achieve significant energy savings and therefore extend the sensor network operational lifetime.

Appendix: End-to-End Delay Distribution in Q-IPA Algorithm Under Poisson Report Arrivals

We assume that the report arrival at each sensor node follows the Poisson process with arrival rate λ, therefore, the report interarrival time X is exponentially distributed, that is, $p_X(t) = \lambda e^{-\lambda t}$, $t \geq 0$. We also assume that $t^{(F)}$ and $t^{(P)}$ are exponentially distributed with parameter

λ_1 and λ_2, respectively. There are two possible operations (cases) to handle this report at a sensor node. Case 1: directly forward the report without using a deferred period (this happens with probability $1 - \gamma$). Case 2: wait for a deferred period τ to perform aggregation (this happens with probability γ). In the following we study the delay distribution by analyzing these two different cases.

Case 1: Sensor Node Does Not Use a Deferred Period τ to Perform Data Aggregation

In this case, when a report arrives, if the system is idle, there is no waiting time for the report to be forwarded; otherwise, it has to wait for the previous report to finish its transmission, and therefore some additional waiting time is introduced. Denote by p_1 the probability that upon its arrival a report finds that another report is still in transmission, then

$$P[t^{(R)} \le t] = P[t^{(R)} \le t|\text{idle}](1 - p_1) + P[t^{(R)} \le t|\text{busy}]p_1$$

where

$$p_1 = \int_0^\infty P[t^{(F)} > x]p_X(x)\,dx$$

$$= \int_0^\infty e^{-\lambda_1 x}\lambda e^{-\lambda x}\,dx = \frac{\lambda}{\lambda_1 + \lambda}$$

Since

$$P[t^{(R)} > t|\text{busy}] = P[t^{(R)} > t|t^{(F)} > X]$$

$$= \int_0^\infty P[t^{(F)} > t]p_X(x)\,dx = e^{-\lambda_1 t}$$

and $P[t^{(R)} \le t|\text{idle}] = 1$, we can obtain

$$P[t^{(R)} \le t] = \begin{cases} 0 & t < 0 \\ (1 - p_1) + p_1\left[1 - e^{-\lambda_1 t}\right] & t \ge 0 \end{cases}$$

and the corresponding pdf is given by

$$f^{(R)}(t) = (1 - p_1)\delta(t) + p_1\lambda_1 e^{-\lambda_1 t}u(t). \quad \text{(without deferred period)} \tag{7.23}$$

Case 2: Sensor Node Uses Deferred Period τ to perform data aggregation

In this case, when a report after the deferred period τ finds the system idle, the waiting time is 0; otherwise, it has to wait for the end of the transmission of the previous report. Let us denote by p_2 the probability that after deferred period τ, a report finds the previous report still in transmission, then:

$$p_2 = P[\text{busy after deferred period } \tau]$$

$$= P[t^{(F)} > X + \tau] = \frac{\lambda}{\lambda_1 + \lambda}e^{-\lambda_1 \tau}$$

Similar to case 1, we can obtain

$$P[t^{(R)} > t|\text{busy after deferred period } \tau] = e^{-\lambda_1(t-\tau)}$$

Therefore the CDF of the reporting (which includes the waiting time) is given by

$$P[t^{(R)} \leq t] = \begin{cases} 0 & t < \tau \\ (1 - p_2) + p_2 \left[1 - e^{-\lambda_1(t-\tau)}\right] & t \geq \tau \end{cases}$$

and the pdf is

$$f^{(R)}(t) = (1 - p_2)\delta(t - \tau) + p_2\lambda_1 e^{-\lambda_1(t-\tau)}u(t - \tau) \quad \text{(with deferred period } \tau\text{)}. \quad (7.24)$$

Since in the proposed Q-IPA approach a node chooses to defer for a period τ with probability γ, the Laplace transform of $f_i(t)$ can be obtained as follows:

$$\begin{aligned}
F_i(s) &= (1 - \gamma)\frac{\lambda_1\lambda_2}{(s + \lambda_1)(s + \lambda_2)}\left[(1 - p_1) + p_1\frac{\lambda_1}{s + \lambda_1}\right] \\
&\quad + \gamma\frac{\lambda_1\lambda_2}{(s + \lambda_1)(s + \lambda_2)}\left[(1 - p_2) + p_2\frac{\lambda_1}{s + \lambda_1}\right]e^{-s\tau} \\
&= \frac{B_1 + B_4e^{-s\tau}}{(s + \lambda_1)(s + \lambda_2)} + \frac{B_2 + B_3e^{-s\tau}}{(s + \lambda_1)^2(s + \lambda_2)}
\end{aligned} \quad (7.25)$$

where

$$\begin{aligned}
B_1 &= (1 - \gamma)(1 - p_1)\lambda_1\lambda_2 & B_4 &= \gamma(1 - p_2)\lambda_1\lambda_2 \\
B_2 &= (1 - \gamma)p_1\lambda_1^2\lambda_2 & B_3 &= \gamma p_2\lambda_1^2\lambda_2
\end{aligned}$$

Thus, the Laplace transform of the pdf of the end-to-end delay for l-hop transmission is

$$\begin{aligned}
\prod_{i=1}^{l} F_i(s) &= \left[\frac{B_2 + B_3e^{-s\tau}}{(s + \lambda_1)^2(s + \lambda_2)} + \frac{B_1 + B_4e^{-s\tau}}{(s + \lambda_1)(s + \lambda_2)}\right]^l \\
&= \sum_{k=0}^{l}\binom{l}{k}\frac{(B_2 + B_3e^{-s\tau})^k(B_1 + B_4e^{-s\tau})^{l-k}}{(s + \lambda_1)^{l+k}(s + \lambda_2)^l} \\
&= \sum_{k=0}^{l}\binom{l}{k}\sum_{h=0}^{k}\binom{k}{h}\sum_{n=0}^{l-k}\binom{l-k}{n}\frac{B_1^{l-k-n}B_4^n B_2^{k-h}B_3^h e^{-s\tau(n+h)}}{(s + \lambda_1)^{l+k}(s + \lambda_2)^l}
\end{aligned} \quad (7.26)$$

Using the partial fraction decomposition [71], we can rewrite (7.26) as

$$\begin{aligned}
\prod_{i=1}^{l} F_i(s) &= \sum_{k=0}^{l}\binom{l}{k}\sum_{h=0}^{k}\binom{k}{h}\sum_{n=0}^{l-k}\binom{l-k}{n}B_1^{l-k-n}B_4^n B_2^{k-h}B_3^h \\
&\quad \times e^{-s\tau(n+h)}\left[\sum_{j=0}^{l+k-1}\frac{A_{1j}}{(s + \lambda_1)^{l+k-j}} + \sum_{j=0}^{l-1}\frac{A_{2j}}{(s + \lambda_2)^{l-j}}\right]
\end{aligned}$$

where

$$A_{1j} = (-1)^j \binom{l+j-1}{j} \frac{1}{(\lambda_2 - \lambda_1)^{l+j}}$$

$$A_{2j} = (-1)^j \binom{l+k+j-1}{j} \frac{1}{(\lambda_1 - \lambda_2)^{l+k+j}}$$

Therefore, the pdf of the end-to-end delay when the number of hops is l can be obtained by taking the inverse Laplace transform of (7.26):

$$
F_{T_l}(t) \triangleq \mathcal{L}^{-1}\left\{\prod_{i=1}^l F_i(s)\right\}
$$

$$
= \sum_{k=0}^l \binom{l}{k} \sum_{n=0}^{l-k} \binom{l-k}{n} B_1^{l-k-n} B_4^n \sum_{h=0}^k \binom{k}{h} B_2^{k-h} B_3^h
$$

$$
\times \left[\sum_{j=0}^{l+k-1} \frac{A_{1j}(t - \tau(n+h))^{l+k-j-1}}{(l+k-j-1)!} e^{-\lambda_1(t-\tau(h+n))} \right.
$$

$$
\left. + \sum_{j=0}^{l-1} \frac{A_{2j}(t - \tau(n+h))^{l-j-1}}{(l-j-1)!} e^{-\lambda_2(t-\tau(h+n))} \right] u[t - \tau(n+h)] \quad (7.27)
$$

From (7.27) we can obtain the pdf of T_L

$$
f_{T_L}(t) = \sum_{l=1}^N p_L(l) F_{T_l}(t) \quad (7.28)
$$

7.5 PROGRESSIVE APPROACH TO DISTRIBUTED MULTIPLE-TARGET DETECTION IN SENSOR NETWORKS

Xiaoling Wang, Hairong Qi, and Steve Beck

In this section, we study the multiple-target detection problem in sensor networks. Due to the resource limitations of sensor networks, the multiple-target detection approach has to be energy efficient and bandwidth efficient. The existing centralized solutions cannot satisfy these requirements. We present a *progressive* detection approach based on the classic Bayesian estimation algorithm. In contrast to the centralized scheme in which data from all the sensors need to be transmitted to a central processing unit, the progressive approach sequentially estimates the number of targets based on only the local observation and the partial estimation result transmitted from previous sensors. In order to further improve the performance, we present a cluster-based distributed estimation approach where local detection within each cluster can adopt either the centralized or the progressive approach, and the estimated probabilities of each possible source number hypothesis are fused between clusters. We thoroughly evaluate the performance of different estimation approaches using data collected in a field demo. The experimental results show that the cluster-based distributed estimation schemes achieve higher detection probability. In addition, the progressive intracluster estimation approach can reduce data transmission by 83.22% and conserve energy by 81.64% compared to the centralized estimation scheme.

7.5.1 Background Information

As the direct interconnection between human being and the physical world, sensor networks have drawn a great deal of attention in recent years and have been employed in a wide variety of applications, ranging from military surveillance to civilian and environmental monitoring [72–76]. Multiple-target detection is a fundamental concern in a different context of sensor network applications that remains a challenging problem. On top of that, the limited wireless communication bandwidth, the large amount of sensor nodes, and the energy efficiency requirement of sensor networks all present further difficulties to the multiple-target detection problem.

If we consider different targets in the field as *sources* emitting signal energy and assume the signals they generate to be independent, multiple-target detection in sensor networks can be solved using traditional blind source separation (BSS) algorithms where the signal captured by the individual sensor is a linear/nonlinear weighted mixture of the signals generated by the sources. The "blind" qualification in BSS refers to the fact that there is no a priori information available on the number of sources, the probabilistic distribution of source signals, or the mixing model [77,78]. However, for conceptual and computational simplicity, the majority of BSS algorithms are developed based on a fundamental assumption: *The number of sources equals the number of sensors* [79,80].

Although the equality assumption is reasonable in small-size, well-controlled sensor array processing, it is not the case in sensor networks since the number of sensors is usually far more than the number of sources. Therefore, there is an immediate need to estimate the number of sources, a problem also referred to as *source number estimation* or *model order estimation* in [81]. Several techniques have been proposed trying to tackle this problem, some heuristic, others based on more principled approaches. As discussed in [81], techniques of the latter category are clearly superior, and heuristic methods at best may be seen as approximations to a more detailed underlying principle. Some examples of principled estimation approaches include the Bayesian estimation method [82], Markov chain Monte Carlo (MCMC) method [83], and variational learning approximation [84]. The Bayesian estimation method uses a set of Laplace approximations to estimate the marginal integrals when calculating the posterior probability of source number hypotheses. The basic idea of the MCMC method is to construct a Markov chain that generates samples from the hypothesis probability and uses the Monte Carlo method to estimate the posterior probability of specific hypotheses. Variational learning provides another approximation of the hypotheses posterior probability by evaluating a so-called negative variational free energy that is inversely proportional to the K-L divergence between signals.

The source number estimation approaches mentioned above have been successfully implemented in several applications such as music mixture separation and EEG (human brain activity) detection [81, 82]. However, their usage in sensor networks is hindered by their *centralized structure* where the raw data from a large amount of sensors have to be transmitted to a central processing unit, which will generate significant data traffic over the network, occupying communication bandwidth, and consuming energy.

To reduce the amount of data transmission and conserve energy, we presented a cluster-based distributed source number estimation algorithm in [85] where sensor networks are divided into clusters of small number of sensors, and the centralized source number estimation is performed within each cluster using the Bayesian estimation approach. The estimated posterior probabilities of possible source number hypotheses from different clusters are then combined using a probability fusion rule. The cluster-based estimation approach can reduce the amount of data transmission to some extent. However, the centralized processing within each cluster still requires raw data transmission and in turn consumes a certain amount of network bandwidth and energy. Moreover, within each cluster, the centralized estimation has to be performed at a central processing unit (the cluster head), which would place the burden of computation on the cluster head. This is contradictory to the property of sensor networks that all sensor nodes should have similar functionality, and they should take turns to be elected as the cluster head [86].

To *completely* eliminate raw data transmission in solving the multiple-target detection problem, we present a progressive approach in this section. The progressive approach carries out the estimation process within each cluster in a sequential manner. Instead of sending data from all sensors to a central unit as in the centralized scheme, the determination of the possible number of targets at a sensor is only based on its local observation and the estimation result received from its previous sensor. Therefore, raw data transmission is avoided and only small packets of partial estimation results are transmitted through the networks. The progressive approach is developed based on the Bayesian estimation method.

7.5.2 Terminologies and Problem Definition

To make our presentation more clear, we first define several terminologies and symbols used in this section that describe different structures of source number estimation.

Terminologies

Definition 7.5.1 A *centralized source number estimation scheme* is a processing structure that all the sensors send their data directly to a central processing unit where source number estimation is performed.

In the centralized scheme, the information transmitted through the network is *raw data* collected by the sensors. The structure of the centralized scheme is shown in Figure 7.35a.

Definition 7.5.2 A *progressive source number estimation scheme* is a processing structure that a group of sensors update the source number estimation result sequentially based on each sensor's local observation and the partial estimation result from its previous sensors in the sequence.

In this scheme, the information transmitted through the network is the *estimation result* or *partial decision*. The structure of the progressive scheme is shown in Figure 7.35b, where sensor i generates its partial decision based on its own observation and the partial decision received from sensor $i - 1$.

Definition 7.5.3 A *distributed* or *cluster-based source number estimation scheme* is a structure including two levels of processing: source number estimation within each cluster and decision fusion between different clusters.

In this scheme, sensor nodes are divided into clusters with each sensor belonging to one and only one cluster, as illustrated in Figure 7.35c. Source number estimation is performed within each cluster in either the centralized or the progressive manner. The estimation results from different clusters are then sent to a fusion center to generate a final decision.

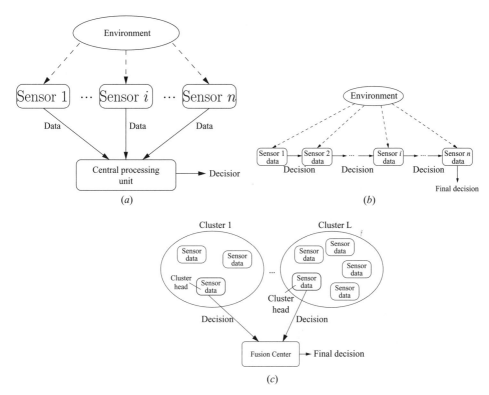

Figure 7.35 Structure of the (*a*) centralized scheme, (*b*) the progressive scheme, and (*c*) the distributed scheme.

Problem Definition Suppose there are m targets in the field generating independent source signals $s_i^{(t)}$, $i = 1, \ldots, m$ and n sensors recording signals $x_j^{(t)}$, $j = 1, \ldots, n$, where $t = 1, \ldots, T$ indicates the time index of the discrete-time signals. Then the sources and the observed mixtures at time t can be denoted as vectors $\mathbf{s}^{(t)} = [s_1^{(t)}, \ldots, s_m^{(t)}]^T$ and $\mathbf{x}^{(t)} = [x_1^{(t)}, \ldots, x_n^{(t)}]^T$, respectively. If we assume the mixing process is linear, the observations can then be represented as $\mathbf{x}^{(t)} = \mathbf{A}\mathbf{s}^{(t)}$ and the sources as $\hat{\mathbf{s}}^{(t)} = \mathbf{W}\mathbf{x}^{(t)}$, where $\mathbf{A}_{n \times m}$ is the unknown nonsingular scalar mixing matrix and the unmixing matrix $\mathbf{W}$ is calculated as the Moore–Penrose pseudoinverse, $\mathbf{W} = (\mathbf{A}^T\mathbf{A})^{-1}\mathbf{A}^T$.

Based on this linear, instantaneous mixing model, the source number estimation can be considered as a hypothesis testing problem, where $\mathcal{H}_m$ denotes the hypothesis on a possible number of targets, and the goal is to find $\hat{m}$ whose corresponding hypothesis $\mathcal{H}_{\hat{m}}$ maximizes the posterior probability given only the observation $\mathbf{x}^{(t)}$. According to Bayes' theorem, the posterior probability of hypotheses can be written as

$$P(\mathcal{H}_m | \mathbf{x}^{(t)}) = \frac{p(\mathbf{x}^{(t)} | \mathcal{H}_m) P(\mathcal{H}_m)}{\sum_{\text{all } \mathcal{H}} p(\mathbf{x}^{(t)} | \mathcal{H}) P(\mathcal{H})} \tag{7.29}$$

Assume the hypothesis $\mathcal{H}_m$ has a uniform distribution. Then the measurement of posterior probability can be simplified to the calculation of likelihood $p(\mathbf{x}^{(t)} | \mathcal{H}_m)$. In other words, the multiple-target detection problem can be defined as finding $\hat{m}$ that maximizes the likelihood function (the objective function).

7.5.3 Classic Centralized Bayesian Estimation Scheme

We briefly review the classic Bayesian source number estimation algorithm. In order to estimate the true number of sources m, Roberts proposed a centralized Bayesian estimation approach [82]. As discussed in [82], maximizing the informativeness of the set of estimated sources may be achieved by making $\mathbf{W}$ as large as possible, which requires some form of constraint. An alternative approach is to linearly map the observations $\mathbf{x}^{(t)}$ to a set of latent variables, $\mathbf{a}^{(t)}$, of the form $\mathbf{a}^{(t)} = \mathbf{W}\mathbf{x}^{(t)}$, followed by a nonlinear transform from this latent space to the set of source estimations, $\hat{\mathbf{s}}^{(t)} = \phi(\mathbf{a}^{(t)})$. Generally speaking, the choice of the nonlinear transform is not critical and can be defined as:

$$\phi(\mathbf{a}^{(t)}) = -\tanh(\alpha \mathbf{a}^{(t)}) \tag{7.30}$$

where α is a scaling factor.

The log-likelihood, $\log p(\mathbf{x}^{(t)}|\mathcal{H}_m)$, only depends on the calculation of marginal integrals of $p(\mathbf{x}^{(t)}|\mathbf{A}, \phi, \beta, \mathbf{a}^{(t)})$, which is the likelihood of $\mathbf{x}^{(t)}$ conditioned on the mixing matrix $\mathbf{A}$, the choice of nonlinearity ϕ, the noise variance β, and the latent variables $\mathbf{a}^{(t)}$. By applying Laplace approximations on the marginal integrals, the log-likelihood function can be estimated as:

$$
\begin{aligned}
L(m) &= \log p(\mathbf{x}^{(t)}|\mathcal{H}_m) \\
&= \log \pi(\hat{\mathbf{a}}^{(t)}) + \frac{n-m}{2} \log\left(\frac{\hat{\beta}}{2\pi}\right) - \frac{1}{2}\log|\hat{\mathbf{A}}^T\hat{\mathbf{A}}| - \frac{\hat{\beta}}{2}(\mathbf{x}^{(t)} - \hat{\mathbf{A}}\hat{\mathbf{a}}^{(t)})^2 \\
&\quad - \frac{mn}{2}\log\left(\frac{\hat{\beta}}{2\pi}\right) - \frac{n}{2}\left(\sum_{j=1}^{m}\log \hat{a}_j^2\right) - mn\log\gamma
\end{aligned}
\tag{7.31}
$$

where $\pi(\cdot)$ is the marginal distribution of the latent variables, $1/\hat{\beta}$ is the noise variance on each component of sensor observations, and it can be optimized using the maximum-likelihood estimation method as

$$\hat{\beta} = \frac{1}{n-m}\|(\mathbf{x}^{(t)} - \hat{\mathbf{A}}\hat{\mathbf{a}}^{(t)})^2\|_2,$$

$\hat{a}_j$ is the jth component of $\hat{\mathbf{a}}^{(t)}$, and $\gamma = 2\|\hat{\mathbf{A}}\|_\infty$. As we can see from Eq. (7.31) that in order to calculate the log-likelihood $L(m)$, the raw data $\mathbf{x}^{(t)}$ from all participating sensor nodes are needed.

The transmission of raw data consumes energy and bandwidth, which is contradictory to the requirement of sensor networks. Therefore, the most challenging problem of source number estimation in sensor networks is to be able to estimate the log-likelihood $L(m)$ locally without direct usage of other sensor observations. This is the motivation of the progressive estimation approach presented in Section 7.5.4.

7.5.4 Progressive Bayesian Source Number Estimation

We present the progressive source number estimation approach developed based on the classic centralized Bayesian estimation algorithm. To accommodate the unique energy efficiency requirement of sensor networks in performing target number estimation, a progressive approach is derived to evaluate the objective function of Eq. (7.31) based on the iterative relationship between sensors, that is, each sensor i updates the log-likelihood evaluation only based on its local observation $x_i^{(t)}$ and the information I_{i-1} transmitted from its previous sensor $(i-1)$ as shown in Figure 7.36a. To carry out the updating process successfully, two problems must be solved: (1) How to update the mixing matrix $\mathbf{A}$ iteratively and keep the independence of source signals, and (2) how to decompose the log-likelihood estimation into different components that depend on the previous log-likelihood estimation and the local observation, respectively.

In the progressive scheme shown in Figure 7.36a, upon receiving a partial estimation result I_{i-1} from its previous sensor $(i-1)$, sensor i will update the log-likelihood $L_i(m)$ and the mixing matrix $\mathbf{A}$ corresponding to different source number hypotheses and then transmit the updated results I_i to its next sensor $(i+1)$. The updating rules at sensor i is denoted as $\mathcal{D}_i(x_i^{(t)}; I_{i-1})$, where $x_i^{(t)}$ is the observation of sensor i at time t and I_{i-1} denotes the information received from sensor $(i-1)$. As shown in Figure 7.36b, the information transmitted by sensor i, I_i, includes the updated mixing matrix $[\mathbf{A}]_i$, the estimated latent variable $[\mathbf{a}^{(t)}]_i$, the accumulated estimation error ϵ_i, and all seven terms in Eq. (7.31) updated at sensor i, $[Term1]_i, [Term2]_i, \ldots, [Term7]_i$.

Progressive Estimation of the Mixing Matrix A   At the first sensor, the mixing matrix $\mathbf{A}$ is initialized randomly. During the progressive implementation, sensor $(i-1)$ modifies matrix $\mathbf{A}_{(i-1)\times m}$ locally and sends the resulting matrix to sensor i. After sensor i receives the information, it first adds one more dimension (an extra row) to $\mathbf{A}$ with random numbers, and then finds the optimal estimate of $\mathbf{A}_{i\times m}$ using the BFGS optimization method, which is one special type of quasi-Newton methods, [87, 88].

Progressive Estimation of the Accumulated Error   We refer to the component

$$\sum_{j=1}^{i-1}\left(x_j^{(t)} - \sum_{k=1}^{m} A_{j,k}\hat{a}_k^{(t)}\right)^2$$

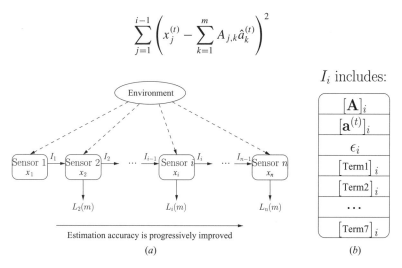

Figure 7.36 Progressive source number estimation scheme: (*a*) structure and (*b*) transmitted information.

as the *accumulated estimation error* ϵ_{i-1} at sensor $(i-1)$, which is the squared difference between the sensor observation and the mixture of estimated source signals. Basically, it originates from the estimation error of the mixing matrix $\mathbf{A}$. This item needs to be carried from sensor to sensor, and its updating rule at sensor i is

$$\epsilon_i = \sum_{j=1}^{i} \left(x_j^{(t)} - \sum_{k=1}^{m} A_{j,k} \hat{a}_k^{(t)} \right)^2 = \epsilon_{i-1} + \left(x_i^{(t)} - \sum_{k=1}^{m} A_{i,k} \hat{a}_k^{(t)} \right)^2 \tag{7.32}$$

Progressive Estimation of the Log-Likelihood Function In order to estimate the log-likelihood function in Eq. (7.31) progressively, we employ the iterative relationship between sensors, that is, the estimation at sensor i, $L_i(m)$, is a function of the estimation at sensor $(i-1)$, $L_{i-1}(m)$, and the local observation $x_i^{(t)}$. The updating rule can be formulated as

$$L_i(m) = L_{i-1}(m) + f\left(x_i^{(t)} \right) \tag{7.33}$$

We derive the updating function f by evaluating the seven terms of the log-likelihood function separately.

Term 1 The first term in Eq. (7.31), $\log \pi(\hat{a}_k^{(t)})$, accounts for the marginal distribution of the latent variables. In normal cases, the distribution of one latent variable $\hat{a}_k$ can be assumed to have the form [82]

$$\pi\left(\hat{a}_k^{(t)} \right) = \frac{1}{Z(\alpha) \left[\cosh(\alpha \hat{a}_k^{(t)}) \right]^{1/\alpha}} \tag{7.34}$$

where $\log Z(\alpha) = a \log(c/\alpha + 1) + b$, α is a scaling factor, $a = 0.522$, $b = 0.692$, and $c = 1.397$. Suppose the mixing matrix at sensor i is $\mathbf{A}^i$; then $\mathbf{W}^i = [(\mathbf{A}^i)^T \mathbf{A}^i]^{-1} (\mathbf{A}^i)^T$, and $\mathbf{w}_k^i$ is the kth row of $\mathbf{W}^i$. Hence, at sensor $(i-1)$, the latent variable $\hat{a}_k^{(t)}$ is of the form

$$\hat{a}_k^{(t)} = \sum_{j=1}^{i-1} w_{k,j} x_j^{(t)} \tag{7.35}$$

and the first term in Eq. (7.31) can be calculated as:

$$\begin{aligned}
[\text{Term1}]_{i-1} &= \left[\log \pi \left(\hat{a}_k^{(t)} \right) \right]_{i-1} = - \log Z(\alpha) - \frac{1}{\alpha} \log \left[\cosh \left(\alpha \hat{a}_k^{(t)} \right) \right] \\
&= - \log Z(\alpha) - \frac{1}{\alpha} \log \left[\frac{\exp \left(\alpha \hat{a}_k^{(t)} \right) + \exp(-\alpha \hat{a}_k^{(t)})}{2} \right] \\
&= - \log Z(\alpha) - \frac{1}{\alpha} \log \left[\exp \left(\alpha \hat{a}_k^{(t)} \right) \left(1 + \exp \left(-2\alpha \hat{a}_k^{(t)} \right) \right) \right] + \frac{1}{\alpha} \log 2
\end{aligned} \tag{7.36}$$

We can restrict the initial choice of $\mathbf{A}^{i-1}$ such that $|\exp(-2\alpha \hat{a}_k^{(t)})| < 1$. According to the Taylor expansion,

$$\log \left[1 + \exp \left(-2\alpha \hat{a}_k^{(t)} \right) \right] = \sum_{p=1}^{\infty} \frac{(-1)^{p-1}}{p} \left[\exp \left(-2\alpha \hat{a}_k^{(t)} \right) \right]^p \tag{7.37}$$

Then Eq. (7.36) becomes

$$[\text{Term1}]_{i-1} = -\log Z(\alpha) - \hat{a}_k^{(t)} - \frac{1}{\alpha} \sum_{p=1}^{\infty} \frac{(-1)^{p-1}}{p} \left[\exp\left(-2\alpha \hat{a}_k^{(t)}\right)\right]^p + \frac{1}{\alpha} \log 2$$

(7.38)

If only we consider the calculation to the first-order precision, that is, $p = 1$, and substitute Eq. (7.35) into Eq. (7.38), term 1 calculated at sensor $(i - 1)$ is

$$[\text{Term1}]_{i-1} \approx -\log Z(\alpha) - \sum_{j=1}^{i-1} w_{k,j} x_j^{(t)} - \frac{1}{\alpha} \exp\left(-2\alpha \sum_{j=1}^{i-1} w_{k,j} x_j^{(t)}\right) + \frac{1}{\alpha} \log 2$$

(7.39)

Similarly, term 1 calculated at sensor i can be written as:

$$[\text{Term1}]_i = \left[\log \pi\left(\hat{a}_k^{(t)}\right)\right]_i = -\log Z(\alpha) - \frac{1}{\alpha} \log\left[\cosh\left(\alpha \sum_{j=1}^{i} w_{k,j} x_j^{(t)}\right)\right]$$

$$\approx -\log Z(\alpha) - \sum_{j=1}^{i} w_{k,j} x_j^{(t)} - \frac{1}{\alpha} \exp\left(-2\alpha \sum_{j=1}^{i} w_{k,j} x_j^{(t)}\right) + \frac{1}{\alpha} \log 2$$

(7.40)

Compare between Eqs. (7.39) and (7.40), the updating rule for term 1 at sensor i can be derived as:

$$[\text{Term1}]_i = [\text{Term1}]_{i-1} - w_{k,i} x_i^{(t)} - \frac{1}{\alpha} \exp\left(-2\alpha \left[\hat{a}_k^{(t)}\right]_{i-1}\right)\left[\exp\left(-2\alpha w_{k,i} x_i^{(t)}\right) - 1\right]$$

(7.41)

Term 2   The second term in Eq. (7.31), $n - m/2 \log(\hat{\beta}/2\pi)$, takes into account the noise variance β, which is estimated by the squared errors between the real observations and their estimated counterparts. At sensor $(i - 1)$, the noise variance is calculated as:

$$\hat{\beta} = \frac{1}{i - 1 - m} \sum_{j=1}^{i-1} \left(x_j^{(t)} - A_j \hat{\mathbf{a}}^{(t)}\right)^2$$

(7.42)

where A_j denotes the jth row of the mixing matrix $\mathbf{A}$ and m is the number of sources. Therefore, term 2 at sensor $(i - 1)$ is

$$[\text{Term2}]_{i-1} = \left[\frac{i - 1 - m}{2} \log\left(\frac{\hat{\beta}}{2\pi}\right)\right]_{i-1}$$

$$= -\frac{i - 1 - m}{2} \log(i - 1 - m) + \frac{i - 1 - m}{2} \log\left[\sum_{j=1}^{i-1} \left(x_j^{(t)} - A_j \hat{\mathbf{a}}^{(t)}\right)^2\right]$$

$$- \frac{i - 1 - m}{2} \log 2\pi$$

(7.43)

By using the Taylor expansion on component $\log[\sum_{j=1}^{i}(x_j^{(t)} - A_j\hat{\mathbf{a}}^{(t)})^2]$ with the first-order precision, the updating rule of term 2 at sensor i is derived as:

$$
[\text{Term2}]_i = -\frac{i-m}{2}\log(i-m) + \frac{i-m}{2}\log\left[\sum_{j=1}^{i}\left(x_j^{(t)} - A_j\hat{\mathbf{a}}^{(t)}\right)^2\right] - \frac{i-m}{2}\log 2\pi
$$

$$
\approx -\frac{i-m}{2}\log(i-m) + \frac{i-m}{2}\log\left[\sum_{j=1}^{i-1}\left(x_j^{(t)} - A_j\hat{\mathbf{a}}^{(t)}\right)^2\right]
$$

$$
+\frac{i-m}{2}\frac{\left(x_i^{(t)} - A_i\hat{\mathbf{a}}^{(t)}\right)^2}{\sum_{j=1}^{i-1}\left(x_j^{(t)} - A_j\hat{\mathbf{a}}^{(t)}\right)^2} - \frac{i-m}{2}\log 2\pi
$$

$$
= \frac{i-m}{i-1-m}[\text{Term2}]_{i-1} + \frac{i-m}{2}\log\left(\frac{i-1-m}{i-m}\right)
$$

$$
+\frac{i-m}{2}\frac{\left(x_i^{(t)} - \sum_{k=1}^{m} A_{i,k}\hat{a}_k^{(t)}\right)^2}{\epsilon_{i-1}}
\tag{7.44}
$$

Term 3   The third term in Eq. (7.31), $-\frac{1}{2}\log|\hat{\mathbf{A}}^T\hat{\mathbf{A}}|$, only depends on the updating rule of mixing matrix $\mathbf{A}$, which has been discussed previously.

Term 4   The fourth term in Eq. (7.31) can be estimated at sensor $(i-1)$ as:

$$
[\text{Term4}]_{i-1} = \left[-\frac{\hat{\beta}}{2}\left(\mathbf{x}^{(t)} - \hat{\mathbf{A}}\hat{\mathbf{a}}^{(t)}\right)^2\right]_{i-1} = -\frac{1}{2(i-1-m)}\left[\sum_{j=1}^{i-1}\left(x_j^{(t)} - A_j\hat{\mathbf{a}}^{(t)}\right)^2\right]^2
\tag{7.45}
$$

Since the updating rule for term 4 at sensor i involves the estimation of noise variance β, it is similar to that of term 2, which is derived as:

$$
[\text{Term4}]_i = -\frac{1}{2(i-m)}\left[\sum_{j=1}^{i}\left(x_j^{(t)} - A_j\hat{\mathbf{a}}^{(t)}\right)^2\right]^2
$$

$$
= -\frac{1}{2(i-m)}\left[\sum_{j=1}^{i-1}\left(x_j^{(t)} - A_j\hat{\mathbf{a}}^{(t)}\right)^2\right]^2 + 2\left(x_i^{(t)} - A_i\hat{\mathbf{a}}^{(t)}\right)^2\sum_{j=1}^{i-1}\left(x_j^{(t)} - A_j\hat{\mathbf{a}}^{(t)}\right)^2
$$

$$
+\left(x_i^{(t)} - A_i\hat{\mathbf{a}}^{(t)}\right)^4
$$

$$
= \frac{i-1-m}{i-m}[\text{Term4}]_{i-1} + 2\epsilon_{i-1}\left(x_i^{(t)} - \sum_{k=1}^{m} A_{i,k}\hat{a}_k^{(t)}\right)^2 + \left(x_i^{(t)} - \sum_{k=1}^{m} A_{i,k}\hat{a}_k^{(t)}\right)^4
\tag{7.46}
$$

Term 5 The fifth term is also affected by the noise variance $\hat{\beta}$. Using the Taylor expansion formula as shown in Eq. (7.37), the updating rule of term 5 at sensor i is

$$
\begin{aligned}
[\text{Term5}]_i &= \left[\frac{im}{2} \log \left(\frac{\hat{\beta}}{2\pi} \right) \right]_i \\
&= \frac{i}{i-1} [\text{Term5}]_{i-1} + \frac{im}{2} \log \left(\frac{i-1-m}{i-m} \right) \\
&\quad + \frac{im}{2} \frac{\left(x_i^{(t)} - \sum_{k=1}^{m} A_{i,k} \hat{a}_k^{(t)} \right)^2}{\epsilon_{i-1}}
\end{aligned}
\tag{7.47}
$$

Term 6 This term in Eq. (7.31) accounts for the estimated latent variables, and its updating rule at sensor i employing the Taylor expansion with the first-order precision in Eq. (7.37) can be written as:

$$
\begin{aligned}
[\text{Term6}]_i &= \left[\frac{i}{2} \sum_{k=1}^{m} \log \left(\hat{a}_k^{(t)} \right)^2 \right]_i = \frac{i}{2} \sum_{k=1}^{m} \log \left(\sum_{j=1}^{i-1} w_{k,j} x_j^{(t)} + w_{k,i} x_i^{(t)} \right)^2 \\
&\approx \frac{i}{2} \sum_{k=1}^{m} \left[\log \left(\sum_{j=1}^{i-1} w_{k,j} x_j^{(t)} \right)^2 + \frac{\left(w_{k,i} x_i^{(t)} \right)^2 + 2 w_{k,i} x_i^{(t)} \sum_{j=1}^{i-1} w_{k,j} x_j^{(t)}}{\left(\sum_{j=1}^{i-1} w_{k,j} x_j^{(t)} \right)^2} \right] \\
&= \frac{i}{i-1} [\text{Term6}]_{i-1} + \frac{i}{2} \sum_{k=1}^{m} \frac{\left(w_{k,i} x_i^{(t)} \right)^2 + 2 w_{k,i} x_i^{(t)} \left[\hat{a}_k^{(t)} \right]_{i-1}}{\left(\left[\hat{a}_k^{(t)} \right]_{i-1} \right)^2}
\end{aligned}
\tag{7.48}
$$

Term 7 The last term in Eq. (7.31) is of the form $im \log \gamma$, where $\gamma = 2 \| \hat{\mathbf{A}} \|_\infty$. This term only depends on the updating rule of matrix $\mathbf{A}$.

Algorithm According to the update rules of the mixing matrix, the log-likelihood, and the estimation error at sensor i, the information to be transmitted from sensor i to sensor $i + 1$ includes the estimated mixing matrix $[\hat{\mathbf{A}}]_i$, the estimated latent variables $[\hat{\mathbf{a}}^{(t)}]_i$, the estimation error ϵ_i, and each term in Eq. (7.31), that is, $I_i = \{ [\hat{\mathbf{A}}]_i, [\hat{\mathbf{a}}^{(t)}]_i, \epsilon_i, [\text{Term1}]_i, \ldots, [\text{Term7}]_i \}$. We denote this term as the *updating information I* hereafter. The progressive estimation algorithm is summarized as:

Algorithm 7.5.7 **Progressive Source Number Estimation Algorithm** {/*Initialization*/ At sensor $i = 1$, for each possible m: Initialize $\mathbf{A}_{1 \times m}$ using random numbers; Compute $\mathbf{W}$ and $\mathbf{a}$; Compute estimation error ϵ; Compute each term in Eq. (7.31); Compute $L(m)$; /* Progressive Update */ $i = 2$; *While* ($\max L(m) < threshold$) {Send $\mathbf{A}$, latent variables $\mathbf{a}$, estimation error ϵ and the seven terms in Eq. (7.31) to sensor i; At sensor i, for each possible m: Add one row to $\mathbf{A}$ with random numbers; *While* (!converge) {Update $\mathbf{A}$ using BFGS method;} Update the accumulated estimation error ϵ; Update each term in Eq. (7.31); Compute $L(m)$; Increase i by 1; } /* Generate the final estimation */ Decide $m = \arg \max L(m)$; Output m; }

7.5.5 Cluster-Based Distributed Source Number Estimation Scheme

The proliferation of low-cost sensors enables large amounts of sensor deployment. As more sensor nodes are put into the field, more data is captured that can enhance decision making. However, the risk of the scalability issue is large data transfer and information overloading. Furthermore, in sensor networks, sensors communicate through low bandwidth and unreliable wireless links compared to wired communication. An individual sensor may suffer intermittent connectivity due to the high bit error rate of the wireless link, and it can be further deteriorated by environmental hazard [89]. In a multihop wireless network, the data traffic between two sensors far away needs to involve a large amount of intermediate "forwarding" sensors that consume extra energy. To tackle these problems, we develop a cluster-based distributed algorithm for multiple-target detection by dividing the sensor nodes into different clusters.

Since clustering is not the focus of this section, in the development of the distributed source number estimation scheme, we assume that a clustering protocol has been applied and the sensor nodes have organized themselves into clusters with each node assigned to one and only one cluster. Local nodes can communicate with other nodes within the same cluster, and different clusters communicate through a cluster head elected from each cluster. An example of a clustered sensor network is illustrated in Figure 7.37.

The distributed scheme is accomplished in two levels: First, source number estimation is conducted within each cluster using either the centralized or the progressive scheme. In the centralized scheme, each sensor within a cluster sends its observation to the cluster head where the log-likelihood of each source number hypothesis is estimated. While in the progressive scheme, the cluster head initiates the local estimation of the log-likelihood, which is then transmitted to the next node closest to it in the cluster such that the initial estimate of the log-likelihood gets updated. This process continues until all sensors within a cluster have updated the log-likelihood estimation. Second, a fusion rule is applied to combine the posterior probability of each source number hypothesis estimated by each cluster. Since sensor nodes may fail or be blocked due to lack of power, physical damage, or environmental inference, sensor fusion provides an efficient way to achieve fault tolerance. At the same time, the fusion algorithm needs to be energy efficient in order to meet the requirement of sensor networks. However, the fault tolerance objective and the energy

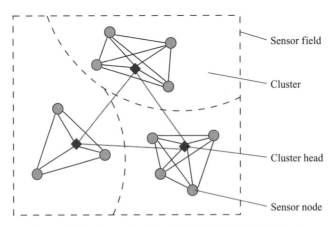

Figure 7.37 Example of clustered sensor network model.

efficiency objective always conflict with each other. In general, energy-efficient approaches try to limit the redundancy in the algorithm so that a minimum amount of energy is required for fulfilling a certain task. On the other hand, redundancy is needed for providing fault tolerance [89]. In this sense, a balance has to be struck between these two objectives in the fusion algorithm.

Suppose the sensor nodes in a sensor network are divided into L clusters. The observations from cluster l, $l = 1, \ldots, L$, are denoted as $x_l^{(t)} = x_1^{(t)}, \ldots, x_{K_l}^{(t)}$ where K_l is the number of sensors in cluster l. Assume the posterior probability derived from cluster l is proportional to the log-likelihood estimation $\log p(x_l^{(t)}|\mathcal{H}_m)$, we then design the fusion rule as a weighted summary of the estimations from all clusters:

$$\log P(\mathcal{H}_m|\mathbf{x}^{(t)}) \propto \sum_{l=1}^{L} c_l \log p(\mathbf{x}_l^{(t)}|\mathcal{H}_m) \tag{7.49}$$

where c_l is the weight of cluster l in determining the fused posterior probability. It reflects the physical characteristic of the clustering in the sensor network. For example, in the case of distributed multiple-target detection using acoustic signals, the propagation of acoustic signals follows the energy decay model that the detected energy is inversely proportional to the square of the distance between the source and the sensor node, that is, $E_{\text{sensor}} \propto 1/d^2 E_{\text{source}}$. Therefore, the weight c_l can be considered as the relative detection sensitivity of the sensor nodes in cluster l and is proportional to the average energy captured by the sensor nodes:

$$c_l = \frac{1}{K_l} \sum_{k=1}^{K_l} E_k \propto \frac{1}{K_l} \sum_{k=1}^{K_l} \frac{1}{d_k^2} \tag{7.50}$$

7.5.6 Performance Evaluation

To compare the performance of the three multiple-target detection schemes presented in this section, we design four experiments and use the real data collected in a field demo (held at BAE Systems, Austin, Texas, August 2002) for the evaluation. In these experiments, two civilian vehicles, a Harley-Davidson motorcycle and a heavy diesel truck as shown in Figure 7.38, are used as the targets. They travel along the north–south road from opposite

 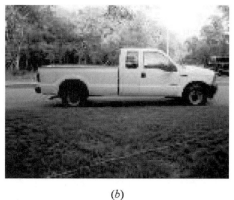

(a)   (b)

Figure 7.38   Vehicles used in the experiments: (a) motorcycle and (b) diesel truck.

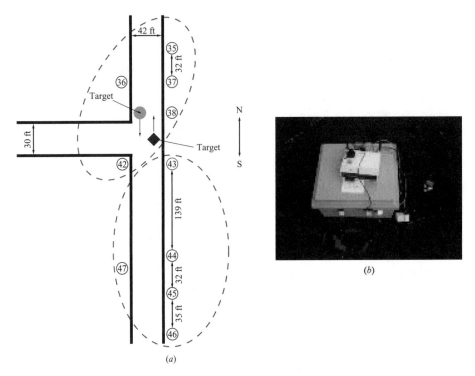

Figure 7.39 (*a*) Sensor laydown and (*b*) the Sensoria sensor node.

directions and intersect at the T-junction. There are 10 nodes deployed along the road, which are divided into 2 clusters of 5 sensor nodes to implement the distributed scheme, as illustrated in Figure 7.39*a*. The Sensoria WINS NG-2.0 sensor nodes are used (as shown in Fig. 7.39*b*, which consist of a dual-issue SH-4 processor running at 167 MHz with 300 MIPS of processing power, radio frequency modem for wireless communication, and up to 4 channels of sensing modalities, including the acoustic, seismic, and infrared.

The four experiments are elaborated as follows:

- Experiment 1: Apply the centralized Bayesian source number estimation scheme on data collected from all the 10 sensors.
- Experiment 2: Apply the progressive source number estimation scheme on all the 10 sensors.
- Experiment 3: Apply the distributed estimation scheme using the centralized Bayesian source number estimation within each cluster.
- Experiment 4: Apply the distributed estimation scheme using the progressive source number estimation scheme within each cluster.

Performance Metrics We use five metrics to evaluate the performance of developed source number estimation schemes, including the average log-likelihood, the detection probability, the amount of data transmission, and the energy consumption.

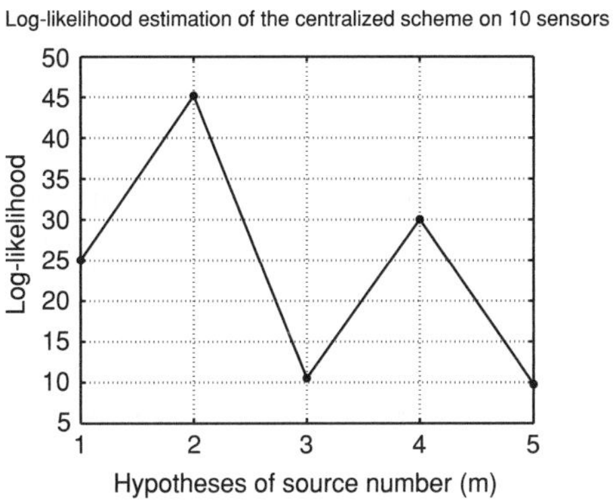

Log-likelihood estimation of the centralized scheme on 10 sensors

Figure 7.40 Average log-likelihood.

Average Log-likelihood   Source number estimation is basically an optimization problem in which an optimal hypothesis $\mathcal{H}_m$ is pursued that maximizes the posterior probability $P(\mathcal{H}_m|\mathbf{x}^{(t)})$ given the observation matrix. A different initialization condition might affect the optimization process at different levels. In order to evaluate the performance using the log-likelihood objectively, we run the experiment 20 times and use the *average log-likelihood* as one of the performance metrics. Figure 7.40 gives an example of the average log-likelihood calculated using the centralized scheme. The x axis refers to the different source number hypotheses $\mathcal{H}_m, m = 1, \ldots, 5$ and the y axis records the corresponding average log-likelihood. From this figure, we can see that on average the correct number of targets ($m = 2$) has the largest log-likelihood compared to other source number hypotheses.

Detection Probability   The *detection probability* ($P_{\text{detection}}$) is defined as the ratio between the number of correct source number estimations and the total number of estimations, that is,

$$P_{\text{detection}} = \frac{N_{\text{correct}}}{N_{\text{total}}} \tag{7.51}$$

where N_{correct} denotes the number of correct estimations and N_{total} is the total number of estimations.

Amount of Data Transmission   In all the experiments, we assume each real number being represented as a floating-point number that is 32 bits long. The *amount of data transmission* is defined as the number of bits transmitted to perform the estimation algorithm. For example, in the centralized estimation scheme, suppose there are 10 sensors deployed along the road and the algorithm is performed over 1-s segments with 500 samples each, then 144,000 bits of data need to be transmitted. While in the progressive scheme under the same conditions, 37,472 bits of data (the updating information) are transmitted.

Energy Consumption   Since in a large-scale sensor network, data communication consumes most of the energy, we only consider the energy consumed on data transmission in the evaluation of *energy consumption*. According to [89], the energy consumed in data transmission can be modeled using a linear equation:

$$E_{\text{tran}} = c \times \text{size} + d \tag{7.52}$$

where c is a coefficient indicating the amount of energy consumed by transferring 1 byte of data, $size$ is the size of data being transferred, and d is a fixed component associated with device state changes and channel acquisition overhead. The values for c and d are different between data transmission and data receiving. Normally, we choose $c = 1.9$ and $d = 454$ for transmission while $c = 1.425$ and $d = 356$ for receiving that are measured based on a Lucent IEEE 802.11 WaveLAN PC Card using a 2.4-GHz direct-sequence spread spectrum.

Decision Delay   In many cases, target detection in sensor networks is very time critical, especially in military applications. It is important that the decision takes place as quickly as possible. Therefore, the measurement of decision delay plays an important role in performance evaluation of multiple-target detection approaches. In our experiments, we only consider decision delay as the time spent on estimating the most possible number of targets in the field.

Experiments and Result Analysis   In our experiments, we test five hypotheses on the number of targets, that is, $m = 1, 2, \ldots, 5$. All algorithms are performed on 1-s acoustic signals with a sampling rate of 500 samples per second, that is, the time instance $t = 1, 2, \ldots, 500$. The observations from sensor nodes are preprocessed component-wise to be zero-mean, unit-variance distributed. The results of the four experiments are shown in Figures 7.41 to 7.43, illustrating the average log-likelihood, the detection probability, the amount of data transmission, and the energy consumption, respectively.

The centralized scheme in experiment 1 is performed at a fusion center using all the 10 sensor observations. Figure 7.41*a* shows the average log-likelihoods corresponding to different source number hypotheses. The progressive scheme in experiment 2 is conducted by sending an initial log-likelihood estimation from sensor number 38, which is at one corner of the T-junction. Figure 7.41*b* shows the average log-likelihoods corresponding to the updated estimations at each local sensor node. Figure 7.41*c* shows the average log-likelihoods according to different hypotheses of the distributed scheme using the centralized intracluster approach. Figure 7.41*d* illustrates the average log-likelihoods using the distributed scheme with the progressive intracluster approach. The first five points along the x axis correspond to the updated result at each sensor within cluster 1 and the last point is the fused result between cluster 1 and cluster 2. Figure 7.41*e* shows the average log-likelihood estimations from cluster 2 in the same manner. It can be seen that all schemes give the strongest support to the true number of sources ($m = 2$). Although in Figure 7.41*e* cluster 2 gives an estimation that $m = 5$, after fusing the estimations from the two clusters, the fusion rule elects the correct source number. In this way, the fusion actually increases the fidelity of the estimation.

The detection probabilities are shown in Figure 7.42. The distributed scheme using the centralized intracluster estimation (experiment 3) provides the highest accuracy, while the distributed scheme using the progressive intracluster estimation (experiment 4) has comparable performance. However, a significant advantage of the progressive source number estimation scheme is the reduction of data transmission within the network and in turn the conservation of energy.

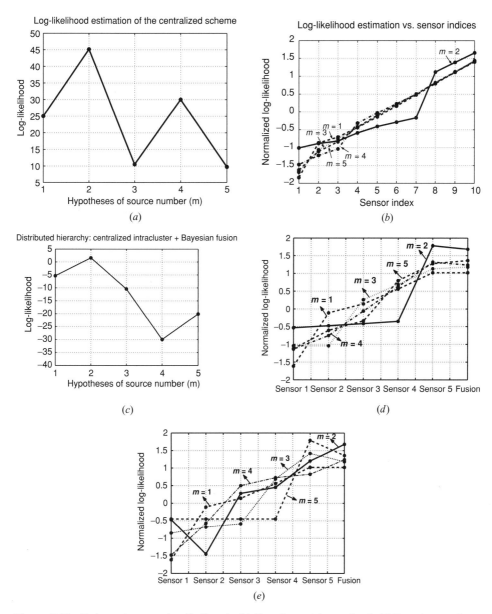

Figure 7.41 Estimated average log-likelihoods: (*a*) Experiment 1 (centralized), (*b*) Experiment 2 (progressive), (*c*) Experiment 3 (cluster-based, centralized within cluster), (*d*) Experiment 4 (cluster-based, progressive in cluster 1), and (*e*) Experiment 4 (cluster-based, progressive in cluster 2.

Figure 7.43*a* shows the amount of data transmitted through the network in all the four experiments. As we can see, in the classic centralized estimation approach (experiment 1), 144,000 bits of data need to be transmitted, while in the progressive approach (experiment 2), only 37,472 bits of data need to be transmitted (73.98% reduction), and in the distributed schemes using the progressive intracluster estimation, 24,160 bits of data transmitted (83.22% reduction). Figure 7.43*b* illustrates the corresponding energy consumption

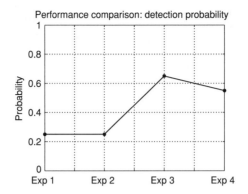

Figure 7.42 Performance comparison: detection probability.

of the four approaches. We can see again that the distributed scheme using the progressive intracluster estimation consumes the least energy among all approaches. The examination of decision delays among the four approaches is shown in Figure 7.43c. Compared to the centralized scheme, the cluster-based distributed approach reduces the processing time by half, while the progressive approaches spend more time on processing when data from all the sensor nodes are employed. However, an advantage of the progressive approach is that

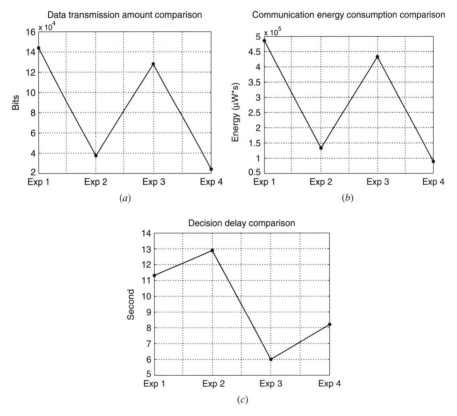

Figure 7.43 Performance comparison: (a) amount of data transmission, (b) energy consumption, and (c) decision delay.

not all the sensor nodes need to be visited. Whenever a predefined accuracy is achieved, the process can terminate and return results. This feature enables saving both computation time and network bandwidth since unnecessary node visits and information transmission are avoided.

Discussion   As demonstrated in the experiments, the cluster-based distributed approach using the progressive intracluster estimation (experiment 4) has the best performance in the sense that it can provide much higher detection probability than the centralized approaches (experiments 1 and 2), while at the same time occupies the least amount of network bandwidth and consumes the least amount of energy. It seems counterintuitive that the distributed approach performs better than the centralized approach. Here, we provide some explanations on this phenomena through experimental study.

First, in our experiments, the sensor nodes are spread in an area 300 long and 42 ft wide, as shown in Figure 7.39a. Based on the energy decay model of acoustic signals, the signal generated by a target degrades during propagation with a factor that is the square of the distance between the target and the sensor node. Therefore, in a large field, some sensors may be out of range for detecting the target, making the information they provide irrelevant to the task. In this case, the fusion among all the sensors would deteriorate the detection accuracy.

Second, in the centralized or the progressive approach, the source number estimation is carried out based on observations from all the sensor nodes. If some sensors are faulty or the signals are corrupted by noise, the estimation result will be significantly affected. However, in the cluster-based distributed algorithm, the clustering approach makes sure that sensor nodes within the same cluster are close to each other. Therefore, the estimation from sensors within a cluster generally has a higher fidelity.

Finally, in the derivation of the posterior probability fusion method, the physical characteristics of sensor networks, such as the signal energy captured by each sensor node versus its geographical position, are considered, making this method more adaptive to real applications.

7.5.7 Conclusion

This section studied the problem of source number estimation in sensor networks for multiple-target detection. Due to the sheer amount of sensors deployed, the limited wireless communication bandwidth, and the battery-powered fact of each sensor node, the classic centralized approach would not provide a satisfactory solution. In this section, a progressive source number estimation algorithm is presented based on the classic Bayesian estimation algorithm, using the iterative relationship between sensors.

Since each sensor node has a limited capability in sensing and can communicate only within a certain range, a cluster-based distributed estimation scheme is further developed in which a local source number estimation is performed within each cluster using either the centralized or the progressive estimation approach; and a probability fusion method is then applied to combine the local estimations to generate a global decision. A distance-based posterior probability fusion method is used to fuse the estimation results from each cluster. Experiments are conducted on multiple civilian vehicle detection using acoustic signals to evaluate the performance of different approaches. It is concluded that the cluster-based distributed algorithm provides higher detection probability. In addition, the distributed scheme using the progressive intracluster estimation can reduce the amount of data transmission by 83.22% and conserves 81.64% of energy compared to the centralized scheme.

REFERENCES

1. W. Ye, J. Heidemann, and D. Estrin, "An energy-efficient mac protocol for wireless sensor networks," in *Proceedings of INFOCOM 2002*, vol. 3, June 2002, pp. 1567–1576.

2. A. Perrig, R. Szewczyk, J. Tygar, V. Wen, and D. Culler, "Spins: Security protocols for sensor network," *Wireless Networks*, vol. 8, no. 5, pp. 521–534, 2002.

3. A. Sinha and A. Chandrakasan, "Dynamic power management in wireless sensor networks," *IEEE Design and Test of Computers*, vol. 18, no. 2, pp. 62–74, Mar.–Apr. 2001.

4. C. Intanagonwiwat, D. Estrin, R. Govindan, and J. Heidemann, "Impact of network density on data aggregation in wireless sensor networks," in *Impact of Network Density on Data Aggregation in Wireless Sensor Networks, Proceedings of the Twenty-Second International Conference on Distributed Computing Systems*, Vienna, July 2002, pp. 575–578.

5. E. Cayirci, "Data aggregation and dilution by modulus addressing in wireless sensor networks," *IEEE Communications Letters*, vol. 7, no. 8, pp. 355–357, 2003.

6. C. Intanagonwiwat, R. Govindan, D. Estrin, J. Heidemann, and F. Silva, "Directed diffusion for wireless sensor networking," *IEEE/ACM Transactions on Networking*, vol. 11, no. 1, Feb. 2003.

7. M. Khan, B. Bhargava, S. Agarwal, L. Lilien, and Pankaj, "Self-configuring node clusters, data aggregation and security in micro sensor networks," http://raidlab.cs.purdue.edu/papers/self-config.pdf, 2003.

8. H. Çam, S. Özdemir, D. Muthuavinashiappan, and P. Nair, "Energy-efficient security protocol for wireless sensor networks," in *Proceedings of IEEE VTC Fall 2003 Conference*, Orlando, FL, Oct. 2003, pp. 2981–2984.

9. H. Çam, S. Özdemir, P. Nair, and D. Muthuavinashiappan, "Espda: Energy-efficient and secure pattern-based data aggregation for wireless sensor networks," in *Proceedings of IEEE Sensors— The Second IEEE Conference on Sensors*, Toronto, Canada, Oct. 22–24, 2003, pp. 732–736.

10. W. Heinzelman, A. Chandrakasan, and H. Balakrishnan, "Energy-efficient protocol for wireless micro sensor networks," in *Proceedings of Thirty-Third Hawaii International Conference on System Sciences*, Maui, 2000, pp. 4–7.

11. L. Eschenauer and V. D. Gligor, "A key-management scheme for distributed sensor networks," in *Proceedings of the Ninth ACM conference on Computer and Communications Security*, Washington, DC, Nov. 2002, pp. 41–47.

12. W. Du, J. D. Y. Han, and P. K. Varshney, "A key management scheme for wireless sensor networks using deployment knowledge," in *IEEE INFOCOM'04*, Hong Kong, China, March 7–11, 2004.

13. H. Chan, A. Perrig, and D. Song, "Random key pre-distribution schemes for sensor networks," in *2003 IEEE Symposium on Research in Security and Privacy*, Berkeley, 2003, pp. 197–213.

14. D. Liu and P. Ning, "Establishing pairwise keys in distributed sensor networks," in *Proceedings of the Tenth ACM Conference on Computer and Communications Security*, Washington, DC, Oct. 27–31, 2003, pp. 52–61.

15. X. Zeng, R. Bagrodia, and M. Gerla, "Glomosim: A library for parallel simulation of large scale wireless networks," in *Proceedings of Twelfth Workshop on Parallel and Distributed Simulations*, Banff, Alberta, Canada, May 1998, pp. 154–161.

16. W. Stallings, *Cryptography and Network Security—Principles and Practices*, 3rd ed., Pearson Education, 2003.

17. A. D. Wood and J. A. Stankovic, "Denial of service attacks in sensor networks," *IEEE Computer Magazine*, pp. 54–62, 2002.

18. W. R. Heinzelman, J. Kulik, and H. Balakrishnan, "Adaptive protocols for information dissemination in wireless sensor networks," in *Proceedings of the Fifth Annual ACM/IEEE International Conference on Mobile Computing and Networking*, Seattle, Aug. 1999, pp. 174–185.

19. A. Sinha and A. Chandrakasan, "Joule track: A web based tool for software energy profiling," in *Proceedings of the Thirty-Eigth Conference on Design Automation*, Las Vegas, NV, pp. 220–226.

20. A. Chandrakasan, R. Amirtharajah, S. Cho, and J. Goodman, "Design considerations for distributed microsensor systems," in *Proceedings of IEEE Customs Integrated Circuits Conference (CICC)*, San Diego, CA, May 1999, pp. 279–286.

21. B. Schneier and D. Whiting, "Fast software encryption: Designing encryption algorithms for optimal software speed on the Intel Pentium processor," in *Fast Software Encryption, Fourth International Workshop Proceedings*, Jan.1997, Springer-Verlag, pp. 242–259.

22. I. F. Akyildiz, W. Su, Y. Sankarasubramaniam, and E. Cayirci, "Wireless sensor networks: A survey," *Computer Networks*, vol. 38, pp. 393–422, 2002.

23. A. Mainwaring, J. Polastre, R. Szewczyk, D. Culler, and J. Anderson, "Wireless sensor networks for habitat monitoring," in *Proceedings of ACM Workshop on Wireless Sensor Networks and Applications*, Atlantia, 2002, pp. 88–97.

24. S. K. Das, D. J. Cook, A. Bhattacharya, E. O. H. III, and T. Y. Lin, "The role of prediction algorithms in the mavhome smart home architecture," *IEEE Wireless Communications*, vol. 9, no. 6, pp. 77–84, 2002.

25. G. J. Pottie and W. J. Kaiser, "Wireless integrated network sensors," *Communications of the ACM*, vol. 43, no. 5, pp. 51–58, 2000.

26. S. Tilak, N. B. Abu-Ghazaleh, and W. Heinzelman, "A taxonomy of wireless micro-sensor networks models," *Mobile Computing and Communications Review*, vol. 6, no. 2, pp. 28–36, 2002.

27. S. Lindsey, C. Raghavendra, and K. M. Sivalingam, "Data gathering algorithm in sensor networks using energy metrics," *IEEE Transactions on Parallel and Distributed Systems*, vol. 13, no. 9, pp. 924–932, 2002.

28. W. Choi, P. Shah, and S. K. Das, "A framework for energy-saving data gathering using two-phase clustering in wireless sensor networks," in *Proceedings of the International Conference on Mobile and Ubiquitous Systems: Networking and Services (MobiQuitous)*, Boston, 2004, pp. 203–212.

29. C. Intanagonwiwat, R. Govindan, and D. Estrin, "Directed diffusion: A scalable and robust communication paradigm for sensor networks," in *Proceedings of MOBICOM*, Boston, 2000, pp. 56–67.

30. A. Manjeshwar, Q. A. Zeng, and D. P. Agrawal, "An analytical model for information retrieval in wireless sensor networks using enhanced apteen protocol," *IEEE Transactions on Parallel and Distributed Systems*, vol. 13, no. 12, pp. 1290–1302, 2002.

31. B. Krishnamachari, D. Estrin, and S. Wicker, "The impact of data aggregation in wireless sensor networks," in *Proceedings of the Twenty-Second International Conference on Distributed Computing Systems Workshops*, Vienna, 2002, pp. 575–578.

32. D. Tihan and N. D. Georganas, "A coverage-preserving node scheduling scheme for large wireless sensor networks," in *Proceedings of ACM WSNA*, Atlantia, 2002, pp. 32–41.

33. H. Gupta, S. R. Das, and Q. Gu, "Connected sensor cover: Self-organization of sensor networks for efficient query execution," in *Proceedings of ACM MOBIHOC*, Annapolis, MD, 2003, pp. 189–199.

34. X. Wang, G. Xing, Y. Zhang, C. Lu, R. Pless, and C. Gill, "Integrated coverage and connectivity configuration in wireless sensor networks," in *Proceedings of SenSys*, 2002, pp. 28–39.

35. Y. Sankarasubramaniam, O. B. Akan, and I. F. Akyildiz, "Esrt: Event-to-sink reliable transport in wireless sensor networks," in *Proceedings of MOBIHOC*, Annapolis, MD, 2003, pp. 177–188.

36. W. Choi and S. K. Das, "Trade-off between coverage and data reporting latency for energy-conserving data gathering in wireless sensor networks," in *Proceedings of IEEE International Conference on Mobile Ad Hoc and Sensor Systems (MASS)*, Fort Lauderdale, FL, 2004.

37. F. Garwood, "The variance of the overlap of geometrical figures with reference to a bombing problem," *Journal of Biometrika*, vol. 34, pp. 1–17, 1947.

38. N. Bulusu, J. Heidemann, and D. Estrin, "Gps-less low cost outdoor localization for very small devices," *IEEE Personal Communications Magazine*, vol. 7, no. 5, pp. 28–34, 2000.

39. A. Nasipuri and K. Li, "A directionality based location discovery scheme for wireless sensor networks," in *Proceedings of ACM Workshop on WSNA*, Atlanta, 2002, pp. 105–111.

40. D. Ganesan, R. Govindan, S. Shenker, and D. Estrin, "Highly-resilient, energy-efficient multipath routing in wireless sensor networks," *Mobile Computing and Communications Review*, vol. 5, no. 4, pp. 11–25, 2001.

41. X. Hong, M. Gerla, and H. Wang, "Load balanced, energy-aware communications for mars sensor networks," in *Proceedings of IEEE Aerospace*, vol. 3, Big Sky, MT, 2002, pp. 1109–1115.

42. W. Choi, S. K. Das, and K. Basu, "Angle-based dynamic path construction for route load balancing in wireless sensor networks," in *Proceedings of IEEE Wireless Communications and Networking Conference*, Atlanta, 2004.

43. G. Xing, C. Lu, R. Pless, and Q. Huang, "On greedy geographic routing algorithms in sensing-covered networks," in *Proceedings of ACM Mobile Adhoc Network Symposium (MOBIHOC)*, Tokyo, 2004, pp. 31–42.

44. A. M. Law and W. D. Kelton, *Simulation Modeling and Analysis*, 3rd ed., New York: McGraw-Hill, 2000.

45. P. Juang, H. Oki, Y. Wang, M. M. L. S. Peh, and D. Rubenstein, "Energy-efficient computing for wildlife tracking: Design tradeoffs and early experiences with zebranet," in *Proceedings of the Tenth International Conference on Architectural Support for Programming Languages and Operating Systems*, San Jose, Oct. 2002, pp. 96–107.

46. A. Mainwaring, D. Culler, J. Polastre, R. Szewczyk, and J. Anderson, "Applications and os: Wireless sensor networks for habitat monitoring," in *Proceedings of the First ACM International Workshop on Wireless Sensor Networks and Applications*, Atlanta, Sept. 2002, pp. 88–97.

47. D. Estrin, R. G. J. Heidemann, and S. Kumar, "Next century challenges: Scalable coordination in sensor networks," in *Proceedings of the Fifth ACM/IEEE International Conference on Mobile Computing and Networking (Mobicom)*, Seattle, Aug.1999, pp. 263–270.

48. E. Shih, S. H. Cho, N. Ickes, R. Min, A. S. A. Wang, and A. Chandrakasan, "Physical layer driven protocol and algorithm design for energy-efficient wireless sensor networks," in *Seventh Annual International Conference on Mobile Computing and Networking*, Rome, July 2001, pp. 272–287.

49. P. Bauer, M. Sichitiu, R. Istepanian, and K. Premaratne, "The mobile patient: Wireless distributed sensor networks for patient monitoring and care," in *Proceedings of 2000 IEEE EMBS International Conference on Information Technology Applications in Biomedicine*, Arlington, VA 2000, pp. 17–21.

50. L. Schwiebert, S. Gupta, and J. Weinmann, "Research challenges in wireless networks of biomedical sensors," in *Proceedings of the ACM Seventh Annual International Conference on Mobile Computing and Networking*, Rome, July 2001, pp. 151–165.

51. V. Shankar, "Energy-efficient protocols for wireless communication in biosensor networks," in *Proceedings of the Twelfth IEEE International Symposium on Personal, Indoor and Mobile Radio Communications*, San Diego, Sept. 2001, pp. D.114–D.118.

52. W. Heinzelman, A. Chandrakasan, and H. Balakrishnan, "Energy efficient communication protocols for wireless microsensor networks," in *Proceedings of the Thirty-Third Hawaii International Conference on System Sciences*, Maui, 2000.

53. C. C. Shen, C. Srisathapornphat, and C. Jaikaeo, "Sensor information networking architecture and applications," *IEEE Personal Communications*, vol. 8, no. 4, pp. 52–59, Aug. 2001.

54. Y. Yao and J. Gehrke, "The cougar approach to in-network query processing in sensor networks," *ACM SIGMOD Record*, vol. 31, no. 3, pp. 9–18, Sept. 2002.

55. J. Zhu and S. Papavassiliou, "On the energy-efficient organization and the lifetime of multi-hop sensor networks," *IEEE Communications Letters*, vol. 7, no. 11, pp. 537–539, Nov. 2003.

56. A. Manjeshwar, Q. A. Zeng, and D. P. Agrawal, "An analytical model for information retrieval in wireless sensor networks using enhanced apteen protocol," *IEEE Transactions on Parallel and Distributed Systems*, vol. 13, no. 12, pp. 1290–1302, Dec. 2002.

57. B. Krishnamachari, D. Estrin, and S. Wicker, "The impact of data aggregation in wireless sensor networks," in *Proceedings of IEEE Twenty-Second International Conference on Distributed Computing Systems Workshop*, Vienna, July 2002, pp. 575–578.

58. J. Chou, D. Petrovic, and K. Ramchandran, "A distributed and adaptive signal processing approach to reducing energy consumption in sensor networks," in *Proceedings of INFOCOM*, San Francisco, vol. 2, 2003, pp. 1054–1062.

59. C. Intanagonwiwat, R. Govindan, D. Estrin, J. Heidemann, and F. Silva, "Directed diffusion for wireless sensor networking," *IEEE/ACM Transactions on Networking*, vol. 11, no. 1, pp. 2–16, 2003.

60. C. Intanagonwiwat, D. Estrin, R. Govindan, and J. Heidemann, "Impact of network density on data aggregation in wireless sensor networks," in *Proceedings of the Twenty-Second International Conference on Distributed Computing Systems*, Vienna, 2002, pp. 457–458.

61. S. R. Madden, M. J. Franklin, J. M. Hellerstein, and W. Hong, "Tag: A tiny aggregation service for ad-hoc sensor networks," in *Proceeings of the Fifth Symposium on Operating Systems Design and Implementation*, Boston, Dec. 2002, pp. 131–146.

62. T. He, B. Blum, J. Stankovic, and T. Abdelzaher, "Aida: Adaptive application independent data aggregation in wireless sensor networks," *ACM Transactions on Embedded Computing Systems*, vol. 3, no. 2, pp. 426–457, May 2004.

63. K. Akkaya and M. Younis, "An energy-aware qos routing protocol for wireless sensor networks," in *Proceedings of the Twenty-Third International Conference on Distributed Computing Systems Workshops*, Providence, 2003, pp. 710–715.

64. C. L. Barret et al., "Routing, coverage, and topology control: Parametric probabilistic sensor network routing," in *Proceedings of the Second ACM International Conference on Wireless Sensor Networks and Applications*, San Diego, Sept. 2003, pp. 122–131.

65. W. Heinzelman, J. Kulik, and H.Balakrishnan, "Adaptive protocols for information dissemination in wireless sensor networks," in *Proceedings of Fifth ACM/IEEE International Conference on Mobile Computing and Networking (MOBICOM)*, Seattle, Aug. 1999, pp. 174–185.

66. J. Mirkovic, G. Venkataramani, S. Lu, and L. Zhang, "A self-organizing approach to data forwarding in large-scale sensor networks," in *Proceedings of IEEE International Conference Communications (ICC 2001)*, Helsinki, vol. 5, 2001, pp. 1357–1361.

67. "Ash transceiver designer's guide," http://www.rfm.com/products/data/tr000.pdf.

68. E. Welsh, W. Fish, and J. P. Frantz, "A testbed for low power heterogeneous wireless sensor networks," in *Proceedings of the 2003 International Symposium on Circuits and Systems*, Bangkok, Thailand, vol. 4, May 2003, pp. IV.836–IV.839.

69. T. J. Kwon, M. Gerla, V. K. V. M. Barton, and T. R. Hsing, "Efficient flooding with passive clustering—An overhead-free selective forward mechanism for ad hoc/sensor networks," *Proceedings of the IEEE*, vol. 91, no. 8, pp. 1210–1218, 2003.

70. R. Krishnan and D. Starobinski, "Message-efficient self-organization of wireless sensor networks," in *Proceedings of WCNC 2003*, New Orleans, vol. 3, 2003, pp. 1603–1608.

71. I. S. Gradshteyn and I. M. Ryzhik, *Table of Integrals, Series, and Products*, New York: Academic, 1980.

72. I. F. Akyildiz, W. Su, Y. Sankarasubramaniam, and E. Cayirci, "A survey on sensor networks," *IEEE Communications Magazine*, vol. 40, no. 8, pp. 102–114, Aug. 2002.

73. A. Cerpa, J. Elson, M. Hamilton, and J. Zhao, "Habitat monitoring: Application driver for wireless communications technology," in *2001 ACM SIGCOMM Workshop on Data Communications in Latin America and the Caribbean*, San Diego, Apr. 2001.

74. S. Kumar, D. Shepherd, and F. Zhao, "Collaborative signal and information processing in micro-sensor networks," *IEEE Signal Processing Magazine*, vol. 19, no. 2, pp. 13–14, Mar. 2002.

75. G. J. Pottie and W. J. Kaiser, "Wireless integrated network sensors," *Communications of the ACM*, vol. 43, no. 5, pp. 51–58, May 2000.

76. A. M. Sayeed, "UW-CSP for detection, localization, tracking and classification," www.ece.wisc.edu/ sensit/presentations/csp_apr_15.ppt, Apr. 2001.

77. A. J. Bell and T. J. Sejnowski, "An information-maximisation approach to blind separation and blind deconvolution," *Neural Computation*, vol. 7, no. 6, pp. 1129–1159, 1995.

78. J. Herault and J. Jutten, "Space or time adaptive signal processing by neural network models," in J. S. Denker (Ed.), *Neural Networks for Computing: AIP Conference Proceedings 151*, New York: American Institute of Physics, 1986.

79. P. Comon, "Independent component analysis, a new concept," *Signal Processing*, vol. 36, no. 3, pp. 287–314, Apr. 1994.

80. A. Hyvarinen and E. Oja, "A fast fixed-point algorithm for independent component analysis," *Neural Computation*, vol. 9, pp. 1483–1492, 1997.

81. S. Roberts and R. Everson, (Eds.), *Independent Component Analysis: Principles and Practice*, Cambridge University Press, 2001.

82. S. J. Roberts, "Independent component analysis: Source assessment & separation, a bayesian approach," *IEE Proceedings on Vision, Image, and Signal Processing*, vol. 145, no. 3, pp. 149–154, 1998.

83. S. Richardson and P. J. Green, "On Bayesian analysis of mixtures with an unknown number of components," *Journal of the Royal Statistical Society, Series B*, vol. 59, no. 4, pp. 731–758, 1997.

84. H. Attias, "Inferring parameters and structure of latent variable models by variational Bayes," in *Proceedings of the Fifteenth Conference on Uncertainty in Artificial Intelligence*, Stockholm, 1999, pp. 21–30.

85. X. Wang, H. Qi, and H. Du, "Distributed source number estimation for multiple target detection in sensor networks," in *IEEE Workshop on Statistical Signal Processing*, St. Louis, Sept. 2003.

86. W. R. Heinzelman, A. Chandrakasan, and H. Balakrishnan, "Energy-efficient communication protocol for wireless micro sensor networks," in *Proceedings of the Thirty-Third Annual Hawaii International Conference on System Sciences*, Maui, 2000, pp. 3005–3014.

87. D. F. Shanno, "Conditioning of Quasi-Newton methods for function minimization," *Mathematics of Computation*, vol. 24, pp. 647–656, 1970.

88. W. H. Press, S. A. Teukolsky, W. T. Vetterling, and B. P. Flannery, *Numerical Recipes in C: The Art of Scientific Computing*, 2nd ed., Cambridge University Press, 1992.

89. H. Qi, Y. Xu, and X. Wang, "Mobile-agent-based collaborative signal and information processing in sensor networks," *Proceedings of IEEE*, Special Issue on Sensor Networks and Applications, vol. 91, no. 8, pp. 1172–1183, Aug. 2003.

8

NETWORK SECURITY

8.1 INTRODUCTION

Sensor networks are expected to support a wide variety of applications, many of which will have at least some requirements for security. For example, some applications may only request that data be authenticated to ensure that is has not been modified on its path from a source to a sink. Others may further require that the source of the data be authenticated to ensure that the data is coming from a valid and trusted source. Still others may require that data be encrypted so that it is kept private from potential intruders. The nature of sensor networks, the lack of predeployed infrastructure, the and reliance on nodes limited in terms of processing resources and power makes it extremely challenging to create a secure environment.

Cryptographic techniques for authentication and encryption can be done in two ways using public keys or private keys. When using public keys, the key value of every node is public information and is, therefore, known by all other nodes. When wishing to communicate privately with a node, the source node simply encrypts data using the public key of the sink node. Due to properties of public key algorithms, only the sink node can correctly decrypt the data. This method uses so-called asymmetric keys because the two communicating nodes use different keys during the session. When using private keys, nodes must first agree on a key before they can communicate securely. One possibility is to use public keys to encrypt information from which private keys can be derived. Private key algorithms use so-called symmetric keys because both communicating nodes use the same keys for encrypting and decrypting data.

In wired data networks, nodes rely on predeployed trusted servers to help establish trust relationships. For public key algorithms, certificate authorities (CAs) are used to verify that a specific public key belongs to a specific destination. The use of CAs prevents nodes from impersonating another node by using a false public key. For private key systems, key distribution centers (KDCs) are sometimes used to securely disseminate keys to nodes

Sensor Network Operations, Edited by Phoha, LaPorta, and Griffin
Copyright © 2006 The Institute of Electrical and Electronics Engineers, Inc.

establishing a secure session. In wireless sensor networks, these trusted authorities do not exist, and hence new methods of establishing trust relationships must be developed. In addition, because sensor nodes have limited memory and central processing unit (CPU) power, cryptographic algorithms must be selected carefully. For instance, public key algorithms require intense processing and, therefore, cannot be used in sensor networks for performance reasons. Likewise, complex algorithms that use long keys may not be useful because they cannot execute on memory-limited devices.

One promising line of research for security in sensor networks has been to predeploy keys in sensor nodes. Care must be taken in how these keys are distributed because if a node is compromised the impact on the network must be minimized. Several research efforts propose clever ways to distributed key information so that networks are robust against compromised nodes, and so that additional key information, for example, group keys, may be derived by the nodes once they are in the field.

In this chapter we present five sections on security in sensor networks. In Section 8.2, by Hodjat and Verbauwhede, the power consumption of two different cryptographic techniques are explored. This discussion points out the necessity of choosing algorithms that are suitable for low-power devices. In Section 8.3, by Di Pietro, Mancini, and Mei, a random key predeployment scheme is proposed to address issues of authentication and confidentiality. In Section 8.4, by Weimerskirch, Westhoff, Lucks, and Zenner, the authors propose to use hash chains to prove authenticity. They then show how, with some additional network infrastructure, identities can be proven. Section 8.5, by Dimitriou, Krontiris, and Nikakis, proposes a nonprobabilistic method of deploying keys in a sensor network. This method organizes nodes into clusters to assist in the generation of keys. Section 8.6, by Lee, Wen, and Hwang, presents a method of generating group keys using partial key disclosure by groups of nodes.

8.2 ENERGY COST OF EMBEDDED SECURITY FOR WIRELESS SENSOR NETWORKS

Alireza Hodjat and Ingrid Verbauwhede

The energy consumption of cryptographic algorithms and security protocols is a crucial factor in wireless networks. Unfortunately, the energy cost of embedded security is only considered as an afterthought. This work presents the energy cost of two well-known public key and secret key cryptographic algorithms when executed on a typical wireless sensor node. It also presents the energy consumption of a key agreement process between two parties that communicate over wireless links. The elliptic curve public key and Rijndael Advanced Encryption Standard (AES) secret key algorithms are chosen to explore the energy cost of a basic Diffie–Hellman and a Kerberos key agreement protocol on a Wireless Integrated Network Sensor (WINS) sensor network. The results show that the total energy cost, including computation and communication energy, of a Diffie–Hellman key agreement process using the elliptic curve public key is between one to two orders of magnitude larger than the key exchange process based on the AES secret key algorithm.

8.2.1 Background Information

Almost every embedded application requires security. This is also the case for wireless sensor networks, where eavesdroppers can easily intercept the transmitted data or introduce false data in the air. Confidentiality and authentication are both requirements of secure data transmission. Public key and private key cryptography algorithms are used to provide data confidentiality, authentication, and key exchange. In the case of wireless sensor networks, energy consumption is a critical design issue as well. As a result, the energy cost of public key and private key cryptography algorithms is an important factor for the implementation of security protocols on wireless sensor nodes.

For low-energy implementation, it is desired that both the hardware and the embedded software consume an extremely low energy budget. Yet, the energy consumption of the embedded software implementation of the cryptography algorithms is not usually considered at design time. This section presents the energy cost of two well-known public key and secret key cryptography algorithms. Moreover, these algorithms are evaluated in the context of protocols.

Key setup protocols are either based on public key algorithms or secret key algorithms. Public key algorithms can be used for a direct node-to-node communication (assuming the public keys are authenticated) in an ad hoc fashion. Secret key algorithms require the use of a trusted central server. It requires that all nodes are within radio range and share a secret key with a trusted base station. The energy cost of these two is compared in this work.

The public key encryption algorithm chosen for this project is the elliptic curve cryptography algorithm based on the IEEE P1363/D1 standard [1]. The secret key encryption algorithm is the Rijndael AES [2]. Using these underlying encryption algorithms and their energy costs, the energy consumption of the Diffie–Hellman and Kerberos key agreement protocols are compared. The basic version of the Diffie–Hellman key agreement protocol is adapted from protocol 12.47 of [4]. We base it on the elliptic curve algorithm because it is the most promising public key algorithm in terms of low power. The second protocol is the key agreement part of the Kerberos protocol [3] listed as protocol 12.24 in [4]. It assumes the availability of a central trusted server, and it uses a secret key algorithm, in this experiment the AES algorithm.

The platform used for this project is the Wireless Integrated Network Sensor. WINS is a StrongARM-based wireless sensor network developed at Rockwell Scientific [5]. In a first step, the energy consumption for Rijndael AES and the elliptic curve encryption algorithms is measured for different data and key lengths. In the next step we explore the energy cost of a public key and a secret key based key exchange protocol and examine their suitability for low-power wireless networks. In the implementation of the whole security protocol, the cost of communication between nodes must be considered as well. Therefore, the whole energy cost consists of the energy consumption of computation and communication. The number of packets exchanged between parties and the energy cost of radio transmission influences the overall communication cost, and the type of encryption algorithm affects the energy cost of computation in each protocol.

8.2.2 Energy Result for AES and ECC

In this section we explore the energy cost for the underlying encryption algorithms—Rijndael AES [6] and elliptic curve cryptography (ECC) [1, 7]. AES is the secret key encryption standard most recently accepted by the National Institute of Standards and Technology (NIST) [7]. Other choices for secret key techniques are DES and RC4. AES was chosen due to the fact that it is more secure and more complex compared to DES and RC4. Moreover, it is useful in exploring the upper bound of energy for secret key applications. As for public key algorithms, the choices are RSA and ECC. ECC has recently gained prominence among public key algorithms. Furthermore, RSA requires much larger word lengths (at least a modulus of 1024 bits) and will require much more energy than ECC and, hence, is not suitable for energy constraint devices. Also, it should be noticed that all the energy numbers that are reported in the figures of this section are derived from the actual measurement of the WINS node's supply voltage, the current drained by the node, and the time that it takes to run the security algorithms.

8.2.2.1 Rijndael Secret Key Algorithm   This algorithm consists of key scheduling, encryption, and decryption primitives [2]. Key scheduling produces a long array of subkeys by key expansion and round key selection routines. This long array of subkeys is used in each round of the encryption or decryption phase to be added to the data. Depending on the key and data size, encryption and decryption algorithms are repeated between 10 to 14 rounds. In the encryption primitive, byte substitution transformation, shift row transformation, mix column transformation, and round key addition are performed [2]. In decryption, the steps are the same as those performed in encryption. However, they are called in reverse order and the substitute table, shift indexes, and fixed polynomials used in the decryption steps are the reverse of those used in encryption.

Figure 8.1 shows the energy consumption for the three different Rijndael parts (key scheduling, encryption, and decryption) on the WINS node. In the original Rijndael algorithm, the input data and key blocks range from 128, 192, to 256 bits, and the output has the corresponding size. Energy for all nine different possible combinations is shown. The energy consumption for key scheduling varies from 110.2 to 315.4 μJ depending on the choice of data and key size. For encryption the variation is from 197.6 to 535.8 μJ and for decryption from 247 to 695.4 μJ. On average, energy consumption for 256-bit data and key is about 2.8 times more than the case where data and key are 128-bit size. The variations in the energy consumption in Figure 8.1 depend on different factors. The most important factor that defines these variations is the number of rounds the algorithm is repeated. The number of rounds for different cases is also shown in Figure 8.1. The key scheduling part differs for a different number of rounds because the key scheduler has to generate one subkey for each round. Each subkey is the size of the data. Hence for larger data sizes and larger number of rounds, the key scheduling algorithm will consume more energy. One sees the same trend in the encryption and decryption algorithms: That is, the energy goes up with the number of

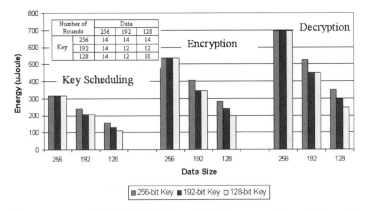

Figure 8.1   The energy cost of Rijndael AES for key scheduling, encryption, and decryption algorithms.

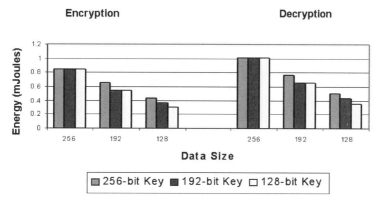

Figure 8.2 The energy cost of Rijndael AES algorithm (key scheduling included in encryption or decryption routines).

rounds and the data size. The main difference between encryption and decryption algorithm is in the individual blocks that make up the algorithm. Most important is the difference in the mix-column operation. This operation is much more energy efficient in the encryption algorithm than the decryption algorithm. This difference is also described by the authors of the Rijndael algorithm [2].

Figure 8.2 shows the energy consumption of the Rijndael AES encryption and decryption algorithms when the key-scheduling algorithm is executed before the encryption and decryption routines. This is usually the case when a node wants to use the AES algorithm in the cryptographic and key agreement protocols. Therefore, the energy cost of making a key schedule and performing the encryption and/or decryption algorithm is measured together. The energy consumption for encryption and key scheduling varies from 0.31 to 0.85 mJ depending on the choice of data and key size. For decryption and key scheduling the variation is from 0.36 to 1.01 mJ. The energy consumption of decryption is 30% more than encryption. The main reason for this difference is the number of shifts performed in the shift row routine and the larger $GF(2^8)$ elements used in mix-column transformation routine.

8.2.2.2 Elliptic Curve Public Key Algorithm

For the elliptic curve public key algorithm, we use the efficient double–add–subtract point multiplication algorithm defined by the IEEE P1363/D1 standard [1]. Elliptic curve point multiplication refers to calculating kP where k is an integer and P is a point on the elliptic curve. The theory of the elliptic curve cryptography is based on the mathematical mapping of an elliptic curve on a Galois field. The elliptic curve points on $GF(2^n)$ and the point in infinity form a group with a specific addition operation [1]. With this definition, kP is equivalent to adding P to itself k times by the group operation.

Figure 8.3 shows the hierarchy of operations in calculating the point multiplication for this implementation. At the highest level there is a point multiplication algorithm to calculate kP using the initial elliptic curve point P and the secret integer key k. This algorithm uses double, add, and subtract operations of the elliptic curve group and uses k and $h = 3k$ (is part of the calculations) to decide the number of times double, add, or subtract operations are required. In the level below, there are double, add, and subtract operations defined using the Galois field operations. These operations are specified exactly in the algorithm specification of the projective coordinates for $GF(2^n)$ following the standard steps and are based on

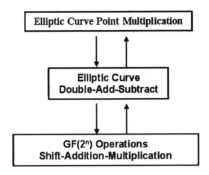

Figure 8.3   Elliptic Curve Implementation Hierarchy.

the multiplication and addition operations of the GF(2^n) field. The lowest level is the implementation of the multiplication and addition operation of GF(2^n). Addition and subtract are just the simple bitwise XOR of the field elements. Multiplication is implemented using shift and XOR of the inputs in an iterative manner to calculate the multiplication of the two polynomials that are the representations of the GF(2^n) elements (n is equal to 128, 192, or 256).

Figure 8.4 shows the pseudocode of the double–add–subtract algorithm. In this algorithm each bit position of k and h are compared. First, the leftmost nonzero bit of h is found. Then from its next bit position onwards, down to the second least significant bit position of h, the comparison repeats. In each iteration one double is performed. In addition, if $h_i = 1$ and $k_i = 0$ one add is performed and if $h_i = 0$ and $k_i = 1$ one subtract is performed. To calculate the lower bound of energy, there should be only doubles in the point multiplication calculation. An example is $k = 1,000,000$ and $h = 11,000,000$. Similarly, to calculate the upper bound, there should be a full number of doubles and the maximum number of additions and/or subtractions. The result shows that the energy consumption of using add and subtract are identical. Also from evaluating the pseudocode, one can derive that the maximum number of adds and/or subtracts that might be called in n-bit point multiplication is $n/2$. So the upper limit of point multiplication is performed with full double range and $n/2$ add/subtract range.

Figures 8.5, 8.6, and 8.7 show the energy consumption of a range of point multiplications for key lengths of 128, 192, and 256 bits, respectively. The execution time and the energy

Input: An integer k and an Elliptic Curve point P
Output: The Elliptic Curve point $k P$

1. Calculate $h = 3k$
2. Let $h_l \, h_{l-1} \, ... \, h_1 \, h_0$ be the binary representation of h
3. Let $k_l \, k_{l-1} \, ... \, k_1 \, k_0$ be the binary representation of k
4. Set $S = P$
5. For i from $l-1$ downto 1 do
    Set $S =$ Elliptic Curve Double Operation (S)
    If $h_i = 1$ & $k_i = 0$ then
        Compute $S =$ Elliptic Curve Add Operation (S, P)
    If $h_i = 0$ & $k_i = 1$ then
        Compute $S =$ Elliptic Curve Subtract Operation (S, P)
6. Output S

Figure 8.4   Pseudo code for the Double-Add-Subtract point multiplication algorithm.

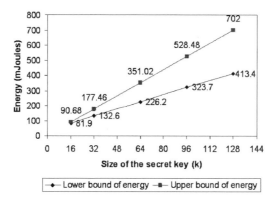

Figure 8.5   The energy cost of point multiplication for 128-bit Galois Field.

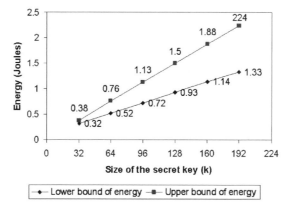

Figure 8.6   The energy cost of point multiplication for 192-bit Galois Field.

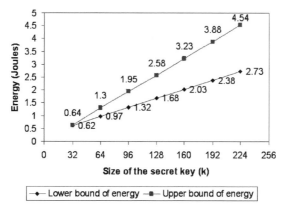

Figure 8.7   The energy cost of point multiplication for 256-bit Galois Field.

for the ECC algorithm strictly depend on the bit pattern of the initial random key k. Therefore, the upper bound and lower bound of energy was measured. In these figures the X axis shows the size of the initial random key k of point multiplication. This is identical to the number of doubles for each experiment. Also, in the case of the upper bound, this axis shows twice the number of adds and/or subtracts. The Y axis shows the energy consumption in terms of mjoules or joules. For instance, if the size of k is 64 for 128-bit point multiplication (Fig. 8.5), then the energy consumption will be between 226.2 and 351.02 mJ depending on the 0 and 1 pattern in k. In this case 64 doubles are performed for the lower limit, and 64 doubles and 32 adds or subtracts are performed for the upper limit.

The results show that the energy consumption of one elliptic curve point multiplication for the 128-bit Galois field varies from 0.082 to 0.702 J. For the 192-bit Galois field the variation is between 0.32 to 2.24 J. In the case of the 256-bit Galois field the energy cost range is between 0.62 to 5.19 J. The average energy cost is 0.30, 1.07, and 2.34 J for the 128-, 192-, and 256-bit key, respectively. All of the above measurements are done on StrongARM-based WINS node running at 133 MHz.

8.2.3 Communication Costs

Radio transmission and reception can cost a lot of energy. In this discussion, we assume that the communication nodes use a radio that is suited for the relative short distance, multihopping sensor network. The power consumption numbers used in this discussion are for Rockwell's WINS node [5], which has a radio module from Conexant Systems. The power numbers for different radio transmission modes are taken from [8].

The energy per transmitted bit depends on the packet size, the header size, a fixed overhead associated with the startup of the radio, the symbol rate, and the modulation scheme [9]. Assuming that the radio is well tuned for low-power operation, it will consume between 5 and 20 μJ per transmitted bit. Reference [10] presents the power consumption of the radio transmission on WINS nodes. Radio's power consumption varies from 396 to 711 mW depending on the transmission power level. This corresponds to a consumption of 771 to 1080 mW for the node. The power consumption of the receive mode is 376 mW for the radio and 751 mW for the node. All these numbers are at a transmission rate of 100 kbits/s.

8.2.4 Key Exchange Energy Consumption

Diffie–Hellman and the basic Kerberos are two solutions for the key distribution and key agreement problem [3, 4]. The Diffie–Hellman key agreement protocol is implemented using a public key encryption algorithm. It is applicable to ad hoc networks due to the fact that it does not require a trusted third party in the key agreement process. Using this protocol, any two nodes in an ad hoc network can agree on their common session key without any trusted node or secure link. In our application, we assume that two nodes that are in the range of each other's radio reception will start a secure communication. On the other hand, basic Kerberos is a key distribution protocol that is based on a secret key encryption technique. It requires that the sensor nodes are within the radio frequency range of a base station and that they share predeployed secret keys with the base station. In the following paragraphs these protocols are described and their energy consumption is explored. It is assumed that the two nodes performing the key agreement process are called *Alice* and *Bob*, and if needed, their trusted party is called *Trent*.

Summary: A and B each send the other one message over and open channel
Result: Shared secret key K known to both parties A and B

- One time set-up : An appropriate prime p and generator α ($2 \leq \alpha \leq p-1$)
- Protocol messages:

$$A \rightarrow B : \alpha^x \bmod p \quad (1)$$
$$A \leftarrow B : \alpha^y \bmod p \quad (2)$$

- Protocol actions:
    1. A chooses a random secret x ($2 \leq x \leq p-1$), and sends B message (1)
    2. B chooses a random secret y ($2 \leq y \leq p-1$), and sends A message (2)
    3. B receives α^x and computes the shared secret key $K = (\alpha^x)^y \bmod p$
    4. A receives α^y and computes the shared secret key $K = (\alpha^y)^x \bmod p$

Figure 8.8 Basic version of the Diffie–Hellman key agreement protocol.

Figure 8.8 shows the basic version of the Diffie–Hellman key agreement protocol, which is from [4]. In the Diffie–Hellman protocol based on elliptic curve point multiplication, there is a common elliptic curve and a specific point on it that is publicly known. Alice and Bob each generate a random initial key, called a and b. They apply point multiplication on the common elliptic curve point P. They exchange their results. Then, they apply point-multiplication on the received data (aP or bP) with their own random initial key again and generate abP. This is their common session key. Both can generate abP, but no one else can generate it without knowing a and b. It can be shown that it is computationally infeasible to generate abP from aP, bP, and P.

Figure 8.9 shows the complete protocol. In this protocol Alice does two elliptic curve point multiplications (calculating aP and abP), once transmits her results to Bob (aP), and once listens to receive Bob's results (bP). Bob also follows the same process. Therefore, this protocol requires four elliptic curve point multiplications as the computation part and two data transmissions and two data receptions for the communication part.

Figure 8.10 shows the energy exploration of the Diffie–Hellman protocol using elliptic curve point multiplication. A header of 256 bits wide is used for each data transmission and the data is either 128, 192, or 256 bits wide. The average energy consumption of point multiplication is used in each case based on the results of Section 8.2.2. Transmission is at the rate of 100 kbits/s, and the energy is shown for different transmission power levels. The radio transmission power level varies form 0.16 to 36.3 mW. As shown in Figure 8.10,

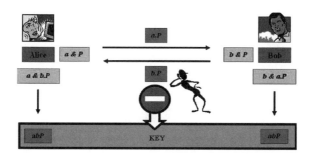

Figure 8.9 Diffie–Hellman key agreement protocol using Elliptic Curve point multiplication.

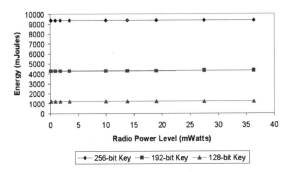

Figure 8.10 The energy cost of one Diffie–Hellman key agreement based on the Elliptic Curve Cryptography algorithm.

the average energy cost of one Diffie–Hellman key agreement protocol in an ad hoc network is 1212.4, 4294.4, and 9376.5 mJ for the key lengths of 128, 192, and 256 bits, respectively. The energy consumption is completely dominated by the computation energy. The communication energy for sending a 512-bit packet is only around 5.12 μJ [10].

In the basic Kerberos protocol based on the AES secret key encryption technique Alice and Bob agree on a common session key with the help of their trusted third party, Trent. Here, we focus on the basic Kerberos key agreement protocol using secret key encryption technique [3] and [4]. We do not discuss the Kerberos network authentication protocol that is used in client/server application. Figure 8.11 shows the basic version of the Kerberos key agreement protocol, which is from [4].

Alice and Bob want to agree on a common session key. At the setup, both have a secret key to communicate securely with Trent. For key agreement, one of them, suppose Alice, sends a message to Trent asking for a secret session key. This message includes Alice's and Bob's identities. Trent generates the random session key K and encrypts it along with time stamp T and lifetime L and Alice or Bob's identity separately. The message for Alice is encrypted with Alice's secret key and Bob's identity and vice versa. The results are $E_A(K, B, T, L)$

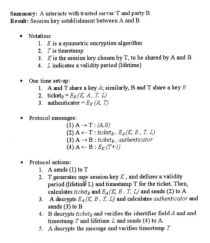

Figure 8.11 Basic version of the Kerberos key agreement protocol.

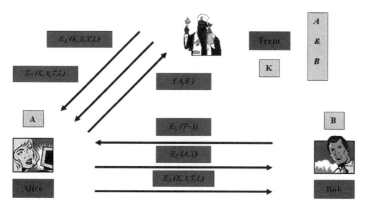

Figure 8.12   Kerberos key agreement protocol using a secret-key encryption algorithm.

and $E_B(K, A, T, L)$. These messages are sent to Alice. Now Alice can decrypt her received message to recover K. Then she makes a message including her identity and time stamp T and encrypts it with $K[E_K(A, T)]$. She sends both $E_B(K, A, T, L)$ and $E_K(A, T)$ to Bob. Now Bob can decrypt the received messages and recover K. He also verifies Alice's identity and then forms a message with $T + 1$ and encrypts it with K and sends it again to Alice. Alice decrypts this message and verifies the timestamp T. Now Alice and Bob have a secret common shared key K. This process is shown in Figure 8.12.

In this protocol there are four data encryptions on the transmitted data and four decryptions on the received data. The total number of transmissions and receptions is six each. Figure 8.13 shows the energy consumption of this protocol using the Rijndael AES secret key algorithm for different transmission power levels. Similar to the Diffie–Hellman experiment, the radio transmission power level varies form 0.16 to 36.3 mW. The packets of size 1 kbit include the header, Alice and Bob's identity, the timestamp, the lifetime, and the generated random session key K, and K can be 128, 192, or 256 bits wide. The three curves show the energy cost for the key agreement process when the Rijndael encryption/decryption algorithm is used for block lengths of 128, 192, and 256 bits. As mentioned in Section 8.2.2, in the case of 128-bit data block, the maximum energy cost is 0.85 and

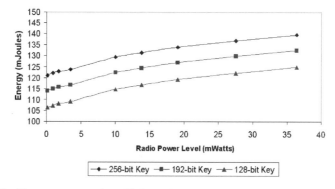

Figure 8.13   The energy cost of one Kerberos key agreement based on the Rijndael algorithm.

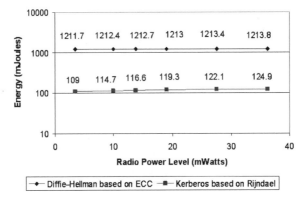

Figure 8.14 The energy comparison of the key agreement protocol for 128-bit key between Diffie–Hellman and Kerberos Protocols.

1.01 mJ for encryption and decryption. Similarly, these numbers are 0.65 and 0.77 mJ for 192-bit data block and 0.44 and 0.5 mJ for 256-bit data block. Therefore, when the Kerberos key agreement protocol is used for a 128-bit key, the energy consumption varies from 106.4 to 124.9 mJ. This variation is between 114.1 and 132.6 mJ and between 121.1 and 139.6 mJ for 192- and 256-bit key lengths, respectively. The total energy consumed varies with the radio transmission power. The fraction of the energy associated with communication varies from 75% for the lowest radio transmission level to 78% to the highest transmission level. This shows that for this protocol the energy cost of the communication is about three times more than the energy cost of computation.

Using the results shown in Figures 8.10 and 8.13, the comparison is done for the key agreement process for the key lengths of 128-, 192-, and 256-bit keys. Figures 8.14, 8.15, and 8.16 show the energy comparison of the Diffie–Hellman key agreement protocol based on the elliptic curve public key algorithm and the Kerberos key agreement protocol based on the Rijndael secret key algorithm for the key length of 128, 192, and 256 bits, respectively.

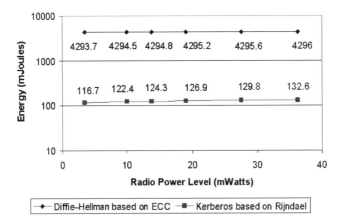

Figure 8.15 The energy comparison of the key agreement protocol for 192-bit key between Diffie-Hellman and Kerberos Protocols.

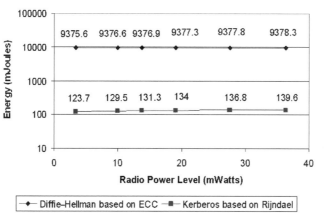

Figure 8.16   The energy comparison of the key agreement protocol for 256-bit key between Diffie–Hellman and Kerberos Protocols.

As shown in these figures, the energy difference is between one to two orders of magnitude depending on the key length.

8.2.5   Related Work

To our knowledge, a detailed energy measurement of neither the ECC algorithm nor the AES algorithm on the StrongARM processor have been published. In related work, the authors of the technical report [12] have made estimates for the modular exponentiation algorithm on several processors used in embedded devices and sensor nodes. The StrongARM processor showed the lowest energy of the set of processors they considered. In [11], the Diffie–Hellman with modular exponentiation needs almost 15 mJ for a modulus of 1024 bits and an exponent of 256 bits on the StrongARM processor. These results are obtained by measuring the number of clock cycles to perform one 128-bit multiplication operation, multiplied by an estimated number of 128-bit multiplications in the algorithm. To obtain similar computational security, the ECC algorithm requires much smaller word lengths: RSA with a 1024-bit modulus has a computational security equivalent to 163 bits of ECC [12].

To obtain an estimate for the AES algorithm, an estimated execution time of 1200 clock cycles is taken for a 128-bit data block, 128-bit key execution of an "average" AES finalist on the StrongARM processor. This results in 2 μJ for one 128-bit block encryption [11]. This is two orders of magnitude lower than our accurate measurement of the Rijndael AES implementation because our result is the energy cost of the sensor node, which includes the energy consumption of the memory, the processor board, and the power supply board in addition to the StrongARM processor.

Several studies describe the energy consumption of hardware/software or ASIC implementations of ECC [13, 14] and AES. Reference [15] shows several orders of magnitude difference between ASIC, assembly code, compiled code, and compiled Java code.

In terms of protocols, different approaches are proposed for security of wireless sensor networks. However, there has not been any report on the energy cost of implementation for the proposed security protocols. SPINS [16] is one of the first studies that proposed the security protocols for wireless sensor networks. SNEP and μTESLA are the two protocols that are proposed for confidentiality, two-party data authentication, integrity, and authentication for data broadcasting. In the key negotiation protocol of SPINS, two nodes can establish a shared session key with the help of a trusted third party. This is similar to the basic Kerberos key agreement process using a secret key cryptography algorithm that was presented in this discussion. On the other hand, Broadcast Session Key (BROSK) [17] is a fully ad hoc key negotiation protocol that constructs link-dependent keys by broadcasting key negotiation messages. This is similar to the Diffie–Hellman key agreement protocol using a public key cryptography algorithm presented earlier in this discussion.

Another related work is [18] by Pottie and Kaiser, which presents the WINS nodes in general. Although this work is not in the field of security or cryptography, it explores the computation–communication trade-off of the energy consumption of the WINS nodes.

8.2.6 Conclusions

Our measurements on WINS nodes show an energy cost of maximum 1 mJ for the AES symmetric-key cryptography algorithm. For the elliptic curve public key cryptography algorithm, the average energy consumption of one point multiplication varies between 300 and 2340 mJ. This cost is between two to three orders of magnitude larger than the AES algorithm. This ratio influences the energy cost of the key agreement protocol. Diffie–Hellman key agreement protocol using elliptic curve point multiplication in a typical ad hoc network costs between 1211 and 9378 mJ. On the other hand it costs less than 140 mJ to exchange keys with an AES-based Kerberos key distribution protocol. This shows that on WINS nodes, the energy cost of a key agreement process in an ad hoc network using a public key encryption technique such as ECC is between one to two orders of magnitude larger than key agreement in regular networks with secret key encryption algorithms such as AES. Therefore, providing security for ad hoc wireless networks with public key algorithms is not only harder than traditional networks based on secret key techniques, but also it might cost between 10 and 100 times more energy. Only when the radio transmission levels become very large, such that the energy per bit transmitted is around 1 mJ/bit, does it make sense to use public key algorithms. This is, however, not suited for sensor networks since the designers of these networks use multihop strategies to avoid the long-distance power-hungry radios. More information on low power embedded software can be found in [19, 20, 21].

This energy evaluation study has shown the need for developing new key distribution protocols specific for distributed ad hoc network applications, such as sensor networks, that address the resource limitations of sensor nodes and that are scalable.

8.3 INCREASING AUTHENTICATION AND COMMUNICATION CONFIDENTIALITY IN WIRELESS SENSOR NETWORKS

Roberto Di Pietro, Luigi V. Mancini, and Alessandro Mei

A distributed wireless sensor network (WSN) is a collection of low-end devices with wireless message exchange capabilities. Due to the scarcity of hardware resources, the lack of network infrastructures, and the threats to security, implementing secure pairwise communications among any pair of sensors is a challenging problem in distributed WSNs.

In particular, memory and energy consumption as well as resilience to sensor physical compromise are the most stringent requirements. In this section, we highlight the fact that pseudorandom key distribution enforces sensor-to-sensor authentication, besides confidentiality. Next, we present a multiparty cooperative protocol to amplify all the security properties provided by the pseudorandom key distribution strategy. The results show that the cooperative protocol (1) requires a limited overhead and it is incurred only once, during the channel setup, (2) is dynamically tunable depending on the level of threat in WSN, and (3) can be combined with any of the pseudorandom key predeployment schemes. We provide extensive simulations of the security properties of the cooperative protocol.

8.3.1 Background Information

A WSN is a collection of sensors whose size can range from a few hundred sensors to a few hundred thousand or possibly more. The sensors do not rely on any predeployed network architecture, thus they communicate via an ad hoc wireless network. The power supply of each individual sensor is provided by a battery, whose consumption for both communication and computation activities must be optimized. Distributed in irregular patterns across remote and often hostile environments, sensors should autonomously aggregate into collaborative, peer-to-peer networks. Sensor networks must be robust and survivable to overcome individual sensor failure and intermittent connectivity (due, e.g., to a noisy channel or a shadow zone).

Establishing secure pairwise communications could be useful for many applications. In particular, it is a prerequisite for the implementation of secure routing, and can be useful for secure group communications. The key establishment problem has been widely studied in general network environments. There are three types of general key establishment schemes: *trusted-server* schemes, *self-enforcing* schemes, and *key predeployment* (or key predistribution) schemes. Trusted-server schemes depend on a trusted server for key agreement among nodes, for example, Kerberos [22]. This type of scheme is not suitable for sensor networks because of the lack of trusted infrastructures. Self-enforcing schemes rely on asymmetric cryptography, such as the use of public key certificates for home banking transactions. However, limited computation and energy resources of low-end devices such as sensors often make it undesirable to use public key algorithms, as is pointed out in [23,24]. In key predistribution schemes, the keys are distributed to all sensors before deployment on the ground. If we knew which sensors would be in the same neighborhood before sensor deployment, keys could be assigned to sensors a priori. However, such a priori knowledge often does not exist.

Yet, a number of key predistribution schemes that do not rely on a priori sensor deployment knowledge do exist [25, 26]. A naïve solution is to let all the sensors carry a master secret key. Any pair of sensors can use this global master secret key to achieve key agreement and obtain a new pairwise key. This scheme does not exhibit desirable network resilience: If one sensor is compromised, the security of the whole network is compromised. Some existing studies suggest storing the master key in tamper-resistant memory to reduce the risk [27,28]. However, a few drawbacks do exist: There is an increase in per-sensor cost and in the energy consumption required to manage tamper-resistant modules; tamper-resistant hardware might not always actually be safe [29]. A second, simple key predistribution scheme for implementing secure pairwise communications is to assign a set of secret keys to each sensor, each key of the set being shared with only one other sensor. This solution requires each member to store $n - 1$ keys, where n is the size of the group, with $n(n - 1)/2$ different keys in the group. In this case, the resilience is perfect since a compromised node does not affect the security of other nodes. However, this scheme is impractical for sensors as n can be very large and memory requirements do not scale. Moreover, adding new sensors to a preexisting sensor network is not easy. To overcome the above-mentioned limitations, random key predeployment schemes have recently been proposed [30–35] and are listed in detail in Section 8.3.2.

This section deals with confidentiality and authentication in WSN. In particular, we point out that pseudorandom key predeployment schemes, besides confidentiality, can also assure node-to-node authentication. Starting from this observation, we introduce the cooperative protocol that amplifies the security properties of confidentiality and authentication. The cooperative protocol enables us to trade-off the desired level of confidentiality and authentication assurance with a moderate increase in communication overhead. Note that the cooperative protocol (1) requires a limited overhead and this overhead is incurred only once, during the channel setup, (2) is dynamically tunable depending on the level of threat in WSN, and (3) can be combined with any of the pseudorandom key predeployment schemes.

8.3.2 Preliminaries and Related Work

We assume a WSN is composed of n sensors. A sensor is denoted by s. To address a specific sensor, we write s_a, where a is the sensor's ID. Each sensor's memory can hold m symmetric encryption keys. The center assigns keys to sensors following a random key predeployment strategy:

1. A *pool* P of random keys $\{k_1^P, \ldots, k_{|P|}^P\}$ is generated.
2. For each sensor s_a in the WSN, a set $M_a = \{k_1^a, \ldots, k_m^a\}$ (called *key ring*) of m distinct keys is randomly drawn from the pool and assigned to sensor s_a.

When two sensors want to communicate securely, they must first find out which keys of the pool they share by executing a *key discovery phase*. Then, they compute a common key as a function of the shared keys. The key is used to secure the channel by using a symmetric key encryption algorithm. $E_k(\cdot)$ denotes the encryption with key k and $E_k^{-1}(\cdot)$ denotes the decryption with key k.

The above-stated idea of probabilistic key sharing for WSN is introduced by Eschenauer and Gligor [30]. The authors also provide a simple and centralized algorithm for rekeying in a distributed WSN. Later, in [31], three mechanisms are described in the framework of random key predistribution. First of all, the *q-composite* random key predistribution scheme, a modification of the basic scheme in [30], achieves better security under small-scale attack while trading off increased vulnerability in the face of a large scale physical attack on the network sensors. Second, the multipath key reinforcement protocol substantially increases the security of the channel by leveraging the security of other links. Lastly, the random pairwise key scheme assigns private pairwise keys to randomly selected pairs of sensors so as to guarantee that the rest of the network remains fully secure even when some of the sensors have been compromised. Moreover, this latter scheme supports node-to-node authentication. However, enforcing the random pairwise key scheme requires a much larger number of keys in each sensor's memory. Indeed, to guarantee sensor-to-sensor visibility with probability p, each sensor must carry np keys. Chan et al. [31] claim that p may grow slowly with n when n is large, depending on the model of connectivity. Still, this means that memory requirement is $\Omega(n)$, which is too large to make the scheme scalable.

In [36], the main idea is that the set of keys assigned to each sensor is computed by using a pseudorandom generator with the ID of the sensor as the seed. This way it is possible for any sensor to know the set of keys stored by any other sensor. This feature supports a very energy-efficient key discovery phase consisting of local computations alone and no message exchange. The protocol admits two operations: group rekeying and threshold-based sensor revocation. In [35], it is shown that the protocol in [36] is weak against the novel smart attacker model.

Two recent schemes build up a secure pairwise channel that combines a deterministic technique with a predistribution random scheme. The first scheme is proposed in [32]. The authors use a deterministic protocol proposed by Blom [37] that allows any pair of nodes in a network to find a pairwise secret key. As a salient feature, Blom's scheme guarantees a so-called λ-*secure property*: As long as no more than λ nodes are compromised, the network is perfectly secure. A λ-secure data structure built this way is called a key space. The authors in [32] create a set $\mathcal{W}$ composed of ω key spaces, and randomly assign up to τ spaces per sensor. Two nodes can find a common secret key if they have picked a common key space. This phase, called the key agreement phase, is carried out by making each node disclose some information about the key spaces it has been assigned, thus either disclosing the indexes of the spaces each node carries or using the challenge response technique in [31]. In the former case, the scheme shows the same weaknesses as [36] against the smart attacker, while, in the latter, the scheme is energy consuming since it requires at least τ messages for each key set up. The second scheme is proposed in [33]. In principle, this work is very similar to [32], where Blundo et al.'s polynomial scheme [38] is used instead of Blom's. Finding the polynomial shared by any two sensors is a crucial phase of the protocol. Again, the authors propose to choose either the weaker disclosing index techniques or the energy demanding challenge response technique.

Finally, in [35] the same authors of this section introduce: (1) a new threat model to communications confidentiality in WSNs (the smart attacker model); under this new, more realistic threat model, the security features of the previous schemes proposed in the literature drastically decrease; (2) a new pseudorandom key predeployment strategy that assures (a) a key discovery phase that requires no communications and (b) high resilience against the smart attacker model. In particular, they propose an *efficient and secure predeployment (ESP) scheme* that works as follows. Consider a new sensor s_a. For every key k_i^P of the pool, compute $z = f_y(a \parallel k_i^P)$; then, put k_i^P into M_a, the key ring of s_a, if and only if $z \equiv 0 \mod (|P|/m)$. This way ESP fills M_a with m keys on average. Further, it is proved that ESP requires a key ring of only $m + o(m)$ size to have m valid keys with high probability. ESP supports a very efficient key discovery procedure. Consider a sensor s_b that is willing to know which keys it shares with sensor s_a. For every key $k_j^b \in M_b$, sensor s_b computes $z = f_y(a \parallel k_j^b)$. Then, by testing $z \equiv 0 \mod (|P|/m)$, s_b discovers whether sensor s_a also has key k_j^b or not.

Note that our cooperative protocol can be implemented on top of any of the above-mentioned schemes to further enhance them by providing an independent degree of security.

8.3.3 Direct Protocol

Key Discovery Phase All the key deployment schemes reviewed in Section 8.3.2 require a key discovery phase. In particular, three different schemes for key discovery have been proposed so far:

1. Key index notification
2. Challenge response
3. Pseudorandom key index transformation

Key index notification [30] requires sensor s_a to send sensor s_b the indexes of the keys in its key ring. Sensor s_b, in turn, notifies sensor s_a of the indexes of the keys that will be used to secure the channel.

Challenge response is also introduced in [30] and is later used in [31–33]. It allows us to find out which keys are shared by two sensors without sending the indexes of the keys in the key ring, albeit at the price of a higher energy cost. Indeed, sensor s_a is supposed to broadcast messages $(\alpha, E_{k_i^a}(\alpha))$, $i = 1, \ldots, m$, where α is a challenge and k_i^a is the ith key assigned to sensor s_a from the pool. The decryption of $E_{k_i^a}(\alpha)$ with the proper key by sensor s_b would reveal the challenge α and the information that s_b shares k_i^a with sensor s_a. As observed in [34, 36], *pseudorandom key index transformation* allows for a more efficient key discovery procedure than key index notification or challenge response. Given a sensor s_a, the idea is to generate the indexes of the keys that will be loaded onto the key ring of s_a by using a pseudorandom generator initialized with a public seed dependent on s_a. Since the seed is known, the m indexes of the keys assigned to s_a can be computed by anyone. Note that the key discovery procedure that is supported by pseudorandom key transformation reveals only the indexes of the keys given to sensor s_a and does not leak any information concerning the keys themselves.

Both key index notification and pseudorandom key index transformation have been criticized for revealing the key indexes of the keys in each sensor's key ring to the attacker [30, 31]. It is argued, on the basis of intuition, that knowing these indexes allows the attacker to choose and to tamper with the sensors whose key rings are most useful for compromising a given target channel. However, neither mathematical nor experimental justifications have been provided to support this intuition. Simply, all previous works in the field used the oblivious attacker, which is not able to take advantage of any of the information leaked by key index notification or by pseudorandom key index transformation. The smart attacker introduced in [35] is able to exploit this information.

Regardless of the key predeployment and key discovery protocol that is used, the strategy of the oblivious attacker does not change. It is basically a random attacker. Conversely, the strategy of the smart attacker does depend on the key discovery protocol. Both key index notification and pseudorandom key index transformation reveal the indexes of the keys in the key ring of sensor s during the key discovery phase. If all the sensors are willing to communicate securely, they have to perform a key discovery phase. Thus, we can assume that the attacker knows the indexes of the keys in each key ring of the WSN. Therefore, for every sensor s, the smart attacker strategy is to compromise the sensor that supplies the largest number of "new" keys in the target key set T.

Let us now take the challenge response into consideration. During one of its key discovery phases, every sensor s leaks enough information to let the attacker know: (1) key ring size m, (2) exactly which keys in the key ring of s are in W, where W is the set of keys that the attacker has already collected, and hence (3) $|M_s \setminus W|$, the number of keys of sensor s outside W. Indeed, anyone that already knows a specific key can discover whether sensor s has that key or not by analyzing the messages exchanged in the key discovery phase of the challenge response. The smart attacker strategy is now clear: At each step of the attack sequence, compromise the sensor that stores the largest number of keys that are not in W, in the hope of maximizing the ones in $T \setminus W$.

Finally, as for ESP [35], the complexity of key discovery phase is comparable to pseudorandom key index transformation: no message exchange and m applications of the pseudorandom function. Moreover, this mechanism reveals to third parties exactly the same information as challenge response. Indeed, whoever already knows key k_i^P is the only one who can know whether k_i^P is in M_a or not. This is computationally impossible for all other entities, since f_y, being a pseudorandom function, is also one-way and thus hard to invert [39].

Channel Setup Assume that sensor s_a wants to communicate securely with s_b and $M_a \cap M_b$ is not empty. After the key discovery phase, sensors s_a and s_b agree on an ordered tuple $\langle k_1^{a,b}, \ldots, k_{|M_a \cap M_b|}^{a,b} \rangle$ containing all the keys in $M_a \cap M_b$. The *direct protocol* builds a secure channel based on key $k^{a,b}$, which is computed by both sensors as follows:

$$k^{a,b} = \mathcal{H} \left(k_1^{a,b} || \ldots || k_{|M_a \cap M_b|}^{a,b} \right) \tag{8.1}$$

where $\mathcal{H}$ is a hash function. This key will be used to secure all messages between sensors s_a and s_b. Basically, the direct protocol is implicitly defined by the key discovery phase and inherits its security properties. Similar protocols are used in [30], [31], and [36].

In the construction of key $k^{a,b}$, the direct protocol uses all the keys shared by sensors s_a and s_b. Intuitively, $k^{a,b}$ is as strong as the strongest key sensors s_a and s_b can agree on by themselves. Since we assume that cryptography is computationally strong, the attacker has to know all of the keys shared by the two sensors to corrupt the channel. Herein, we will often refer to key $k^{a,b}$ as the channel (indeed $k^{a,b}$ provides effective implementation of the secure channel). Expressions such as "the channel has been corrupted" mean that the attacker has managed to collect all the shares of key $k^{a,b}$.

Channel Confidentiality and Probabilistic Node-to-Node Authentication in the Direct Protocol

The channel confidentiality of the direct protocol directly depends on the key discovery phase on which it is implemented. The direct protocol guarantees that a computationally bounded attacker cannot corrupt the confidentiality of a given channel without collecting all the pool keys of which it is made up. The difficulty of this task fully depends on how keys are predeployed and how much information is disclosed by sensors during the key discovery phase. In our case, the key property that challenge response and ESP alone can enforce is *index hiding*: Whoever knows a particular key is the only one who can possibly know whether another sensor has that key or not.

The second security property we focus on is node-to-node authentication. Authentication is a key property. It is crucial for supporting a number of security functionalities. Any scheme that is used to revoke misbehaving nodes is based on the certainty that node identities are correctly detected. A number of powerful attacks, such as cloning a large number of malicious sensors with different IDs and the same stolen key ring, are mitigated if authentication is enforced.

Assume that sensor s_a successfully sets up a channel with sensor s_b by using the direct protocol. To what extent is sensor s_a sure of the identity of sensor s_b, and vice-versa? In principle, sensor identities could be faked since multiple copies of the same keys are distributed in the network. However, a fundamental property of pseudorandom key index transformation and ESP is that, given a pool, the key ring of each sensor fully depends on its identity. The same does not hold for index notification and challenge response. This property supports the following authentication method: Sensor s_b proves its identity by proving it knows which keys sensor s_b is supposed to know. In case the above-mentioned property does hold, the attacker can pretend to be sensor s_a during a channel set up with s_b only if it has all the keys in $M_a \cap M_b$. Hence, the identity of s_a can be faked only if $M_a \cap M_b \subseteq W$, where W is the set of keys that attacker collected by corrupting the sensors. This also means that channel confidentiality between s_a and s_b is corrupted. Consequently, as long as the system is built to guarantee high confidentiality of node-to-node channels, it also implicitly guarantees node-to-node authentication with high probability, at no additional cost.

Finally, note that our scheme based on pseudorandom key index transformation defends also against the threat posed by the Sybil attack [40] to wireless sensor networks, where a node illegitimately claims multiple identities.

8.3.4 Cooperative Protocol

The Protocol The *cooperative protocol* involves multiple cooperating sensors when setting up a channel between any two sensors s_a and s_b. The core idea is to use the key rings of some cooperators to enlarge the set of keys sensor s_a, which can be use to secure the channel. Herein, $k^{a,b}$ denotes the channel key between sensors s_a and s_b built by using the direct protocol with ESP. When sensor s_a wants to establish a secure channel with sensor s_b, sensor s_a chooses a set $\mathcal{C}$ of cooperating sensors such that $a, b \notin \mathcal{C}$, $|\mathcal{C}| \geq 0$, and s_a has a secure channel with all sensors in $\mathcal{C}$. Then, s_a sends a request for cooperation to each sensor $s_c \in \mathcal{C}$. The request carries the ID of s_b and is sent to s_c over the secure channel $k^{a,c}$. Next, each cooperating sensor s_c builds a share $f^{c,a,b}$ by hashing $k^{c,b}$, its channel key with s_b, together with the ID of s_a:

$$f^{c,a,b} = HMAC(a, k^{c,b}) \tag{8.2}$$

In case sensor s_c does not have a secure channel with s_b, s_c ignores the request for cooperation. After a timeout, sensor s_a realizes that s_c does not respond and removes it from the set of cooperators. Next, sensor s_c encrypts the share with $k^{c,a}$ and sends it back to s_a. Moreover, sensor s_c also sends a notification to sensor s_b in order to tell s_b that s_c itself is one of the cooperators for the channel that is being built; the message is encrypted with $k^{c,b}$. When all the shares are received, sensor s_a computes $k^{a,b}$ and combines it with the shares in order to obtain a cooperative channel key $k^{a,b,\mathcal{C}}$:

$$k^{a,b,\mathcal{C}} = k^{a,b} \oplus \left(\bigoplus_{s_c \in \mathcal{C}} f^{c,a,b} \right) \tag{8.3}$$

Sensor s_a sends to sensor s_b message $\mathcal{H}(k^{a,b,\mathcal{C}})$. Once sensor s_b has received this message and all the notifications from the cooperators, sensor s_b has all the information it needs to locally compute $k^{a,b,\mathcal{C}}$, without sending or receiving any other messages and can double-check

Cooperative_Protocol(b :sensorID)
Input: the receiving sensor;
Output: $k^{a,b,C}$

1. Generate set C;
2. Set timeout Δ;
3. $k^{a,b,C} = 0$;
4. **for all** $c \in C$ **do begin**
5. $k^{a,c} = $ Direct_Protocol(c);
6. $a \to c :< a, c, E_{k^{a,c}}(\text{req\_coop}||b) >$
7. **end**;
8. $C' = C$;
9. **while** ($C' \neq \emptyset$ **and** (**not** elapsed(Δ))) **do begin**
10. $a \leftarrow c :< c, a, E_{k^{c,a}}(f^{c,a,b}) >$;
11. $tmp = E_{k^{a,c}}^{-1}(E_{k^{a,c}}(f^{c,a,b}))$;
12. $C' = C' - \{c\}$;
13. $k^{a,b,C} = k^{a,b,C} \oplus tmp$
14. **end**
15. $k^{a,b,C} = k^{a,b} \oplus k^{a,b,C}$;
16. $a \to b :< a, b, \mathcal{H}(k^{a,b,C}) >$;

Figure 8.17 Cooperative protocol: pseudocode for sensor s_a.

the resulting key with $\mathcal{H}(k^{a,b,C})$ as well. Note that, when C is empty, the direct protocol with ESP and the cooperative protocol are equivalent.

Figure 8.17 shows the pseudocode of the cooperative protocol for sensor s_a in detail. In Figure 8.18, we show a simplified instance of the messages that are exchanged by the cooperative protocol in case there is only one cooperating sensor (c). In particular, Table 8.1 summarizes the changes introduced in the key discovery phase and the channel setup phase in the cooperative protocol with respect to the direct protocol introduced in Section 8.3.3.

It is important to note that the set C can be chosen on the basis of several policies. A first energy-preserving option is to include only relatively nearby sensors in C, for example, only sensors within one or two hops. This choice yields a more efficient key setup phase, although the protocol can be weaker against geographically localized attacks. A second option is to choose C randomly, thus providing more security but at the price of a larger communication overhead, especially in large networks. By choosing this second option, the protocol would support both random and geographically localized attacks. Third, sensors can be chosen according to individual properties, such as tamper resistance, therefore supplying a potentially more secure channel.

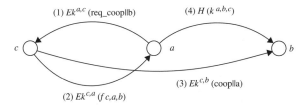

Figure 8.18 Example of cooperative approach with one cooperating sensor.

Table 8.1 Comparison of Phases Involved by Direct Protocol and Cooperative Protocol

Phase	Direct Protocol	Cooperative Protocol
Key discovery	Performed by the sender	Triggered by the sender and performed by the sender and all the sensors in $\mathcal{C}$.
Channel setup	The channel is built composing all the keys the sender shares with the receiver.	The channel is built composing the keys the sender, and each sensor in $\mathcal{C}$ shares with the receiver.

The cooperative protocol shows the following features:

- *Sensor Failure Resistance* If a sensor in $\mathcal{C}$ is not available for any reason (e.g., the sensor is destroyed or its battery has run out), the protocol will not fail or deadlock. Moreover, if any sensor does not reply to the request for cooperation within a certain timeout, sensor s_a can decide to add other sensors to $\mathcal{C}$ to achieve a satisfactory level of security.

- *Adaptiveness* If no information is available on the set of corrupted sensors but an upper bound on its cardinality, set $\mathcal{C}$ can be chosen in such a way to secure all channels with the desired probability.

- *Load Balance* The workload generated by the cooperating protocol is equally distributed among all the sensors in $\mathcal{C}$, that is, two messages per sensor in $\mathcal{C}$. Only sensor s_a has to send $|\mathcal{C}| + 1$ messages (one for each of the cooperating sensors in $\mathcal{C}$ and one to s_b). The total cost for the WSN is thus $3|\mathcal{C}| + 1$ potentially multihop messages. However, this cost is incurred only once, that is, during the channel setup. Finally, note that sensor s_b does not need to send any message to build up the cooperative channel.

- *Node-to-Node Authentication* Exactly like in the direct protocol, ESP provides the cooperative protocol with implicit node-to-node authentication. Hence, as far as the parameters of the protocol are chosen to guarantee confidentiality with high probability, the cooperative protocol also guarantees node authentication with high probability.

Denial of Service Attacks of Malicious Cooperators In general, corrupted sensors may be chosen as cooperators in the cooperative protocol. This is explicitly taken into consideration in the experiments, which show that channel confidentiality can be probabilistically guaranteed. However, it is interesting to explore all possible attacks from a cooperating sensor to other properties of the protocol.

Assume that $s_c \in \mathcal{C}$ is a corrupted sensor included in the set of cooperators for the channel set up from s_a to s_b. When s_c is asked by s_a to provide its share, s_c can behave as follows:

1. Sensor s_c does not send any message.
2. Sensor s_c sends a correct key $k_{c,b}$ and a correct notification to sensor s_b.
3. Either sensor s_c sends a bogus share $\tilde{f}^{c,a,b}$ or sends a bogus notification to sensor s_b.

Case 1 is equivalent to the case of a cooperating sensor that is no longer available, for instance, it has been destroyed or its battery has run out. After the timeout Δ defined in Figure 8.17, sensor s_a can decide to involve other sensors if the number of cooperating sensors is too low to guarantee its security requirements.

Case 2 is the best choice if the attacker is attempting to break the channel confidentiality. Indeed, the channel is set up, and the attacker may have enough keys to recover $k^{a,b,\mathcal{C}}$. Since this section focuses on confidentiality, in the experiments we assume that this is the default behavior of corrupted sensors. Note that the presence of malicious cooperators does not imply that the channel is corrupted.

Case 3 may result in a denial of service (DoS) on sensor s_a. Indeed, the channel built by sensor s_a with a bogus share does not match the channel that is locally computed by sensor s_b. The same happens if a bogus notification is sent to sensor s_b. Even though the aim of this section is to address confidentiality, in what follows we sketch a few countermeasures that can be taken to mitigate this sort of DoS attack. A first countermeasure is to randomly select another set C of cooperating sensors and to reapply the cooperative protocol. In order to be an effective solution, set C should be small enough to have Pr[every cooperator in C is not corrupted] $> 1/t$, for some small positive integer t. This way, after t iterations on the average, a good set of cooperators can be found. Note that other subsets of cooperators can be added later on to strengthen the same final cooperative channel. A second countermeasure is possible if a trusted channel leading to a base station is available. Each sensor can store an individual secret key that is shared with the base station, which also knows all the secret keys of the pool. When sensor s_a finds that a channel key $\tilde{k}^{a,b,C}$ might contain a bogus share, sensor s_a sends it to the base station along with all the shares that it used to build the key. Assuming that s_a is not corrupted, the base station has all the information it needs to efficiently track down the cheating sensors among the ones in C and s_b. Then, this is sent back to s_a. Note that the base station is used only when a cheater is detected, and corrupted sensors are discouraged from giving bogus shares in this setting. Finally, even when a channel to the center is not available, the presence of a bogus share provides important information, and lets us know that the WSN is under attack.

Overhead Introduced by the Cooperative Protocol   In this subsection we evaluate the overhead introduced by the cooperative protocol and by the direct protocol on both the sender and the cooperators. We take into consideration the amount of computations and the number of multihop messages required by each of them. The actual energy consumption of the computation depends on the specific implementation, while the energy consumption for message transmission depends on parameters such as the transmission radius. The interested reader can find in [26, 41] an analytic quantification of the energetic cost of each of these primitives. In particular, [26] analyzes different sensor prototypes and algorithm implementations. In the following, we count the number of messages sent, as well as the number of times we invoke operations such as encryption E, decryption E^{-1}, and hashing $\mathcal{H}$. As seen in Section 8.3.3, the cost of the key discovery phase is equal to m invocations of the hash function. The total overhead for both protocols, including the key discovery phase, is shown in Table 8.2.

The computational cost of the direct protocol is due to the fact that the sender has to perform a key discovery phase [which costs $(m + 1)\mathcal{H}$] and one encryption before sending the message. The multihop message accounts for the message sent to the receiver. The computational cost of the cooperative protocol is the following: (1) The sender has to perform

Table 8.2   Overhead Incurred by Cooperative and Direct Protocols

Protocol	Computation	Multihop Messages
Direct	$(m + 1)\mathcal{H} + E$	1
Cooperative	$\lvert C \rvert(m\mathcal{H} + E^{-1} + E) + \lvert C \rvert(2m\mathcal{H} + 2E + E^{-1}) + (m + 1)\mathcal{H}$	$3\lvert C \rvert + 1$

$|\mathcal{C}|$ key discovery phases to send the encrypted request of cooperation to cooperators; once a cooperator sends the share back to the sender, it must be decrypted by the sender. This accounts, for the sender, to $|\mathcal{C}|(m\mathcal{H} + E + E^{-1})$ computations and $|\mathcal{C}|$ messages; (2) each cooperator has to perform a key discovery phase on receiving the request for the share from the sender; then the cooperator has to decrypt the received message, to perform a key discovery phase to find out the keys it shares with the intended receiver, and then sends (a) the encrypted share to the sender and (b) an encrypted message to notify the receiver that it is participating with its share as a cooperator. These operations account for $(|\mathcal{C}|(2m\mathcal{H} + 2E + E^{-1}))$ computations and $2|\mathcal{C}|$ messages; (3) the sender finds out the keys it shares with the receiver, combines these keys with the keys provided by the cooperators, and then sends an image of $k^{a,b,\mathcal{C}}$ to the receiver. These operations account for $(m + 1)\mathcal{H}$ computations and one message.

Between computations and communications, communications are the more energy demanding operations [26]. Hence, it might seem that the cooperative protocol, requiring $3|\mathcal{C}|$ times the communications required by the direct protocol, is very energy consuming. However, it is worth noting that: (1) the effective cost of a single multihop message is strictly related to the policy according to which sensors in $\mathcal{C}$ are selected, as discussed earlier. For instance, when the cooperators are selected within the neighbors of the sender, the number of hops of all multihop messages between the sender and the cooperators is just one; and (2) the overhead is incurred only once by the sender; when the cooperative channel is established, it can be used for multiple messages. Finally, as will be shown in the next section, the slight increase in overhead of the cooperative protocol guarantees an excellent resilience against network link corruption when compared to the resilience of the direct protocol.

8.3.5 Experiments on the Cooperative Protocol Against the Smart Attacker

Threat Model   A common assumption is that the attacker is compliant with the Dolev–Yao model [42]. This means that the attacker can perform the following actions: (1) intercept and learn any message and (2) introduce forged messages into the system using all the available information. However, a distinguishing feature of WSNs is that sensors may be unattended. Hence, we assume there is a third more powerful action: (3) The attacker can capture a sensor and acquire all the information stored within it. To cope with the latter action, it could be possible to assume that sensors are tamper-proof [27, 29].

However, a very large number of sensors can be built only if sensors are low-cost devices, and this makes it difficult for manufacturers to make them tamper-proof. Therefore, in the rest of this section we will assume that sensors do not have tamper-proof components and that they can be captured. Moreover, we assume that the attacker's goal is to violate the confidentiality of the channel between two arbitrarily chosen sensors s_a and s_b. In order to achieve its goal, the attacker can compromise any subset of the sensors in the WSN but s_a and s_b. To formalize the attacker in a random predeployment scheme, we refer to the work in [35], which is summarized in the following.

Definition 8.3.1   Assume that the attacker's goal is to collect a subset T of the keys in the pool. The attacker has already compromised a number of sensors and has collected all their keys in a set W. For every sensor s in the WSN, the *key information gain* $G(s)$ is a random variable equal to the number of keys in the key ring of s that are in T and are not in W.

For example, if the attacker's goal is to compromise the channel between sensors s_a and s_b, subset T in the above definition is equal to $M_a \cap M_b$, that is, it contains all the keys that are in the key ring of both s_a and s_b. Assuming that the attacker has collected a set W of keys, the value of the random variable $G(s)$ is equal to $|(M_s \cap M_a \cap M_b) \setminus W|$.

The authors in [35] define two attackers, both of which tamper with sensors sequentially and collect all the keys in a set W.

1. *The Oblivious Attacker* At each step of the attack sequence, the next sensor to be tampered with is chosen randomly among the ones that have yet to be compromised.
2. *The Smart Attacker* At each step of the attack sequence, the next sensor to be tamper with is sensor s, where s maximizes $E[G(s)|\ I(s)]$, the expectation of the key information gain $G(s)$ given the information $I(s)$ that the attacker knows on sensor s key ring.

Intuitively, the oblivious attacker does not take advantage of any information leaked by compromised sensors. Conversely, the smart attacker greedily uses this information to choose which sensor to corrupt in order to maximize the number of useful keys it is expected to collect. Note that information $I(s)$ may also depend on the protocol used by sensor s and, in particular, on the key discovery phase.

Simulation Model We set the stage for our simulations as follows: The network is supposed to be connected. Note that our simulations are not dependent on the communication range of sensors. The objectives of our simulations will be: (a) to assess, as for the cooperative protocol, the impact of both cooperating sensors and the key ring size toward the resilience in link network corruption and (b) to compare the resilience to link network corruption of the direct protocol and the cooperative protocol.

To perform our experiments, we simulated the cooperative protocol with 4 and 8 cooperators in a sensor network of size 1000. Pool size is set to 10, 000. ESP is used as the key predeployment and discovery phase. The attacker is smart. Key ring size ranges from 70 to 220 keys. Herein, s_a is the sensor that starts the protocol to establish a secure channel with sensor s_b.

The attacker tampers with sensors until it collects enough keys to compromise the cooperative channel. This means that it has corrupted channel $k^{a,b}$ and, for every cooperator s_c, either channel $k^{a,c}$ or channel $k^{c,b}$. First of all, note that the simplified approach of simulating the protocol by assuming that each of the above-mentioned links is independently compromised with a certain probability is not correct. Indeed, even if it may seem counterintuitive, the security of $k^{a,c}$ is not independent from the security of $k^{c,b}$ and from the security of $k^{a,b}$, although the key rings of sensors s_a, s_b, and s_c are drawn independently. Therefore, in order to get accurate and realistic results, we performed a detailed simulation of every step of the key predeployment phase, of the cooperative protocol, of the direct protocol with ESP as discovery phase, and of the attack as described in the previous sections.

Figure 8.19 shows how many sensors the smart attacker has to corrupt in order to compromise an arbitrary channel with 0 (i.e., the direct protocol), 4, and 8 cooperators. Take into consideration key ring size of 100. While the channel based on the direct protocol is compromised after tampering with 96 sensors, as many as 131 sensors are necessary if 4 cooperators are used, and 150 sensors if 8 cooperators are used. Not only does the cooperative protocol improve the average number of sensors required to corrupt an arbitrary channel, it also provides a much stronger resilience against small-scale attacks. Figure 8.20 shows the average number of sensors the smart attacker has to corrupt in order to compromise 10% of the network links based on the cooperative protocol. Here, when the key ring is set to

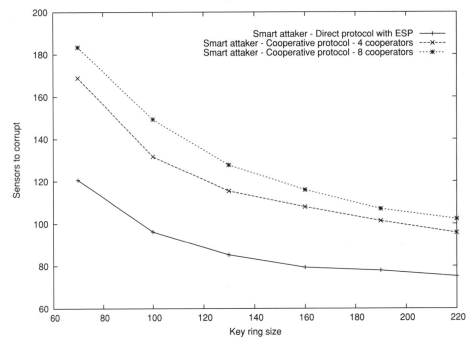

Figure 8.19 Experimental results on the direct protocol and the cooperative protocol. Average number of sensors the smart attacker have to corrupt in order to compromise an arbitrary channel.

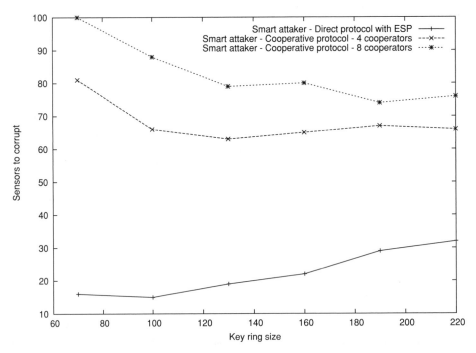

Figure 8.20 Experimental results on the direct protocol and the cooperative protocol. Average number of sensors the oblivious and smart attacker have to corrupt in order to compromise 10% of the network links.

100 keys, the direct protocol tolerates 15 sensors. The cooperative protocol greatly enhances the security of the channel even when a small number of cooperators is used. Indeed, it tolerates 66 sensors when 4 cooperators are used and 88 sensors when 8 cooperators are used. This is more than a factor of 4 improvement over the direct protocol.

8.3.6 Concluding Remarks

In this section we point out that the pseudorandom key predeployment schemes, besides confidentiality, implicitly enforce an authentication mechanism among sensors. Next, our main contribution is to introduce the cooperative protocol that amplifies both the above security properties. In particular, our scheme enhances the resilience against the Sybil attack, where a node illegitimately claims multiple identities.

The overhead required by the cooperative protocol is limited and must be paid only once, during the channel setup. Note that the cooperative protocol is dynamically tunable depending on the level of threat in WSN and can be adopted by any of the cited pseudorandom key predeployment schemes.

An area that is open to research could be to enhance the resilience of the cooperative protocol against malicious but silent nodes that can be selected as cooperators and attempt to perform DoS.

8.4 EFFICIENT PAIRWISE AUTHENTICATION PROTOCOLS FOR SENSOR AND AD HOC NETWORKS

André Weimerskirch, Dirk Westhoff, Stefan Lucks, and Erik Zenner

Sensor networks face huge security lacks. In the most general case nodes need to build up a well-defined security association without any preestablished secret or common security infrastructure. In this section we present and analyze two protocols: (1) a protocol that we call zero common-knowledge (ZCK) recognition protocol to provide a basic form of authentication, namely recognition, and (2) a proof of identity at the cost of some external infrastructure that we call identity certified (IC) authentication protocol. The latter serves as an extension to the ZCK protocol. We analyze the ZCK and IC protocols with respect to their overhead, including operational complexity and energy consumption. The ZCK protocol is an extremely efficient authentication protocol for sensor networks as it uses only hash chains, whereas IC still is an efficient identification protocol closing the gap to asymmetric solutions. From an energy point of view the protocols are efficient in networks with a dynamic topology and in networks where the average distance between a node to its neighborhood is moderate.

8.4.1 Background Information

Wireless sensor networks (WSN) are a particular class of ad hoc networks that attract more and more attention both in academia and industry. The sensor nodes themselves are preferably inexpensive and tiny consisting of an application-specific sensor, a wireless transceiver, a simple processor, and an energy unit that may be battery or solar driven. Such sensor nodes are envisioned to be spread out over a geographical area to form a multihop network in a self-organizing manner. Most frequently, such WSNs are stationary although also mobile WSNs are possible. For instance, potential applications for such a scenario—besides obviously military ones—can be found in monitoring environmental data such as the detection of fire in huge forest areas.

In this section we consider security solutions for WSNs to provide integrity and authentication. We do not assume tamper-resistant units as it is unrealistic for the envisioned ultrarestricted devices [43]. Consequently, one has to face the possibility that a subset of nodes is corrupted and that all their secrets are revealed to an attacker. Hence, pairwise security associations between these nodes is advantageous. Concepts such as the TinySec approach [44], which is based on a single shared group key, require nearly no overhead but cannot fulfill the above security requirements. A pairwise authentication based on predistributed keys comes at the cost of storage overhead and does not allow network extendibility except by using asymmetric key agreement schemes such as Diffie–Hellman, which are too demanding for sensor devices if frequently used. Promising approaches are the pairwise key predistribution schemes for wireless sensor networks in [45] and [46], which save considerable memory storage. Unfortunately, both approaches are "closed shop" in a twofold meaning. First, due to the predistribution of some secrets, nodes must originate from one manufacturer. Second, once the network is spread over an area, it is subsequently only extendible by introducing more nodes using the same predistributed share set at the cost of security. Such an extension may be valuable in cases where a larger region needs to be monitored, or in cases where due to an empty battery of some nodes the network's connectivity is not guaranteed anymore.

Authentication Schemes   It is possible to perform entity as well as message authentication with a public key scheme and a symmetric key scheme, respectively. Both can be done in a unilateral or mutual fashion. In a public key scenario each entity has a certificate ⟨PK⟩ issued by a certificate authority (CA) and an assigned public/private key pair. A unilateral message authentication works as presented in Algorithm 8.4.1. Let SIG (m, SK) be a signature of the message m by the private key SK, and VER(S, m, PK) be the verification of the signature S to the message m by the public key PK. It is VER(S, m, PK) valid, if S is the signature of m by the corresponding secret key of PK, that is, $S = \text{SIG}(m, \text{SK})$, and it is invalid otherwise.

Algorithm 8.4.1   Public Key Message Authentication

1. A signs a message m as $S := \text{SIG}(m, \text{SK})$ and sends $m, S, \langle \text{PK} \rangle$ to B.

2. B verifies whether $\langle \text{PK} \rangle \stackrel{?}{=} valid$ and whether VER$(S, m, \text{PK}) \stackrel{?}{=} valid$.

This can easily be converted into an entity authentication process by introducing a random challenge r and using a challenge–response method [47].

In practice, often hybrid schemes are used. Here a key agreement scheme such as a Diffie–Hellman (DH) [4] is used followed by a symmetric MAC scheme for each authentication. This is presented in Algorithm 8.4.2.

Algorithm 8.4.2   Hybrid Message Authentication

1. A and B agree on a shared symmetric key K, for example, by DH.

2. A computes $M := \text{MAC}(m, K)$ and sends m, M to B.

3. B checks whether $M \stackrel{?}{=} \text{MAC}(m, K)$.

Contributions In this section we present two new protocols for a pairwise mutual authentication that support different levels of authentication and that are more efficient in many scenarios than the above approaches. Both protocols are robust against a set of corrupted nodes in the sense that only a limited number of trust associations is corrupted if a set of nodes is compromised. The first protocol is neither closed shop in its manufacturer related meaning nor with respect to the extendibility of the network. The second protocol, which provides a proof of identity, assumes devices from a single manufacturer (or a well-defined consortium of manufacturers) but is still open in the sense that it is extendible at any time with additional nodes from the consortium.

We now explain the main idea of our protocols. Imagine two strangers that meet in the real world. Usually, and obviously strongly related to the considered level and type of interaction, these people do not explicitly identify each other, for example, by showing their passports. Assuming there is nobody else available to ask about the opposite's reputation, the best both can do is to establish step-by-step a trust relationship built on their bilateral, personal experiences. To do so, it is mandatory that these two people recognize each other the next time they meet. In our model we do not assume geographical proximity though. We call this form of authentication recognition and present a protocol that we called *zero common-knowledge* (ZCK) protocol to provide recognition.

This approach aims at sensor and ad hoc networks without any kind of infrastructure. However, it lacks the possibility of establishing a trust relationship based on the identity (which probably is not possible to provide at all without any infrastructure or previous knowledge). In the second protocol, which we call *identity certified (IC) authentication*, we assume that there are some storage resources available to provide identification, and we assume that nodes are loosely time synchronized. We use the term identification in the sense that an entity A is able to prove its identity to an entity B assuming that both entities trust some third party, usually the common manufacturer.

We introduced these protocols in [48,49]. They did, however, contain a security flaw that we described in [50]. In the same study, we introduced an improved version that is provably secure and that is the basis of the following protocols. The main contribution of this section is a corrected and extended suite of two new authentication protocols especially suited to pervasive networks and an analysis of these protocols. The first scheme only requires one-way hash functions and is suited to nearly any computationally weak device since it is more efficient than any public key scheme by some order of magnitude. The scheme provides entity and message recognition and is provably secure. The second scheme, IC authentication, enhances the ZCK recognition protocol by identification at the cost of some infrastructure. Furthermore we present a performance analysis of the ZCK and the IC protocols for stationary WSNs where a major cost metric is energy consumption.

8.4.2 Zero Common-Knowledge Recognition

We follow a fully infrastructure-less approach of establishing trust relationships in pervasive networks that have a frequently changing topology. We define that a network is of a *pure* kind (or just pure) if there exist neither central services nor a fixed infrastructure, and if there is no preestablished knowledge between two entities. We assume that the network has no special topology, specifically it might be a wireless multihop network that is exposed to several attacks such as a man-in-the-middle attack. We define a *weak device* to be a simple device without any further assumptions. Such assumptions would be tamper resistance or a unique identification number. We assume the most general case in which there is no common trusted third party (TTP), and the ad hoc network is indeed pure. We define our security objective, which we call entity recognition, as follows:

Definition 8.4.1 *Entity recognition* is the process whereby one party, after having initially met, can be assured in future recognitions that it is communicating with the same second party involved in a protocol and that the second party has actually participated (i.e., is active at, or immediately prior to, the time the evidence is acquired).

Note that entity recognition is a weaker form of entity authentication. Hence, a protocol that performs entity authentication has, implicitly, performed entity recognition. The reverse of this is not true. The accumulation of trust, gained in several recognition processes, can bring two communicating parties close to—but not fully attain—entity authentication. During the first recognition A cannot be sure that it is communicating with the correct entity B; however, the more information that A receives that it expected B to have, the more certain A can be that it is correct. However, trust accumulation is very situational dependent and requires further methods to evaluate trust that we do not focus on here. Similar to entity recognition, Definition 8.4.2 defines message recognition.

Definition 8.4.2 *Message recognition* provides data integrity with respect to the original message source and assures the data origin is the same in future recognitions but does not guarantee uniqueness and timeliness.

As for entity and message authentication, the difference between message recognition and entity recognition is the lack of a timeliness guarantee in the former with respect to when a message was created.

Now it is clear what we want to achieve: recognition in a pure pervasive network consisting of weak devices where there are no preestablished secrets. We do not require any assumptions such as tamper-resistant devices, devices that are able to frequently perform expensive public key operations, special access structures for deploying a distributed PKI, and so on. It seems that entity authentication is impossible in a pure pervasive network. However, entity recognition can approach the security of entity authentication. We believe that such an approach is more practical and realistic, yet still sufficient. Our approach focuses on problems that are inherent to pure pervasive and ad hoc networks. For example, our solution is able to speed up cooperation and motivation-based schemes [51–53] as well as secure routing methods. Our scheme also makes sense for all kinds of client–server and peer-to-peer relations, respectively.

Note that entity recognition prevents a man-in-the-middle attack after the first recognition, hence if C was modifying information, C would have to have been in the middle from the beginning (and hence, A has been recognizing C, not B). A *reliable relay channel* for the first recognition process is a requirement in a formal model. Such a reliable relay channel provides integrity of messages and ensures that all messages are received in order to prevent manipulation or interception of messages. In many applications, the initial contact and with it the requirement of a reliable relay channel might become meaningless, though. Here, A is never able to distinguish whether the other party is B or C. A will accumulate trust based on the interaction with its communication partner such that the initial contact, that is, the question whether A is talking to B or C, becomes irrelevant. For instance, if A is communicating with C, and C is able to provide A with all services needed, it is not important for A whether C was a man-in-the-middle at the initial contact that was supposed to happen between A and B.

General Recognition Protocols   Consider the case where an entity A wants to be able to recognize an entity B after an initial contact. Obviously, the communication channel they use is insecure. In this scenario, B might be a service provider and A the client. In the first step B provides A with some data that allows the later one to recognize B. Furthermore, A can verify that a message origins of B. A simple example is a private/public key pair of a digital signature scheme. Here, B sends his public key PK to A. Then B signs his messages, which A can verify. For entity recognition, a challenge–response protocol as described in [47] can be performed. We consider as the main objective of this scheme the capability to ensure that entities are able to re-recognize each other in order to receive a service they requested. Hence the public key PK always has to be sent together with the offered service, that is, service and key have to be bound together to ensure that a malicious entity cannot inject his public key to a service that he did not offer at all.

In contrast to PKI scenarios where there is a logical central certificate directory, A has to store B's public key (together with B's ID string) to be able to recognize B. After A deletes B's public key from her memory, A is not able anymore to build a connection to a previous relationship with B. Note that in many applications a mutual authentication process is required. The above protocol can easily be extended for this case.

We now consider the traditional security objectives, namely authentication, confidentiality, integrity, and nonrepudiation [47]. The above scheme ensures message recognition, that is, it is possible to establish integrity of messages by authenticating these messages as shown above. Obviously, the scheme does not provide nonrepudiation. However, for some type of scenario a weaker form of nonrepudiation may be appropriate. We define *recognition nonrepudiation* to be the service that prevents an entity from denying a commitment or action chain. In our case this means that an entity A is able to prove to a third party that a number of actions or commitments were taken by the same (probably unknown) entity B. Obviously, a scheme that provides signatures satisfies the recognition nonrepudiation objective.

The presented protocol is as secure as the underlying security scheme, that is, to break the protocol an adversary has to construct an authenticated message SIG(m, SK) for given m. Note that a man-in-the-middle attack is always possible, and consider what this means as an example. There is an entity B that offers a service, an entity A that seeks this service, and a malicious entity M. The entity B sends PK_B to A. M interrupts this message and sends PK_M to A. Then M satisfies the needs of A by offering its services. All that A is interested in was that service. She does not care who provided the service, but she wants to receive the service in the same quality again. It becomes now clear that a man-in-the-middle attack is no threat to our protocol.

Two protocols to provide recognition that immediately arise are based on traditional signature schemes as well as symmetric ciphers. In the case of digital signatures as shown above, the scheme ensures entity recognition and nonrepudiation, and also message recognition. In most cases public key operations overstrain the capabilities of weak devices. Thus the frequent usage of RSA and even ECC is too slow and too resource consuming. In the case of a symmetric cipher a secret key has to be shared a priori, that is, the protocol requires a secret channel to exchange the shared secret. It is suited for applications where devices are geographically close, for example, where devices can exchange the keys by infrared channel, or where there is only a single trust authority. Clearly, a secure channel for key exchange could also be established by a secure key exchange such as Diffie–Hellman. Such a key exchange has to be performed for each communication pair. Since frequent key exchanges demand computationally powerful devices and the heavily changing topology of pervasive networks of mobile nodes induces many key exchanges such a solution is not suited to weak devices. We now introduce our new scheme that overcomes these limitations.

Zero Common-Knowledge Protocol Our new ZCK protocol for recognition only requires one-way hash functions but no expensive public key operations. Doing so the scheme is extremely efficient and orders of magnitude faster than any public key scheme.

Consider the case where Alice wants to send authenticated messages to Bob. We define a hash chain, which is also known as Lamport's hash chain [47], as $x_{i+1} = h(x_i)$ with x_0 being the anchor and h being an unkeyed one-way hash function. Alice chooses randomly an anchor a_0 and computes the final element a_n of the hash chain. We call a_n the public key and a_0 the secret key of Alice. Bob is doing likewise to obtain b_0 and b_n. We use a chain value as key to generate an authenticated message by a MAC. The core idea of the protocol is as follows: First exchange a value a_i, which the receiver will tie together with some experience. Then prove knowledge of the preimage of a_i, that is, a_j with $j < i$, in order to authenticate by establishing a relationship to a_i and the past experience. In order to repeat the authentication step arbitrarily many times, a hash chain based on a one-way hash function is used. The protocol works as illustrated in Figure 8.21. Here, $\mathcal{P}(.)$ represents the storage of some data. If any comparison $\overset{?}{=}$ fails, the protocol flow is interrupted. The protocol can then only be resumed at the same position.

Remarks

- Steps 1 and 2 in Figure 8.21 ensure the exchange of public keys, which is done only once per pair A and B.
- Steps 3 to 7 are done for each recognition.
- For each recognition of each communication pair we assume that A stores B's key b_i and that B stores A's key a_j (step 3). Hence, after the initial key exchange A stores b_n and B stores a_n. After each successful recognition, A and B replace a_j and b_i by a_{j-1} and b_{i-1}, respectively. Furthermore, A stores the private key a_0 that she created especially for Bob, and Bob is doing likewise.
- The exchanged message is guaranteed to be fresh since both parties are involved actively.
- The message sent in step 4 can be read but is not authenticated at this moment. Messages can only be checked after the keys were opened in step 6, that is, there is a message buffer required.

Transmitting:	Processing:
For key exchange, do Steps 1–2 only once:	
1. $A \rightarrow B : a_n$	$B : \mathcal{P}(a_n)$
2. $B \rightarrow A : b_n$	$A : \mathcal{P}(b_n)$
For each recognition process, repeat Steps 3–7:	
3.	A knows b_i
	B knows a_j
4. $A \rightarrow B : m, M := \mathrm{MAC}(m, a_{j-1})$	
5. $B \rightarrow A : b_{i-1}$	$A : h(b_{i-1}) \overset{?}{=} b_i$
6. $A \rightarrow B : a_{j-1}$	$B : h(a_{j-1}) \overset{?}{=} a_j, \mathrm{MAC}(m, a_{j-1}) \overset{?}{=} M$
7.	$A : \mathcal{P}(b_{i-1})$
	$B : \mathcal{P}(a_{j-1})$

Figure 8.21 Zero common-knowledge message recognition protocol.

- If A opens a key that B does not receive due to network faults, A can send the key again without endangering the security of the scheme. The same holds for keys opened by B.

- A man-in-the-middle attack at initialization time is possible. However, once the public keys a_n and b_n were exchanged, messages cannot be forged.

- The scheme is not resistant to denial-of-service attacks. A malicious entity can try to overflow the message buffer. Thus, an implementation must take care of an appropriately sized message buffer.

- The scheme is provably secure if its building blocks h and MAC are secure [50].

- When a key a_0 is compromised, only the security of the communication channel between Alice and Bob is affected.

- To save memory and to avoid extensive computations, short hash chains can be used. These can be renewed by transmitting a new final element as an authenticated message in order to keep the trust association.

- The public key is only sent once. A message recognition requires three messages, each of them a t-bit string. Using an efficient hash-chain algorithm as presented in [54], an element of the hash chain can be computed by $\frac{1}{2} \log_2(n)$ hash iterations with storage of $\log_2(n)$ hash-chain elements. In the following, we let $n = 100$, which should suffice most applications. Then there is storage required for seven elements each t bits in size, and four hash iterations are needed to obtain an element of the chain. Note that the execution of a hash function has very low running time compared to any asymmetric operation.

- For each communication pair of A with any B, she needs to store B's public key b_i of t bits. Thus altogether, for each communication partner there is storage needed of seven hash-chain elements as well as the public key resulting in t bytes.

- ZCK entity recognition is performed by replacing the message m by a random value r. Timeliness in our protocol is introduced by the active involvement of Bob and Alice. Hence, Alice can choose r, and neither a challenge by Bob nor a timestamp is required to introduce timeliness. Hence, we save the initial message here such that the ZCK entity recognition and ZCK message recognition protocol have the same operational complexity.

For a better understanding, we illustrate our protocol again in Figure 8.22. Here, the initial key exchange as well as two recognition processes are depicted. First, Alice and Bob exchange keys b_n and a_n. In the first recognition, Alice authenticates message m to Bob by key a_{n-1}. After a successful message recognition, Alice and Bob store keys b_{n-1} and a_{n-1}, respectively. At any time later on, if Alice wishes to authenticate another message m' to Bob she computes the MAC over m' by key a_{n-2}, and follows the ZCK recognition protocol. After the initial key exchange, any number of recognition processes can be performed between Alice and Bob.

Let t be the output size of the MAC and the hash function. It is widely believed that computing a collision to a given message (target collision resistance) in a one-way hash function that maps strings to $t = 80$ bits is approximately as hard as factoring an RSA modulus of 1024 bits [55]. We believe that target collision resistance is sufficient for our application, whereas collision resistance (finding any pair of colliding messages) requires twice the number of bits. An attack that finds any preimage of an opened key a_i has a complexity of around 2^t. Hence we assume $t = 80$ in the following. We believe that an average hash chain of length $n = 100$ should meet the life span demands of most security associations.

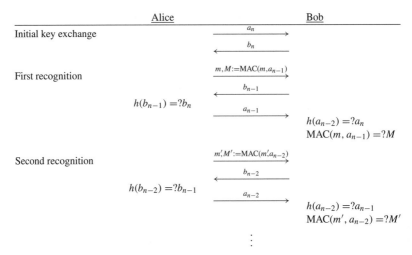

Figure 8.22 Zero common-knowledge recognition protocol example.

The above scheme provides unilateral recognition. A mutual recognition can also be provided as presented in Figure 8.23. The number of exchanged messages increases to four. Here, Alice must start each round of the authentication.

8.4.3 Identity Certified Authentication Protocol

We now extend our previous scheme to provide identification. We assume that the devices are able to perform a signature verification that is reasonable in the case of RSA with short exponent. However, the devices do not need to perform signature verifications frequently, but only once in a while. We further assume that devices are loosely time synchronized to a global time. Our goal here is to extend the ZCK recognition protocol in such a way that it provides identification to entities that do not know each other nor had any contact in the past. This can be seen as similar to the exchange of a certificate in the public key scenario. Once the identity proof was performed the ZCK recognition protocol can be used, which

Transmitting:	Processing:
For key exchange, do Steps 1–2 only once:	
1. $A \to B : a_n$	$B : \mathcal{P}(a_n)$
2. $B \to A : b_n$	$A : \mathcal{P}(b_n)$
For each recognition process, repeat Steps 3–8:	
3.	A knows b_i
	B knows a_j
4. $A \to B : m_A, M_A := \mathrm{MAC}(m_A, a_{j-1})$	
5. $B \to A : m_B, M_B := \mathrm{MAC}(m_B, b_{i-2}), b_{i-1}$	$A : h(b_{i-1}) \stackrel{?}{=} b_i$
6. $A \to B : a_{j-1}$	$B : h(a_{j-1}) \stackrel{?}{=} a_j, \mathrm{MAC}(m_A, a_{j-1}) \stackrel{?}{=} M_A$
7. $B \to A : b_{i-2}$	$A : h(b_{i-2}) \stackrel{?}{=} b_{i-1}, \mathrm{MAC}(m_B, b_{i-2}) \stackrel{?}{=} M_B$
8.	$A : \mathcal{P}(b_{i-2})$
	$B : \mathcal{P}(a_{j-1})$

Figure 8.23 Zero common-knowledge mutual message recognition protocol.

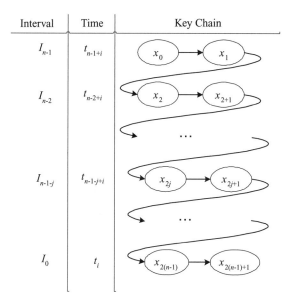

Figure 8.24 Time intervals for hash chains.

ensures that this level of trust in the other's identity is maintained.[1] Thus an exchange of keys that can be used for the ZCK recognition protocol for future identification or message authentication processes is included in our new scheme. For each communication pair A and B, the certificate exchange only needs to be performed once for proving identity, whereas later on only the ZCK recognition protocol needs to be executed for identification based on the identity proven level of trust.

The idea of the IC protocol is to divide time into intervals and let a set of keys of a hash chain only be valid for one time interval, similar to the idea of TESLA [56]. Let $a_{i+1} = h(a_i)$ be a hash chain, a_0 be the anchor of the chain, and $a_{2(n-1)+1}$ be the final element of the hash chain. We let two keys of the chain be valid in a time interval. Let t_i be a point in time that we assume to be the starting time, L be the length of the time intervals, and I_{j-1} be the jth time interval. We assume that time is divided into time intervals of length L, that is, there are time values t_i, t_{i+1}, t_{i+2}, and so on, such that the time difference between t_j and t_{j+1} is L, and the time difference between t_j and t_{j+m} is mL. The kth time interval I_{k-1} lasts from t_{i+k-1} to t_{i+k}. Let d_s be the time difference measured in interval lengths L between the current time t_c and the time t_s, that is, $d_s = c - s$. Assume the starting point is t_i and the keys valid during the corresponding interval I_0 are $a_{2(n-1)}$ and $a_{2(n-1)+1}$ such that there are keys for n time intervals in the hash chain. This fact is depicted in Figure 8.24. Then $a_{2(n-1)}$ and $a_{2(n-1)+1}$ are keys for the first time interval I_0 between time t_i and t_{i+1}, a_{2j} and a_{2j+1} are keys valid for the time interval I_{n-1-j}, and a_0 and a_1 are the keys valid in the final time interval I_{n-1}. Furthermore, the interval difference between the current time t_c and starting time t_i is $d_i = c - i$ such that the current time interval is I_{d_i}, and the keys valid for this interval are $A_c = a_{2(n-1-d_i)}$ as well as $A_{c+1} = a_{2(n-1-d_i)+1}$.

Assume Alice wants to identify herself to Bob. Then Alice holds a secret anchor a_0 and the public value $PK_A = a_{2(n-1)+1}$. Bob likewise holds b_0 and $PK_B = b_{2(n-1)+1}$. The public

---

[1] The ZCK recognition scheme improves or maintains the level of trust between two entities. Since the proof of identity is the highest trust level we can achieve here, in this case the ZCK recognition maintains this level.

Transmitting	Processing
Certificate Exchange	
0. $A \rightarrow B$: hello	
1. $B \rightarrow A$: r_B	
2. $A \rightarrow B$: $r_A, \langle t_u, id_A, r_B, PK_A \rangle$	B: verify certificate
3. $B \rightarrow A$: $\langle t_v, id_B, r_A, PK_B \rangle$	A: verify certificate
Compute Time Differences	
4.	A: determine t_i^A with $d_u^A = i - u, d_v^A = i - v$
	B: determine t_j^B with $d_v^B = j - v, d_u^B = j - u$
Key Exchange	
5. $A \rightarrow B$: $a_n', MAC(a_n', A_c)$	
6. $B \rightarrow A$: $b_n', MAC(b_n', B_c), B_{c+1}$	A: $h^{2d_v^A \pm \{0,1\}}(B_{c+1}) \stackrel{?}{=} PK_B$
Knowledge Proof	
7. $A \rightarrow B$: A_c	B: $h^{2d_u^B + 1 \pm \{0,1\}}(A_c) \stackrel{?}{=} PK_A$
8. $B \rightarrow A$: B_c	A: $h(B_c) \stackrel{?}{=} B_{c+1}$
Store ZCK-Key	
9.	A: $\mathcal{P}(b_n')$
	B: $\mathcal{P}(a_n')$

Figure 8.25 Identity certified authentication protocol.

values are signed by the manufacturer MF at some time interval t_u and t_v, respectively, such that Alice and Bob obtain certificates $\langle t_u, PK_A \rangle$ and $\langle t_v, PK_B \rangle$. Now, if Alice wants to prove her identity to Bob, she sends Bob her certificate. Assume Bob, determines the current time as t_j^B and Alice as t_i^A. These two timings might be different since we do not require a strict time synchronization. Then Alice and Bob determine the time difference between the current time they measure and the certificate time so that they obtain the interval differences $d_u^A = i - u$, $d_v^A = i - v$, $d_v^B = j - v$, and $d_u^B = j - u$. Alice then opens the key $A_{c+1} = a_{2(n-1-d_u^A)+1}$. Bob now checks whether $h^{2d_u^A \pm \{0,1\}}(A_{c+1}) = PK_A$ where $h^i(x)$ is the ith iteration of the hash function. We allow that Bob's determined time t_j^B differs from Alice's determined time t_i^A by one interval length L, that is, Alice's and Bob's determined times may each differ from the actual global time by L.[2] Hence, Bob will tolerate that he got a chain element that is one position behind or before the element he expected. We express this as a correction term "$\pm\{0, 1\}$" in the formula.

We now want to reduce the risk that Bob uses key values he just obtained to impersonate Alice (or vice-versa). To prevent such an attack where Bob obtains the hash-chain value for some time interval and then uses it to negotiate with a third entity, we introduce a random seed. Alice and Bob select a random seed r_A and r_B. Both Alice and Bob have a set of precomputed certificates in the form $\langle t_A, r, PK_A \rangle$ and $\langle t_B, r, PK_B \rangle$ for all possible r, which they obtain from the manufacturer. Now Bob can only impersonate Alice to a third party John if John uses for some time interval the same random seed as Alice. Since this is an interactive proof system, we expect that a small set for the seeds r suffices, for example, r being an 8-bit number. In that case each entity has to store 2^8 precomputed certificates. If we would set the time interval to be 1 min, Bob had to ask on average 2^7 entities in 2 min to find an entity using the same random seed as Alice. Further possibilities to increase the security level are given below.

Assume Alice wants to authenticate to Bob and starts the authentication process. The mutual ZCK protocol of Figure 8.23 is used as a basis here in order to pairwise exchange

---

[2] Note that the required level of time synchronization only depends on the size of the time intervals.

a certified hash chain. The IC protocol is illustrated in Figure 8.25. The core idea here is that for each time interval there is a different key. One has to prove knowledge of the appropriate key for a given time interval. However, a third party cannot use this captured key for an impersonation attack since time moves forward and the captured key is not valid anymore. The protocol has several phases. First, there is the certificate exchange (steps 0 to 3 in Fig. 8.25). Then Alice and Bob determine the difference between current time and issuing time of the certificates (step 4), and perform the ZCK key exchange (steps 5 and 6). Now Alice and Bob prove knowledge of a previous hash-chain element by opening keys (steps 7 and 8). Finally, if the knowledge proof was successful, Alice and Bob store the exchanged ZCK key for further identification processes performed by the ZCK protocol. We observe the following characteristics of the IC protocol:

- Once a session key of a hash chain is exchanged and authenticated, this key is used as described in the basic ZCK protocol, that is, it is used for all further identification and message authentication procedures for the entity pair Alice and Bob.
- For the public key certificate an arbitrary signature scheme might be used. We assume an RSA-like scheme because it performs signature verification very efficiently.
- The scheme is vulnerable to a denial-of-service attack by sending invalid certificates. To the authors' knowledge there is no solution known when using asymmetric cryptography for this problem in general, though.

If we set the interval length L to be infinitely small and the bit size of the challenge r to an appropriate security level, it is clear that this scheme is as secure as the ZCK protocol. However, the loose time synchronization as well as a small security parameter l_r for r introduce weaknesses. We see the following two attacks. An adversary Mallory starts an identification process with Alice to obtain her certificate and a key of the hash chain for the seed r_A. He then uses these values for Bob to obtain his certificate and one of Bob's keys for the seed r_B. Now he uses Bob's key and certificate against Alice who has to use the same seed r_A again such that the adversary obtains the second key for the current time interval. Mallory is then able to impersonate Alice against Bob in the current time interval. In the second attack, Mallory eavesdrops an identification process between Alice and Bob to obtain a certificate and the keys for the current time interval. He then tries to impersonate Alice against a third entity John with these values in that same time interval. In both cases Mallory uses the fact that he knows keys that are valid in the current time interval. Hence he has at most two time intervals to impersonate Alice because of the time tolerance we allow. However, in both cases Mallory has to find an entity that uses the same random seed r_A as Alice since the keys he obtained are only valid for this seed. It becomes clear that by increasing the bit size of the seed r and shortening the time intervals the security level can be set arbitrarily high.

For a higher security level it is possible to perform several authentication processes in parallel. In such a case, all messages of the parallel processes are sent at once. The number of exchanged messages does not increase, whereas the data, computation, and storage overhead increases linearly. Note that the used certificates need to be distinguishable, that is, for two parallel processes there need to be two distinguishable sets of certificates for each r. For instance, when performing three processes in parallel, an adversary Mallory has 2 min to find an entity requiring the proper nonce that on average takes him $2^{23} = 8388608$ identification processes. Then there have to be three distinguishable sets of certificates.

Hence there is memory space required for $p \times 2^r$ certificates, where p is the number of parallel instances.

The certificate lifetimes depend on the application and the available infrastructure. If certificates can be renewed once in a while, the lifetime can be chosen quite short. In many applications a temporary network will be available to support certificate renewal. Other applications may require that the certificate lifetime exceeds the device's lifetime. In the following, we consider a sensor network scenario. We assume that the IC hash chain has $n = 1000$ elements at a time interval length of 5 min such that the certificate lifetime is more than 3 days. Using the methods of [54], such a hash chain can be computed with $\frac{1}{2} \log_2(n)$ computations at the storage of $\log_2(n)$ elements, that is, if 10 elements are stored, each element can be recovered by 5 hash function iterations. The verifier has to iterate the hash chain to the public key, that is, on average he has to run $n/2$ iterations. Since an element is 10 bytes in size, there is storage needed of 100 bytes for each hash chain. Furthermore, each 1024-bit RSA certificate requires memory of $128 + 10 = 138$ bytes. Thus, using a certificate set of 2^8, there is memory space required of $256/(138 + 100)$ bytes ≈ 60 kbytes. When using the IC protocol for a static sensor network, it is reasonable to limit the number of successful IC authentications to one per time interval.

Thus an adversary Mallory has less than 5 min time left to impersonate Alice when there are 2^8 different challenges r, that is, Mallory's probability of a successful impersonation is 2^{-7}, which for some applications might be sufficient for an interactive proof system in the considered network class. However, it is not appropriate for applications such as financial transactions of significant value. Clearly, if there is more memory space and computational power available, for example, in a more powerful device such as a sensor aggregator or a more powerful device in an ad hoc network, there could be used 6 parallel sessions with a certificate set of $r = 5$. Thus the complexity of an attack would increase to 2^{30}, which is considered to be a high security level for interactive proof systems. In ad hoc networks with more powerful devices the certificate lifetime can be chosen far higher by lengthening the hash chains. Furthermore, recent research by M.Fischlin suggests including further information with the public key to reduce the effort for the verifier [57]. Hence, we believe that chain lengths of up to $n = 1,000,000$ can be chosen to support a certificate lifetime of several years in ad hoc networks consisting of powerful devices.

8.4.4 Energy Consumption

We introduce a simple energy model that suffices for our purpose. We hereby consider the energy consumption that results of radio transmission and processor activity due to the cryptographic overhead. A more sophisticated energy model was introduced in [11]. Actual comparisons of the energy consumption of standard cryptographic algorithms can be found in [58,59].

Device Architecture For the subsequent estimation of the rising energy cost we consider a highly constrained homogeneous network consisting of devices equipped with an 8-bit CPU and small memory and bandwidth capacity. For our analysis, we use the Mica Motes of Crossbow [60] as a reference platform. The Mica2 has 128 kbytes of programmable flash memory, 4 kbytes SRAM, and runs at 8 MHz. Hence, it is already a relatively high powered sensor device. A less costly sensor might have a 4-bit CPU at less than 4 MHz, and might thus be an order of magnitude slower. The packet size is 36 bytes of raw data, and 28 bytes of data without header at 38.4 kbaud and a transmission range of at most 150 m. The Mica Motes run under TinyOS [61].

Asymmetric cryptographic primitives, for example, digital signatures, require quite some execution time and thus cause much energy consumption. In some cases, asymmetric methods might be too demanding at all. A 134-bit ECC operation on an 8-bit 8051 processor, which is roughly comparable to our considered hardware platform and our security level, takes 2998 ms for a point multiplication [62]. Thus, the execution time for a signature verification is about 6 s (two-point multiplications). We can estimate that a 1024-bit RSA signature verification is more than 10 times faster than an ECC verification at a similar security level [63]. Thus we expect an RSA verification to run in 0.5 s [64]. For the subsequent estimation of the protocols' energy consumption we chose RC5 for both, computing the MAC and the hash chain since it is already part of TinySec, the security module of TinyOS. An additional scheme would cost additional storage space. On the Mica Motes an RC5-based hash function $\mathcal{H}$ has a running time of 2.22 ms and a *CBC–MAC* based on RC5 takes 4.18 ms [65].

Energy Model There are two operation modes of sensor nodes that largely reduce the battery charge, namely processor activities and the transmission of data. The number of CPU operations for the IC and the ZCK is fixed and so is the resulting battery consumption for processing. However, the energy cost for data transmissions is variable for both protocols. They increase linearly with the packet size, and proportionally to the square of the transmitting distance. In particular, the latter means that with a varying node density in the network, the energy cost for transmission will significantly vary.

Our subsequent estimation of the total energy consumption for the ZCK protocol and the IC protocol is based on the following assumptions. All values are derived of the Mica2 Motes specification [60]. At a speed of 38.4 kbaud, the transmission of one packet of 36 bytes takes about 10 ms. The Motes are powered by 2 AA batteries, which supply around 2200 mAh [66]. For computation, we assume that the energy consumption requires a current consumption of 8000 nA such that computation of 1 ms costs 0.0022 nAh. Receiving a data packet also has a current of 8000 nA, that is, receiving of a packet that takes 10 ms costs 0.022 nAh. Transmitting at maximum power requires a current of 25 mA, that is, transmitting a packet that takes 10 ms costs 0.07 nAh. The Mica Motes specification gives a maximum distance of 150 m. As this value is for outdoors in a perfect environment, we assume a distance of 50 m here at maximum transmission power. We now give an overview of our assumptions.

Processing Energy E_P

- The energy consumption E_P for processing a basic operation ζ is linear to its execution time t_ζ.
- $E_P(t_\zeta) := t_\zeta \cdot E_P(1)$.

Transmitting Energy E_T

- Let $p(\eta)$ be the number of packets required to send a message η. The energy consumption E_T for transmitting a message η for a given distance d is linear to the number of packets $p(\eta)$. Furthermore, the energy consumption E_T for transmitting a given message η is quadratic to its transmission distance d.
- $E_T(d, \eta) := d^2 p(\eta) E_T(1, 1)$.

Receiving Energy E_R

- The energy consumption E_R for receiving a message η is linear to the number of packets $p(\eta)$.
- $E_R(p(\eta)) = p(\eta)E_R(1)$.

Note that for the device architecture introduced before it holds that

- $E_P(r \text{ ms}) = 0.0022r \text{ nAh}$.
- $E_T(d \text{ m}, p(\eta) \text{ packets}) = 0.07(d/50)^2 p(\eta) \text{ nAh}$.
- $E_R(p(\eta)) = p(\eta)0.022 \text{ nAh}$.

We are aware that the Mica Motes do not allow dynamic adjustment of the transmission power. However, as future motes might allow this, we include this possibility in our model. A node's energy consumption in idle, sleep, or receive mode is negligible compared to transmission and processing energy. In the reminder, we use the following values:

Operations ζ	t_ζ	$E_P(\zeta)$
VER$_{RSA}$	0.5 s	1.1 nAh
$\mathcal{H}$	2.22 ms	0.0049 nAh
$\mathcal{P}$	Negligible	—

Messages η	Size (bytes)	$p(\eta)$	$E_T(d \text{ m}, \eta)$	$E_R(\eta)$
$\mathcal{H}$	10	1	0.07 $(d/50)^2$ nAh	0.022 nAh
cert$_{RSA}$	138	5	0.35 $(d/50)^2$ nAh	0.11 nAh
r	1	1	0.07 $(d/50)^2$ nAh	0.022 nAh

Clearly, our model assumes an ideal environment. One may argue that our model simplifies unrealistically. Nevertheless, we believe that it is well suited as a basic metric to decide under which circumstances protocols such as the ZCK and the IC protocol are applicable to sensor networks.

Energy Map of the ZCK Protocol   We now list the energy consumption of the ZCK protocol of Figure 8.21 per each protocol step and per each involved sensor node. We are using the notations of the previous section. Steps 1 and 2 are done only once and include the computation of the public keys a_n and b_n, respectively. Steps 4 and 5 include the authentication of the message for which the right key of the hash chain needs to be computed. For hash chains of length $n = 100$, there are four iterations of the hash function necessary.

In Table 8.3 we present a node's total energy consumption when running the ZCK protocol. We only consider the energy overhead here and not the energy consumption for normal messages or the MAC as this needs to be transmitted in every scheme to obtain a basic security level. Obviously, the energy consumption depends on a node's role within the authentication process. If a node is in the role of Alice, Bob, or an intermediate node,

Table 8.3   Energy Map of ZCK Protocol per Protocol Step and per Involved Entity

	Alice	$N_i(\forall i = 1, \ldots, n)$	Bob
1.	$100 E_P(\mathcal{H}) + E_T(d, \mathcal{H})$	$E_R(\mathcal{H}) + E_T(d, \mathcal{H})$	$E_R(\mathcal{H}) + E_P(\mathcal{P})$
2.	$E_R(\mathcal{H}) + E_P(\mathcal{P})$	$E_R(\mathcal{H}) + E_T(d, \mathcal{H})$	$100 E_P(\mathcal{H}) + E_T(d, \mathcal{H})$
3.	—	—	—
4.[a]	$4 E_P(\mathcal{H})$	—	—
5.[a]	$E_R(\mathcal{H}) + E_P(\mathcal{H})$	$E_R(\mathcal{H}) + E_T(d, \mathcal{H})$	$4 E_p(\mathcal{H}) + E_T(d, \mathcal{H})$
6.	$E_T(d, \mathcal{H})$	$E_R(\mathcal{H}) + E_T(d, \mathcal{H})$	$E_R + E_P(\mathcal{H})$
7.	$E_P(\mathcal{P})$	—	$E_P(\mathcal{P})$

[a] : = message exchange.

its energy consumption is E_A, E_B, or E_N. For the key exchange, the following energy cost is necessary only initially (steps 1 and 2):

$$E'_A = E'_B = E_R(\mathcal{H}) + E_T(d, \mathcal{H}) + E_P(\mathcal{P}) + 100 E_P(\mathcal{H})$$
$$= 0.506 + 0.07(d/50)^2 \text{ nAh} \qquad (8.4)$$

$$E'_N = 2 E_R(\mathcal{H}) + 2 E_T(d, \mathcal{H})$$
$$= 0.044 + 0.14(d/50)^2 \text{ nAh} \qquad (8.5)$$

For each authentication process, there is the following energy necessary (steps 3 to 7):

$$E_A = E_B = E_R(\mathcal{H}) + E_T(d, \mathcal{H}) + E_P(\mathcal{P}) + 5 E_P(\mathcal{H})$$
$$= 0.0462 + 0.07(d/50)^2 \text{ nAh} \qquad (8.6)$$

$$E_N = 2 E_R(\mathcal{H}) + 2 E_T(d, \mathcal{H})$$
$$= 0.044 + 0.14(d/50)^2 \text{ nAh} \qquad (8.7)$$

Let x be the number of message authentication processes that are performed between a pair Alice and Bob. Thus the total energy can be computed as $\hat{E} = E' + xE$. Hence we obtain

$$\hat{E}_A = \hat{E}_B = (x + 1) E_R(\mathcal{H}) + (x + 1) E_T(d, \mathcal{H}) + (x + 1) E_P(\mathcal{P}) + (100 + 5x) E_P(\mathcal{H})$$
$$= x \, 0.0462 + 0.506 + (x + 1) 0.07(d/50)^2 \text{ nAh} \qquad (8.8)$$

$$\hat{E}_N = 2(x + 1) E_R(\mathcal{H}) + 2(x + 1) E_T(d, \mathcal{H})$$
$$= (x + 1) \, 0.044 + (x + 1) 0.14(d/50)^2 \text{ nAh} \qquad (8.9)$$

Energy Map of the IC Protocol   We now derive the energy consumption of the IC protocol of Figure 8.25 as presented in Table 8.4. Again, the energy that is required by Alice, Bob, and the intermediate nodes is divided into the different columns. The precomputation that is done by a workstation is not part of this analysis. We assume that the hash chains have $n = 1000$ elements. On average there are 10 hash iterations computed by Alice and 500 by Bob to verify the public key (steps 6 and 7). Furthermore, we assume that the ZCK hash

Table 8.4   Energy Map of IC Protocol per Protocol Step and per Involved Entity

	Alice	$N_i(\forall i = 1 \dots, n)$	Bob
1.	$E_R(r)$	$E_T(d, r) + E_R(r)$	$E_T(d, r)$
2.	$E_T(d, r + \text{cert}_{\text{RSA}})$	$E_T(d, r + \text{cert}_{\text{RSA}}) +$ $E_R(r + \text{cert}_{\text{RSA}})$	$E_R(r + \text{cert}_{\text{RSA}}) +$ $E_P(\text{VER}_{\text{RSA}})$
3.	$E_R(\text{cert}_{\text{RSA}}) + E_P(\text{VER}_{\text{RSA}})$	$E_T(d, \text{cert}_{\text{RSA}}) + E_R(\text{cert}_{\text{RSA}})$	$E_T(\text{cert}_{\text{RSA}})$
4.	—	—	—
5.	$E_T(d, 2\mathcal{H}) + 110 E_P(\mathcal{H})$	$E_T(d, 2\mathcal{H}) + E_R(2\mathcal{H})$	$E_R(2\mathcal{H})$
6.	$E_R(3\mathcal{H}) + 500 E_P(\mathcal{H})$	$E_T(d, 3\mathcal{H}) + E_R(3\mathcal{H})$	$E_T(d, 3\mathcal{H}) + 110 E_P(\mathcal{H})$
7.	$E_T(\mathcal{H})$	$E_T(d, \mathcal{H}) + E_R(\mathcal{H})$	$E_R(d, \mathcal{H}) + 500 E_P(\mathcal{H})$
8.	$E_R(\mathcal{H}) + E_P(\mathcal{H})$	$E_T(d, \mathcal{H}) + E_R(\mathcal{H})$	$E_T(d, \mathcal{H})$
9.	$E_P(\mathcal{P})$	—	$E_P(\mathcal{P})$

chain, which is created and transmitted in steps 5 and 6, has a length of $n' = 100$ elements such that altogether $500 + 100 + 10 = 610$ hash iterations are performed both by Alice and Bob in steps 5 to 7. Note that the 100-element chain is used for the subsequent ZCK protocol, whereas the 1000-element chain is used for the IC protocol only. When using the ZCK protocol afterwards, the initial phase of the ZCK protocol, does not need to be redone.

From this energy map we derive the following equations, which represent the sensor node's energy consumption of the IC protocol for the different roles.

$$E_A'' = E_T(d, r + \text{cert}_{\text{RSA}} + 3\mathcal{H}) + E_R(r + \text{cert}_{\text{RSA}} + 4\mathcal{H}) + E_P(611\mathcal{H} + \text{VER}_{\text{RSA}} + \mathcal{P})$$
$$= 0.63(d/50)^2 + 4.3072 \text{ nAh} \tag{8.10}$$

$$E_N'' = E_T(d, 7\mathcal{H} + 2r + 2\text{cert}_{\text{RSA}}) + E_R(7\mathcal{H} + 2r + 2\text{cert}_{\text{RSA}})$$
$$= 1.33(d/50)^2 + 0.418 \text{nAh} \tag{8.11}$$

$$E_B'' = E_T(d, r + \text{cert}_{\text{RSA}} + 4\mathcal{H}) + E_R(r + \text{cert}_{\text{RSA}} + 3\mathcal{H}) + E_P[610\mathcal{H} + \text{VER}_{\text{RSA}} + \mathcal{P}]$$
$$= 0.7(d/50)^2 + 4.2804 \text{ nAh} \tag{8.12}$$

After the identification phase and the exchange of a ZCK key, the ZCK protocol can be used for message authentication. Again, let x be the number of message authentications that are performed afterwards. Then we can compute the total energy as $\hat{E} = E'' + x \times E$ by combining Eqs. (8.6) and (8.7) with Eqs. (8.10) to (8.12).

$$\hat{E}_A = E_T[d, r + \text{cert}_{\text{RSA}} + (3 + x)\mathcal{H}] + E_R[r + \text{cert}_{\text{RSA}} + (4 + x)\mathcal{H}]$$
$$+ E_P[(111 + 5x)\mathcal{H} + \text{VER}_{\text{RSA}} + (1 + x)\mathcal{P}]$$
$$= (x\, 0.07 + 0.63)\, (d/50)^2 + x\, 0.0462 + 4.3072 \text{ nAh} \tag{8.13}$$

$$\hat{E}_N = E_T[d, (7 + 2x)\mathcal{H} + 2r + 2\text{cert}_{\text{RSA}}] + E_R[(7 + 2x)\mathcal{H} + 2r + 2\text{cert}_{\text{RSA}}]$$
$$= (x\, 0.14 + 1.33)\, (d/50)^2 + x\, 0.044 + 0.418 \text{ nAh} \tag{8.14}$$

$$\hat{E}_B = E_T[d, r + \text{cert}_{\text{RSA}} + (4 + x)\mathcal{H}] + E_R[r + \text{cert}_{\text{RSA}} + (3 + x)\mathcal{H}]$$
$$+ E_P[(110 + 5x)\mathcal{H} + \text{VER}_{\text{RSA}} + (1 + x)\mathcal{P}]$$
$$= (x\, 0.07 + 0.7)\, (d/50)^2 + x\, 0.0462 + 4.2804 \text{ nAh} \tag{8.15}$$

Energy Map of Traditional Protocols With respect to the following comparisons we now want to present an energy consideration of alternative traditional protocols using hybrid methods as presented in Section 8.4.1. We presented the ZCK and the IC protocols to provide recognition and identification, respectively. Alternatively, one can use a single shared key for all devices, which does not cause any overhead. However, then all the devices had to be under the control of the same authority, and the entire network depends on the security of each of the devices, that is, if one device gets broken or tampered with, all the device are insecure. A pairwise predistribution of symmetric keys overcomes some of these problems, but it does not allow the network to be extended and thus is limited to only a few authorities. Thus we consider the following two protocols, which are comparable to the ZCK and IC protocol:

1. *Uncertified DH (UDH)* Uncertified Diffie–Hellman key exchange with subsequent symmetric authentication ↔ ZCK
2. *Certified DH (CDH)* Certified Diffie–Hellman key exchange with subsequent symmetric authentication ↔ IC + ZCK

The first one, an uncertified DH key exchange, is used to exchange a symmetric key that in the following is used to authenticate messages or entities by a symmetric MAC scheme. The second one, a certified DH key exchange, implements a key exchange using certificates to provide identification of the participating parties. Both schemes work as presented in Algorithm 8.4.2. However, in UDH a public key is first computed and then exchanged, whereas in CDH the public key is exchanged as part of a certificate. The UDH protocol provides recognition and is equivalent to using the ZCK protocol, and the CDH protocol provides the functionality of the IC + ZCK protocol, that is, using IC for a ZCK key exchange and then using ZCK for message authentication. To derive the energy consumption, we extend our previous assumptions. We assume that elliptic curve cryptography (ECC) is used for the key exchange (ECDH) as it is far more efficient for this task than RSA. An ECC certificate $cert_{ECC}$ consists of an Elliptic Curve Digital Signature Algorithm (ECDSA) signature of 40 bytes and a public key PK_{ECC} of 21 bytes when using point compression such that 3 packets are required to transmit an ECC certificate. A point multiplication OP_{ECC}, which is the core operation on an elliptic curve, takes 3 s as described before, such that a signature generation SIG_{ECC} takes 3 s and a signature verification VER_{ECC} takes 6 s. Altogether we obtain:

Operations ζ	t_ζ	$E_P(\zeta)$
OP_{ECC}	$3s$	6.6 nAh
SIG_{ECC}	$3s$	6.6 nAh
VER_{ECC}	$6s$	13.2 nAh

messages η	size (bytes)	$p(\eta)$	$E_T(d \text{ ft}, \eta)$	$E_R(\eta)$
$cert_{ECC}$	61	3	$0.21(d/50)^2$ nAh	0.066 nAh
PK_{ECC}	21	1	$0.07(d/50)^2$ nAh	0.022 nAh

We now derive the energy equations for the UDH protocol. To be able to act anonymously such as it is provided by the ZCK protocol, that is, to change identity, we cannot assume that the public key is precomputed but needs to be computed once in a while. Here, for the ECDH each party performs an ECC operation OP_{ECC} to obtain a public key, submits this value and then performs another ECC operation to obtain the shared secret. Hence we obtain the energy consumption of the protocol flow for each of the involved entities.

$$E_A = E_B = 2E_P(\text{OP}_{\text{ECC}}) + E_T(\text{PK}_{\text{ECC}}) + E_R(\text{PK}_{\text{ECC}})$$
$$= 0.07(d/50)^2 + 13.222\,\text{nAh}$$
$$E_N = 2\,E_T(d, \text{PK}_{\text{ECC}}) + 2\,E_R(\text{PK}_{\text{ECC}})$$
$$= 0.14(d/50)^2 + 0.044\,\text{nAh}$$

If the CDH protocol is used, the certificate also needs to be checked, thus requiring another signature verification on each side. Contrary to the uncertified version, here the public key is precomputed as it is part of the certificate. Thus, the signature step is omitted, but another verification step for the certificate is included. Therefore, we can derive the energy consumption of the protocol flow for each of the involved entities as follows:

$$E_A = E_B = E_T(\text{cert}_{\text{ECC}}) + E_R(\text{cert}_{\text{ECC}}) + E_P(\text{OP}_{\text{ECC}} + \text{VER}_{\text{ECC}})$$
$$= 0.21(d/50)^2 + 19.866\,\text{nAh}$$
$$E_N = 2E_T(d, \text{cert}_{\text{ECC}}) + 2E_R(\text{cert}_{\text{ECC}})$$
$$= 0.42(d/50)^2 + 0.132\,\text{nAh}$$

Since we only consider the overhead caused in addition to a basic security level by using a MAC when using UDH or CDH, there is no additional overhead for the following message authentications. Hence for the total energy it is $\hat{E} = E$ in this case.

8.4.5 Efficiency Analysis

The previously introduced energy maps help us to compare the ZCK and IC protocols to the hybrid protocols CDH and UDH, respectively. In the following we first compare the operational complexity in detail, and then we analyze how the ZCK and IC protocols compare to the UDH and the CDH protocols regarding the energy consumption.

Operational Complexity We now consider the operational complexity of the above protocols for message recognition and message authentication, respectively. We assume the same parameters as above. Table 8.5 gives an overview of the complexity of the ZCK and

Table 8.5 Operational Complexity of ZCK and UDH

		ZCK	UDH
Initialization	Public key size (bytes)	10	21
	Exchanged messages	2	2
	Exchanged bytes	20	42
	Computational effort	—	4 PK Op.
Authentication	Exchanged messages	3	1
	Exchanged bytes	30	10
	Computational effort	—	—
	Code size	35 kbytes	—

Table 8.6 Operational Complexity of IC and CDH

		IC	CDH
Initializaton	Certificate size (bytes)	138	61
	Exchanged messages	7	2
	Exchanged bytes	348	122
	Computational effort	2 PK Op.	4 PK Op.
Authentication	Exchanged messages	3	1
	Exchanged bytes	30	10
	Computational effort	—	—
	Code size	45 kbytes	—

UDH protocols. It is distinguished in the initialization phase and the recognition phase. Note that contrary to the previous considerations here the transmission of the MAC for each recognition is also considered. We implemented the ZCK protocol on the Mica2 Motes. A preliminary version has a code size of around 35 kbytes. Since a hash function has negligible complexity compared to a public key operation, we omit it here. Note that for both the ZCK and the UDH protocols a full distribution of all the public keys requires $n(n - 1)$ keys as there is no central directory of public keys as there is for a public key infrastructure.

We now present the operational complexity of the IC and the CDH protocols in Table 8.6. Note that the IC protocol needs quite some memory storage for precomputation. An implementation of the IC protocol is to be expected at around a further 10 kbytes for the RSA verification such that we estimate a code size of 45 kbytes. Again, we distinguished the initialization phase and the authentication phase. The number of exchanged bytes for the IC can be computed as 2×138 bytes for the certificates, a further 70 bytes for the hash and MAC values, as well as 2 bytes for the seeds, such that we obtain 348 bytes altogether.

Device Lifetime We define the lifetime of a device as the duration of time the node works with a given power supply (e.g., battery) once it is deployed until the energy resources are exhausted and the device needs to be recharged or replaced. For simplicity we are considering a flat and homogeneous network, that is, each node has the same hardware characteristics, and there is no clustering structure within the network. Note that for the energy efficiency of the ZCK and the IC protocols, a flat homogeneous network is the most challenging architecture since here the total initial energy load of the network is most restricted. We now consider the lifetime of a network when using our new protocols. Let n be the average number of forwarding nodes, x be the average number of message authentication processes between each pair of parties that established a communication channel, and p be the average number of channels that a party establishes to other parties. Then the average energy consumption of an involved node $\widetilde{E}$ can be computed as follows:

$$\widetilde{E} = p(\hat{E}_A + n\hat{E}_N + \hat{E}_B)/(n + 2) \tag{8.16}$$

For the ZCK protocol, by applying Eqs. (8.8) and (8.9) we obtain

$$\widetilde{E}_{\text{ZCK}} = p/(n + 2)\,[n(x + 1)0.14(d/50)^2 + n(x + 1)0.044 \tag{8.17}$$
$$+ (x + 1)\,0.14(d/50)^2 + x\,0.0924 + 1.012]\quad[\text{nAh}]$$

Note that for simplicity we do not consider additional costs resulting from retransmission over the wireless. Let D be the average distance between Alice and Bob over several hops, and d be the distance of a node to its neighbor node, that is, $D = (n + 1)d$. For a given D, $\widetilde{E}$ will obviously be minimal for small d as this goes into Eq. (8.17) with square complexity. The battery of a sensor node has a capacity of around 2200 mAh of energy. For instance, if $D = 50$ m, $d = 10$ m, $n = 4$, and $x = 1000$, a sensor node can establish trust associations to $p > 40,000$ different parties before the energy is exhausted. If $D = 100$ m, $d = 50$ m, $n = 1$, and $x = 1000$, this decreases to around $p = 15,000$.

We now consider the lifetime of a sensor network when using the IC protocol to establish identified channels and afterwards ZCK to authenticate messages. The average energy consumption of an involved node can be computed by Eqs. (8.13) to (8.15) as follows:

$$
\begin{aligned}
\widetilde{E}_{IC} &= p(\hat{E}_A + n\hat{E}_N + \hat{E}_B)/(n + 2) \\
&= p/(n + 2)\{[(n + 1)0.14x + (n + 1)1.33](d/50)^2 \\
&\quad + nx\,0.044 + n\,0.418 + x\,0.0924 + 8.5876\} \quad \text{[nAh]} \quad (8.18)
\end{aligned}
$$

Since the IC protocol requires more computational resources than the ZCK protocol, we assume that a node only establishes identified relationships to its neighborhood. Thus we pick a random node Alice and consider the neighborhood of this node. All nodes that are reachable by at most l hops are in the l-hop neighborhood of our node. We denote by c the connectivity of nodes, that is, the number of nodes that a node can directly reach without any intermediate nodes. The number $p(l, c)$ of l-hop neighbors in a c-connected network for $3 \leq c \leq 6$ is then computed as

$$
p(l, c) = c \sum_{i=1}^{l+1} i \quad (8.19)
$$

Next, we need to know how many intermediate nodes are on average involved on the route from our node to a node of its neighborhood. All nodes that are reachable by one hop do not involve any intermediate nodes. All nodes that are reachable by two hops involve one intermediate node, and so on. Thus, altogether the number of intermediate nodes $n(l, c)$ involved in pairwise authentication in a c-connected network within an l-hop neighborhood of our node Alice can be computed as follows for $3 \leq c \leq 6$:

$$
n(l, c) = \left[c \sum_{i=1}^{l+1} (i - 1)i \right] \Big/ p(l, c) \quad (8.20)
$$

Note that with respect to their position nodes need to be counted multiple times to take into account that their involvement in multiple authentications with varying communication partners from the l-hop neighborhood.

Thus, we can compute the total average energy of a node within its neighborhood by plugging in values p and n of Eqs. (8.19) and (8.20) into (8.16). For instance, consider a homogeneous network that is 4-connected with $d = 20$ m. When using the IC protocol to establish an identified channel to the 1-hop neighborhood, that is, to $p(1, 4) = 12$ nodes, and on average there are $x = 1000$ messages exchanged, then $n(1, 4) = \frac{2}{3}$. By inserting these values into Eq. (8.18) we obtain an energy consumption of $\widetilde{E} = 757$ nAh, which is less than 0.04% of the total battery capacity of 2200 mAh. Clearly, the energy consumption of the CDH and UDH protocols also depends on the overall distance D and the distances d between each two nodes. As argued above, the energy consumption is minimal for a small distance d as it goes into the energy consumption with a square complexity. The average total energy can easily be obtained for the UDH protocol as

$$\widetilde{E}_{\mathrm{UDH}} = p/(n+2)\left[(n+1)0.14(d/50)^2 + n\,0.044 + 26.444\right] \quad [\mathrm{nAh}] \qquad (8.21)$$

and for the CDH as

$$\widetilde{E}_{\mathrm{CDH}} = p/(n+2)\left[(n+1)0.42(d/50)^2 + n\,0.132 + 39.732\right] \quad [\mathrm{nAh}] \qquad (8.22)$$

Comparison of the Protocols   Finally, we want to compare our new protocols to the hybrid ones. We first emphasize the benefits of the ZCK protocol: (1) it is computationally extremely efficient; (2) it is provably secure; (3) it works between two low-power devices; and (4) it allows a party easily to start fresh (anonymity). First, it seems clear that for recognition our ZCK protocol is computationally far more efficient than any asymmetric scheme, and it is nearly as efficient as a message authentication scheme using a MAC. Hence, for time-critical applications, or when there is the need of establishing new channels, often ZCK should be the first choice. If the number of intermediate nodes is small, ZCK is also from an energy perspective a good solution. If identification is needed, IC performs well in some scenarios. If the network is reliable and packets do not get lost frequently, the large certificate size only has moderate impact. Furthermore, if very low power devices are used, for example, sensors that are equipped with a 4-bit CPU at less than 4 MHz, IC might be the only possibility to perform an identification since an ECC key agreement is too demanding for these devices. Again, if there are time-critical applications, it might be a good idea to use IC since it is operationally more efficient than an ECC solution.

We now want to compare the energy consumption in more detail. Thus we compare the energy consumption of the ZCK to the UDH protocol, and of the IC+ZCK to the CDH protocol. Namely, we compare Eq. (8.17) to (8.21) and Eq. (8.18) to (8.22). Hence we are interested when

$$\mathrm{ZCK} \leftrightarrow \mathrm{UDH}\colon n(x+1)0.14(d/50)^2 + n(x+1)0.044 + (x+1)0.14(d/50)^2$$
$$+ x\,0.0924 + 1.012 < (n+1)0.14(d/50)^2 + n\,0.044 + 26.444$$

and

$$\mathrm{IC} + \mathrm{ZCK} \leftrightarrow \mathrm{CDH}\colon [(n+1)0.14x + (n+1)1.33](d/50)^2 + nx\,0.044$$
$$+ n\,0.418 + x\,0.0924 + 8.5876 < [(n+1)0.42(d/50)^2] + n\,0.132 + 39.732$$

Figure 8.26 illustrates the trade-off between the ZCK and the UDH protocols. There are five graphs for fixed $n = 0, \ldots, 4$ that determine the trade-off of messages x for different distances d, that is, the number of messages x where the ZCK and UDH protocols are equivalent regarding their energy consumption. Below the curve, the ZCK protocol requires less energy than UDH. Figure 8.27 illustrates the trade-off between the IC + ZCK and the CDH protocols in the same manner. For instance, let us consider the first case for a scenario where $d = 20$ m and $n = 4$. If on average there are at most $x = 66$ message authentications performed, the ZCK protocol requires less energy then the UDH protocol, otherwise it

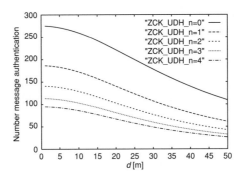

Figure 8.26   Trade-off between the ZCK and UDH protocol.

requires more energy. For a network where $d = 50$ m and $n = 0$, ZCK is more energy efficient for $x \leq 109$. Now let us compare the IC + ZCK to the CDH protocol. Then for $d = 20$ m and $n = 4$, IC + ZCK is more energy efficient for $x \leq 77$, and for $d = 50$ m and $n = 0$ it is for $x \leq 131$.

It now becomes clear that the ZCK protocol and also the IC protocol are very energy efficient for a small average number of hops n in the network and a small number of message exchanges x per communication pair. Hence, we foresee applications in highly mobile networks. Furthermore, when computational efficiency is of importance, for example, in time-critical applications, our protocols are by far more efficient than any solution using intensive asymmetric cryptography. On the other hand our ZCK protocol requires caching of a message before it can be verified. Thus the scheme is vulnerable to a message overflow attack. Due to the increased number of messages compared to other schemes our ZCK as well as our IC protocol might be less powerful in unreliable networks. If data packets get lost frequently, the number of resent messages will decrease network performance.

8.4.6 Conclusions and Outlook

We presented two pairwise authentication protocols for WSNs that mainly require symmetric primitives such as key chains and avoid exhaustive computations. Whereas the ZCK protocol provides pairwise rerecognition, the initial usage of the IC protocol within the bootstrapping phase also provides pairwise identification of sensor nodes.

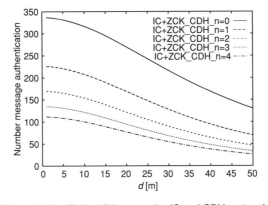

Figure 8.27   Trade-off between the IC and CDH protocol.

We analyzed both protocols with respect to their energy consumption by applying a simple energy model that considers processing activities as well as transmitting activities. Our results show that our protocols are very efficient regarding the computational complexity that makes them an appropriate fit for time-critical applications. However, regarding energy efficiency our protocols are only applicable if a relatively small number of messages is exchanged between two parties after the initial phase, and if the distance between devices is moderate. This might be the case if the network topology changes steadily such that new pairs of communicating parties are established all the time. Furthermore, our protocols are useful if only a communication channel to a node's neighborhood is established. If the distance between a node to its neighborhood is rather large, it is clear that the additional transmissions required for our protocol cost too much energy.

The considered class represented by the Mica2 Motes cannot be considered to be an ultra-low-powered device anymore as it has a similar computational power as a smart card. Its price is also not in a range such that the motes could be deployed by the thousands. If devices are to be used that are in the range of $1, our protocols might be an interesting choice as they enable authentication, whereas asymmetric protocols might be too demanding at all. However, in such cases a static solution with predistributed keys might be the default solution except where the devices need to be more flexible. It seems to be clear that the IC protocol for sensor networks only has a weak security and thus is better suited to devices in ad hoc networks that are more powerful such as PDAs or cell phones. The IC protocol should only be rarely used in sensor networks, for example, when a sensor needs to be replaced once its battery is exhausted.

The energy model we use reflects reality only very simplistic. Furthermore, we were not able yet to verify the energy estimations for computation and transmission that we use. We are currently working on improving our energy model and measure energy values using a test bed of Mica Motes. We are also evaluating the characteristics of recognition by comparing it to authentication to derive possible application scenarios of our protocols.

We believe that our schemes shorten the gap to digital signature schemes that are still more flexible and powerful but also unrealistic with respect to their energy consumption. We think that for sensor networks, which consist of computationally weak devices with restricted energy capacity, protocols have to be designed that are largely based on symmetric primitives while getting close to the functionality of public key schemes. Furthermore, these protocols should also be provably secure. Only such protocols will allow that sensor networks are deployed with a sufficient level of security and flexibility. We believe that our ZCK and IC protocols, which are extremely efficient and also provably secure, are a good start in this direction.

8.5 FAST AND SCALABLE KEY ESTABLISHMENT IN SENSOR NETWORKS

Tassos Dimitriou, Ioannis Krontiris, and Fotios Nikakis

We present a protocol for key establishment in sensor networks. Unlike random key predistribution schemes, our protocol does not rely on probabilistic key sharing among sensors. During a quick bootstrapping phase, nodes use their predeployed key material to form groups or clusters of small size that share a common key. Intercluster communication is then achieved by nodes sharing cluster keys. Our scheme is *scalable* and provides *resiliency* against node capture and replication. This is due to the fact that keys are localized; keys that appear in some part of the network are not used again. So, even if a node is compromised and its keys exposed, an adversary can have access only to a very small portion of the network centered around the compromised node. What is more important, however, is that our protocol is optimized for *message broadcast*; when a node has to broadcast a message, it does not have to encrypt it each time with a key targeted for a specific neighbor. This saves energy and makes retransmissions unnecessary. Furthermore, our scheme is suited for *data fusion* and aggregation processing; if necessary, nodes can "peak" at encrypted data using their cluster key and decide upon forwarding or discarding redundant information.

8.5.1 Background Information

Sensor networks have attracted much scientific interest during the past few years. These networks use hundreds to thousands of inexpensive wireless sensor nodes over an area for the purpose of monitoring certain phenomena and capture geographically distinct measurements over a long period of time (see [67] and [68] for a survey). Nodes employed in sensor networks are characterized by limited resources such as storage, computational, and communication capabilities. The power of sensor networks, however, lies exactly in the fact that their nodes are so small and cheap to build that a large number of them can be used to cover an extended geographical area. In most sensor networks there are one or more base stations to which data collected by sensor nodes is relayed for uploading to external systems.

Many critical applications require that the data must be exchanged in a secure way. Establishing secure communication between sensor nodes becomes a challenging task, given the limited processing power, storage, bandwidth, and energy resources. Public key algorithms, such as RSA, are undesirable, as they are computationally expensive. Instead, symmetric encryption/decryption algorithms and hashing functions are between two to four orders of magnitude faster [69] and constitute the basic tools for securing sensor network communication.

For these algorithms a shared key between the two communicating parties is required. Since we do not have prior knowledge of which nodes will be neighbors of each other after deployment, a solution would be for every pair of sensor nodes in the network to share a unique key. However, this is not feasible due to memory constraints. A more scalable solution is the use of a key common to all sensor nodes in the network [70]. The problem with this approach is that if a single node is compromised then the security of the whole network is disrupted. Furthermore, refreshing the key becomes too expensive due to communication overhead.

There exist several schemes [71–74] proposed in the literature that suggest random key predistribution: Before deployment each sensor node is loaded with a set of symmetric keys that have been randomly chosen from a key pool. Then nodes can communicate with each other by using one or more of the keys they share according to the model of random key predistribution used. These schemes offer network resilience against node capture but they are not "infinitely" scalable. As the size of the sensor network increases, the number of symmetric keys needed to be stored in sensor nodes must also be increased in order to provide sufficient security of links. Unfortunately, the more keys are stored in a network, the more links become compromised (even not neighboring ones) in case of node capture. Hence these schemes offer only "probabilistic" security as other links may be exposed with certain probability. A desirable feature is therefore that compromise of nodes should result in a breach of security that is *constrained* within a small, localized part of the network.

Another problem that is not handled well by current key predistribution schemes is that of simple message broadcast. Many sensor network protocols require a node to broadcast a message to a subset or all of its neighbors. If a random key predistribution scheme is used, it is highly unlikely that a node can share the *same* key (or set of keys) with *all* its neighbors. Hence nodes have to make multiple transmissions of messages, encrypted each time with a different key. Thus such schemes suffer from this message broadcast problem and focus only on establishing node-to-node security. In general, we believe that transmissions must be kept as low as possible because of their high energy consumption rate.

Finally, a closely related problem to that of broadcasting encrypted messages is the ability to perform aggregation and data fusion processing [75]. This, however, can be done only if intermediate nodes have access to encrypted data to (possibly) discard extraneous messages reported back to the base station. The use of pairwise shared keys effectively hinders data fusion processing.

Our Contribution In this work we present a security protocol that has the following properties:

- *Resilience to Node Capture* Our scheme offers deterministic security as a single compromised node disrupts only a *local* portion of the network while the rest remains fully secured. We designed our protocol without the assumption of tamper resistance. Once an adversary captures a node, key materials can be revealed. Building tamper resistance into sensor nodes can substantially increase their cost, and, besides, trusting such nodes can be problematic [29].
- *Resistance to Node Replication* Even if a node is compromised and can be used to populate the network with its clones, an adversary cannot gain control of the network as key material from one part of the network cannot be used to disrupt communications to some other part of it.
- *Energy Efficiency* We enable secure communication between a node and its neighbors by requiring only *one* transmission per message. Thus messages do not have to be encrypted multiple times with different keys to reach all neighbors. This saves energy as transmissions are among the most expensive operations a sensor can perform [76].
- *Intermediate Node Accessibility of Data* An effective technique to extend sensor network lifetime is to limit the amount of data sent back to reporting nodes since this reduces communications energy consumption. This can be achieved by some processing of the raw data to discard extraneous reports. However, this cannot be done unless intermediate sensor nodes have access to the protected data to perform *data fusion* processing. Although existing random key predistribution schemes provide a secure path between a source and a destination, nearby nodes cannot have access to this information as it is highly unlikely they possess the right key to decrypt data.
- *Scalability* The number of keys stored in sensor nodes is *independent* from the network size and the security level remains unaffected.
- *Easy Deployment* Our protocol enables a newly deployed network to establish a secure infrastructure quickly using only local information and total absence of coordination.

8.5.2 Security Protocol

In this section we first describe a localized algorithm for key management in wireless sensor networks and then provide a scheme that utilizes the established keys in order to provide secure communication between a source node and the base station. In general, our scheme can provide secure communication between *any* pair of nodes by building transitive keying relationships, but we demonstrate our protocol with respect to the "node-to-base station" data delivery model, since this is the most commonly used in sensor networks. The protocol is divided into the following phases:

1. Initialization phase, which is performed before sensor nodes are deployed.
2. Cluster key setup phase, which splits the network into disjoint sets (clusters) and distributes a unique key to each cluster. That key is shared between all the cluster members, as well as the nodes that are one-hop away from the cluster.
3. Secure communication phase, which provides confidentiality, data authentication, and freshness for messages relayed between nodes toward the base station.

In what follows, we describe in detail the three phases of our protocol. In our discussion the following notation is used:

$M_1|M_2$ Concatenation of messages M_1 and M_2.

$E_{K,IV}(M)$ Encryption of message M, with key K and the initialization vector IV, which is used in counter mode (CTR).

$MAC_K(M)$ MAC of message M using key K.

Initialization Sensor nodes are assigned a unique ID that identifies them in the network, as well as three symmetric keys. Since wireless transmission of this information is not secure, it is assigned to the nodes during the manufacturing phase, before deployment. In particular the following keys are loaded into sensor nodes:

- K_i Shared between each node i and the base station. This key will be used to secure information sent from node i to the base station. It is not used in establishing the security infrastructure of the network but only to encrypt the sensed data D that must reach the base station in a secure manner. If we are interested in data fusion processing, this key should not be used to encrypt D as otherwise intermediate nodes will not be able to evaluate and possibly discard the data.

- K_c^i Shared between each node i and the base station. This key will be used only by those nodes that will become cluster heads and it will be the cluster key. These are the keys used to forward information to the base station in a hop-by-hop manner.

- K_m A master key shared among all nodes, including the base station. This key will be used to secure information exchanged during the cluster key setup phase. Then it is erased from the memory of the sensor nodes.

The base station is then given all the ID numbers and keys used in the network before the deployment phase. Since the base station stores information used to secure the entire network, it is necessary to include it in our trusted computing base.

Cluster Key Setup In this section we describe how sensor nodes use the predeployed key material in order to form a network where nodes can communicate with each other using a set of trusted keys. The cluster key setup procedure is divided into two phases: organization into clusters and secure link establishment. During the first phase the sensor nodes are organized into clusters and agree on a common cluster key, while in the second phase, secure links are established between clusters in order to form a connected graph.

An implicit assumption here is that the time required for the underlying communication graph to become connected (through the establishment of secure links) is smaller than the time needed by an adversary to compromise a sensor node during deployment. As security protocols for sensor networks should *not* be designed with the assumption of tamper resistance [29], we must assume that an adversary needs more time to compromise a node and discover the master key K_m. In the experimental section we give evidence that this is indeed the case.

Organization into Clusters   In this phase, the creation of clusters happens in a probabilistic way that requires the nodes to make at most one broadcast. Each node i waits a random time (according to an exponential distribution) before broadcasting an HELLO message to its neighbors declaring its decision to become a cluster head. This message is encrypted using K_m and contains the ID_i of the node, its key K_c^i and an authentication tag:

$$E_{K_m}(ID_i | K_c^i | MAC_{K_m}(\langle ID_i | K_c^i \rangle))$$

Upon receiving an HELLO message, a node decrypts and authenticates the message. Then reacts in the following way:

1. If the node has not made any decision about its role yet, it joins the cluster of the node that sent the message and cancels its timer. No transmission is required for that node. The key that it is going to be used to secure traffic is $K_c = K_c^i$.
2. If the node has already decided its role, it rejects the message. This will happen if the node has already received an HELLO message from another node and became a cluster member of the corresponding cluster, or the node has sent an HELLO message being a cluster head itself.

Upon termination of the first phase, the network will have been divided into clusters. All nodes will be either cluster heads or cluster members, depending on whether they sent an HELLO message or received one. We are assuming here that collisions are resolved at a lower level, otherwise acknowledgments must be incorporated in this simple protocol. There is, however, the possibility that two neighboring nodes send HELLO before they receive the same message from their neighbor, thus becoming cluster heads of themselves. Furthermore, there is a case for a node to send an HELLO message after all its neighboring nodes have decided their role, and thus become a head of a cluster with no members. Although these possibilities can be minimized by the right exponential distribution of the time delays that nodes send the HELLO messages, they do not affect the proper running of the protocol. In Figure 8.28 we show the distribution of nodes to clusters for densities (average number of neighbors per sensor) equal to 8 and 20. As it can be seen in the figure, for smaller densities a larger percentage of nodes forms clusters of size one. However, the probability of this event decreases as the density becomes larger.

At the end of this phase each cluster will be given an identifier CID, which can be the cluster head's ID. All nodes in a cluster will be sharing the same key, K_c, which is the key K_c^i of the cluster head. From this point on, cluster heads turn to normal members, as there is no more need for a hierarchical structure. This is important since cluster-based approaches usually create single points of failure as communications must usually pass

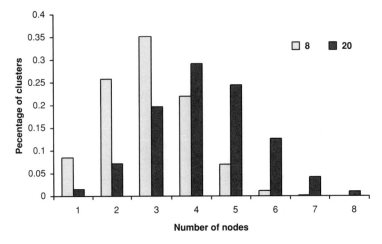

Figure 8.28 Distribution of nodes to clusters.

through a clusterhead. Figure 8.29 shows an example topology where three clusters have been created with CIDs 13, 9, and 19.

As seen from Figure 8.29, the maximum distance between two nodes in a cluster is two hops. Since all nodes in a cluster share a common key K_c, we need to keep the size of the clusters as small as possible in order to minimize the damage done by the compromise of a single node. In the experimental section, we give evidence that indeed clusters contain on average a small number of nodes that is independent of the size of the network.

Secure Link Establishment   In the second phase, all nodes get informed about the keys of their neighboring clusters. We need this phase in order to make the whole network connected since up to this point it is only divided into clusters whose nodes share a common key. We say that a node is *neighbor* of a cluster CID when that node has within its communication range

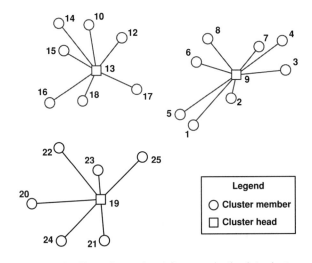

Figure 8.29 Example topology after organization into clusters.

at least one member of that cluster. This phase is executed with a simple local broadcast of the cluster key by all nodes. The message sent contains the tag and the CID, encrypted using K_m:

$$E_{K_m}(CID_i|K_c|MAC_{K_m}(\langle CID_i|K_c\rangle))$$

Nodes of the same cluster simply ignore the message, while any nodes from neighboring clusters will store the tuple $\langle CID, K_c \rangle$ and use it to decrypt traffic coming from that cluster, as explained in the next section. If the message has been sent from a member of the same cluster, then that message should be ignored.

We must emphasize again that the total time of both steps is too short for an adversary to capture a node and retrieve the key K_m (see also Fig. 8.36 for a justification of this claim.) Nevertheless an adversary could have monitored the key setup phase and by capturing a node at a later time it could retrieve all cluster keys. Therefore after the completion of the key setup phase, all nodes erase key K_m from their memory.

At this point, each node i of the sensor network will have its key K_i and a set S of cluster keys that includes its own cluster key and the keys of its neighboring clusters. The total number of the keys that a node will have to store depends on the number of its neighboring clusters, thus not all cluster members store the same number of keys. (In the experimental section we give evidence that each node needs to store on average a handful of cluster keys.) Most importantly, however, the number of keys that each node gets is *independent* of the network size, and therefore there is no upper limit on the number of sensor nodes that can be deployed in the network.

We illustrate the operations of the cluster key setup phase with the following example. Consider the sensor network depicted in Figure 8.30. Three clusters with CIDs 13, 9, and 19 have been formed from the first step. The figure also shows the transmission radius of nodes 25, 17, and 5. As can be seen, node 25 has two neighboring clusters, since node 17 from cluster 13 and nodes 5 and 1 from cluster 9 are within its communication range. Therefore, node

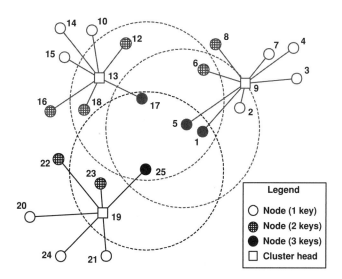

Figure 8.30 Example topology during the key setup phase.

17 will store 3 cluster keys. Likewise, nodes 17, 5, and 1 also have 2 neighboring clusters and will store 3 cluster keys each. On the other hand, node 6 is within the range of node 17 but outside the radius of any node from cluster 19; therefore, it will only store 2 cluster keys.

Secure Message Forwarding   In this section we describe how information propagating toward a base station can be secured to guarantee confidentiality, data authentication, and freshness. *Confidentiality* guarantees that sensor readings and other sensitive information (e.g., routing information) are protected from disclosure to unauthorized parties. This can be achieved by encrypting data with a secret key. *Data authentication* allows the base station to verify that data was really sent by the claimed sender and not from an adversary that injected it in the network. Even if messages cannot be inferred or forged, an adversary can capture them and replay them at a later time (message replay attack). *Data freshness* ensures that messages are fresh, meaning that they obey a message ordering and have not been reused.

Here we make the assumption that sensor readings must first be encrypted (step 1 in the description below) and then authenticated in a hop-by-hop manner (step 2) as data is forwarded to the base station through intermediate nodes. If we are interested in data fusion processing, then step 1 should be omitted. It is only used when we want to make sure that sensor readings can only be seen by the base station.

It is worth mentioning here that while we describe our protocol for the case that a sensed event must be securely forwarded to a base station, *the same ideas can be used with any routing protocol that requires secure communication between one-hop neighbors*. The key translation character of our protocol is very important in establishing a transitive chain of trust.

Step 1 (Optional)   To achieve the security requirements for the data D that will be exchanged between the source node and the base station, we use a SNEP [76] like exchange, as shown in Figure 8.31. A good security practice is to use different keys for different cryptographic operations; this prevents potential interactions between the operations that might introduce weaknesses in a security protocol. Therefore, we use independent keys for the encryption and authentication operations, K_{encr} and K_{MAC}, respectively, which are derived from the unique key K_i that node shares with the base station. For example we may take $K_{encr} = F_{K_i}(0)$ and $K_{MAC} = F_{K_i}(1)$, where F is some secure pseudorandom function.

As can be seen from Figure 8.31, we use the encrypt-then-authenticate method to construct a "secure channel" between source node and base station. It is shown in [77] that this is the most secure method for symmetric encryption and authentication. Encryption is performed through the use of a counter C that is shared between the source node and the base station. We do this in order to achieve semantic security; an adversary will not be able to obtain partial information about a plaintext, even if it is the same plaintext that is encrypted multiple times. This can also be achieved through randomization, but then the random value used in the encryption of the message must also be transmitted. The counter approach results in less transmission overhead as the counter is maintained in both ends. If counter synchronization is a problem (usually the receiver can try a small window of

$$
\begin{aligned}
y_1 &\leftarrow E_{K_{encr}, C}(D) \\
t_1 &\leftarrow MAC_{K_{MAC}}(y_1) \\
c_1 &\leftarrow y_1 | t_1
\end{aligned}
$$

Figure 8.31 Step 1 for secure communication between source node and base station. This step is applied by the source node alone.

counter values to recover the message) then the counter or the random value used can be sent alongside the message. We leave the choice to the particular deployment scenario as one alternative may be better than the other.

Step 2 (Required)   Since the encrypted data must be forwarded by intermediate nodes in order to reach the base station, we need to further secure the message so that an adversary cannot disrupt the routing procedure. Thus, no matter what routing protocol is followed, *intermediate* nodes need to verify that the message is not tampered with, replayed, or revealed to unauthorized parties, before forwarding it.

To secure the communication between one-hop neighbors, we use the protocol described in Figure 8.32. Each node (including the source node) uses its cluster key to produce the encryption key K'_{encr} and the MAC key K'_{MAC}. These keys are used to secure the message produced by step 1, before it is further forwarded. (As we emphasized previously, if we are only interested in hop-by-hop encryption and authentication, step 1 should be omitted in which case c_1, in message 2 below, is simply the data D.) Since the nodes that will receive that message do not know the sender and therefore the key that the message was encrypted with, the cluster ID is included in c_2. This way intermediate sensors will use the right key in their set S to authenticate the message.

If authentication is not successful, the message should be dropped since it is not a legitimate one. Otherwise, each node will apply step 2 with its own cluster key to further forward the message. The fact that this key is shared with all of its neighbors, allows the node to make only *one* transmission per message. Notice that this is the point where our protocol differs from random key predistribution schemes. To broadcast a message in such a scheme the transmitter must encrypt the message multiple times, each time with a key shared with a specific neighbor. And this, of course, is extremely energy consuming.

To continue the example shown in Figure 8.30, assume that node 14 must send a message m toward the base station that lies in the direction of node 4. It first encrypts and tags the message to produce a ciphertext c_1 according to the protocol shown in Figure 8.31 and then wraps this to produce an encrypted block c_2 according to the specifications shown in Figure 8.32. When ready, it broadcasts c_2 to its neighbors. Eventually, an encapsulation of c_1 will reach node 12, maybe through node 10. This node will decrypt and authenticate the message since it shares the same cluster key as node 14 and once all the checks are passed, it will reencrypt c_1 and forward it to its neighbors. One of them is node 8, which is a member of cluster with CID $= 9$. This node will look at its set of cluster keys S and use the one that it shares with node 12 (the one corresponding to CID $= 13$). Upon success it will reencrypt the message with its cluster key and forward it toward its neighbors. Thus nodes that lie at the edge of clusters will be able to "translate" messages that come from neighboring clusters and be able to *authenticate* them in a hop-by-hop manner.

In summary, every node decrypts and authenticates the packets it receives by using the keys it derived from each cluster key. If the node belongs to a different cluster than the

$$\begin{aligned}
\tau &\leftarrow \text{time}() \\
y_2 &\leftarrow E_{K'_{encr}}(c_1, \tau, CID) \\
t_2 &\leftarrow \text{MAC}_{K'_{MAC}}(y_2) \\
c_2 &\leftarrow CID | y_2 | t_2
\end{aligned}$$

Figure 8.32 Step 2 for secure communication between source node and base station. This step is applied by all intermediate nodes, besides the source node.

transmitter of the message, it will find the right cluster key from those stored in its memory and use it to reencrypt the message and pass it along to its neighbors. Thus messages get authenticated as they traverse the network. However, while this approach defends against adversaries who do not possess the required cluster keys, it falls prey to insider attacks since an adversary can easily inject spurious messages after it has compromised a node and discovered its cluster key. Unfortunately this kind of attack is not only hard to prevent but is also hard to detect.

To increase security and avoid sending too much traffic under the same keys, cluster keys may be refreshed periodically. To support such functionality, sensor nodes can repeat the key setup phase with a predefined period in order to form new clusters and new cluster keys. Since K_m is no longer available to the nodes, the current cluster key may be used by the nodes instead. The fact that each node can communicate with all of its neighbors using the current cluster key makes it possible to broadcast an HELLO message in a secure way.

The message will contain the new cluster key, created by a secure key generation algorithm embedded in each node. Since the key setup phase requires very low communication overhead (as it will be showed in the next section) and takes only a short time to complete, the refreshing period can be as short as needed to keep the network safe. Alternatively, if we do not like the fact that certain nodes are assigned the task of creating new keys (as they may be the compromised ones), we can renew the cluster keys by periodically hashing these keys at fixed time intervals.

8.5.3 Experimental Analysis

We simulated a sensor network to determine some parameter values of our scheme. We deployed several thousand nodes (2500 to 3600) in a random topology and ran the key setup phase. Of particular interest is the scalability, the communication overhead, and memory requirements of our approach.

The storage requirements of our approach are determined by the number of cluster keys stored in each node. We performed experiments with various network sizes, and we found that the curves matched exactly (modulo some small statistical deviation). So, we experimented with respect to network *density*. Figure 8.33 shows the average number of cluster keys that each node stores as a function of the average number of neighbors per node

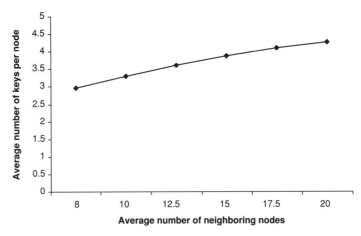

Figure 8.33 Average number of cluster keys held by sensor nodes as a function of network density.

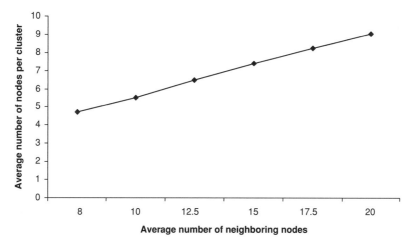

Figure 8.34 Average number of nodes in clusters as a function of network density.

(density of network). The number of cluster keys also indicates the number of neighboring clusters that each node has. As is obvious from the figure, the number of stored keys is very small and increases with low rate as the number of neighbors increases, requiring negligible memory resources from the sensor node. The point to be made is that the number of required keys remains *independent* of the actual network size. Thus our protocol behaves the same way in a network with 2000 or 20,000 nodes.

In Figure 8.34 we further show the average number of nodes per cluster for various network densities. Nodes of the same cluster share a common cluster key, and thus an adversary, upon compromising such a node, can also control the communication links of the rest of cluster nodes. Thus, having small clusters, as is indicated in the figure, minimizes the damage inflicted by the compromised node and prevents its spreading to the rest of the network.

The communication traffic required by the key setup phase is partly due to the number of messages sent by the cluster heads to their cluster members during phase one, and partly due to the messages sent by all nodes of the network during the link establishment phase. The former quantity depends on the number of cluster heads and is shown in Figure 8.35. The second quantity is always constant and equal to n, the number of nodes in the network. Bearing in mind that the key setup phase is executed only once, the total communication overhead due to that phase is kept very low. Further evidence to this fact is given in Figure 8.36, where the average number of messages required *per node* to set up the keys is shown. Thus the overall time needed to establish the keys is a little more than transmission of one message plus the time to decrypt the material sent during this phase.

8.5.4 Security Analysis

We now discuss one by one some of the general attacks [78] that can be applied to routing protocols in order to take control of a small portion of the network or the entire part of it.

- *Spoofed, Altered, or Replayed Routing Information* As sensor nodes do not exchange routing information, this kind of attack is not an issue.
- *Selective Forwarding* In this kind of attack an adversary selectively forwards certain packets through some compromised node while it drops the rest. Although such an attack

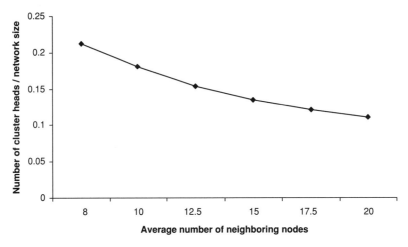

Figure 8.35   Percentage of cluster heads with respect to total sensor nodes in the network.

is always possible when a node is compromised, its consequences are insignificant since nearby nodes can have access to the same information through their cluster keys.

• *Sinkhole and Wormhole Attacks*   Since all nodes are considered equal and there is not a distinction between more powerful and weak nodes, an adversary cannot launch attacks of this kind. Furthermore, in our protocol such an attack can only take place during the key establishment phase. But the authentication that takes place in this phase and its small duration, as we described in the previous section, makes this kind of attack impossible.

• *Sybil Attacks*   Since every node shares a unique symmetric key with the trusted base station, a single node cannot present multiple identities. An adversary may create clones of a compromised node and populate them into the same cluster or the node's neighboring

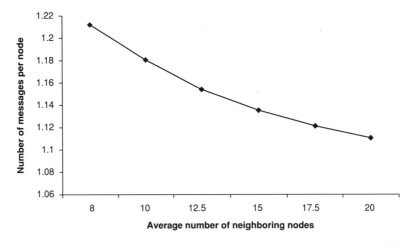

Figure 8.36   Number of messages exchanged *per node* for organization into clusters and link establishment in a network of 2000 nodes and various densities.

clusters, but this does not offer any advantages to the adversary with respect to the availability of the information to the base station.

*   *Hello Flood Attacks* In our protocol, nodes broadcast an HELLO message during the cluster key setup phase in order to announce their decision to become cluster heads and distribute the cluster key. Since, however, messages are authenticated this attack is not possible. (A necessary assumption for all key establishment protocols is, of course, that the duration of this phase is small so that an adversary cannot compromise a node and obtain the key K_m. In the previous section we presented evidence that this is indeed the case.)

    However, this kind of attack is possible during key refresh. If we assume that a laptop-class attacker has compromised a node and retrieved its cluster keys, then she could broadcast such an HELLO message during a key refresh phase and could attract nodes belonging to neighboring clusters as well and form a new larger cluster with himself as a cluster head. One way to defend against this is to constraint the key-refresh phase within clusters, that is, not allow new clusters to be created. Therefore, cluster keys will be refreshed within the same clusters, and an adversary cannot take control of more nodes than she already has, that is, the nodes within the same cluster. A better way, however, which makes this kind of attack useless, is to refresh the keys by hashing instead of letting nodes generate new ones.

*   *Acknowledgment Spoofing* Since we do not rely on link layer acknowledgments, this kind of attack is not possible in our protocol.

The power of our proposal lies exactly in its simplicity. No location information, path reinforcements, or routing updates are used. By strengthening security with encryption and authentication, it becomes really hard for an adversary to disrupt the routing procedure. Not many sensor network routing protocols currently proposed in the literature resist so well (or at all) to such security attacks as [78] demonstrated, and their design remained up to now an open problem.

8.5.5 Related Work

We have already mentioned Basagni et al.'s pebblenets architecture [70] that uses a global key shared by all nodes. Having network-wide keys for encrypting information is very good in terms of storage requirements and energy efficiency as no communication is required among nodes to establish additional keys. It suffers, however, from the obvious security disadvantage that compromise of even a single node will reveal the universal key. Since one cannot have keys that are shared pairwise between all nodes in the network, a key predistribution scheme must be used.

Random key predistribution schemes [71–74] offer a trade-off between the level of security achieved using shared keys among nodes and the memory storage required to keep at each node a set of symmetric keys that have been randomly chosen from a key pool. Then, according to the model used, nodes use one or more of these keys to communicate securely with each other. A necessary requirement of a good key distribution scheme is that compromise of nodes should *not* result in a breach of security that is spread in the whole network. Our approach guarantees this since it is not a probabilistic scheme; compromised nodes cannot expose keys in another part of the network. In our work, however, we emphasize on another desirable characteristic: When a node wants to broadcast a message to a subset or all of its neighbors it should *not* have to make multiple transmissions of the same message, encrypted each time with a different key. We now review some other proposals that use security architectures similar to ours.

In LEAP [79] *every* node creates a cluster key that distributes to its immediate neighbors using pairwise keys that shares with each one of them. In this case, however, clusters highly overlap so every node has to apply a different cryptographic key before forwarding the message. While this scheme offers deterministic security and broadcast of encrypted messages, it has a more expensive bootstrapping phase and increased storage requirements as each node must set up and store a number of pairwise *and* cluster keys that is proportional to its actual neighbors.

Slijepcevic et al. [80] propose dividing the network into hexagonal cells, each having a unique key shared between its members. Nodes belonging to the bordering region between neighboring cells store the keys of those cells, so that traffic can pass through. The model works under the assumption that sensor nodes are able to discover their exact location, so that they can organize into cells and produce a location-based key. Moreover, the authors assume that sensor nodes are tamper resistant; otherwise the set of master keys and the pseudorandom generator, preloaded to all sensor nodes, can be revealed by compromising a single node, and the whole network security collapses. Those assumptions are usually too demanding for sensor networks.

8.5.6 Conclusions

We have presented a key establishment protocol that is suitable for sensor network deployment. The protocol provides security against a large number of attacks and guarantees that data securely reaches the base station in an energy-efficient manner. Our protocol is based on hop-by-hop encryption, allowing nodes to share keys only with neighboring nodes. The protocol has a number of important characteristics among which are:

- *Resiliency Against Node Capture and Replication* This is due to the fact that keys are localized. After a deployment phase, nodes share a handful of keys to securely communicate with their neighbors. Thus compromised keys in one part of the network do not allow an adversary to obtain access in some other part of it.

- *Efficient Broadcasting of* Encrypted Messages When a node wants to broadcast a message to its neighbors, it does not have to make multiple transmissions encrypted each time with a different key. We achieve this by encrypting messages with a cluster key that is shared between neighboring nodes. This makes our scheme very energy efficient.

- *Intermediate Node Accessibility of Data* When multiple nodes receive the same message, some of them may decide not to forward it. However, this is not possible unless nodes can have access to encrypted data. Using our approach, nodes can "peak" at encrypted information using their cluster key and decide upon forwarding or discarding redundant messages, thus enabling data aggregation processing.

- *Scalability* Our protocol scales very well as the key establishment phase requires only local information and no global coordination. Furthermore the keys that need to be stored at each node do not depend on the size of the sensor network but only on its *density* (the average number of neighbors per node). Thus our protocol behaves similarly in networks of 2000 or 20,000 nodes as long as the density is the same.

8.6 WEIL PAIRING-BASED ROUND, EFFICIENT, AND FAULT-TOLERANT GROUP KEY AGREEMENT PROTOCOL FOR SENSOR NETWORKS

Tian-Fu Lee, Hsiang-An Wen, and Tzonelih Hwang

A group key agreement protocol enables a group of participants to securely establish a common key that can be used to encrypt all communications over an insecure channel. This investigation proposes a Weil pairing-based round, efficient, and fault-tolerant group key agreement protocol for sensor networks. In the proposed protocol, all sensor nodes just broadcast their contributions of the common key one time, and then they can calculate the common key simultaneously. The proposed protocol provides the properties of fault tolerance and fairness. All good sensor nodes can certainly obtain a common key, and any failed nodes will be automatically excluded from the sensor network.

8.6.1 Background Information

A sensor network is composed of a large amount of sensor nodes that can communicate to one another by broadcasting. Sensor nodes can be randomly deployed in inaccessible place so that the sensor network is useful for many applications. For example, the deployment of sensor nodes is helpful for medical purposes. This way it is possible to monitor dangerous areas for humans. On the other hand, the military can rapidly deploy sensor nodes in critical areas for reconnaissance and communication. However, intruders can easily intercept the content of the communication by eavesdropping since the sensor nodes broadcast messages without applying any security mechanism. Therefore, how to construct a secure communication environment is an important task for sensor networks [81–83].

A group communication protocol can assist a group of participants to establish a common key for securely communicating over an insecure channel. In general, the group communication protocol can be divided into two categories: the group key distribution protocol and the group key agreement protocol. In group key distribution protocol a chairman or a server selects a common key and distributes the key to all participants. In a group key agreement protocol all participants cooperate to compute a common key without help of a chairman or a server. Since the sensor nodes form a distributed environment, the group key agreement protocol is more suitable to be applied in sensor networks.

Recently, several group key schemes have been presented, such as the group key distribution protocols [84–90] and the group key agreement protocols [91–94]. In 1995, Klein et al. [95] presented a fault-tolerant group key agreement protocol. Kim et al. [96] in 2000 proposed a simple and fault-tolerant key agreement for dynamic collaborative groups. Besides providing secure communication, a fault-tolerant group key agreement protocol should satisfy the following property: All good participants who work correctly cannot be excluded from the participant set, but any failed participant will be detected and excluded from the set.

Subsequently, Burmester and Desmedt [97] proposed a round-efficient group key agreement protocol over the broadcast channel in the same year. In 2000, Tzeng et al. [98] proposed a round-efficient group key agreement protocol that provides the fault tolerance. Tzeng [99] proposed an interactive and fault-tolerant group key agreement protocol in 2002. Although the conference key agreement protocols [98, 99] proposed by Tzeng have already provided the property of fault tolerance, these protocols are not very practical because they require large amounts of random numbers and modular exponential computations during the execution time.

In these years, several new cryptosystems based on elliptic curves have been developed. Comparing the elliptic curve cryptosystem (ECC) with the discrete logarithm cryptosystem (DLC), the ECC offers the same level security for a far smaller bit size, thereby reducing processing overhead [100]. Since, in sensor networks, sensor nodes are limited in power, computational capacities, and memory [101], using ECC is more suitable for sensor networks than using Diffie–Hellman key agreement [102] or RSA [103]. Besides, Joux [104] discovered some new cryptographic properties of Weil pairing over elliptic curves. For example, the decisional Diffie–Hellman (DDH) problem can be easily solved in Weil pairing while the computational Diffie–Hellman (CDH) problem remains hard. Since then, several variant cryptographic systems based on Weil pairing [104–107] have been proposed.

This investigation proposes a secure and round and efficient group key agreement protocol using Weil pairing for sensor networks. Applying the outstanding properties of ECC based on Weil pairing, we are able to design and implement a group key agreement protocol, which provides the properties of authentication, privacy, and fault detection, in a round. In the proposed scheme, each sensor node generates one random number and computes and broadcasts the subkey and signature files over an insecure channel. Then, on receiving messages from other sensor nodes, all sensor nodes can verify the validity of the messages and compute a common group key. In addition, since all good sensor nodes have contributions to generate the common group key and all failed sensor nodes will be detected and excluded from the sensor network, the proposed protocol exhibits the properties of fairness and fault tolerance. Furthermore, the security of the proposed protocol is also carefully analyzed so that the authenticity and the privacy of the proposed protocol can be guaranteed.

8.6.2 Preliminaries

We review the properties of the Weil pairing, Diffie–Hellman assumptions, and the short signature scheme using Weil pairing proposed by Boneh et al. [106].

Weil Pairing   Let G be a prime-order subgroup of an elliptic curve E over the field F_q, l be the order of G, and k be the smallest integer such that $l | q^k - 1$. The modified Weil pairing $\hat{e}$ is defined as $\hat{e} : G \times G F_{q^k}^*$ and satisfies the following properties [103, 104]:

1. Bilinear:

$$\hat{e}(P_1 + P_2, Q) = \hat{e}(P_1, Q)\hat{e}(P_2, Q)$$

$$\hat{e}(P, Q_1 + Q_2) = \hat{e}(P, Q_1)\hat{e}(P, Q_2)$$

2. Nondegenerate: There exists a $P \in G$ such that $\hat{e}(P, P) \neq 1$.
3. Computability: There exists an efficient algorithm for calculating $\hat{e}(P, Q)$ in polynomial time.

Discrete Logarithm Problem in Elliptic Curve   Let G be a prime-order subgroup of an elliptic curve E and l be the order of G. The elliptic curve discrete logarithm problem (ECDLP) is: Given $(P, aP) \in G$, if l is large enough, then calculating $a \pmod{l}$ is computationally infeasible [103].

Diffie–Hellman Assumptions Let G be a prime-order subgroup of an elliptic curve E and l be the order of G. Let $\hat{e} : G \times G \times F_{q^k}^*$ be a Weil pairing. The Diffie–Hellman assumptions over G is stated as follows:

1. *Computational Diffie–Hellman Problem 1 (CDHP-1)* Given $(P, aP, bP) \in G$, if l is large enough, then calculating $(ab)P \in G$ is computationally infeasible.
2. *Computational Diffie–Hellman Problem 2 (CDHP-2)* Given $(P, aP, bP, cP) \in G$, if l is large enough, then calculating $\hat{e}(P, P)^{abc} (\bmod l)$ is computationally infeasible.
3. *Decisional Diffie–Hellman Problem (DDHP)* Given $(P, aP, bP, cP) \in G$, then we can easily determine whether $c = ab (\bmod l)$ as follows. We first compute

$$\hat{e}(aP, bP) = \hat{e}(P, P)^{ab} \quad \text{and} \quad \hat{e}(P, cP) = \hat{e}(P, P)^c$$

If the equation

$$\hat{e}(aP, bP) = \hat{e}(P, cP)$$

holds, then we output $c = ab (\bmod l)$. Else, we output $c \neq ab (\bmod l)$. Therefore, DDHP is easy to solve in G.

Short Signature Scheme Using Weil Pairing In 2001, Boneh et al. [107] proposed an efficient signature scheme using Weil pairing, called the short signature scheme. This scheme is suitable for systems where signatures are sent over a lower bandwidth channel. In [107], the authors use relatively small values of l to obtain short signatures, and the security is based on the discrete log problem in a large finite field. The following will briefly review the lemmas for describing the behavior of these curves and the short signature scheme.

Lemma 8.6.1 (107) The curve E^+ defined by $y^2 = x^3 + 2x + 1$ over F_{3^l} satisfies

$$\frac{\#E^+}{F_{3^l}} = \begin{cases} 3^l + 1 + \sqrt{3 - 3^l} & \text{when } l = \pm 1 \bmod 12 \\ 3^l + 1 - \sqrt{3 - 3^l} & \text{when } l = \pm 5 \bmod 12 \end{cases}$$

The curve E^- defined by $y^2 = x^3 + 2x + 1$ over F_{3^l} satisfies

$$\frac{\#E^-}{F_{3^l}} = \begin{cases} 3^l + 1 - \sqrt{3 - 3^l} & \text{when } l = \pm 1 \bmod 12 \\ 3^l + 1 + \sqrt{3 - 3^l} & \text{when } l = \pm 5 \bmod 12 \end{cases}$$

Lemma 8.6.2 (107) Let E be an elliptic curve defined by $y^2 = x^3 + 2x + 1$ over F_{3^l}, where $l \bmod 12$ equals ± 1 or ± 5. Then $\#(E/F_{3^l})$ divides $3^{6l-1} - 1$.

Definition 8.6.1 (107) Automorphism of E^+, $E^-/F_{3^{6l}}$: For l such that $l \bmod 12$ equals ± 1 or ± 5, compute three elements of $F_{3^{6l}}$, u, r^+, and r^-, satisfying $u^2 = -1, (r^+)^3 + 2r^+ + 2 = 0$, and $(r^-)^3 + 2r^- - 2 = 0$. Now consider the following maps over $F_{3^{6l}}$:

$$\phi^+(x, y) = (-x + r^+, uy) \quad \text{and} \quad \phi^-(x, y) = (-x + r^-, uy)$$

Lemma 8.6.3 (107)   Let l mod 12 equal ± 1 or ± 5. Then ϕ^+ is an automorphism of $E^+/F_{3^{6l}}$ and ϕ^- is an automorphism of $E^-/F_{3^{6l}}$. Moreover, if P is a point of order q on E^+/F_{3^l} (or E^-/F_{3^l}), then $\phi^+(P)$ [or $\phi^-(P)$] is a point of order q that linearly independent of P.

Assume that E/F_{p^l} is an elliptic curve of order m defined by $y^2 = f(x)$ and $P \in E/F_{p^l}$ is a point of prime order q, where m is not divided by q^2. In [107], the deterministic algorithm, called *MapToGroup*$_{h'}$, is used to hash messages in $\{0, 1\}^*$ onto G^*. That is, h: $\{0, 1\}^* \to F_{p^l} \to \{0, 1\}$, where the hash functions h' are built from standard cryptographic hash functions.

Lemma 8.6.4 (107)   Suppose the GDH signature scheme is (t, q_H, q_S, E) secure in the subgroup G when using a random hash function $h: \{0, 1\}^* \to G^*$. Then it is $(t - 2^I q_H lgm, q_H, q_S, E)$ secure when the hash function h is computed with MapToGroup$_{h'}$ where h' is a random hash function $h' : \{0, 1\}^* \to F_{p^l} \times \{0, 1\}$, where $I = \lceil \log_2 \log_2(1/\delta) \rceil$ is a small and fixed parameter and δ is some desired bound on the probability of failure.

The short signature scheme comprises the following three algorithms:

1. *The Key Generation Algorithm*   Given a security parameter 1^k as input, this algorithm produces a public/private key pair (pk, sk) as output. The key generation algorithm is probabilistic.
2. *The Signing Algorithm*   Given a private key sk and a message M as input, this algorithm produces a signature σ as output. The key generation algorithm is deterministic.
3. *The Verification Algorithm*   Given (pk, M, σ) as input, this algorithm verifies the validity of the signature. If the signature is valid, the algorithm outputs accept; otherwise, the algorithm outputs reject.

The short signature scheme of Boneh et al. [107] is described as follows:

1. *Key Generation*   Let a point $P \in G$. Pick a random number $x \in Z_q^*$ and set $R = xP$. Then (l, q, P, R) is a public key and x is private key.
2. *Signing*   Let $h : \{0, 1\}^* \to G$ denote a cryptographic hash function and M be the message to be signed. Compute

$$P_M = h(M) \quad \text{and} \quad S_M = x P_M$$

The signature σ is the x coordinate of S_M.
3. *Verification*   Given the public key (l, q, P, R), a message M and its signature σ, a verifier has to do the following steps:
   a. Find a point S whose x coordinate is σ and y coordinate y for some y. If no such point exists, then signature σ is regarded as invalid and rejected.
   b. Set $u = \hat{e}(P, \phi(S))$ and $v = \hat{e}(R, \phi(h(M)))$, where $\hat{e} : G \times G \to F_{q^k}^*$ is a Weil pairing map and $\phi: E \to E$ is the automorphism of the curve [107].
   c. Check whether $u = v$ or $u^{-1} = v$. If successful, then accept the signature; otherwise, reject.

8.6.3 Proposed Group Key Agreement Protocol for Sensor Networks

We propose a round and efficient and fault-tolerant group key agreement protocol for sensor networks. We first define the notation used and then describe the proposed group key agreement protocol in detail.

Notation

U_i: The participant (sensor node) of sensor networks for $i = 1, 2, \ldots, n$.

$(x_i, x_i P)$: The private key/public key pair of the participant U_i for $i = 1, 2, \ldots, n$.

$H(.): \{0, 1\}^* \rightarrow \{0, 1\}^l$: A one-way hash function.

Γ: A symmetric encryption/decryption scheme consisting of the following two algorithm:

$$E_k(m) = C \qquad \text{a symmetric encryption scheme}$$

$$D_k(C) = m \qquad \text{a symmetric decryption scheme, where } m \text{ is the plaintext,}$$

C is the ciphertext, and k is the secret key

δ_i: The candidate set belonging to the participant U_i.

Proposed Group Key Agreement Protocol for Sensor Networks   The proposed group key agreement protocol is suitable for sensor networks environment and comprises four phases:

1. *Message Broadcasting Phase*   In this phase, all participants compute and broadcast messages containing the signature files and the encrypted subkey of the common group key.
2. *Verification Phase*   In this phase, each participant verifies the messages from the other participants in message broadcasting phase. Besides, the subkeys will be also decrypted in this phase.
3. *Fault Detection Phase*   In this phase, any failed participant who cannot work correctly will be detected and excluded from the participant set.
4. *Group Key Computing Phase*   In this phase, all good participants who work correctly compute the common group key using the subkeys obtained in verification phase.

The proposed group key agreement protocol for sensor networks is described as follow:

1. *Message Broadcasting*   Each participant U_i performs the following steps:
   a. Select a random number a_i.
   b. Compute $Q_i = h(a_i, t_i)$ and the subkey $\kappa_i = x_i Q_i$, where t_i is a time stamp.
   c. Compute $s_{ij} = \hat{e}(\kappa_i, x_j P)$ and $u_{ij} = E_{s_{ij}}(\kappa_i)$, for $j = 1, 2, \ldots, n, j \neq i$.
   d. Broadcast $(a_i, t_i, u_{i1}, \ldots, u_{i(i-1)}, u_{i(i+1)}, \ldots, u_{in})$.
2. *Verification*   Each participant U_j performs the following steps:
   a. Setup the candidate set $\delta_j = \{1, 2, \ldots, n\}$.

    b. Check the validity of the time stamp t_i. If t_i is invalid, then remove i from the candidate set δ_j, for $i = 1, 2, \ldots, n, i \neq j$.

    c. Calculate $Q'_i = h(a_i, t_i)$, $s'_{ij} = \hat{e}(x_j Q_i, x_i P)$, and $\kappa'_i = D_{s_{ij}}(u_{ij})$, for $i \in \delta_j, i \neq j$.

    d. Check whether $\hat{e}(Q'_i, x_i P) = \hat{e}(\kappa'_i, P)$. If the equation does not hold, then remove i from the candidate set δ_j, for $i \in \delta_j, i \neq j$.

3. *Fault Detection* Each participant U_j performs the following steps:

    a. Calculate $s_{im} = \hat{e}(\kappa'_i, x_m P)$, and $\kappa_{im} = D_{s'_{im}}(u_{im})$ for $i \in \delta_j$, $i \neq j$ and $m = 1, 2, \ldots, n, m \neq i$.

    b. Check whether $\kappa_{im} = \kappa'_i$. If the equation does not hold, then remove i from the candidate set δ_j.

4. *Group Key Computing* Each participant U_j computes the common key

$$K = H\left(\kappa'_{j_1}, \kappa'_{j_2}, \ldots, \kappa'_{j_n}.\right)$$

where the current candidate set is $\delta_j = \{j_1, j_2, \ldots, j_{n'}\}$.

8.6.4 Security Analysis

We analyze the security properties of the proposed scheme in detail. First, we verify the correctness, fault tolerance, and fairness of the proposed group key agreement protocol for sensor networks. Second, we guarantee that the proposed scheme can achieve the requirements of authenticity and privacy. Finally, we show that the proposed scheme can prevent several possible attacks such as the known-key attacks and replay attacks.

Correctness For correctness, we show that all participants can calculate a common group key if they follow the proposed group key agreement protocol.

Claim 1. *All participants can calculate a common group key if they follow the proposed protocol.*

PROOF   Assume that all participants follow the proposed protocol. Then, on receiving

$$(a_i, t_i, u_{i1}, \ldots, u_{i(i-1)}, u_{i(i+1)}, \ldots, u_{in})$$

from participant U_i, the participant U_j can obtain the correct subkey κ'_i of U_i since U_j can derive the temporary session key

$$s'_{ij} = \hat{e}(x_j Q'_i, x_i P) = \hat{e}(Q_i, P)^{x_i x_j} = \hat{e}(x_i Q_i, x_j P) = s_{ij}$$

and then compute

$$\kappa'_i = D_{s'_{ij}}(u_{ij}) = D_{s'_{ij}}[E_{s_{ij}}(\kappa_i)] = D_{s_{ij}}[E_{s_{ij}}(\kappa_i)] = \kappa_i$$

Moreover, since κ'_i is a short signature signed by U_i, U_j can also verify the validity of κ'_i by checking

$$\hat{e}(Q'_i, x_i P) = \hat{e}(x_i Q_i, P) = \hat{e}(\kappa_i, P) = \hat{e}(\kappa'_i, P)$$

On the other hand, U_j can confirm the subkey κ_i'' (derived by $U_m, m = 1, 2, \ldots, n$ and $m \neq i, j$) is equivalent to κ_i' in fault detection phase because U_j can calculate

$$s_{im}' = \hat{e}(\kappa_i', x_m P) = \hat{e}(x_i Q_i, x_m P),$$

$$\kappa_i'' = D_{s_{im}'}(u_{im})$$

and check

$$\kappa_i'' = \kappa_i'$$

Hence, all participants can obtain the same $\kappa_1, \kappa_2, \ldots, \kappa_n$ and calculate the common key as

$$K = H(\kappa_1, \kappa_2, \ldots, \kappa_n)$$

∎

Fault Tolerance   For fault tolerance property, we show that the proposed group key agreement protocol provides the following two properties. First, any failed participant that cannot work correctly will be excluded from the participant set. Second, all good participants that work correctly will not be excluded from the participant set.

Claim 2.   *All good participants who work correctly will not be excluded from the current participant set, but any failed participant will be detected and excluded from the participant set consisting of all good participants.*

PROOF   Assume that a failed participant U_i broadcasts the message

$$(a_i, t_i, u_{i1}, \ldots, u_{i(m-1)}, \overset{*}{u}_{im}, u_{i(m+1)}, \ldots, u_{in})$$

where $\overset{*}{u}_{im}$ is a bad ciphertext, so that participant U_m has a faulty verification in verification phase and sets U_i as a failed participant. Then, U_m removes i from the candidate set.

On the other hand, each participant U_j calculates

$$sk_{im}' = \hat{e}(\kappa_i', x_m P) \quad \text{and} \quad \kappa_i'' = D_{sk_{im}'}(u_{im})$$

for $m = 1, 2, \ldots, n$ and $j \neq m$ in fault detection phase. But he cannot successfully check $\kappa_i'' = \kappa_i'$. Then, all other participants can discover the inconsistency of the subkey κ_i and remove i from the candidate set simultaneously. Therefore, U_i is also set as a failed participant and excluded from the participant set of all good participants. In addition, the good participant who works correctly can still compute the same common key.   ∎

Fairness   For fairness, we show that all participants fairly have contributions to produce the common group key in the proposed protocol.

Claim 3.   *The group key is fairly determined by all participants together.*

PROOF   Note that the group key is

$$K = H(\kappa_1, \kappa_2, \ldots, \kappa_n)$$

where $\kappa_1, \kappa_2, \ldots, \kappa_n$ are the subkeys of all participants that can successfully broadcast messages. Therefore, all participants have contributions to produce the group key.   ■

Authentication   For authentication, we show that the proposed protocol enables participants mutually to verify each other's signature file in order to check the private key.

Claim 4.   *No intruder can impersonate a legal participant unless he/she knows the private key of the participant.*

PROOF   Assume that an adversary who impersonates a legal participant U_i chooses a random number a_i, a time stamp t_i, and computes

$$Q_i^* = h(a_i, t_i)$$

Then he is able to successfully sign Q_i^* with a nonnegligible probability and constructs a signature (Q_i^*, sk_{ij}^*) that satisfy

$$\hat{e}(Q_i^*, x_i P) = \hat{e}(\kappa_i^*, P)$$

in verification phase and

$$\sigma_{ij}^* = \hat{e}(\kappa_i^*, x_j P) = \hat{e}(Q_i^*, P)^{x_i x_j}$$

in fault detection phase if $\kappa_i^* = x_i Q_i^*$. Let $Q_i^* = y P$ for some constants $y \in Z/l$. The coefficient y is difficult to be obtained since DLP in the elliptic curve cannot be solved in polynomial time. So, the equations

$$\kappa_i^* = x_i Q_i^* \quad \text{and} \quad \sigma_{ij}^* = \hat{e}(P, P)^{x_i x_j y} .$$

hold.

The private key x_i is unknown; then, given $(P, x_i P, y P)$, $x_i y P = x_i Q_i^*$ cannot be calculated since CDHP-1 in the Diffie–Hellman assumption cannot be solved in polynomial time. Furthermore, given $(P, x_i P, x_j P, y P)$, the term $\hat{e}(P, P)^{x_i x_j y}$ cannot be determined since CDHP-2 in the Diffie–Hellman assumption cannot be solved in polynomial time. Then, x_i, x_j, and y are unknown:

$$sk_{ij}^* = \hat{e}(P, P)^{x_i x_j y}$$

cannot be determined. Thus, the adversary is capable of computing the private key x_i or x_j. It is a contradiction and the proof is concluded.   ■

Privacy   For privacy, we show that an eavesdropper cannot learn any valuable information about the group key established by the legal participants even though he/she collects all communication messages.

Claim 5. *No eavesdropper can get any information about the group key established by the legal participants.*

PROOF   Since the group key is equal to

$$K = H(\kappa_1, \kappa_2, \ldots, \kappa_{n\kappa})$$

and $H(\cdot)$ is considered as a random oracle [101], the eavesdropper must know each κ_i (for $i = 1, 2, \ldots, n'$) to derive the common key K. In the following, we show that the eavesdropper cannot get any κ_i by eavesdropping the network.

CASE 1   Compute κ_i directly. Assume $Q_i = y_i P$ for some $y_i \in Z/l$. Then we have

$$\kappa_i = x_i Q_i = x_i y_i P$$

Given $(P, x_i P, Q_i)$, without knowing x_i or y_i, the eavesdropper must solve the Diffie–Hellman problem (CDHP-1) to obtain κ_i. Thus, the eavesdropper cannot compute κ_i directly.

CASE 2   Get κ_i by deciphering $u_{ij} = E_{s_{ij}}(\kappa_i)$. Since each κ_i is encrypted by the keys s_{ij} for some $1 \le j \le n \, j \ne i$ and $E_k(m)$ is considered as an ideal cipher [108, 109], the eavesdropper must know one of the decryption keys s_{ij} to obtain κ_i. Assume $Q_i = y_i P$ for some $y_i \in Z/l$. Then we have

$$s_{ij} = \hat{e}(x_i Q_i, x_j P) = \hat{e}(x_i y_i P, x_j P) = \hat{e}(P, P)^{x_i x_j y_i}$$

Given $(P, x_i P, x_j P, Q_i)$, without knowing x_i, x_j, or y_i, the eavesdropper must solve the Diffie–Hellman problem (CDHP-2) to obtain s_{ij}. Thus, the eavesdropper cannot get the decryption key s_{ij} to derive κ_i.   ∎

Security Against Known Key Attacks   For known key attacks, we show that the compromise of one session group key does not reveal other session group keys in the proposed protocol.

Claim 6.   *The proposed group key agreement protocol prevents the known key attacks.*

PROOF   In the proposed protocol, each participant selects a random number a_i and computes $Q_i = h(a_i, t_i)$ and the subkey $\kappa_i = x_i Q_i$. So, the subkeys $\kappa_1, \kappa_2, \ldots, \kappa_n$ may be regarded as random numbers. Since these random numbers are independent among protocol executions and the common group key is equal to

$$K = H(\kappa_1, \kappa_2, \ldots, \kappa_n)$$

where the current candidate set is $\delta = \{1, 2, \ldots, n\}$, the group keys generated in different runs are independent. Hence, the proposed protocol resists known key attacks.   ∎

Security Against Replay Attacks   A replay attack is performed by an unauthorized participant that records previous messages transmitted through an insecure channel between the sender and the receiver and uses them to trick the receiver in later transactions.

Claim 7. *Replay attacks are unsuccessful against the proposed group key agreement protocol.*

PROOF   Assume an intruder tries to impersonate a legal participant and cheats honest participants by sending out a previous message. A failed attack must be detected since the message contains the sender's time stamp and time stamp t_i' is out of the expected legal time interval. Hence, replay attacks are unsuccessful against the proposed protocol. ∎

8.6.5 Conclusions

The sensor network can be easily deployed so that it can be applied to a wide range of applications. This study proposes a Weil pairing-based round, efficient, and fault-tolerant group key agreement protocol for sensor networks. In the proposed protocol, each sensor node generates one random number and computes and broadcasts his subkey over an unsecured channel. After receiving broadcasted messages, a secure common key can be calculated by each legal node. Thus, the sensor nodes can encrypt all communications using the common key. In addition, the proposed protocol also provides the properties of fairness and fault tolerance. Moreover, passive attacks, impersonate attacks, known key attacks, and replay attacks are unsuccessful against this protocol.

REFERENCES

1. "Standard specification for public-key cryptography," IEEE Standard p1363/d1, Nov. 1999.
2. J. Daemen and V. Rijmen, "Aes proposal: Rijndael," http://csrc.nist.gov/encryption/aes/rijndael/Rijndael.pdf, NIST AES Proposal, June 1998.
3. D. R. Stinson, *Cryptography Theory and Practice*, 1st ed., Boca Raton, FL: CRC Press, 1995.
4. A. Menezes, P. van Oorschot, and S. Vanstone, *Handbook of Applied Cryptography*, Boca Raton, FL: CRC Press, 1996.
5. http://wins.rockwellscientific.com/.
6. "Advanced encryption standard," http://csrc.nist.gov/publication/drafts/dfips-AES.pdf, National Institute of Standards and Technology (U.S.).
7. "Recommended elliptic curves for federal government use," http://csrc.nist.gov/CryptoToolkit/dss/ecdsa/NISTReCur.pdf, National Institute of Standards and Technology, May 1998.
8. V. Raghunathan, C. Schurgers, S. Park, and M. Srivastava, "Energy-aware wireless microsensor networks," *IEEE Signal Processing Magazine*, pp. 40–50, Mar. 2002.
9. C. Schurgers, O. Aberthorne, and M. B. Srivastava, "Modulation scaling for energy aware communication systems," in *International Symposium on Low Power Electronics and Design*, Huntington Beach, CA, Aug. 6–7, 2001, pp. 96–99.
10. A. Savvides, S. Park, and M. B. Srivastava, "On modeling networks of wireless microsensors," in *ACM Sigmetrics*, 2002.
11. D. Carman, P. Kruus, and B. Matt, "Constraints and approaches for distributed sensor network security (final)," Technical Report 00-010, NAI Labs, 2000.
12. I. F. Blake, G. Seroussi, and N. P. Smart, *Elliptic Curves in Cryptography*, Cambridge, MA: Cambridge University Press, 1999.
13. J. Goodman and A. P. Chandrakasan, "An energy efficient reconfigurable public-key cryptography processor," *IEEE Journal of Solid State Circuit*, vol. 36, no. 11, Nov. 2001.

14. S. Janssens, J. Thomas, W. Borremans, P. Gijsels, I. Verbauwhede, F. Vercauteren, B. Preneel, and J. Vandewalle, "Hardware/software co-design of an elliptic curve public-key cryptosystem," in *SIPS*, 2001.

15. P. Schaumont and I. Verbauwhede, "Domain-specific co-design for embedded security," *IEEE Computer*, Apr. 2003.

16. A. Perrig, R. Szewczyk, V. Wen, D. Culler, and J. D. Tygar, "Spins: Security protocols for sensor networks," in *MobiCom*, July 2001.

17. B. Lai, S. Kim, and I. Verbauwhede, "Scalable session key construction protocol for wireless sensor networks," in *IEEE Workshop on Large Scale Real-Time and Embedded Systems (LARTES)*, Dec. 2002.

18. G. Pottie and W. Kaiser, "Wireless integrated network sensors," *Communications of the ACM*, vol. 43, no. 5, pp. 51–58, May 2000.

19. A. Sinha and A. Chandrakasan, "Jouletrack—A web based tool for software energy profiling," in *DAC 2001*, Las Vegas, NV, 2001.

20. V. Tiwari, S. Malik, and A. Wolfe, "Power analysis of embedded software: A first step towards software power minimization," *IEEE Transactions on VLSI Systems*, vol. 2, no. 4, Dec. 1994.

21. H. Mehta, R. M. Owens, M. J. Irwin, R. Chen, and D. Ghosh, "Techniques for low energy software," in *Proceedings of the 1997 ISLPED*, Monterey, CA, Aug. 18–20, 1997.

22. B. C. Neuman and T. Tso, "Kerberos: An authentication service for computer networks," *IEEE Communications Magazine*, vol. 32, no. 9, pp. 33–38, Sept. 1994.

23. A. Perrig, R. Szewczyk, V. Wen, D. Culler, and J. D. Tygar, "Spins: Security protocols for sensor networks," in *Proceedings of the Seventh Annual International Conference on Mobile Computing and Networking (MOBICOM'01)*, Rome, ACM Press, 2001, pp. 189–199.

24. I. Akyildiz, Y. S. W. Su, and E. Cayirci, "Wireless sensor networks: A survey," *Journal of Computer Networks*, vol. 38, pp. 393–422, 2002.

25. R. Di Pietro, "Security issues for wireless sensor networks," Ph.D. dissertation, Università di Roma "La Sapienza," Dipartimento di Informatica, 2004.

26. D. W. Carman, P. S. Kruus, and B. J. Matt, "Constraints and approaches for distributed sensor network security (final)," Technical Report 00-010, NAI Labs, 2000.

27. R. Di Pietro, L. V. Mancini, and S. Jajodia, "Providing secrecy in key management protocols for large wireless sensors networks," *Journal of AdHoc Networks*, vol. 1, no. 4, pp. 455–468, 2003.

28. S. Basagni, K. Herrin, D. Bruschi, and E. Rosti, "Secure pebblenets," in *Proceedings of the 2001 ACM International Symposium on Mobile Ad Hoc Networking and Computing*, Long Beach, CA, ACM Press, 2001, pp. 156–163.

29. R. Anderson and M. Kuhn, "Tamper resistance—A cautionary note," in *Second USENIX Workshop on Electronic Commerce Proceedings*, San Diego, 1996, pp. 1–11.

30. L. Eschenauer and V. D. Gligor, "A key-management scheme for distributed sensor networks," in *Proceedings of the Ninth ACM Conference on Computer and Communications Security*, Washington, DC, ACM Press, 2002, pp. 41–47.

31. H. Chan, A. Perrig, and D. Song, "Random key predistribution schemes for sensor networks," in *Proceedings of the IEEE Symposium on Security and Privacy*, Oakland, CA, May 11–14, 2003, pp. 197–213.

32. W. Du, J. Deng, Y. S. Han, and P. K. Varshney, "A pairwise key predistribution scheme for wireless sensor networks," in *Proceedings of the Tenth ACM Conference on Computer and Communications Security (CCS '03)*, Washington, DC, Oct. 27–31, 2003.

33. D. Liu and P. Ning, "Establishing pairwise keys in distributed sensor networks," in *Proceedings of the Tenth ACM Conference on Computer and Communications Security (CCS '03)*, Washington, DC, Oct. 27–31, 2003.

34. R. Di Pietro, L. V. Mancini, and A. Mei, "Random key-assignment for secure wireless sensor networks," in *Proceedings of the First ACM Workshop on Security of Ad Hoc and Sensor Networks (SASN'03)*, ACM Press, 2003, pp. 62–71.

35. R. Di Pietro, L. V. Mancini, and A. Mei, "Efficient and resilient key discovery based on pseudo-random key pre-deployment," in *Proceedings of the IEEE Fourth International Workshop on Algorithms for Wireless, Mobile, Ad Hoc and Sensor Networks (WMAN '04)*, Santa Fe, NM, Apr. 2004, pp. 26–30.

36. S. Zhu, S. Xu, S. Setia, and S. Jajodia, "Establishing pairwise keys for secure communication in ad hoc networks: A probabilistic approach," in *Proceedings of the Eleventh IEEE International Conference on Network Protocols*, IEEE Computer Society, Atlanta, 2003, pp. 326–335.

37. R. Blom, "An optimal class of symmetric key generation systems," in *Advances in Cryptology: Proceedings of EUROCRYPT '84*, Lecture Notes in Computer Science, vol. 338, Springer-Verlag, 1985.

38. C. Blundo, A. D. Santis, A. Herzberg, S. Kutten, U. Vaccaro, and M. Yung, "Perfectly-secure key distribution for dynamic conferences," in *Advances in Cryptology: Proceedings of CRYPTO '92*, Lecture Notes in Computer Science, vol. 740, Springer-Verlag, 1993.

39. O. Goldreich, *Foundations of Cryptography: Basic Tools*, Cambridge University Press, 2001.

40. J. R. D. J. R. Douceur, "The sybil attack," in *Revised Papers from the First International Workshop on Peer-to-Peer Systems*, 2002, pp. 251–260.

41. H. Çam, S. Ozdemir, P. Nair, D. Muthuavinashiappan, and H. O. Sanli, "Energy-efficient secure pattern based data aggregation for wireless sensor networks," *Journal of Computer Communications on Sensor Networks*, in press.

42. D. Dolev and A. C. Yao, "On the security of public key protocols," *IEEE Transactions on Information Theory*, vol. 29, no. 2, pp. 198–208, 1983.

43. R. Anderson and M. Kuhn, "Tamper resistance—A cautionary note," in *Second Usenix Workshop on Electronic Commerce*, San Diego, Nov. 1996, pp. 1–11.

44. C. Karlof, N. Sastry, and D. Wagner, "Tinysec: Link layer security for tiny devices," http://www.cs.berkeley.edu/ nks/tinysec/, 2003.

45. W. Du, J. Deng, and Y. Han, "A pairwise key pre-distribution scheme for wireless sensor networks," in *Proceedings of the Tenth ACM Conference on Computer and Communications Security (CCS 2003)*, Washington, DC, 2003.

46. D. Liu and P. Ning, "Establishing pairwise keys in distributed sensor networks," in *Proceedings of the Tenth ACM Conference on Computer and Communications Security (CCS 2003)*, Washington, DC, 2003.

47. A. J. Menezes, P. C. van Oorschot, and S. A. Vanstone, *Handbook of Applied Cryptography*, Boca Raton, FL: CRC Press, 1997.

48. A. Weimerskirch and D. Westhoff, "Zero common-knowledge authentication for pervasive networks," in *Selected Areas in Cryptography—SAC, 2003*, 2003.

49. A. Weimerskirch and D. Westhoff, "Identity certified authentication for ad-hoc networks," in *ACM Workshop on Security of Ad Hoc and Sensor Networks in Conjunction with the Tenth ACM SIGSAC Conference on Computer and Communications Security (ACM SASN'03)*, Fairfax, VA Oct. 2003.

50. S. Lucks, E. Zenner, A. Weimerskirch, and D. Westhoff, "Efficient entity recognition for low-cost devices," submitted for publication.

51. B. Lamparter, K. Paul, and D. Westhoff, "Charging support for ad hoc stub networks," *Journal of Computer Communication*, vol. 26, pp. 1504–1514, Aug. 2003, special Issue on Internet Pricing and Charging: Algorithms, Technology and Applications.

52. L. Buttyán and J.-P. Hubaux, "Nuglets: A virtual currency to stimulate cooperation in self-organized mobile ad hoc networks," Technical Report DSC/2001/001, Swiss Federal Institute of Technology—Lausanne, Department of Communication Systems," 2001.

53. S. Zhong, Y. R. Yang, and J. Chen, "Sprite: A simple, cheat-proof, credit-based system for mobile ad-hoc networks," in *Proceedings of IEEE INFOCOM '03*, San Francisco, Mar. 2003.

54. D. Coppersmith and M. Jakobsson, "Almost optimal hash sequence traversal," in *Proceedings of the Fourth Conference on Financial Cryptography (FC '02)*, Springer-Verlag, 2002.

55. National Institute of Standard and Technology, "Recommendation for key management," Special Publication 800-57, Federal Information Processing Standards, National Bureau of Standards, U.S. Department of Commerce, Jan. 2003.

56. A. Perrig, R. Szewczyk, V. Wen, D. Cullar, and J. D. Tygar, "SPINS: Security protocols for sensor networks," in *Proceedings of MOBICOM 2001*, Rome, 2001.

57. M. Fischlin, "Fast verification of hash chains," in *RSA Security Cryptographer's Track 2004*, Lecture Notes in Computer Science, vol. 2964, Springer-Verlag, 2004, pp. 339–352.

58. N. Potlapally, S. Ravi, A. Raghunathan, and N. K. Jha, "Analyzing the energy consumption of security protocols," in *IEEE International Symposium on Low-Power Electronics and Design (ISLPED)*, Seoul, Aug. 2003.

59. A. Hodjat and I. Verbauwhede, "The energy cost of secrets in ad-hoc networks," in *IEEE Circuits and Systems Workshop on Wireless Communications and Networking*, Orlando, FL Sept. 2002.

60. Crossbow, "Webpage," http://www.xbow.com, 2004.

61. T. Project, "TinyOS: A component-based OS for the networked sensor regime," http://webs.cs.berkeley.edu/tos/download.html.

62. S. Kumar, M. Girimondo, A. Weimerskirch, C. Paar, A. Patel, and A. Wander, "Embedded end-to-end wireless security with ECDH key exchange," in *IEEE Midwest Symposium on Circuits and Systems*, San Juan, Puerto Rico, Dec. 2003.

63. M. Wiener, "Performance comparison of public-key cryptosystems," Technical Report, RSA Laboratories, 1998.

64. J. López and R. Dahab, "Performance of elliptic curve cryptosystems," Technical Report IC-00-08, ICUniCamp, May 2000.

65. J. Deng, R. Han, and S. Mishra, "Security support for in-network processing in wireless sensor networks," in *Proceedings of First ACM Workshop on Security of Ad Hoc and Sensor Networks (SASN'03)*, Oct. 2003, pp. 83–93.

66. A. Mainwaring, J. Polastre, R. Szewczyk, D. Culler, and J. Anderson, "Wireless sensor networks for habitat monitoring," in *ACM International Workshop on Wireless Sensor Networks and Applications (WSNA'02)*, Atlanta, GA, Sept. 2002.

67. I. F. Akyildiz, W. Su, Y. Sankarasubramaniam, and E. Cayirci, "Wireless sensor networks: A survey," *Computer Networks*, vol. 38, pp. 393–422, Mar. 2002.

68. C.-Y. Chong and S. Kumar, "Sensor networks: Evolution, opportunities and challenges," *Proceedings of the IEEE*, vol. 91, pp. 1247–1256, Aug. 2003.

69. D. Carman, P. Kruus, and B. J. Matt, "Constraints and approaches for distributed sensor network security (final)," Technical Report 00-010, NAI Labs, 2000.

70. S. Basagni, K. Herrin, D. Bruschi, and E. Rosti, "Secure pebblenet," in *Proceedings of the 2001 ACM International Symposium on Mobile Ad Hoc Networking and Computing, MobiHoc 2001*, Long Beach, CA, Oct. 2001, pp. 156–163.

71. L. Eschenauer and V. D. Gligor, "A key-management scheme for distributed sensor networks," in *Proceedings of the Ninth ACM Conference on Computer and Communications Security*, 2002, Washington, DC, pp. 41–47.

72. H. Chan, A. Perrig, and D. Song, "Random key predistribution schemes for sensor networks," in *IEEE Symposium on Security and Privacy*, Berkeley, May 2003, pp. 197–213.

73. D. Liu and P. Ning, "Establishing pairwise keys in distributed sensor networks," in *Proceedings of the Tenth ACM Conference on Computer and Communication Security*, Fairfax, VA, Oct. 2003, pp. 52–61.

74. W. Du, J. Deng, Y. S. Han, and P. K. Varshney, "A pairwise key pre-distribution scheme for wireless sensor networks," in *Proceedings of the Tenth ACM Conference on Computer and Communication Security*, Washington, DC, Oct. 2003, pp. 42–51.

75. C. Intanagonwiwat, R. Govindan, D. Estrin, J. Heidemann, and F. Silva, "Directed diffusion for wireless sensor networking," *ACM/IEEE Transactions on Networking*, vol. 11, Feb. 2002.

76. A. Perrig, R. Szewczyk, V. Wen, D. E. Culler, and J. D. Tygar, "SPINS: Security protocols for sensor networks," in *Seventh Annual International Conference on Mobile Computing and Networks (MobiCOM)*, Rome, 2001, pp. 189–199.

77. H. Krawczyk, "The order of encryption and authentication for protecting communications (or: How secure is SSL?)," in Lecture Notes in Computer Science, vol. 2139, 2001, pp. 310–331.

78. C. Karlof and D. Wagner, "Secure routing in wireless sensor networks: Attacks and countermeasures," *Ad Hoc Network Journal*, Special Issue on Sensor Network Applications and Protocols, vol. 1, pp. 293–315, Sept. 2003.

79. S. Zhu, S. Setia, and S. Jajodia, "LEAP: Efficient security mechanisms for large-scale distributed sensor networks," in *Proceedings of the Tenth ACM Conference on Computer and Communication Security*, Washington, DC, Oct. 2003, pp. 62–72.

80. S. Slijepcevic, M. Potkonjak, V. Tsiatsis, S. Zimbeck, and M. Srivastava, "On communication security in wireless ad-hoc sensor networks," in *Eleventh IEEE International Workshops on Enabling Technologies: Infrastructure for Collaborative Enterprises*, Pittsburgh, June 2002, pp. 139–144.

81. N. Asokan and P. Ginzboorg, "Key agreement in ad-hoc networks," *Computer Communications*, vol. 23, 2000.

82. S. Capcun, L. Buttyan, and J.-P. Hubaux, "Self-organized public-key management for mobile ad hoc networks," *IEEE Transactions on Mobile Computing*, pp. 52–64, Jan.–Mar. 2003.

83. W. Du, J. Deng, Y. S. H. S. Chen, and P. Varshney, "A key management scheme for wireless sensor networks using deployment knowledge," in *IEEE INFOCOM'04*, Hong Kong, China, Mar. 2004.

84. S. Berkovits, "How to broadcast a secret," in *Proceedings of Advances in Cryptology—Eurocrypt '91*, 1991, pp. 535–541.

85. C. C. Chang, T. C. Wu, and C. P. Chen, "The design of a conference key distribution system," in *Proceedings of Advances in Cryptology—Asiacrypt '92*, 1992, pp. 459–466.

86. K. Koyama, "Secure conference key distribution schemes for conspiracy attack," in *Proceedings of Advances in Cryptology—Eurocrypt '92*, 1993, pp. 447–453.

87. C. Blundo, A. D. Santis, A. Herzberg, S. K. U. Vaccaro, and M. Tung, "Perfectly-secure key distribution for dynamic conferences," in *Proceedings of Advances in Cryptology—Crypto '92*, 1993, pp. 471–486.

88. C. C. Chang and C. H. Lin, "How to converse securely in a conference," in *Proceedings of IEEE Thirtieth Annual International Carnahan Conference*, 1996, pp. 42–45.

89. A. Perrig, D. Song, and J. Tygar, "Elk, a new protocol for efficient large-group key distribution," in *Proceedings of IEEE Security and Privacy Symposium*, Oakland, CA, May 2001.

90. A. Perrig and J. D. Tygar, *Secure Broadcast Communications in Wired and Wireless Networks*, Boston, MA: Kluwer Academic, 2002.

91. T. C. Wu, "Conference key distribution system with user anonymity based on algebraic approach," *IEEE Proceedings on Computers and Digital Techniques*, vol. 144, no. 2, pp. 145–148, 1997.

92. A. Perrig, Y. Kim, and G. Tsudik, "Communication-efficient group key agreement," in *International Federation for Information Processing IFIP SEC*, Paris, 2001.

93. A. Perrig, "Tree-based group key agreement," *ACM Transactions on Information and System Security (TISSEC)*, vol. 7, no. 1, 2004.

94. Y. Kim, A. Perrig, and G. Tsudik, "Group key agreement efficient in communication," *IEEE Transactions on Computers*, vol. 53, no. 7, pp. 905–921, July 2004.

95. B. Klein, M. Otten, and T. Beth, "Conference key distribution protocols in distributed systems," in *Proceedings on Codes and Ciphers—Cryptography and Coding IV*, 1995, pp. 225–242.

96. Y. Kim, A. Perrig, and G. Tsudik, "Simple and fault-tolerant key agreement for dynamic collaborative groups," in *Proceedings of the Seventh ACM Conference on Computer and Communications Security*, Athens, Greece, 2000, pp. 235–244.

97. M. Burmester and Y. Desmedt, "A secure and efficient conference key distribution system," in *Proceedings of Advances in Cryptology—Eurocrypt '94*, 1995, pp. 275–286.

98. W. G. Tzeng and Z. J. Tzeng, "Round-efficient conference key agreement protocols with provable security," in *Proceedings of Advances in Cryptology—Asiacrypt 2000*, 2001, pp. 614–627.

99. W. G. Tzeng, "A secure fault-tolerant conference-key agreement protocol," *IEEE Transactions on Computers*, vol. 51, no. 4, pp. 373–379, 2002.

100. W. Stallings, *Cryptography and Network Security: Principles and Practice*, 2nd ed., Upper Saddle River, NJ: Prentice-Hall, 1999.

101. I. F. Akyildiz, W. Su, Y. Sankarasubramaniam, and E. Cayirci, "A survey on sensor networks," *IEEE Communications Magazine*, pp. 102–114, Aug. 2002.

102. W. Diffie and M. E. Hellman, "New directions in cryptography," *IEEE Transactions on Information Theory*, vol. 22, pp. 644–654, Nov. 1976.

103. R. L. Rivest, A. Shamir, and L. Adleman, "A method for obtaining digital signatures and public-key cryptosystems," *Communications of the ACM*, vol. 21, no. 2, pp. 120–126, Feb. 1978.

104. A. Joux, "A one round protocol for tripartite diffie-hellman," in *Algorithmic Number Theory Symposium*, Leiden, Netherlands, 2000, pp. 385–394.

105. D. Boneh. and M. Franklin, "Identity-based encryption from the weil pairing," in *Advances in Cryptology—CRYPTO 2001*, 2001, pp. 213–229.

106. J. C. Cha and J. H. Cheon, "An identity-based signature from gap Diffie–Hellman groups," in *Proceedings of PKC '03*, Miami, FL, 2003, pp. 18–30.

107. D. Boneh, B. Lynn, and H. Shacham, "Short signatures from the weil pairing," in *Proceedings of Advances in Cryptology—Asiacrypt 2001*, 2001, pp. 514–532.

108. M. Bellare, D. Pointcheval, and P. Rogaway, "Authenticated key exchange secure against dictionary attacks," in *Proceedings of Advances in Cryptology—Eurocrypt 2000*, 2000, pp. 139–155.

109. M. Steniner, P. B. T. Eirich, and M. Waidner, "Secure password-based cipher suite for tls," *ACM Transactions on Information Systems and Security*, vol. 4, no. 2, 2001.

SENSOR NETWORK APPLICATIONS

PURSUER–EVADER TRACKING IN SENSOR NETWORKS

Murat Demirbas, Anish Arora, and Mohamed Gouda

In this chapter we present a self-stabilizing program for solving a pursuer–evader problem in sensor networks. The program is a hybrid between two orthogonal programs, an evader-centric program and a pursuer-centric program, and can be tuned for tracking speed or energy efficiency. In the program, sensor nodes close to the evader dynamically maintain a tracking tree of depth R that is always rooted at the evader. The pursuer, on the other hand, searches the sensor network until it reaches the tracking tree and then follows the tree to its root in order to catch the evader.

9.1 INTRODUCTION

Due to its importance in military contexts, pursuer–evader tracking has received significant attention [1–4] and has been posed by the DARPA network-embedded software technology (NEST) program as a challenge problem. Here, we consider the problem in the context of wireless sensor networks. Such networks comprising potentially many thousands of low-cost and low-power wireless sensor nodes have recently became feasible, thanks to advances in microelectromechanical systems technology and are being regarded as a realistic basis for deploying large-scale pursuer–evader tracking.

Previous work on the pursuer–evader problem is not directly applicable to tracking in sensor networks since these networks introduce the following challenges: First, sensor nodes have very limited computational resources [e.g., 8K random-access memory (RAM) and 128K flash memory]; thus, centralized algorithms are not suitable for sensor networks due to their larger computational requirements. Second, sensor nodes are energy constrained; thus, algorithms that impose an excessive communication burden on nodes are not acceptable since they drain the battery power quickly. Third, sensor networks are fault-prone: Message

Sensor Network Operations, Edited by Phoha, LaPorta, and Griffin
Copyright © 2006 The Institute of Electrical and Electronics Engineers, Inc.

losses and corruptions (due to fading, collusion, and hidden node effect), and node failures (due to crash and energy exhaustion) are the norm rather than the exception. Thus, sensor nodes can lose synchrony and their programs can reach arbitrary states [5]. Finally, on-site maintenance is not feasible; thus, sensor networks should be self-healing. Indeed, one of the emphases of the NEST program is to design low-cost fault-tolerant, and more specifically self-stabilizing services for the sensor network domain.

In this chapter we present a tunable and self-stabilizing program for solving a pursuer–evader problem in sensor networks. The goal of the pursuer is to catch the evader (despite the occurrence of faults) by means of information gathered by the sensor network. The pursuer can move faster than the evader. However, the evader is omniscient—it can see the state of the entire network—whereas the pursuer can only see the state of one sensor node (say the nearest one). This model captures a simple, abstract version of problems that arise in tracking via sensor networks.

Tunability We achieve tunability of our program by constructing it to be a hybrid between two orthogonal programs: an evader-centric program and a pursuer-centric program. In the evader-centric program, nodes communicate periodically with neighbors and dynamically maintain a tracking tree structure that is always rooted at the evader. The pursuer eventually catches the evader by following this tree structure to the root: The pursuer asks the closest sensor node who its parent is; then it proceeds to that node and, thus, reaches the root node (and hence the evader) eventually.

In the pursuer-centric program, nodes communicate with neighbors only at the request of the pursuer: When the pursuer reaches a node, the node resets its recorded time of a detection of an evader to zero and directs the pursuer to a neighboring node with the highest recorded time.

The evader-centric program converges and tracks the evader faster, whereas the pursuer-centric program is more energy-efficient. In the hybrid program we combine the evader-centric and pursuer-centric programs:

1. We modify the evader-centric program to limit the tracking tree to a bounded depth R to save energy.
2. We modify the pursuer-centric program to exploit the tracking tree structure.

The hybrid program is tuned for tracking speed or energy efficiency by selecting R appropriately. In particular, for the extended hybrid program in Section 9.6, the tracking time is $3(D - R) + R\alpha/(1 - \alpha)$ steps, and at most n communications take place at each program step, where D denotes the diameter of the network, α is the ratio of the speed of the evader to that of the pursuer, and n is the number of sensor nodes included in the tracking tree.

Self-Stabilization In the presence of faults, our program recovers from arbitrary states to states from where it correctly tracks the evader; this sort of fault tolerance is commonly referred to as stabilizing fault tolerance. In particular, starting from any arbitrary state, the tracking time is $2R + 3(D - R) + R\alpha/(1 - \alpha)$ steps for our extended hybrid program.

9.2 THE PROBLEM

System Model A sensor network consists of a (potentially large) number of *sensor nodes*. Each node is capable of receiving/transmitting messages within its field of communication. All nodes within this communication field are its neighbors; we denote this set for node j as $nbr.j$. We assume the nbr relation is symmetric and induces a connected graph. (Protocols for maintaining biconnectivity in sensor networks are known [6, 7].)

Problem Statement   Given are two distinguished processes, the *pursuer* and the *evader*, that each reside at some node in the sensor network. Each node can immediately detect whether the pursuer and/or the evader are resident at that node.

Both the pursuer and the evader are mobile: Each can atomically move from one node to another, but the speed of the evader movement is less than the speed of the pursuer movement. The strategy of evader movement is unknown to the network. The strategy could in particular be intelligent, with the evader omnisciently inspecting the entire network to decide whether and where to move. By way of contrast, the pursuer strategy is based only on the state of the node at which it resides.

Required is to design a program for the nodes and the pursuer so that the pursuer can "catch" the evader, that is, guarantee in every computation of the network that eventually both the pursuer and the evader reside at the same node.

Programming Model   A *program* consists of a set of variables, node actions, pursuer actions, and evader actions. Each variable and each action resides at some node. Variables of a node j can be updated only by j's node actions. Node actions can only read the variables of their node and the neighboring nodes. Pursuer actions can only read the variables of their node. The evader actions can read the variables of the entire program; however, they cannot update any of these variables.

Each action has the form:

$$\langle guard \rangle \longrightarrow \langle assignment\ statement \rangle$$

A *guard* is a Boolean expression over variables. An assignment statement updates one or more variables. A *state* is defined by a value for every variable in the program, chosen from the predefined domain of that variable. An action whose guard is true at some state is said to be *enabled* at that state.

We assume maximal parallelism in the execution of node actions. At each state, each node executes all actions that are enabled in that state. (Execution of multiple enabled actions in a node is treated as executing them in some sequential order.) Maximal parallelism is not assumed for the execution of the pursuer and evader actions. Recall, however, that the speed of execution of the former exceeds that of the latter. For ease of exposition, we assume that evader and pursuer actions do not occur strictly in parallel with node actions.

A *computation* of the program is a maximal sequence of program steps: In each step, actions that are enabled at the current state are executed according to the above operational semantics, thereby yielding the next state in the computation. The maximality of a computation implies that no computation is a proper prefix of another computation.

We assume that each node has a clock that is synchronized with the clocks of other nodes. This assumption is reasonably implemented at least for Mica nodes [8]. Later, in Section 9.7, we show how our programs can be modified to work without the synchronized clocks assumption.

Notation   In this chapter, we use j, k, and l to denote nodes. We use $var.j$ to denote the variable var residing at j. We use $[]$ to separate the actions in a program and $x :\in A$ to denote that x is assigned to an element of set A.

Each *parameter* in a program ranges over the *nbr* set of a node. The function of a parameter is to define a set of actions as one parameterized action. For example, let k be a parameter whose value is 0, 1, or 2; then an action *act* of node j parameterized over k abbreviates the following set of actions:

$$act \backslash (k = 0) \quad \square \quad act \backslash (k = 1) \quad \square \quad act \backslash (k = 2)$$

where $act \backslash (k = i)$ is *act* with every occurrence of k substituted with i.

We describe certain conjuncts in a guard in English: {Evader resides at j} and {Evader detected at j}. The former expression evaluates to true at all states where the evader is at j, whereas the latter evaluates to true only at the state immediately following any step where the evader moves to node j, and evaluates to false in the subsequent states even if the evader is still at j.

We use N to denote the number of nodes in the sensor network, D the diameter of the network, and M the distance between the pursuer and evader. We use distance to refer to hop distance in the network; unit distance is the single-hop communication radius. Finally, we use α to denote the ratio of the speed of the evader to that of the pursuer.

Evader Action   In each of the programs that we present in this chapter, we use the following evader action:

$$\{\text{Evader resides at } j\} \quad \longrightarrow \quad \text{Evader moves to } l, \ l :\in \{k \mid k \in nbr.j \cup \{j\}\}$$

When this action is executed, the evader moves to an arbitrary neighbor of j or skips a move. This notion of nondeterministic moves suffices to capture the strategy of an omniscient evader.

Recall from the discussion in the problem statement that, when the evader moves to a node, the node immediately detects this fact (i.e., the detection actions have priority over normal node actions and are fired instantaneously).

Fault Model   Transient faults may corrupt the program state. Transient faults may also fail-stop or restart nodes (in a manner that is detectable to their neighbors); we assume that the connectivity of the graph is maintained despite these faults.

A program P is *stabilizing fault-tolerant* iff starting from an arbitrary state, provided that no other faults occur during recovery, P eventually recovers to a state from where its specification is satisfied.

9.3 EVADER-CENTRIC PROGRAM

In this section we present an evader-centric solution to the pursuer–evader problem in sensor networks. In our program every sensor node, j, maintains a value, $ts.j$, that denotes the latest time stamp that j knows for the detection of the evader. Initially, for all $j, ts.j = 0$. If j detects the evader, it sets $ts.j$ to its clock's value. Every node j periodically updates its $ts.j$ value based on the ts values of its neighbors: j assigns the maximum time-stamp value it is aware of as $ts.j$. We use $p.j$ (read parent of j) to record the node that j received the maximum time-stamp value. Initially, for all j, $p.j$ is set to null, that is, $p.j = \perp$.

As the information regarding the evader propagates through the network via gossipping of the neighbors, the parent variables at these nodes are set accordingly. Note that the parent relation embeds a tree rooted at the evader on the sensor network. We refer to this tree as the *tracking tree*.

In addition to the above variables, we maintain a variable $d.j$ at each node j. Initially, for all j, $d.j = \infty$. When the evader is detected at a node j, $d.j$ is set to 0. Otherwise, $d.j$ is updated by setting it to be the parent's d value plus 1, that is, $d.j := d.(p.j) + 1$. This way, $d.j$ at each node corresponds to the distance of j from the evader. In the case where $ts.j$ is equal to ts values of j's neighbors, j uses the d values of its neighbors to elect its parent to be the one offering the shortest distance to the evader.

Thus, the actions for j (parameterized with respect to neighbor k) in the evader-centric program is as follows:

$$\{\text{Evader resides at } j\} \longrightarrow p.j := j; \; ts.j := clock.j; \; d.j := 0$$

$$[]$$

$$ts.k > ts.j \vee (ts.k = ts.j \;\wedge\; d.k + 1 < d.j)$$
$$\longrightarrow \; p.j := k; \, ts.j := ts.(p.j);$$
$$d.j := d.(p.j) + 1$$

Once a tracking tree is formed, the pursuer follows this tree to reach the evader simply by querying its closest node for its parent and proceeding to the parent node. Thus, the pursuer action is as follows:

$$\{\text{Pursuer resides at } j\} \; \longrightarrow \; \text{Pursuer moves to } p.j$$

9.3.1 Proof of Correctness

As Figure 9.1 illustrates, if the evader is moving, it may not be possible to maintain a minimum distance spanning tree. Note that this is a worst-case scenario and occurs when the evader speed is as fast as the communication speed of the nodes. (Tracking is not achievable if the evader is faster than the communication speed of the nodes.) In practice, node communication (25 ms) is faster than the evader movement and construction of a minimum distance spanning tree is possible. However, even in this worst-case scenario, we can still prove the following properties in the absence of faults.

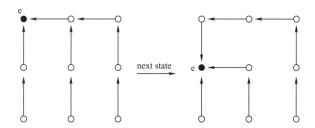

Figure 9.1 Minimum spanning tree is not maintained when evader is moving.

Theorem 9.3.1   The tracking tree is a spanning tree rooted at the node where the evader resides and is fully constructed in at most D steps.

PROOF   From the synchronized clocks assumption and the privileged detection action, {Evader resides at j}, it follows that the node j where the evader resides has the highest time-stamp value in the network. Observe from the second node action that the $p.k$ variable at every node k embeds a logical tree structure over the sensor network. Cycles cannot occur since $(\forall k : ts.k > 0 : d.(p.k) < d.k).$[1] Since $(\forall k : ts.k > 0 : ts.(p.k) > ts.k)$, the network is connected, and the node j where the evader resides has the highest time-stamp value in the network, it follows that there exists only one tree in the network and it is rooted at j. Within at most D steps all the nodes in the network receive a message from a node that is already included in the tracking tree (due to the maximal parallelism model and the second node action), and a tracking tree covering the entire network is constructed.   ■

Lemma 9.3.1   The distance between the pursuer and evader does not increase once the constructed tree includes the node where the pursuer resides.

PROOF   Once the constructed tree includes the node k_x where the pursuer resides, there exists a path $k_1, k_2, \ldots, k_x$ such that $(\forall i : 1 < i \leq x : p.k_i = k_{i-1})$ and the evader resides at k_1. At any program step, if the evader moves to a neighboring node, the pursuer, being faster than the evader, also moves to the next node in the path.

Note that at each program step, any node k_i in this path may choose to change its parent, rendering a different path between the pursuer and the evader. However, observe from the second node action that k_i changes its parent to be the neighbor that has a shorter path to the evader (higher time stamp implies shorter path since the node where the evader resides has the highest time stamp and nodes execute under maximal parallelism model). Thus, the net effect is that the path length can only decrease but not increase.   ■

Theorem 9.3.2   The pursuer catches the evader in at most $M + 2M\lceil \alpha/(1 - \alpha) \rceil$ steps.

PROOF   Since the initial distance between the evader and the pursuer is M, after M program steps the tracking tree includes the node at which the pursuer resides. Since the evader's speed is below unit time step of the protocol execution, within this period the evader can move to at most M hops away, potentially increasing the distance between the evader and pursuer to $2M$. From Lemma 9.3.1, it follows that this distance cannot increase in the subsequent program steps. Since the pursuer is faster than the evader, it catches the evader in at most $2M\lceil \alpha/(1 - \alpha) \rceil$ steps [follows from solving $\alpha = X/(X + 2M)$ for X].   ■

9.3.2 Proof of Stabilization

In the presence of faults, variables of a node j can be arbitrarily corrupted. However, for the sake of simplicity we assume that even in the presence of faults the following two conditions hold:

1. **always** $ts.j \leq clock.j$.
2. **always** $\{p.j \in \{nbr.j \cup \{j\} \cup \{\bot\}\}$.

---

[1] The predicate $(\forall i : R.i : X.i)$ may be read as "for all i that satisfy $R.i$, $X.i$ is also true."

The first condition states that the time stamp for the detection of evader at node j is always less than the local clock at j (i.e., $ts.j$ cannot be in the future). The second condition states that the domain of $p.j$ is restricted to the set $\{nbr.j \cup \{j\} \cup \{\bot\}\}$ where $p.j = \bot$ denotes that j does not have any parent. These are both locally checkable and enforceable conditions; in order to keep the program simple, we will not include the corresponding correction actions in our presentation.

Lemma 9.3.2   The tracking tree stabilizes in at most D steps.

PROOF   Since we have **always** $ts.j \leq clock.j$, even at an arbitrary state (which might be reached due to transient faults) the node where the evader resides has the highest time-stamp value in the network. From Theorem 9.3.1 it follows that a fresh tracking tree is constructed within at most D steps, and this tracking tree is a spanning tree rooted at the node where the evader currently resides.   ■

Theorem 9.3.3   Starting from an arbitrarily corrupted state, the pursuer catches the evader in at most $D + 2D\alpha/(1 - \alpha)$.

PROOF   The proof follows from the proofs of Lemma 9.3.2 and Theorem 9.3.2.   ■

9.3.3   Performance Metrics

The evader-centric program is not energy efficient since every node communicates with its neighbor at each step of the program. That is, $\omega \times N$ communications occur each step, where ω denotes the average degree of a node. The communications can be treated as broadcasts, and, hence, the number of total communications per step is effectively N. On the other hand, the tracking time and the convergence time of the evader-centric program is fast: Starting from an arbitrarily corrupted state it takes at most $D + 2D\alpha/(1 - \alpha)$ steps for the pursuer to catch the evader.

9.4   PURSUER-CENTRIC PROGRAM

In this section we present a pursuer-centric solution to the pursuer–evader problem in sensor networks. Here, similar to the evader-centric program, every sensor node, j, maintains a value, $ts.j$, that denotes the latest time stamp that j knows for the detection of the evader. Initially, for all j, $ts.j = 0$. If j detects the evader, it sets $ts.j$ to its clock's value.

In this program, nodes communicate with neighbors only at the request of the pursuer: When the pursuer reaches a node j, j resets $ts.j$ to zero and directs the pursuer to a neighboring node with the highest recorded time (we use $next.j$ to denote this neighbor). Note that if all ts values of the neighbors are the same (e.g., zero), the pursuer is sent to an arbitrary neighbor. Also, if there is no pursuer at j, $next.j$ is set to $\bot$ (i.e., undefined).

Thus, the actions for node j in the pursuer-centric program are as follows:

$$\{\text{Evader detected at } j\} \longrightarrow ts.j := clock.j$$

$$\square$$

$$\{\text{Pursuer detected at } j\} \longrightarrow next.j :\in \{k \mid k \in nbr.j \wedge$$

$$ts.k = \max(\{ts.l \mid l \in nbr.j\})\};$$

$$ts.j := 0$$

The pursuer's action is as follows:

$$\{\text{Pursuer resides at } j\} \longrightarrow \text{Pursuer moves to } next.j$$

9.4.1 Proof of Correctness

In the absence of faults, our pursuer-centric program satisfies the following properties.

Lemma 9.4.1 If the pursuer reaches a node j where $ts.j > 0$, the pursuer catches the evader in at most $N\alpha/(1 - \alpha)$ steps.

PROOF   If the pursuer reaches a node j where $ts.j > 0$, then there exists a path between the pursuer and the evader that is at most of length N. This distance does not increase in the following program steps (due to maximal parallel execution semantics and the program actions).   ∎

In [9], it is proven that during a random walk on a graph the expected time to find N distinct vertices is $O(N^3)$. However, a recent result [10] shows that by using a local topology information (i.e., degree information of neighbor vertices) it is possible to achieve the cover time $O(N^2 \log N)$ for random walk on any graph. Thus, we have:

Lemma 9.4.2 The pursuer reaches a node j where $ts.j > 0$ within $O(N^2 \log N)$ steps.

Theorem 9.4.1 The pursuer catches the evader within $O(N^2 \log N)$ steps.

9.4.2 Proof of Stabilization

Since each node j resets $ts.j$ to zero upon a detection of the pursuer, arbitrary $ts.j$ values eventually disappear, and, hence, the pursuer-centric program is self-stabilizing.

Theorem 9.4.2 Starting from an arbitrary state, the pursuer catches the evader within $O(N^2 \log N)$ steps.

9.4.3 Performance Metrics

The pursuer-centric program is energy efficient. At each step of the program only the node where the pursuer resides communicates with its neighbors. That is, ω communications occur at each step. On the other hand, the tracking and the convergence time of the pursuer-centric program is slow: $O(N^2 \log N)$ steps.

9.5   HYBRID PURSUER–EVADER PROGRAM

In the hybrid program we combine the evader-centric and pursuer-centric approaches:

1. We modify the evader-centric program to limit the tracking tree to a bounded depth R to save energy.
2. We modify the pursuer-centric program to exploit the tracking tree structure.

We limit the depth of the tracking tree to R by means of the distance, d, variable.

$$\{\text{Evader resides at } j\} \quad \longrightarrow \quad p.j := j;\ ts.j := clock.j;\ d.j := 0$$

$$\square$$

$$d.k < R \wedge (ts.k > ts.j \ \vee \ (ts.k = ts.j \wedge d.k+1 < d.j))$$
$$\longrightarrow \quad p.j := k;\ ts.j := ts.(p.j);$$
$$d.j := d.(p.j) + 1$$

By limiting the tree to a depth R, we lose the advantages of soft-state stabilization: There is no more a flow of fresh information to correct the state of the nodes that are outside the tracking tree. To achieve stabilization, we add explicit stabilization actions. Next we describe these two actions.

Starting from an arbitrarily corrupted state where the graph embedded by the parent relation on the network has cycles, each cycle is detected and removed by using the bound on the length of the path from each process to its root process in the tree. To this end, we exploit the way that we maintain the d variable: j sets $d.j$ to be $d.(p.j)+1$ whenever $p.j \in nbr.j$ and $d.(p.j) + 1 \leq R$. The net effect of executing this action is that if a cycle exists then the $d.j$ value of each process j in the cycle gets "bumped up" repeatedly.

Within at most R steps, some $d.(p.j)$ reaches R, and since the length of each path in the adjacency graph is bounded by R, the cycle is detected. To remove a cycle that it has detected, j sets $p.j$ to $\perp$ (undefined) and $d.j$ to ∞, from which the cycle is completely cleaned within the next R steps. Note that this action also takes care of pruning the tracking tree to height R (e.g., when the evader moves and as a result a node j with $d.j = R$ becomes $R + 1$ away from the evader).

Node j also sets $p.j$ to $\perp$ (undefined) and $d.j$ to ∞ if $p.j$ is not a valid parent [e.g., $d.j \neq d.(p.j)+1$ or $ts.j > ts.(p.j)$ or $(p.j = j \wedge d.j \neq 0)$]. We add another action to correct the fake tree roots. If a node j is spuriously corrupted to $p.j = j \wedge d.j = 0$, this is detected by explicitly asking for a proof of the evader at j. Thus, the stabilization actions for the bounded length tracking tree are as follows:

$$p.j \neq \perp \wedge ((p.j = j \wedge d.j \neq 0) \vee ts.j > ts.(p.j)$$
$$\vee\, d.j \neq d.(p.j) + 1 \vee d.(p.j) \geq R)$$
$$\longrightarrow \quad p.j := \perp;\ d.j := \infty$$

$$\square$$

$$p.j = j \wedge d.j = 0 \wedge \neg\{\text{Evader resides at } j\}$$
$$\longrightarrow \quad p.j := \perp;\ d.j := \infty$$

We modify the node action in the pursuer-centric program only slightly so as to exploit the tracking tree structure.

$$\{\text{Pursuer detected at } j\} \ \longrightarrow$$
$$\text{if}\,(p.j \neq \perp) \text{ then } next.j := p.j$$
$$\text{else}$$

$$next.j :\in \{k \mid k \in nbr.j \wedge ts.k = \max(\{ts.l \mid l \in nbr.j\})\};$$

$$ts.j := 0$$

Finally, the pursuer action is the same as that in Section 9.4.

9.5.1   Proof of Correctness

In the absence of faults, the following lemmas and theorem follow from their counterparts in Sections 9.3 and 9.4:

Lemma 9.5.1   The tracking tree is fully constructed in at most R steps.

Below n denotes the number of nodes included in the tracking tree.

Lemma 9.5.2   The pursuer reaches the tracking tree within $O((N - n)^2 \log(N - n))$ steps.

Theorem 9.5.1   The pursuer catches the evader within $O((N - n)^2 \log(N - n))$ steps.

Since the evader is mobile, the number of nodes in the tracking tree of depth R varies over time depending on the location of the evader and the density of nodes within the R-hop neighborhood. However, the number n we use in the $O(\,)$ notation depends only on the number of nodes included in the first tracking tree constructed. More specifically, in the $O(\,)$ notation we use $N - n$ to denote the number of nodes that the pursuer needs to perform a random walk to reach a node that was once involved in the tracking tree, that is, $ts > 0$. Even though the maximum number of nodes that the pursuer needs to visit monotonously decreases as the evader moves and new tracking trees are constructed, in our analysis we still use $N - n$, which resulted from the construction of the first tracking tree, to capture the worst-case scenario.

9.5.2   Proof of Stabilization

Lemma 9.5.3   The tracking tree structure stabilizes in at most $2R$ steps.

PROOF   Stabilization of the nodes within the tracking tree follows from Lemma 9.3.2. The discussion above about the stabilization actions of the hybrid program states that a cycle outside the tracking tree is resolved within $2R$ steps. These two occur in parallel; thus system stabilization is achieved within $2R$ steps.   ■

Theorem 9.5.2   Starting from an arbitrary state, the pursuer catches the evader within $O((N - n)^2 \log(N - n))$ steps.

9.5.3   Performance Metrics

The hybrid program for the nodes can be tuned to be energy efficient by decreasing R since it decreases n. At each step of the program at most $n + \omega$ communications take place. The hybrid program can also be tuned to track and converge faster by increasing R since it increases

n, and the time, $O((N - n)^2 \log (N - n))$ steps, a random walk takes to find the tracking tree. From that point on it takes only $R\alpha/(1 - \alpha)$ steps for the pursuer to catch the evader.

Note that there is a trade-off between the energy efficiency and the tracking time. In Section 9.7, we provide an example where we choose a suitable value for R to optimize both energy efficiency and tracking time concurrently.

9.6 EFFICIENT VERSION OF HYBRID PROGRAM

In this section we present a communication- and, hence, energy-efficient version of the hybrid program. We achieve this by replacing the random walk of the pursuer with a more energy-efficient approach, namely that of constructing a search tree rooted at the pursuer. To this end, we first present the extended and energy-efficient version of the pursuer-centric program, and then show how this extended pursuer-centric program can be incorporated into the hybrid program.

Extended Pursuer-Centric Program   In the extended version of the pursuer-centric program, instead of the random walk prescribed in Section 9.4, the pursuer uses *agents* to search the network for a trace of the evader. The pursuer agents idea can be implemented by constructing a (depth-first or breadth-first) tree rooted at the node where the pursuer resides. If a node j with $ts.j > 0$ is included in this *pursuer tree*, the pursuer is notified of this result along with a path to j. The pursuer then follows this path to reach j. From this point on, due to Lemma 9.4.1, it will take at most $N\alpha/(1 - \alpha)$ steps for the pursuer to catch the evader.

This program can be seen as an extension of the original pursuer-centric program in that instead of a 1-hop tree construction (i.e., the node k where the pursuer resides contacts $nbr.k$) embedded in the original pursuer-centric program, we now employ a D-hop tree construction. To this end we change the original pursuer program as follows. The node k where the pursuer resides sets $next.j$ to $\perp$ if none of its neighbors has a time-stamp value greater than 0, instead of setting $next.j$ to point to a random neighbor of j. The pursuer upon reading a $\perp$ value for the $next$ variable, starts a tree construction to search for a trace of the evader. Note that by using a depth D, the pursuer tree is guaranteed to encounter a node j with $ts.j > 0$.

Several extant self-stabilizing tree construction programs [11–13] suffice for constructing the pursuer tree in D steps and to complete the information feedback within another D step. Also since the root of the pursuer tree is static (root does not change dynamically unlike the root of the tracking tree), it is possible to achieve self-stabilization of a pursuer tree within D steps in an energy efficient manner. That is, in contrast to the evader-centric tracking tree program where all nodes communicate at each program step, in the pursuer tree program only the nodes propagating a (tree construction or information feedback) wave need to communicate with their immediate neighbors.

Extended Hybrid Program   It is straightforward to incorporate the extended version of the pursuer-centric program into the hybrid program. The only modification required is to set the depth of pursuer tree to be $D - R$ hops instead of D hops. Note that $D - R$ hops is enough for ensuring that the pursuer will encounter a trace of the evader (i.e., pursuer tree will reach a node included in the tracking tree). After a node that is/has been in the tracking tree is reached, the pursuer program in Section 9.5 applies as is.

9.6.1   Performance Metrics

The extension improves the tracking and the convergence time of the pursuer-centric program from $O(N^2 \log N)$ steps to $3D + N\alpha/(1 - \alpha)$ steps ($2D$ steps for the pursuer tree construction and information feedback, and D steps for the pursuer to follow the path returned by the pursuer tree program). The extended pursuer-centric program remains energy efficient; the only overhead incurred is the one-time invocation of the pursuer tree construction.

In the extended hybrid program, it takes at most $3(D - R)$ steps for the pursuer to reach the tracking tree. [Compare this to $O((N - n)^2 \log (N - n))$ steps in the original hybrid program.] From that point on it takes $R\alpha/(1 - \alpha)$ steps for the pursuer to catch the evader. At each step of the extended hybrid program at most n communications take place. Due to the pursuer tree computation, a one-time cost of $(N - n)$ is incurred.

1-Pursuer 0-Evader Scenario   The evader-centric program is energy efficient in a scenario where there is no evader but there is a pursuer in the system: No energy is spent since no communication is needed. On the other hand, the pursuer-centric program performs poorly in this case: At each step the pursuer queries the neighboring nodes incurring a communication cost of w. The hybrid program, since it borrows the pursuer action from the pursuer-centric program, also performs badly in this scenario.

The extended pursuer-centric program fixes this problem by modifying the pursuer tree construction to require that an answer is returned only if the evader tree is encountered. That is, if there is no evader in the network, the pursuer tree program continues to wait for the information feedback wave to be triggered, and, hence, it does not waste energy.

0-Pursuer 1-Evader Scenario   By enforcing that pursuers authenticate themselves when they join the network and notify the network when they leave, we can ensure that a tracking tree is maintained only when there is a pursuer in the system and achieve energy efficiency.

9.7   IMPLEMENTATION AND SIMULATION RESULTS

In this section, we present implementation and simulation results and show an example of tuning our tracking program to optimize both energy efficiency and tracking time concurrently.

9.7.1   Implementation

We have implemented an asynchronous version of the evader-centric program on the Berkeley's Mica node platform [8] for a demonstration at the June 2002 DARPA–NEST retreat held in Bar Harbor, Maine.

Asynchronous Program   Even though we assumed an underlying clock synchronization service for our presentation, it is possible to modify the evader-centric program slightly (only one line is changed) to obtain an asynchronous version. The modification is to use, at every node j, a counter variable $val.j$ that denotes the number of detections of the evader that j is aware of, instead of $ts.j$ that denotes the latest times tamp that j knows for the

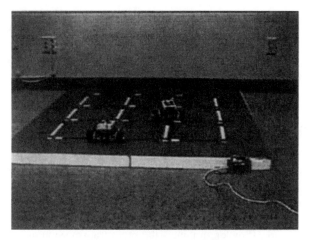

Figure 9.2 Snapshot from our demo.

detection of the evader. When j detects the evader, instead of setting $ts.j$ to $clock.j$, j increases $val.j$ by 1.

The extended pursuer program is also made asynchronous in a straightforward manner, since the idea of pursuer agents (a tree rooted at the pursuer) is readily implemented in the asynchronous model [11–13]. In our demonstration, a Lego Mindstorms robot serving as a pursuer used our program to catch another Lego Mindstorms robot serving as an evader, in a 4×4 grid of nodes subject to a variety of faults. Figure 9.2 shows a snapshot from our demo.

The sensor nodes are embedded in a foam panel under the board. There are small rectangular holes in the board corresponding to sensor locations. A node detects an evader via its optic sensor. When the evader reaches a sensor location, its body closes the hole and triggers a "darkness" reading. The pursuer avoids this detection thanks to the glow sticks attached under its body.

The colored lines on the board encode the four directions, e.g., "long yellow" followed by a "short black" indicates north. The pursuer calculates which direction it is heading after a complete line traversal.

We have soldered infrared (IR) light-emitting diodes (LEDs) on the nodes. Nodes blink IR LEDs at four different frequencies to communicate the four directions. On reaching a node, the pursuer detects the frequency of emitted IR signals and decides how to turn.

The evader is remotely controlled by a human playing the role of an omniscient adversary. We showed, in our demo, that despite node failures or transient corruption of the nodes, the tracking tree stabilizes to a good state, and the pursuer catches the evader by following the tree.

The pursuer robot could get disoriented on the grid or lose track of the grid lines, so we also built in stabilization in the robot program to find the grid lines from an arbitrary state. The pursuer converges to the grid from any point within the board, without falling off the edge. Upon converging, the pursuer regains its sense of direction within one complete line traversal and starts to track the tree by following direction signals from the nodes.

Due to incorrect interaction between the motes and the pursuer robot, the robot could be driven into bad states. The stabilizing robot program was designed to tolerate such bad interactions. For example, if the pursuer reaches a failed mote, it cannot get a direction signal. The pursuer then chooses a random direction to follow. If this direction leads it off the grid, it backs up and retries till it finds a grid direction. On reaching the next nonfailed node, the pursuer gets a proper direction to follow and catches the evader eventually.

We have recently ported our TinyOS sensor node code to nesC [14]; the source code for the sensors, the pursuer, and evader robots (written in NQC) and video shots for the demo are available at www.cis.ohio-state.edu/~demirbas/peDemo/.

9.7.2 Simulation Results

In the preceding sections we have presented analytical worst-case bounds on the performance of our tracking service. In this section we consider a random movement model for the evader and compare and contrast the average case performances of our tracking programs through simulation.

For our simulations, we use Prowler [15], a MATLAB-based, event-driven simulator for wireless sensor network. Prowler simulates the radio transmission/propagation/reception delays of Mica2 motes [8], including collisions in ad hoc radio networks and the operation of the MAC layer. The average transmission time for a packet is around 25 ms. Our implementations are per node and are message-passing distributed programs. Our code for the simulations is also available from www.cis.ohio-state.edu/~demirbas/peDemo. In our simulations, the network consists of 100 nodes arranged in a grid topology of 10×10. The distance between two neighboring nodes on the grid is a unit distance. The node communication radius is approximately 2 units. The evader makes a move every 2 s and the pursuer every 1 s. The average move distance for both the pursuer and evader is 2 units.

Table 9.1 shows the number of total messages and the catching times for the evader-centric program, the pursuer-centric program, hybrid program with $R = 1$, and hybrid program with $R = 2$. These averages are calculated using 30 runs of these programs with random starting locations for the pursuer and the evader.

Figure 9.3 shows a snapshot from the simulation of the evader-centric program. The direction of the arrows denotes the parent pointers at the motes. The two numbers next to a node correspond to val and d (in hops) variables at that node. The tracking tree is rooted at the evader; the evader happens to reside at the middle of the grid in that run. The tracking tree spans the entire network and may have up to 4 hops. It is costly to maintain this tree: on average a total of 256 messages are sent before the evader is captured.

Since the transmission time of a message (25 ms) is much smaller than the speed of the evader, the tracking tree is effectively a minimum spanning tree. (The irregularities and long links are due to the nondeterministic nature of the radio model, fading effects, and collusions.) Hence, the pursuer catches the evader within about 3 to 4 moves: average catching time is 3.3 s.

Figure 9.4 shows a snapshot from the pursuer-centric program. In this run, the evader started in the middle of the network and moved to the upper-left corner within 16 moves. The pursuer started at the lower-left corner, randomly wandered around for a while, found a node

Table 9.1 Number of Messages and Catching Times

	Evader-Centric	Pursuer-Centric	Hybrid $R = 1$	Hybrid $R = 2$
Total number of messages	256	21	286	208
Catch time (s)	3.3	15.7	10.7	3.9

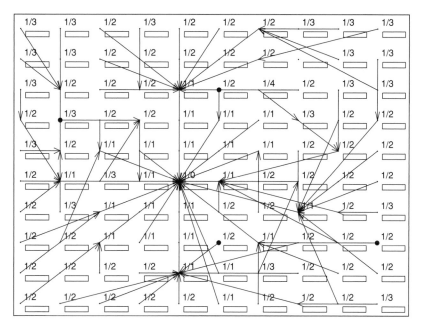

Figure 9.3 Simulation for evader-centric program.

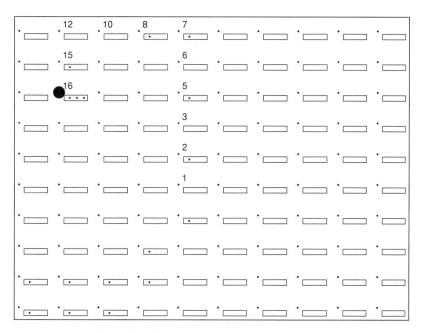

Figure 9.4 Simulation for pursuer-centric program.

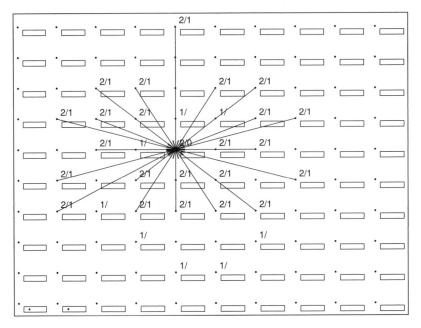

Figure 9.5   Simulation for hybrid program for $R = 1$.

that was visited by the evader, and followed these tracks to catch the evader at the upper-left corner (a dot in the rectangle denotes that the pursuer has visited the corresponding node).

Since the pursuer-centric program does not maintain a tracking tree, the total number of messages sent is low (21). On the other hand, it takes more time for the pursuer to find a track of the evader, hence catching time is high (15.7 s).

Figure 9.5 shows a snapshot from the hybrid program with $R = 1$ (hybrid1), and Figure 9.6 shows the hybrid program with $R = 2$ (hybrid2). Our simulations show that hybrid2 performs better than hybrid1: Both the total number of messages sent and the catching time of hybrid2 is significantly smaller than those of hybrid1.

Hybrid1 cannot provide good coverage over the network with its one-hop tracking tree. Hence the pursuer wanders around for 5 to 10 moves before it can encounter a node that had some information about the evader (a node that is/has been part of the tracking tree). Due to this wandering around time, the catching time increases, and energy is wasted for maintaining a tracking tree for an elongated time. (Note that the maintenance of a 1-hop tree is not achievable only by a broadcast of the root node. The leaf nodes also broadcast messages; these broadcasts are required for informing the nodes that are to be pruned, that is, the nodes that were included in the previous tracking tree but that are outside the new tracking tree as a result of evader movement and accompanying root node change.)

Hybrid2, on the other hand, provides a reasonable coverage over the network; hence the pursuer discovers the track of the evader earlier than that of hybrid1. The tracking tree is maintained to $R = 2$, and, thus, after each evader move hybrid2 sends more messages than hybrid1. But since the catching time is significantly shortened in hybrid2, the 2-hop tracking tree is maintained only for this short time. As a result, the total number of messages sent by hybrid2 is less than that of hybrid1.

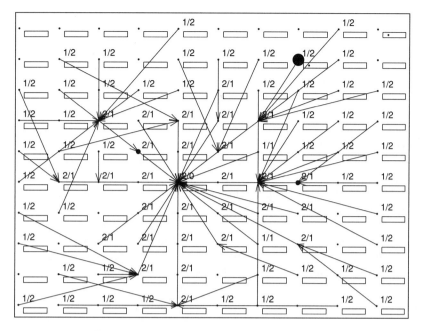

Figure 9.6 Simulation for hybrid program for $R = 2$.

Hybrid2 optimizes both energy efficiency and tracking time concurrently. The total number of messages sent by hybrid2 is less than that of the evader-centric program (208 versus 256), and the catching time of hybrid2 is comparable to that of the evader-centric program (3.9 s versus 3.3 s). (Hybrid program with $R = 3$ gives similar results to the evader-centric program and is omitted from our discussion.)

9.8 DISCUSSION AND RELATED WORK

In this chapter we have investigated a pursuer–evader game for sensor networks. More specifically, we have presented a hybrid, tunable, and self-stabilizing program to solve this problem. We proved that the pursuer catches the evader even in the presence of faults. For the sake of simplicity, we have adopted a shared-memory model in our presentation; our results are still valid for a message-passing memory model. We have provided message-passing implementations of our programs in Section 9.7. Note that the semantics of the message-passing program is event-based execution (e.g., upon receiving a message or detecting an evader/pursuer), rather than maximal parallelism.

Energy Efficiency We have demonstrated that our program is tunable for tracking speed or energy efficiency. Our program is also tunable for stabilization speed or energy efficiency. The periodicity of soft-state updates for stabilization should be kept low if the faults are relatively rare in the network. For example, in the absence of faults, the first action (i.e., {Evader resides at j} action) need not be executed unless the evader moves to a different node. Similarly, the stabilization actions (actions 3 and 4 of the hybrid program) can be executed with low frequency to conserve energy.

Another way to improve the energy efficiency is to maintain the tracking tree over a small number of nodes. For example, hierarchical structuring can be employed to maintain tracking information with accuracy proportional to the distance from the evader. Also maintaining the tracking tree in a directional manner and only up to the location of the pursuer will help conserve energy.

Related Work   Several self-stabilizing programs exist for tree construction [11–13] to name a few. However, our evader-centric program is unique in the sense that a spanning tree is maintained even though the root changes dynamically.

A self-stabilizing distributed directory protocol based on path reversal on a network-wide, Fixed spanning tree is presented in [16]. The spanning tree is initialized to guarantee a reachability condition: Following the links from any node leads to the evader. When the evader moves from a node j to another node k, all the links along the path from j to k in the spanning tree are reversed. This way, the tree always guarantees the reachability condition. This protocol suffers from a nonlocal update problem because it is possible to find at least two adjacent nodes j, k in the network such that the distance between j and k in the overlayed spanning tree structure is twice the height of the tree (i.e., equal to the diameter of the network). An evader that is dithering between these two nodes may cause the protocol to perform nonlocal updates for each small move and would result in a scenario where the pursuer is never able to catch the evader. In contrast, our protocol maintains a dynamic tree and does not suffer from the nonlocal update problem.

In our program, we choose to update the location of the evader immediately. In [17], three strategies for when to update the location the evader (time-based, number of movements-based, and distance-based) are evaluated with respect to their energy efficiency.

Relating to the idea of achieving energy efficiency by using a small number of nodes, Awerbuch and Peleg [1] present a local scheme that maintains tracking information with accuracy proportional to the distance from the evader. They achieve this goal by maintaining a hierarchy of $\log D$ regional directories (using the graph-theoretic concept of *regional matching*) where the purpose of the ith-level regional directory is to enable a pursuer to track the evader residing within 2^i distance from it. They show that the communication overhead of their program is within a polylogarithmic factor of the lower bound. Loosely speaking, their regional matching idea is an efficient realization of our pursuer-centric program, and their forwarding pointer structure is analogous to our tracking tree structure.

By way of contrast, their focus is on optimizing the complexity during the initialized case, whereas we focus on optimizing complexity during stabilization as well. That is, we are interested in (a) tracking that occurs while initialization is occuring; in other words, soon after the evader joins the system, and (b) tracking that occurs from inconsistent states; in other words, if the evader moves in an undetectable/unannounced manner for some period of time yielding inconsistent tracks. Their complexity of initialization is $O(E \log^4 N)$ where E is the number of edges in the graph and N is the number of nodes. Thus, brute force stabilization of their structure completes in $O(E \log^4 N)$ time as compared with the $2R$ steps it takes in our extended hybrid program.

We have recently found that [18] if we restrict the problem domain to tracking in planar graphs, it is possible to optimize the tracking time in the presence of faults as well as the communication cost and tracking time in the absence of faults. A topology change triggers a global initialization in Awerbuch and Peleg's program since their m-regional matching structure depends on a nonlocal algorithm that constructs sparse covers [19]. Assuming that the graph is planar (neither [1] nor this chapter assumes planarity), a local and self-stabilizing clustering algorithm [20] for constructing the m-regional matching structure is achievable, and hence, it is possible to deal with topology changes locally.

The concept of self-stabilization is particularly useful for dealing with unanticipated and undetectable faults [21]. To achive such an ambitious goal, self-stabilization assumes for convenience that no further faults occur within the stabilization period. It is possible to improve stabilizing programs by adding masking fault tolerance for common and detectable faults; this way occurrence of trivial, common faults during stabilization can be masked immediately and does not affect the stabilization time. The design of this type of fault tolerance, known as multitolerance, is discussed in [22].

Moreover, for the type of faults for which masking is impossible or infeasible, preventing them from spreading is useful for achieving scalability of stabilization for large-scale networks. To this end, several fault containment techniques [23–26]. have been proposed in the self-stabilization literature.

Furthermore, by choosing a weaker invariant, it is possible to show that the stabilization of our tracking programs are unaffected by common faults such as message losses or node fail-stops. That is, by accepting a degraded tracking performance in the presence of these faults, we can show that stabilization to a weaker invariant—for example, a tracking tree, albeit not the optimal tree—is still achievable under message losses and node fail-stops.

Future Work   We have found several variations of the pursuer–evader problem to be worthy of study, where we change, for instance, the communication time between nodes, the number of pursuers and evaders, and the range of a move. Especially of interest to us are general forms of the tracking problem where efficient solutions can be devised by hybrid control involving traditional control theory and self-stabilizing distributed data structures (such as tracking trees and regional directories).

REFERENCES

1. B. Awerbuch and D. Peleg, "Online tracking of mobile user," *Journal of the Association for Computing Machinery*, vol. 42, pp. 1021–1058, 1995.

2. E. Pitoura and G. Samaras, "Locating objects in mobile computing," *Knowledge and Data Engineering*, vol. 13, no. 4, pp. 571–592, 2001.

3. A. P. Sistla, O. Wolfson, S. Chamberlain, and S. Dao, "Modeling and querying moving objects," in *ICDE*, 1997, pp. 422–432.

4. A. Bar-Noy and I. Kessler, "Tracking mobile users in wireless communication networks," in *INFOCOM*, San Francisco, 1993, pp. 1232–1239.

5. M. Jayaram and G. Varghese, "Crash failures can drive protocols to arbitrary states," paper presented at the *ACM Symposium on Principles of Distributed Computing*, Philadelphia, 1996.

6. Y. Choi, M. Gouda, M. C. Kim, and A. Arora, "The mote connectivity protocol," in *Proceedings of the International Conference on Computer Communication and Networks (ICCCN-03)*, Dallas, 2003.

7. A. Woo, T. Tong, and D. Culler, "Taming the underlying challenges of reliable multihop routing in sensor networks," in *Proceedings of the First International Conference on Embedded Networked Sensor Systems*, Los Angeles, 2003, pp. 14–27.

8. J. Hill, R. Szewczyk, A. Woo, S. Hollar, D. Culler, and K. Pister, "System architecture directions for network sensors," in *ASPLOS*, 2000, pp. 93–104.

9. G. Barnes and U. Feige, "Short random walks on graphs," *SIAM Journal on Discrete Mathematics*, vol. 9, no. 1, pp. 19–28, 1996.

10. S. Ikeda, I. Kubo, N. Okumoto, and M. Yamashita, "Local topological information and cover time," research manuscript, 2002.

11. A. Arora and M. G. Gouda, "Distributed reset," *IEEE Transactions on Computers*, vol. 43, no. 9, pp. 1026–1038, 1994.

12. N. Chen and S. Huang, "A self-stabilizing algorithm for constructing spanning trees," *Information Processing Letters (IPL)*, vol. 39, pp. 147–151, 1991.

13. A. Cournier, A. Datta, F. Petit, and V. Villain, "Self-stabilizing PIF algorithms in arbitrary networks," in *International Conference on Distributed Computing Systems (ICDCS)*, Phoenix, 2001, pp. 91–98.

14. D. Gay, P. Levis, R. von Behren, M. Welsh, E. Brewer, and D. Culler, "The NESC language: A holistic approach to network embedded systems," submitted to the ACM SIGPLAN(PLDI), San Diego, June 2003.

15. G. Simon, P. Volgyesi, M. Maroti, and A. Ledeczi, "Simulation-based optimization of communication protocols for large-scale wireless sensor networks," paper presented at IEEE Aerospace Conference, Big Sky, MT, Mar. 2003.

16. M. Herlihy and S. Tirthapura, "Self-stabilizing distributed queueing," in *Proceedings of Fifteenth International Symposium on Distributed Computing*, Oct. 2001, pp. 209–219.

17. A. Bar-Noy, I. Kessler, and M. Sidi, "Mobile users: To update or not to update?" in *INFOCOM*, Ontario, 1994, pp. 570–576.

18. M. Demirbas, A. Arora, T. Nolte, and N. Lynch, "STALK: A self-stabilizing hierarchical tracking service for sensor networks," Technical Report OSU-CISRC-4/03-TR19, The Ohio State University, Apr. 2003.

19. B. Awerbuch and D. Peleg, "Sparse partitions (extended abstract)," in *IEEE Symposium on Foundations of Computer Science*, 1990, pp. 503–513.

20. V. Mittal, M. Demirbas, and A. Arora, "LOCI: Local clustering in large scale wireless networks," Technical Report OSU-CISRC-2/03-TR07, The Ohio State University, Feb. 2003.

21. A. Arora and Y.-M. Wang, "Practical self-stabilization for tolerating unanticipated faults in networked systems," Technical Report OSU-CISRC-1/03-TR01, The Ohio State University, 2003.

22. A. Arora and S. S. Kulkarni, "Component based design of multitolerant systems," *IEEE Transactions on Software Engineering*, vol. 24, no. 1, pp. 63–78, Jan. 1998.

23. S. Ghosh, A. Gupta, T. Herman, and S. V. Pemmaraju, "Fault-containing self-stabilizing algorithms," in *ACM PODC*, 1996, pp. 45–54.

24. M. Nesterenko and A. Arora, "Local tolerance to unbounded byzantine faults," in *IEEE SRDS*, 2002, pp. 22–31.

25. Y. Azar, S. Kutten, and B. Patt-Shamir, "Distributed error confinement," in *ACM PODC*, 2003, pp. 33–42.

26. A. Arora and H. Zhang, "LSRP: Local stabilization in shortest path routing," in *IEEE-IFIP DSN*, June 2003, pp. 139–148.

10

EMBEDDED SOFT SENSING FOR ANOMALY DETECTION IN MOBILE ROBOTIC NETWORKS

Vir V. Phoha, Shashi Phoha, Asok Ray, Kiran S. Balagani, Amit U. Nadgar, and Raviteja Varanasi

We present a framework for detecting and mitigating software anomalies using soft sensors in mobile robotic networks. The experience obtained in this study is applicable to a wide class of mobile networks of complex systems, such as computer networks and networks of electromechanical components. This work develops an integrated hierarchical control mechanism for detection and mitigation of anomalies in mechanical, electrical, and software systems on a network of three mobile robots interconnected through a wireless network. The test environment consisted of two platforms: (1) a player/stage robotic simulation system and (2) three networked autonomous robots, with an on-board Linux operating system, where we built fault injection, detection, and mitigation mechanisms controlled by a centralized computer. We developed soft sensors to detect memory leak and mutex lock. Experiments show that, if unchecked, both memory leak and mutex lock result in erratic robot behavior (e.g., gripper failure). This erratic behavior was successfully detected and corrected. There are four major contributions of this work: (1) formalizing protocols for reporting anomalous behavior by mobile units to a centralized controller, (2) giving examples of software-induced abnormal behavior by robots, (3) developing soft sensors to detect software anomalies, and (4) developing control protocols and mechanisms to mitigate the effect of software anomalies.

10.1 INTRODUCTION

Present-day human-engineered complex systems, such as robots, computer networks, and aircrafts, are mostly software-driven and incorporate dynamic interactions of many components. Faults in one component may affect the performance of other interrelated components.

Sensor Network Operations, Edited by Phoha, LaPorta, and Griffin
Copyright © 2006 The Institute of Electrical and Electronics Engineers, Inc.

Figure 10.1   Two robots equipped with grippers and sensors in a wireless mobile robotic network.

Even with most sophisticated software engineering methods and checks [1–7], the system software is prone to contain bugs and errors, which can lead to degraded performance or catastrophic failures [8]. A fundamental understanding of the causal dynamics intrinsic to the software system that permits the evolution and propagation of anomalous behavior is essential to diagnose and mitigate emerging faults and anomalies. In this study, we report our experience of developing methods to mitigate software-instigated pathological behavior of mobile robots in an experimental environment containing three autonomous mobile robots interacting with each other through wireless connections. We present a hierarchical control mechanism to detect and mitigate faults, introduce the concept of soft sensors, develop methods to sense software faults and anomalies, and present simulation results on the Player/Stage [9, 10] virtual robotic system. The developed methods have been implemented as a proof of concept on real robots in the Robotics Laboratory of Penn State University. The environment in which these robots operate includes sensors, effectors, obstacles, grippers, and different colored pucks. Figures 10.1 and 10.2 show photographs of two robots equipped with sensors (color blob detectors, laser range finders, and sonar obstacle finders)

Figure 10.2   The robots communicate with the mission command and control unit to pick a colored puck amid obstacles.

and grippers. The goal of the robots is to pick a preselected color puck placed amid obstacles without colliding with the obstacles obstructing their navigation path.

10.1.1   Hierarchical Control

The hierarchical control and mitigation architecture in the robotic network consists of three levels of control: (1) local control, (2) coordination, and (3) mission control. Each mobile robot has its own local control which is provided by the local discrete-event supervisory (DES) controller. Figure 10.3 shows the hierarchical robotic network control architecture.

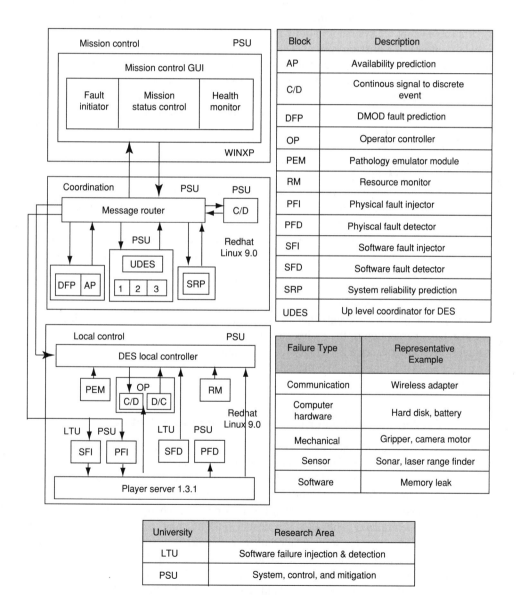

Block	Description
AP	Availability prediction
C/D	Continous signal to discrete event
DFP	DMOD fault prediction
OP	Operator controller
PEM	Pathology emulator module
RM	Resource monitor
PFI	Physical fault injector
PFD	Phyiscal fault detector
SFI	Software fault injector
SFD	Software fault detector
SRP	System reliability prediction
UDES	Up level coordinator for DES

Failure Type	Representative Example
Communication	Wireless adapter
Computer hardware	Hard disk, battery
Mechanical	Gripper, camera motor
Sensor	Sonar, laser range finder
Software	Memory leak

University	Research Area
LTU	Software failure injection & detection
PSU	System, control, and mitigation

Figure 10.3   Integrated hierarchical robotic network control architecture for software anomaly detection and mitigation in complex systems. The architecture consists of three levels of control: (1) local control, (2) coordination level of control, and (3) mission control.

The implementation of the architecture is a joint effort of four universities: (1) Penn State University (PSU), Louisiana Tech University (LTU), Duke University, and Carnegie Mellon University. In Figure 10.3 we show the components of the architecture most relevant to this work, realized at Penn State and Louisiana Tech Universities. The local DES controller takes inputs from various sensors such as a physical fault detector (PFD) to detect electromechanical faults, a software fault detector (SFD) to detect software anomalies, and a resource monitor (RM) to monitor the status of resources (software resources and battery life) and operation (OP) modules to monitor operational status. The local DES controller passes the sensor input to the upper level message router (MR), which is a part of the coordination level control. This control level coordinates the activities of the mobile robots in a robotic network. In this level of control, the sensor input is used to predict the operational reliability of a robot using the system reliability prediction (SRP) modules. The third and highest level of control is the mission control. The mission control contains three subcomponents: (1) mission status control, (2) fault initiator, and (3) health monitor. The mission status control component sends command messages to mobile robots and receives their status through the MR. The fault initiator component injects different types of electromechanical, electrical (hardware), and software faults into robotic systems through the MR in the coordination level that makes use of the recently developed theory of optimal discrete-event supervisory control [11] for decision making. The key concept is briefly outlined below.

Discrete-event dynamic behavior of robot operation is modeled as regular languages that can be readily realized by finite-state automata [7]. The sublanguage of a controlled robot is likely to be different under different supervisors. These controlled sublanguages form a partially ordered set and hence require a quantitative measure for total ordering of their respective performance. The real signed measure of regular languages [12, 13] serves as a common quantitative tool for comparison performance of different supervisors; a brief review of the language measure theory is presented in Appendix A. The language measure is assigned an event cost $\tilde{\Pi}$ matrix and a state characteristic X vector. Event costs (i.e., elements of the $\tilde{\Pi}$ matrix) based on the plant states, where they are generated, are physical phenomena dependent on the plant behavior and are similar to conditional probabilities of the respective events. On the other hand, the X vector is chosen based on the designer's perception of the individual state's impact on the system performance. The discrete-event supervisory algorithm of robot operations is built upon the theory of optimal control of regular languages [11]; the optimal control synthesis procedure is summarized in Appendix B.

10.1.2 Message Handling in Hierarchical Control Architecture

In this section we describe the basic message structure used for communication among different components of the hierarchical anomaly detection and control architecture. Communication between the centralized controller (i.e., the base station for command and control operations) and mobile hosts (i.e., mobile robots) uses a transmission control protocol (TCP). Figure 10.4 illustrates the message format for information exchange between the mobile robots and the centralized controller. The header field has nine subfields: (1) the source field (SRC) that identifies the software module in the hierarchical architecture initiating the message; (2) the sequence field (SEQ) to identify the message sequence; (3) the length field (LEN) to identify the length of the message in bytes; (4) the TYPE field to identify the structure of payload; (5) the robot identifier field (RID) to identify the mobile robots; (6) the subsystem field (SYS) to identify the subsystem (e.g., mechanical, electrical, software);

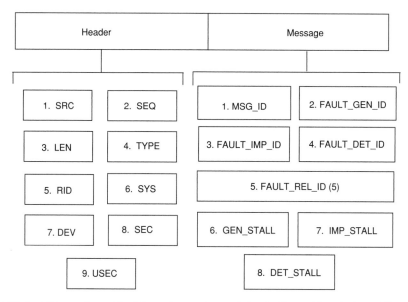

Figure 10.4 Message format for information exchange between the mobile robots and the mission control and command center. The message format has two fields: (1) the header and (2) the message. The header field is used to identify subsystems and devices of the mobile robots. The message field is used for identifying and describing the faults that are to be injected/detected/released into mobile robot subsystems or devices.

(7) the device field (DEV) to identify the device in the subsystem (e. g., gripper in mechanical subsystem); (8) the SEC field containing the time stamp in seconds; and (9) the USEC field containing the time stamp in microseconds. The message field consists of eight subfields: (1) the message identifier field (MSG_ID) carries the identification number for the fault message; (2) the fault generation command identifier field (FAULT_GEN_ID) identifies the fault generation command; (3) the fault implementation identifier field (FAULT_IMP_ID) identifies the status (successful/unsuccessful) of fault implementation at the target subsystem; (4) the fault detection identifier field (FAULT_DET_ID) indicates fault detection status (detected/undetected); (5) the fault release identifier field (FAULT_REL_ID) is set to stall the release of a fault in a subsystem; (6) the fault abort field (GEN_STALL) is set when a generated fault is stalled; (7) the fault implementation abort field (IMP_STALL) is set when the implementation of a fault on a target subsystem is stalled; and (8) the fault detection abort field (DET_STALL) is set to stall the detection of a fault in a subsystem.

Figure 10.5 illustrates the exchange of messages between the centralized controller and the mobile robots in different memory leak injection, detection, and release scenarios. The mission control graphic user interface (GUI) issues a memory leak injection message as shown in Figure 10.5b and releases the fault as shown in Figure 10.5e. Figure 10.5 also illustrates different states of mission control GUI during fault (i.e., memory leak) injection and release. In the idle state, there is no exchange of messages between the robot and the mission control. The mission control GUI injects a memory leak and moves to the inject state. The robot acknowledges the memory leak inject command by issuing the fault implemented message, thereby changing the state of the mission control to IMP_ACK. Once the memory leak fault is detected by the memory leak detection mechanism, the robot

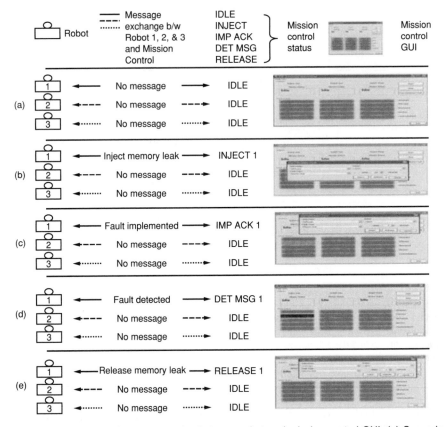

Figure 10.5 Message exchange scenarios between robot and mission control GUI. (*a*) Scenario in which all the robots are idle with no message exchange between the robots and the mission control GUI. (*b*) The mission control GUI issues a memory leak inject command to robot 1. (*c*) Robot 1 issues a fault implemented acknowledgment to mission control GUI. (*d*) The fault (memory leak) detected message is issued by robot 1. (*e*) The mission control GUI issues a memory leak release command to the robot 1.

issues a fault-detected message to the mission control GUI changing its state to DET_MSG followed by a release-memory-leak message and moves to the release state.

10.2 MOBILE ROBOT SIMULATION SETUP

The Player/Stage [10] software has been used to simulate a real mobile robot environment (shown in Figs. 10.1 and 10.2) in Louisiana Tech University's Anomaly Detection and Mitigation (ADAM) Laboratory. The virtual environment has three autonomous robots that interact in a wireless setup to achieve a common goal. The Player is a multithreaded robot device server that gives the user simple and complete control over the physical sensors and actuators on the mobile robot. The Player program runs on the mobile robot. A client program [14] (e.g., the robot control program) connects to the Player via standard TCP sockets. The Player and client programs communicate with each other by sending and receiving messages. The Player program runs on the mobile robot, but in the absence of an

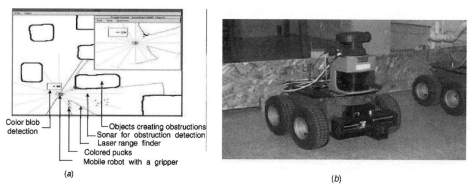

Color blob detection

Objects creating obstructions
Sonar for obstruction detection
Laser range finder
Colored pucks
Mobile robot with a gripper

(a)

(b)

Figure 10.6 Virtual and real mobile robot simulation environments. (*a*) A virtual robotic environment using Player/Stage 1.3.1 software was set up at Louisiana Tech University's ADAM Laboratory. Also shown are the virtual robot, sensors, gripper, goals (as colored pucks), and obstacles as rectangle-like objects. (*b*) Experimental setup with actual robots operating in real environments at Penn State University's Robotic Laboratory.

actual robot the Player software uses Stage, which simulates a population of mobile robots, sensors, and objects in a two-dimensional bitmapped environment. If an actual robot is not available, then both Stage and Player can run on a computer system, where Stage provides virtual devices for Player.

Figure 10.6*a* shows the interactive interface exported by the Player/Stage package. This screenshot shows a scenario where there are obstacles, colored pucks and a robot equipped with a gripper and sensors such as a color blob detector, laser range finder, and the sonar for obstacle detection. The goal of the mobile robot is to identify and pick a user-specified colored puck amid obstructions. Figure 10.6*b* shows a picture of a real robot colliding with an obstacle in Penn State University's Robotic Laboratory.

10.3 SOFTWARE ANOMALIES IN MOBILE ROBOTIC NETWORKS

We consider two software anomalies in mobile robot applications: (1) memory leak and (2) mutex lock. The first anomaly, a memory leak [15–17], also referred to as software aging [18], develops due to software errors in programs that fail to release memory after allocation and usage. Over a period of time, the amount of free memory for the software execution becomes so low that it leads to outages such as a system crash. The free physical memory available for applications under execution reduces with time because of memory leaks and other software faults. This leads to the degradation of performance or the failure of the applications, especially in those developed in C and C++, where the programmer is required to manage dynamic memory allocation and deallocation. In such applications, there is a high possibility for memory leaks to go undetected, despite the fact that the application is developed using advanced software development methodologies by highly skilled C/C++ programmers [15,16,19]. In mobile robot applications with limited memory, there is a high possibility for memory leaks to occur. The unavailability of memory caused by memory leaks can increase the robots' response time and in extreme cases would crash the robots.

The second anomaly is a mutex lock that is highly probable in applications using multi-threaded programs [20–22]. The mutex lock arises in classical interprocess communication problems like the producer–consumer [23] and reader–writer [24]. A mutex lock is used when two threads attempting to access a shared mutual-exclusive region lock a mutex variable before entering the region and unlock it after its usage. In a mobile robotic application it is highly possible for a thread to run for an indefinite period of time in the shared region without unlocking the mutex variable. The blocked thread waiting for the mutex variable to be unlocked may starve leading to erroneous operation of the robot.

10.4   SOFT SENSOR

A soft sensor is defined as a software entity that senses the state of a software system. In a mobile robotic environment, we implement soft sensors as C/C++ programs that monitor the system resource utilization status of an executing software application. Examples of soft-sensor percepts are the amount of memory used by a process, the number of instruction cycles executed per unit time, the number of files open, the amount and frequency of disk swaps made by the process, the status of mutex lock, and other statistics that an operating system maintains for a program in execution. Overutilization of resources (on known workloads generated by test application programs) is an indicator of anomalies, errors, or bugs in the program. Therefore, by monitoring the system resource utilization, one can ensure that the software execution is in control.

In our experiments we use two soft sensors: (1) a memory leak sensor and (2) a mutex lock sensor to monitor memory leak and mutex lock, respectively. The Player server is a process on the robot for which the/proc file system maintains information such as the total physical memory (in pages) that the Player is using and the amount of time the Player spends in user mode and system mode. This information is available in a file that is indexed in the /proc file system by Player's process identification number (PID).

Similarly, information pertaining to total available physical memory and free memory for the entire system is maintained in a /proc/meminfo file. The memory leak sensor reads data from /proc/pid(player)/stat and /proc/meminfo files periodically (e.g., every second) and passes the memory information (free and total physical memory available in the robot) to the memory leak detector (see Fig. 10.7). The mutex lock sensor reads data from /proc/pid(player)/stat periodically (i.e., every second) and passes the information (user mode time and system mode time of a process) to the mutex lock detector (see Fig. 10.7).

10.5   SOFTWARE ANOMALY DETECTION ARCHITECTURE

This section presents the architecture for software anomaly detection. Our simulation environment consists of three robots (see Section 10.2). Figure 10.7 shows the components of our software fault detection architecture built on each mobile robot operating in the Redhat Linux 7.3 environment. The proposed architecture monitors the robotic system for software faults due to memory leaks and mutex locks and performs on-line detection of such faults by analyzing the collected resource usage data while continuously monitoring the Player server software on the mobile robot. The Linux operating system provides the /proc file system [25, 26] that holds overall system information as well as data relevant to all processes running on the system. Figure 10.7 shows the software fault detection architecture

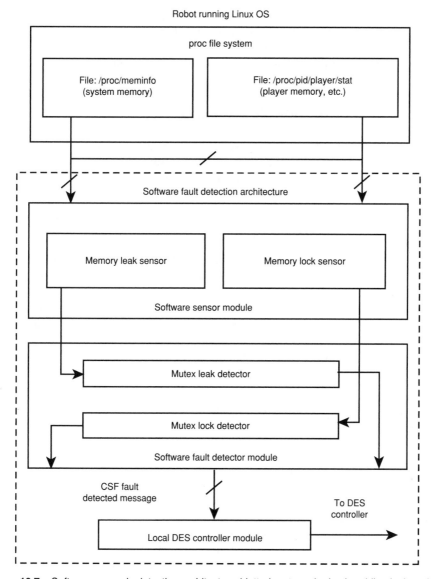

Figure 10.7 Software anomaly detection architecture (dotted rectangular box) residing in the robot's Linux operating system. The architecture consists of three components: (1) sensor software module, (2) SFD module, and (3) local DES controller module. Also shown is the /proc file system containing information on the state of a process under execution.

that consists of three modules: (1) the soft-sensor module, (2) the SFD module, and (3) the local DES controller module (discussed in Section 10.1.1). The soft-sensor module contains software sensors that supply information relevant to detection of memory leaks and mutex locks. The SFD module is the heart of our architecture containing two subcomponents: (i) the memory leak detector (MemLD) and (ii) the mutex lock detector (MuLD). Data that are relevant to memory leak detection is analyzed by the MemLD. Similarly, the timing information related to mutex lock is analyzed by the MuLD. The local DES controller module receives fault detection messages from SFD and sends it to the DES coordinator (discussed in Section 10.1.1). The following section discusses the software fault detection mechanisms used by the SFD module to detect conditions of low memory availability due to memory leak and thread blockage due to a locked mutex variable.

10.6   ANOMALY DETECTION MECHANISMS

This section presents mechanisms for detecting memory leaks and mutex locks in mobile robot applications. The memory leak detection mechanism resides in the MemLD component and the mutex lock detection mechanism resides in the MuLD component of the SFD module presented in Section 10.5.

10.6.1   Memory Leak Detection Mechanism

The memory leak detection mechanism uses two variables: (1) Resident set size (RSS) [25] of a process, which gives the number of physical pages a process has in physical memory, and (2) an adaptive threshold variable τ, which limits the growth of RSS. When a memory leak occurs, the process starts allocating memory in excess. Due to such an allocation, the address space of the process grows, thereby increasing the value of the RSS variable. A variable logical limit is maintained by using threshold variable τ on the growth of RSS, which is updated depending on the available amount of free memory. The algorithm for memory leak detection follows.

10.6.2   Mutex Lock Detection Mechanism

For an active process/thread (an active process is a process execution), the kernel updates the time the process or thread spends in the user mode and system mode. The mutex lock detection mechanism receives the system and the user mode times periodically (every second) from the mutex lock sensor. These values are available in the /proc/x/stat file system associated with the process, where x is the process identifier. The mutex lock detection scheme uses two variables: (1) variable threshold $\tau_{\text{mutexlock}}$ and (2) total time Γ. The sum of the system and user mode times remains unchanged in a process waiting for a mutex release. The sum of system mode time and user mode time is represented by Γ. A counter keeps track of the number of program iterations for which Γ remains unchanged. The user-specified variable threshold $\tau_{\text{mutexlock}}$ limits this counter to a predefined value (in our experiments we assign a value of 50 to $\tau_{\text{mutexlock}}$). Finally, a detection message for a probable case of mutex lock is sent to the local DES controller.

```
Input:   Memory leak counter variable η and RSS variable.
         In our experiments we set η to 81.
         RSS values are provided by memory_leak_sensor.
Output:  Memory leak detection message.
Initialization Set probable_leak_counter to 0 and Δ^old to RSS.
Procedure Detect_memory_leak(RSS, η)
{
     /* If probable_leak_counter variable exceeds η, the algorithm
     issues a memory leak detection message. */
```

```
while(probable_leak_counter < η) {
    if(RSS > τ) {
            /* The reduction variable δ is determined by the
            level of severity in which the memory leak detection
            mechanism operates. We used 5 levels of adaptation
            before sending a memory leak detection message.
            The threshold update factor Δ is reduced by
            reduction variable δ as free memory starts falling
            below a fixed percentage of total physical
            memory. */
            /* Begin LEVEL 1 */
            if(AFM ≤ 0.1 × TFM&AFM > 0.07 × TFM) {
                    δ ← 50; /* Assign 50 to δ */
                    Δ^new ← Δ^old − δ; /* Update the value of Δ */
            }
            /* End LEVEL 1. Begin LEVEL 2 */
            elseif(AFM ≤ 0.07 × TFM&AFM > 0.05 × TFM) {
                    δ ← 40; /* Assign 40 to δ */
                    Δ^new ← Δ^old − δ; /* Update the value of Δ */
            }
            /* End LEVEL 2. Begin LEVEL 3 */
            elseif(AFM ≤ 0.05 × TFM&AFM > 0.03 × TFM) {
                    δ ← 10; /* Assign 10 to δ */
                    Δ^new ← Δ^old − δ; /* Update the value of Δ */
            }
            /* End LEVEL 3. Begin LEVEL 4 */
            elseif(AFM ≤ 0.03 × TFM&AFM > 0.01 × TFM) {
                    δ ← 8; /* Assign 8 to δ */
                    Δ^new ← Δ^old − δ; /* Update the value of Δ */
            }
            /* End LEVEL 4. Begin LEVEL 5 */
            elseif(AFM ≤ 0.01 × TFM) {
                    δ ← 3; /* Assign 3 to δ */
                    Δ^new ← Δ^old − δ; /* Update the value of Δ */
            } /* End LEVEL 5 */
            probable_leak_counter++;
            τ^new ← τ^old + Δ; /* Update threshold with update
            factor Δ */
    } /* End If */
} /* End While */
} /* End Procedure Detect_memory_leak */
```

10.7   TEST BED FOR SOFTWARE ANOMALY DETECTION IN MOBILE ROBOT APPLICATION

This section describes a test bed for evaluating the software anomaly detection architecture (see Section 10.5) for mobile robot applications. The memory leak and mutex lock detection mechanisms are tested in a mobile robot application simulated using the Player/Stage [9,10] software.

Figure 10.8 illustrates the fault detection test-bed architecture used for fault injection and detection. The architecture has (1) the mobile robot environment, (2) fault injection and

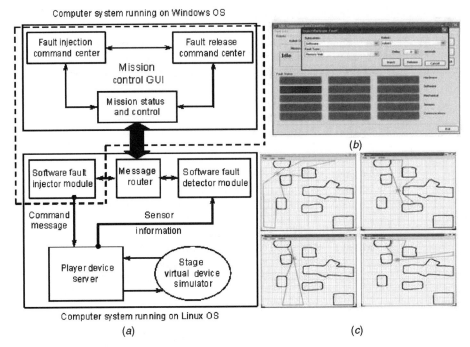

Figure 10.8 Implementation of mobile robot environment, fault detection architecture, and fault test-bed architecture on two computer systems. (*a*) The fault test-bed architecture (enclosed in the dashed box) consists of two components: (1) mission control GUI and (2) software fault injector module. The mission control GUI (on the Windows operating system) is an interface to inject and release faults into the player device server. The software fault injector module injects faults whenever it receives an injection command from the mission control GUI through the MR. (*b*) A screenshot of mission control GUI showing a "memory leak detected" status. (*c*) The scenario in which the robot under normal operation hits an obstacle due to injection of a memory leak (to be viewed clockwise).

detection modules, and (3) mission control GUI, all operating on two computer systems. Mobile robot simulation involves creating the "scenario" or the "world" in which the robot operates, where the world consists of a bitmapped image of the obstacle positions. The initial position of the robot in this scenario and the ancillary devices are specified in a world file. Stage interprets this world file to simulate various devices that the robot can use. Player which is the device server, provides the interface between the robot control program and Stage. A Reader thread in Player continuously reads status data from various sensor devices that were created according to the world file and sends the data to the C++ Client [14].

The test bed consists of two components: (1) the mission control GUI and (2) the fault injector (FI). The mission control GUI consists of a fault injection command sender, a fault release command sender, and a mission status and control unit. The fault injection command sender sends a fault injection command to the MR. The mission control GUI maintains information about the faults injected, monitors the status of robots, and shows the fault injection/detection status to the user. The fault release command sender sends the fault release command to the MR. The fault injector consists of a MR and a fault injection unit. The MR broadcasts the received message to all the components. The software fault injector module, upon receiving the fault injection message from the MR, injects faults in the player device server. The software fault detector detects the faults and sends the message to the MR, which in turn sends it to the mission control GUI.

10.7.1 Software Fault Injection

The software fault injector (SFI) operates on each mobile robot. Injection of a fault is triggered by a user command that comes through the fault sender in the mission control GUI. We instrumented the Player server code with algorithms to inject memory leaks or mutex locks. The fault injection code in the Player server is activated when the SFI receives a fault injection command (shown as a switch in Fig. 10.9) from the MR. Due to the

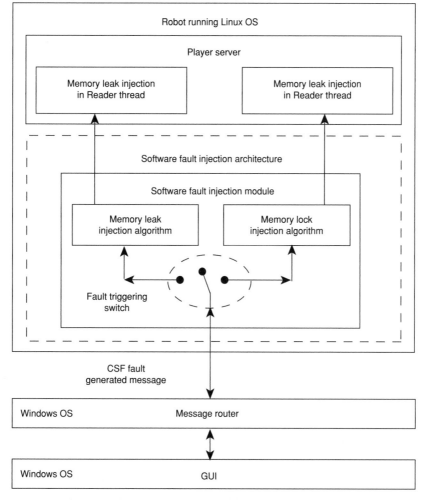

Figure 10.9 Software fault injection architecture (enclosed in dashed rectangle). The architecture consists of two components: (1) the software fault injection module and (2) the fault triggering switch. The memory leak injection and mutex lock injection algorithms are implemented in the software injection module. The triggering switch allows sequential injection of faults. The fault injection architecture communicates with the mission control GUI through the MR.

switching mechanism of the fault injector, faults can be injected sequentially. The details of our memory leak and mutex lock fault injection methods are discussed in the next section.

Memory Leak Injection Mechanism   The memory leak injection algorithm is implemented within Player. The objective of the injection mechanism is to reduce the amount of free physical memory and increase the amount of memory used by the Player process. Player operates in an infinite loop checking continuously for sensor input. The following algorithm has been implemented in this loop so that a fixed block of memory (size of the block is equal to page size, i.e., 4 kbytes) is allocated for every iteration. For every block of memory that is allocated, a corresponding linked list node is created that holds the address of this block. The algorithm for memory leak injection follows.

```
Algorithm: Memory Leak Injection
Input:      'Memory Leak Inject command' from mission control GUI.
Output:     'Memory Leak injected' message to mission control GUI.

/* Structure of a singly linked list with two variables,
(1) mem_allocated  to allocate fixed amount of memory, and
(2) a pointer *next to keep track of allocated memory */
structure leak_list
{
      char *mem_allocated;
      structure leak_list *next;
}


Procedure Memory_leak_inject()
{
      if(lh = 0) {
            /* Initialize the variables in the 'leak_list'
            structure */
            lh =  allocate(size(char)× 4096);
            lh → next = 0;
            lp = lh;
      }
      elseif(lh ≠ 0) {
            /* Insert nodes in leak_list to track memory
            allocation */
            lp → next = allocate(size(char) × 4096);
            lp = lp → next;
            lp → next = 0;
      } /* end if */
} /* End Procedure Memory_leak_inject */
```

Mutex Lock Fault Injection Mechanism   The purpose of a mutex lock fault injection mechanism is to block a thread indefinitely until the mutex is released by the thread that has locked the mutex. To inject this fault in the Player software, the software fault injection algorithm continuously locks a mutex. The Reader thread in Player checks if the mutex variable mutex_lock is free. If free, it locks the mutex and later releases it. However, if a fault has been injected by the SFI, the Player thread indefinitely waits for the mutex to be unlocked. This introduces the mutex lock fault in the player. The algorithm for mutex lock injection follows.

```
Algorithm: Mutex Lock Injection
Input:      Mutex variable mutex_lock.
Output:     'Mutex Lock injected' message to mission control GUI.
Procedure
/* Mutex lock injection thread continuously checks the status
(locked/unlocked) of a mutex variable. If the variable is unlocked,
a mutex lock is injected by locking the mutex variable */

/* These two threads run concurrently */
Mutex_Lock_Injection_Thread(Mutex variable mutex_lock)
{
    Do {
        if(!mutex_lock) {
            lock(mutex_lock);
        }
    }while(true);
}


Reader Thread(Mutex variable mutex_lock)
{
    Do {
        if(!mutex_lock) {
            lock(mutex_lock);
        }
    }while(true);
    ⋮
    /* statements to control the robot */
    ⋮
    unlock(mutex_lock)
}
```

10.8 RESULTS AND DISCUSSION

Using the experimental setup discussed in Section 10.2, we conduct tests to evaluate the performance of the fault detection algorithms (discussed in Section 10.6). Faults are created in the system using the fault injection algorithms given in Section 10.7.1.

10.8.1 Observation on Memory Leak Injection

The memory leak injection command is given through the fault injection command sender in the mission control GUI. The Player server is instrumented to accommodate the memory leak injection algorithm in Section 10.7.1. When the command is routed to the fault injection unit of a robot, it activates the memory leak fault injection algorithm.

The memory leak injection algorithm allocates 4 kbytes of memory, equivalent to a page size in each iteration. Figure 10.10 shows that by allocating a page of memory and by not

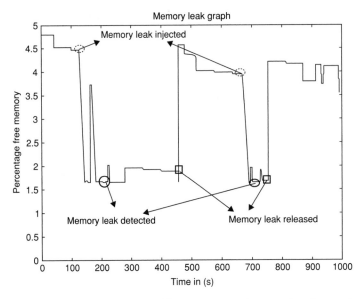

Figure 10.10 Variations in the percentage of free memory under normal and anomalous operating conditions. The point where the memory leak detection mechanism identifies a memory leak is shown. The percentage of free memory decreases rapidly on injection of a memory leak.

deallocating it, the amount of free memory in the system reduces at a higher rate. Figure 10.10 also shows that for the simulated robot the percentage of free memory (PFM) has reached between 1 and 2% when the fault detection message is sent. When the PFM in this region reaches a limit, the fault detection algorithm increases its threshold according to the level of severity and then increments probable_leak_counter. In our experiment the threshold for probable_leak_counter was set to 81, which meant that the memory leak was detected after player had allocated $50 \times 4096 \times 81$ bytes of physical memory. The effects of memory leak on three real robots (experiments conducted at PSU Robotic Laboratory) are the missing of goals (picking color pucks) and gripper failures.

10.8.2 Observation on Mutex Lock Injection

We demonstrate the result of a mutex lock fault injection by considering the case where the robot tries to avoid an obstacle that is in its path. Player consists of a Reader thread which continuously checks for input from the DES controller. The robot avoids an obstacle in its path when the Reader thread receives an obstacle avoidance command from the Player server. Figures 10.11a and 10.11b show the proper working of the robot as it avoids the obstacle in its path. This is the normal operation of the robot when the mutex lock fault is not injected in the robot.

Figures 10.11c and 10.11d show the working of the robot after injecting a mutex lock fault, where the robot has not detected the obstacle in its path due to a blocked Reader thread. The mutex lock fault is injected by the SFI operating on a robot by locking the mutex variable which the Reader thread probes in each iteration. Once the mutex variable is locked, the Reader thread waits for its release indefinitely, resulting in erroneous behavior of the robot, shown in Figure 10.11d.

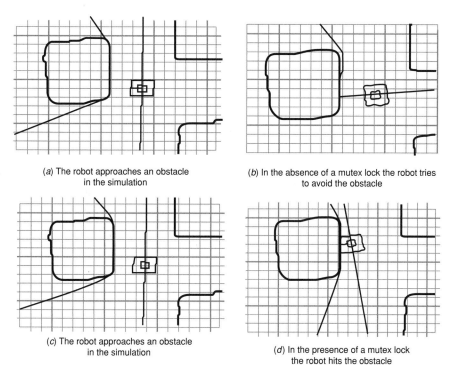

(a) The robot approaches an obstacle in the simulation

(b) In the absence of a mutex lock the robot tries to avoid the obstacle

(c) The robot approaches an obstacle in the simulation

(d) In the presence of a mutex lock the robot hits the obstacle

Figure 10.11   Scenario showing the anomalous behavior of the robot on injecting a mutex lock. (a),(b) Normal behavior of the robot before injecting a mutex lock. Here the robot safely avoids the obstacles in its path to achieve its goal. (c),(d) Erroneous behavior of the robot on injecting a mutex lock. Here the robot hits the obstacle on its navigation path.

When the mutex lock fault is injected, the Reader thread is blocked and therefore shows no activity. Due to its blockage, the time it spends in the user and system mode remains unchanged. Figure 10.12 shows that once the mutex lock fault is injected, the sum of user mode time and system mode time remains constant. The mutex lock detection algorithm (see Section 10.6.2) reads the system mode and the user-mode times of the Reader thread

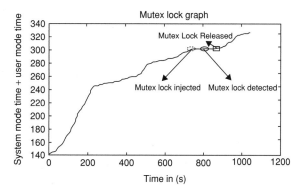

Figure 10.12   Variations in the total time variable (sum of system and user mode times). Also shown is the detection of the mutex lock by the mutex lock detection mechanism.

checks if their sum is constant. If this sum does not change for a user-defined period, which was set to a safe threshold value of 50, a fault detection message is sent to the local DES controller.

10.9 CONCLUSIONS AND FUTURE WORK

This chapter has introduced a class of sensors, called soft sensors, that monitor the status of software characteristics. Using these soft sensors, a framework of software anomaly detection and mitigation is developed for interconnected robotic networks. The underlying principles are valid for a wide class of complex systems of interconnected components. We experimentally validated the percolation of software anomalies resulting in erroneous behavior of electrical and mechanical components, such as a gripper failure in the robot, which occurred as a result of memory leak. We are exploring scaling issues related to performance of soft sensors and our integrated architecture as the complexities of networks and processes in software systems increase and detection of other software anomalies such as extensive memory page swaps and their effects on interconnected components of complex systems.

APPENDIX A

This appendix reviews the previous work on language measure [12, 13]. It provides the background information necessary to develop a performance index and an optimal control policy.

Let the dynamical behavior of a physical plant be modeled as a deterministic finite-state automaton (DFSA) $G_i \equiv (Q, \Sigma, \delta, q_i, Q_m)$ with $|Q| = n$ and $|\Sigma| = m$.

Definition 10.9.1   A DFSA G_i initialized at $q_i \in Q$ generates the language $L(G_i) \equiv \{s \in \Sigma^* : \delta^*(q_i, s) \in Q\}$ and its marked sublanguage $L_m(G_i) \equiv \{s \in \Sigma^* : \delta^* (q_i, s) \in Q_m\}$.

Definition 10.9.2   The language of all strings that start at $q_i \in Q$ and terminate at $q_j \in Q$ is denoted as $L(q_i, q_j)$.

Definition 10.9.3   The characteristic function that assigns a signed real weight to state-partitioned sublanguages $L(q_i, q_j)$ is defined as: $\chi : Q \to [-1, 1]$ such that

$$\chi_i \equiv \chi(q_j) \in \begin{cases} [-1, 0) & \text{if } q_j \in Q_m^- \\ \{0\} & \text{if } q_j \notin Q_m \text{ independent of } q_j \\ (0, 1] & \text{if } q_j \in Q_m^+ \end{cases}$$

The $n \times 1$ characteristic vector is denoted as $\bar{X} = [\chi_1, \chi_2 \cdots \chi_n]^T$

Definition 10.9.4   The event cost is defined as $\tilde{\pi} : \Sigma^* \times Q \to [0, 1)$ such that, $\forall q_j \in Q, \forall \sigma_k \in \Sigma, \forall s \in \Sigma^*,$

- $\tilde{\pi}[\sigma_k | q_j] = 0$ if $\delta(q_j, \sigma_k)$ is undefined; $\tilde{\pi}[\epsilon | q_j] = 1$;
- $\tilde{\pi}[\sigma_k | q_j] \equiv \tilde{\pi}_{jk} \in [0, 1); \sum_k \tilde{\pi}_{jk} < 1$;
- $\tilde{\pi}[\sigma_k s | q_j] = \tilde{\pi}[\sigma_k | q_j \tilde{\pi}[s | \delta(q_j, q_k)]$.

The event cost matrix is denoted as $\tilde{\Pi} \equiv \tilde{\pi}_{ij}$

Definition 10.9.5   The state transition cost of the DFSA is defined as a function $\pi : Q \times Q \to [0, 1)$ such that $\forall q_j, q_k \in Q$, $\pi(q_k|q_j) = \sum_{\sigma \in \Sigma : \delta(q_j, \sigma) = q_k} \tilde{\pi}(\sigma|q_j) \equiv \pi_{jk}$ and $\pi_{jk} = 0$ if $\{\sigma \in \Sigma : \delta(q_j, \sigma)\} = \varnothing$. The state $n \times n$ transition cost matrix, denoted as Π, is defined as

$$
\Pi = \begin{bmatrix}
\pi_{11} & \pi_{12} & \cdots & \pi_{1n} \\
\pi_{21} & \pi_{22} & \cdots & \pi_{2n} \\
\vdots & & \ddots & \vdots \\
\pi_{n1} & \pi_{n2} & \cdots & \pi_{nn}
\end{bmatrix}
$$

Definition 10.9.6   The signed real measure μ of a singleton string set $\{s\}$ is defined as

$$
\mu(s) \equiv \chi(q_j)\tilde{\pi}(s|q_j) \qquad \forall s \in L(q_i, q_j) \subseteq L(G_i)
$$

The signed real measure of $L(q_i, q_j)$ is defined as

$$
\mu(L(q_i, q_j)) \equiv \left(\sum_{s \in L(q_i, q_j)} \mu(\{s\}) \right)
$$

The signed real measure of a DFSA G_i, initialized at the state $q_i \in Q$, is defined as

$$
\mu_i \equiv \mu(L(G_i)) = \sum_j \mu(L(q_i, q_j))
$$

The $n \times 1$ real signed measure vector is denoted as

$$
\tilde{\pi} \equiv [\mu_1 \mu_2 \cdots \mu_n]^T
$$

Wang and Ray [13] have shown that the measure of the language $L(G_i)$, where $G_i = (Q, \Sigma, \delta, q_i, Q_m)$, can be expressed as $\mu = \sum_j \pi_{ij}\mu_j + \chi_j$. Equivalently, in vector notation, $\tilde{\mu} = \Pi\tilde{\mu} + \tilde{X}$. Since Π is a contraction operator, the measure vector $\tilde{\mu}$ is uniquely determined as $\tilde{\mu} = [I - \Pi]^{-1}\tilde{X}$.

APPENDIX B

Fu et al. [11] have introduced the concept of unconstrained optimal control of regular languages based on a specified measure. The state-based optimal control policy is obtained by selectively disabling controllable events to maximize the measure of the controlled plant language without any constraints. In each iteration, the optimal control algorithm attempts to disable all controllable events leading to "bad marked states" and enable all controllable events leading to "good marked states." It is also shown that computational complexity of the control synthesis is polynomial in the number of plant states [11]. Let G be the DFSA plant model without any constraint of operational specifications. Let the state transition cost matrix of the open loop plant be $\Pi^{\text{plant}} \in \mathfrak{R}^{n \times n}$ and the characteristic vector be $\chi \in \mathfrak{R}^n$.

Starting with $k = 0$ and $\Pi^0 \equiv \Pi^{\text{plant}}$, the control policy is constructed by the following two-step procedure [11]:

Step 1: For every state q_j for which $\mu_j^0 < 0$, disable controllable events leading to q_j. Now, $\Pi^1 = \Pi^0 - \Delta^0$, where $\Delta^0 \geq 0$ is composed of event costs corresponding to all controllable events that have been disabled at $k = 0$.

Step 2: Starting with $k = 1$, if $\mu_j^k \geq 0$, reenable all controllable events leading to q_j, which were disabled in step 1. The cost matrix is updated as $\Pi^{k+1} = \Pi^k - \Delta^k$ for $k \geq 1$, where $\Delta^k \geq 0$ is composed of event costs corresponding to all currently reenabled controllable events. The iteration is terminated if no controllable event leading to q_j remains disabled for which $\mu_j^k \geq 0$. At this stage, the optimal performance is $\mu^* = [I - \Pi^*]^{-1} X$.

REFERENCES

1. J. Carreira, H. Madeira, and J. Silva, "Xception: A technique for the experimental evaluation of dependability in modern computers," *IEEE Transactions on Software Engineering*, vol. 24, no. 2, pp. 125–136, 1998.

2. M. Hsueh, T. Tsai, and R. K. Iyer, "Fault injection techniques and tools," *IEEE Transactions on Computers*, vol. 44, no. 2, pp. 248–260, 1997.

3. G. A. Kanawati, N. A. Kanawati, and J. A. Abraham, "Ferrari: A flexible software-based fault and error injection system," *IEEE Transactions on Computers*, vol. 44, no. 2, pp. 248–260, 1995.

4. W. Kao, R. K. Iyer, and D. Tang, "FINE: A fault injection and monitoring environment for tracing the unix system behavior under faults," *IEEE Transactions on Software Engineering*, vol. 19, no. 11, pp. 1105–1118, 1993.

5. S. W. Keckler, A. Chang, W. S. Lee, S. Chatterjee, and W. J. Dally, "Concurrent event handling through multithreading," *IEEE Transactions on Computers*, vol. 48, no. 9, pp. 903–916, 1999.

6. W. D. Pauw and G. Sevitsky, "Visualizing reference patterns for solving memory leaks in java," *Concurrency: Practice and Experience*, vol. 12, no. 14, pp. 1431–1454, 2000.

7. M. Ricardo, L. Henriques, D. Costa, and H. Madeira, "Xception - enhanced automated fault-injection environment," in *Proceedings of the International Conference Dependable Systems and Networks (2002)*, Washington, DC, 2002, pp. 547–550.

8. The Space Shuttle Columbia Disaster, CNN, 2003.

9. B. Gerkey and R. T. V. A. Howard, "The player/stage project: Tools for multi-robot and distributed sensor systems," in *Proceedings of the Eleventh International Conference on Advanced Robotics*, Coimbia, Portugal, 2003, pp. 317–323.

10. B. Gerkey, R. Vaughan, K. Stoy, A. Howard, G. Sukhatmeand, and M. Mataric, "Most valuable player: A robot device server for distributed control," in *Proceedings of the IEEE/RSJ International Conference on Intelligent Robots and Systems (IROS 2001)*, Maui, 2001, pp. 1226–1231.

11. J. Fu, A. Ray, and C. Lagoa, "Unconstrained optimal control of regular languages," *Automatica*, vol. 40, no. 4, pp. 639–646, 2003.

12. A. Surana and A. Ray, "Signed real measure of regular languages," in *Proceedings of the Forty-second IEEE Conference on Decision and Control (CDC)*, Maui, 2003, pp. 3233–3238.

13. X. Wang and A. Ray, "A language measure for performance evaluation of discrete event supervisory control systems," *Applied Mathematical Modelling*, 2004.

14. B. P. Gerkey, R. T. Vaughan, and A. Howard, "Player c++ client library version 1.4," Techical Rep., USC Robotics Lab, 2003.

15. R. Hastings and B. Joyce, "Fast detection of memory leaks and access errors," in *Proceedings of the Winter '92 USENIX Conference*, Seattle, 1992, pp. 125–136.

16. B. Willard and O. Frieder, "Autonomous garbage collection: Resolving memory leaks in long running network applications," in *Proceedings of the International Conference on Computer Communications and Networks*, Lafayette, LA, 1998, pp. 886–896.

17. J. Xu, X. Wang, and C. Pham, "Less intrusive memory leak detection inside kernel," in *Proceedings of the Fault-Tolerant Computing Symposium*, 2003.

18. Y. Hong, D. Chen, L. Li, and K. Trivedi, "Closed loop design for software rejuvenation," *SHAMAN Journal of the International Society for Shamanistic Research*, June 2002.

19. P. J. Ramadge and W. M. Wonham, "Supervisory control of a class of discrete event processes," *SIAM Journal of Control and Optimization*, vol. 25, no. 1, pp. 206–230, 1987.

20. H. Madeira, M. Vieira, and D. Costa, "On the emulation of software faults by software fault injection," in *Proceedings of the IEEE International Conference on Dependable Systems and Networks*, New York, 2000, pp. 417–426.

21. N. Tuck and D. M. Tullsen, "Initial observations of the simultaneous multithreading pentium 4 processor," in *Proceedings of the Twelfth International Conference on Parallel Architectures and Compilation Techniques (PACT'03)*, 2003, pp. 26–34.

22. P. Watcharawitch and S. Moore, "Jma: The java-multithreading architecture for embedded processors," in *Proceedings of the 2002 IEEE International Conference on Computer Design: VLSI in Computers and Processors (ICCD'02)*, Austin, 2002, pp. 527–529.

23. J. S. Gray, *Interprocess Communications in UNIX: The Nooks and Crannies*, 2nd ed., Upper Saddle River, NJ: Prentice-Hall, 1998.

24. A. Silberschatz, G. Gagne, and P. Galvin, *Operating System Concepts*, 6th ed., New York: John Wiley & Sons, 2002.

25. D. Bovet and M. Cesati, *Understanding the Linux Kernel*, 2nd ed., O'Reilly & Associates, 2002.

26. A. Rubini and J. Corbet, *Linux Device Drivers*, 2nd ed., O'Reilly & Associates, 2001.

11

MULTISENSOR NETWORK-BASED FRAMEWORK FOR VIDEO SURVEILLANCE: REAL-TIME SUPERRESOLUTION IMAGING

Guna Seetharaman, Ha V. Le, S. S. Iyengar, N. Balakrishnan, and R. Loganantharaj

A network of multimodal sensors with distributed and embedded computations is considered for a video surveillance and monitoring application. Practical factors limiting the video surveillance of large areas are highlighted. A network of line-of-sight sensors and mobile-agents-based computations are proposed to increase the effectiveness. CMOS digital cameras in which both sampling and quantization occur on the sensor focal plane are more suitable for this application. These cameras operate at very high video frame rates and are easily synchronized to acquire images synaptically across the entire network. Also, they feature highly localized short-term memories and include some SIMD parallel computations as an integral part of the image acquisition. This new framework enables distributed computation for piecewise stereovision across the camera network, enhanced spatiotemporal fusion, and superresolution imaging of steadily moving subjects. A top-level description of the monitor, locate, and track model of a surveillance and monitoring task is presented. A qualitative assessment of several key elements of the mobile-agents-based computation for tracking persistent tokens moving across the entire area is outlined. The idea is to have as many agents as the number of persons in the field of view and perform the computations in a distributed fashion without introducing serious bottlenecks. The overall performance is promising when compared against that of a small network of cameras monitoring large corridors with a human operator in the loop.

11.1 INTRODUCTION

Increased access to inexpensive fabrication of CMOS circuits has vitalized research in intelligent sensors with embedded computing and power-aware features. Video image sensors [1, 2] and infrared detectors have been built using standard CMOS processes, with embedded digital signal processing in the pixel planes. They indicate an emerging trend in modeling the flow of information in an image processing system. The old "acquire, plumb, and process—in chain" model of image processing systems would be replaced. The chain would most likely evolve into a top-sorted graph comprised of acquire-and-fuse, macroassimilate, and meta-process stages in which several nodes within each stage may be connected via lateral parallelism (fusion at that level). In this new paradigm, the data—at each stage of the chain—would be subjected to appropriately designed intelligent processing, giving rise to a pipeline of incrementally inferred knowledge. Insight into data representation and modeling of data flow in this framework could have a profound impact on making large surveillance systems more tractable through simpler distributed and parallel computing. The new approach would require sensor and data fusion at all levels, in both time and space. Smart CMOS pixel planes with simple and networked computational features could make video surveillance more effective and tractable. This chapter is an effort to introduce a framework and examine at least one of the newly enabled benefits of such a basic multisensor network system.

11.2 BASIC MODEL OF DISTRIBUTED MULTISENSOR SURVEILLANCE SYSTEM

We present the basic factors influencing the analysis and performance of the data flow and computation in a large-scale multisensor network, in the context of a airport surveillance system. The relevant factors are captured in Figure 11.1, including the dynamic computational states of the embedded software agents. The basic model assumes that some mechanism is available to distinctly locate multiple objects (persons) within its field of view, as they enter. Let m be the number of people that can check in simultaneously and pass through the

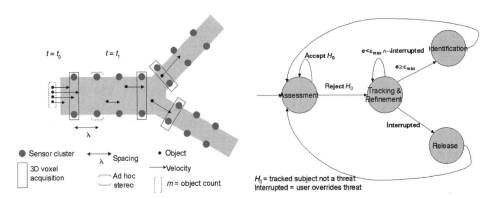

Figure 11.1 Central theme is to observe video image of a busy corridor, with multiple lanes and branches. The principal paradigm is to track every moving entity until a threshold has been reached to raise an alarm or abandon tracking.

predetermined points of entry. Their exact locations are picked up by fast three-dimensional (3D) sensors. Several multisensors, such as X-ray and infrared cameras, may acquire additional data and tag the data to the voxel. This is in fact the origin of a spatiotemporal thread associated with the events triggered by the moving person. Each person moves at an arbitrary pace. Let v be the average velocity. They are monitored by sensor clusters downstream. The local agents at these sensor clusters will need a speed and power proportional to $m \times v$. The local memory required and the monitoring complexity are $m\lambda v$. Network data flow is $m \times v$. The spatiotemporal registration between two stations separated by a distance λ will be proportional to m^2. It is envisioned that large variations in velocity can complicate the matter. In that case, simple dynamic programming approach using the last recently known location as the index of search space could be exploited to reduce the complexity of the problem. Along these lines, we estimate the tracking complexity to be proportional to $m^2(1 + \alpha|v_{max} - v_{min}|)$, for the purpose of spatiotemporal registration. Further inspection reveals that the peak load on the agents would also be influenced by a factor proportional to the variation. The complexity of local agent processing power and memory requirements will scale up by a factor $(1 + \alpha|v_{max} - v_{min}|)$. In all these cases, the value of α would assume different values, one for each context.

Given a voxel and a camera whose field of view covers the same, the location of the image of the voxel in the captured video image is trivially determined if the cameras have been fully calibrated. We assume this to be the case. In a sense, this abstraction treats some clusters of sensors to be more adept at identifying distinct events, whereas others downstream are more efficient in tracking them. Also, we do not preclude the possibility of any pair of adjacent sensors in forming ad hoc means to resolve 3D locations should there be a need triggered by local temporal events. That is, more rigorous voxel acquisition sensors are placed sporadically, and loosely coupled video cameras are widespread.

11.2.1 Sensors: Design, Deployment, Data Acquisition, and Low-Level Fusion

Rapid acquisition of the 3D location (voxel occupancy) of people in their field of view is essential. A very high speed three-dimensional sensor published in the literature [3] may be used to acquire the 3D image of physical scenes. The sensor is constructed with an array of smart analog-pixel sensors. Each smart pixel is made of a photo cell, an analog comparator, and a sample-and-hold circuit. The 3D data acquisition requires a planar laser beam to sweep through the scene. When the laser sweeps through the scene, it would produce an event of significance at various pixels at different times—easily detected by an increase in intensity. The exact time of the event is captured, which amounts to sensing the depth.

It is possible to acquire 3D data by a set of two video cameras and not require a laser beam, or use laser minimally when necessary, such as a hybrid range and passive video sensor [4]. The sensor consists of two video cameras with significant overlap in their fields of view and similarity in their parallax. It seeks to detect a number of feature points in each view, in an attempt to establish point correspondence [9] and compute 3D data. It involves spatial search for concurring observation(s) across the views. The spatial search has a well-defined geometric pattern known as epipolar lines. Then, a SIMD parallel computing array makes it possible to acquire up to 15,000 voxels per second [4]. In addition two image sequences are delivered by the cameras.

The computation described above is an example of low-level fusion, facilitated by a pair of networked sensors with mutual access to very low level data of each other. A number

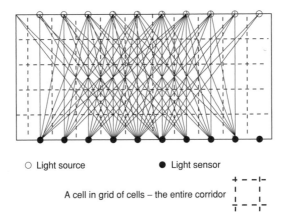

○ Light source   ● Light sensor

A cell in grid of cells – the entire corridor

Figure 11.2   Line-of-sight sensor network comprised of light emitters and light sensors shown above effectively eliminates the bottleneck of locating people (points of interest) in video frame rate.

of higher-level approaches exist for stereovision; and they differ in terms of the feature space used to detect the points of interest, the methods used for matching them across two views, and the controllability of the observed space. Such computations, in our view, fall in the classification of macro and mid-level fusion. Often they use scene knowledge and object knowledge and not the temporal signatures. That is, mutual access to the raw data of neighboring sensors and spatiotemporal fusion at low level, we believe, would offer speed, albeit with a need for postprocessing.

A set of omnidirectional (isotropic) light-emitting diodes driven by a multiphase clock and a set of omnidirectional light sensors can be deployed in large numbers along the corridor. These would be packaged in easily installed and networked strips. The spatiotemporal signals acquired through these rudimentary sensor nets would help reduce the camera count and increase their separation. Some of them come packaged with local microcontrollers for communication purposes. Such a sensor network is illustrated in Figure 11.2.

11.2.2 Overview of Cooperative Agents That Monitor and Track

An agent plays a pivotal role in integrating raw sensor inputs and information coming from sensor fusion and image analysis so as to achieve effective monitoring and tracking objects by collaborating with other agents. A software agent is a program that perceives an environment through sensors and acts on the environment [5–8] so as to achieve the intended purposes, which is monitoring and tracking all the objects with its monitoring boundaries. Agents are autonomous within the context of its intended purposes and thus an ideal choice to perform continuous monitoring and tracking purposes. A functional architecture of our agent is given in Figure 11.3, which is inspired by the architecture of the remote agent that was successfully deployed by NASA. The agent has five functional components, namely execution monitor, planning and scheduler, knowledge base, detector of unusual behavior, and a communicator.

We envision a set of agents are collaborating together to achieve the overall purpose. The communicator module of an agent is responsible for maintaining all relevant information of all other agents in the system and respond to the request from the execution monitor to send

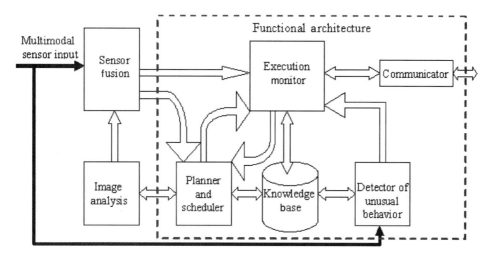

Figure 11.3 Functional architecture of the intelligent agent.

a message to another agent. Here we assume that the agents do have a global knowledge of other agents in the system, their capabilities, and their layout along with their physical monitoring boundaries. If for some reason an agent fails and another agent becomes active in covering the void left by the failing agent, the new agent communicates with all other agents, specially the one in its neighborhood to inform about its existence, its coverage area, and its capability. Steps have been taken to safeguard against some malicious agents pretending to be part of the cooperating agents.

To monitor and to track an object, each object entering into the monitoring space is given a unique identification (ID) number, which in our case is a time stamp followed by a predefined number of digits, say a two-digit number, generated randomly. If many objects enter at the same time through different entrances, there may be a possibility (1/1000 in the case of two-digit random numbers) of assigning the same ID for two objects. If such a violation occurs, it will be corrected appropriately. In addition to assigning a unique ID for a 3D voxel, other features of the object is also stored, which will help to identify the voxel and reassign the ID after the object temporarily leaves the monitoring area, such as a restroom, and the like.

For tracking and monitoring purposes, spatial temporal history of an object, that is, its location at different time points, is maintained by the module for planning and scheduling. From the information of location over the time period, the current speed and the direction is inferred, and it is being used to predict the future location in the next time slot. The approximate locations in the next time slot of objects are fed into the image analyzers. The image analyzer along with the sensors confirm or obtain the new locations of the requested objects and update the planning and scheduling module. For some reason no object is identified in the region; then the area is gradually increased and the identification process continues. This is the basis of tracking. If many objects appeared in an area, it is resolved by matching with the candidates who may have moved from their previous positions. The projected path of an object is being used to predict whether it will leave the monitoring boundary of the current agent. The ID and the coordinates of the objects that are predicted to cross the boundary are handed over to the agents who will subsequently monitor those objects. The knowledge base has a predefined template of features and the expected behaviors of the objects that matched the features.

The agility of the application demands immediate recognition of the unusual behavior of an object. To realize such recognition, we implement reactive behavior of an agent using the sensor input. The abnormal detecting module is trained to recognize classes of abnormal behaviors from the raw sensor input and thereby avoiding processing time. Based on the thread levels of abnormal behavior, the execution module of the agent will take appropriate action. For example, if the sensor detects smoke, the agent will immediately enabled the fire alarm and inform all security personnel who are trained to deal with the physical situation. In the context of mobile sensor networks, such as unmanned aerial vehicles (UAVs) equipped with video cameras and wireless commucnication, the agents should also take into account the cost of communication with the other nodes. The optimal computation of local temporal computations, such as image sequence analysis, should be designed to exploit local data first and use stereovision sparingly.

11.3   SUPERRESOLUTION IMAGING

The objective of superresolution imaging is to synthesize a higher resolution image of objects from a sequence of images whose spatial resolution is limited by the operational nature of the imaging process. The synthesis is made possible by several factors that effectively result in subpixel-level displacements and disparities between the images.

Research on superresolution imaging has been extensive in recent years. Tsai and Huang were the first trying to solve the problem. In [9], they proposed a frequency-domain solution that uses the shifting property of the Fourier transform to recover the displacements between images. This as well as other frequency-domain methods such as [10] have the advantages of being simple and having low computational cost. However, the only type of motion between images that can be recovered from the Fourier shift is the global translation; therefore, the ability of these frequency-domain methods is quite limited.

Motion-compensated interpolation techniques [11, 12] also compute displacements between images before integrating them to reconstruct a high-resolution image. The difference between these methods and the frequency-domain methods mentioned above is that they work in the spatial domain. Parametric models are usually used to model the motions. The problem is, most parametric models are established to represent rigid motions such as camera movements, while in the real-world motions captured in image sequences are often nonrigid, too complex to be described by a parametric model. Model-based superresolution imaging techniques such as back-projection [13] also face the same problem.

More powerful and robust methods such as the *projection onto convex sets* (POCS)-based methods [14], which are based on set theories, and stochastic methods such as *maximum a posteriori* (MAP)-based [15] and *Markov random field* (MRF)-based [16] algorithms are highly complex in terms of computations; hence, they are unfit for applications that require real-time processing.

We focus our experimental study [17] on digital video images of objects moving steadily in the field of view of a camera fitted with a wide-angle lens. These assumptions hold good for a class of video-based security and surveillance systems. Typically, these systems routinely perform MPEG analysis to produce a compressed video for storage and offline processing. In this context, the MPEG subsystem can be exploited to facilitate superresolution imaging

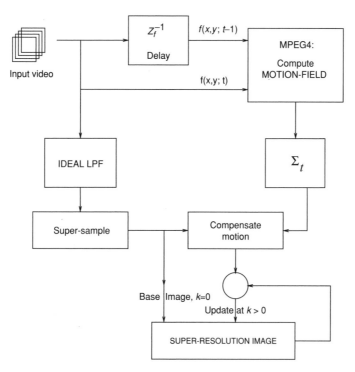

Figure 11.4 Schematic block diagram of the proposed superresolution imaging method.

through a piecewise affine registration process that can easily be implemented with the MPEG-4 procedures. The method is able to increase the effectiveness of camera security and surveillance systems.

The flow of computation in the proposed method is depicted in Figure 11.4. Each moving object will be separated from the background using standard image segmentation techniques. Also, a set of feature points, called the points of interest, will be extracted. These points include places were the local contrast patterns are well defined, exhibit a high degree of curvature, and demonstrate such geometric features. We track their motions in the 2D context of a video image sequence. This requires image registration or some variant of point correspondence matching. The net displacement of the image of an object between any two consecutive video frames will be computed with subpixel accuracy. Then, a rigid coordinate system is associated with the first image, and any subsequent image is modeled as though its coordinate system has undergone a piecewise affine transformation.

We recover the piecewise affine transform parameters between any video frame with respect to the first video frame to a subpixel accuracy. Independently, all images will be enlarged to a higher resolution using a bilinear interpolation [18] by a scale factor. The enlarged image of each subsequent frame is subject to an inverse affine transformation to help register it with the previous enlarged image. Given K video frames, then, in principle, it will be feasible to synthesize $K - 1$ new versions of the scaled and interpolated and inverse-motion-compensated image at the first frame instant. Thus, we have K high-resolution images from which to assimilate. (See Figure 11.5 for some results.)

Mean square errors

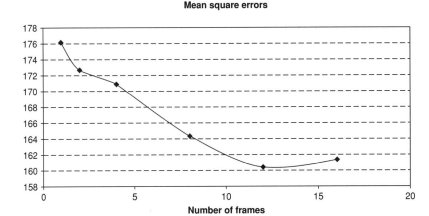

Figure 11.5   Graph of mean square errors between reconstructed images and the original frame.

11.4   OPTICAL FLOW COMPUTATION

We follow a framework proposed by Cho et al. [19] for optical flow computation based on a piecewise affine model. A surface moving in the 3D space can be modeled as a set of small planar surface patches. Then, the observed motion of each 3D planar patch in the 2D image plane can be described by an affine transform. Basically, this is a mesh-based technique for motion estimation, using 2D content-based meshes. The advantage of content-based meshes over regular meshes is their ability to reflect the content of the scene by closely matching boundaries of the patches with boundaries of the scene features [20], yet finding feature points and correspondences between features in different frames is a difficult task. A multiscale coarse-to-fine approach is utilized in order to increase the robustness of the method as well as the accuracy of the affine approximations. An adaptive filter is used to smooth the flow field such that the flow appears continuous across the boundary between adjacent patches, while the discontinuities at the motion boundaries can still be preserved. Many of these techniques are already available in MPEG-4.

Our optical flow computation method includes the following phases:

1. *Feature Extraction and Matching:* In this phase the feature points are extracted and feature matching is performed to find the correspondences between feature points in two consecutive image frames.

2. *Piecewise Flow Approximation:* A mesh of triangular patches is created whose vertices are the matched feature points. For each triangular patch in the first frame there is a corresponding one in the second frame. The affine motion parameters between these two patches can be determined by solving a set of linear equations formed over the known correspondences of their vertices. Each set of these affine parameters define a smooth flow within a local patch.

Finding the correspondences between feature points in consecutive frames is the key step of our method. We devised a matching technique in which the cross correlation, curvature, and displacement are used as matching criteria. The first step is to find an initial estimate for the motion at every feature point in the first frame. Some matching techniques described in [21] would consider all of $M \times N$ pairs where M and N are the number of feature points in the first and second frames, respectively. Some others assume the displacements are small to limit the search for a match to a small neighborhood of each point. By giving an initial estimate for the motion at each point, we are also able to reduce the number of pairs to be examined without having to constrain the motion to small displacements. We have devised a multiscale scheme in which the initial estimation of the flow field at one scale is given by the piecewise affine transforms computed at the previous level. At the starting scale, a rough estimation can be made by treating the points as if they are under a rigid 2D motion. It means the motion is a combination of a rotation and a translation. Compute the centers of gravity, C_1 and C_2, the angles of the principal axes, α_1 and α_2, of the two sets of feature points in two frames. The motion at every feature points in the first frame can be roughly estimated by a rotation around C_1 with the angle $\phi = \alpha_2 - \alpha_1$, followed by a translation represented by the vector $\mathbf{t} = \mathbf{x}_{C_2} - \mathbf{x}_{C_1}$, where $\mathbf{x}_{C_1}$ and $\mathbf{x}_{C_2}$ are the vectors representing the coordinations of C_1 and C_2 in their image frame.

Let i^t and j^{t+1} be two feature points in two frames t and $t + 1$, respectively. Let i'^{t+1} be the estimated match of i^t in frame $t + 1$, $d(i', j)$ be the Euclidean distance between i'^{t+1} and j^{t+1}, $c(i, j)$ be the cross correlation between i^t and j^{t+1}, $0 \leq c(i, j) \leq 1$, and $\Delta\kappa(i, j)$ be the difference between the curvature measures at i^t and j^{t+1}. A *matching score* between i^t and j^{t+1} is defined as follows:

$$
\begin{aligned}
d(i', j) &> d_{\max} : \\
&\quad s(i, j) = 0 \\
d(i', j) &\leq d_{\max} : \\
&\quad s(i, j) = w_c c(i, j) + s_k(i, j) + s_d(i, j)
\end{aligned}
\tag{11.1}
$$

where

$$
\begin{aligned}
s_k(i, j) &= w_k[1 + \Delta\kappa(i, j)]^{-1} \\
s_d(i, j) &= w_d[1 + d(i', j)]^{-1}
\end{aligned}
\tag{11.2}
$$

The quantity $d_{\max}$ specifies the maximal search distance from the estimated match point; and w_c, w_k, and w_d are the weight values determining the importance of each of the matching criteria. The degree of importance of each of these criteria changes at different scales. At a finer scale, the edges produced by a Canny edge detector become less smooth, meaning the curvature measures are less reliable. Thus, w_k should be reduced. On the other hand, w_d should be increased, reflecting the assumption that the estimated match becomes closer to the true match. For each point i^t, its optimal match is a point j^{t+1} such that $s(i, j)$ is maximal and exceeds a threshold value t_s. Finally, interpixel interpolation and correlation matching are used in order to achieve subpixel accuracy in estimating the displacement of the corresponding points.

Using the constrained Delaunay triangulation [22] for each set of feature points, a mesh of triangular patches is generated to cover the moving part in each image frame. A set of line segments, each of which connects two adjacent feature points on a same edge, is used to constrain the triangulation, so that the generated mesh closely matches the true content of the image. Each pair of matching triangular patches results in six linear equations made of piecewise local affine motion parameters, which can be solved to produce a dense velocity field inside the triangle.

11.4.1 Evaluation of Optical Flow Computation Technique

We conducted experiments with our optical flow estimation technique using some common image sequences created exclusively for testing optical flow techniques and compared the results with those in [23] and [24]. The image sequences used for the purpose of error evaluation include the translating tree sequence (Fig. 11.6), the diverging tree sequence

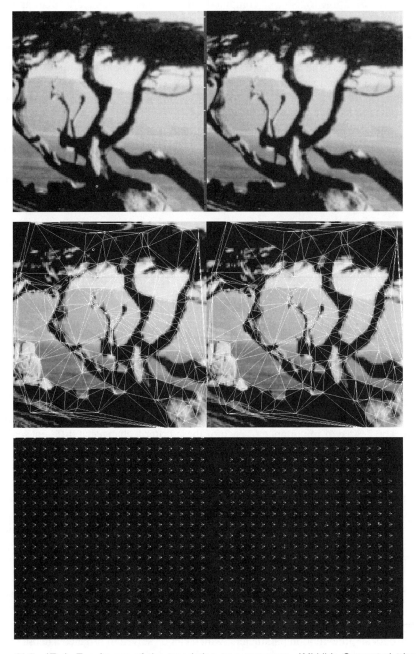

Figure 11.6 (*Top*): Two frames of the translating tree sequence. (*Middle*): Generated triangular meshes. (*Bottom*): The correct flow (left) and the estimated flow (right).

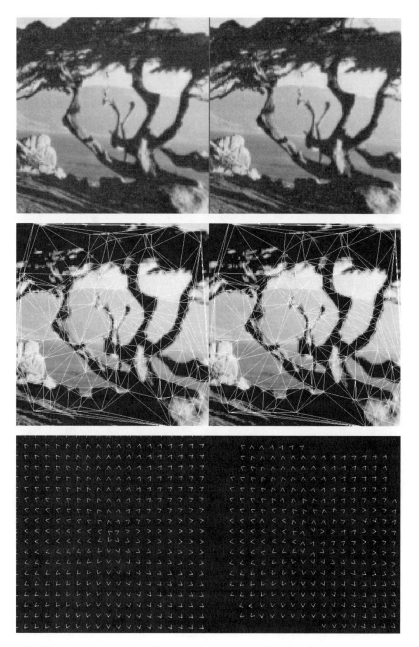

Figure 11.7 (*Top*): Two frames of the diverging tree sequence. (*Middle*): Generated triangular meshes. (*Bottom*): The correct flow (left) and the estimated flow (right).

(Fig. 11.7), and the Yosemite sequence (Fig. 11.8). These are simulated sequences for which the ground truth is provided.

As in [23] and [24], an angular measure is used for error measurement. Let $\mathbf{v} = [u \quad v]^T$ be the correct 2D motion vector and $\mathbf{v}_e$ be the estimated motion vector at a point in the image plane. Let $\tilde{\mathbf{v}}$ be a 3D unit vector created from a 2D vector $\mathbf{v}$:

$$\tilde{\mathbf{v}} = \frac{[\mathbf{v} \quad 1]^T}{|[\mathbf{v} \quad 1]|} \tag{11.3}$$

Figure 11.8 (*Top*): Two frames of the Yosemite sequence. (*Middle*): Generated triangular meshes. (*Bottom*): The correct flow (left) and the estimated flow (right).

The angular error ψ_e of the estimated motion vector $\mathbf{v}_e$ with respect to the correct motion vector $\mathbf{v}$ is defined as follows:

$$\psi_e = \arccos(\tilde{\mathbf{v}}\tilde{\mathbf{v}}_e) \qquad (11.4)$$

Using this angular error measure, bias caused by the amplification inherent in a relative measure of vector differences can be avoided.

To verify if the accuracies are indeed subpixel, we use the distance error $d_e = |\mathbf{v} - \mathbf{v}_e|$. For the translating tree sequence (Table 11.1), the mean distance error is 11.40% of a pixel and the standard deviation of errors is 15.69% of a pixel. The corresponding figures for the diverging tree sequence (Table 11.2) are 17.08 and 23.96%, and for the Yosemite sequence (Table 11.3) they are 31.31 and 46.24%. It is obvious that the flow errors at most points of the images are subpixel.

Table 11.1 Performance of Various Optical Flow Techniques on the Translating Tree Sequence

Techniques	Average Errors (deg)	Standard Deviations (deg)	Densities (%)
Horn and Schunk (original)	38.72	27.67	100.0
Horn and Schunck (modified)	2.02	2.27	100.0
Lucas and Kanade	0.66	0.67	39.8
Uras et al.	0.62	0.52	100.0
Nagel	2.44	3.06	100.0
Anandan	4.54	3.10	100.0
Singh	1.64	2.44	100.0
Heeger	8.10	12.30	77.9
Waxman et al.	6.66	10.72	1.9
Fleet and Jepson	0.32	0.38	74.5
Piecewise affine approximation	2.83	4.97	86.3

Table 11.2 Performance of Various Optical Flow Techniques on the Diverging Tree Sequence

Techniques	Average Errors (deg)	Standard Deviations (deg)	Densities (%)
Horn and Schunk (original)	12.02	11.72	100.0
Horn and Schunck (modified)	2.55	3.67	100.0
Lucas and Kanade	1.94	2.06	48.2
Uras et al.	4.64	3.48	100.0
Nagel	2.94	3.23	100.0
Anandan	7.64	4.96	100.0
Singh	8.60	4.78	100.0
Heeger	4.95	3.09	73.8
Waxman et al.	11.23	8.42	4.9
Fleet and Jepson	0.99	0.78	61.0
Piecewise affine approximation	9.86	10.96	77.2

Table 11.3 Performance of Various Optical Flow Techniques on the Yosemite Sequence

Techniques	Average Errors (deg)	Standard Deviations (deg)	Densities (%)
Horn and Schunk (original)	32.43	30.28	100.0
Horn and Schunck (modified)	11.26	16.41	100.0
Lucas and Kanade	4.10	9.58	35.1
Uras et al.	10.44	15.00	100.0
Nagel	11.71	10.59	100.0
Anandan	15.84	13.46	100.0
Singh	13.16	12.07	100.0
Heeger	11.74	19.04	44.8
Waxman et al.	20.32	20.60	7.4
Fleet and Jepson	4.29	11.24	34.1
Black and Anandan	4.46	4.21	100.0
Piecewise affine approximation	7.97	11.90	89.6

11.5 SUPERRESOLUTION IMAGE RECONSTRUCTION

Given a low-resolution image frame $\mathbf{b}_k(m, n)$, we can reconstruct an image frame $\mathbf{f}_k(x, y)$ with a higher resolution as follows [18]:

$$\mathbf{f}_k(x, y) = \sum_{m,n} \mathbf{b}_k(m, n) \frac{\sin \pi(x\lambda^{-1} - m)}{\pi(x\lambda^{-1} - m)} \frac{\sin \pi(y\lambda^{-1} - n)}{\pi(y\lambda^{-1} - n)} \tag{11.5}$$

where $\sin\theta/\theta$ is the ideal interpolation filter, and λ is the desired resolution step-up factor. For example, if $\mathbf{b}_k(m, n)$ is a 50×50 image and $\lambda = 4$, then, $\mathbf{f}_k(x, y)$ will be of the size 200×200.

Each point in the high-resolution grid corresponding to the first frame can be tracked along the video sequence from the motion fields computed between consecutive frames, and the superresolution image is updated sequentially:

$$x^{(1)} = x, \, y^{(1)} = y, \mathbf{f}_1^{(1)}(x, y) = \mathbf{f}_1(x, y) \tag{11.6}$$

$$x^{(k)} = x^{(k-1)} + u_k\left(x^{(k-1)}, y^{(k-1)}\right), \, y^{(k)} = y^{(k-1)} + v_k\left(x^{(k-1)}, y^{(k-1)}\right) \tag{11.7}$$

$$\mathbf{f}_k^{(k)}(x, y) = \frac{k-1}{k}\mathbf{f}_{k-1}^{(k-1)}(x, y) + \frac{1}{k}\mathbf{f}_k\left(x^{(k)}, y^{(k)}\right) \tag{11.8}$$

for $k = 2, 3, 4, \cdots$. The values u_k and v_k represent the dense velocity field between $\mathbf{b}_{k-1}$ and $\mathbf{b}_k$. This sequential reconstruction technique is suitable for online processing in which the superresolution images can be updated every time a new frame comes.

11.6 EXPERIMENTAL RESULTS

In the first experiment we used a sequence of 16 frames capturing a slow-moving book (Fig. 11.9). Each frame was down-sampled by a scale of four. High-resolution images were reconstructed from the down-sampled ones, using $2, 3, \ldots, 16$ frames, respectively. The graph in Figure 11.5 shows errors between reconstructed images and their corresponding original frame, decreasing when the number of low-resolution frames used for reconstruction is increased, until the accumulated optical flow errors become significant. Even though this is a simple case because the object surface is planar and the motion is rigid, it nevertheless presented the characteristics of this technique.

The second experiment was performed on images taken from a real surveillance camera. In this experiment we tried to reconstruct high-resolution images of faces of people captured by the camera (Fig. 11.10). Results show obvious improvements of reconstructed superresolution images over original images.

For the time being, we are unable to conduct a performance analysis of our superresolution method in comparison with others' because (1) there has been no study on quantitative evaluation of the performance of superresolution techniques so far; and (2) there are currently no common metrics to measure the performance of superresolution techniques (in fact, most of the published works on this subject did not perform any quantitative performance

Figure 11.9 (*Top*): Parts of an original frame (left) and a down-sampled frame (right). (*Middle*): Parts of an image interpolated from a single frame (left) and an image reconstructed from 2 frames (right). (*Bottom*): Parts of images reconstructed from 4 frames (left) and 16 frames (right).

analysis at all). The number of superresolution techniques are so large that a study on comparison of their performances could provide enough contents for another study.

11.7 CONCLUSION

We have presented a method for reconstructing superresolution images from sequences of low-resolution video frames, using motion compensation as the basis for multiframe data fusion. Motions between video frames are computed with a multiscale piecewise affine model that allows accurate estimation of the motion field even if the motion is nonrigid. The reconstruction is sequential—only the current frame, the frame immediately

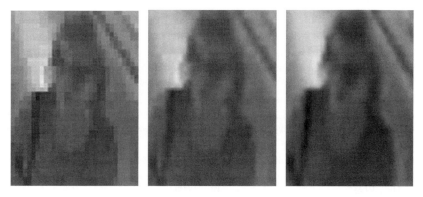

Figure 11.10 (*Left*): Part of an original frame containing a human face. (*Center*): Part of an image interpolated from a single frame. (*Right*): Part of an image reconstructed from four frames.

before it, and the last reconstructed image are needed to reconstruct a new superresolution image. This makes it suitable for applications that require real-time operations like in surveillance systems. The proposed superresolution is one example of a number of new computations made feasible in a distributed multisensor network comprised of many video cameras and line-of-sight sensors. Coarser measurement of people in 3D field of view is made available by the line-of-sight sensors. This in fact helps decouple the tasks of detecting points of interest in images and recognizing/monitoring the dynamically moving objects giving rise to those points of interest. The decoupling facilitates high frame rate of computation. The task of interpreting the superresolution images, and its dynamics, is currently in progress.

REFERENCES

1. E. R. Fossum, "CMOS image sensors: Electronic camera-on-a-chip," *IEEE Transactions on Electronic Devices*, vol. 44, no. 10, 1997.

2. S. Kleinfelder et al., "A 10000 frames/sec CMOS digital pixel sensor," *IEEE Journal of Solid-State Circuits*, vol. 36, no. 12, 2001.

3. L. R. C. A. Gruss and T. Kanade, "Integrated sensor and range-finding analog signal processor," *IEEE Journal of Solid-State Circuits*, vol. 26, no. 3, pp. 184–191, Mar. 1991.

4. G. Seetharaman, M. Bayoumi, K. Valavanis, and M. Mulder, "A VLSI architecture for stereo image sensors," in *Proceedings of the Workshop on Computer Architecture for Machine Perception*, Paris, Dec. 1991.

5. S. Russell and P. Norvig, *Artificial Intelligence: A Modern Approach*, Upper Saddle River, NJ: Prentice-Hall, 2003.

6. J. M. Bradshaw, *Software Agents*, Cambridge, MA: MIT Press, 1997.

7. K. Rajan, M. Shirley, W. Taylor, and B. Kanefsky, "Ground tools for the 21st centrury," in *Proceedings of the IEEE Aerospace Conference*, Big Sky, MT, 2000.

8. D. Bernard, G. Dorias, E. Gamble, B. Kanefsky, et al., "Spacecraft autonomy flight experience: The DS1 remote agent experiment," in *Proceedings of the AIAA 1999 Annual Conference*, Albuquerque, NM, 1999.

9. R. Y. Tsai and T. S. Huang, "Multiframe image restoration and registration," in R. Y. Tsai and T. S. Huang (Eds.), *Advances in Computer Vision and Image Processing*, vol. 1, JAI Press, 1984, pp. 317–339.

10. S. P. Kim and W.-Y. Su, "Recursive high-resolution reconstruction of blurred multiframe images," *IEEE Transactions on Image Processing*, vol. 2, no. 10, pp. 534–539, Oct. 1993.

11. A. M. Tekalp, M. K. Ozkan, and M. I. Sezan, "High resolution image reconstruction from low resolution image sequences, and space varying image restoration," in *Proceedings of the IEEE Conference on Acoustics, Speech, and Signal Processing*, vol. 3, San Francisco, CA, Mar. 1992, pp. 169–172.

12. M. Elad and Y. Hel-Or, "A fast super-resolution reconstruction algorithm for pure translational motion and common space-invariant blur," *IEEE Transactions on Image Processing*, vol. 10, no. 8, pp. 1187–1193, Aug. 2001.

13. M. Irani and S. Peleg, "Motion analysis for image enhancement: Resolution, occlusion and transparency," *Journal of Visual Communications and Image Representation*, vol. 4, pp. 324–335, Dec. 1993.

14. A. J. Patti, M. I. Sezan, and A. M. Tekalp, "Superresolution video reconstruction with arbitrary sampling lattices and nonzero aperture time," *IEEE Transactions on Image Processing*, vol. 6, no. 8, pp. 1064–1076, Aug. 1997.

15. M. Elad and A. Feuer, "Restoration of a single superesolution image from several blurred, noisy and undersampled measured images," *IEEE Transactions on Image Processing*, vol. 6, no. 12, pp. 1646–1658, Dec. 1997.

16. R. R. Schultz and R. L. Stevenson, "Extraction of high-resolution frames from video sequences," *IEEE Transactions on Image Processing*, vol. 5, no. 6, pp. 996–1011, June 1996.

17. H. Le and G. Seetharaman, "A method of super-resolution imaging based on dense subpixel accurate motion fields," in *Proceedings of the International Workshop on Digital Computational Video DCV2002*, Clearwater Beach, FL, Nov. 2002.

18. E. Meijering, "A chronology of interpolation: From ancient astronomy to modern signal and image processing," *Proceedings of the IEEE*, vol. 90, no. 3, pp. 319–344, Mar. 2002.

19. E. C. Cho, S. S. Iyengar, G. Seetharaman, R. J. Holyer, and M. Lybanon, "Velocity vectors for features of sequential oceanographic images," *IEEE Transactions on Geoscience and Remote Sensing*, vol. 36, no. 3, pp. 985–998, May 1998.

20. Y. Altunbasak and M. Tekalp, "Closed-form connectivity-preserving solutions for motion compensation using 2-D meshes," *IEEE Transactions on image processing*, vol. 6, no. 9, pp. 1255–1269, Sept. 1997.

21. R. N. Strickland and Z. Mao, "Computing correspondences in a sequence of non rigid images," *Pattern Recognition*, vol. 25, no. 9, pp. 901–912, 1992.

22. S. Guha, "An optimal mesh computer algorithm for constrained Delaunay triangulation," in *Proceedings of the International Parallel Processing Symposium*, Cancun, Mexico, Apr. 1994, pp. 102–109.

23. J. L. Barron, D. J. Fleet, and S. S. Beauchemin, "Performance of optical flow techniques," *International Journal of Computer Vision*, vol. 12, no. 1, pp. 43–77, 1994.

24. M. J. Black and P. Anandan, "The robust estimation of multiple motions: Parametric and piecewise-smooth flow field," *Computer Vision and Image Understanding*, vol. 63, no. 1, pp. 75–104, Jan. 1996.

USING INFORMATION THEORY TO DESIGN CONTEXT-SENSING WEARABLE SYSTEMS

Holger Junker, Paul Lukowicz, and Gerhard Troester

The development of low-power, short-range wireless communication technologies and the advances in sensor miniaturization have opened up new fields of applications for sensor networks such as context-aware computing. Major challenges that have to be resolved concern the management and the design of such networks.

This chapter focuses on design issues and shows how feature selection methods using mutual information can be applied to help in the design process of sensor networks with a particular focus on the design of wearable systems for context recognition. After a review of suitable methods, we describe how they can be adapted to meet the specific requirements of our target problem. Based on a standard recognition task for wearable computers, we derive an optimized design for our own wearable sensor platform using the adapted feature selection methods.

12.1 INTRODUCTION

Recently, context awareness has emerged as an important application of mobile sensor networks. The idea is to use simple sensors that are seamlessly integrated in the user's outfit and distributed in his environment to provide a computer system with a degree of "understanding" of what the user is doing, where he is, and what is happening around him. Using such information, the system can automatically configure itself to best suit the user's needs or even proactively retrieve, deliver, and/or record relevant information. In addition, the interaction between the system and the user can be simplified, reducing the cognitive load on the user. This, in turn, makes the system more appropriate for mobile applications. Examples of this concept range from simple extensions for mobile phones (e.g., to make the phone switch off the ringer when the user is in a meeting [1]),

Sensor Network Operations, Edited by Phoha, LaPorta, and Griffin

through automatic reminder systems [2], to elaborate assembly and maintenance support systems where the progress of the assembly task is being automatically followed by the computer [3].

The design of context-aware systems is a complex and as yet not fully solved problem. It involves a number of issues to be resolved, ranging from situation modeling and application design to the actual recognition. This chapter focuses on the latter. Specifically, we address the question of what information sources are best suited for a given recognition task. This selection problem includes the selection of sensors, their optimal placement, and the choice of suitable processing algorithms to extract relevant features from the raw sensor data. In the chapter, we investigate this issue with an emphasis on body-worn sensors, often associated with wearable systems.

12.1.1 Feature Selection Problem in Wearable Systems

The feature selection problem plays a key role in the overall design of a context-aware system. For one, relevant information must be provided to the classifier to enable reliable and efficient context recognition. This implies the use of relevant features with discriminative power. On the other hand, the design of the system must aim to minimize cost. This is particularly important in wearable systems where a certain sensor location might be inconvenient to the user and where power consumption issues impose strict limits on the amount of processing and communication that can be performed. Thus, it is essential to find a minimum set of features that provide sufficient and reliable information for the classification task, while considering the given cost constraints.

Finding those features allows the identification of relevant sensors to be integrated into the sensor network. Hence, the feature selection problem is crucial to the overall system design. In general, the existing feature selection techniques only focus on the selection of features. Thus, they do not consider any network design issues. Some techniques, however, have been reviewed for the dynamic and efficient management of network resources (see Section 12.2). The work presented here proposes to use feature selection techniques to derive design decisions for a context-aware sensor system, which is an important issue.

12.1.2 Chapter Contributions

Today, in most context recognition applications, the feature selection process is performed heuristically by the system designer. It is often a trial-and-error process based on using as many sensors as available, which in turn leads to system architectures that are more costly and cumbersome than necessary and sometimes perform worse than architectures that rely on a subset of well-selected sensors and features.

The contribution of this chapter is to demonstrate how information-theoretic-based feature selection methods can be adapted to perform this task in a systematic way. This includes:

1. The review of existing feature selection methods with respect to their suitability for the targeted problem domain
2. Appropriate adaptation of the methods
3. Experimental demonstration of the method on a standard context recognition problem using a wearable sensor platform developed by our group.

12.2 RELATED WORK

Many feature selection techniques for classification tasks have been proposed. An excellent overview of existing techniques and their characteristics is provided by Dash and Liu [4]. According to this work, a typical feature selection method consists of a generation procedure that generates a candidate subset of features, an evaluation function that evaluates the candidate subset, a stopping criterion that is used to terminate the selection process, and a validation step that validates the selected subset. In [4], the generation procedures are categorized as either complete, heuristic, or random. Popular evaluation functions are distance measures [5], classifier performance, and information-based measures such as mutual information [6–10]. The choice of the stopping criterion can be influenced by the generation procedure or by the evaluation function used. In the first case this can simply be the number of features to be selected, in the second case it can be a specific value a subset must achieve according to some evaluation function.

The feature selection techniques cannot only be used to improve classification performance but also as a tool to dynamically manage resources in classification systems. This issue has been investigated by several researchers [5, 11–13]. In this respect, general objective functions have been introduced to consider trade-offs between the usefulness of particular resources and the costs associated with using those resources [5].

12.3 THEORETICAL BACKGROUND

The feature selection scheme applied here uses mutual information as an evaluation criterion. It selects features according to the information gain they contribute to the recognition task. This makes the selection process independent of a specific induction algorithm, thus providing classifier-independent, comparable results.

One of the main problem with feature selection techniques that also applies to the technique used here is the fact that finding the optimal feature subset with respect to a given evaluation criterion requires in general an exhaustive search within the full feature set. This is, however, often computationally too expensive since the number of possible feature subsets increases exponentially with the total number of features. For our evaluation, we therefore use a standard forward feature selection method that selects and also ranks features according to the following scheme:

Step 1: Initialization: Start with an empty feature set $S = \{\ \}$.

Step 2: Choose a feature that has maximum mutual information with respect to the class.

Step 3: Add best feature to feature set according to some evaluation criterion.

Step 4: Repeat step 3 until stopping criterion is reached.

In the following, the theoretical background of the evaluation function will be reviewed, and the choice of our stopping criterion explained.

12.3.1 Mutual Information

The mutual information (MI) between two continuous random variables X and Y is defined as

$$\mathrm{MI}(X; Y) = \int_{x \in X} \int_{y \in Y} p(x, y) \log \frac{p(x, y)}{p(x)p(y)} dx\, dy \qquad (12.1)$$

$MI(X; Y)$ measures the statistical dependencies between the two variables X and Y and provides a measure of how much information one variable contains about the other. Assuming that Y is a discrete random variable, (12.1) can be written in terms of the Shannon entropy $H(Y)$[1] and the conditional entropy $H(Y|X)$ as shown in (12.2).

$$MI(X; Y) = H(Y) - H(Y|X) \tag{12.2}$$

where $H(Y)$ measures the uncertainty about Y, and $H(Y|X)$ the uncertainty of Y when X is observed. $MI(X; Y)$ quantifies the reduction in uncertainty about Y when X is known. MI is also referred to the information gain provided by X.

Assuming Y to be a class vector c and X a feature vector f, then the mutual information $MI(c; f)$ provides a measure of how much information feature f contains about the class c. Mutual information can also be calculated between multiple variables, allowing to evaluate the information gained about the class using an N-dimensional feature set F.

The mutual information is therefore a suitable measure to evaluate the relevance of individual features and feature combinations. The usefulness of mutual information as an evaluation criterion is also stated by Fano's [14] inequality, which provides a lower bound on the error probability. Maximizing $MI(c; F)$ lowers this bound [14]. The calculation of the mutual information between a feature vector and the output class requires the estimation of the joint probability density function $p(c; f)$. In this chapter, we use two different density estimation techniques. One is based on common histograms, the other on kernel functions [15]. Given these two density estimation techniques, two different evaluation criteria will be reviewed. The first one was proposed by Battiti [6], the second one by Kwak and Chong-Ho [15]. Since the two different approaches lead to different feature selectors, we refer to them as feature selector 1 (Battiti) and feature selector 2 (Kwak and Chong-Ho).

12.3.2 Feature Selector 1

The calculation of the evaluation criterion value suggested by Battiti uses a histogram technique to estimate the joint mutual information between random variables. However, the calculation of the joint probability density function (jpdf) between multiple features, and the class suffers from the curse of dimensionality [16]. An N-dimensional joint histogram, with $N - 1$ features and one class vector with N_c classes, consists of $N_c n_{bins}(f_1) \times \cdots \times n_{bins}(f_{N-1})$ bins. The estimation of the jpdf between 4 features and a class vector with 3 different classes using 20 bins for each feature, requires already 3×20^4 bins. Storing the frequency counts for all bins may require a lot of memory, which already becomes a problem with a small number of features. Battiti introduced the following evaluation function (12.3) where the mutual information only between a single feature and the class and between two features need to be calculated. This avoids the problem described above.

$$f_{Eval1} = \operatorname{argmax}_{f_i \in F} \left\{ MI(c; f_i) - \beta \sum_{f_s \in S} MI(f_i; f_s) : f_i \in F \right\} \tag{12.3}$$

With each call of step 3 in the selection process, the feature that maximizes this expression will be selected.

---

[1] For a continuous random variable Y, $H(Y)$ is called differential entropy.

In (12.3), $f_i \in F$ is a candidate feature that has not yet been selected, f_s are already selected features, where $f_s \in S$. The parameter β regulates the relative importance of the MI between the candidate feature and the already-selected features with respect to MI$(c; f_i)$ [6]. If $\beta = 0$, statistical dependencies between features are not considered, which may lead to subsets with redundant features. If $\beta > 0$, redundancies between a candidate feature and the already-selected features are considered. Increasing β reduces those redundancies. However, for large values of β, the evaluation function only considers the relation between features and not the relation between the features and the class [7].

Battiti's algorithm provides a simple method to select a subset of features from a large feature space that is based on estimation techniques that are well understood. However, his approach has some major shortcomings, one of which is to find a suitable value for β. Although Battiti [6] reported that values of β between 0.5 and 1 are appropriate for many classification tasks, the influence of β with respect to the classification performance is not well understood. The second shortcoming deals with the fact that the evaluation function is only based on mutual information between two variables rather than between multiple variables. Therefore, the approach does not allow the calculation of the total information gain provided by a specific feature subset. This would, however, be very useful for the derivation of an appropriate stopping criterion.

12.3.3 Feature Selector 2

The estimation technique proposed by Nojun et al. [7] uses Gaussian kernels and provides a means to calculate the joint mutual information between the class and multiple features without encountering the problems inherent to the histogram technique. Furthermore, the proposed estimation technique allows the use of an evaluation criterion (12.4) that is based on the overall information gain of the investigated feature subset.

$$f_{\text{Eval2}} = \text{argmax}_{f_i \in F}\{\text{MI}(c; S; f_i) : f_i \in F\} \qquad (12.4)$$

MI$(c; S; f_i)$ is calculated by estimating

$$H(\hat{C}|F) = \sum_{j=1}^{n} \frac{1}{n} \sum_{c=1}^{N_c} p(\hat{c}|f_j) \log(p(\hat{c}|f_j)) \qquad (12.5)$$

with

$$p(\hat{c}|f) = \frac{\frac{1}{n_c} \sum_{i \in I_c} \phi(f - f_i, h_c)}{\sum_{k=1} \frac{1}{n_k} \sum_{i \in I_k} \phi(f - f_i, h_k)} \qquad (12.6)$$

where N_c is the number of classes, n the number of samples, h_c and h_k class-specific window width parameters, n_k the number of samples belonging to class c_k, I_k the set of indices belonging to class c_k, and $\phi(z, h)$ a Gaussian window function. Further details on the calculation can be found in [15]. Since the approach allows the calculation of the joint mutual information between multiple variables, the information gain added with each selected feature can be calculated.

12.3.4 Stopping Criterion

One stopping criterion that is meaningful in this context would be a minimum information gain that has to be achieved. Such a stopping criterion is, however, only feasible for the second feature selector. For reasons of comparison, we therefore use a predefined number of features to select as a common stopping criterion for both feature evaluators.

12.4 ADAPTATIONS

The two feature selectors reviewed in the previous section provide a means to identify relevant features. However, the selection process depends on a specific class distribution reflected in the joint probability density functions $p(c; f)$. Considering other class distributions may therefore lead to different feature subsets. In the following, we explain this problem and propose suitable adaptations to the existing feature selectors.

12.4.1 Problem Formulation

The calculation of mutual information requires the estimation of joint probability density functions. In classification tasks, the estimation is normally based on experimental data that consists of labeled feature vectors with fixed length. This data has an inherent class distribution. For illustrating the problem with the ranking of features that is based on a particular class distribution, we consider the following simple supervised learning task that is closely related to our case study in Section 12.6: Given a set of four sensors integrated into a person's outfit, we would like to recognize three different predefined activities (classes). Since we do not know the likelihood functions $p(f|c)$ that would be used to calculate the joint probability density functions $p(c; f)$, we carry out experiments and estimate the required density functions from the recorded data. A possible outcome of such an experiment is shown in Figure 12.1. Illustrated are four artificially generated features with the corresponding classes (top graph). Note that the data set is heavily biased toward the occurrence of c_1. The underlying class distribution was set to $p(c_1) = 0.7$, $p(c_2) = 0.15$, and $p(c_3) = 0.15$. Despite the imbalance in the class distribution, we assume that the data represents the underlying likelihood functions. The likelihood functions that were used to generate the artificial feature values for feature f_1, f_2, f_3, and f_4 are given in Table 12.1.

As can be seen, f_1 can be used to identify c_1 but cannot distinguish between c_2 and c_3. With feature f_2 we can recognize c_2 but not differentiate between c_1 and c_3. Feature f_3 discriminates c_3 from the two other classes, but c_1 and c_2 are indistinguishable using that feature. Feature f_4 can distinguish between all three classes but with some error.

If we calculate the mutual information between each feature and the class based on the artificial generated data, we receive the results given in Table 12.2.[2] Feature f_1 provides on average the maximum information gain since the class distribution is heavily biased toward c_1, and f_1 is very good in detecting that class. The change of information gain for the individual features when other class distributions are considered (see Table 12.3) is shown in Figure 12.2. The required probability densities $p(c; f)$ were calculated by estimating the

---

[2] The results are obtained using the histogram technique. The bin size of the histogram is set to $\log 2(n) + 1$, where n is the total number of data samples $n = 3000$.

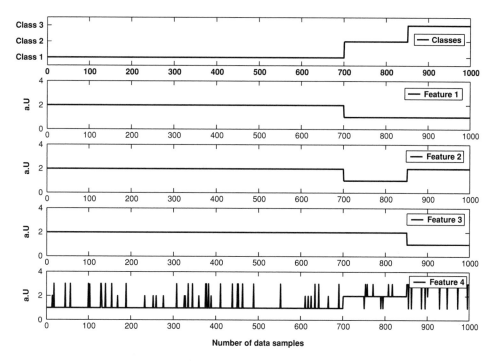

Figure 12.1 Artificial data.

Table 12.1 Likelihoods

$p(f_1\|c_i)$	$f_1 = 1$	$f_1 = 2$	$f_1 = 3$
Class 1	0	1	0
Class 2	1	0	0
Class 3	1	0	0

$p(f_2\|c_i)$	$f_2 = 1$	$f_2 = 2$	$f_2 = 3$
Class 1	0	1	0
Class 2	1	0	0
Class 3	0	1	0

$p(f_3\|c_i)$	$f_3 = 1$	$f_3 = 2$	$f_3 = 3$
Class 1	0	1	0
Class 2	0	1	0
Class 3	1	0	0

$p(f_4\|c_i)$	$f_4 = 1$	$f_4 = 2$	$f_4 = 3$
Class 1	$\frac{14}{15}$	$\frac{1}{30}$	$\frac{1}{30}$
Class 2	$\frac{1}{30}$	$\frac{14}{15}$	$\frac{1}{30}$
Class 3	$\frac{1}{30}$	$\frac{1}{30}$	$\frac{14}{15}$

Table 12.2 Mutual Information of Features Derived from Artificial Data[a]

$MI(f_1; c)$	$MI(f_2; c)$	$MI(f_3; c)$	$MI(f_4; c)$
0.881	0.610	0.610	0.8457

[a] $p(c_1) = 0.7$, $p(c_2) = 0.15$, $p(c_3) = 0.15$.

Table 12.3 Class Probabilities Used to Investigate the Mutual Information of the Features

Distribution	$p(c_1)$	$p(c_2)$	$p(c_3)$
1	0.7	0.15	0.15
2	0.1	0.8	0.1
3	0.1	0.1	0.8
4	0.33	0.33	0.33

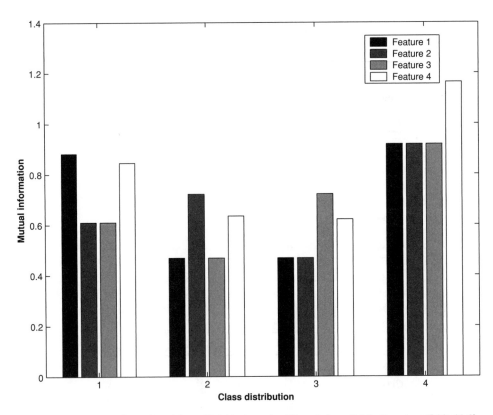

Figure 12.2 Mutual information of the artificial features for different class distributions (see Table 12.3).

likelihood functions[3] using the histogram-based technique and applying Bayes rule (12.7):

$$p(f; c_i) = p(f|c_i)p(c_i) \tag{12.7}$$

As can be seen in Figure 12.2, the relevance of individual features depends on the class distribution. If the probability of a specific class $p(c_k)$ with $k \in \{1, 2, 3\}$ is sufficiently high, the feature that can distinguish between c_k and the other two classes provides the highest information gain. Figure 12.3a illustrates the information gain of the four features for various probability distributions that are representatives of all possible class distributions. The distributions are taken from $p(c_i) \in (0 : 1/29 : 1)$ with $\sum_{c_i} p(c_i) = 1; i \in \{1, 2, 3\}$. Figure 12.3$b$ shows a top view of Figure 12.3a. For reasons of comparison, we also calculate the mutual information based on the kernels. We apply a stratified sampling method to the artificial data and obtain samples from each class randomly in such a way that the class distribution to be investigated is achieved. Figures 12.4a and 12.4b show the results obtained by this method.

The four regions that are given in Figure 12.3b allow the identification of the feature that provides maximum information gain for a specific class distribution. A comparison of the region borders between Figures 12.3b and 12.4b indicates that both methods provide similar results on the ranking of features. The fact that the information gain of individual features varies depending on the underlying class distribution will also influence the selection of multiple features. Table 12.4 shows different rankings of features[4] for three specific class distributions.

In dynamic sensor management, which can be formulated as a sequential Bayesian estimation problem [5], suitable sensors can be dynamically selected and deselected, for example, with respect to the current estimated class distribution. However, we are interested in identifying those sensors that are needed to perform robust recognition not only for specific class distributions.

12.4.2 Solution

To use the described feature selection techniques as a design tool, we propose to choose those sensors for the system that are not only relevant for some but for representatives of all possible class distributions. Since the sensors are derived from relevant features, we introduce the ranking scheme of individual features first, then describe the feature subset selection, and finally the scheme to identify the sensors.

Ranking of Individual Features   The ranking of features is based on the average relevance of each individual feature. By average relevance we mean the relevance of a feature with respect to all possible class distributions. Note, however, that if the classification is only based on present observations of features and if the classifier does not incorporate any knowledge about estimated class distributions, we would rather rank features with respect to the distribution where classes have equal probability.

The following scheme describes the ranking procedure:

Step 1: Set $F \leftarrow$ Initial set of all candidate features, $S \leftarrow$ Empty set.

Step 2: $\forall f_i \in F$, compute $\text{ORF}_\alpha(f_i) = n_{f_i}/N_{tot}$, where n_{f_i} is the number of class distributions where $(1 + \alpha)\text{MI}(f_i; c) > \text{MI}(f_j; c)$ for $\forall f_j \in F \setminus \{f_i\}$, and N_{tot} is the total

---

[3] The likelihoods provided in Table 12.1 were only used to generate the data.
[4] The ranking was obtained using feature selector 1 with $\beta = 0.5$.

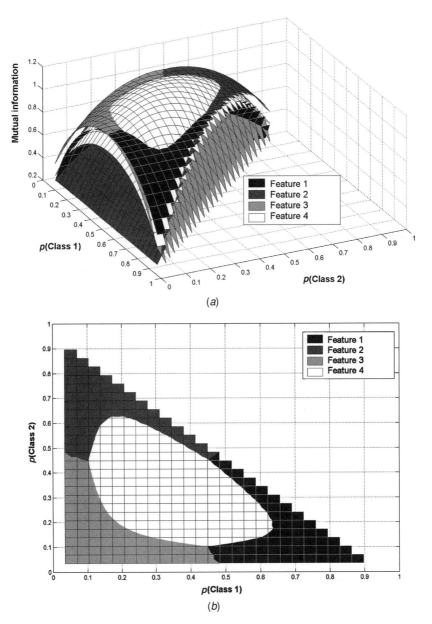

Figure 12.3 Mutual information of artificial features using likelihoods estimated with histogram technique: (*a*) 3D perspective and (*b*) top view.

number of considered class distributions. $\text{ORF}_\alpha(f_i)$ is denoted the overall relevance of feature f_i.

Step 3: Find the feature that maximizes $\text{ORF}_\alpha(f_i)$, set $F \leftarrow F \setminus \{f_i\}$, $S \leftarrow \{f_i\}$ and $\text{LF} \leftarrow \{f_i\}$, where LF is a list that stores the sequence of ranked features, and $\text{LF}(1)$ denotes the feature with the highest overall relevance. Repeat step 2 until all features are ranked.

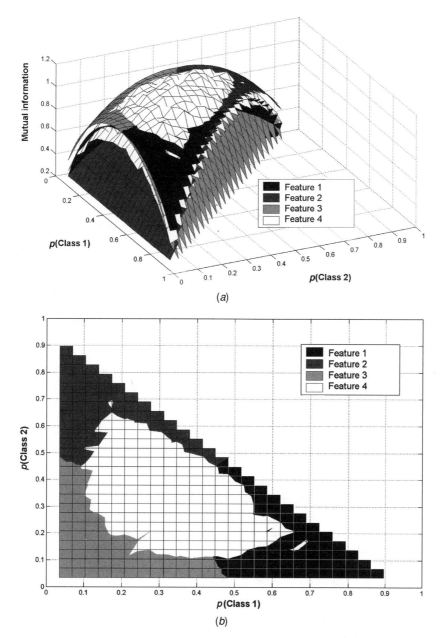

Figure 12.4 Mutual information using stratified sampling method and kernel density estimation technique: (*a*) 3D perspective and (*b*) top view.

Table 12.4 Selection Order for Specific Class Distributions

$p(c_1)$	$p(c_2)$	$p(c_3)$	List
0.2069	0.3103	0.4828	$[f_4, f_3, f_2, f_1]$
0.3103	0.0690	0.6207	$[f_3, f_1, f_4, f_2]$
0.1034	0.5517	0.3448	$[f_2, f_3, f_4, f_1]$

Table 12.5   Ranking of Individual Features Using Estimated Likelihoods[a]

Rank	Features ($\alpha = 0$)		Features ($\alpha = 0.1$)	
1	f_4	[39.15]	f_4	[67.72]
2	f_1	[34.66]	f_1	[50.26]
3	f_2	[50.79]	f_2	[60.85]
4	f_3	[100.00]	f_3	[100.00]

[a] The numbers in parenthesis correspond to the ORF values.

The parameter α is introduced to account for the following: Assuming we have to choose between two features in step 3 of the selection process, f_A and f_B, where $\text{ORF}_{\alpha=0}(f_A) > \text{ORF}_{\alpha=0}(f_B)$. We would like to assign more relevance to f_B than to f_A if the information gain of f_B is insignificantly smaller for those class distributions where $\text{MI}(f_A, c) \geq \text{MI}(f_B, c)$, while being significantly larger for distributions where $\text{MI}(f_B, c) \geq \text{MI}(f_A, c)$. This can be achieved with $\alpha > 0$.

Tables 12.5 and 12.6 show the ranking of the four artificial features for $\alpha = \{0; 0.1\}$ using the histogram-based and the kernel density estimation technique, respectively.[5] Both approaches provide the same ranking sequence with f_4 being most relevant. Note that in this cases the calculated sequences for both α values are identical but that the individual $\text{ORF}_\alpha(f_i)$ values for $\alpha = 0.1$ are much higher than those for $\alpha = 0$.

Feature Subset Selection   Based on the selection of features for the individual class distributions, we identify those features that are often used in feature combinations. We therefore introduce a score metric for each feature. This metric represents the number of occurrences of a feature in all feature selection lists LF_d, weighted with their position of the feature in the corresponding list. A more formal representation of that scheme is given below:

Step 1: Set $F \leftarrow$ Initial set of all candidate features, $S \leftarrow$ Empty set; $\forall f_i \in F$, $\text{ScF}(f_i) = 0$, where $\text{ScF}(f_i)$ is the score of feature f_i.

Step 2: For each class distribution d obtain the feature subset FSub_d from the feature selection process together with list LF_d that contains the order in which the features have been selected, where d denotes a particular class distribution and N_c the number of classes.

Step 3: Repeat this step for all class distributions d: $\forall f_j \in \text{FSub}_d$, calculate $\text{ScF}(f_j) = \text{ScF}(f_j) + 1/\text{order}(f_j, LF_d)$, where $\text{order}(f_j, LF_d)$ returns the position of feature f_j in the sorted feature list LF_d with respect to class distribution d.

Step 4: Rank all features f_i according to their scores $\text{ScF}(f_i)$ and output LF_{FC}, where LF_{FC} is the sorted list of features derived from feature combinations.

Applying this ranking scheme to the artificial data using feature selector 1 with $\beta = 0.5$ and $k = 4$ as stopping criterion ranks the four test features according to Table 12.7. Feature selector 1 has selected f_4 first in the feature selection process. That is not very surprising

---

[5] α was set empirically to 0.1.

Table 12.6   Ranking of Individual Features Using Stratified Sampling Method and Kernel Density Estimation Technique[a]

Rank	Features ($\alpha = 0$)		Features ($\alpha = 0.1$)	
1	f_4	[39.68]	f_4	[66.40]
2	f_1	[34.66]	f_1	[50.26]
3	f_2	[51.85]	f_2	[60.85]
4	f_3	[100.00]	f_3	[100.00]

[a] The numbers in parenthesis correspond to the ORF values.

because f_4 provides on the average more information to the problem than any other feature with respect to all class distributions. Feature selector 2, however, assigns f_4 the lowest rank. Here, we see one of the strengths of selector 2. For all considered class distributions where f_4 was not selected first, f_4 was selected last. The reason is that any of the following feature combinations, $\{f_1; f_2\}, \{f_1; f_3\}$, and $\{f_2; f_3\}$, already provides maximum achievable information gain.

Sensor Selection   The selection of relevant sensors to be incorporated into the design of a system works in almost the same way as the selection of relevant features described in the previous paragraph. For the artificial data we assumed that the features were derived from four different sensors. In this case, the feature selection and the sensor selection problem are identical. However, we would also like to consider cases where two or more features are extracted from the same sensor. Similar to the score ScF that provides a ranking criterion for features, we introduce a score ScS to order sensors according to their relevance. The calculation of those scores follows the procedure below:

Step 1: Initialization $\forall S_i$, $\mathrm{ScS}(S_i) = 0$, where $\mathrm{ScS}(S_i)$ is the score of sensor S_i.

Step 2: For each class distribution obtain the sorted list of features LF_d. Based on LF_d obtain LS_d, LS_d is a list of the sensors S_j required to extract the features in LF_d. (Example: $\mathrm{LF}_d = [f_4, f_2, f_1, f_3]$) Assuming that f_3 and f_4 are extracted from sensor 1, and f_1 and f_2 from sensor 2, LS_d becomes $\mathrm{LS}_d = [S_1, S_2, S_2, S_1]$.

Step 3: Repeat this step for all class distributions $d : \forall S_j | S_j \in \mathrm{LS}_d, \mathrm{ScS}(S_j) = \mathrm{ScS}(S_j) + \sum_m 1/m_{S_j}$, where m are the positions in LS_d where S_j occurs.

Step 4: Rank all sensors S_i according to their scores ScS_{S_i} and output $\mathrm{LS}_{\mathrm{FC}}$, where $\mathrm{LS}_{\mathrm{FC}}$ is the sorted list of sensors.

The suggested adaptations provide a means to identify relevant sensors for the recognition task. With respect to the implementation of a system, the authors would like to address two

Table 12.7   Feature Subsets Derived Using Feature Selector 1 and Feature Selector 2[a]

Rank	1	2	3	4
Selector 1	f_4 (240.25)	f_3 (238.00)	f_1 (235.35)	f_2 (231.50)
Selector 2	f_1 (275.75)	f_2 (244.75)	f_3 (217.50)	f_4 (207.00)

[a] The numbers in parenthesis correspond to the ORF values.

design considerations that might affect the decisions of sensors to be incorporated into the design.

12.5 DESIGN CONSIDERATIONS

This section briefly addresses two specific considerations that may affect the design decisions.

12.5.1 Within-Class Problem

The first consideration covers the *within-class problem*. In certain context applications, it might suffice to distinguish between a specific class and all other classes (c_i or not c_i), rather than recognizing one of the possible classes. To illustrate the within-class problem, we refer to the example with the artificial data that has been used in the previous sections. If we want to distinguish between c_1 and all other classes, f_1 is the feature that will provide maximum mutual information for any class distribution, followed by f_4. If c_2 is to be distinguished between the other classes, f_2 has maximum information gain. Concerning the within-class problem for c_3, we should use f_3. Compared to the ranking of features from above (see Tables 12.5 and 12.6) that assigned maximum relevance to f_4, other features provide more information gain with respect to the within-class problem. In applications where the within-class problem is of certain relevance it might be useful to expand the feature analysis accordingly. Figure 12.5 shows the mutual information for various class

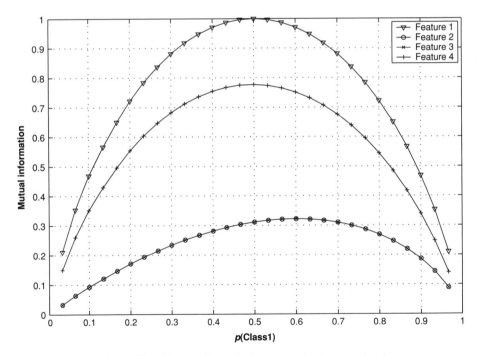

Figure 12.5 Mutual information of features of within-class problem for c_1.

Table 12.8 Ranking of Individual Features for Different Within-Class Problems (c_i or Not c_i) Using the Histogram Technique[a]

Rank	c_1 or Not c_1	c_2 or Not c_2	c_3 or Not c_3
1	f_1 [100]	f_2 [100]	f_3 [100]
2	f_4 [100]	f_4 [100]	f_4 [100]
3	f_2 [100]	f_1 [100]	f_1 [100]
4	f_3 [100]	f_3 [100]	f_2 [100]

[a] The number in parenthesis represent the scores of the features.

distributions $p(c_1)$ for the within-class problem of c_1. The results of the individual rankings are given in Table 12.8. Note that α will not affect the order of features since none of the lines in Figure 12.5 intersect with any other line.

12.5.2 System Reliability

In the special case of wearable systems, where the required resources should be kept small, it might, however, be necessary to incorporate redundancy into a system for reliability issues. This depends of course on the targeted application. It would therefore be useful to provide a design tool that allows the identification of the vulnerable parts of a system while providing information on how to add redundancy efficiently. The very same technique that has been presented for the identification of relevant features could be used for this task since it will not only identify relevant but also redundant sensors.

12.6 CASE STUDY

With the given feature selectors, we evaluate a context-sensing, wearable platform that targets the recognition of three simple human activities (level walking and ascending and descending stairs). This specific problem has been investigated by many researchers [17–23], and there is still ongoing research. Despite the apparently simple recognition task, there are still many unsolved issues, concerning for example, the search for optimal sensor placement, suitable sensing modalities, and relevant and efficient features. Therefore, this specific classification task has become a quasi-standard classification problem within the context community. Applying our proposed method to that specific problem allows to incorporate and analyze existing approaches with their introduced features and also contributes new findings to that problem.

12.6.1 Experimental Platform

To conduct an analysis of the described problem, we have implemented an experimental platform that allows easy distribution of sensors over the human body and simple integration of different sensing modalities into the system. A specific hardware interface has been designed to interface the sensors and to handle the communication between the different nodes of the sensor network. Figure 12.6 depicts this interface module. It incorporates an MSP430F149 low-power 16-bit mixed signal microprocessor from Texas Instruments.

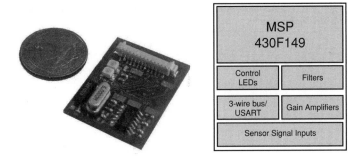

Figure 12.6   Interface module: (*left*) printed circuit board and (*right*) block diagram.

Our platform is based on a master–slave architecture, where the slaves represent the sensor nodes and the master the central processing unit that controls the overall communication and collects the data of the sensors connected to the network. The master also provides a global system clock used to synchronize the sensor nodes.

For our experiments, we evaluated different sensing modalities and features that have been introduced in recent publications. Table 12.9 gives an overview of the sensors that have

Table 12.9   Overview of Sensing Modalities, Sensor Locations, and Features[a]

Sensing Modalities	Sensor Location	Features
2 two-axis accelerometers	First accelerometer: above right knee, with axes aligned to antero-posterior—Acc knee forward—Sensor 1 and vertical body axis—Acc knee down—Sensor 2; Second accelerometer: Back of body, mounted on belt (same axes orientation as first accelerometer: Acc cent forward—Sensor 3, Acc cent down—Sensor 4,	Mean, variance, rms, number of zero crossings, interquartile range (IQR), low-frequency power, high-frequency power, Frequency Entropy, Number of zero-crossings of differentiated signals from vertical axes of belt and knee accelerometers.
1 air pressure sensor	Inside right, inner pocket of jacket—Air pressure—Sensor 5	Slope*
1 one-axis gyroscope	Above right knee, sensitive axis aligned with lateral axis—gyro knee lateral—Sensor 6	Mean, variance, rms, number of zero crossings, IQR, low-frequency power, high-frequency power, frequency entropy.
2 force-sensitive resistors (FSR)	1st FSR: mounted under heel of right shoe sole; 2nd FSR: mounted under ball of right shoe sole (in the evaluation process the two FSRs are combined to a single sensor (*Foot Switch—Sensor 7*)	Lag,* gait cycle time*

[a] All features listed in the table are calculated using a sliding window approach, except those that are marked with the symbol *, which are derived using other processing schemes.

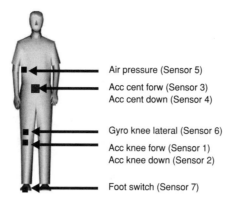

Figure 12.7 Placement of sensors on the body.

been integrated into the platform, describes their placement, and summarizes the features used for our evaluation. In total 43 features were extracted from 7 sensors.[6]

12.6.2 Data Acquisition

To acquire experimental data, test subjects were equipped with the sensor platform (see Fig. 12.7). They were asked to walk repeatedly a predefined path in a building without further instructions, such as speed of walking. The path included level walking and ascending and descending stairs. During the experiments, the sensor data was manually labeled and recorded on a remote laptop using a wireless-based transmission. The analog signals of all sensors were sampled synchronously at 100 Hz and digitally converted with 12-bit resolution.

12.6.3 Feature Extraction

In a postprocessing step, features are extracted from the recorded data. Table 12.9 lists the specific features derived from the raw signals of the individual sensors. These features have been recently proposed for the described recognition task (see [18–20, 23, 25–29]). In the following, the less common features are briefly described: Low-frequency power describes the signal power in the frequency band between 0.1 and 5 Hz. High-frequency power represents the signal power in the high-frequency band 5 to 50 Hz. The frequency entropy is a measure of the distribution of frequency components in the frequency band 7 to 50 Hz. Slope describes the gradient of a signal. Lag represents the time difference between initial heel strike and corresponding ball strike [30] and correlates with the foot orientation angle at initial ground contact, which can be used to distinguish between the different activities [31]. The gait cycle time represents the duration of a gait cycle.

All features—except those features marked with an asterisk (∗) in Table 12.9—are calculated using a sliding-window-based method with fixed window width and overlap parameters.[7] The extraction of the other features rely on different signal processing schemes. A discussion on those schemes is, however, beyond the scope of this chapter.

[6] The two force-sensitive resistors account for one sensor.

[7] In this chapter, those parameters were set to 200 and 199 samples, respectively.

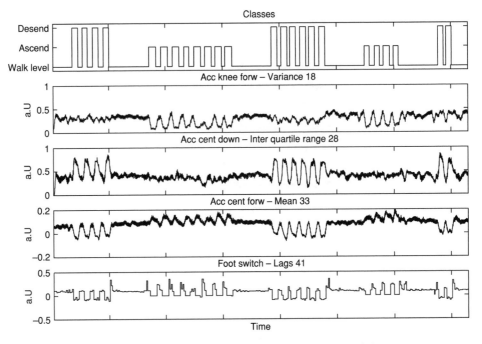

Figure 12.8   Selected features extracted from experimental data.

Figure 12.8 shows four different features that have been extracted from recorded data of one test subject. The upper graph of Figure 12.8 shows the corresponding activities performed over time.

In the following, the features that are extracted from the experimental data are analyzed with respect to their relevance for the targeted recognition task. The final goal, however, is to identify the sensors that have a significant influence on the system's classification performance.

12.6.4 Evaluation Procedure

The evaluation of the system is divided into three steps. In a first step, we investigate the importance of individual features and identify the most relevant features according to the ranking scheme presented in Section 12.4. Then, we evaluate the best ranked features derived in the first step with respect to their classification performance using a standard maximum-likelihood (ML) detector. Finally, we analyze and identify important feature combinations using both feature selectors. We compare the results and derive the relevant sensors.

12.7 RESULTS

The results of the feature evaluation are presented according to the evaluation steps provided in Section 12.6.4.

Table 12.10 Ranking of Features Using Histogram-Based Technique[a]

Rank	$\alpha = 0$		$\alpha = 0.1$	
1	f_{41}	[89.68]	f_{41}	[99.47]
2	f_{33}	[96.83]	f_{33}	[100.00]
3	f_{38}	[90.48]	f_{38}	[96.03]
4	f_2	[78.84]	f_2	[89.68]
5	f_3	[70.63]	f_3	[84.66]
6	f_{21}	[50.26]	f_{30}	[63.23]
7	f_{30}	[61.64]	f_{21}	[78.31]
8	f_{18}	[68.52]	f_{18}	[78.57]
9	f_{28}	[100.00]	f_{28}	[100.00]

[a] Numbers in parenthesis represent ORF values.

12.7.1 Ranking of Individual Features

The ranking of individual features was carried out with both feature selectors and for two different α values. The results of the rankings are summarized in Tables 12.10 and 12.11 for the histogram-based and the kernel density estimation technique, respectively. For the sake of clarity, only the 9 best ranked features out of 43 are listed. As can be seen, both approaches provide the same ranking of features. The ranking is not influenced if α is set to 0.1 except for rank 6 and 7 (see Table 12.10). If we look at the ORF-value of the first selected feature (f_{41}), we see that it provides maximum information gain for approximately 90% of all class distributions and therefore clearly dominates all other features. The dominance of f_{41} is illustrated in Figures 12.9a and 12.9b. Both figures show the mutual information of f_{41}, f_{33}, f_{18}, and f_{28}. The ORF value of feature f_{33} indicates a similar dominance. Note that in the selection process, the ORF values of unselected features are always recalculated without considering already selected features. Therefore, it is possible that an ORF value of one feature might be larger compared to the ORF value of another feature although it has a lower rank.

Classification Performance of Individual Features To evaluate the usefulness of the features with respect to the classification task we use a maximum-likelihood (ML)

Table 12.11 Ranking of Features Using Kernel Density Estimation Technique[a]

Rank	$\alpha = 0$		$\alpha = 0.1$	
1	f_{41}	[88.10]	f_{41}	[97.62]
2	f_{33}	[91.01]	f_{33}	[99.74]
3	f_{38}	[88.62]	f_{38}	[95.50]
4	f_2	[80.16]	f_2	[88.10]
5	f_3	[70.63]	f_3	[83.07]
6	f_{21}	[53.70]	f_{21}	[65.87]
7	f_{30}	[49.47]	f_{30}	[82.80]
8	f_{18}	[68.25]	f_{18}	[78.57]
9	f_{28}	[94.44]	f_{28}	[99.74]

[a] Numbers in parenthesis represent ORF values.

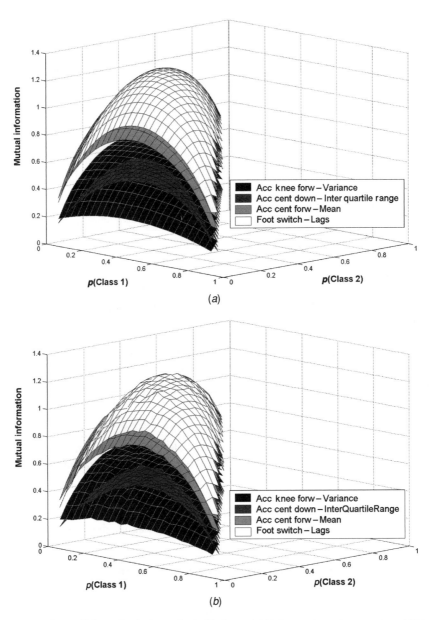

Figure 12.9   Mutual information for four selected features: (*a*) histogram-based approach and (*b*) kernel density estimation-based approach.

detector and calculate the individual recognition rates. We used the recorded data sets of a single test subject for this evaluation, divided the recordings into 10 subsets and carried out a 10-fold cross validation. Table 12.12 summarizes the recognition results sorted from best to worst performance.

From Table 12.12 we see that f_{41}, f_{33}, and f_{38} are very useful for the recognition task; especially f_{41} with which recognition rates of around 95% are achieved. Thus, sensor 7,

Table 12.12   Recognition Performance of ML Detector with Different Inputs[a]

Feature	Recognition Rate (%)	Rank
f_{41}	94.69	1
f_{33}	86.35	2
f_{38}	83.97	3
f_2	80.82	4
f_3	79.23	6
f_{30}	79.07	5
f_{18}	71.31	7
f_{28}	66.81	8
f_{21}	65.70	9

[a] Results obtained using 10-fold cross validation.

from which f_{41} is derived, should be considered for our design. As a second choice, we would take sensor 3 from which f_{33} and f_{38} are extracted. If we compare the performance of the features with the ranking provided in Tables 12.10 and 12.11, we see a strong correlation considering the first 4 features, which could be expected according to Fano's inequality.

12.7.2   Feature Combinations

The primary goal of this step is to identify relevant feature combinations that will further enhance the classification performance. The selection process considers 28 class probability distributions covering the full distribution space [$p(c_i) \in 0 : 1/10 : 1$], where $\sum_{c_i} p(c_i) = 1; i \in [1, 2, 3]$). For the selection process we considered all 43 features (see Table 12.17).

The rankings based on feature selector 1 and feature selector 2 are shown in Tables 12.13 and 12.14, respectively. Not surprisingly, the feature selectors chose different features due

Table 12.13   Feature Ranking with Feature Selector 1[a]

Rank	Feature	Name	ScF
1	f_{41}	Foot switch—Lags	28.0
2	f_{10}	Acc knee down—Variance	12.6
3	f_{38}	Acc cent forw—lowFreqPower	11.2
4	f_{15}	Acc knee down—freqEntropy	6.8
5	f_{14}	Acc knee down—highFreqPower	5.6
6	f_4	Gyro knee lateral—NumZeroCrossings	4.8
7	f_7	Gyro knee lateral—highFreqPower	3.8
8	f_{40}	Acc cent forw—freqEntropy	3.6
9	f_{21}	Acc knee forw—InterQuartileRange	3.2
10	f_6	Gyro knee lateral—lowFreqPower	1.6
11	f_{37}	Acc cent forw—InterQuartileRange	1.4
12	f_8	Gyro knee lateral—freqEntropy	0.8
13	f_{31}	Acc cent down—freqEntropy	0.2
14	f_{11}	Acc knee down—RmsSig	0.2
15	f_1	Gyro knee lateral—Mean	0.2

[a] $\beta = 0.75$.

Table 12.14   Feature Ranking with Feature Selector 2

Rank	Feature	Name	ScF
1	f_{41}	Foot switch—Lags	27.8
2	f_{42}	Foot switch—Cycletime	11.8
3	f_{21}	Acc knee forw—InterQuartileRange	10.4
4	f_2	Gyro knee lateral—Variance	9.6
5	f_{12}	Acc knee down—InterQuartileRange	7.8
6	f_{25}	Acc cent down—Mean	4.4
7	f_{43}	Air pressure—Slope	3.2
8	f_{19}	Acc knee forw—RmsSig	2.4
9	f_3	Gyro knee lateral—RmsSig	1.6
10	f_{22}	Acc knee forw—lowFreqPower	1.6
11	f_1	Gyro knee lateral—Mean	1.4
12	f_{38}	Acc cent forw—lowFreqPower	1.0
13	f_{32}	Diff of acc cent down—NumZeroCrossings	0.6
14	f_{18}	Acc knee forw—Variance	0.2
15	f_9	Acc knee down—Mean	0.2

to the different evaluation criteria. The influence on the relevance of individual sensors can be seen in Table 12.15. To compare the two feature/sensor combinations, we calculated the average mutual information for a single feature, and for combinations with 2, 3, 4, and also with all 5 features using the feature combinations obtained for the individual class distributions. Figure 12.10 illustrates the average mutual information gain when using 1, 2, 3, 4, and 5 features. The gain values given in Figure 12.10 were obtained by first normalizing the individual mutual information values with the corresponding class entropy $H(C)$ of the underlying class distribution. To compare the results from both feature selectors, we calculated the average information gain of the features provided by feature selector 1 using the kernel density estimation method that is part of feature selector 2. As can be seen, the features from selector 1 do not provide as much information compared to the features of selector 2. Furthermore, concerning the features selected by selector 1 using more than two features does not provide more information gain.

Classification Performance of Selected Feature Combinations   Figure 12.11 shows the recognition performance of the features that have been selected by selector 1

Table 12.15   Scores of Individual Sensors[a]

Sensor Number	Score: Selector 1	Score: Selector 2
Sensor 1 (Acc knee forw)	3.2	**14.6**
Sensor 2 (Acc knee down)	**25.2**	8.0
Sensor 3 (Acc cent forw)	16.2	1.0
Sensor 4 (Acc cent down)	0.2	5.0
Sensor 5 (Air pressure)	0.0	3.2
Sensor 6 (Gyro lateral)	11.2	12.6
Sensor 7 (Foot switch)	**28.0**	**39.6**

[a] The two highest scores are marked in bold numbers.

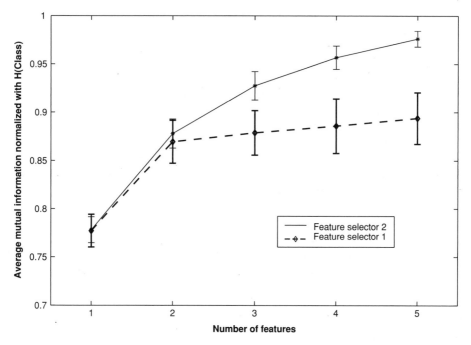

Figure 12.10 Average information gain with respect to the number of features.

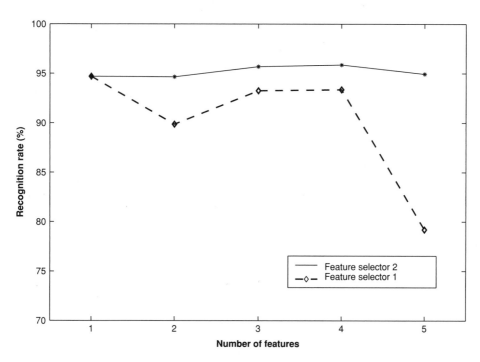

Figure 12.11 Recognition performance of selected features.

and selector 2. The results were obtained using a Naive Bayes classifier and a 10-fold cross validation. The recognition results indicate that the features from selector 2 perform better than those chosen by selector 1. However, we also see that the recognition performance is not enhanced by using multiple features.

Although the results indicate that selector 2 is superior to selector 1, a detailed analysis for this specific classification task is limited due to the fact that a single feature already provides very high information gain and allows for recognition rates of approximately 95%. Adding further features may therefore only lead to a small increase in recognition performance. Hence, the analysis of performance improvements using multiple features is not very significant. The main goal of providing design decisions to a system engineer has been achieved. From a set of 7 sensors, we identified a single sensor to facilitate robust recognition of the user's modes of locomotion.

For reasons of comparison, we carried out our evaluation procedure with a reduced feature set that contained all features from Table 12.17 except the features f_{41} and f_{42} obtained from the foot switch.

The results of the evaluation with feature selector 2 are shown in Figures 12.12 and 12.13. From Figure 12.12 we see that the selected features from the reduced feature set provide less information gain than the selected features from our initial evaluation where features f_{41} and f_{42} were included in the feature set. The difference in information gain decreases, however, with the number of features used.

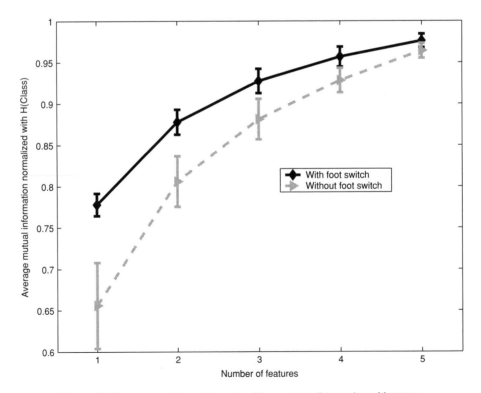

Figure 12.12   Average information gain with respect to the number of features.

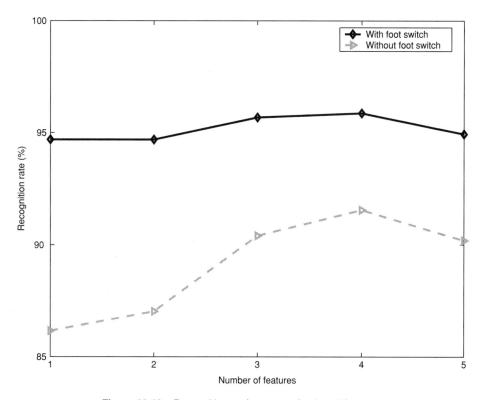

Figure 12.13   Recognition performance of selected features.

Figure 12.13, shows the recognition performance of the features that have been selected by feature selector 2 from the reduced feature set (see Table 12.16 for the feature names). When only f_{33} is considered for the recognition, the performance is around 86% compared to 94.7% when the lag feature f_{41} from the foot switch is used instead (see also Table 12.12). Although, in this particular case, the use of multiple features will not result in recognition rates higher than 91%, the increase in performance from 86 to 91% recognition rate (when the 4 features f_{33}, f_{12}, f_{22}, and f_{25} are used) is much more significant compared to our first investigation where the maximum recognition performance was around 95.9% (with four features) and already 94.7% with a single feature. If for any reason, the foot switch could not be used or if redundancy is required, we would integrate the acceleration sensors into our system from which the features f_{33}, f_{12}, f_{22}, and f_{25} are derived.

Table 12.16   Selected Features from Reduced Feature Set

Rank	Feature	Name
1	f_{33}	Acc cent forw—Mean
2	f_{12}	Acc knee down—InterQuartileRange
3	f_{22}	Acc knee forw—lowFreqPower
4	f_{25}	Acc cent down—Mean
5	f_{21}	Acc knee forw—InterQuartileRange

12.8   CONCLUSION

We investigated the use of two feature selection methods based on mutual information as a design tool for context-sensing wearable systems. We proposed certain adaptations to those methods to derive relevant sensing modalities for a given classification task and applied them to an existing context-sensing platform. We identified a single sensor that provides robust recognition, thus reducing the amount of necessary resources to a bare minimum. However, for an in-depth validation of our proposed approach, more complex classification tasks are required where only a combination of multiple features will lead to a reasonable recognition performance.

Future work will focus on design issues considering the trade-offs between classification performance and costs associated with the use of selected sensors. Cost functions may reflect power consumption, computational load, hardware complexity, and a measure for the unobtrusiveness of the system, which is of special interests for wearable, body-worn systems.

APPENDIX

The appendix comprises Table 12.17.

Table 12.17   List of Evaluated Features

No.	Name	No.	Name
f_1	Gyro knee lateral—Mean	f_{23}	Acc knee forw—highFreqPower
f_2	Gyro knee lateral—Variance	f_{24}	Acc knee forw—freqEntropy
f_3	Gyro knee lateral—RmsSig	f_{25}	Acc cent down—Mean
f_4	Gyro knee lateral—NZC[a]	f_{26}	Acc cent down—Variance
f_5	Gyro knee lateral—InterQuart.Range	f_{27}	Acc cent down—RmsSig
f_6	Gyro knee lateral—lowFreqPower	f_{28}	Acc cent down—InterQuartileRange
f_7	Gyro knee lateral—highFreqPower	f_{29}	Acc cent down—lowFreqPower
f_8	Gyro knee lateral—freqEntropy	f_{30}	Acc cent down—highFreqPower
f_9	Acc knee down—Mean	f_{31}	Acc cent down—freqEntropy
f_{10}	Acc knee down—Variance	f_{32}	Diff(acc cent down)—NZC[a]
f_{11}	Acc knee down—RmsSig	f_{33}	Acc cent forw—Mean
f_{12}	Acc knee down—InterQuart.Range	f_{34}	Acc cent forw—Variance
f_{13}	Acc knee down—lowFreqPower	f_{35}	Acc cent forw—RmsSig
f_{14}	Acc knee down—highFreqPower	f_{36}	Acc cent forw—NZC[a]
f_{15}	Acc knee down—freqEntropy	f_{37}	Acc cent forw—InterQuartileRange
f_{16}	Diff(acc knee down)—NZC[a]	f_{38}	Acc cent forw—lowFreqPower
f_{17}	Acc knee forw—Mean	f_{39}	Acc cent forw—highFreqPower
f_{18}	Acc knee forw—Variance	f_{40}	Acc cent forw—freqEntropy
f_{19}	Acc knee forw—RmsSig	f_{41}	Foot switch—Lags
f_{20}	Acc knee forw—NZC[a]	f_{42}	Foot switch—Cycletime
f_{21}	Acc knee forw—InterQuartileRange	f_{43}	Air pressure—Slope
f_{22}	Acc knee forw—lowFreqPower		

[a] NZC: Number of zero crossings.

REFERENCES

1. D. Siewiorek, A. Smailagic, J. Furukawa, A. Krause, N. Moraveji, K. Reiger, J. Shaffer, and F. L. Wong, "Sensay: A context-aware mobile phone," in *Proceedings of the Sixth International Symposium on Wearable Computers IS WC 2002*, White Plains, NY, 2003.

2. T. Starner, B. Schiele, and A. Pentland, "Visual contextual awareness in wearable computing," in *Proceedings of the Second International Symposium on Wearable Computers IS WC 1998*, Pittsburgh, PA 1998.

3. M. Bauer, T. Heiber, G. Kortuem, and Z. Segall, "A collaborative wearable system with remote sensing," Proc. 2nd International Symposium on Wearable Computers, Pittsburg, PA 1998.

4. M. Dash and H. Liu, "Feature selection for classification," in *Intelligent Data Analysis*, vol. 1, no. 3, Aug. 1997.

5. Z. Feng, S. Jaewon, and J. Reich, "Information-driven dynamic sensor collaboration," *IEEE Signal Processing Magazine*, vol. 19, no. 2, pp. 61–72, Mar. 2002.

6. R. Battiti, "Using mutual information for selecting features in supervised neural net learning," *IEEE Transactions on Neural Networks*, vol. 5, no. 4, pp. 537–550, July 1994.

7. N. Kwak and C. Chong-Ho, "Input feature selection for classification problems," *IEEE Transactions on Neural Networks*, vol. 13, no. 1, pp. 143–159, Jan. 2002.

8. G. D. Tourassi, E. D. Frederick, M. K. Markey, and C. E. Floyd, "Application of the mutual information criterion for feature selection in computer-aided diagnosis," *Medical Physics*, vol. 28, no. 12, pp. 2394–2402, Dec. 2001.

9. A. Al-Ani and M. Deriche, "Feature selection using a mutual information based measure," in R. Kasturi, D. Laurendeau, and C. Suen (Eds.), *Proceedings of the Sixteenth International Conference on Pattern Recognition*, Quebec, vol. 4, 2002.

10. Y. Howard-Hua and J. Moody, "Feature selection based on joint mutual information," in *Proceedings of the International ICSC Congress on Computational Intelligence Methods and Applications*, Rochester, NY, June 1999.

11. P. Greenway and R. Deaves, "An information filter for decentralized sensor management," in *Proceedings of the Sixth International Symposium on Wearable Computers IS WC 2002*, Seattle, vol. 2232, 2002, pp. 70–78.

12. K. Kastella, "Discrimination gain for sensor management in multitarget detection and tracking," in *Symposium on Control, Optimization and Supervision*, Lille, France vol. 1, 1996, pp. 167–172.

13. J. Manyika and H. Durrant-Whyte, *Data Fusion and Sensor Management: A Decentralized Information-Theoretic Approach*. Englewood Cliffs, NJ: Prentice-Hall, 1994.

14. R. Fano, "Class notes for transmission of information, course 6.574," MIT, Cambridge, MA, 1952.

15. N. Kwak and C. Chong-Ho, "Input feature selection by mutual information based on parzen window," *IEEE Transactions on Pattern Analysis and Machine Intelligence*, vol. 24, no. 12, pp. 1667–1671, Dec. 2002.

16. D. Scott, *Multivariate Density Estimation: Theory, Practice and Visualization*, New York: John Wiley & Sons, 1992.

17. P. H. Veltink, H. B. J. Bussmann, W. de Vries, W. L. J. Martens, and R. C. Van-Lummel, "Detection of static and dynamic activities using uniaxial accelerometers," *IEEE Transactions on Rehabilitation Engineering*, vol. 4, no. 4, pp. 375–385, Dec. 1996.

18. K. Sagawa, T. Ishihara, A. Ina, and H. Inooka, "Classification of human moving patterns using air pressure and acceleration," in *IECON '98*, Aachen, Germany, vol. 2, 1998.

19. K. Van-Laerhoven and O. Cakmakci, "What shall we teach our pants?" in *Digest of Papers, Fourth International Symposium on Wearable Computers*, 2000, pp. 77–83.

20. C. Randell and H. Muller, "Context awareness by analysing accelerometer data," in *Digest of Papers*, 2000.

21. M. Sekine, T. Tamura, T. Fujimoto, and Y. Fukui, "Classification of walking pattern using acceleration waveform in elderly people," *Engineering in Medicine and Biology Society*, vol. 2, pp. 1356–1359, July 2000.

22. L. Seon-Woo and K. Mase, "Recognition of walking behaviors for pedestrian navigation," in *Proceedings of the 2001 IEEE International Conference on Control Applications (CCA'01)*, Tampa, FL, 2001, pp. 1152–1155.

23. J. Mantyjarvi, J. Himberg, and T. Seppanen, "Recognizing human motion with multiple acceleration sensors," in *2001 IEEE International Conference on Systems, Man and Cybernetics*, Tucson, AZ, vol. 3494, 2001, pp. 747–752.

24. L. Bao, "Physical activity recognition from acceleration data under semi-naturalistic conditions," Master's thesis, Massachusetts Institute of Technology, Cambridge, MA, 2003.

25. A. Schmidt, K. A. Aidoo, A. Takaluoma, U. Tuomela, K. V. Laerhoven, and W. V. de Velde, *Advanced Interaction in Context*, Lecture Notes in Computer Science, vol. 1707, Springer-Verlag, 1999, pp. 89–93.

26. N. Kern, B. Schiele, and A. Schmidt, "Multi-sensor activity context detection for wearable computing," in *European Symposium on Ambient Intelligence*, Eindhoven, The Netherlands, Nov. 2003.

27. I. Pappas, I.-P. R. Popovic, M. T. Keller, V. Dietz, and M. Morari, "A reliable gait phase detection system," *IEEE Transactions on Neural Systems and Rehabilitation Engineering*, vol. 9 no. 2, pp. 113–125, June 2001.

28. N. Kern, B. Schiele, H. Junker, P. Lukowicz, and G. Tröster, "Wearable sensing to annotate meeting recordings," in *Proceedings of the Sixth International Symposium on Wearable Computers IS WC 2002*, Seattle, Oct. 2002.

29. K. Aminian, P. Robert, E. Jequier, and Y. Schutz, "Estimation of speed and incline of walking using neural network," *IEEE Transactions on Instrumentation and Measurement*, vol. 44, no. 3, pp. 743–746, June 1995.

30. H. Junker, P. Lukowicz, and G. Troester, "Locomotion analysis using a simple feature derived from force sensitive resistors," in *Proceedings of the Second IASTED International Conference on Biomedical Engineering, 2004*, Innsbruck, Feb. 2004.

31. R. Riener, M. Rabuffetti, and C. Frigo, "Stair ascent and descent at different inclinations," *Gait and Posture*, vol. 15, no. 1, pp. 32–44, 2002.

13

MULTIPLE BIT STREAM IMAGE TRANSMISSION OVER WIRELESS SENSOR NETWORKS

Min Wu and Chang Wen Chen

In this chapter, we proposed a novel scheme for error-robust and energy-efficient image transmission over wireless sensor networks. The innovations of the proposed scheme are twofold: multiple bit stream image encoding to achieve error-robust transmission and small-fragment burst transmission to achieve energy-efficient transmission. In the case of image encoding, a multiple bit stream image encoding is developed based on the decomposition of images in the wavelet domain. We first group wavelet coefficients into multiple trees according to parent–child relationship and then code them separately by the set partitioning in hierarchical trees (SPIHT) algorithm to form multiple bit streams. We show that such decomposition is able to reduce error propagation in transmission. To further enhance the error robustness, unequal error protection strategy with a rate-compatible punctured convolutional and a cyclic redundancy check (RCPC/CRC) is implemented. The coding rate is designed based on the bit error sensitivity of different bit planes. In the stage of image transmission, we divide the multiple bit streams into small fragments and transmit them in burst. We demonstrate that such transmission strategy matches well with the media access control layer protocol as well as the link layer protocol currently employed in wireless sensor networks. We achieve energy-efficient transmission by saving energy consumed on control overhead and device switching from sleep to active. Experimental results show that the proposed scheme not only has high energy efficiency in transmission but also graceful degradation in peak signal-to-noise ratio (PSNR) performance over high bit error rate (BER) channels in terms of image reconstruction quality at the base station.

13.1 INTRODUCTION

Research and development in wireless sensor networks are becoming increasingly widespread because of their significance in homeland security and defense applications

Sensor Network Operations, Edited by Phoha, LaPorta, and Griffin
Copyright © 2006 The Institute of Electrical and Electronics Engineers, Inc.

[1, 2]. In general, each sensor node contains sensing, processing, and communication elements and is designed to monitor events specified by the deployer of the network. Imaging sensors are able to provide intuitive visual information for quick recognition, real-time monitoring, and live surveillance. However, image sensors usually generate vast amounts of data for transmission from sensor node to the central station [3]. For those battery-powered sensors, energy-efficient transmission of the images collected in the sensor network presents the most challenging problem.

A number of research efforts are currently under way to address the issues on different layers of energy-efficient transmission. In network layer, several types of energy-efficient protocols, including data-centric and hierarchical protocols, have been proposed. In data-centric routings, such as directed diffusion [4], the base station first sends queries to certain areas and waits for the data to return from the sensor in those areas. Routes are dynamically formed when data of interest are transported. In hierarchical routings, such as LEACH [5], the clusters are formed to perform data aggregation and fusion. The data collected in each cluster are reduced before sending to a central base station. In medium access control (MAC) layer, Ye et al. proposed a energy-efficient MAC protocol [6]. The RTS (request to send)/CTS (clear to send) mechanism is adopted to eliminate collisions of medium access. Periodic listen and sleep is adopted to avoid overhearing while a sensor is idling. In the link layer, many researchers address the issues of error control and frame length adaptation to the circuit features and channel conditions. Lettieri et al. [7] proposed an adaptive radio model designed for wireless multimedia communications over ATM. Frame length and forward error correction parameters can be adaptive to channel conditions to obtain low energy consumption of the radio and to improve overall data throughput.

There are also numerous research efforts on robust image transmission over wireless channel [8–10]. Sherwood and Zeger [10] proposed a variable-length joint source and channel coding scheme for robust wireless image transmission. Ruf and Modestino [8] proposed a fixed-length joint source and channel coding scheme for image transmission over additive white Gaussian noise channels. One important characteristic of fixed-length coding scheme is the development of the explicit rate distortion function. However, comparing with variable-length coding, fixed-length image coding generally has lower coding efficiency and therefore requires more bit budget to code an image of the same quality.

In this chapter, we consider a sensor network that monitors an environment with cameras. Each node in the network takes pictures of its surrounding area, compresses each image using the SPIHT algorithm, and transmits the data to the base station. SPIHT, a refinement of embedded zero-tree coding approach (EZW) [11], is among the best wavelet coders. It generates an embedded bit stream that has nice properties for bit rate adaptation in time-varying network conditions. However, SPIHT bit stream is very vulnerable to the channel errors. The first uncorrected bit error may cause the decoder to discard all subsequent bits whether or not the subsequent bits are received correctly. This is why SPIHT bit stream needs nearly perfect channel error protection in transmission.

Various channel protection schemes have been proposed for transmitting the SPIHT-coded bit stream over error-prone channels. Sherwood and Zeger [10] proposed a scheme with a concatenated coder consisting of a CRC as inner coder and a RCPC coder as outer coder. The whole bit stream is equally protected with a fixed coding rate during the transmission. However, it is well known that the importance of each bit decreases as it moves toward the end of the bit stream. Li and Chen [12] recognized this issue and developed an unequal error protection scheme by providing different protection levels to different bit planes. When channel BER increases beyond the designed value, the PSNR of the reconstructed image degrades more gracefully than that of the Sherwood and Zeger [10] equal protection scheme. Cao and Chen [13] proposed a multiple-substream unequal protection scheme to reduce error propagation. The wavelet coefficients are split into multiple substreams. Each substream is SPIHT coded; hence the error propagation is limited within a single substream.

Unlike conventional error-robust schemes for wireless image transmission, energy efficiency has the highest priority in the design of the image transmission in wireless sensor networks. In this chapter, we proposed a cross layer design that considers both energy efficiency and robustness in transmission. In link layer, the design of the proposed scheme is based on the basic idea that it is more efficient to transmit a very long bit stream by dividing it into small fragments and transmit them in a burst [6]. We divide a long bit stream into many small fragments and transmit in a burst for two reasons. First, when automatic retransmission request (ARQ) mode is adopted, the cost for retransmission of a very long packet is high if only a few bits have been corrupted in the transmission. In forward error correction (FEC) coding mode, the cost is high because the whole long packet has to be discarded when errors in transmission cannot be corrected. Second, if we fragment a long bit stream into many small fragments and transmit them in a burst, we can save energy by reducing control overhead and the device switching from sleep to active. In the application layer, to achieve error-robust image transmission, we first generate multiple SPIHT bit streams to reduce error propagation, the unequal error protection similar to the one developed in [12] is then employed to ensure graceful degradation in PSNR over a change of channel bit error rates. Experimental results show that the proposed scheme exhibits not only high energy efficiency in transmission but also error robust in terms of graceful degradation in PSNR over high channel BERs.

13.2 SYSTEM DESCRIPTION

The proposed scheme is shown in Figure 13.1 First, we group the wavelet coefficients into multiple trees according to the parent–child relationship after wavelet decomposition of the image. Each tree is then encoded by the SPIHT algorithm independently. Unequal error protection strategy is implemented by the RCPC/CRC channel coding in order to combat time-varying channel errors. The channel coding rate is designed based on the bit error sensitivity of different bit planes. To achieve energy-efficient transmission, we further divide the multiple bit streams into small fragments and transmit them in bursts so as to

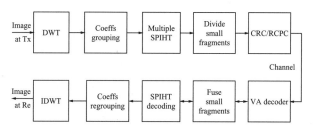

Figure 13.1 Diagram of proposed system.

save energy consumed on control overhead and device switching. At the receiving end, a list Viterbi algorithm is applied for channel decoding. Within each decoded bit stream, when the first uncorrected error has been detected, the subsequent bits have to be discarded because of the nature of progressive coding in SPIHT. The decoded data fragments are then assembled for SPIHT decoding. The decoded wavelet trees are finally regrouped for inverse discrete wavelet transform to reconstruct the image.

13.2.1 Generation of Multiple Bit Streams

The well-known SPIHT algorithm [14] is a refinement of zero tree wavelet coding initially proposed by Shapiro [11]. This algorithm generates an embedded bit stream and achieves best trade-off between complexity and performance. Since the bit stream is progressive, rate change can be made easily and instantaneously by dropping packets either at the source or at the intermediate nodes. Hence, SPIHT-encoded images possess a nice property in bit rate adaptation for time-varying network conditions.

However, the SPIHT-encoded bit stream is fragile to the channel errors; one bit error may cause the sequential bits successfully received to be unusable for image decoding. To reduce the error propagation, we first group the wavelet coefficients into multiple blocks according to parent–child relationship after hierarchical wavelet decomposition of the image. Since the coefficients in wavelet transform domain form a natural hierarchy of subbands, the hierarchical structure of these subbands exhibits parent–child relations along with each orientation. Each coefficient at a given scale is related to a set of coefficients at the next finer scale along the same orientation. Those coefficients having such a relationship can be grouped to form a wavelet tree suitable for SPIHT encoding.

Matucci et al. [15] and Creusere [16] showed that a wavelet block can be constructed by reorganizing coefficients of each wavelet tree. This concept of wavelet block provides a direct association between wavelet coefficients and what they represent spatially in the image. The relation between a wavelet block and an image block facilitates the generation of multiple substreams based on their spatial representation. Therefore, we group the wavelet coefficients into blocks according to the parent–child relationship, as shown in Figure 13.2. Each block is then coded independently by the SPIHT algorithm. Multiple bit streams are therefore generated. In this case, each bit stream is only related to its corresponding region of the image. If bit errors occur in one bit stream, only the image contents within the same block are affected. The bit streams from other blocks will not be affected. With such multiple bit stream generation, the error propagation is limited within one image block. Since the

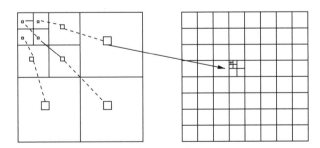

Figure 13.2 Wavelet coefficients grouping.

parent–child relationships are intact within each block, the high coding efficiency of the SPIHT algorithm can still be preserved.

13.2.2 Reduction in Error Propagation

To evaluate the performance of the multiple bit streams for the reduction of error propagation, we test the image transmission over a binary symmetric, memoryless communication channel with channel BER P_{eb}. The image is compressed to B bits and divided into S substreams. Each substream contains B/S bits. The probability of correctly receiving the first k bits in each substream is

$$p(k) = \begin{cases} P_{eb}(1 - P_{eb})^k & 0 \leq k < B/S \\ (1 - P_{eb})^k & k = B/S \end{cases} \tag{13.1}$$

where the first line of (13.1) refers to the probability that the first bit error occurs at the $(k + 1)$th bit, the second line of (13.1) refers to the probability that there is no bit error within a substream. The mean value of k in each substream is therefore

$$m = \sum_{k=0}^{B/S} p(k)k \tag{13.2}$$

$$= \frac{(1 - P_{eb})}{P_{eb}}[1 - (1 - P_{eb})^{B/S}] \tag{13.3}$$

If B/S is much larger than $1/P_{eb}$, then, this means the average number of errors in each substream is much larger than 1. In this case, the mean value m approaches $(1 - P_{eb})/P_{eb}$, which is related only to channel BER. This ideal case means that the first bit error always happens at the same position for a given BER no matter how long the substream is. Therefore, ideally, in a multiple bit stream scheme, the total number of usable bits of S substreams is $S(1 - P_{eb})/P_{eb}$, which is S times the number of usable bits of a single bit stream scheme, if we discard the subsequent bits after we encountered the first bit error.

13.2.3 Error Control from Battery Perspective

To combat the channel bit errors, the SPIHT bit stream needs error protection during transmission. Typically, FEC and ARQ are two common schemes used as error control strategies. Intuitively, one would expect that, as the packet loss probability goes up, which may happen because of increasing bit error rate or because of larger packet size, an FEC code will do increasingly better while an ARQ scheme will do worse from the battery power consumption perspective [7].

However, whether FEC or ARQ is more energy efficient is actually a function of the quality of the service, packet size, and channel characteristics [7]. In FEC, we use RCPC/CRC code because of its flexible structure for unequal error protection. Different coding rates for an input bit stream can be designed using only one encoder/decoder pair [17]. This is especially suitable for the channel coding employed to enable unequal error protection for SPIHT bit layers. The error correcting ability of the RCPC/CRC code can be obtained through simulation.

In the case of ARQ, if we assume P_{eb} is the BER of a binary symmetrical channel, and L is the length of a packet. The packet error probability P_{ep} will be

$$P_{ep} = 1 - (1 - P_{eb})^L \tag{13.4}$$

We can use the correctly received packet ratio at the receiver P_{cp} as a measure of the quality of the service. Suppose T rounds of retransmission of lost packets are required to maintain a given P_{cp}, the following condition should be satisfied:

$$(1 - P_{ep})(1 + P_{ep} + P_{ep}^2 + \cdots + P_{ep}^T) \geq P_{cp} \tag{13.5}$$

That is, the parameter T should be such that the above inequality (13.5) holds. Under this condition, the mean retransmission ratio can be computed as:

$$R = P_{ep}(1 - P_{ep}^T)/(1 - P_{ep}) \tag{13.6}$$

Figure 13.3 shows the mean retransmission ratio vs. channel BER if we set the correctly received packet ratio at the receiver P_{cp} at 10^{-5}. From this figure, we can verify that, at a low BER range (less than 10^{-3}), the mean retransmission ratio is very low (less than 0.1). Hence, ARQ is more energy efficient.

To compare ARQ and RCPC/CRC performance at high channel BER, which is common for wireless sensor network application, we set the retransmission bit ratio equal to the channel coding rate in RCPC/CRC. Such setting implies that both strategies use the same bandwidth resource for transmitting image data of given size. In FEC, if we select a rate of 1/4 memory 4 RCPC code with punctured rate of 10/8, 220 bit packet with 16 CRC bits and 4 flushing bits. The actual coding rate will be 1.375. In ARQ, retransmission does not change BER at the receiver but increases the correctly received packet ratio. We calculate P_{eb} from Eq. (13.1) and compare it with the RCPC/CRC scheme. The result is shown in

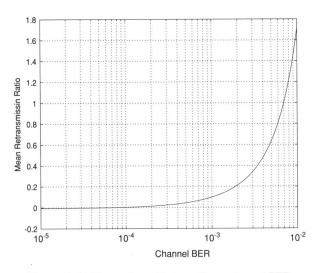

Figure 13.3   Mean retransmission ratio vs. channel BER.

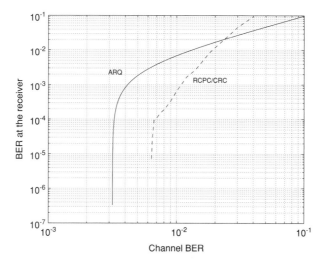

Figure 13.4   BER at receiver for ARQ and RCPC/CRC.

Figure 13.4. RCPC/CRC is better than ARQ when channel BER is less than 0.025. In high BER range, RCPC/CRC scheme requires less redundant bits than ARQ in order to maintain the same quality of the service in P_{cp}. Therefore, RCPC/CRC is more energy efficient in such a channel BER range. However, when channel BER is higher than 0.025, channel error rate has become so high that it is beyond the correcting ability of RCPC/CRC with a coding rate of 1.375. A higher coding rate of RCPC/CRC is required to combat channel error to ensure the quality of service.

A true adaptive scheme would require the transmission strategy to switch between ARQ and FEC. However, the time-varying nature of the wireless channels does not allow frequent switching between these two quite different transmission operations. The hierarchical error sensitivity of the SPIHT-encoded image is able to facilitate the unequal error protection and therefore avoids the required switching for fluctuating wireless channels. In the case of unequal error protection adopted in this research, we apply a different channel coding rate to different bit planes of image data. Top bit planes of image data are protected with much higher coding rate, and therefore we can combat higher rates of channel error. Such strategy guarantees that we receive the top bit planes of the compressed image without error and results in graceful degradation in terms of reconstructed image quality.

13.2.4   Unequal Error Protection Channel Coding

We adopt the unequal error protection scheme developed in [12] to different bit planes of each SPIHT bit stream. Suppose the uniform threshold quantizer is applied to the range $[-Q/2, Q/2]$ with n bits. The quantization step is $Q/2^n$. The magnitude sensitivity for jth bit plane is

$$\epsilon_{n,j} = \left(2^j \frac{Q}{2^n}\right)^2 \qquad \text{for } j = 0, \ldots, n-2 \tag{13.7}$$

where $j = 0$ denotes the least significant bit (LSB), $j = n - 2$ denotes the most significant bit (MSB), and $j = n - 1$ denotes the sign bit. The bit sensitivity for the sign bit is higher than that of the magnitude bits. For example, if the MSB is k, the magnitude is between $2^k Q/2^n \leq A < 2^{k+1} Q/2^n$. However, the sign bit error will be $2A^2$. The bit streams are ordered by bit plane and each bit plane is assigned a different channel coding rate based on the relative bit error sensitivity. Since the importance of each bit decreases as it moves toward the end of the bit stream, high bit plane deserves more protection than low bit planes. Because more channel coding protections are assigned to the bits from a high bit plane, uncorrected bit errors are more likely to occur in the lower bit plane. This reduces image distortions and makes the scheme more error robust in image transmission.

13.2.5 Energy-Efficient Transmission

To void collisions in MAC, we adopt a contention-based scheme using RTS (request to send) and CTS (clear to send). This scheme has been adopted in IEEE 802.11 wireless local area network (LAN) standards. Figure 13.5 shows the timing relationship between a sender and receiver. We use a time division multiple access (TDMA) scheme in image transmission. This scheme requires periodic synchronization among neighboring nodes in order to correct possible clock drifts. Before sending data, a sender first broadcasts a SYNC packet. Then, it sends out the RTS packet to win the medium. If the receiver reply with a CTS packet, the sender begins to send data to the receiver. We also adopt periodic listen and sleep in [6] to avoid overhearing, which is another source of energy waste in wireless sensor network.

To further enhance energy efficiency, we divide the multiple bit streams into small fragments. In our FEC mode, small packet size has the advantage that, if errors in transmission cannot be corrected by FEC code, we only need to discard a small number of bits. If the packet size is large, more bits will have to be discarded and, therefore, will cause more distortion at the receiving end. We then transmit those small fragments in bursts.

Energy saving is achieved in two different aspects. First, since we use RTS and CTS packets in contention for each independent fragment transmission, if we transmit small fragments in burst, only one RTS and one CTS packet are used. This will save energy that would otherwise be wasted for extra RTS and CTS packets. Second, transmitting in bursts can also save energy consumed in frequent device switching because switching the device from sleep to active does not occur instantaneously. For example, the radio device needs time to hop the desired carrier frequency before data transmission and reception. The time the phase-locked loops (PLL) take to settle down within an acceptable error is called the lock time. The lock time could be several hundred microseconds [18]. Data cannot be transmitted or received during the lock time period. If the transmission rate is 1 Mbps,

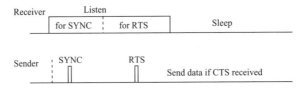

Figure 13.5   Timing relationship between a receiver and a sender.

transmitting a 100-bit packet only needs about 100 μs. If we transmit many fragments in burst, we will save energy on device switching.

13.3 EXPERIMENTAL RESULTS

13.3.1 Experiment Setup

The node model used in our research is based on the Intel StrongARM SA 1110 and National Semiconductor LMX3162. They are used as processor and transceiver, respectively. LMX3162 works in a 2.4-GHz unlicensed band. The transmission power is 80 mJ when the transceiver sends data. The transmission rate is 1 Mbps. If we count the time for RTS and CTS transmission, the switching time from sleep to active is longer than the lock time. The average power during lock time is almost the same as the power when the transceiver sends data. We adopt the switching time of 466 μs measured in [19], which has the same processor and transceiver.

We test image transmission with 512×512 8 bit/pixel gray scale Lena image. We applied a 5-level hierarchical wavelet transform to the image. The lowest frequency subband has 16×16 coefficients. These wavelet coefficients are grouped into 64 trees. SPIHT is adopted to code each tree independently, and each of these coefficient trees corresponds to one image block of 32×32 pixels. We divide each bit stream into packets with the length of 100 bytes. In our experiments, we adopt an unequal error protection scheme developed in [12], and assign {8/24, 8/22, 8/20, 8/18, 8/16, 8/12, 8/12} RCPC rates to packets of different bit planes. The transmission is based on binary symmetric channels.

13.3.2 Results and Analysis

Figure 13.6 shows the results of a reconstructed image with one and multiple bit streams at 0.5 bpp for both source and channel coding. From the experimental results, we find that

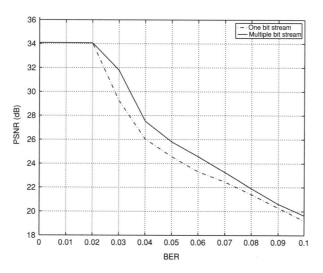

Figure 13.6   PSNR of reconstructed image.

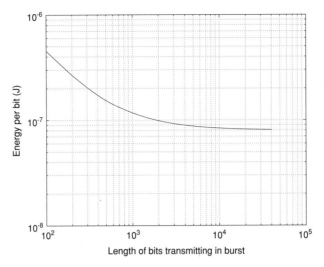

Figure 13.7   Energy per bit in transmission.

when BER ranges from 0 to 0.02, all transmission errors are corrected by RCPC code, and both schemes produce virtually the same PSNR. When BER is higher than 0.02 and errors cannot be completely corrected, the PSNR of the proposed scheme decreases more gracefully. Therefore, the proposed multiple bit streams scheme performs constantly better than the single bit stream scheme for wireless image transmission.

We further divide multiple bit streams into small fragments; each fragment has 100 bits. Figure 13.7 shows the energy per bit based on different lengths of fragments transmitting in bursts. If we transmit each fragment independently, the energy per bit is about 0.45 μJ. If we transmit 100 fragments in bursts, the energy per bit is about 0.084 μJ. The energy could be greatly saved if we choose transmitting fragments in bursts. However, the saving in transmission energy flat out when the length of bursts in transmission reaches beyond 10^4.

13.4   SUMMARY AND DISCUSSION

In this chapter, we have proposed a novel scheme for cross-layer design of energy-efficient image transmission over wireless sensor network. First, we generate multiple bit streams of the compressed image to achieve error-robust image transmission over wireless sensor network. High-coding efficiency is preserved and error propagation is significantly reduced. Unequal error protection strategy is then implemented by RCPC/CRC channel coding to combat the channel errors. The coding rate is designed based on the bit error sensitivity of different bit planes. To achieve energy-efficient transmission by saving energy consumed on control overhead and device switching, we divide the multiple bit streams into small fragments and transmit them in bursts. Experimental results show that the proposed scheme has advantages in two aspects: high-energy efficiency in transmission and graceful degradation in PSNR performance in terms of image reconstruction quality at the base station over a wide range of channel BERs.

REFERENCES

1. D. Estrin, L. Girod, G. Pottie, and M. Srivastava, "Instrumenting the world with wireless sensor network," in *Proceedings of ICASSP '2001*, Salt Lake City, 2001.

2. D. Li, K. D. Wong, Y. H. Hu, and A. M. Sayeed, "Detection, classification, and tracking of targets," *IEEE Signal Processing Magazine*, vol. 19, no. 2, pp. 17–29, Mar. 2002.

3. O. Schrey, J. Huppertz, G. Filimonovic, A. Bussmann, W. Brockherde, and B. J. Hosticka, "A 1K × 1K high dynamic range CMOS image sensor with on chip programmable region-of-interest readout," *IEEE Journal of Solid-State Circuits*, vol. 37, no. 7, pp. 911–915, 2002.

4. C. Intanagonwiwat, R. Govindan, and D. Estrin, "Directed diffusion: A scalable and robust communication paradigm for sensor network," in *Proceedings of ACM MobiCom '00*, Boston, 2000, pp. 56–67.

5. W. Heinzelman, A. Chandrakasan, and H. Balakrishnan, "Energy-efficient communication protocol for wireless sensor networks," in *Proceedings of HICSS 2000*, Maui, 2000.

6. W. Ye, J. Heidemann, and D. Estrin, "An energy-efficient MAC protocol for wireless sensor network," in *Proceedings of INFOCOM 2002*, New York, vol. 3, June 2002, pp. 1567–1576.

7. P. Lettieri, C. Fragouli, and M. B. Srivastava, "Low power error control for wireless links," in *Proceedings of the Third ACM/IEEE International Conference on Mobile Computing and Networking*, Budapest, pp. 139–150, 1997.

8. M. J. Ruf and J. W. Modestino, "Operational rate-distortion performance for joint source and channel coding of images," *IEEE Transactions on Image Processing*, vol. 8, no. 3, pp. 305–320, 1999.

9. A. J. Goldsmith and M. Effros, "Joint design of fixed-rate source codes and multiresolution channel codes," *IEEE Transactions on Communications*, vol. 46, pp. 1301–1312, 1998.

10. P. G. Sherwood and K. Zeger, "Progressive image coding for noisy channels," *IEEE Signal Processing Letters*, vol. 4, no. 7, pp. 189–191, 1997.

11. J. M. Shapiro, "Embedded image coding using zerotrees of wavelet coefficients," *IEEE Transactions on Signal Processing*, vol. 41, no. 12, pp. 3445–3462, 1993.

12. H. Li and C. W. Chen, "Robust image transmission with bidirectional synchronization and hierarchical error correction," *IEEE Transactions on Circuits and Systems for Video Technology*, vol. 11, no. 11, pp. 1183–1187, 2001.

13. L. Cao and C. W. Chen, "Multiple hierarchical image transmission over wireless channels," *Journal of Visual Communication and Image Representation*, vol. 13, no. 3, pp. 386–399, 2002.

14. A. Said and W. A. Pearlman, "A new, fast, and efficient image codec based on set partitioning in hierarchical trees," *IEEE Transactions on Circuits and Systems for Video Technology*, vol. 6, no. 3, pp. 243–250, 1996.

15. S. A. Martucci, I. Sodagar, T. Chiang, and Y. Q. Zhang, "A zerotree wavelet video coder," *IEEE Transactions on Circuits and Systems for Video Technology*, vol. 7, Feb. 1997.

16. C. D. Creusere, "A new method of robust image compression based on the embedded zerotree wavelet algorithm," *IEEE Transactions on Image Processing*, vol. 6, pp. 1436–1442, Oct. 1997.

17. J. Hagenauer, "Rate-compatible punctured convolutional codes (rcpc codes) and their applications," *IEEE Transactions on Communications*, vol. 36, no. 4, pp. 389–400, 1988.

18. *LMX3162 Evaluation Notes and Datasheet*, National Semiconductor Corporation, 1999.

19. E. Shih, S. H. Cho, N. Ickes, R. Min, A. Sinha, A. Wang, and A. Chandrakasan, "Physical layer driven protocol and algorithm design for energy-efficient wireless sensor network," in *Proceedings of MOBICOM 2001*, Rome, 2001.

14

HYBRID SENSOR NETWORK TEST BED FOR REINFORCED TARGET TRACKING

Pratik K. Biswas and Shashi Phoha

We describe some sensor network operational components and illustrate how they can all be combined to construct an integrated surveillance experiment for a hybrid sensor network test bed. The infrastructure for the test bed provides a hybrid platform consisting of real and virtual sensor nodes augmented with a simulated environment. Data from real-world tracking is provided instantaneously to the simulated environment, where it is used to self-organize the randomly deployed sensor network (simulation) through an indigenously developed self-organizing protocol. Results from the simulation are then fed back to the real world to enable the sensor network to reorganize for reinforced tracking. The results indicate that a large number of distributed and interacting sensor nodes, with different capabilities and operating in different environments, can be incorporated in high-fidelity experiments to analyze the challenging aspects of the surveillance problem through realistic application scenarios.

14.1 INTRODUCTION

Recent technological advances have directed research toward distributed sensor networks [1, 2]. The Emergent Surveillance Plexus (ESP) [3] is a Multidisciplinary University Research Initiative (MURI), which has been funded by the Defense Advanced Research Projects Agency (DARPA) to advance the surveillance capabilities of sensor networks. It involves participants from such universities as Penn State University (PSU), University of California at Los Angeles (UCLA), Duke, Wisconsin, Cornell, and (Louisiana State University (LSU). One of its goals is to develop an infrastructure [4] that can support applications [5] adaptable to dynamic environments. Adaptation in sensor networking involves the reconfiguration and reorganization of the sensor nodes and is often limited by distributed sensing inaccuracy, resource constraints, and communication bottlenecks.

Sensor Network Operations, Edited by Phoha, LaPorta, and Griffin
Copyright © 2006 The Institute of Electrical and Electronics Engineers, Inc.

We have set up a test bed at PSU/ARL (Applied Research Laboratory) that provides a heterogeneous platform to support realistic surveillance applications, whose results can be independently validated. To demonstrate the applicability of the test bed, we have constructed an experiment which integrates some of the MURI participants' innovations to provide insights into energy-efficient self-organization of sensor networks for integrated target surveillance [6]. Our eventual goal is to create a hybrid sensor network infrastructure that can be used for independent validation of surveillance-based experiments as well as for building applications that can integrate as many ESP innovations as possible.

The remainder of the chapter is organized as follows: Section 14.2 provides a background to some of the sensor network operational building blocks that have been developed under the ESP MURI. Section 14.3 states the challenge problem and formulates our immediate and long-term goals. Section 14.4 describes an experiment that demonstrates how these operational building blocks can be integrated on a hybrid test bed to build a sensor network application for integrated surveillance. Section 14.5 defines an evaluation metrics for validating the experimental results and presents results from the experiment. Section 14.6 concludes the chapter with an insight to future challenges.

14.2 SENSOR NETWORK OPERATIONAL COMPONENTS

A sensor network can be easily organized for surveillance by a set of core activities and reusable software patterns. These can be viewed as sensor network operational building blocks for surveillance applications. In this section, we provide a background to some of the technologies, developed under this MURI, which we have integrated into our sensor network infrastructure for building realistic surveillance applications.

14.2.1 Beamforming: An Algorithm for Target Localization

The *beamforming algorithm* [7] is a localization algorithm that uses the relative time delays of arrival of an acoustic signal at sensor nodes to estimate the *direction of arrival (DOA)* of a noise source. It is a method of estimating the target location at any time instance on the basis of the time series of the acoustic data collected by each sensor node at that instance. The beamforming algorithm that we have implemented uses the approximate maximum-likelihood version [8]. It is much faster and less power draining than blind beamforming. It utilizes acoustic data collected from microphones attached to a node to track targets. Each sensor node has four (or more) microphones that "receive" acoustic data from the source. The inputs from the microphone are time stamped by the node clock. Each node then uses these data to calculate the DOA and forwards it to the rest of the relevant network. If there is more than one node involved, then the *crossover point* of the DOAs is calculated by a supervisory or sink node (more powerful processing node) to determine the location of the source. This is repeated numerous times to produce a target track. The beamforming algorithm has been developed at UCLA.

14.2.2 Dynamic Space–Time Clustering: A Target Tracking Algorithm

The dynamic space–time clustering algorithm (DSTC) [9, 10] is a *tracking algorithm* that is based on event clustering. It depends upon the *closest point of approach (CPA)* of a target to the nodes in the sensor network. Conceptually, a CPA is a local maximum in time of the intensity of the sensor energy detected at the sensor node. For acoustic and seismic sensors, it is reasonable to assume that the intensity of the signal is highest when the target is closest to the node. As each node detects a CPA event, it transmits it to all nodes in the local neighborhood. The neighborhood defines a space–time window around each sensor node that includes all nodes within a specified radius and whose readings occur within a specified time window. The radius and the time window define the size of the neighborhood. After a set time interval, each node checks its queue of CPA events acquired from the local neighborhood. If a node finds that the local intensity is larger than the others in the queue, then it declares itself as the cluster head. It then estimates the target location as a weighted mean of the CPA positions of all the nodes that detected the target, with the square root of CPA intensity as the weight. In addition the velocity of the target is computed using a weighted regression technique that gives both the magnitude and direction. The cluster head then propagates this track information to the nodes that are in the neighborhood of the target's estimated direction. It may be noted that fewer than five CPA events can seriously bias the result and we need at least two clusters to draw a track. It was developed at PSU/ARL.

14.2.3 AESOP: Self-Organizing Protocol for Integrated Surveillance

An Emergent-Surveillance-Plexus Self-Organizing Protocol (AESOP) is a self-organizing protocol [6] for surveillance. It can be used to self-organize a randomly deployed sensor network, with an arbitrary number of static and mobile sensor nodes, when there is a priori knowledge about a possible target trajectory. A priori knowledge about a possible target trajectory may include a training path from a past track (training pattern) or an estimated target track (from image analysis) or a predicted track (based on extrapolation). AESOP consists of the four phases *self-awareness, self-reconfiguration, self-repositioning*, and *self-adaptation*, each utilizing one or more algorithms developed under ESP MURI. It is based on an innovative integration of location estimation, state reconfiguration, and mobility management algorithms. AESOP can be very useful in self-organizing a sensor network for tracking targets, one target at a time, that traverses a very similar path, as in the case of a convoy of enemy carriers or when the nodes can assume the existence of a target path that is yet realistically unknown. The protocol has been developed at PSU/ARL.

14.2.4 Location Estimation Algorithm for Node Localization

The *location estimation algorithm* [11] allows sensor nodes to exchange messages to establish their local neighborhood and estimate their own locations based on the positions of their static neighbors. It assumes the presence of a number of nodes with Global Positioning System (GPS) capabilities whose locations are accurately known. Each node's location estimate is modeled as a rectangle. It uses connectivity-induced constraints to localize a node and exploits nonneighbor (neighbor of a neighbor) relationships to improve on the accuracy of the location estimation. The algorithm has been developed at Wisconsin.

14.2.5 Self-Configuration and Adaptive Reconfiguration Algorithm for State Management

The self-configuration and adaptive reconfiguration algorithm [12] ensures that the network deploys the minimal number of nodes to provide the same coverage and connectivity of the surveillance area as the original deployment by switching off the redundant nodes with full overlapping coverage. This algorithm runs in two stages: The first one is used for each node's sensing coverage evaluation, while the second is used for node state reevaluation and connectivity checking for the overall network. It employs a token exchange mechanism between any pair of neighbors to loop through the network as each node calculates/recalculates its coverage redundancy and ensures node connectivity. Consequently, some nodes reconfigure their states to *active* while others are set to *inactive/sleeping*. The algorithm is useful in maximizing the lifetime of the network. The algorithm has been developed at Duke.

14.2.6 Distributed Annealing Algorithm for Mobility Strategy

The *distributed annealing algorithm* [13–15] demonstrates node mobility as a trade-off between surveillance and communication tasks when a target follows a preconceived (training pattern) or an evolving trajectory. Some nodes close to the track act as source nodes emitting target information, which gets transmitted through a host of intermediate relay nodes (active nodes from the previous phase) to a more powerful node, assuming the role of a sink node or a base station. During this process, as the origin of communication moves from one source node to the next, some of these intermediate relay nodes reposition themselves to achieve an energy-efficient configuration vis-à-vis the training/evolving path. The algorithm uses a distributed version of simulated annealing. A node moves either when the sum of power consumption at the new position and the moving cost (power required for movement) are less than the power consumption at the old position or when the probability of acceptance (as being globally successful) of the move, based on the local power loss, is greater than a uniformly generated random number. As the local power loss increases with time, the freedom to move gets effectively diminished. If the optimization is successful, the final solution will be a global minimum. The algorithm has been developed at PSU/EE.

14.2.7 C3L: Command, Control and Communication Language for Autonomous Devices

The *Command, Control and Communication Language (C3L)* [3] is a useful tool for the implementation of large systems containing heterogeneous, autonomous devices. It is designed to capture the causal dynamical architecture of autonomous devices conducting operations in complex, time-critical environments. It can be used for synthesizing distributed mission scripts and expressing individual and collective behaviors. It is device-oriented and event-driven.

C3L is capable of being used across many levels of hierarchical command and control in a multidevice structure. It can be used as a communication language to command and query the devices and fuse (aggregate) responses. In addition device controllers written in C3L can be used to control the behavior of individual devices.

In C3L, programming of a device is done at three levels. First, the specifications of the base device are described at the *plant* level. Second, the basic memory structures and the event handlers are defined at the *controller* level. Third, the communication infrastructure consisting of the groupings, information content, and communicative events is detailed at the *communication* level. We summarize here, axiomatically, the key features of the language:

Formality C3L incorporates the formal model of hierarchical discrete-event control.

Event Handling C3L supports handling of controllable and uncontrollable events.

Portability C3L achieves a separation between the program and the hardware through the prespecification of the controllable and uncontrollable events.

Expressiveness C3L includes semantic structures allowing recursive, repetitive, conditional, sequential, and parallel execution as well as nesting of controllable and uncontrollable events (asynchronous behaviors). It provides syntax for authorization direct lists and group constructs.

Extensibility C3L is extendible at all three levels of its programming.

Fidelity The language is complete enough to express a wealth of behaviors, mission scripts, and hierarchical plans.

A sensor network can also be viewed as a network of interacting devices. Each sensor node receives commands and query messages from sink nodes or cluster heads, communicative events from other sensor nodes, and uncontrollable events from its own event handler. It can invoke any controllable event for any peer as well as raise communicative events on other sensor nodes. Besides, it can raise its own controllable events.

We have used C3L as a command, control, and communication language for sensor networks. Supported by a *plant model* for sensor nodes, C3L can also be used to command, control, and query the devices at a lower level. It has been developed at PSU/ARL.

14.2.8 STACS: Interactive Controller Synthesizer

The *Software Tool for Automated Controller Synthesis* (STACS) [16] is a controller synthesizer that will take an application specification and generate a controller for individual sensor nodes. Given a high-level specification of goals specified as behaviors of the network, our aim is to generate local controllers for each node that execute the network behaviors based upon observation of certain events identified in the sensor data.

STACS facilitates the generation and debugging of discrete-event controllers for sensor nodes from application level specifications of sensor network applications. Initially the natural language specifications are translated into finite-state machines and composed to form the controller. STACS facilitates the translation process by guiding the user to phrase the specification in such a way that their structures meet the necessary conditions for controllability. It also provides an interactive error correction system that allows the user to iteratively refine poorly designed specifications. Finally, it outputs the compileable code in common control language (CCL).

STACS helps configure local node behavior that is globally consistent with the general behavior of the sensor network application. The controller can be ported to each node during the initialization phase.

STACS was written in Java, revision 1.4.1 available from Sun Microsystems. It has been developed at PSU/ARL.

14.2.9 Mechanism for Code Mobility

Code mobility refers to the sensor node's ability to move code. It consists of interfaces for code and data migration [10]. A node can push or pull data files, C3L code, and executables on demand. Nodes are deployed with a minimal software suite. If they require additional functionality, the desired software can be easily downloaded from the sink node/base station or other sensor nodes and executed. To conserve memory, different mobile software

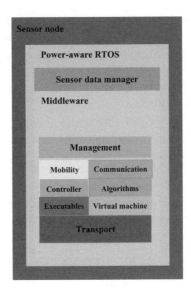

Figure 14.1 Sensor node software architecture.

components can be made *mutually exclusive* using *software regions*, ensuring one component must be removed before another gets downloaded. Code swapping enables sensor nodes to *dynamically reconfigure their roles* and behaviors to *conserve battery power and storage space*.

14.2.10 Sensor Node Software

Sensor node software consists of at least an *operating system*, (OS), a *sensor manager*, and a *middleware* (Fig. 14.1). The OS in the sensor node is a power-aware real-time operating system (RTOS), which can reduce power consumption by coordinating device usage [17]. It allows the node to adjust its processing speeds. The sensor manager manages and measures the local sensor data and provides an interface to the middleware for accessing the sensor information. The sensor node middleware coordinates the integration of the sensor node with the rest of the network. It is made up of the controller(s), management module, communication module, mobility module, task-specific algorithms, downloaded executables, and a virtual machine (VM) that provides a sensor node execution environment (SNEE). The Communication module also provides a C3L interpreter and supports wireless communication. The algorithms determine the behavior of the other modules.

14.3 SENSOR NETWORK CHALLENGE PROBLEM

The sensor network challenge problem addressed in this chapter calls for the construction of a test bed to conduct an experiment that can demonstrate the integration of the various sensor network operational components developed under this MURI on one platform through a single application scenario.

14.3.1 Objectives

Accordingly, our *short-term* objective is to construct a sensor network test bed and then use it to develop and validate an experiment by integrating the aforesaid components (from Section 14.2) to demonstrate the following innovations:

- Implementation of beamforming algorithm on power-aware RTOS
- Creation, implementation, and testing of track data association techniques
- Integration of DOA- and CPA-based tracking
- Self-organized tracking
- Modifications of real-time Linux to show power awareness
- A mobility strategy that compromises between a surveillance task and the goal of minimizing communication energy
- State management of nodes in a dense sensor network
- Location estimation scheme that works in environments in which GPS is unavailable or too expensive in power and/or cost

Our *long-term* objective is to incorporate the contributions of some of the other participants left out in this phase of the experiment and integrate as many uncovered indigenously developed innovations as possible to build realistic surveillance applications on our hybrid test bed that are of practical importance to our sponsor.

14.4 INTEGRATED TARGET SURVEILLANCE EXPERIMENT

We have used our components to construct an integrated target surveillance experiment. Through a single sensor network application, we have tried to demonstrate that we can track a target in real time by an integration of the beamforming algorithm [8] and DSTC algorithm [9, 10] and such tracking can be reinforced by reorganizing nodes based on simulation results from AESOP (Fig. 14.2). Accordingly, the application runs in three stages: (1) real-time tracking of a target using an integration of beamforming and DSTC algorithms, (2) self-organization of the network through simulation of AESOP, and (3) reinforced, real-time tracking using the reorganized network with the DSTC algorithm. In the real sensor network, we have used a combination of static and mobile real sensor nodes, fixed sensors, and virtual sensor nodes (sensing, communicating, and tracking processes) for tracking a target.

14.4.1 Scenario Flow-Through

1. One target enters combat zone.
2. Beamformer(s) (one to two), running on special laptops with AMD chips track targets and identify initial position.
3. Beamformer(s) set their processing speeds.
4. Virtual sensor nodes running the DSTC algorithm receive initial position of the target and track the rest of its path.
5. The track is displayed on the screen.

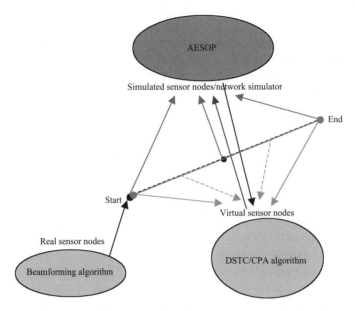

Figure 14.2 Experimental scenario.

6. The initial random deployment configuration along with track data are fed to a network simulator (NS) that can simulate a sensor network.

7. The simulated sensor network (consisting of simulated nodes), employing the initial deployment configuration, self-organizes using AESOP based on the track path.

8. The simulator outputs a self-organized, energy-efficient configuration for tracking a target following a similar path.

9. The real sensor network is reorganized based on the self-organized configuration.

10. Sensor nodes download the DSTC algorithm and track a second target in real time following the same path.

11. Sensor nodes stop the DSTC algorithm.

12. The track is displayed on the screen.

14.4.2 Architecture of the Test-bed

We have constructed a test bed to conduct this experiment (Fig. 14.3).

The following is a brief description of the functional components:

1. *Network Simulator:* NS (2.27) runs algorithms for the self-organizing protocol. The resultant self-organized configuration is fed to a real sensor network through a formatted flat file (ASCII). The information router is responsible for polling that file for reinforced tracking.

2. *Static Sensor Nodes:* One to two beamforming nodes (laptops with AMD chips) are available within the network. They compute their own target localization. This gets transmitted through the information router. The beamforming nodes are equipped with RTOS [18] and the proposed middleware.

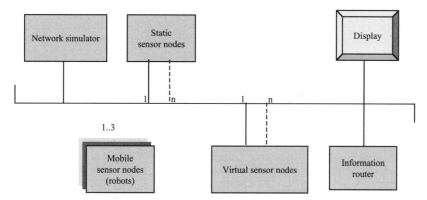

Figure 14.3 High-level architecture of the test bed.

3. *Mobile Robots:* Three mobile robots act as sensor nodes. Each is running an instance of the tracking process (DSTC algorithm) just as a sensor node would. The three sensors receive CPA information just like virtual sensor nodes. Movement instructions are communicated from NS to the robots via the information router.

4. *Virtual Sensor Nodes:* Sensor nodes are realized as separate processes running on a Windows 2000/XP PC. Each sensor node is modeled by a separate instance of the tracking process. These processes receive sensory inputs from the physical sensors (lasers, pressure floor sensors, and passive infrared sensors). Each sensor node has a sensor model and an associated CPA generator for running the DSTC/CPA algorithm [9, 10]. The CPA generator uses inputs from the physical sensors to generate CPAs for the grid of virtual sensor nodes and node positions to be displayed on screen. Virtual sensor nodes communicate through the information router so that packet dropping and other network effects can be simulated.

5. *Information Router:* The information router is a clearinghouse for CPA events (transmitted between the real and virtual sensor nodes), position information, and so on. The information router essentially performs three services: internode communication (real and virtual), CPA/information routing to the screen, and information exchange between the rest of the network simulator and the rest of the network.

6. *Display:* Big screen (9 × 19 in.) displays tracks, coordinates, graphs, CPA events, and so on.

14.4.3 Sensor Network and Node Controllers for Integrated Target Surveillance

Figure 14.4 provides a list of controllers for the surveillance experiment.

Figure 14.5 displays the C3L code snippets for the network and the node controllers corresponding to Figure 14.4.

14.5 EXPERIMENTAL RESULTS AND EVALUATION

In this section, we define an evaluation metrics for our integrated surveillance experiment and then present the experimental results.

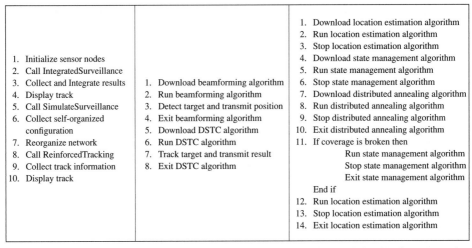

1. Initialize sensor nodes
2. Call IntegratedSurveillance
3. Collect and Integrate results
4. Display track
5. Call SimulateSurveillance
6. Collect self-organized configuration
7. Reorganize network
8. Call ReinforcedTracking
9. Collect track information
10. Display track
	1. Download beamforming algorithm
	2. Run beamforming algorithm
	3. Detect target and transmit position
	4. Exit beamforming algorithm
	5. Download DSTC algorithm
	6. Run DSTC algorithm
	7. Track target and transmit result
	8. Exit DSTC algorithm
		1. Download location estimation algorithm
		2. Run location estimation algorithm
		3. Stop location estimation algorithm
		4. Download state management algorithm
		5. Run state management algorithm
		6. Stop state management algorithm
		7. Download distributed annealing algorithm
		8. Run distributed annealing algorithm
		9. Stop distributed annealing algorithm
		10. Exit distributed annealing algorithm
		11. If coverage is broken then
		        Run state management algorithm
		        Stop state management algorithm
		        Exit state management algorithm
		End if
		12. Run location estimation algorithm
		13. Stop location estimation algorithm
		14. Exit location estimation algorithm

Figure 14.4 Controllers for sensor network, real/virtual sensor nodes, and simulated sensor nodes (AESOP simulation).

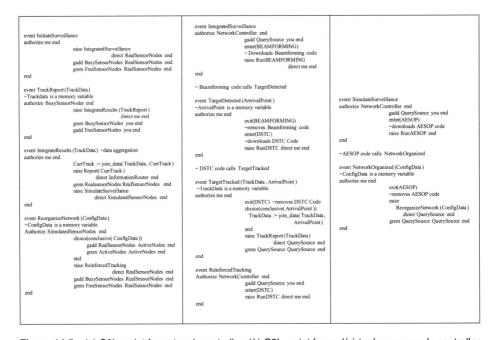

Figure 14.5 (*a*) C3L script for network controller, (*b*) C3L script for real/virtual sensor node controller, and (*c*) C3L script for simulated sensor node controller.

14.5.1 Evaluation Metrics

A distributed sensor network environment lends itself to many different types of metrics. We have used the following metrics for validating the results from our experiment:

- r^2 *Error* The *goodness-of-fit measure* of the estimated track. It is computed as

$$r^2 = 1 - \frac{-d_1^2}{-d_1^{-2}}$$

  where $\dfrac{d_t}{d_t} = $ distance at time t between estimated location and actual location
  $= $ distance at time t between estimated location and center of actual trajectory

- *Variance* (S_e^2) The point variance in x, y, and t for all the points in a track versus the points on actual target path to which they correspond. It is computed as

$$S_e^2 = \frac{1}{n-1} - d_t^2$$

  where $\dfrac{d_t}{d_t} = $ distance at time t between estimated location and actual location
  $= $ distance at time t between estimated location and center of actual trajectory

- *Average Location Error* The average error between the actual location and the estimated location for each simulated sensor node.
- *Percent Redundancy* The Percentage of inactive nodes (simulated) left behind in the sensor network from the initial configuration, after executing the state management algorithm.
- *Energy Saved* The total energy saved by state management of nodes
- *Power Usage* The total power used by the active mobile sensor nodes in the network for communication and mobility.
- *Average Movement* The average distance moved by each node and the average distance per move due to its mobility strategy, which compromises between a surveillance task and the goal of minimizing communication energy.

14.5.2 Experimental Results

Next, we present some results from our integrated surveillance experiment. It includes results from all three stages, that is, target track from real-time integrated tracking, network configurations from AESOP, and target tracks from reinforced tracking.

1. *Real-Time Integrated Target Tracking* We track a single target (mobile robot) in 30×30 (ft^2) grid with 48 sensor nodes: 3 real, mobile sensor nodes (robots) and 45 virtual sensor nodes, using a combination of Beamforming and DSTC algorithms. Figure 14.6 displays a target track from integrated target tracking in real time.

2. *Simulation of AESOP* We start the simulation with the same initial random deployment of 48 nodes mapped to a 30×30-m^2 grid, each node with a sensing range and a communication range of 10 m (scaled). Each node has an initial energy of 50 J and transmission power as well as receiving power of 0.28183815 mW. We have used a motion constant of 0.0005 W/m.

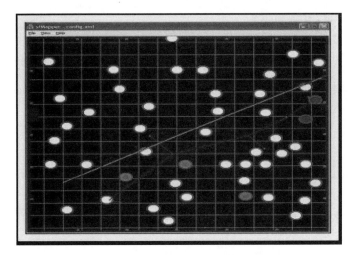

Figure 14.6   Real-time integrated target tracking.

Figure 14.7*a* shows the NAM output for the initial random deployment from the real sensor network configuration. Gray circles indicate GPS nodes while green circles represent the three mobile robots from the real world. Figure 14.7*b* shows the resultant self-organized configuration, with 31 active nodes. Brown squares indicate the *active nodes*, while snow-white squares (with/without gray circles) indicate *sleeping nodes*. Note that the positions of some green mobile nodes have changed in the self-organized configuration due to the *mobility strategy*.

From the simulation results, we can report an *average location error* of 1.802288 m. Seventeen nodes configure their states to *inactive*, saving a *total energy* of approximately 1.735 J through redundancy (35.4%). The three mobile nodes move an *average distance* of 11.36 m *per node* and consume a *total power* of 0.327 W for *mobility* and 38.335 W for *communication*.

3. Reinforced Target Tracking   We track a second target in the original 30×30-ft^2 grid with the self-organized configuration, as obtained from the simulation, now consisting of only 31 nodes. The nodes track using the DSTC/CPA algorithm. Figure 14.8 displays

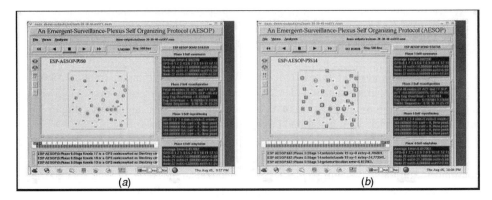

Figure 14.7   (*a*) Initial deployment and (*b*) final self-organized configuration.

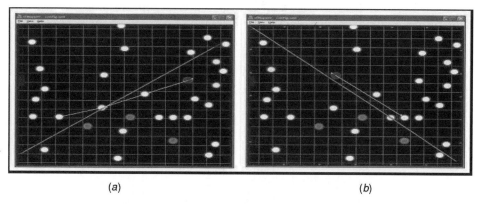

(a)   (b)

Figure 14.8 Screenshots of reinforced target tracking in real time (same and similar path).

tracks for a single target, as tracked by the distributed sensor network in real time with the self-organized configuration. Figure 14.8*a* shows the same path, while Figure 14.8*b* shows a similar path that can also be tracked by this configuration. The green circles represent mobile robots (real sensor nodes), the purple circle indicates the latest cluster head, and red circles represent nodes that have received the latest CPA events.

Figure 14.9 presents statistics from real-time tracking, corresponding to Figures 14.6 and 14.8. Table 1 in Figure 14.9 shows error statistics for the initial and the self-organized configurations, when the same path (Figs. 14.6 and 14.8*a*) was varied up/down by 5 m (in steps of 1 m). Table 2 shows the tracking statistics for a similar yet unrelated path, corresponding to Figure 14.8*b*, for both 48 and 31 nodes.

14.6 CONCLUSION

In this chapter, we have presented some indigenously developed sensor network operational components and demonstrated how they can be integrated to build realistic surveillance applications for a hybrid sensor network. The distributed, hybrid architecture integrates real sensor nodes with processes that sense and track as well as simulated sensor nodes. We have described an experiment that involves real-time tracking, reinforced by results from the simulation of a self-organizing protocol.

Challenges remain in conducting experiments with a larger number of sensor nodes in more complicated scenarios. Further work remains to be done in integrating all the other innovations, accomplished under the ESP initiative, possibly through one application that can then be implemented in our test bed.

Accordingly, we will strive to demonstrate the following in our future endeavors:

- A test bed for a distributed sensor network with a larger number of real, virtual, and simulated sensor nodes
- Tracking, surveillance, and classification of multiple targets (types)

Path Variable (m)	r^2 Value		Point Variance		Track Coverage	
	48nodes	31nodes	48nodes	31nodes	48nodes	31nodes
-5.0	0.99816	099701	0.5703	1.6512	291	48.8
-4.0	0.99852	099832	0.5955	0.8593	679	48.7
-3.0	0.99898	099890	0.9282	1.0082	607	60.7
-2.0	0.99937	099813	0.6106	1.6161	501	60.6
-1.0	0.99879	099768	0.4921	0.5361	406	76.9
0.0	0.99821	099695	0.7911	3.7180	400	87.7
1.0	0.99914	098863	1.0351	5.5605	876	39.6
2.0	0.99881	099067	1.4021	8.3914	868	70.5
3.0	0.99932	098439	0.8754	11.8124	282	89.4
4.0	0.99947	099427	0.7399	6.0536	280	81.6
5.0	0.99947	099838	1.1115	2.1012	440	89.6

Table 1: Same Target Path Shifted up/down by Path Var

Path Variable (m)	r^2 Value		Point Variance		Track Coverage	
	48 nodes	31nodes	48 nodes	31nodes	48 nodes	31nodes
0.0	0.99891	0.99702	1.8262	4.9876	815	815

Table 2: Similar (Unrelated) Target Path (Acid Test)

Figure 14.9   Comparative error statistics for tracking with the initial deployment and self-organized configuration.

- A packet routing strategy that is appropriate for volatile networking conditions and can accommodate both delay-sensitive and delay-insensitive traffic in an energy-efficient manner
- Protocols that enable a seamless, distributed "switch-over" of routing protocols in response to perceived network volatility by the nodes, for example, choosing between pheromone routing and adaptive online distance vector (AODV)
- Robustness/adaptability to network volatility caused by faults, attacks, dwindling energy, high mobility, varying communication channel and traffic demands, terrain conditions, and so on.

Acknowledgment   The authors would like to thank E. Keller, J. Douglas, and J. Koch of PSU/ARL; G. Kesidis and his students from PSU/EE; K. Chakrabarty and Y. Zou from Duke; and P. Ramanathan and his students from Wisconsin for contributing to the implementation of the integrated surveillance experiment.

REFERENCES

1. H. Qi, S. Iyengar, and K. Chakrabarty, "Distributed sensor networks—A review of recent research, short survey, *Journal of the Franklin Institute*, vol. 338, no. 6, pp. 655–668, Sept. 2001.

2. G. J. Pottie, "Wireless sensor networks," in *IEEE Information Theory Workshop*, Killerney, Ireland, June 1998, pp. 139–140.

3. Emergent Surveillance Plexus MURI—ESP TR0201, *Experimental Plan and Validation (Winter 2003)*.

4. P. K. Biswas and S. Phoha, "A hybrid test-bed for surveillance-based sensor network experiments, in *Fourth International IEEE/ACM Conference on Information Processing in Sensor Networks (IPSN 2005)*, Los Angeles, Apr. 25–27, 2005.

5. P. K. Biswas and S. Phoha, "A sensor network test-bed for an integrated target surveillance experiment," in *Proceedings of IEEE Workshop on Embedded Networked Sensors (EmNetS-I)*, Tampa, FL, Nov. 16, 2004.

6. P. K. Biswas and S. Phoha, "Self-organizing sensor networks for integrated target surveillance," in *Wiley European Transactions on Telecommunications*.

7. K. Yao, R. Hudson, C. Reed, D. Chen and F. Lorenzelli "Blind beamforming on a randomly distributed sensor array system," *IEEE Journal on Selected Areas in Communications*, vol. 16, pp. 1555–1567, 1998.

8. J. C. Chen and K. Yao, "Beamforming," in S. Iyengar and R. Brooks (Eds.), *Frontiers in Distributed Sensor Networks*, Boca Raton, FL: CRC Press, 2003.

9. D. Friedlander, C. Griffin, N. Jacobson, S. Phoha, and R. Brooks, "Dynamic agent classification and tracking using an ad hoc mobile acoustic sensor network," *EURASIP Journal on Applied Signal Processing*, vol. 4, pp. 371–377, 2003.

10. R. Brooks, J. Moore, T. Keiser, S. Phoha, D. Friedlander, J. Koch, and N. Jacobson, "Tracking targets with self-organizing distributed ground sensors," in *Proceedings of IEEE Aerospace Conference*, Big Sky, MT, Mar. 10–15, 2003.

11. N. Sundaram and P. Ramanathan, "Connectivity based location estimation scheme for wireless ad hoc networks," *Proceedings of Globecom*, vol. 1, pp.143–147, Taipen Taiwan, Nov. 2002.

12. H. Sabbineni and K. Chakrabarty, "SCARE: A scalable self-configuration and adaptive reconfiguration scheme for dense sensor networks," in S. Phoha et al. (Eds.), *Sensor Network Operations*, Piscataway, NJ: IEEE Press Wiley, 2006.

13. R. Rao and G. Kesidis, "Purposeful mobility for relaying and surveillance in mobile ad-hoc sensor networks," *IEEE Transactions on Mobile Computing*, vol. 3, no. 3, pp. 225–232, 2004.

14. G. Kesidis, T. Konstantopoulos, and S. Phoha, "Surveillance coverage and communication connectivity properties of ad-hoc sensor networks under a random mobility strategy," in *Proceedings of the Second IEEE International Conference on Sensors*, Toronto, Canada, Oct. 22–24, 2003.

15. G. Kesidis and E. Wong, "Optimal acceptance probability for simulated annealing," *Stochastics and Stochastics Reports*, vol. 29, pp. 221–226, 1990.

16. S. Damiani, C. Griffin, and S. Phoha, "Automated generation of discrete event controllers for dynamic reconfiguration of sensor networks," in *Proceedings of the Forty-Eighth Annual SPIE Conference*, San Diego, Aug. 3–8, 2003.

17. V. Swaminathan, C. B. Schweizer, K. Chakrabarty, and A. A. Patel, "Experiences in implementing an energy-driven task scheduler in RT-linux," in *Proceedings of the Real-Time and Embedded Technology and Applications Symposium*, San Jose, Sept. 2002, pp. 229–239.

15

NOISE-ADAPTIVE SENSOR NETWORK FOR VEHICLE TRACKING IN THE DESERT

Shashi Phoha, Eric Grele, Christopher Griffin, John Koch, and Bharat Madan

As part of the Defense Advanced Research Projects Agency (DARPA) SensIT program, an operational system of networked acoustic sensors was fielded at the Marine Corps Air Ground Combat Center, 29 Palms, California. In coordination with multiple universities and industrial partners, we formulated tracking algorithms for identifying mobile targets using a network of low-cost narrow-band acoustic microsensing devices randomly dispersed over the region of interest. In this chapter, we develop adaptive methods to mitigate the time-varying effects of wind, turbulence, and temperature gradients and other environmental factors that decrease the signal-to-noise ratio (SNR) and introduce random errors that cannot be removed through calibration. In our sensors and controls test bed, described in Chapter 14, we added background noise to the signal data collected in the field experiments to achieve controllable SNRs. Generally, effective sensing radii may have to be reduced as the SNR decreases. Hence more sensors are needed for dynamic clustering at the closest point of approach for maintaining desired performance thresholds. By treating the SNR as a control variable, we can define a random variable X that represents the minimum number of nodes needed to achieve a specified performance threshold. The statistical distribution parameters of X are approximated from observed data. The dynamic space–time clustering (DSTC) algorithm is adapted for temporally and spatially distributed signal processing at each cluster head to detect, identify, and track targets in noisy environments with variable SNR. As the track information is passed on, the number of nodes needed to achieve performance thresholds under the current SNR is determined and passed on to the next space–time window along with the track data. The results are compared to "ground truth" data obtained from Global Potioning System (GPS) receivers on the vehicles.

Elsewhere we demonstrated the performance of DSTC to be better than beamforming in noisy environments [4]. Studies performed suggest that hybrid algorithms yield the best

---

Sensor Network Operations, Edited by Phoha, LaPorta, and Griffin

performance in terms of energy and robustness. In this chapter, we analyze and quantify the impact of noise on the subprocedures of the DSTC algorithm. We also compute a lower bound on the confidence a user can place on the DSTC algorithm for given sensor density and SNR.

15.1   INTRODUCTION

When sensor networks are embedded in real environments, ambient noise levels can adversely affect the performance of these networks. The surface layer of the atmosphere is driven by the heat flux from the sun and, at nighttime, its reradiation into space, the latent heat contained in water vapor, the forces of gravity, and the forces of the prevailing geotropic wind. Swanson [1] has studied the effects of atmospheric dynamics to conclude that wind noise is the most significant performance limitation on acoustic sensors as turbulence in the atmosphere tends to scatter and refract sound rays. Attenuation of the emitted acoustic signal from a target, as it propagates toward the sensors, may further decrease the SNR on acoustic sensors. Hence, network performance can be improved if the sensors are positioned closer to a target. Also, improving the SNR using low-power signal processing techniques can significantly improve the overall performance of the sensor network.

Real-time distributed microsensor networks are also very limited in resources. There is often a very limited amount of battery power and limited communications bandwidth. Improvements in performance come at the expense of increased power consumption, which may reduce the active lifetime of a sensor network, in case the battery power is consumed too quickly. Thus sensor network design is often evaluated in terms of trade-offs between performance and resource consumption with the goal of maximizing the effectiveness of the sensor network. The purpose of this work is to obtain dependable performance of distributed sensing algorithms for acoustic-based target localization and tracking in the presence of varying SNR.

15.1.1   29 Palms Field Testing Experiments

The technologies described in this chapter were part of a larger operational demonstration undertaken by DARPA IXO in November 2001 at 29 Palms Marine Base in the Mojave Desert [2], whose goal was to show an end-to-end integrated sensor network application in the field. Sensor deployment is shown in Figure 15.1.

This network located and tracked vehicles using acoustic, seismic, or magnetic sensors over an ad hoc network. The sensors reported the positions of the vehicles and their speed and direction of travel and displayed the results to a base station a few kilometers away. A wide array of vehicles was used, including high-mobility multipurpose wheeled vehicles (HMMWVs), Dragonwagons, 5-ton trucks, light armored vehicles, and tanks. Vehicles included hardware used by American forces and by the former Soviet Union. The technologies described in this chapter performed target parameter estimation, classification, and track maintenance. Network data routing was performed by the diffusion routing methods given elsewhere [3]. Targets were detected and closest-point-of-approach (CPA) events stored in a local data repository.

The operational demonstration took place over a two-week period. During the first week, the sensor network was installed. It was centered at the intersection of two roads and continued for approximately 1 km along the roads. A total of 70 sensor nodes were used.

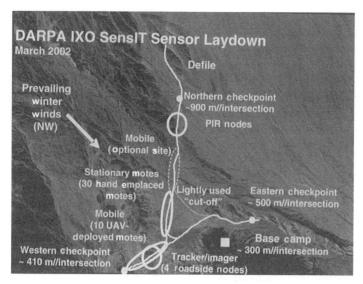

Figure 15.1 Deployment of sensors for the March 2002 SITEX experiments at 29 Palms MCAGCC, CA.

Figure 15.2 provides a diagram of the intersection, roads, and sensor placement. For instrumentation purposes, the network clusters and local connectivity were fixed in advance. Internal routing structures were dynamic.

During the second week, the sensor network was used for live testing and system debugging. Test runs were made using sport utility vehicles to establish the proper parameters to use for cluster formation. Each day there were a number of live vehicle runs through the sensor field. In addition to testing the software suite, each vehicle run was used to record

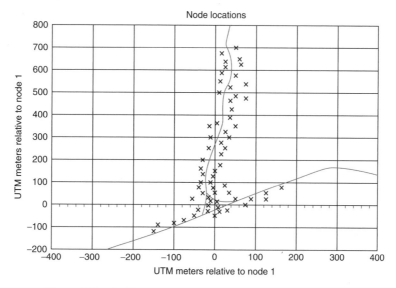

Figure 15.2 Positions of sensor nodes (with roads) during testing.

time series data from all sensors on all the nodes. These data were stored along with ground truth data for postprocessing.

The target tracking software included algorithms for determining CPA, object classification, data association, and track prediction, as described in the next section. The tracking suite ran with the Penn State University/Applied Research Laboratory (PSU/ARL) classifier and the University of Wisconsin classifiers running on separate nodes. Target positioning data and ground truth data collected during these experiments were used in laboratory experiments to formulate the DSTC algorithm [4, 5] and to validate results of this chapter.

15.2   DISTRIBUTED TRACKING

We consider target tracking as the following sequence of problems:

1. *Object detection*: Data features indicate the presence of targets.
2. *Object classification*: Features assign the entity to one of a set of known classes.
3. *Data association*: Detections are associated with tracks.
4. *Track prediction*: Likely future trajectories are used to cue sensor nodes that could continue tracking the entity.

Our approach embeds this sequence of problems into dynamic clusters formed by a network of sensor nodes. Decisions are made using locally available information. Figure 15.3 gives a flowchart of the logical actions performed at each node. Each node executes the same logic.

Initialization invokes subscribe primitives in the diffusion API [6]. Subscribe expresses interest in a data resource. It is invoked three times. Invocations have parameters associating them with "near," "mid," or "far" tracks. The subscribe invocations announce to the network routing substrate the node's intent to receive candidate tracks of entities that may pass through its sensing range. "Near," "mid," and "far" differ in the distance between the node receiving the candidate track and the target's last position in the track. Once this is done, the node starts receiving track information on targets that are moving in its direction. This information is stored in stacks of candidate track data structures. Separate stacks are kept for near, mid, and far tracks.

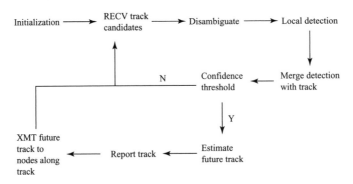

Figure 15.3   Flowchart of processing performed at each node for distributed tracking.

If a clump of nodes detects a target, and the clump head and the list of candidate tracks are empty, then a new track record is initiated. Track records contain the following information:

1. Track ID—time and universal transverse mercator (UTM) coordinates of initial detection
2. Time of last track update
3. Current UTM position estimate (easting and northing)
4. Current estimate of velocity with easting and northing components
5. Codebook value from classification
6. Certainty value of track
7. List of last three detection events (time, UTM coordinates, and velocity estimation) of track

The track record is reported to the user community by insertion in a distributed database.

A set of 12 calls are made to the publish primitive of the network routing API [6]. Each call has an associated region centered on the estimated target position (there are 12 positions in all). The regions consist of northeast, northwest, southeast, and southwest headings for near, mid, and far distances. The headings define rectangular regions with one corner at the estimated target position. The size of the rectangles is determined by multiplying the target's speed by a constant factor. Publish calls announce the arrival of a data resource to the diffusion routing layer. Diffusion routing determines routes between the published data sources and their associated subscriptions. Send primitives [6] are then used to transmit the track record to all nodes in the direction the target is heading.

When a candidate track record is received, it is placed on the appropriate stack: near, mid, or far. A disambiguation procedure looks for duplicate track records and removes them from the stack. Track records are also removed when they are old enough that the track can no longer be valid. When a clump of nodes detect a target and the clump head's list of candidate tracks is not empty, the clump head checks the lists of candidate tracks to find the best match between the current detection event and candidate track. Tracks on the near (mid) stack are favored over tracks on the mid (far) stack.

During our test at 29 Palms, we associated detections with tracks using a Euclidean metric. The estimated target heading and position in the track record were used to estimate the tracked target's location when the detection occurred. The track, whose estimated position was closest to the detection event, was then associated with the detection event. The clump detection is associated with the candidate track with the best fit, as long as the fit is beneath a threshold value. The threshold value was chosen heuristically. If no candidate track is beneath the threshold, a new track is initiated. In many ways, this is a nearest-neighbor data association algorithm.

When detection is associated with a candidate track, it is used to update the track record. The current detection is added to the list of three latest detections. If the track contains fewer than three detections, then the estimates in the candidate track are replaced with the estimates in the current detection. If three or more detections are available, then the extended Kalman filter (EKF) is used to update the position and velocity estimates. The predicted future track determines the regions likely to detect the entity in the future. The algorithm invokes the send method to propagate this information to nodes in those regions, thus completing the processing loop. Figure 15.4 shows how track information is propagated through a distributed network of sensor nodes.

Association of entities with tracks is almost trivial in this approach as long as sampling rates and sensor density are high enough and target distribution is low enough to avoid

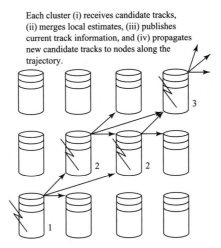

Each cluster (i) receives candidate tracks, (ii) merges local estimates, (iii) publishes current track information, and (iv) propagates new candidate tracks to nodes along the trajectory.

Figure 15.4   Example of track information propagation in a network.

ambiguity. When that is not the case, other established data association techniques may be applicable [7]. We find a fully decentralized approach, like the one proposed here, attractive. In our opinion, it could be more robust and efficient than current centralized methods. We are continuing our research to verify this supposition by instrumented testing of common centralized and decentralized approaches [4, 5, 8, 9, 14].

15.3   ALGORITHMS

15.3.1   CPA Estimation

The CPA in DSTC nodes is determined from a measurement of the acoustic signal power available at any given time, as shown in Figure 15.5. An initial noise reading is taken to determine background noise. It is assumed that no targets are present during the noise-sampling period. If is the signal sampled at frequency for seconds, then the scaled average signal strength is

$$P_n = n_m \frac{\|x\|^2}{f_s t} \tag{15.1}$$

Here x is the signal, f_s the sampling rate, t time.

Signal strength is multiplied by some noise multiplier n_m to determine the base power against which future signal samples will be compared. Once this is complete, the DSTC nodes enter a tracking mode.

In this mode, the acoustic signal is sampled at the given frequency until some time slice samples have been collected. (In our experiments, seconds and Kilohertz, yielding 960 samples). The signal power of these samples is calculated as above and saved as the current signal power. The signal power is then compared to the noise power via

$$Z = \log\left(\frac{P_s}{P_n}\right)$$

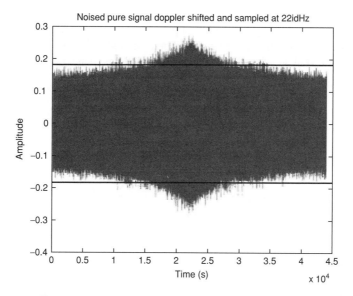

Figure 15.5 Sound samples are analyzed for CPA effects.

If $Z < 0$, then no CPA is detected. Once $Z > 0$, we assume the target is approaching, thus allowing its acoustic signal strength to rise above the background noise. This Z and its corresponding time stamp are saved as the first possible CPA value and a predetermined time window t_w is used to determine whether a true CPA has occurred.

If a new $Z > Z_{max}$ occurs before the t_w seconds, then the new Z and its corresponding time stamp become the new Z_{max}, and the process repeats. This continues until the time window expires before a Z higher than the current Z_{max} occurs. At this point, the currently saved Z_{max} and its time stamp are reported as a CPA.

To prevent any new CPAs being detected on the down slope of the signal peak, the DSTC node calculates the amount of time between the first possible CPA and the reported CPA and does not reenter the tracking model until that amount of time passes.

15.3.2 Position and Velocity Estimation

Models of human perception of motion may be based on the spatiotemporal distribution of energy detected through vision [10, 11]. Similarly, the network detects motion through the spatiotemporal distribution of sensor energy.

We extend techniques found in the literature [12] and adapt them to find vehicle velocity estimates from acoustic sensor signals. The definitions shown below are for one temporal and two spatial dimensions, $\bar{x} = (x, y)$; however their extension to three spatial dimensions is straightforward.

The platform location data from the CPA event cluster can be organized into the following sets of observations: $\{(x_0, t_0), \ldots, (x_n, t_n)\}$ and $\{(y_0, t_0), \ldots, (y_n, t_n)\}$, where (x_0, y_0) is the location of event k_{ij} which contains the largest amplitude CPA peak in the cluster. We redefine the times in the observations so that $t_0 = 0$, where t_0 is the time of CPA event k_{ij}.

We weight the observations based on the CPA peak amplitudes on the assumption that CPA times are more accurate when the target passes closer to the sensor, to give

$\{(x_0, t_0, w_0), \ldots, (x_n, t_n, w_n)\}$ and $\{(y_0, t_0, w_0), \ldots, (y_n, t_n, w_n)\}$, where w_i is the weight of the ith event in the cluster. This greatly improved the quality of the predicted velocities. We defined the spatial extent of the neighborhoods so nodes do not span more than a few square meters and vehicle velocities are approximately linear. Under these assumptions, we can apply least squares linear regression to obtain the following equations [13]:

$$x(t) = v_x t + c_1 \qquad y(t) = v_y t + c_2$$

where

$$v_x = \frac{\sum_i t_i \sum_i x_i - \sum_i w_i \sum_i x_i t_i}{\left(\sum_i t_i\right)^2 - \sum_i w_i \sum_i t_i^2}$$

$$v_x = \frac{\sum_i t_i \sum_i y_i - \sum_i w_i \sum_i y_i t_i}{\left(\sum_i t_i\right)^2 - \sum_i w_i \sum_i t_i^2}$$

and the position $\bar{x}(t_0) = (c_1, c_2)$. The space–time coordinates of the target for this event are $(\bar{x}(t_0), t_0)$.

This simple technique can be augmented to ensure that changes in the target trajectory do not degrade the quality of the estimated track. The correlation coefficients for the velocities in each spatial dimension, r_x, r_y, can be used to identify large changes in vehicle direction and thus limit the CPA event cluster to include only those nodes that will best estimate local velocity. Assume that the observations are sorted as follows:

$$o_i < o_j \Rightarrow |t_i - t_0| < |t_j - t_0|$$

where o_i is an observation containing a time, location, and weight and t_i is the time of the event.

The velocity elements are computed once with the entire event set. After this, the final elements of the list are removed and the velocity is recomputed. This process is repeated while at least five CPAs are present in the set and subsequently the event subset with the highest velocity correlation is used to determine velocity. Fewer than five CPA points could bias the computed velocity and thus render our approximation useless. We summarize the DSTC process in Figure 15.6.

Dynamic space–time clustering has been implemented elsewhere [4, 5]. Data routing is done via the diffusion routing approach described by Heidemann et al. [3], which supports communications based on data attributes instead of node network addresses. Communications can be directed to geographic locations or regions.

15.4 EXPERIMENTAL METHODS

In this section, we describe the experimental setup. All simulations were run on an Intel 686 processor using Redhat Linux. The DSTC algorithm was run with the software that was used in field testing at 29 Palms.

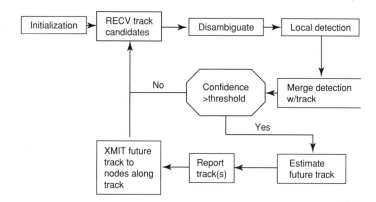

Figure 15.6 Flowchart of the processing performed at any given node to allow distributed target tracking.

15.4.1 CPA Testing

A single simulated node was used to test the effect of SNR on CPA quality. A single sound source passed the node within a distance of r meters ($1 \leq r \leq 20$). The sound source emitted a pure sine wave. To model the sound at frequency f produced by an object moving at velocity v vertically displaced from the sensor node plane and sampled with frequency f_s, we constructed a vector

$$x(t) = \frac{A}{d(t)} \sin[2\pi\omega(t)] \tag{15.2}$$

where $d(t)$ is the node–target distance at time t and

$$\omega(t) = f \frac{v}{v - v_s \left[\sqrt{d(t)^2 - r^2} \Big/ d(t)\right]} \tag{15.3}$$

is the signal frequency corrected for the Doppler shift associated with an off-angle moving vehicle. Dividing by the distance to the sensor node attenuates the signal to account for the spread of the sound energy. A schematic of the modeled behavior is shown in Figure 15.7.

White Gaussian noise (WGN) was added to the signal to create a new signal with a chosen SNR at the CPA. If R is the desired SNR and P_s is the power of the pure signal [computed using Eq. (15.1)], then the variance of the WGN is computed as

$$\sigma_n^2 = \frac{P_s}{10^{R/10}}$$

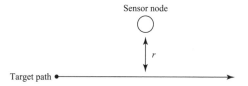

Figure 15.7 A simulated target moves toward a sensor node. The target emits a pure sound that is adjusted to model off-angle Doppler shift and attenuation.

The resulting signal used in the simulation is then

$$x_n = x + \sqrt{\sigma_n^2 w\,(0,\,1)}$$

where $w(0,1)$ is the WGN with mean zero and variance 1. Such a signal is shown in Figure 15.5. The CPA algorithm was used on the constructed signal at various values of r and SNR to determine whether and when a CPA occurred. The noise threshold and time window of the CPA algorithm were varied to determine their impact on the quality of the CPA obtained.

15.5 RESULTS AND DISCUSSION

In this section we summarize and discuss the results from our two experiments.

15.5.1 CPA Confidence Thresholds

To construct a lower bound on the confidence given to a CPA for a given SNR, we computed the maximum distance r at which all 30 experimental runs at a given SNR returned perfect CPA information. We defined a perfect execution of the CPA algorithm as having the following three properties:

1. The CPA generated was not a false positive and no other false-positive CPAs were generated.
2. The CPA run successfully generated a true positive; that is, no false negatives occurred.
3. The CPA was generated at the best possible time—one time window behind the ground truth time of the true CPA.

Using these criteria, we constructed Table 15.1.

Table 15.1 Maximum Range of Trustworthy CPA Information Computed by Fixing SNR and Simulating 30 Target Passes at Varying Distances[a]

SNR (dB)	Lower Bound Ranges of CPA 100% Confidence (m)					
	(2, 2)	(2, 1.75)	(2, 1.5)	(1.5, 2)	(1.5, 1.75)	(1.5, 1.5)
0	0	1	1	0	1	1
5	1	2	2	1	2	2
10	3	3	4	3	3	4
15	5	6	8	5	6	8
20	10	12	7	10	12	4
25	12	14	19	9	12	17
30	20	20	19	18	19	20
35	20	20	20	20	20	20
40	20	20	20	20	20	20

[a] The farthest distance at which each CPA detection was *perfect* was used as the maximum trustworthy distance.

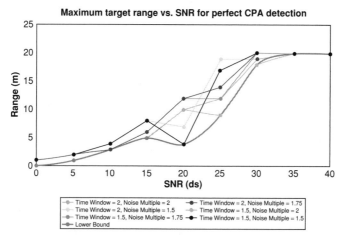

Figure 15.8 Deployed sensor networks may have to deal with various target types traveling at various speeds. Empirically derived lower bound trust curves for CPA detections can be used for determining the quality of results when varying time windows are used. Shorter time windows must be used to track faster and more densely packed targets.

A graphical representation of the results is shown in Figure 15.8. The lower bound on the data is shown as the smoothed boundary line. This represents the lower bound perfect confidence for the CPA estimator. This information can be used for designing sensor fields with sufficient density if the average SNR is known a priori. It can also be used for estimating the confidence in a given CPA if the sensor field density is known and the SNR can be estimated.

15.6 CONCLUSION

In this chapter, we have presented a sensitivity analysis for a CPA algorithm that can be used for the DSTC algorithm. Using these results we constructed a trust bound for results from CPA that can be used to determine confidence at an estimated SNR. These results establish thresholds of confidence for sensor networks that have been tested in mock military conditions. They provide the basis for designing acceptable levels of dependability in sensor network operations by observing the changes in SNR. Due to environmental or other effects, as significant changes in SNR are observed, the distance over which high confidence levels can be maintained for target tracking begins to shrink and hence more sensors must be added, thus increasing the density of sensor coverage.

As sensor networks become more prevalent, practical real-world results such as these will become more and more important for their dependable operation.

REFERENCES

1. D. Swanson, "Environmental effects," in *Frontiers in Distributed Sensor Networks*, Boca Raton, FL: CRC Press, 2004.

2. D. Shephard and S. Kumar, "Microsensor applications," in *Frontiers in Distributed Sensor Networks*, Boca Raton, FL: CRC Press, 2004.

3. J. Heidemann, F. Silva, C. Intanagonwiwat, R. Govindan, D. Estrin, and D. Ganesan, "Building efficient wireless sensor networks with low-level naming," in *Proceedings of the Symposium on Operating Systems Principles*, Chateau Lake Louise, Banff, Alberta, Canada, Oct. 2001, pp. 146–159.

4. S. Phoha, N. Jacobson, and D. Friedlander, "Sensor network based localization and target tracking through hybridization in the operational domains of beamforming and dynamic space-time clustering," in *Proceedings of the 2003 Global Communications Conference*, San Francisco, Dec. 1–5, 2003.

5. D. Friedlander and S. Phoha, "Semantic information fusion of coordinated signal processing in mobile sensor networks," *Journal of High Performance Computing*, Special Issue on Sensing, vol. 16, no. 3, pp. 235–242, 2002.

6. D. Coffin, D. Van Hook, R. Govindan, J. Heidemann, and F. Silva, "Network routing application programmer's interface (api) and walk through 8.0, "Technical Report, USC/ISI, Mar. 2001. Available: http://www.isi.edu/ johnh/PAPERS/Coffin01a.html.

7. Y. Bar-Shalom and T. Fortmann, *Tracking and Data Association*, Orlando, FL: Academic, 1988.

8. R. R. Brooks, C. Griffin, and D. Friedlander, "Self-organized distributed sensor network target tracking," *Journal of High Performance Computing*, Special Issue on Sensing, vol. 16, no. 3, pp. 207–220, 2002.

9. R. R. Brooks and C. Griffin, "Traffic model evaluation of ad hoc target tracking algorithms," *Journal of High Performance Computing*, Special Issue on Sensing, vol. 16, no. 3, pp. 221–234, 2002.

10. E. H. Adelson and J. R. Bergan, "Spatiotemporal energy modes for the perception of motion," *Journal of the Optical Society of America (A)*, vol. 2, no. 2, pp. 284–299, 1985.

11. E. H. Adelson, "Mechanisms for motion perception," *Optics and Photonics*, vol. 46, no. 1, pp. 24–30, 1991.

12. M. Hellebrant, R. Mathar, and M. Scheibenbogen, "Estimating position and velocity of mobiles in a cellular radio network," *IEEE Transactions on Vehicular Technology*, vol. 46, no. 1, pp. 65–71, 1997.

13. W. H. Press, S. Teukolsky, W. Vetterling, and B. Flannery, *Numerical Recipes in C*, Cambridge: Cambridge University Press, 1992.

14. R. R. Brooks, P. Ramanathan, and A. Sayeed, "Distributed target tracking and classification in sensor networks," *Proceedings of the IEEE*, vol. 91, no . 8, pp. 1163–1171, 2003.

ACKNOWLEDGMENTS

Chapter 2, Section 2.2   This research was sponsored in part by the Office of Naval Research (ONR) under grant no. N00014-01-1-0712. It was also supported by DARPA and administered by the Army Research Office under Emergent Surveillance Plexus MURI Award No. DAAD19-01-1-0504. Any opinions, findings, and conclusions or recommendations expressed in this publication are those of the authors and do not necessarily reflect the views of the sponsoring agencies.

Chapter 2, Section 2.5   This work is supported by the State Planning Organization of Turkey under grant number 03K120250 and by the Boğaziçi University Research Projects under grant number 04A105.

Chapter 3, Section 3.5   This work is supported by the Natural Sciences and Engineering Research Council (Canada) and the Bell University Laboratories Program.

Chapter 4, Section 4.2   This work was partially sponsored by NSF CAREER CCR-0092724, DARPA Grant OSURS01-C-1901, ONR Grant N00014-01-1-0744, NSF Equipment Grant EIA-0130724, and a grant from Michigan State University.

Chapter 6, Section 6.3   This work was supported by the U.S. Office of Naval Research (ONR) Young Investigator Award under Grant N00014-03-1-0466.

Chapter 6, Section 6.4   This work was carried out at the National Science Foundations State/Industry/University Cooperative Research Centers (NSFS/IUCRC) Center for Low Power Electronics (CLPE). CLPE is supported by the NSF (Grant #EEC-9523338), the State of Arizona, and an industrial consortium. Any opinions, findings, and conclusions or recommendations expressed in this material are those of the author(s) and do not necessarily reflect the views of the National Science Foundation.

Chapter 7, Section 7.3   This work was supported by NSF ITR grants IIS-0121297 and IIS-0326505.

Chapter 8, Section 8.2   This material is based upon work supported by the National Science Foundation under contract no. CCR-0098361 and by the Rockwell Scientific Company through UC-Micro. The authors would like to acknowledge the funding of this project.

Chapter 8, Section 8.3 This work was partially funded by the EU under the IST-2001-34734 EYES project, by the WEB-MINDS project supported by the Italian MIUR under the FIRB program, and by the "Progetto Giovani Ricercatori" program from the Italian MIUR. All the authors are with the Dipartimento di Informatica, Via Salaria 113, 00198 Roma, Italy, {dipietro,mancini,mei}@di.uniroma1.it. Roberto Di Pietro is partially funded by ISTI-CNR, Pisa, with a postdoctoral grant.

Chapter 8, Section 8.5 The authors wish to thank the anonymous reviewers for the useful comments that helped improve the readability of this work.

Chapter 8, Section 8.6 The authors would like to thank the anonymous referees for their valuable comments. The research was partially supported by the National Science Council of the Republic of China (R.O.C.) under contract no. NSC 92-2213-E-006-056.

Chapter 11 This work was initially conceived when the primary author was with The Center for Advanced Computer Studies, University of Louisiana. The views expressed in this chapter are those of the authors and do not necessarily reflect the official policy or position of the U.S. Air Force, the U.S. Department of Defense, or the U.S. government (AFIT-35-101 pp. 15.5). This chapter has been cleared for public dissemination with a requirement that this disclaimer be included.

Chapter 12 The authors would like to thank Nojun Kwak for providing the source code for the kernel density estimation technique. This research is supported by ETH Zurich under the Polyproject "Miniaturized Wearable Computing: Technology and Applications."

Chapter 14 This material is based upon work supported by in part or in total by DARPA and Army Research Office under Award No. DAAD19-01-1-0504. Any opinions, findings, and conclusions or recommendations expressed in this publication are those of the author(s) and do not necessarily reflect the view of the sponsoring agencies.

Chapter 15 This material is based upon work supported by in part or in total by DARPA and Army Research Office under Award No. DAAD19-01-1-0504. Any opinions, findings, and conclusions or recommendations expressed in this publication are those of the author(s) and do not necessarily reflect the view of the sponsoring agencies.

INDEX

Sensor Network Operations, Edited by Phoha, LaPorta, and Griffin
Copyright © 2006 The Institute of Electrical and Electronics Engineers, Inc.

ABOUT THE EDITORS

Shashi Phoha is the director of the Information Technology Laboratory at the National Institute of Standards and Technology. Since 1991, she has been a professor of Electrical and Computer Engineering at the Pennsylvania State University, and was the director of the Information Sciences and Technology Division of the Applied Research Laboratory of Penn State. Prior to that, she was the director of Information Systems Analysis Division of the Computer Sciences Corporation where she led the development of the Global Transportation Network Architecture for the DoD. She headed the Command, Control, Communications and Intelligence Systems Department and was a senior scientist at ITT Defense Communications from 1984 to 1990. She worked at the MITRE Corporation from 1978 to 1984 where she developed innovative computing architectures, software and protocols for global information systems (WIS and JTIDS), and led the simulation and modeling effort for modernizing the Command and Control System at NORAD.

Dr. Phoha has pioneered new research directions in computational sciences that enable dependable distributed automation of multiple interacting devices. Her work in integrating computation, communications, and control pushes traditional computation beyond the digital abstractions of cyberspace to interact with networks of physical devices in human time frames, with many compelling applications. Many researchers are applying her methods to anomaly detection in complex software systems, multirobot control, internet security, and damage mitigation in electro-mechanical systems. These innovations have had a significant impact on the design and dependable operation of several pragmatic, hitherto computationally infeasible civil and military missions. She has developed real-time data driven sensor fusion and self-organization algorithms for wireless sensor networks for target tracking and for autonomous ocean sampling robotic networks.

Dr. Phoha was the recipient of the 2004 IEEE Technical Achievement Award from the IEEE Computer Society for her work in distributed automation. She has published more than 200 research papers, two books, several book chapters, and has two patent applications in her name. She has served on the board of directors of several international consortia dedicated to technology innovation and international collaboration. Dr. Phoha chaired the Springer-Verlag Technical Advisory Board for publishing the *Dictionary of Internet Security* in May 2002. She was an associate editor of the *IEEE Transactions on Systems, Man, and Cybernetics* for four years and is editor of the *International Journal of Distributed Sensor Networks*.

Thomas F. La Porta received his Ph.D. degree in Electrical Engineering from Columbia University, New York. He joined the Computer Science and Engineering Department at the Pennsylvania State University in 2002 as a full professor. He is the director of the Networking and Security Research Center at Penn State. Prior to joining Penn State, Dr. La Porta was with Bell Laboratories since 1986. He was the director of the Mobile Networking Research Department in Bell Laboratories, Lucent Technologies where he led various projects in wireless and mobile networking. He is an IEEE Fellow, Bell Labs Fellow, received the Bell Labs Distinguished Technical Staff Award in 1996, and an Eta Kappa Nu Outstanding Young Electrical Engineer Award in 1996.

Dr. La Porta was the founding editor-in-chief of the *IEEE Transactions on Mobile Computing*. He served as editor-in-chief of *IEEE Personal Communications Magazine* for three years and is currently a senior advisor. He has published more than 50 technical papers and holds more than 25 patents.

Christopher Griffin has a Masters degree in Mathematics from the Pennsylvania State University and is completing a Ph.D. in Industrial Engineering and Operations Research. He is also a full-time associate research engineer with the Penn State Applied Research Laboratory, the largest independent research lab on Penn State's campus. Mr. Griffin has worked on DoD projects with the Applied Research Laboratory for almost seven years. He was the principal system architect for the ARL *JFACC* Program and the *IEIST* project subcontracted through the Boeing Corporation and the Air Force Research Lab. Mr. Griffin has worked on the Reactive Sensor Networks (RSN) program sponsored by DARPA IXO as a software engineer. RSN explored collaborative signal processing to aggregate information moving through the network, and the use of mobile code for coordination among intelligent sensor nodes. He was mathematical modeling coordinator for the Critical Infrastructure Protection University Research Initiative (CIP/URI), Mobile Ubiquitous Security Environment (MUSE). The MUSE program created survivable network infrastructures by combining peer-to-peer and mobile code technology. Mr. Griffin is currently serving as the test bed coordinator for the Multi-University Research Initiative Emergent Surveillance Plexus (ESP). ESP is developing distributed, power-aware sensor technologies for target tracking and recognition in noisy environments. He is a reviewer for *IEEE Transactions on Mobile Computers*, the *International Journal of Distributed Sensor Networks*, the International Conference on Distributed Computer Systems, and Elsevier's *Information Fusion*. Mr. Griffin's published works span a diverse set of interests including distributed control, game theory, distributed sensing algorithms, bioinformatics and network security.